TECHNOLOGY MANUAL

LINDA M. MEYERS • DIANE L. BENNER
Harrisburg Area Community College
KIRK ANDERSON
Grand Valley State University
DEBRA HYDORN
University of Mary Washington

STATISTICS

THE ART AND SCIENCE OF LEARNING FROM DATA

Agresti • Franklin

Upper Saddle River, NJ 07458

Editor-in-Chief: Sally Yagan
Senior Acquisitions Editor: Petra Recter
Supplements Editor: Joanne Wendelken
Executive Managing Editor: Kathleen Schiaparelli
Assistant Managing Editor: Karen Bosch
Production Editor: Jenelle J. Woodrup
Supplement Cover Manager: Paul Gourhan
Supplement Cover Designer: Christopher Kossa
Manufacturing Buyer: Ilene Kahn
Manufacturing Manager: Alexis Heydt-Long

Pearson Prentice Hall
Pearson Education, Inc.
Upper Saddle River, NJ 07458

Printed in the United States of America

10 9 8 7 6 5 4 3 2 1

ISBN 0-13-149737-5

Pearson Education Ltd., *London*
Pearson Education Australia Pty. Ltd., *Sydney*
Pearson Education Singapore, Pte. Ltd.
Pearson Education North Asia Ltd., *Hong Kong*
Pearson Education Canada, Inc., *Toronto*
Pearson Educación de Mexico, S.A. de C.V.
Pearson Education—Japan, *Tokyo*
Pearson Education Malaysia, Pte. Ltd.

MINITAB™ MANUAL

LINDA M. MEYERS • DIANE L. BENNER
Harrisburg Area Community College

STATISTICS

THE ART AND SCIENCE OF LEARNING FROM DATA

Agresti • Franklin

Table of Contents

CHAPTER 1
INTRODUCTION TO MINITAB

INTRODUCTION: This lab is designed to introduce you to the statistical software MINITAB. During this session you will learn how to enter and exit MINITAB, how to enter data and commands, how to print information, and how to save your work for use in subsequent sessions.

BEGINNING AND ENDING A MINITAB SESSION

To start MINITAB

Double-click on the icon placed on your desktop when you installed the program.

If you didn't choose to have the icon installed, you may choose, from the taskbar, **Start > Programs > MINITAB 14 Student > MINITAB 14 Student.**

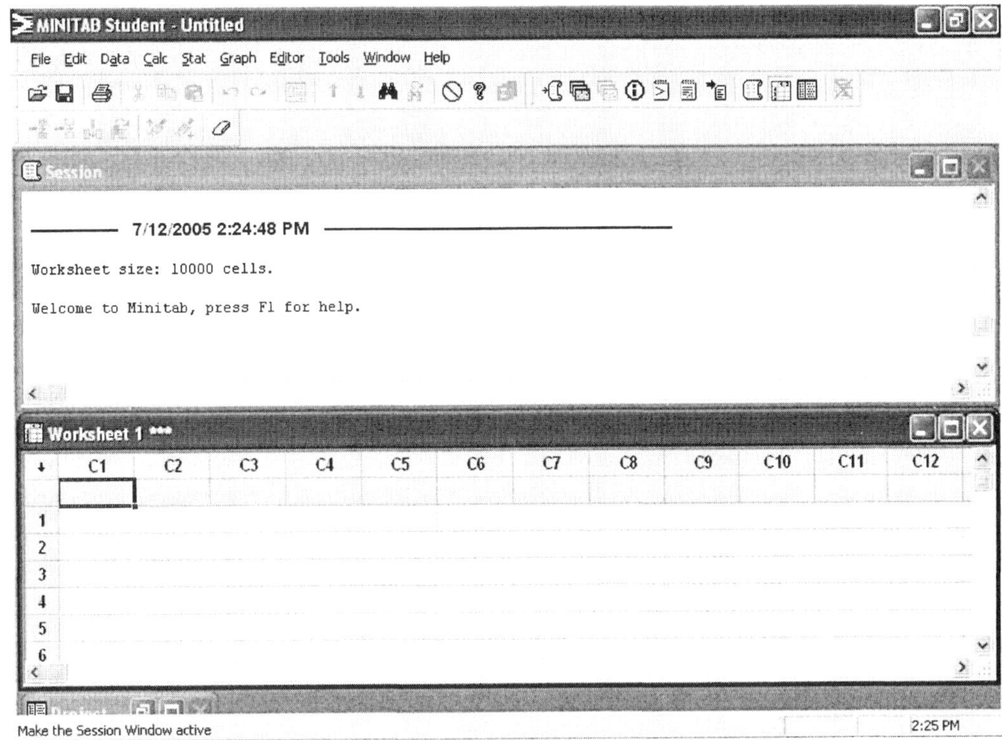

To exit MINITAB

To end a MINITAB session and exit the program, choose **File** from the menu bar and then choose **Exit**. A dialog box will appear, asking if you want to save the changes made to this worksheet. Click **Yes** or **No**.

It is also possible to exit MINITAB by clicking the X in the upper right corner of the window.

In MINITAB commands are executed with menus. When you click a menu selection, MINITAB performs an action or opens a dialog box.

MINITAB WINDOWS

The main MINITAB window opens when you first start MINITAB. You will be in a window titled "MINITAB - Untitled " within which a split window is shown; one titled "Session" and the other titled "Worksheet 1". The Session window displays text output such as tables of statistics. Data windows are where you enter, edit, and view the column data for each worksheet. Another window in the MINITAB environment that can be accessed through the Window menu is the Project Manager. The Project Manager summarizes each open worksheet. Within the Project manager, the History window records all the commands you have used. The graph window display graphs that were generated in the project.

Session Window

The data window is active when you first start MINITAB. To move to the Session Window just point the mouse to the Session Window and click. In older versions of MINITAB, whenever you issue a command from a menu, its corresponding Session command appears in the Session window. In version 14, the command will appear in the History folder within the Project Manager and will only appear in the session window if you have enabled the command language.

The Help Window in MINITAB

Information about MINITAB is stored in the computer. If you forget how to use a command or subcommand, or need general information, you can ask MINITAB for help. There are three methods for accessing Help: choose Help from the menu, select "?" from the toolbar, or press F1. It would be beneficial for you to read "How to use MINITAB Help" the first time you enter the program to help you understand the structures used in MINITAB.

Students: Practice using the HELP command by selecting the following and reading what is presented on the screen:

Choose: **Help > Help**

Select: **Index Help on**
Enter: **MEAN**

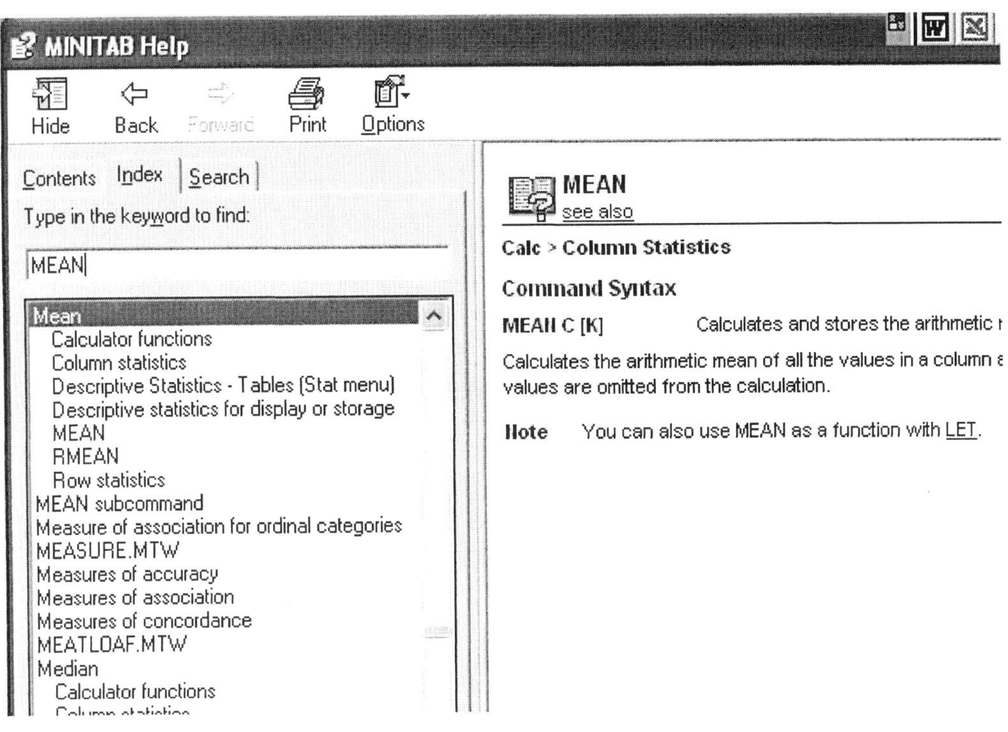

Close Help and click in the worksheet.

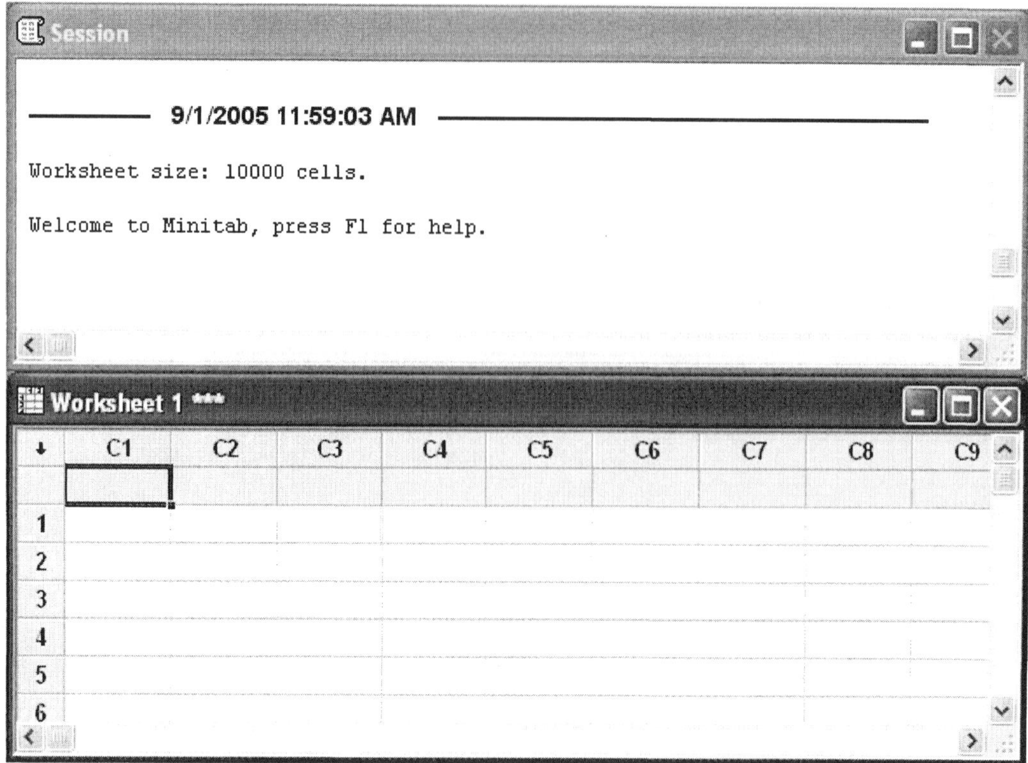

The worksheet is arranged by rows and columns. The columns C1, C2, C3, . . . , correspond to the variables in your data, the rows to observations. In general, a column contains all the data for one variable, and a row contains all the data for an individual subject or observation. You can refer to the columns as C1, C2, or by giving them descriptive names. Click into the column name cell (the blank space below the column number).

ENTERING DATA

As an example, consider a portion of the CEREAL data used in a number of future exercises:

Cereal	Sodium	Sugar	Type
Frosted Mini Wheats	0	7	A
Raisin Bran	210	12	A
All Bran	260	5	A

Now that we are in the data window, let's enter data in the first four columns. Name each column, appropriately.

Worksheet 1 ***

↓	C1-T	C2	C3	C4-T	C5	C6
	Cereal	Sodium	Sugar	Type		
1	Frosted Mini Wheats	0	7	A		
2	Raisin Bran	210	12	A		
3	All Bran	260	5	A		
4						
5						
6						

Changing a value entered

We can edit data directly in the data window. Let's suppose we had incorrectly entered the third data item in the second column. It should have been a 620. Click cell C2 row 3 to make it active. Type in the correct value and press enter. Double-clicking allows insertion of new characters without retyping the entire entry.

Suppose we had inadvertently left out a single value and we wish to enter it in a particular position. Place the cursor in the cell in which you wish to insert the new value. Right click and choose Insert Cells. A blank cell is created and the missing value can be entered . There are buttons on the taskbar that are active when you are in the data window. They can be used to insert a cell, row, or column.

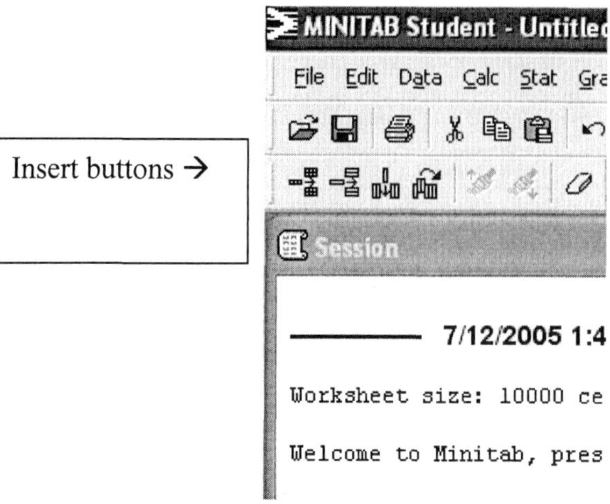

A cell can be deleted by making the cell active, then Choose: **Edit > Delete Cells** (or press the Del key).

Rows of values can also be inserted or deleted in a similar manner. The menu command to insert a row is only functional when the data window is active, and a row is active. To make a row active, click the row header (ie. the row number). An empty row will be added above the active row in the Data window and the remaining rows will be moved down.

Choose: **Editor > Insert Row**

To print your data choose **File > Print Worksheet,** make the appropriate selections and click **OK**

Suppose we wish to copy a column into another column. We can use the **COPY** command instead of reentering the data.

Choose: **Data> Copy> Copy columns to columns**
Enter: Copy from columns: **Cereal**
Select **Store Copied Data in columns** (choose from drop down arrow to select Column)
Click: **OK**

To erase an entire column we use the **ERASE** command.

Choose: **Data > Erase Variables**
Enter: Columns and constants: **select appropriate variable**
Click**: OK**

Entering Patterned Data

Suppose we wish to create a column that contains the integers 1 to 10. Although we could enter these numbers directly into the Data window by typing, there is a much easier way. Open a new worksheet by selecting **File > New > Worksheet**

Choose: **Calc > Make Patterned Data > Simple Set of Numbers**
Enter: Store patterned data in: **C1**
from first value: **1**
to last value: **10**
Click: **OK**

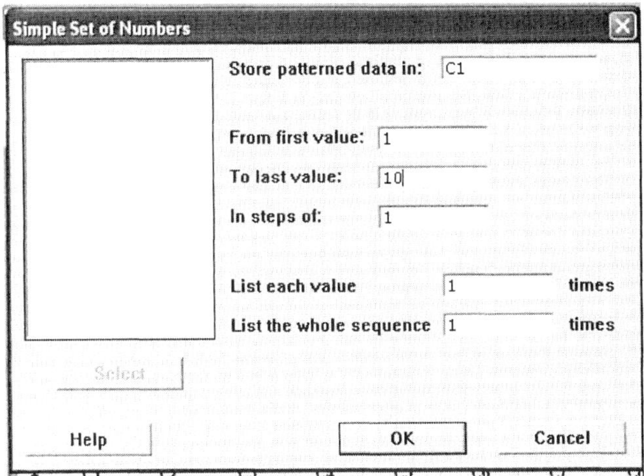

Column 1 should now contain the integers 1 through 10.

SAVING YOUR WORK

A MINITAB project contains all of your work; the data, text output from the commands, graphs, and more. When you save a project, you save all of your work at once. When you open a project, you can pick up right where you left off.
The project's many pieces can be handled individually. You can create data, graphs, and output from within MINITAB. You can also add data and graphs to the project by copying them from files. The contents of most windows can be saved and printed separately from the project, in a variety of formats. You can also *discard* a worksheet or graph, which removes the item from the project without saving it. Let's save the project and name it "Intro". Be sure to note where you are saving it.

To open, save, or close a project
To open a new project, choose **File > New**, click **Project**, and click **OK**.

To open a saved project, choose **File > Open Project**.

To save a project, choose **FILE > Save Project**.

To close a project, you must open a new project, open a saved project, or exit MINITAB.

RETRIEVING A FILE

To retrieve the project that we just saved:

Choose: **File > Open Project**

Click: Look in drop-down list arrow
Locate the file
Double-click: INTRO.MPJ
Click: **OPEN**

The data window now displays the test data you saved previously.

A CD ROM accompanies the Agresti/Franklin text, containing data sets for selected problems. They are all saved as MINITAB worksheets in the MINITAB folder.

Let's do exercise 1.19, which uses data from the CD ROM.

Click **File>Open Worksheet**

You must tell MINITAB where the file is located, usually the D drive.

Click the **appropriate drive > MINITAB > fla_student_survey.mtw**

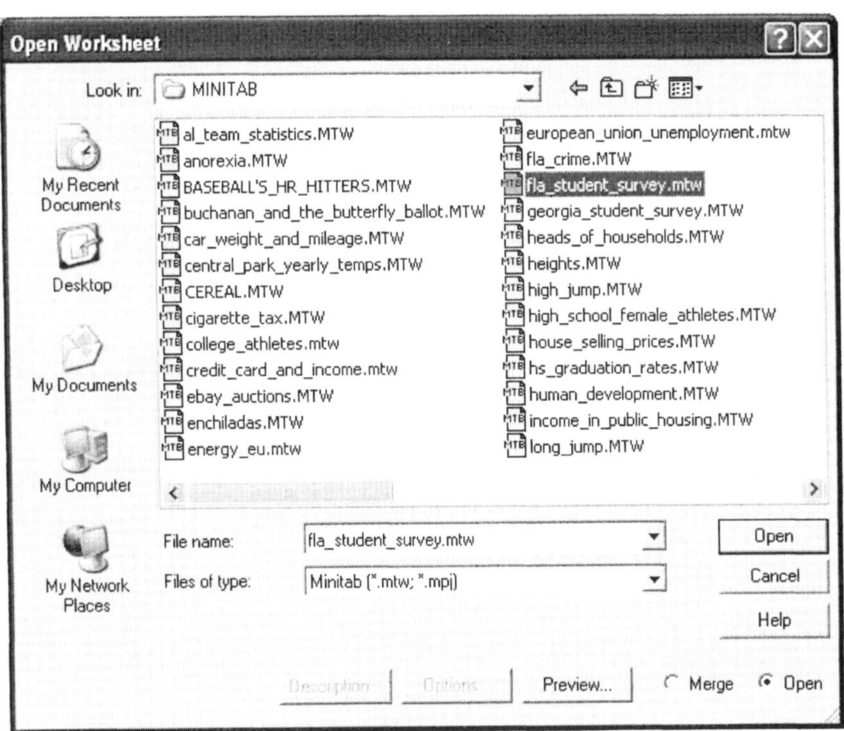

florida_student_survey.mtw ***

↓	C1-T subject	C2-T gender	C3 age	C4 high_sch_GPA	C5 college_GPA	C6 distance_home	C7 distance_residence	C8 TV	s
51	51	m	32	3.0	3.0	2000	5.00	5.0	
52	52	f	41	4.0	4.0	0	8.00	8.0	
53	53	f	29	3.0	3.9	300	3.70	2.0	
54	54	f	50	3.5	3.8	6	6.00	7.0	
55	55	f	22	3.4	3.7	80	7.00	10.0	
56	56	f	23	3.6	3.2	375	1.50	5.0	

Chapter 2 GRAPHIC PRESENTATIONS OF DATA

Graphically representing data is one of the most helpful ways to become acquainted with the sample data. In this lab you will use MINITAB to present data graphically. There are several ways to display a picture of the data. These graphical displays help us get acquainted with the data and to begin to get a feel for how the data is distributed and arranged. To see what is available to you, use the menu bar to select **Graph**. Note the different types of graphs that are listed there. We will use the menu bar to make our selections.

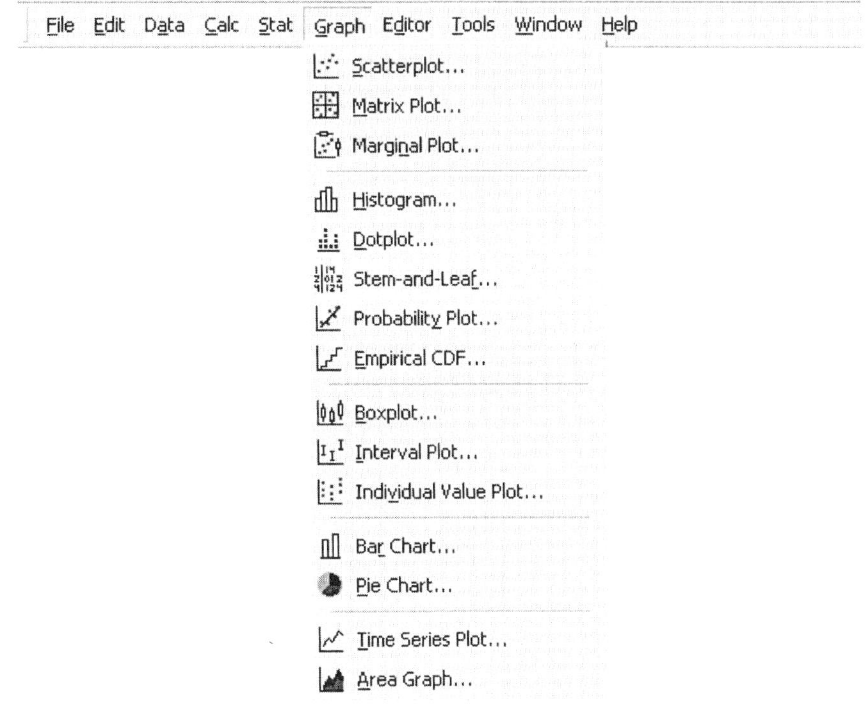

Chapter 2 Example 3 - How Much Electricity Comes From Renewable Energy Sources? Constructing a Pie Chart

Copy the data from the text into the worksheet.

Worksheet 1 ***

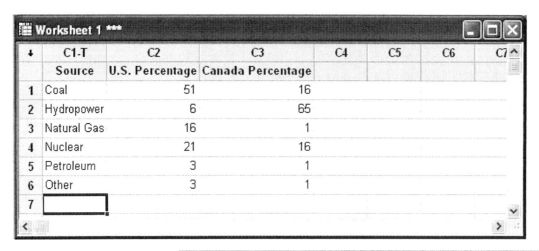

↓	C1-T	C2	C3	C4	C5	C6	C7
	Source	U.S. Percentage	Canada Percentage				
1	Coal	51	16				
2	Hydropower	6	65				
3	Natural Gas	16	1				
4	Nuclear	21	16				
5	Petroleum	3	1				
6	Other	3	1				
7							

Choose:
Graph > Pie Chart

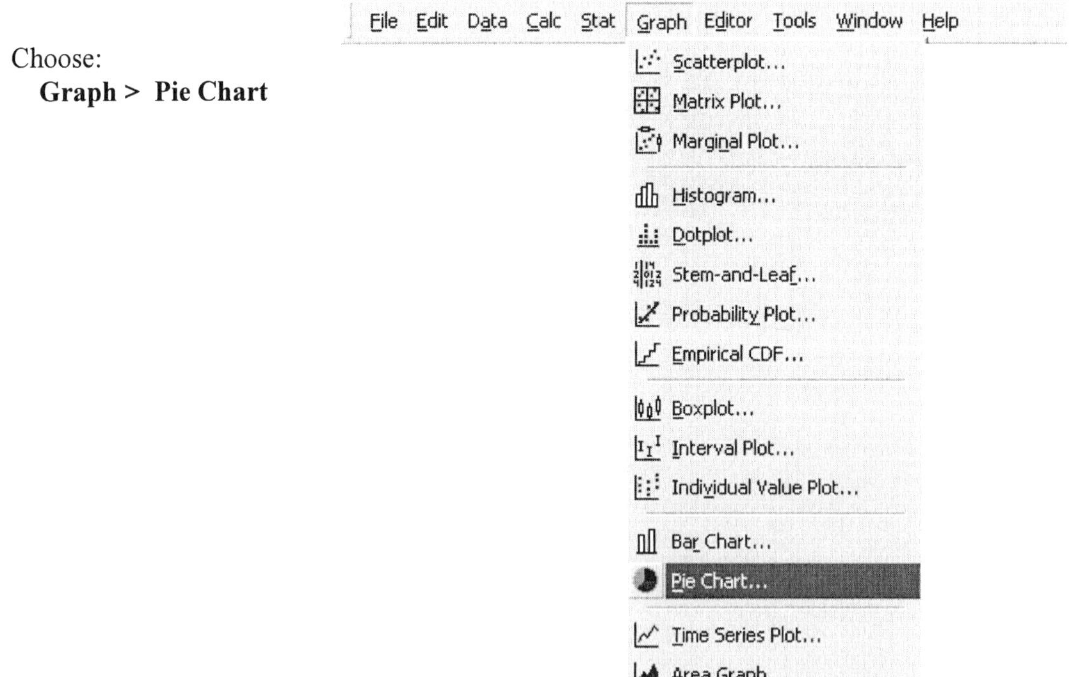

Select: **Chart values from a table**
Enter: Catagorical variable: **Source**
Summary variable: **US percentages**
Choose **labels**: Enter titles
Click: **OK**

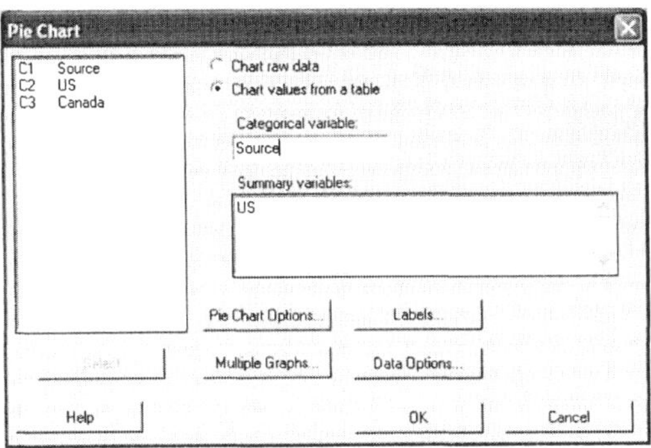

Use the following menu commands to get the Bar chart:

Choose: **Graph > Bar Chart**

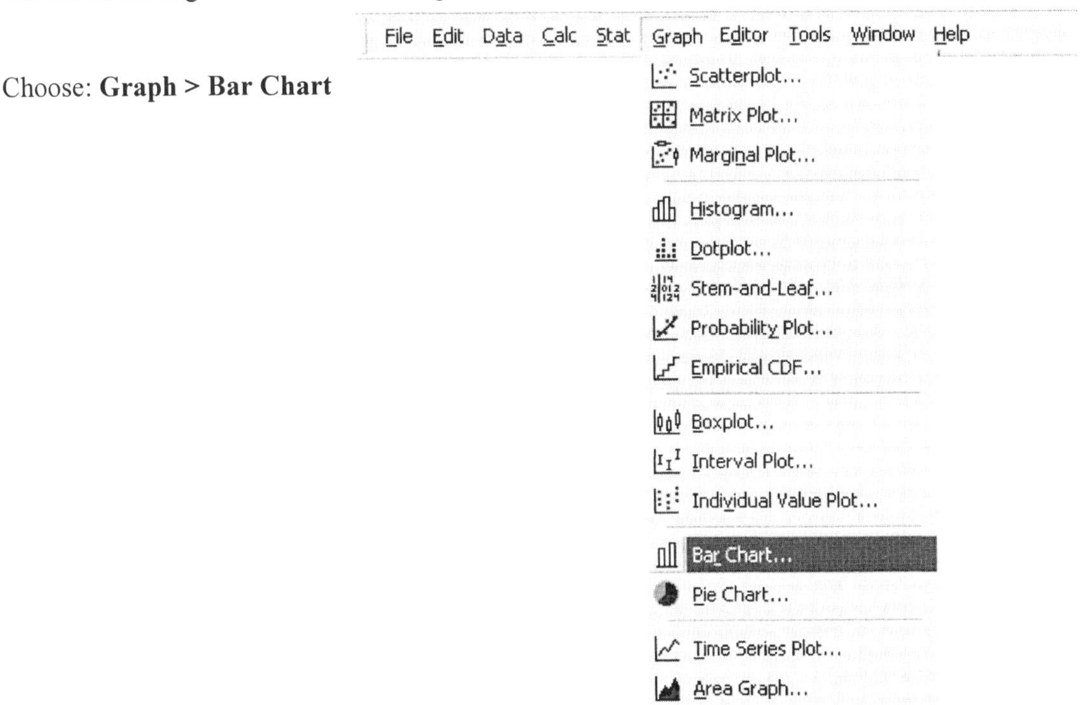

Choose: **Simple > OK**
Select: Bars represent:
Values from a table
Select: **Simple**
Click: **OK**

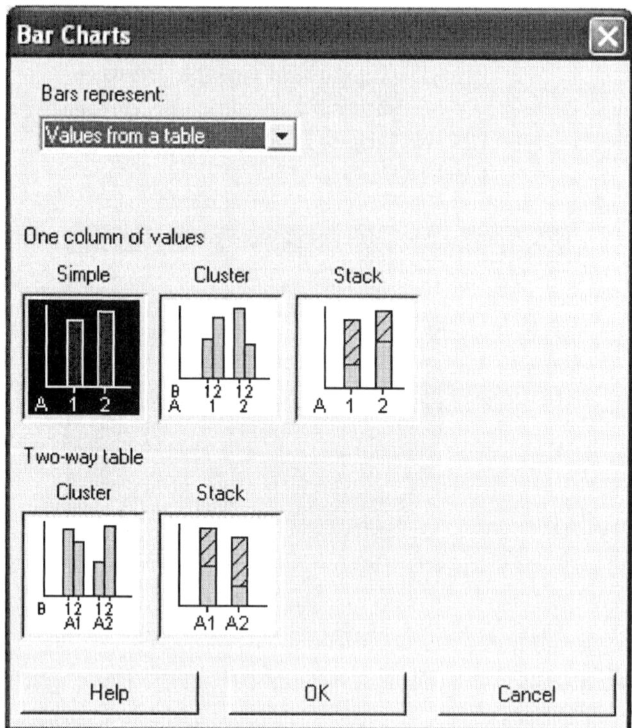

Enter: Graph variables: **C2**
Enter: Categorical variables: **C1**
Choose: **Labels**
Enter: **an appropriate title**
Click: **OK**

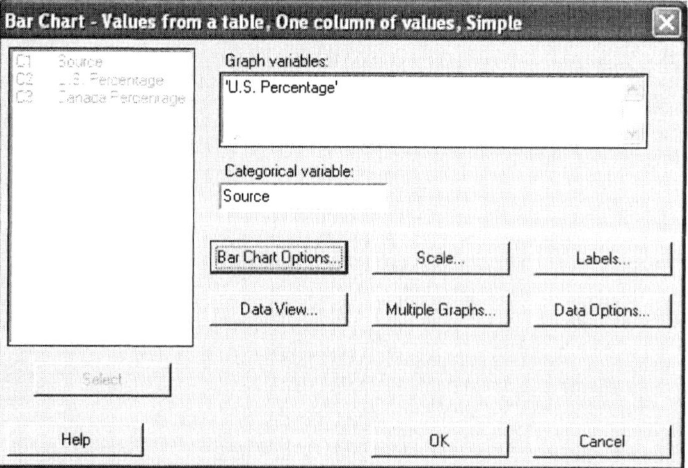

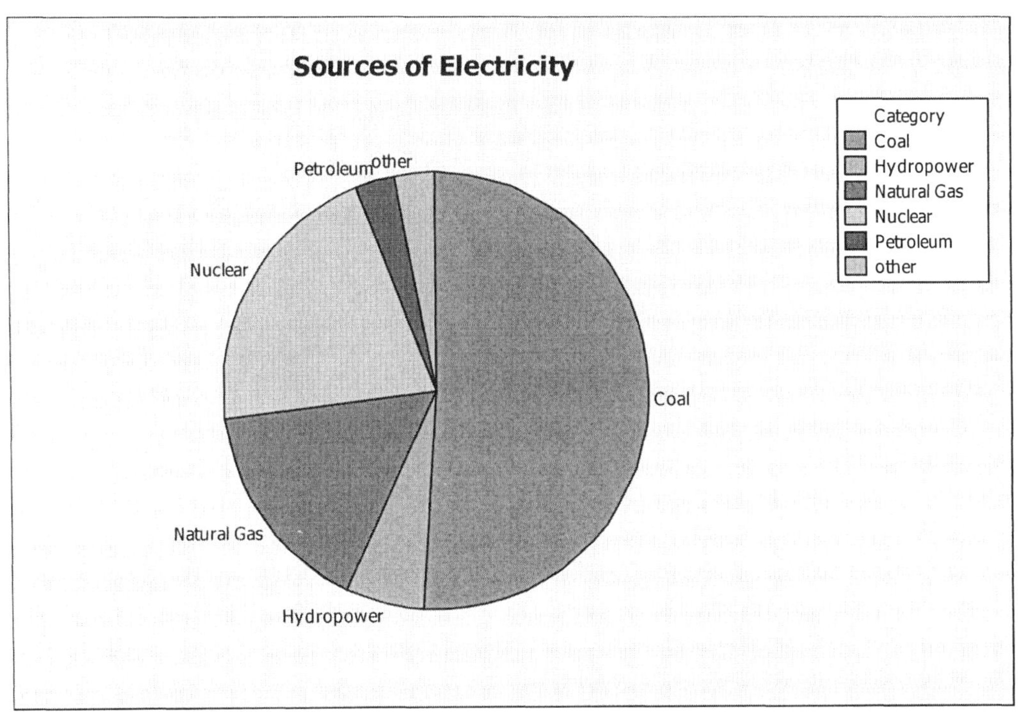
Sources of Electricity
Category
Coal
Hydropower
Natural Gas
Nuclear
Petroleum
other
Petroleum
other
Nuclear
Coal
Natural Gas
Hydropower

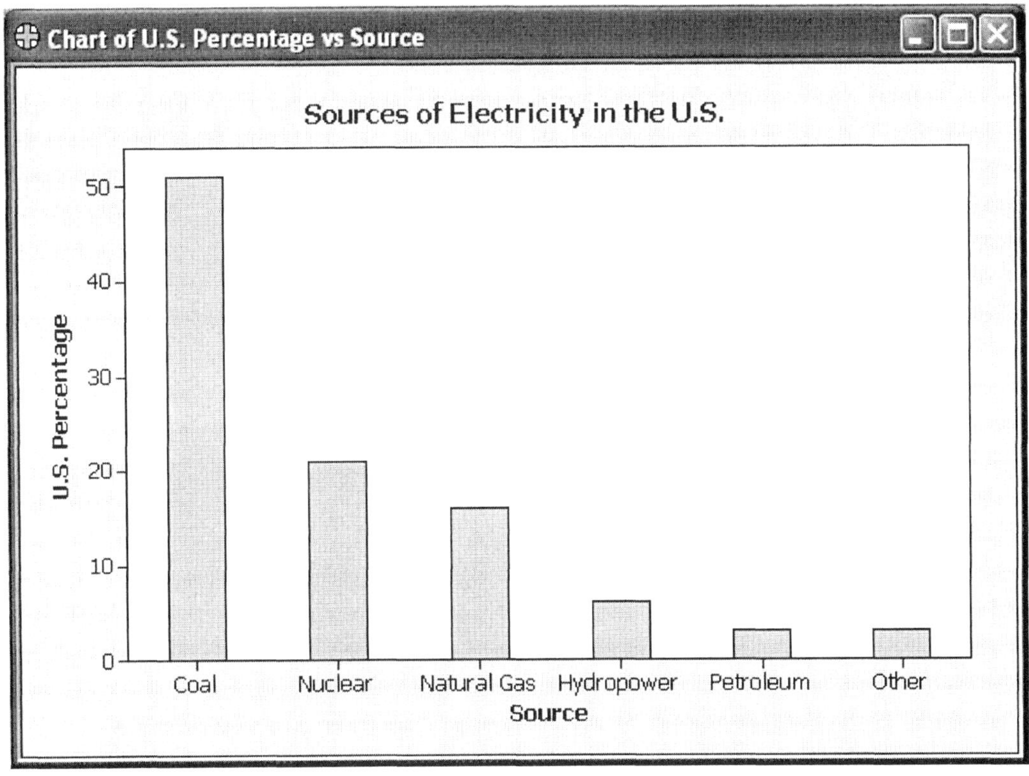
Chart of U.S. Percentage vs Source
Sources of Electricity in the U.S.
U.S. Percentage
50
40
30
20
10
0
Coal
Nuclear
Natural Gas
Hydropower
Petroleum
Other
Source

Chapter 2 Exercise 2.11 Weather Stations
Constructing a Pie Chart

Using the data shown in the pie chart in the text:

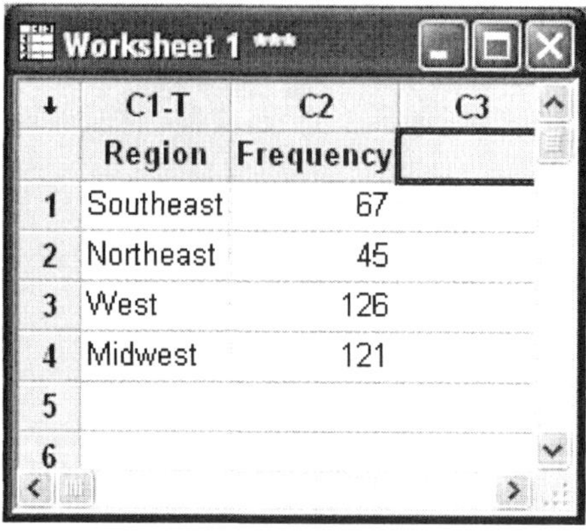

↓	C1-T	C2	C3
	Region	Frequency	
1	Southeast	67	
2	Northeast	45	
3	West	126	
4	Midwest	121	
5			
6			

Choose: **Graph > Pie Chart**

Select**: Chart values from a table**
Enter: Categorical variable: **Region**
Summary variable: **Frequency**
Choose: **Labels**
Enter: **appropriate title**
Click: **OK**

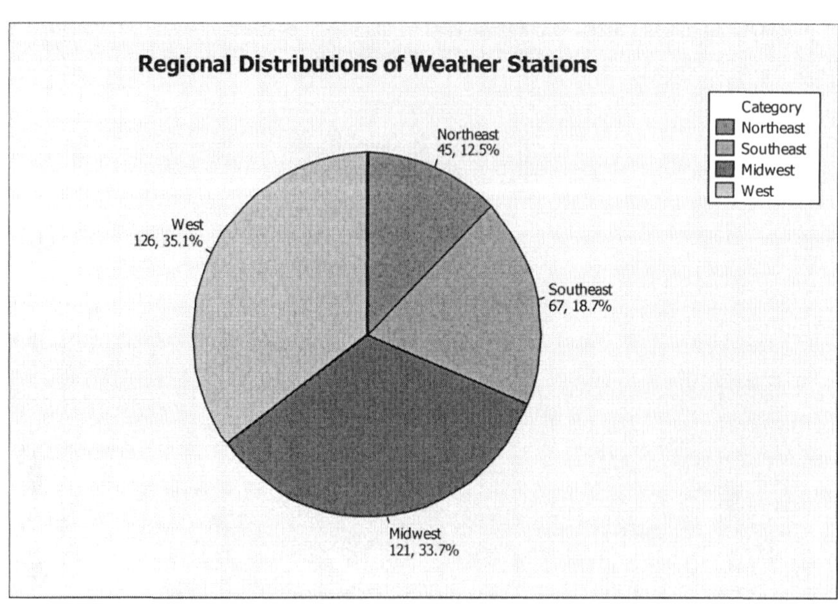

Note that by checking the boxes for category name, frequency and percent (on the Slice Labels tab), we get labels on each slice. If you compare the pie chart to the bar graph below, in which is it easier to identify the mode?

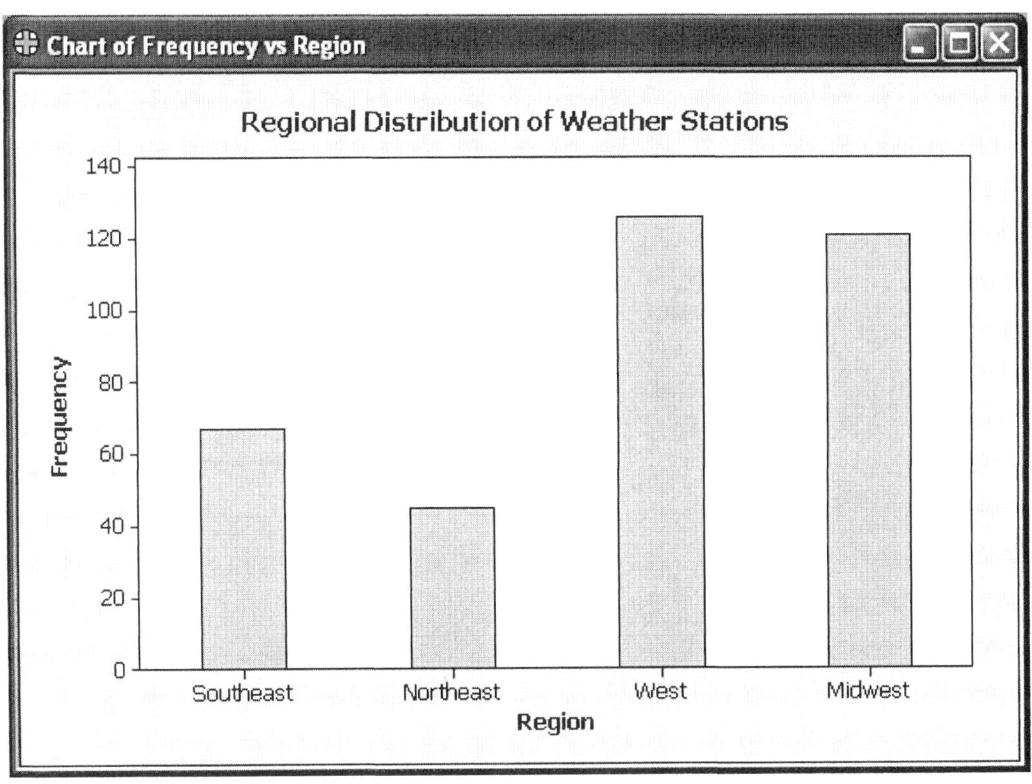

Chapter 2 Exercise 2.13 Shark attacks worldwide
Bar Chart and Pareto Chart

Enter the data from table 2.1 into the worksheet.

Worksheet 1 ***

↓	C1-T	C2	C3	C
	Region	Frequency		
1	Florida	289		
2	Hawaii	44		
3	California	34		
4	Australia	44		
5	Brazil	55		
6	South Africa	64		
7	Reunion Island	12		
8	New Zealand	17		
9	Japan	10		
10	Hong Kong	6		
11	Other	160		

Sort the data alphabetically based on region and store the sorted columns in original columns.

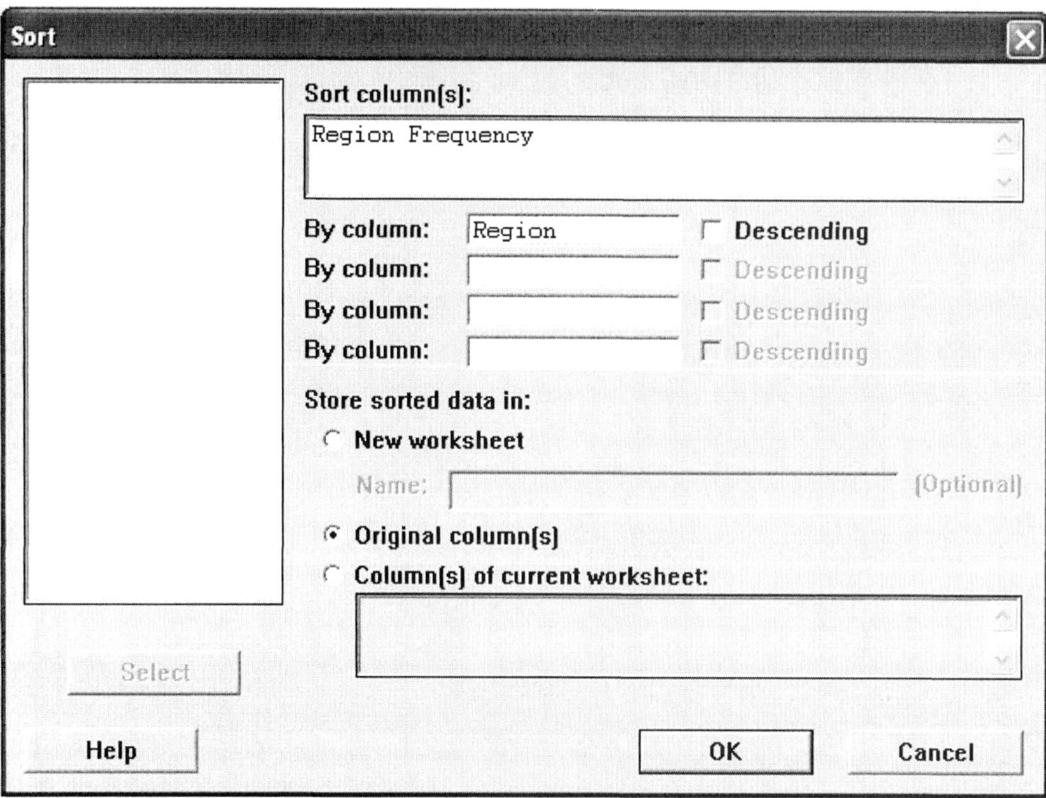

To create the bar chart

Choose: **Graph > Bar Chart**
Select: Bars represent: **Values from a table**
Select: **Simple > OK**
  Enter: Graph variables: **C2**
  Enter: Catagorical variables: **C1**
  Choose: **Labels**
    Enter: **an appropriate title**
    Click: **OK**

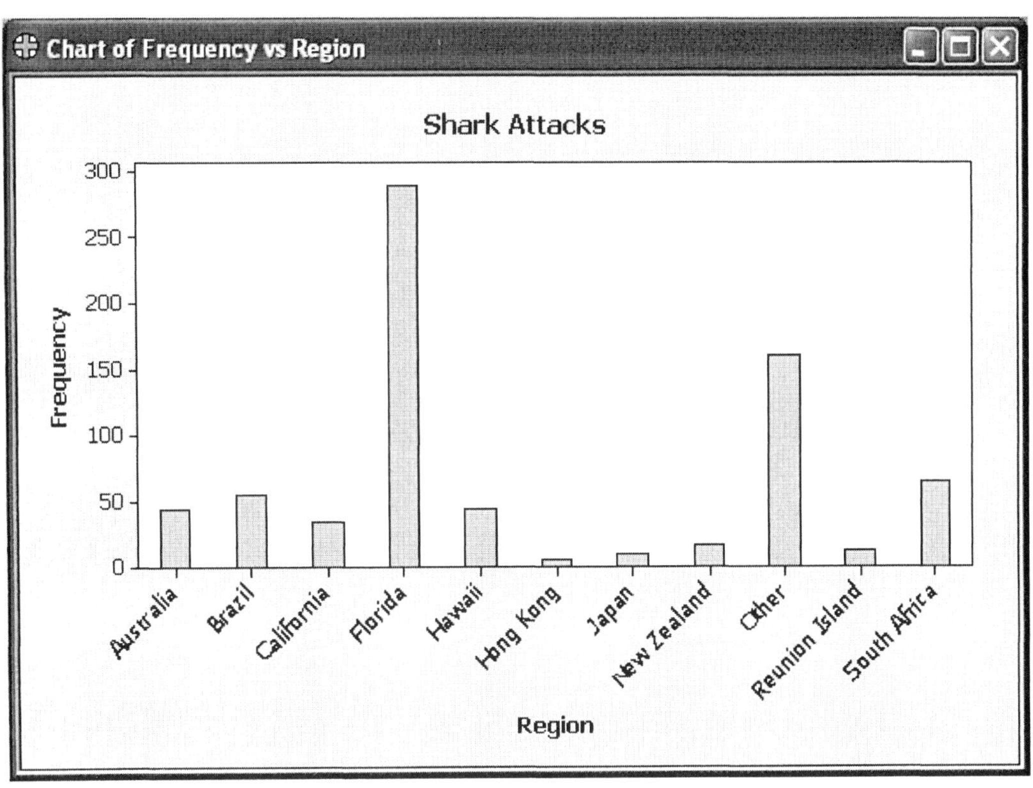

To obtain a Pareto Chart for the same data, use the following menu choices

Choose: **Stat > Quality Tools > Pareto Chart**
Click: Chart defects table
Enter: Labels in: **C1**
Frequencies in: **C2**
Combine all defects after **99** % into one bar
Options: Title: **Shark Attacks**
Click: **OK** **OK**

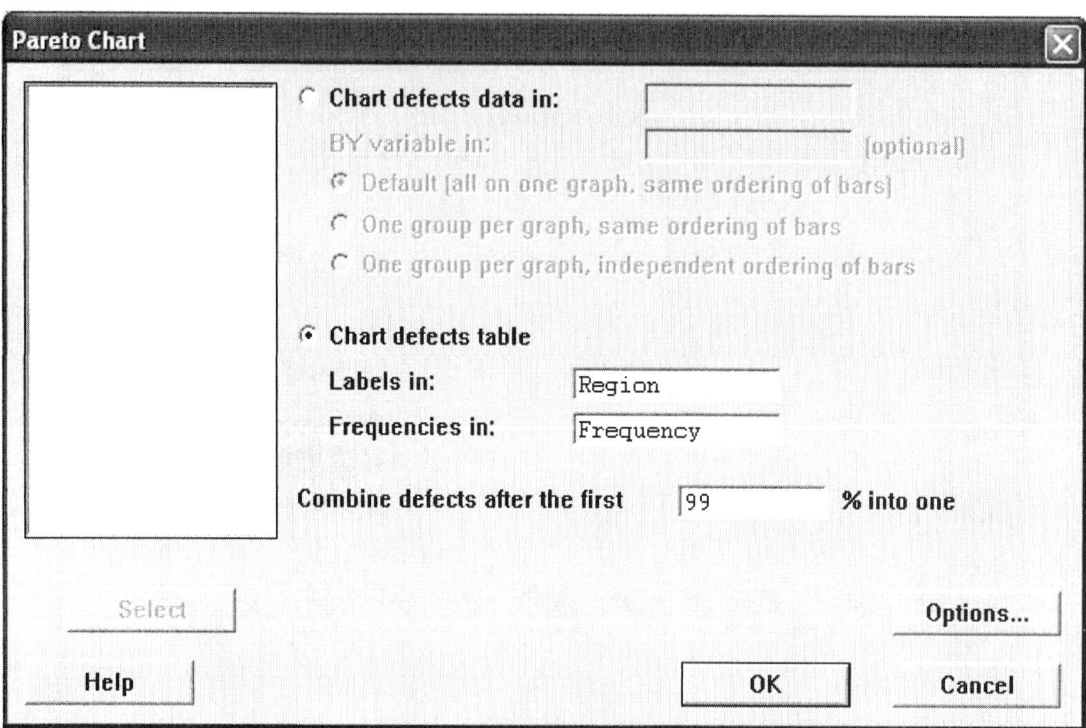
Pareto Chart
Chart defects data in:
BY variable in:
[optional]
Default [all on one graph, same ordering of bars]
One group per graph, same ordering of bars
One group per graph, independent ordering of bars
Chart defects table
Labels in:
Region
Frequencies in:
Frequency
Combine defects after the first
99
% into one
Select
Options...
Help
OK
Cancel

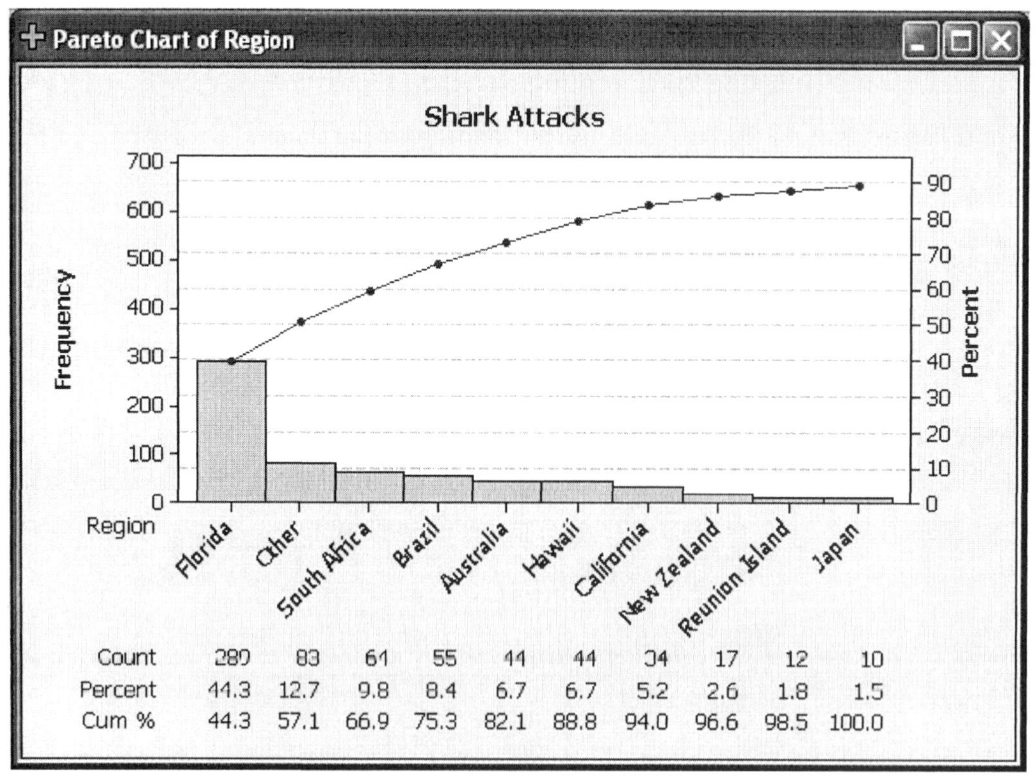
Pareto Chart of Region
Shark Attacks
Frequency
Percent
Region
Florida
Other
South Africa
Brazil
Australia
Hawaii
California
New Zealand
Reunion Island
Japan
Count 289 83 64 55 44 44 34 17 12 10
Percent 44.3 12.7 9.8 8.4 6.7 6.7 5.2 2.6 1.8 1.5
Cum % 44.3 57.1 66.9 75.3 82.1 88.8 94.0 96.6 98.5 100.0

Chapter 2 Example 4 Exploring the Health Value of Cereals
Constructing a Dot Plot

Dot plots are a quick and efficient way to get a preliminary understanding of the distribution of your data. It results in a picture of the data as well as sorts the data into numerical order. Open the worksheet **CEREAL** from the MINITAB folder of the data disk

Choose: **Graph > Dotplot**

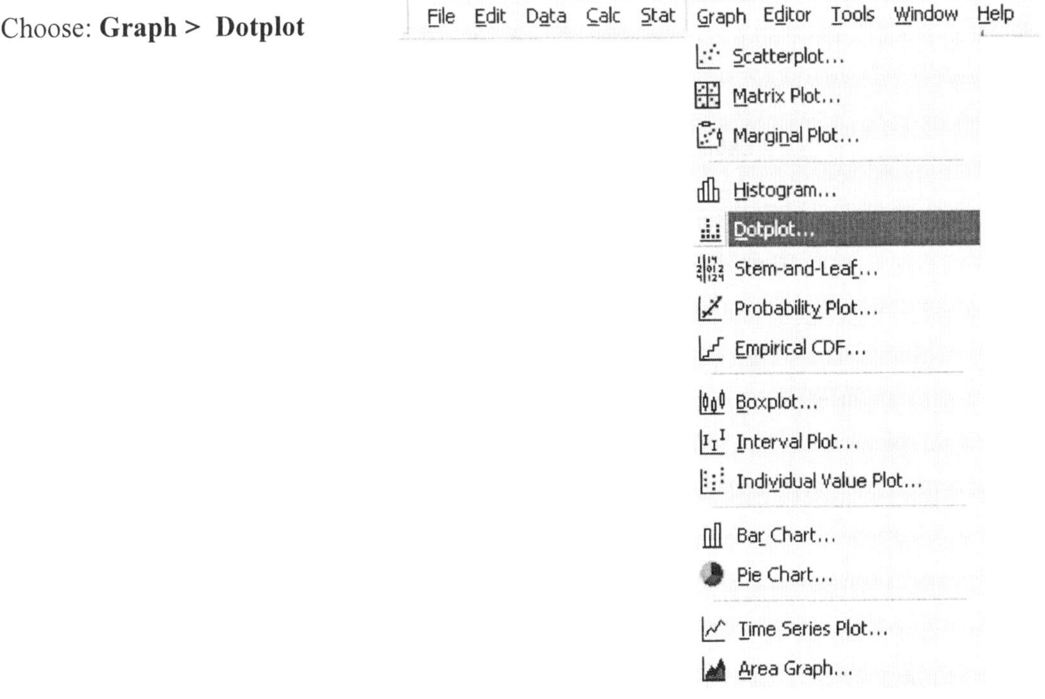

Choose: **One Y/Simple**
Click: **OK**

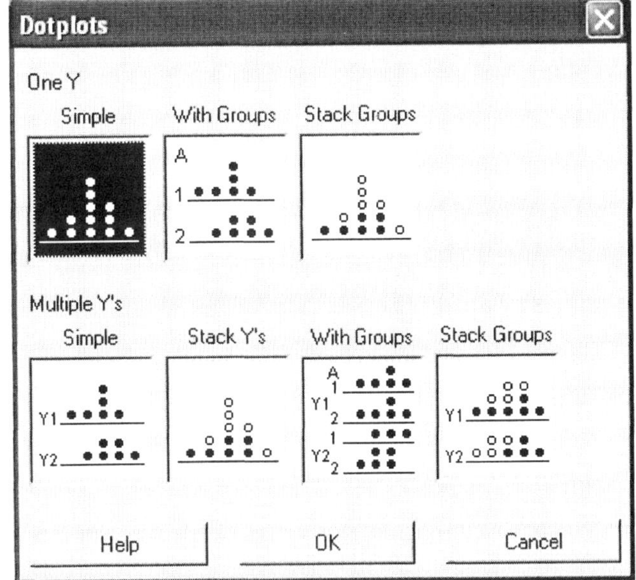

Enter: Graph variables: **C2**
Click: **OK**

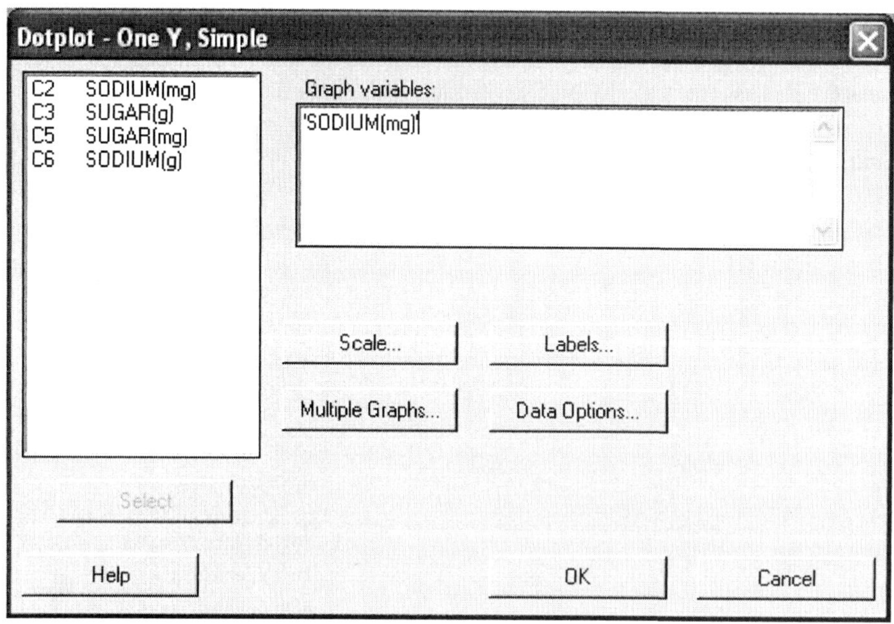

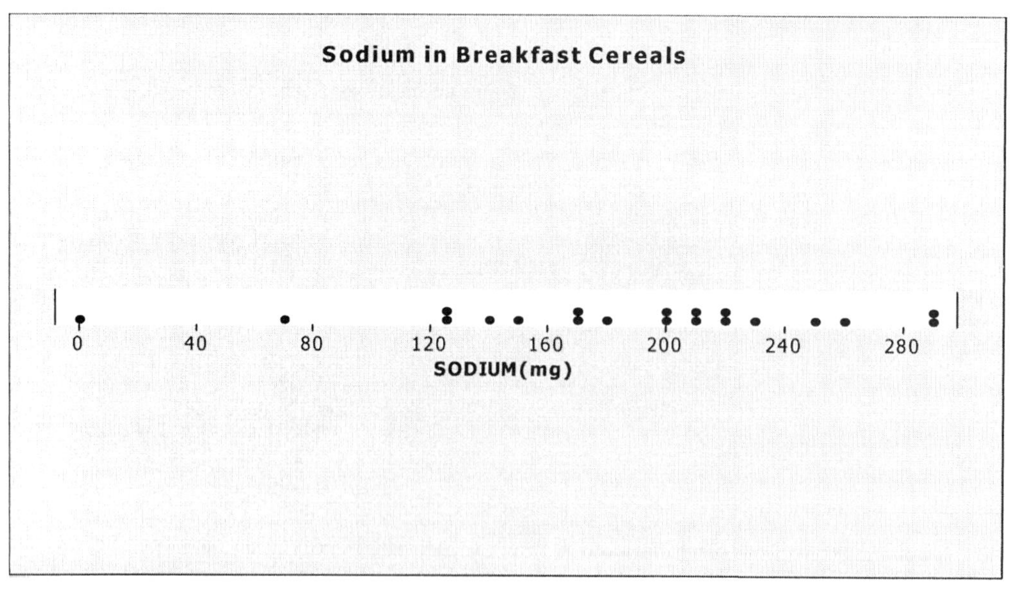

Chapter 2 Exercise 2.14 Sugar Dot Plot
Constructing a Dot Plot

Open the worksheet **CEREAL** from the MINITAB folder of the data disk
Choose: **Graph > Dotplot**
Choose: **One Y/Simple > OK**
Enter: Graph variables: **C3**
Click: **OK**

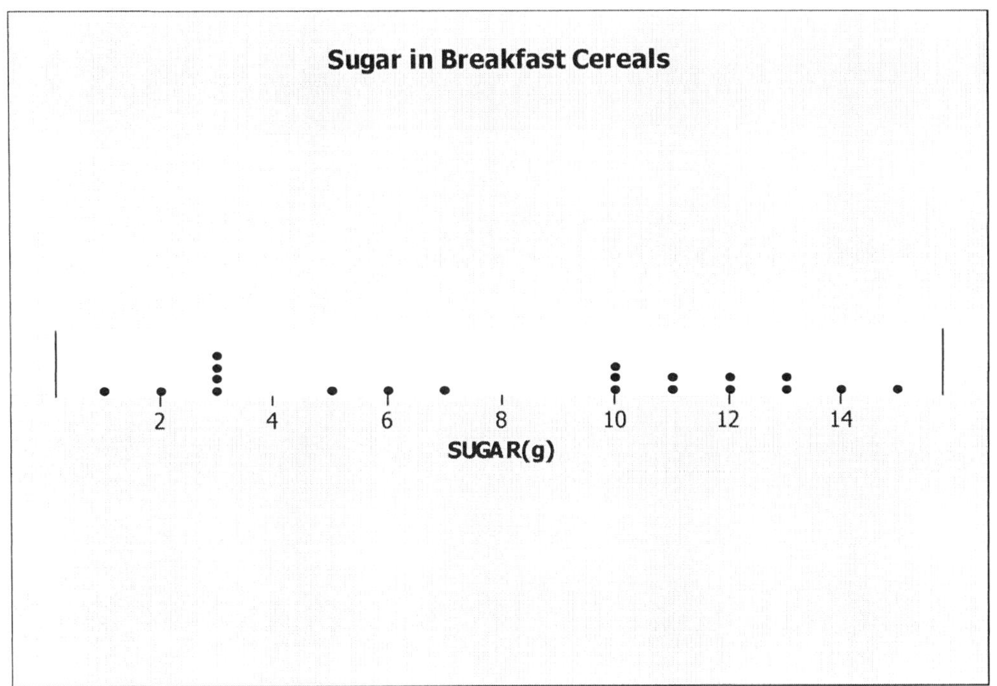

Chapter 2 Example 5 Exploring the Health Value of Cereals
Stem and Leaf Plots

To illustrate the commands necessary to construct a stem-and-leaf display, let's use the data from the **CEREAL** worksheet, sodium values in cereals.

Choose: **Graph >Stem-and-Leaf**

File Edit Data Calc Stat Graph Editor Tools Window Help

Scatterplot...
Matrix Plot...
Marginal Plot...
Histogram...
Dotplot...
Stem-and-Leaf...
Probability Plot...
Empirical CDF...
Boxplot...
Interval Plot...
Individual Value Plot...
Bar Chart...
Pie Chart...
Time Series Plot...
Area Graph...

Enter: Variable: **C2**
Uncheck **"Trim Outliers"**
Click **OK**

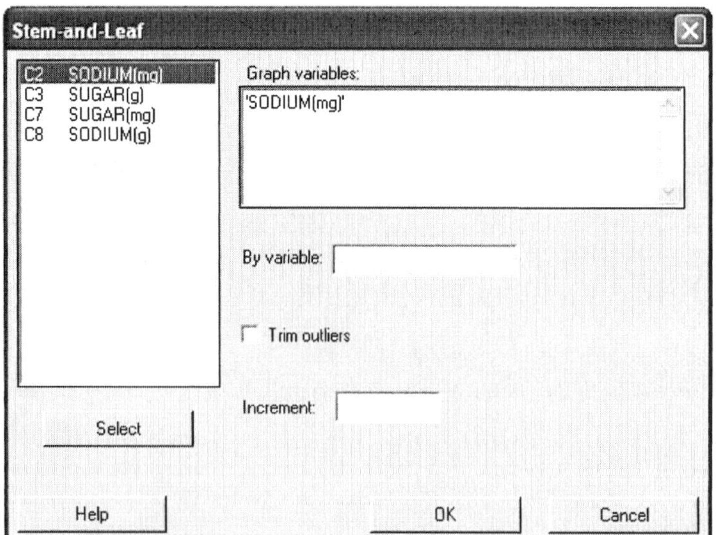

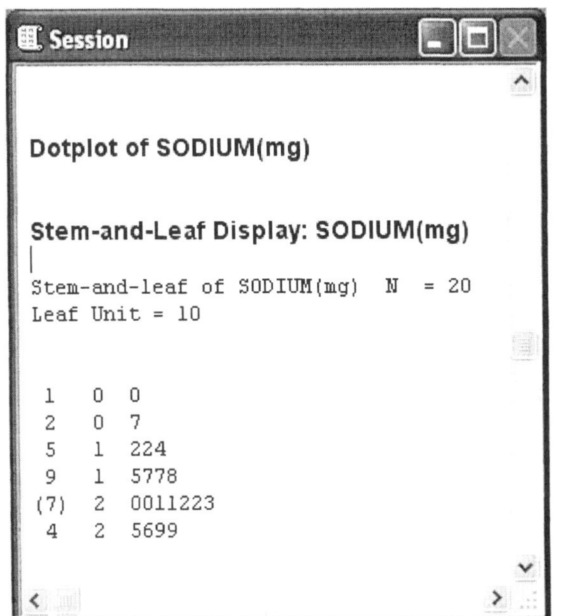

If we were to comment on the shape of the distribution, we can clearly see that it is skewed left.

We allowed MINITAB to choose the increment. We may choose to specify the increment ourselves. Try various values for "increment". Which Stem-and-Leaf display do you prefer?

Chapter 2 Example 7 Exploring the Health Value of Cereals
Constructing a Histogram

Open the **CEREAL** worksheet from the MINITAB folder of the data disk.

Choose: **Graph > Histogram**

Choose: **Simple > OK**

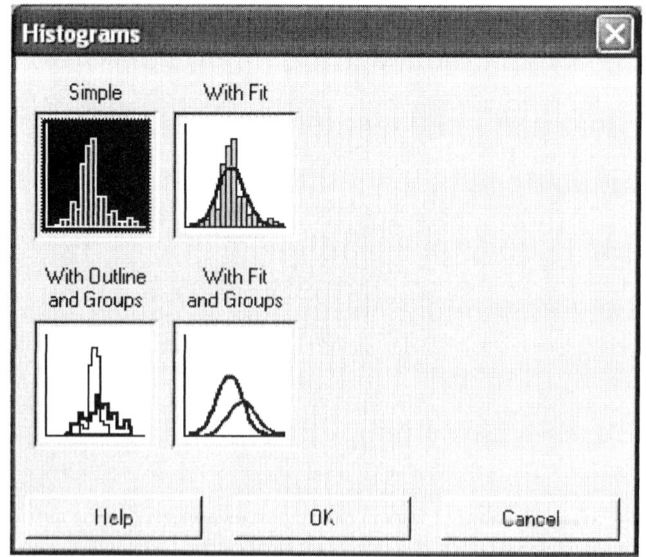

Enter: Graph variables: **C2**

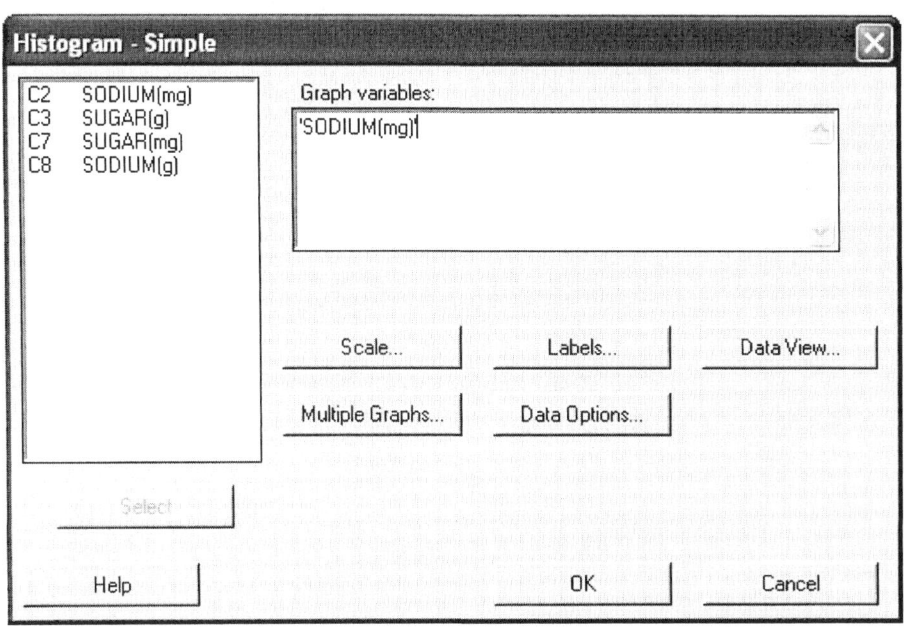

Select: **Scale > Y-Scale**
Select: either **Frequency**
or **Percent**
Click: **OK OK**

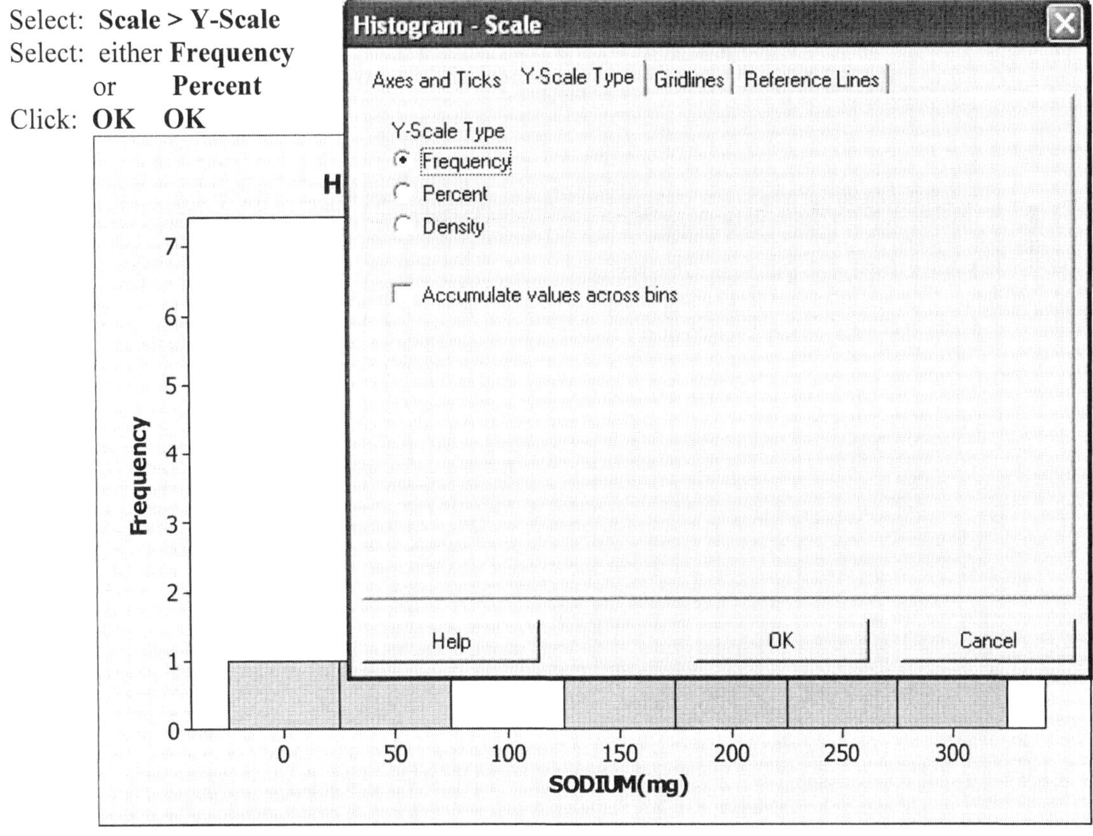

Chapter 2 Exercise 2.20 Sugar plots
Dot Plot, Stem-and-Leaf Plot and Histogram

Open the **CEREAL** worksheet from the MINITAB folder of the data disk.

Making the menu selections as previously demonstrated, we get the following graphical displays.

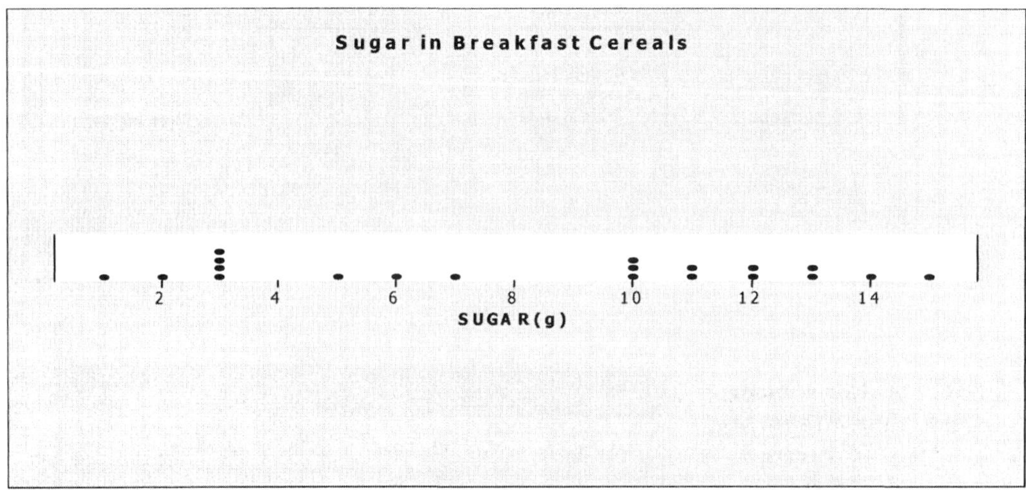

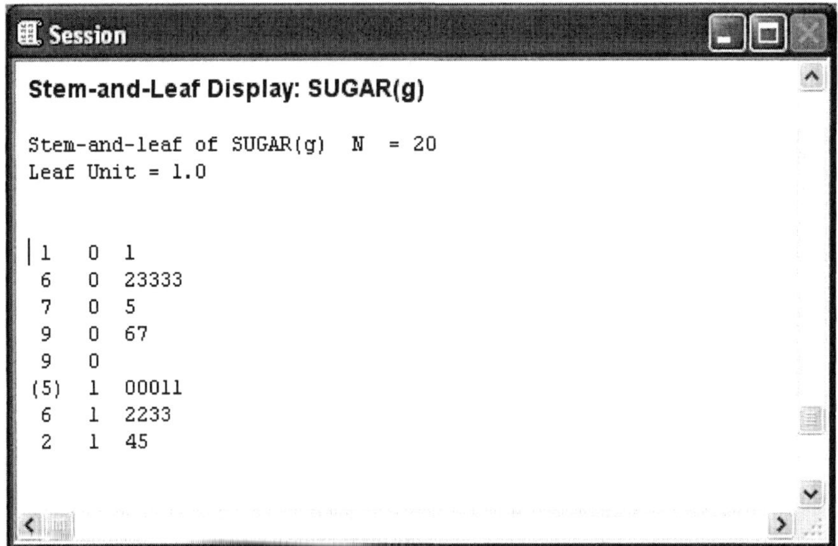

Session

Stem-and-Leaf Display: SUGAR(g)

```
Stem-and-leaf of SUGAR(g)  N  = 20
Leaf Unit = 1.0


 1    0  1
 6    0  23333
 7    0  5
 9    0  67
 9    0
(5)   1  00011
 6    1  2233
 2    1  45
```

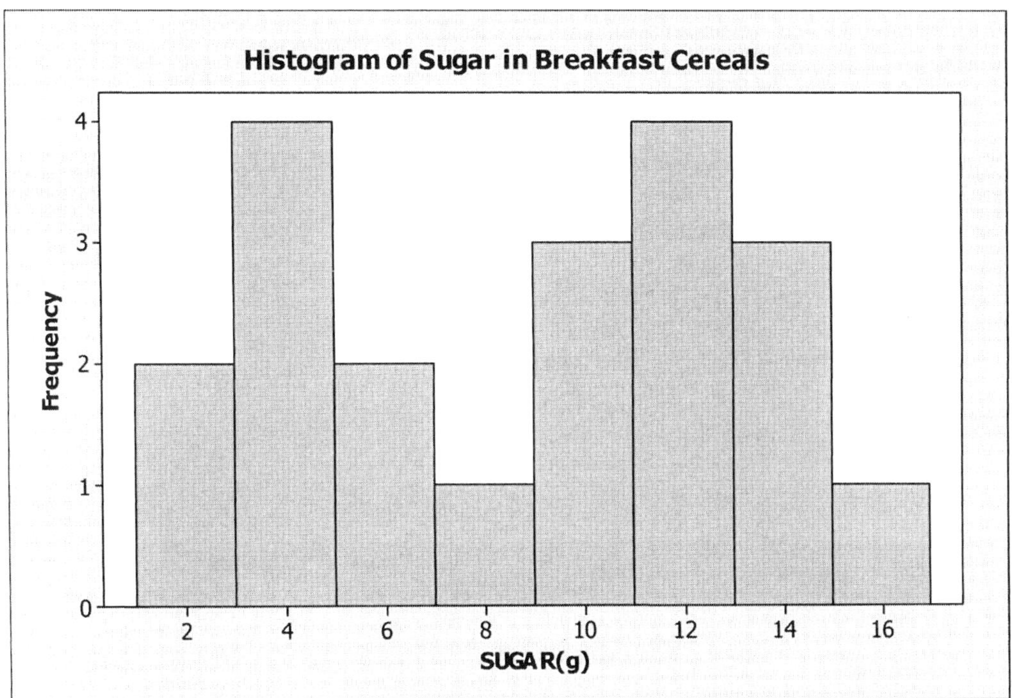

How would you explain each of these?

Chapter 2 Exercise 2.103a - Temperatures in Central Park
Constructing a Histogram

Open the worksheet **central_park_yearly_ temps** from the MINITAB folder of the data disk.

This time we will make the selection "With Fit"

Choose: **Graph > Histogram > With Fit > OK**
Select: **Label**
    Enter: Title: **Central Park Annual Temperatures**
Select: **Frequency**
Click: **OK OK**

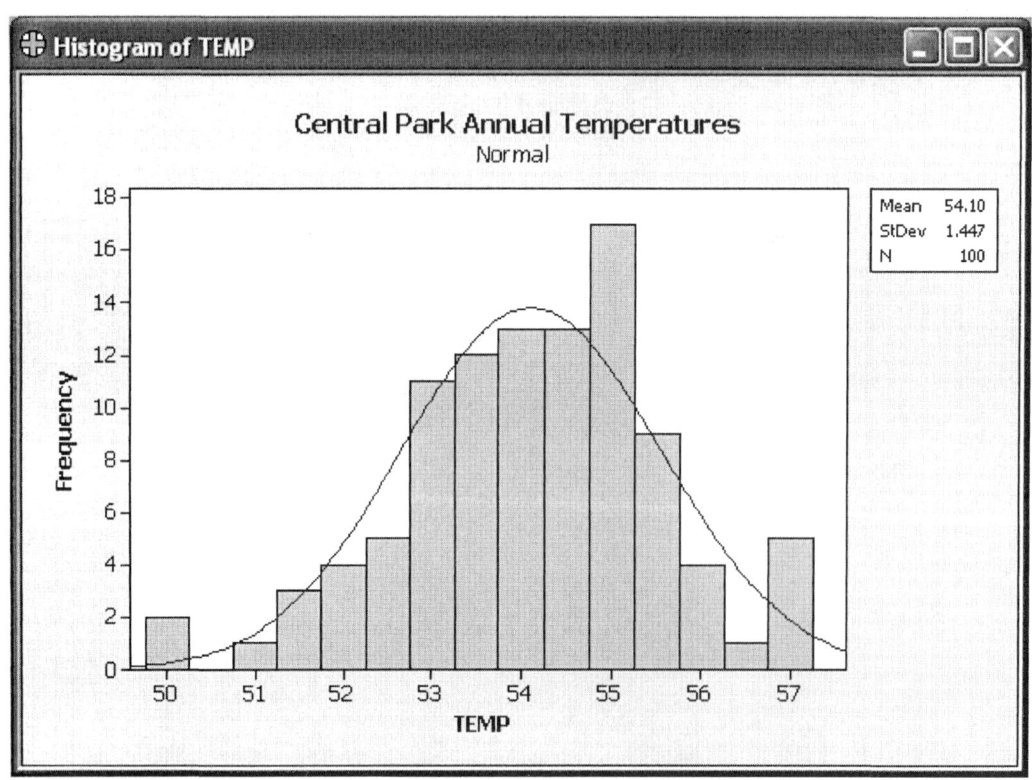

Chapter 2 Example 9 – Is There a Trend Toward Warming in New York City? Constructing a Time Plot

For some variables, observations occur over a period of time. It is useful for showing trends in the data.

Open the worksheet **central_park_yearly_ temps** from the MINITAB folder of the data disk.

Select: **Graph > Time Series Plot**

Select: **Simple**
Click: **OK**

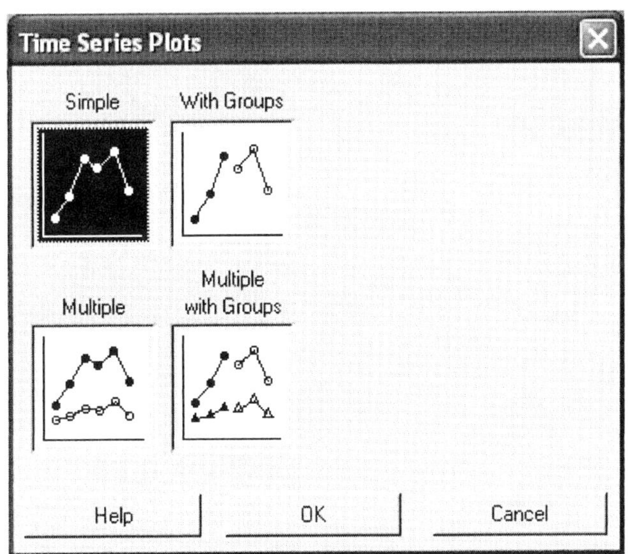

Enter:
Series: **C1**

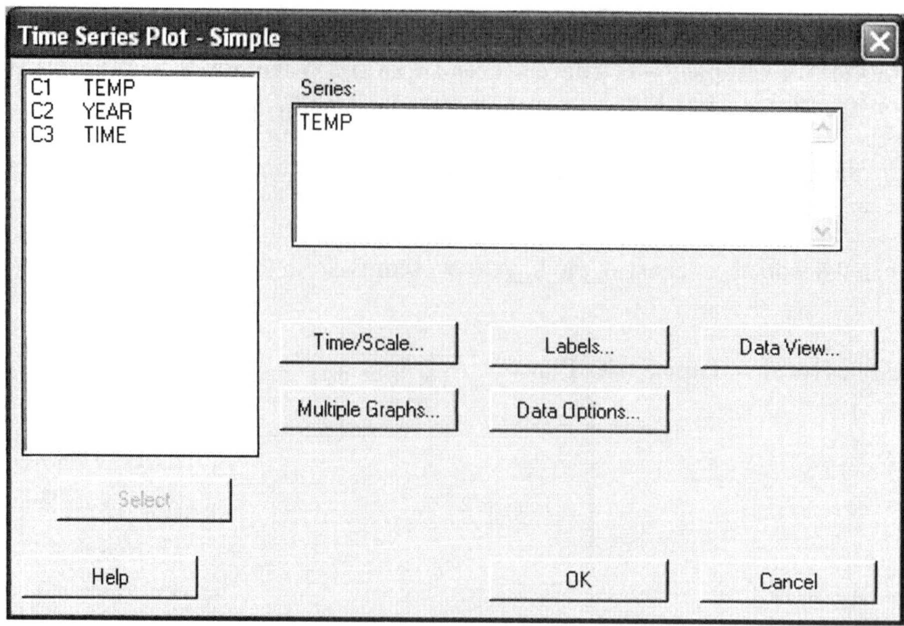

Click **Time/Scale**… **> Calendar > Year > OK**

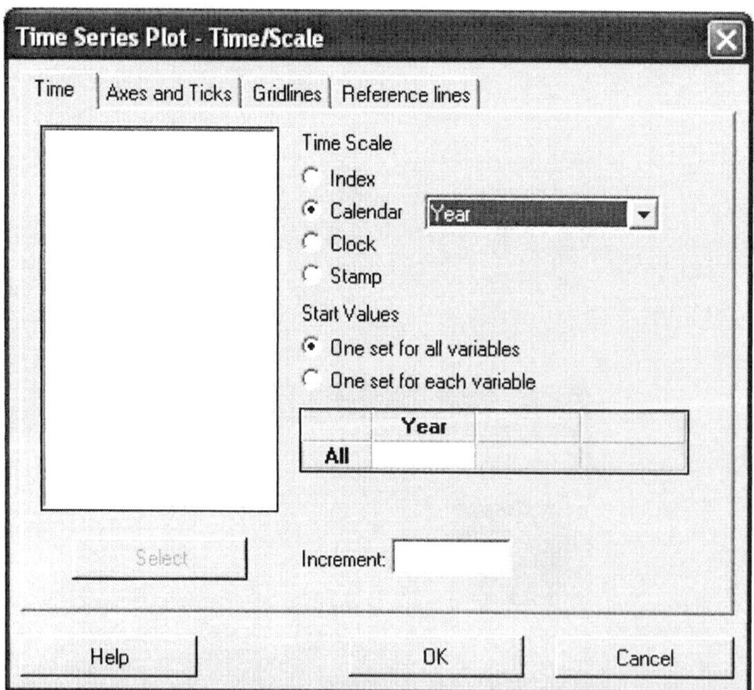

Click **Labels**
Enter: Title: **an appropriate title**
Click: **OK**

Time Series Plot - Labels

Titles/Footnotes | Data Labels

Title:
Average Yearly Temperatures in Central Park
Subtitle 1:
Subtitle 2:
Footnote 1:
Footnote 2:

Help OK Cancel

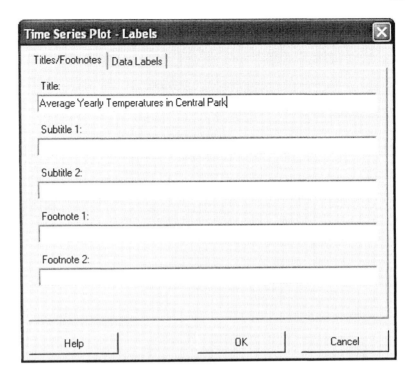

Click **Data View**
Select: **Symbols**
Connect Line
Click: **OK OK**

Select: **Smoother** tab

Select: Smoother: **Lowess**

(This adds a general trend line to the time series plot.)

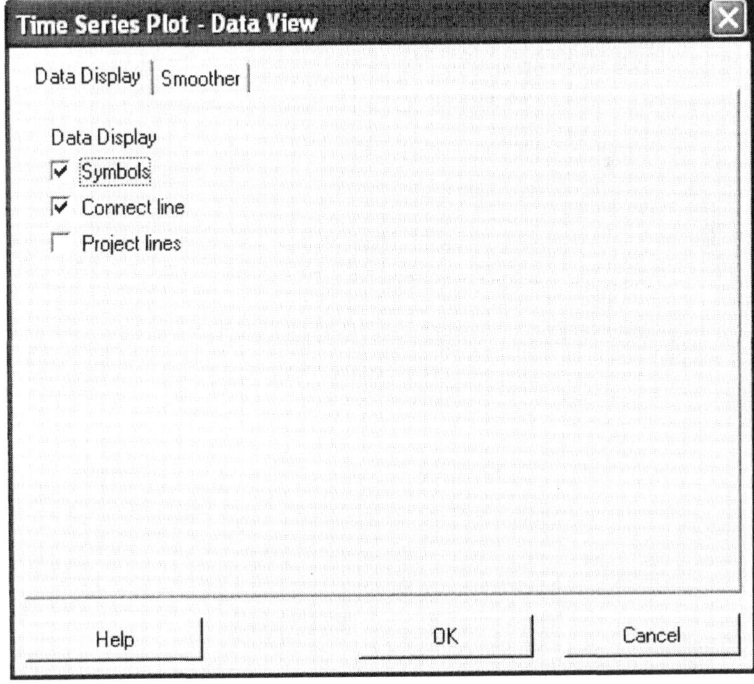

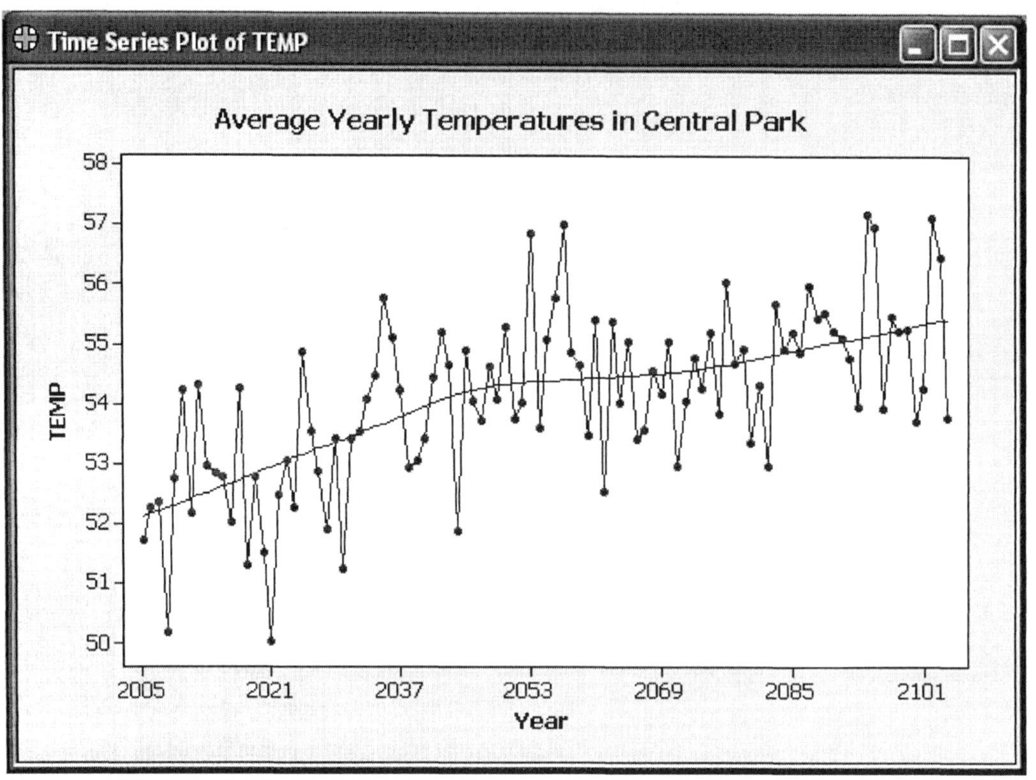

Chapter 2 Exercise 2.27 Warming in Newnan, Georgia?
Constructing a time plot

Open the worksheet **newnan_ga_temps** from the MINITAB folder of the data disk

Select: **Graph > Time Series Plot > Simple > OK**

Enter: Series: **C1**

Click **Time/Scale… > Calendar > Year >** Start Value **One set for all variables > 1901**

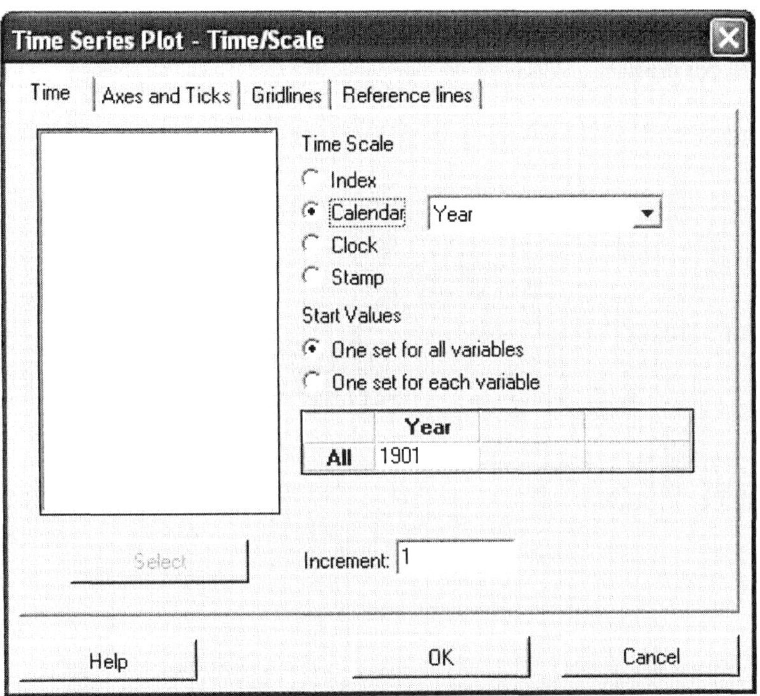

Click **Data View**
Select: **Symbols**
Connect Line
Click: **OK OK**

Select: **Smoother** tab

Select: Smoother: **Lowess**

(Note: These options remain as previously selected.)

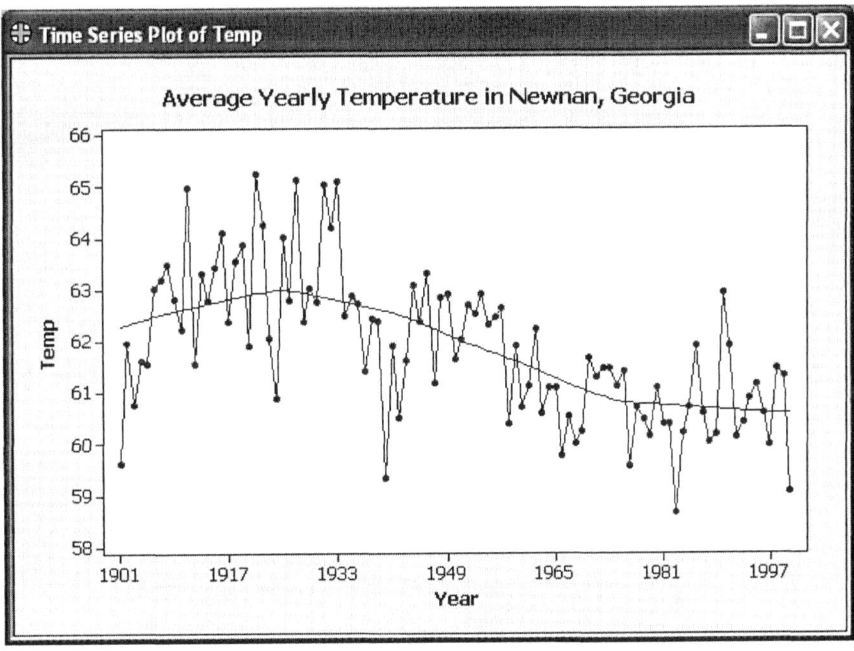

Chapter 2 Example 10 What is the Center of the Cereal Sodium Data?
Determining Mean and Median

Open the Minitab worksheet **cereal** from the MINITAB folder of the student data disk.

Calculate the mean and median using the following commands.

Choose: **Calc > Column Statistics**

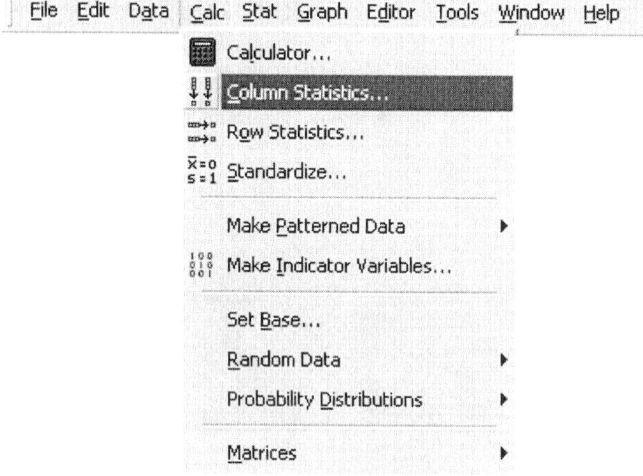

Select: **Median**
Enter: Input variable: **C2**
Click: **OK**

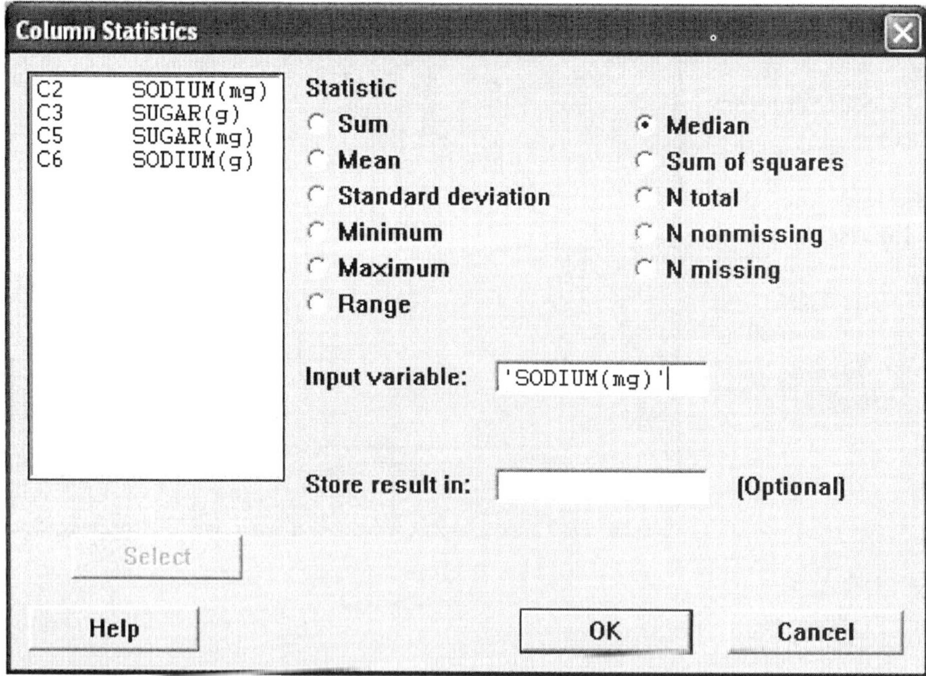

The results are displayed in the Session Window.

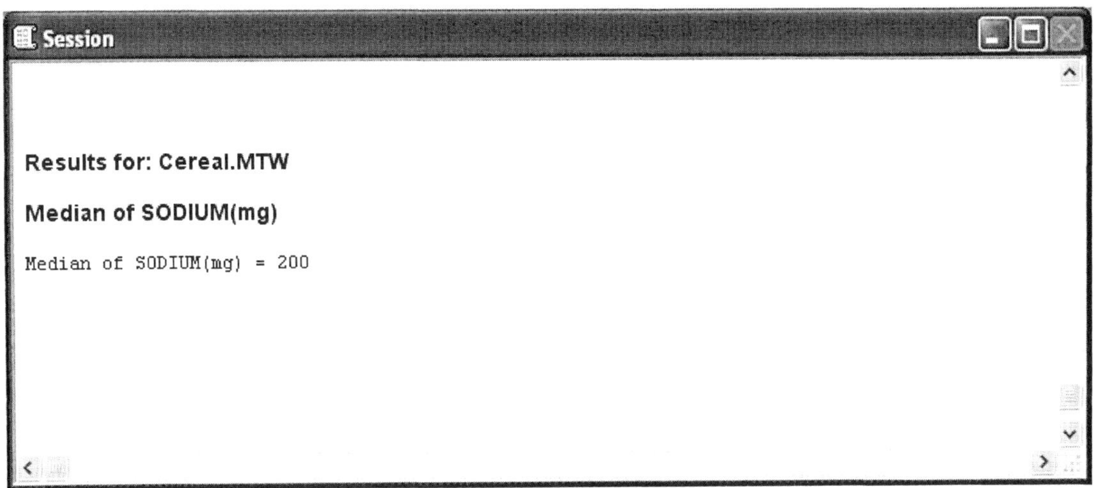

Use similar commands to find the mean:
Choose: **Calc > Column Statistics**
Select: **Mean**
Click: **OK**

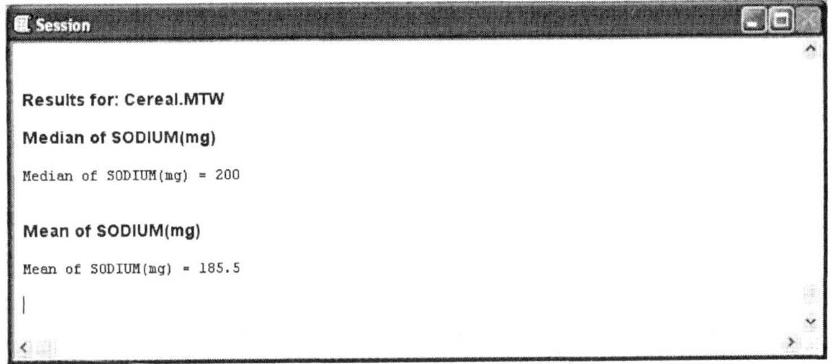

If you are interested in a variety of statistics, including median and mean, these values can be found more easily using the following:

Choose: **Stat > Basic Statistics > Display Descriptive Statistics**

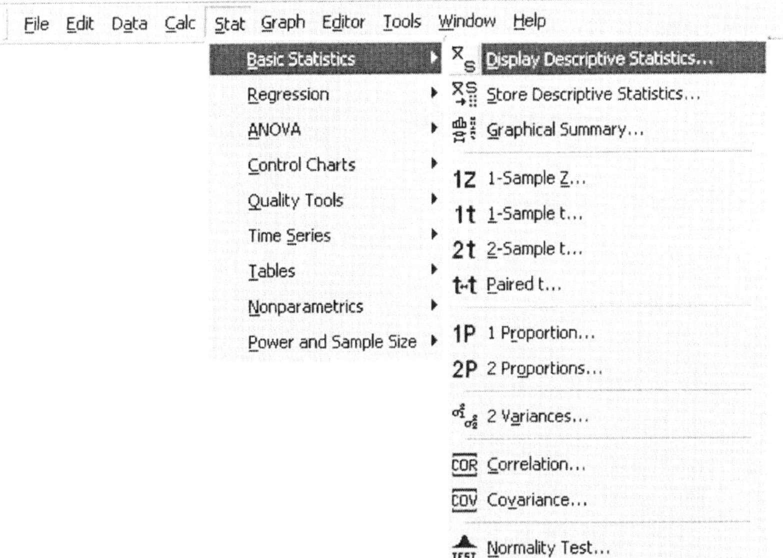

Enter: Variables: **C2**
Click: **OK**

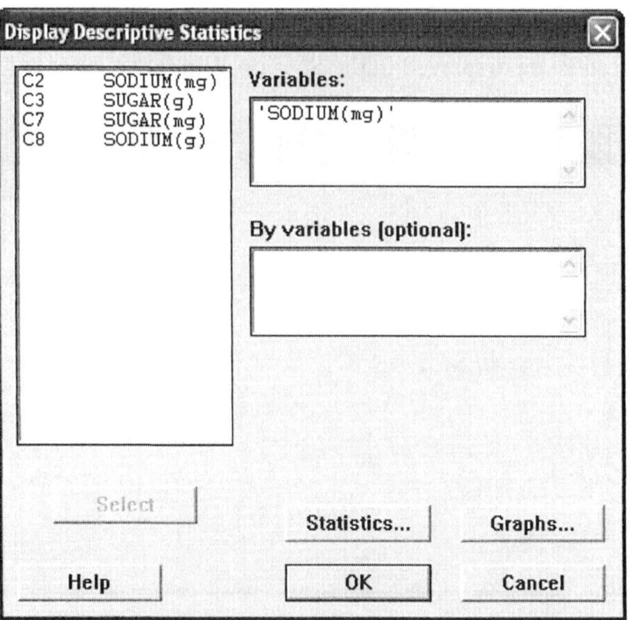

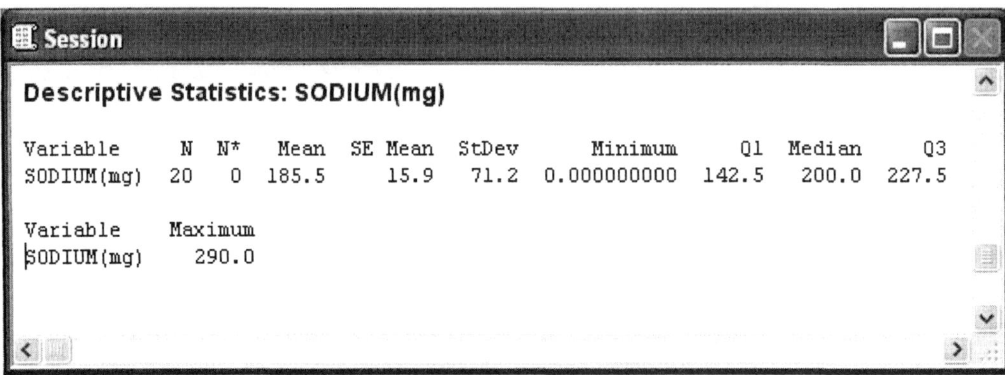

The only values we are interested in, at this point, are mean and median.

We can also find values by entering a formula and storing the result in a column. For example, the midrange = (high + low)/2. To do this in MINITAB we would do the following:

Select: **Calc>Calculator**

Store Result In: **midrange**
Type in the expression: **(MAX(C2) + MIN(C2)) /2**

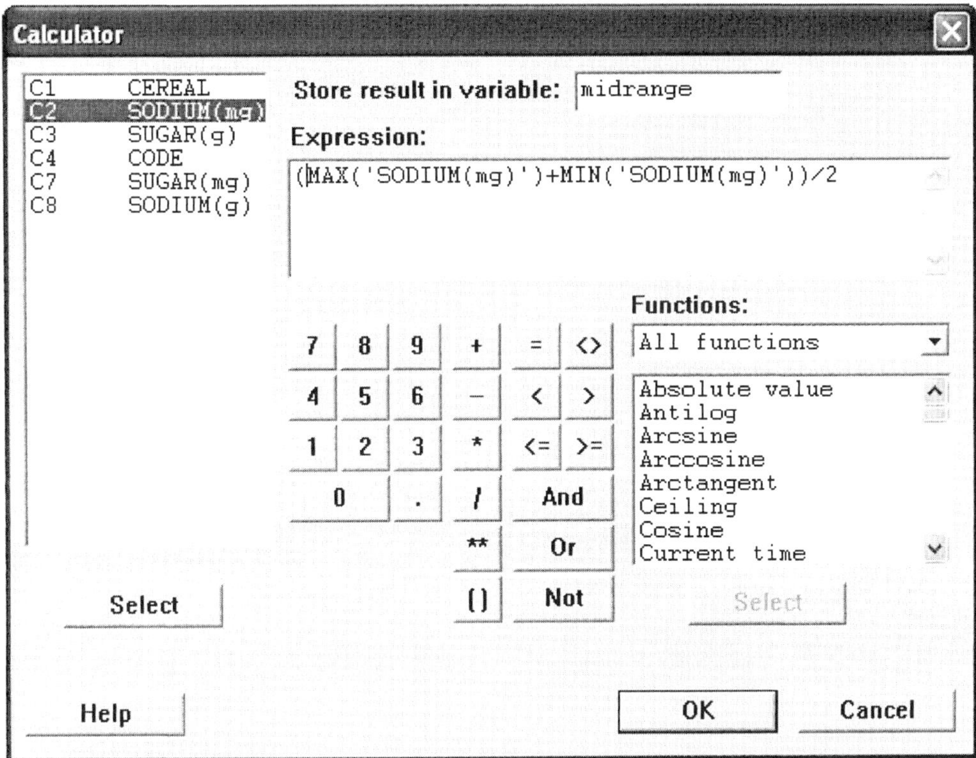

There is now a new column in the worksheet named midrange, which contains the result of the expression.

Cereal.MTW ***

↓	C1-T	C2	C3	C4-T	C5	C6	C7	C8	C9
	CEREAL	SODIUM(mg)	SUGAR(g)	CODE			SUGAR(mg)	SODIUM(g)	midrange
1	FMiniWheats	0	7	A			7000	0.000	145
2	ABran	260	5	A			5000	0.260	
3	AJacks	125	14	C			14000	0.125	
4	CCrunch	220	12	C			12000	0.220	
5	Cheeros	290	1	C			1000	0.290	
6	CTCrunch	210	13	C			13000	0.210	
7	CFlakes	290	2	A			2000	0.290	
8	RBran	210	12	A			12000	0.210	
9	COakBran	140	10	A			10000	0.140	
10	Crispix	220	3	A			3000	0.220	

Chapter 2 Exercise 2.29 More On CO_2 Emissions Mean and Median

Enter the data (country and million metric tons of carbon equivalent) into the worksheet.

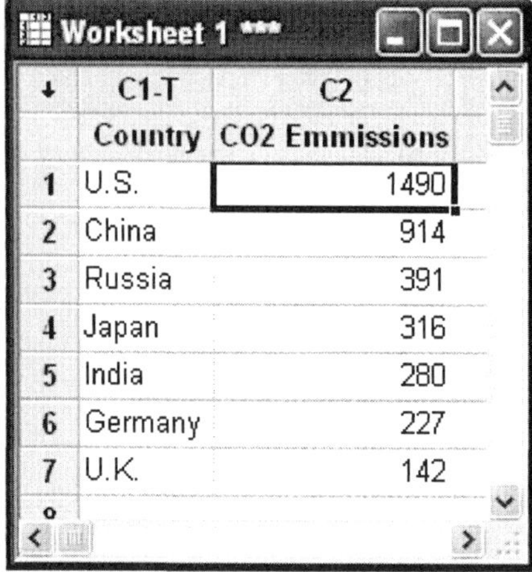

Worksheet 1 ***

↓	C1-T	C2
	Country	CO2 Emmissions
1	U.S.	1490
2	China	914
3	Russia	391
4	Japan	316
5	India	280
6	Germany	227
7	U.K.	142

Calculate the mean and median using the following commands.

Choose: **Calc > Column Statistics**
Select: **Mean** and then **Median**
Enter: Input variable: **C2**
Click: **OK**

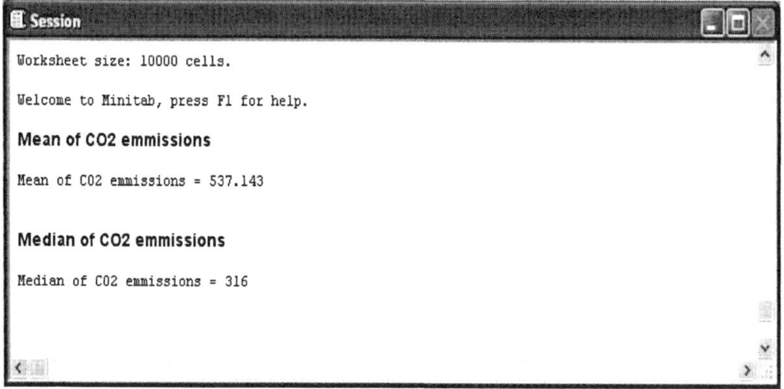

Session

Worksheet size: 10000 cells.

Welcome to Minitab, press F1 for help.

Mean of CO2 emmissions

Mean of CO2 emmissions = 537.143

Median of CO2 emmissions

Median of CO2 emmissions = 316

Chapter 2 Example 15 Describing Female College Student Heights Empirical Rule

When we open the data file **heights** on the text CD, we find that the data includes male (indicated by 0) and females(indicated by 1). In this example we are only concerned with the heights of the females.

heights.MTW ***

→	C1	C2	C3	C4	C5
	HEIGHT	GENDER			
1	70.0	0			
2	75.0	0			
3	67.0	0			
4	67.0	1			
5	73.0	0			
6	65.0	0			
7	73.0	0			
8	70.0	0			

To generate the histogram of female student height data only, we must specify that the data to be used should be taken from only those rows where the gender is a 1.

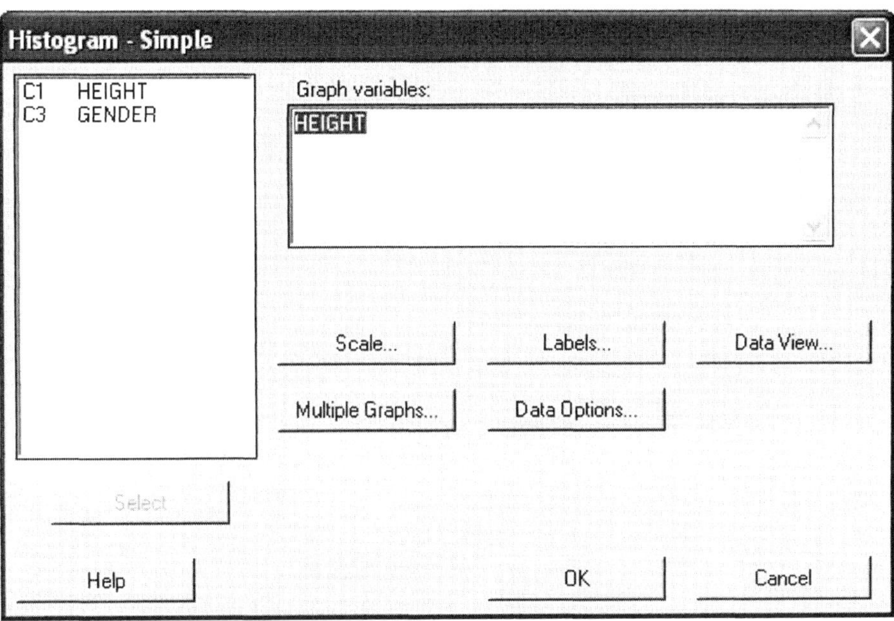

This can be accomplished by clicking the **Data Options** button within the **Graph>Histogram>Simple** task window.

In the **Data Options** window,
Click:**Specify Which Rows to Include > Rows that match > Condition** and within the Condition select **C3** and type "**=1**",
click **OK** for all three windows.

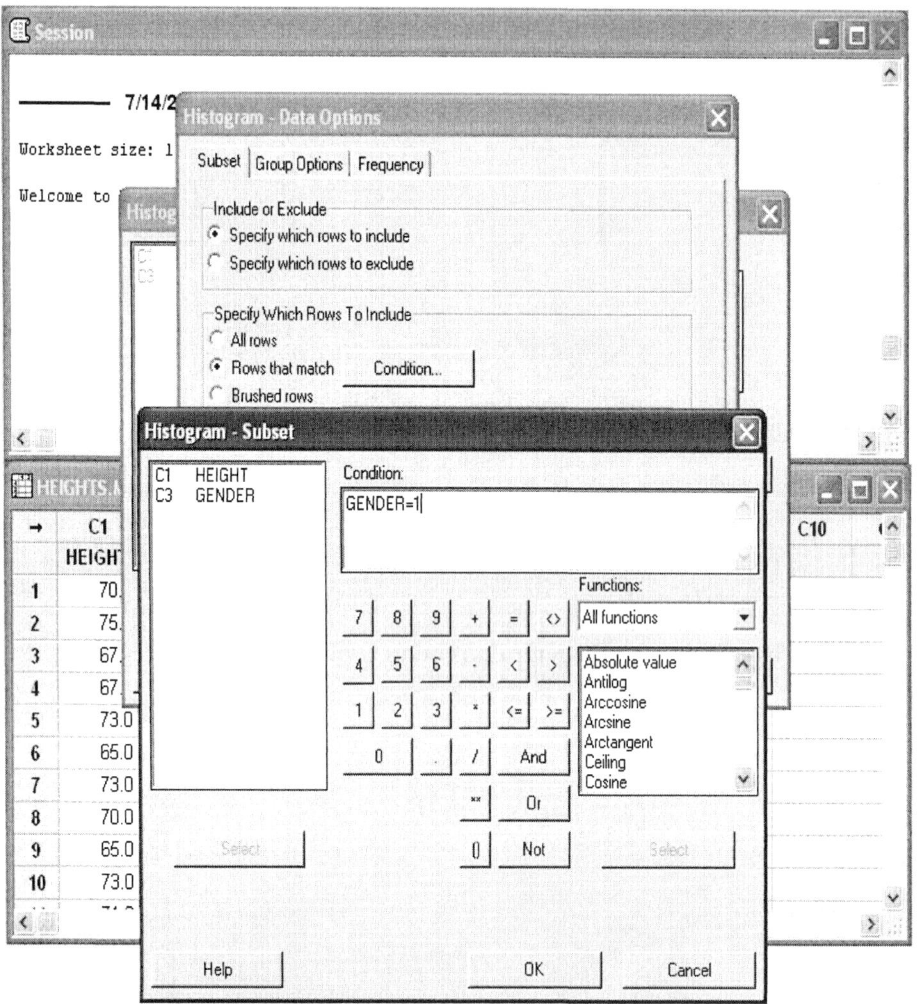
Session
7/14/2
Worksheet size: 1
Welcome to
Histogram - Data Options
Subset
Group Options
Frequency
Include or Exclude
Specify which rows to include
Specify which rows to exclude
Specify Which Rows To Include
All rows
Rows that match
Condition...
Brushed rows
Histogram - Subset
C1 HEIGHT
C3 GENDER
Condition:
GENDER=1
Functions:
All functions
Absolute value
Antilog
Arccosine
Arcsine
Arctangent
Ceiling
Cosine
And
Or
Not
Select
Help
OK
Cancel
C10
HEIGHT
70.
75.
67.
67.
73.0
65.0
73.0
70.0
65.0
73.0

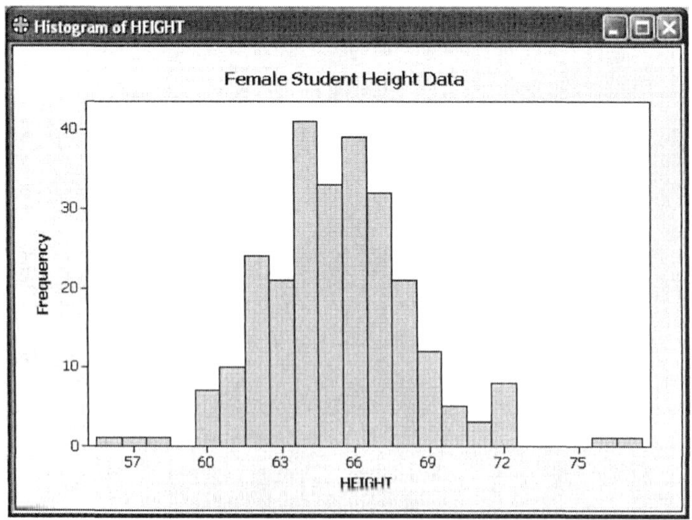
Histogram of HEIGHT
Female Student Height Data
Frequency
HEIGHT
40
30
20
10
0
57
60
63
66
69
72
75

To generate the descriptive statistics for only the females, a similar selection is made from **Stat> Basic Statistics > Display Descriptive Statistics** within the By Variables box.

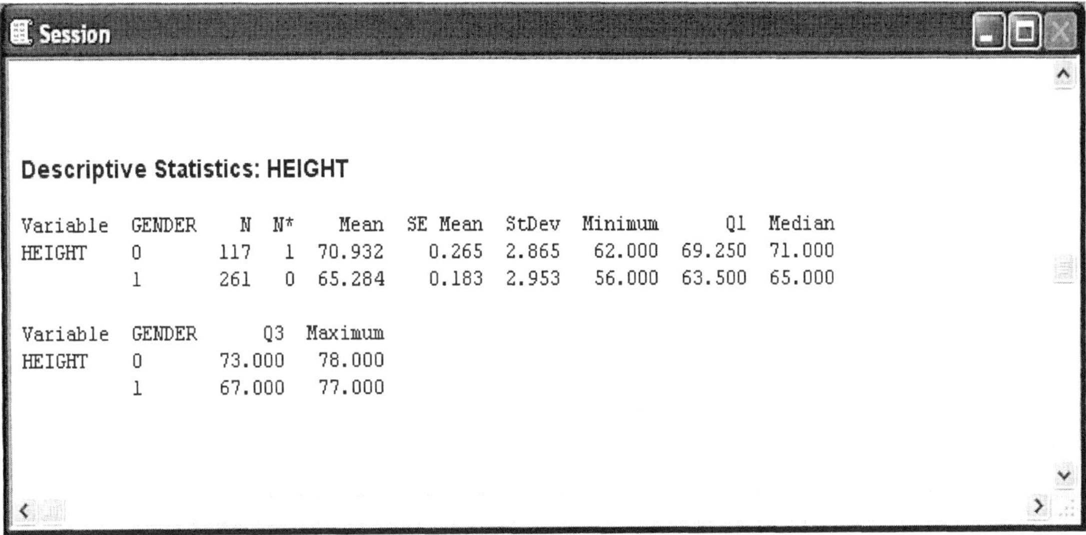
Session

Descriptive Statistics: HEIGHT

Variable	GENDER	N	N*	Mean	SE Mean	StDev	Minimum	Q1	Median
HEIGHT	0	117	1	70.932	0.265	2.865	62.000	69.250	71.000
	1	261	0	65.284	0.183	2.953	56.000	63.500	65.000

Variable	GENDER	Q3	Maximum
HEIGHT	0	73.000	78.000
	1	67.000	77.000

Notice this generates basic statistics for both the males and the females indicated by GENDER.

Now, to use MINITAB to count the number of observations within each of the ranges generated in the example, we will sort the data to facilitate the counting. Click **Data > Sort**

Enter both C1 and C3 into the **Sort Column(s)**, and C1 into the **By Column**. This way the correct gender identifier remains with the associated height. Also indicate into which **Column(s) of the current worksheet** you would like the sorted lists to be placed. Since we wish to handle the females separately from the males, we'll sort by both the Gender and the Height.

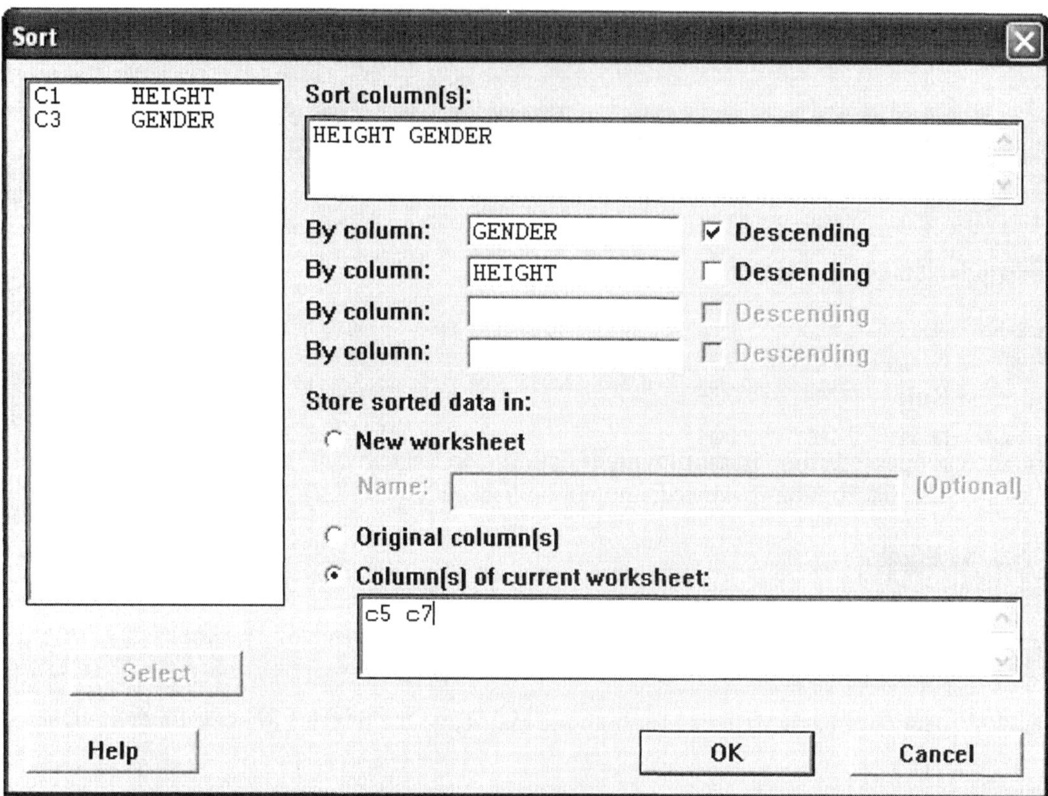

Now, with the lists sorted, counting the number of observations of female heights within each range is easier.

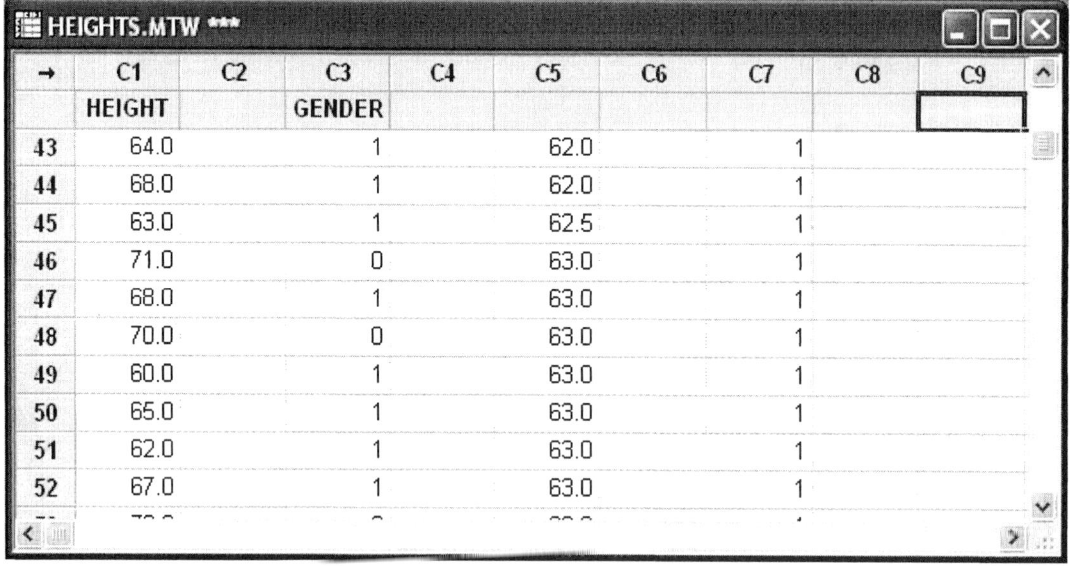

HEIGHTS.MTW ***

→	C1	C2	C3	C4	C5	C6	C7	C8	C9
	HEIGHT		GENDER						
43	64.0		1		62.0		1		
44	68.0		1		62.0		1		
45	63.0		1		62.5		1		
46	71.0		0		63.0		1		
47	68.0		1		63.0		1		
48	70.0		0		63.0		1		
49	60.0		1		63.0		1		
50	65.0		1		63.0		1		
51	62.0		1		63.0		1		
52	67.0		1		63.0		1		

The heights between 62.3 and 68.3 lie from row 45 to row 231, which is 187 values. 72% of all female heights.

The heights between 59.3 and 71.3 lie from row 4 to row 251, which is 248 values. 95% of all female heights.

The heights between 56.3 and 74.3 lie from row 2 to row 259,which is 378 values. 99% of all female heights.

Chapter 2 Exercise 2.55 EU data file
Empirical Rule

Open the **european_union_unemployment** data file on the text CD. Generating the histogram, we see that the distribution is skewed, and not bell shaped.

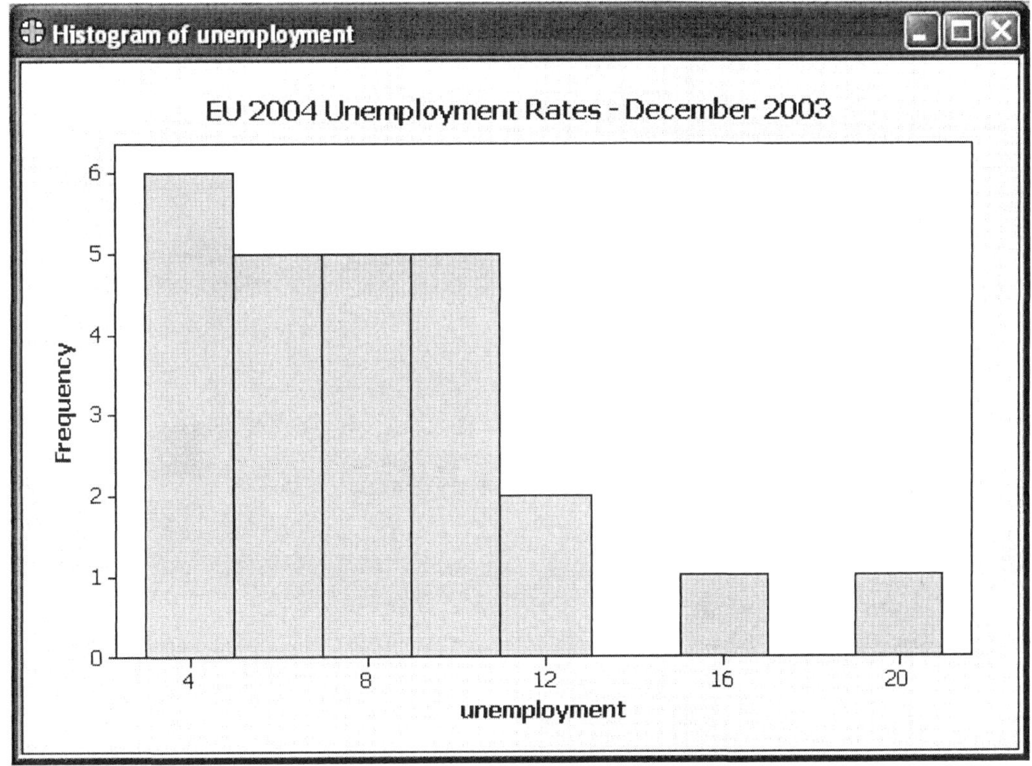

Generating basic statistics

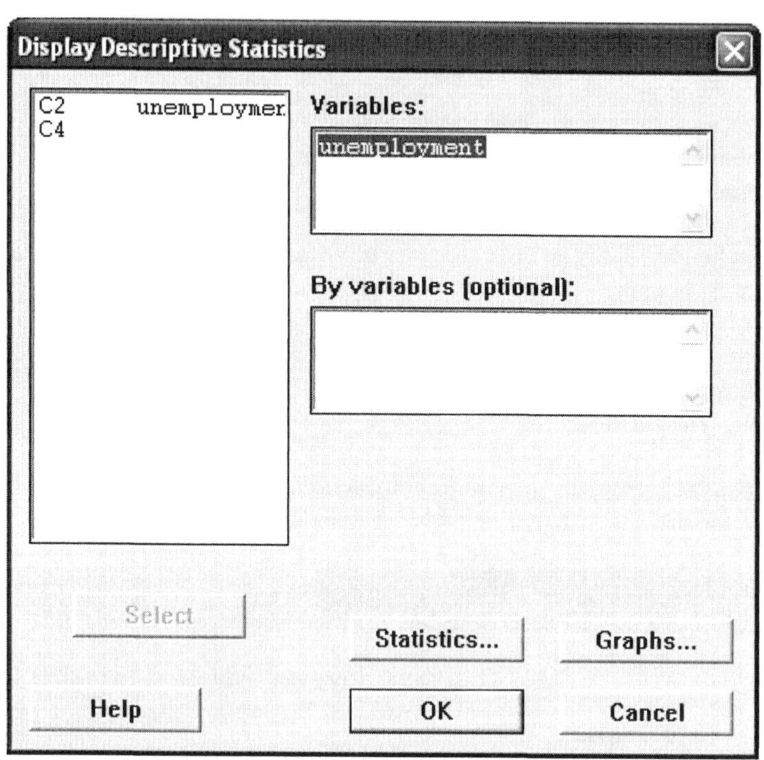

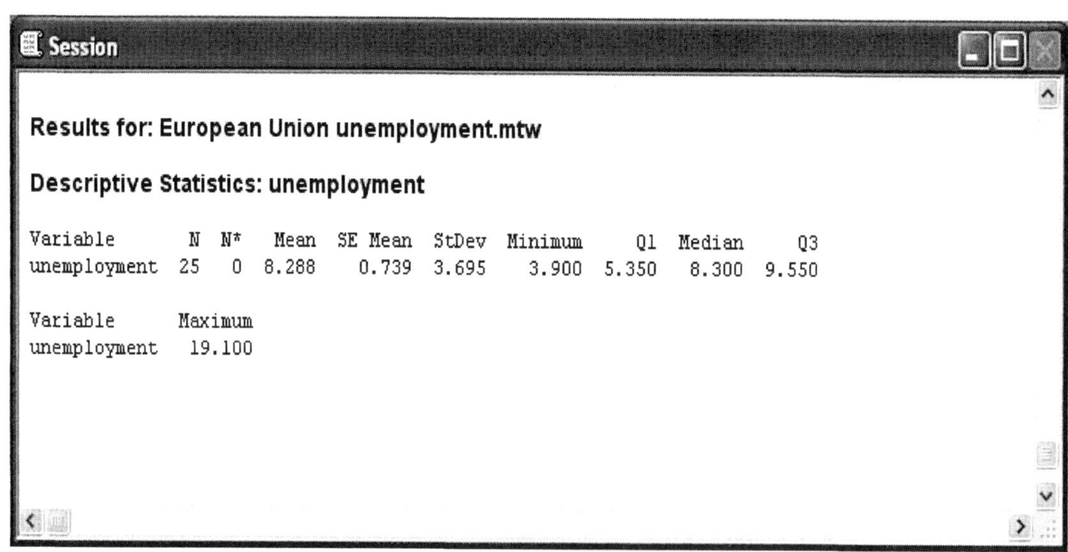

To determine the number of data values that fall in the ranges
$\bar{x} \pm 1 \cdot s = (4.593, 11.983), \bar{x} \pm 2 \cdot s = (.898, 15.678), \bar{x} \pm 3 \cdot s = (-2.797, 19.373)$,
sort the unemployment rates

European Union unemployment.mtw ***

↓	C1-T	C2	C3	C4	C5-T	C6	C7	C8	C9
	country	unemployment							
1	Belgium	8.3			Luxembourg		3.9		
2	Denmark	6.0			Netherlands		4.4		
3	Germany	9.2			Austria		4.5		
4	Greece	9.3			Ireland		4.6		
5	Spain	11.2			Cyprus		4.7		
6	France	9.5			United Kingdom		4.8		
7	Ireland	4.6			Hungary		5.9		
8	Italy	8.5			Denmark		6.0		
9	Luxembourg	3.9			Sweden		6.0		
10	Netherlands	4.4			Slovenia		6.4		

We determine that there are 20, 23, and 25 data values within the three ranges, respectively, which is 80%, 92 %, and 100%. Notice that theses percentages don't even begin to match the 68%, 95% and 99.7% of the Empirical Rule. This provides further evidence that the distribution of unemployment rates is not bell shaped.

Chapter 2 Example 16 What Are the Quartiles for the Cereal Sodium Data? Quartiles

Open the **CEREAL** data file on the text CD.

In this example, we wish to determine the quartiles for the sodium values. Click **Stats > Basic Statistics > Display Descriptive Statistics > Statistics** and select First quartile, Median, and Third Quartile. Click **OK** for each window.

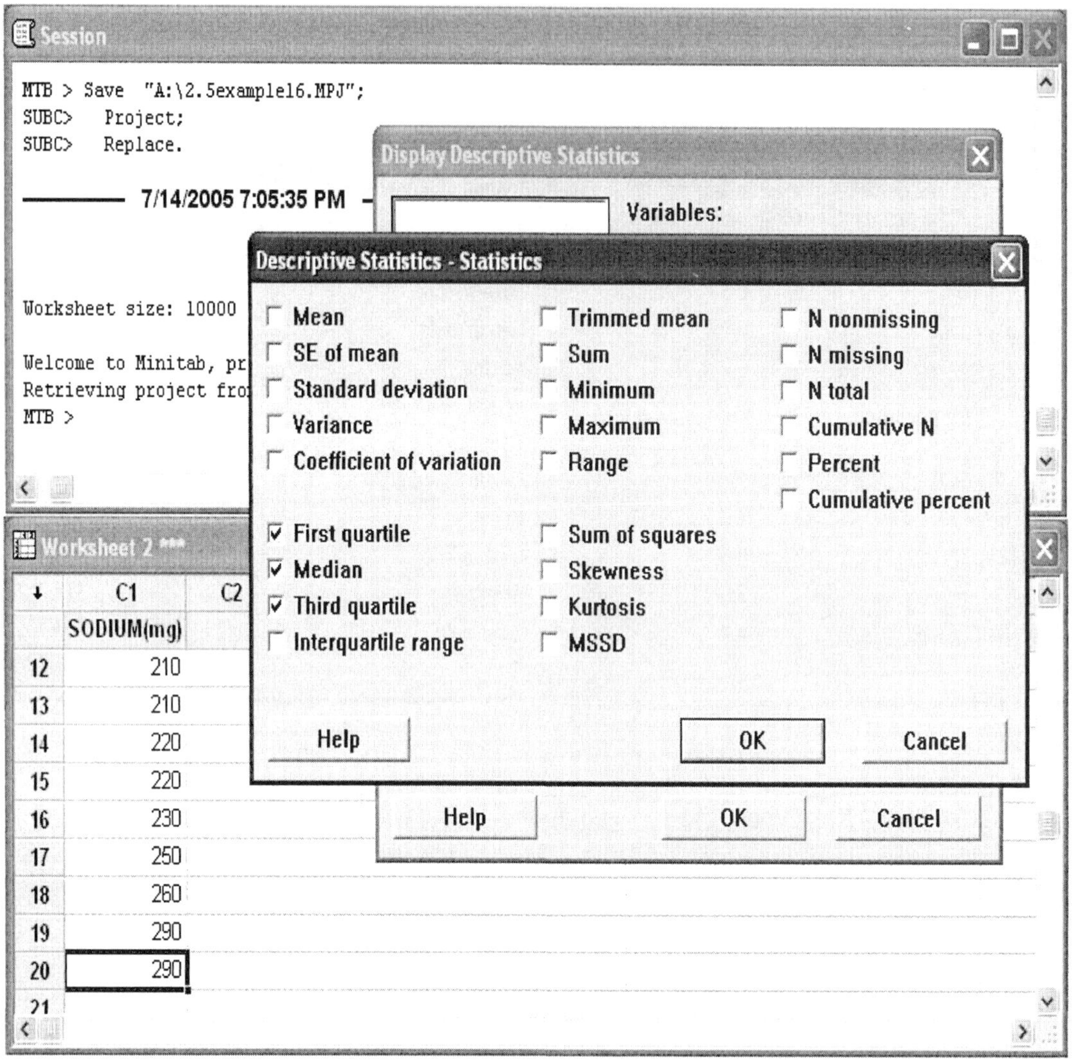

Session

```
MTB > Save "A:\2.5example16.MPJ";
SUBC>   Project;
SUBC>   Replace.
```

—— 7/14/2005 7:05:35 PM ——

Worksheet size: 10000

Welcome to Minitab, pr

Retrieving project fro

MTB >

Display Descriptive Statistics

Variables:

Descriptive Statistics - Statistics

☐ Mean	☐ Trimmed mean	☐ N nonmissing
☐ SE of mean	☐ Sum	☐ N missing
☐ Standard deviation	☐ Minimum	☐ N total
☐ Variance	☐ Maximum	☐ Cumulative N
☐ Coefficient of variation	☐ Range	☐ Percent
		☐ Cumulative percent
☑ First quartile	☐ Sum of squares	
☑ Median	☐ Skewness	
☑ Third quartile	☐ Kurtosis	
☐ Interquartile range	☐ MSSD	

Help OK Cancel

Help OK Cancel

Worksheet 2 ***

↓	C1	C2
	SODIUM(mg)	
12	210	
13	210	
14	220	
15	220	
16	230	
17	250	
18	260	
19	290	
20	290	
21		

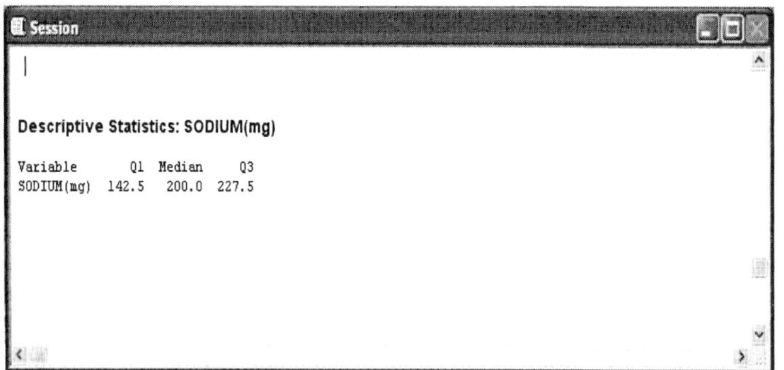

Note that the results differ from those in the text. Not all statistical programs use the same method for determining the quartiles. You should explore how and why MINITAB generates Q1=142.5 and Q3 = 227.5

Chapter 2 Exercise 2.58 European Unemployment
Quartiles

Enter the data into C1 and C2 of the Data Window. Be sure to label the columns.

Click **Stats > Basic Statistics > Display Descriptive Statistics > Statistics** and select First quartile, Median, and Third Quartile. Click **OK** for each window. The selected descriptive statistics will be displayed in the Session Window.

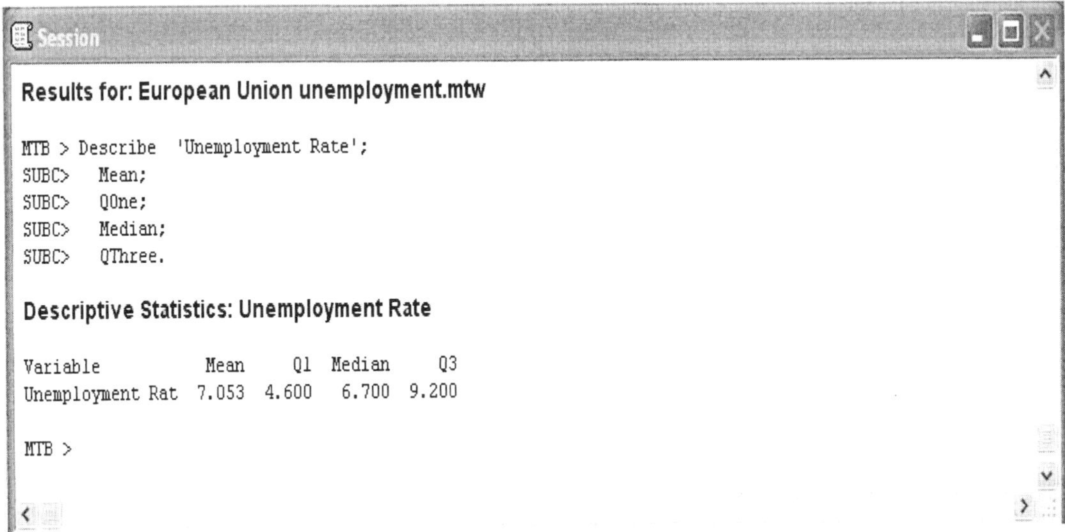

Chapter 2 Example 17 Box Plot for the Breakfast Cereal Sodium Data
Box Plots

Open worksheet **CEREAL** from the MINITAB folder of the data disk. To construct a boxplot for the breakfast cereal sodium data, use the following commands:

Click: **Graph > Boxplot**

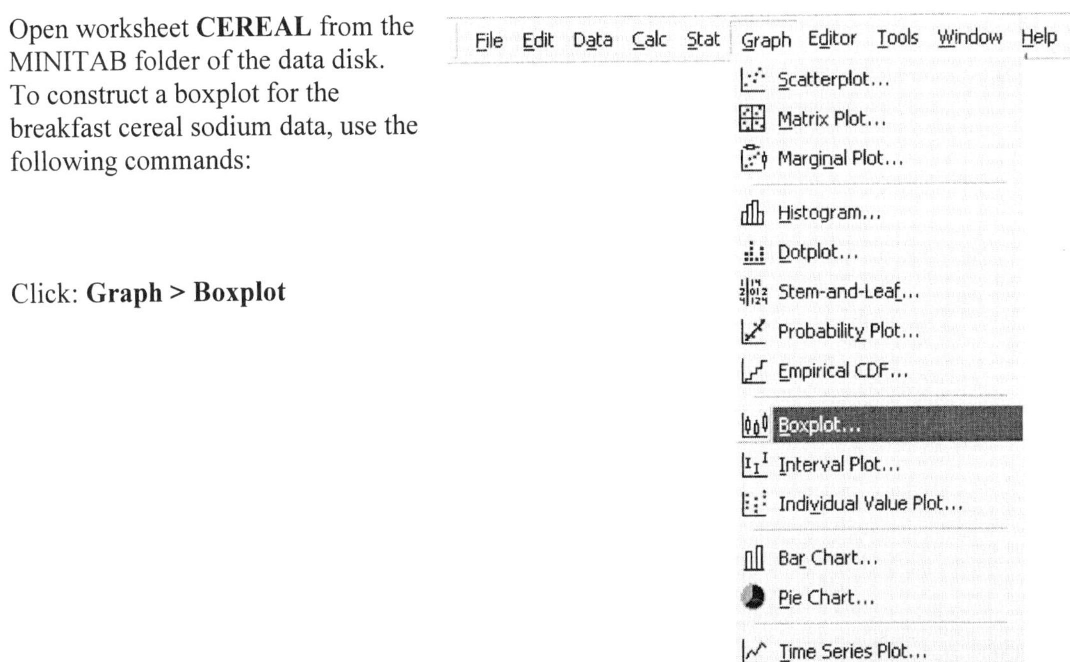

Select: **One Y/Simple > OK**

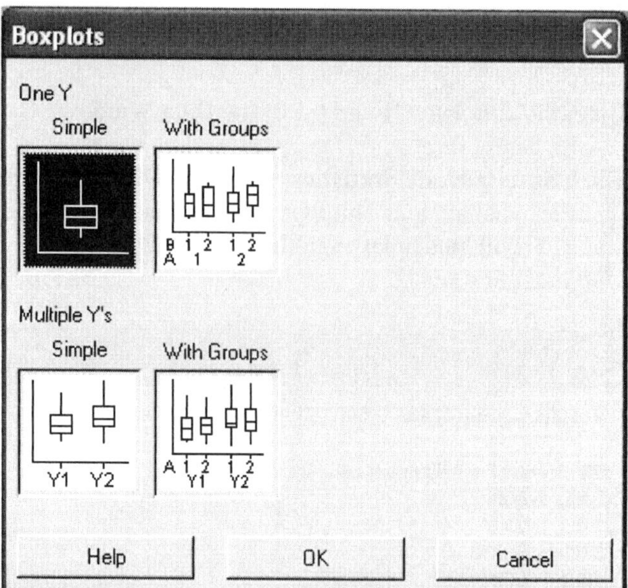

Enter:
Graph variable: **C2**
Click: **OK**

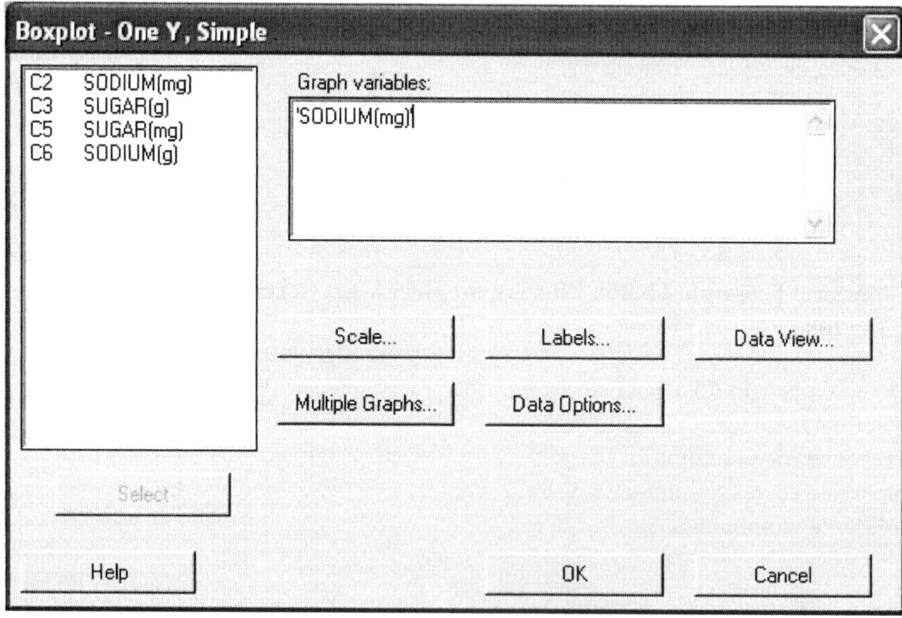

The default MINITAB boxplot plots vertically. This can be changed. Follow the commands above, and before clicking the second OK ,
Click: **Scale**
Select: **Transpose value and category scales**
Click: **OK** twice

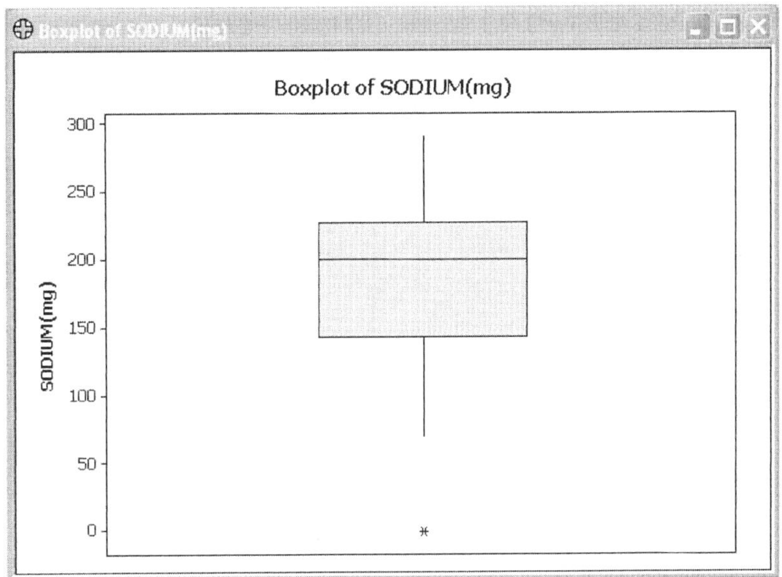

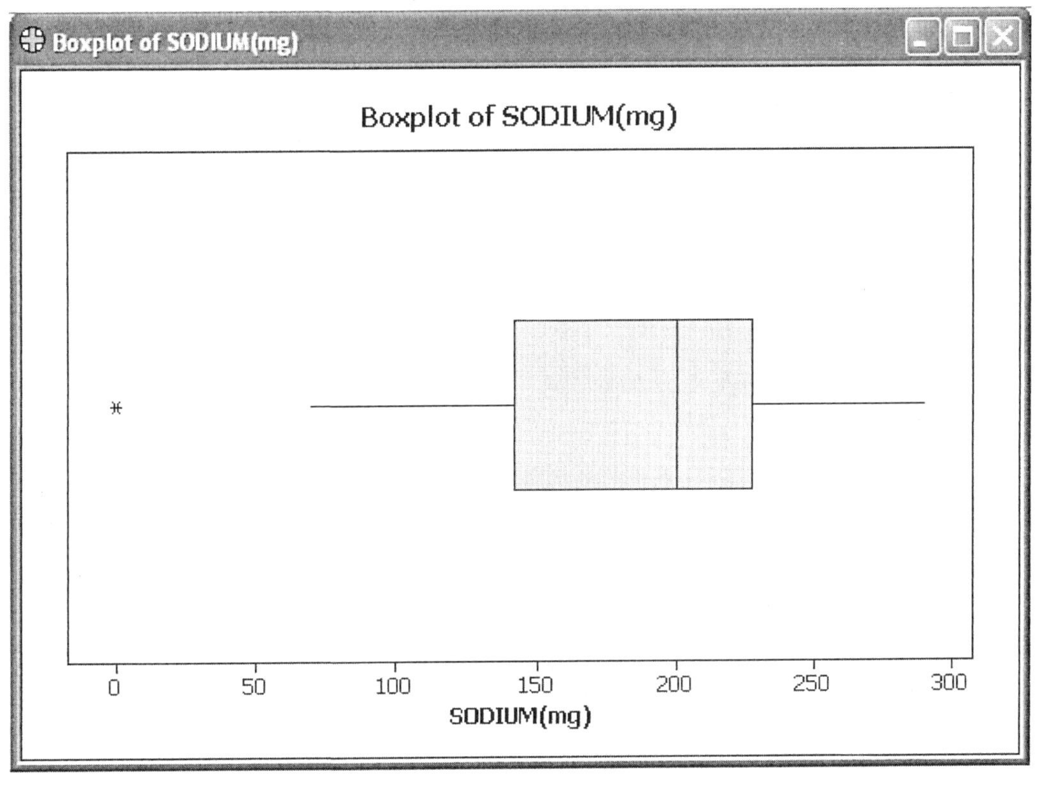

Chapter 2 Exercise 2.69 European Union Unemployment Rates
Box Plots

Enter the data from exercise 2.58 into C1 and C2.

Click: **Graph > Boxplot > Simple**
Click: **OK**
Enter: Graph variable: **C2**
Select: **Scale**
Click: **Transpose value and category scales**
Click: **OK OK**

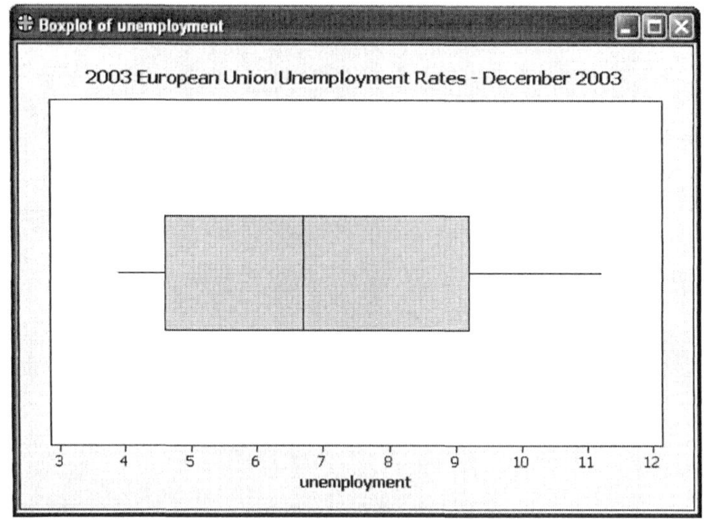

Chapter 2 Exercise 2.73 Florida Students Again
Box Plots

Open worksheet **fla_student_survey** from the MINITAB folder of the data disk.

Click: **Graph > Boxplot > Simple**
Click: **OK**
Enter: Graph variable: **C8**
Select: **Scale**
Click: **Transpose value and category scales**
Click: **OK OK**

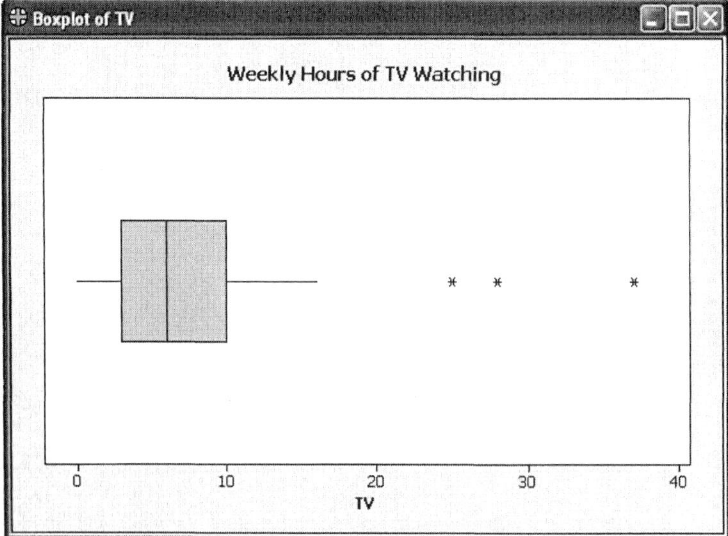

Chapter 2 Exercise 2.103b Temperatures in Central Park
Box Plot

Open worksheet **central_park_yearly_temps** from the MINITAB folder of the data disk.

Click: **Graph > Boxplot > Simple**
Click: **OK**
Enter: Graph variable: **C1**
Select: **Scale**
Click: **Transpose value and category scales**
Click: **OK OK**

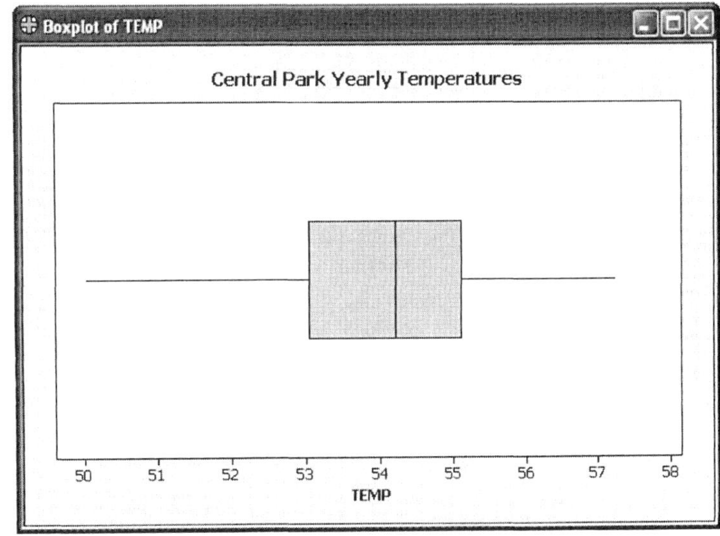

We can compare this to the histogram produced for part (a) of this exercise.

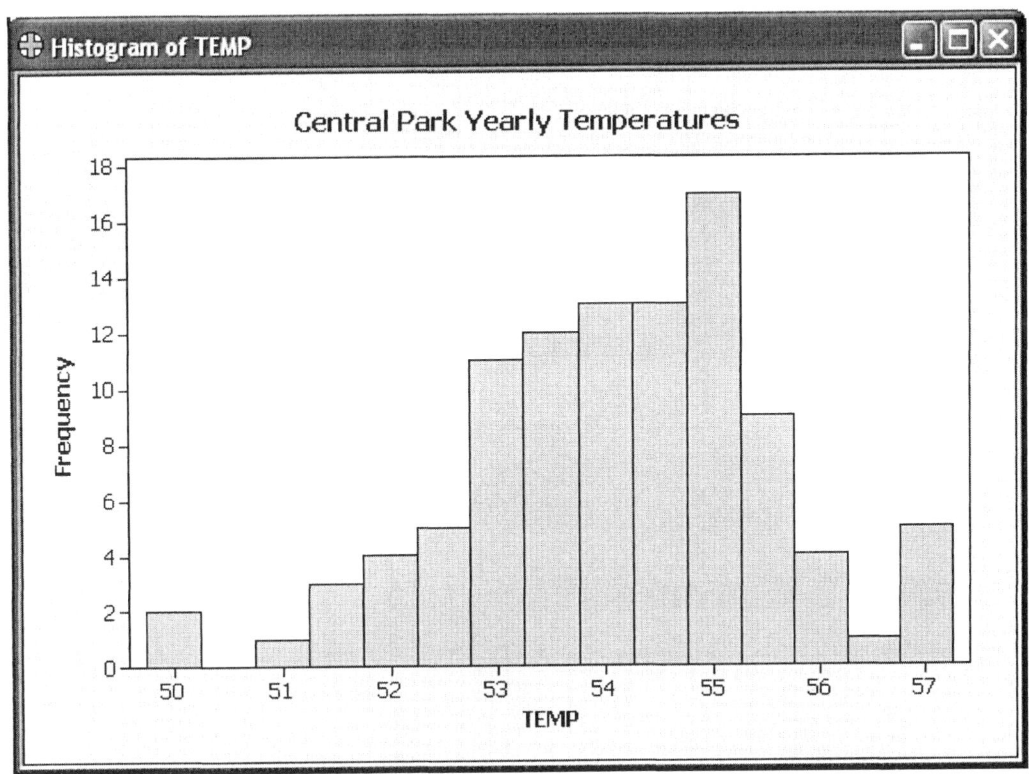

Chapter 2 Exercise 2.113 How Much is Spent on Haircuts?
Side-by-Side Plots

Open worksheet **georgia_student_survey** from the MINITAB folder of the data disk. For this exercise we wish to compare how much males and females spend on a haircut. First note that gender is indicated in C2 as 0 for male and 1 for female. Many of the graphs illustrated in this chapter can be done as side-by-side plots, graphing with a separate plot for the males and the females, using the same scale, so that comparison of the distributions can be made.

georgia_student_survey.MTW ***

↓	C1	C2	C3	C4	C5	C6	C7	C8
	Height	Gender	Haircut	Job	Studytime	Smokecig	Dated	HSGPA
1	65	1	25	1	7.0	0	0	3.90
2	71	0	12	0	2.0	0	1	3.79
3	68	1	4	0	4.0	0	1	3.00
4	64	1	0	1	3.5	0	1	3.90
5	64	1	50	0	4.5	0	1	3.60
6	66	0	10	1	3.0	0	3	3.20

Click: **Graph > Boxplot (or Histogram or Dotplot) > One Y, With Groups**
Click: **OK**

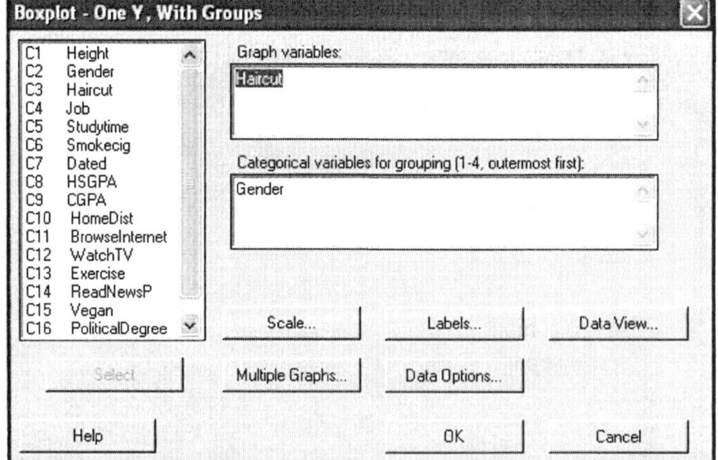

Select: Graph variable: **C3**
Select: Categorical variables for grouping: **G2**
Select: **Scale**
Click: **Transpose value and category scales**
Click: **OK OK**

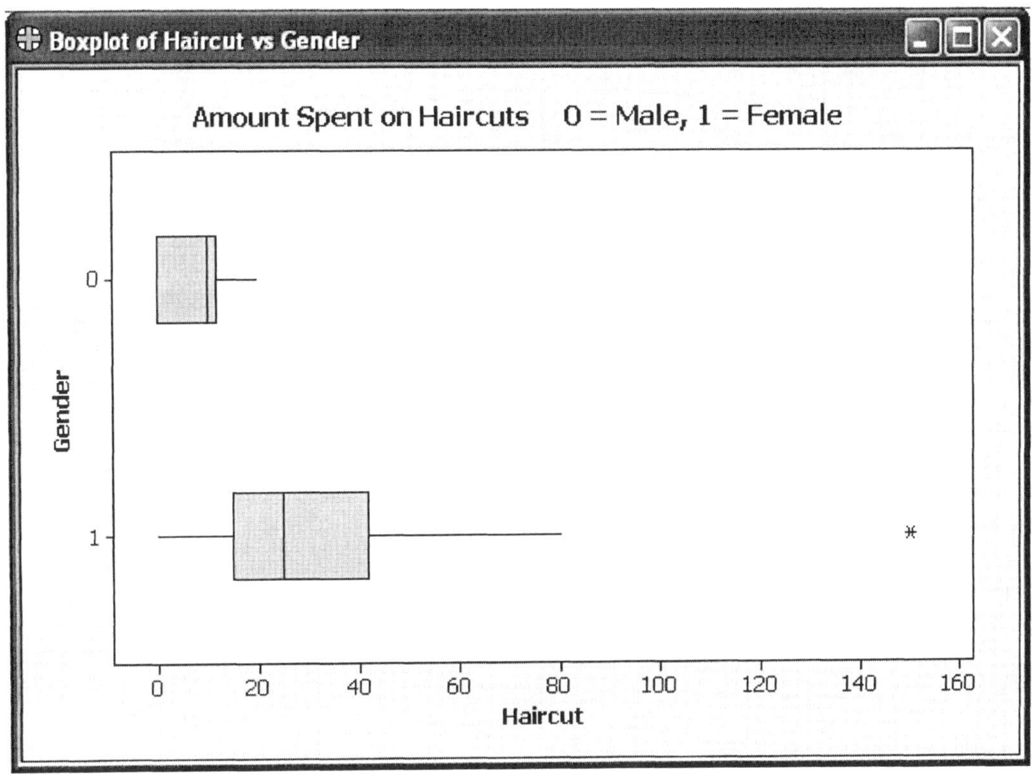

For comparing histograms, the results are more easily readable if we generate histograms based on gender in separate panels.

Click: **Graph > Histogram > Simple**
Click: **OK**
Click: **Labels**
  Enter: Title: **an appropriate title**
  Click: **OK**
  Click: **Multiple Graphs**

On the Multiple Variables tab
Select: **In separate panels of the same graph**

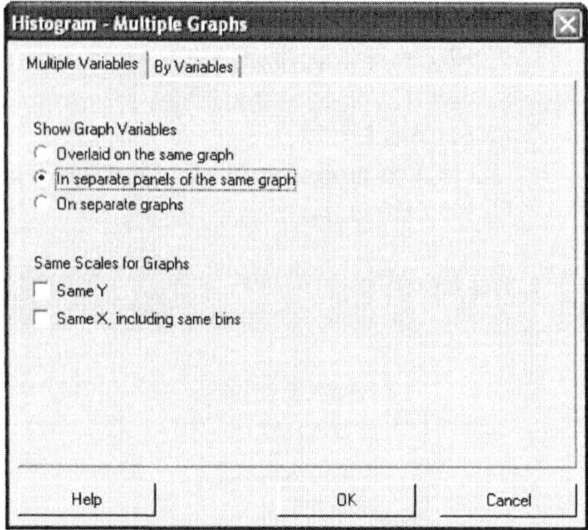

Click: the **By Variables** tab
Enter: By variables with groups in separate panels: **C2**

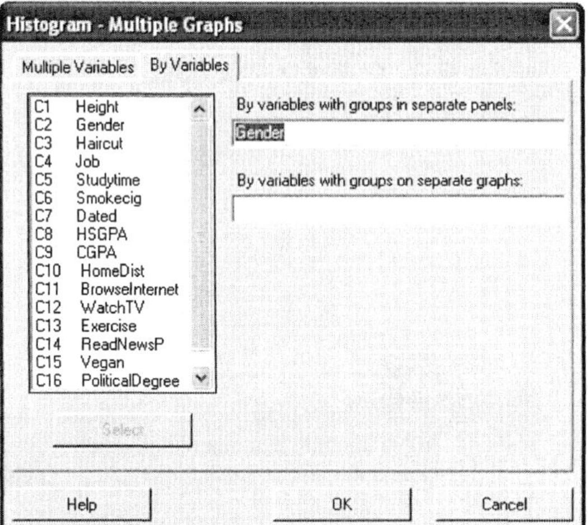

Then click **OK** for each window.

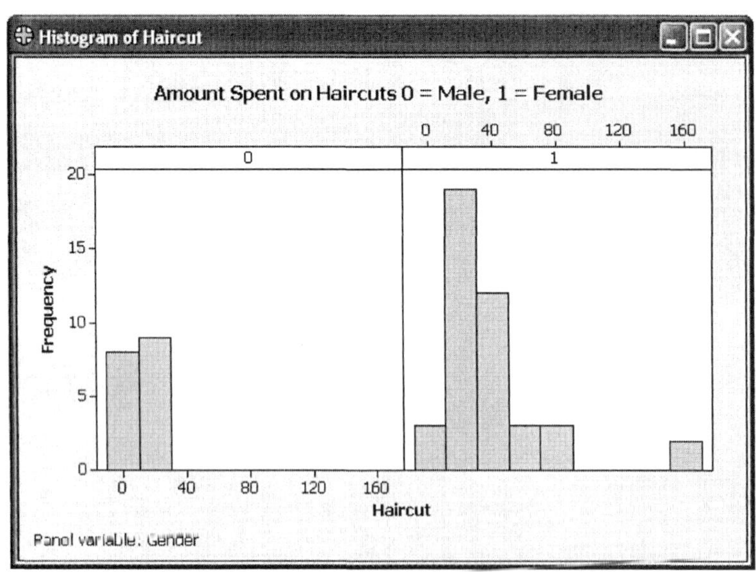

Chapter 3 Example 3 How Can We Compare Pesticide Residues For the Food Types Graphically?
Side-by-Side Bar Chart

Enter the conditional proportions (Table 3.2) into the worksheet.

Worksheet 1 ***

↓	C1-T	C2	C3
	Food Type	Pesticide Present	Pesticide Not Present
1	Organic	0.23	0.77
2	Conventional	0.73	0.27

To generate a single bar graph that show side-by-side bars to compare the conditional proportion of pesticide residues in conventionally grown and organic foods:

Choose: **Graph > Bar Chart**
Select: Bars represent: **Values from a table**
Click: **Two-way table / Cluster > OK**
Enter: Graph Variables : **C2 C3**
Row labels: **C1**
Select: Table Arrangement: **Columns are outermost categories and rows are innermost**
Click: **OK**

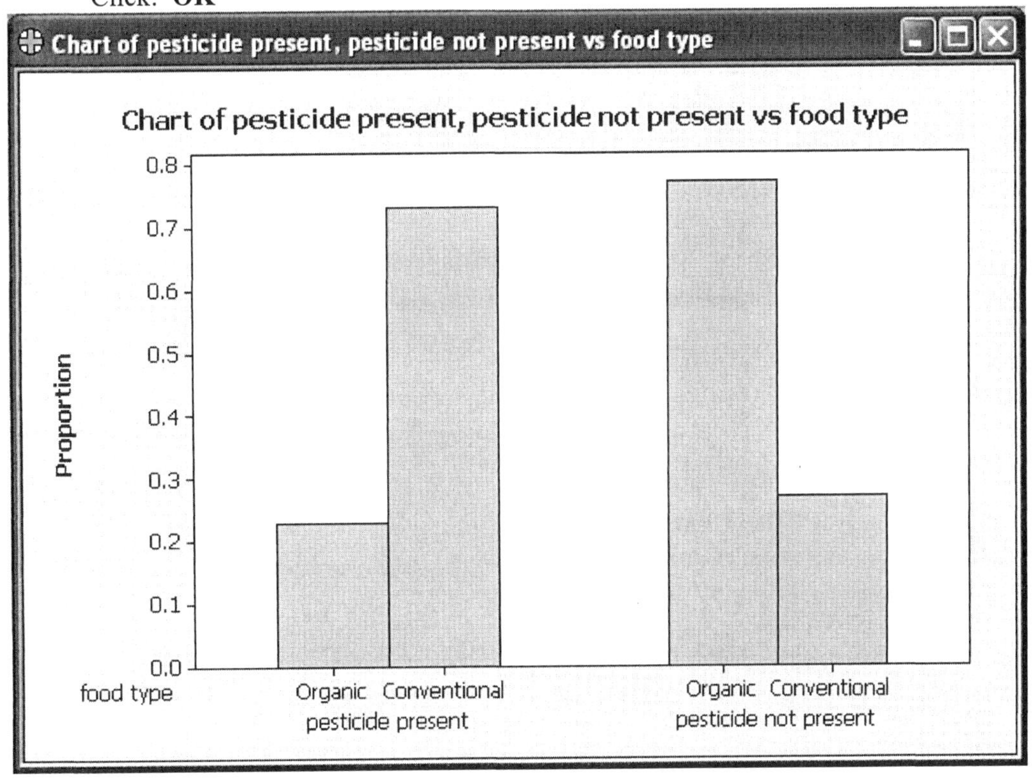

Making the selection of **Rows are outermost categories and columns are innermost** produces the following side-by-side bar chart.

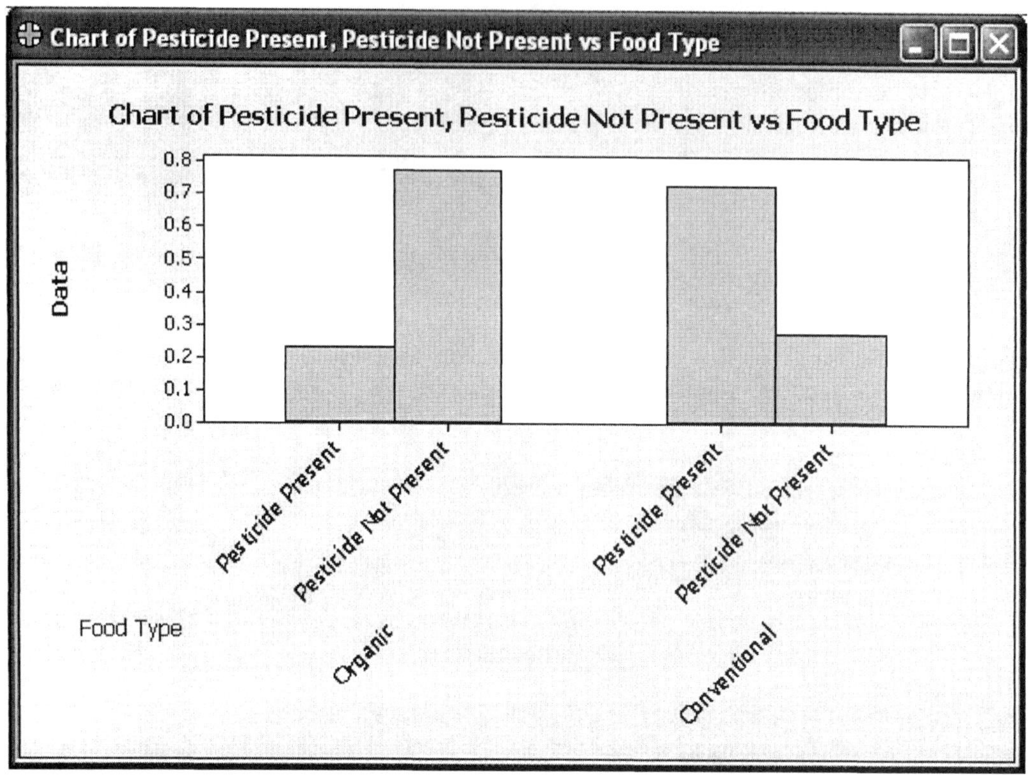

Chapter 3 Exercise 3.4 Religious Activities
Side-by-Side Bar Chart

Enter the data into the worksheet. Since we are comparing unequal numbers of males and females, we need to enter the data as proportions of hours of home religious activity within each category of gender.

↓	C1-T	C2	C3	C4	C5	C6
	Gender	0	1-9	10-19	20-39	40 or more
1	Female	0.29896	0.38773	0.114880	0.13446	0.06397
2	Male	0.43533	0.38328	0.093060	0.06309	0.02524

Choose: **Graph > Bar Chart**
Select: Bars represent: **Values from a table**
Click: **Two-way table / Cluster > OK**
Enter: Graph Variables : **C2 C3 C4 C5 C6**
Row labels: **C1**
Select: Table Arrangement: **Columns are outermost categories and rows are innermost**
Click: **OK**

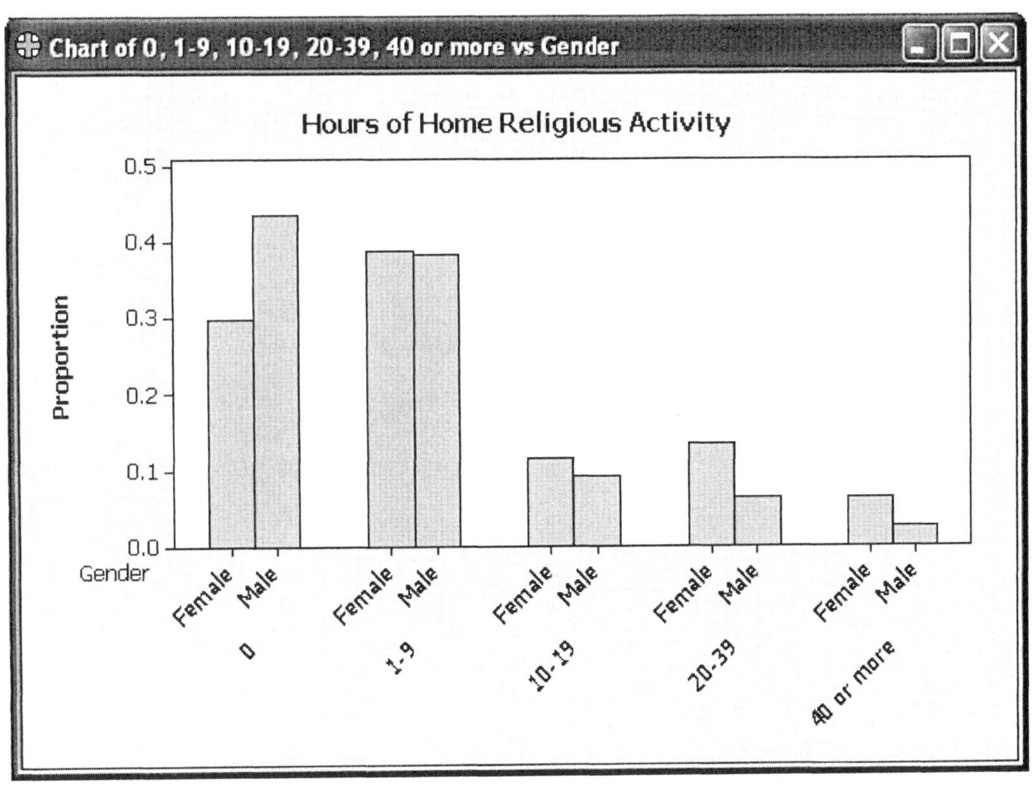

Chapter 3 Example 5 Constructing a Scatterplot for Internet Use and GDP
Constructing a Scatterplot

Open the worksheet **human_development**, which is found in the MINITAB folder on the data disk.

↓	C1-T	C2	C3	C4	C5	C6	C7
		INTERNET	GDP	CO2	CELLULAR	FERTILITY	LITERACY
1	Algeria	0.65	6.09	3.0	0.3	2.8	58.3
2	Argentina	10.08	11.32	3.8	19.3	2.4	96.9
3	Australia	37.14	25.37	18.2	57.4	1.7	100.0
4	Austria	38.70	26.73	7.6	81.7	1.3	100.0
5	Belgium	31.04	25.52	10.2	74.7	1.7	100.0
6	Brazil	4.66	7.36	1.8	16.7	2.2	87.2
7	Canada	46.66	27.13	14.4	36.2	1.5	100.0
8	Chile	20.14	9.19	4.2	34.2	2.4	95.7
9	China	2.57	4.02	2.3	11.0	1.8	78.7
10	Denmark	42.95	29.00	9.3	74.0	1.8	100.0

Since we are interested in how Internet use depends on GDP, C3 GDP is the x-variable, and C2 Internet is the y-variable.

To create a scatterplot of the data

Choose: **Graph > Scatterplot**

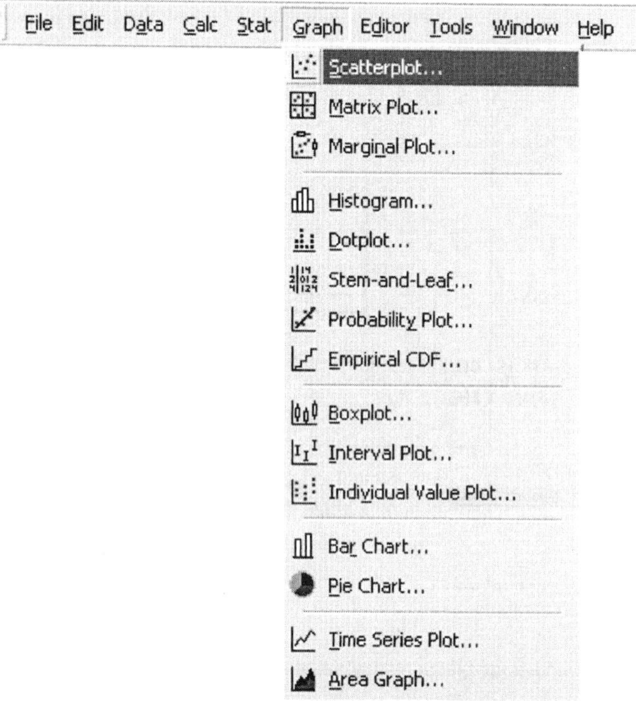

Select: **Simple > OK**

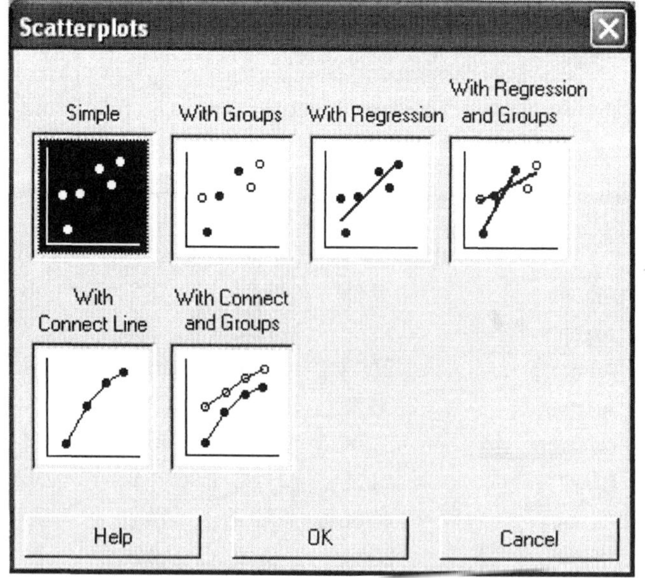

Enter: Y variables **C2**
X variables **C3**

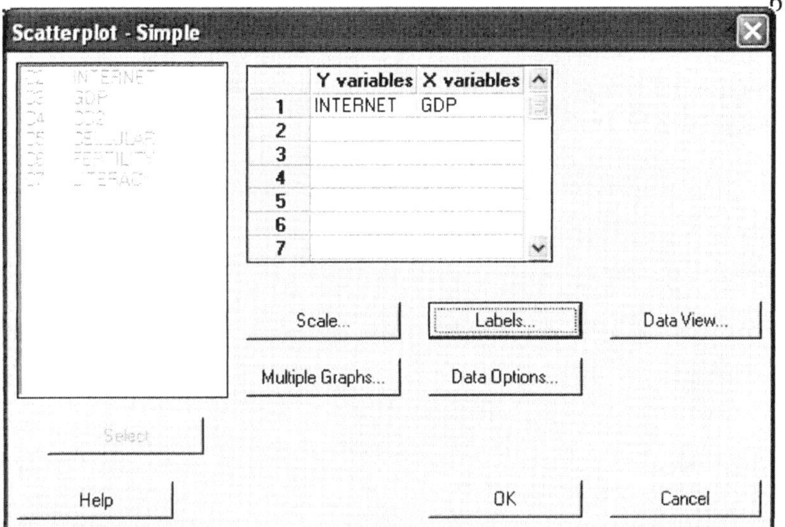

Select: Labels **enter an appropriate title**
Click: **OK OK**

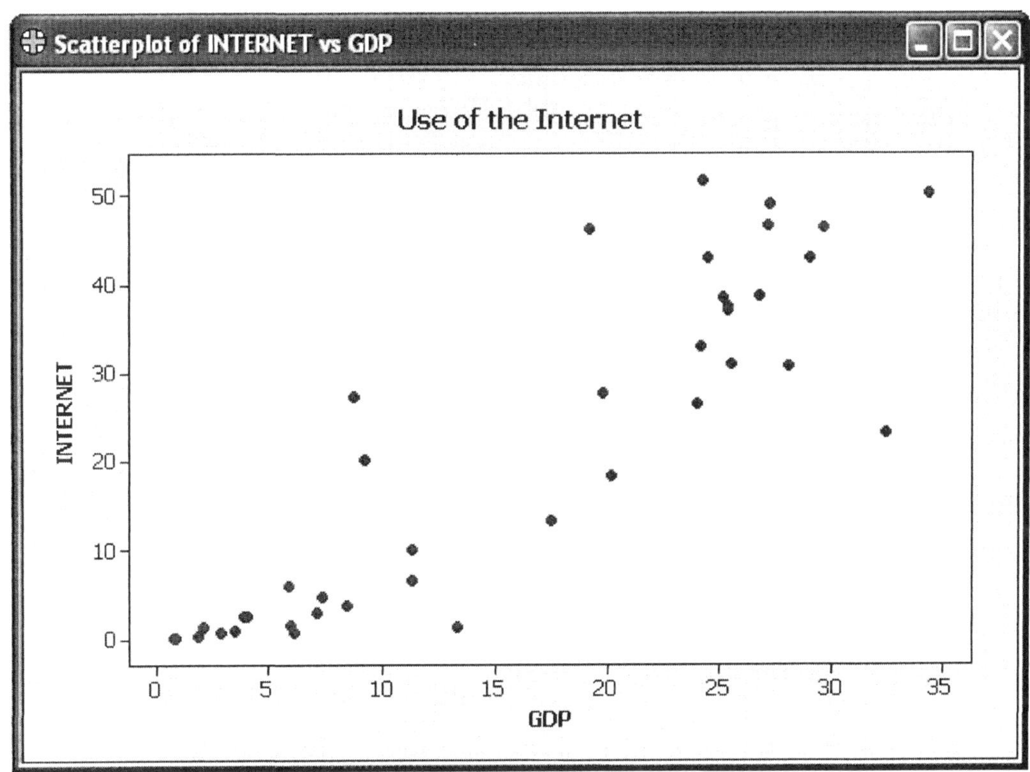

Chapter 3 Example 7 What's the Correlation Between Internet Use and GDP? Computing Correlation

Open the worksheet **human_development**, which is found in the MINITAB folder on the data disk.

↓	C1-T	C2	C3	C4	C5	C6	C7
		INTERNET	GDP	CO2	CELLULAR	FERTILITY	LITERACY
1	Algeria	0.65	6.09	3.0	0.3	2.8	58.3
2	Argentina	10.08	11.32	3.8	19.3	2.4	96.9
3	Australia	37.14	25.37	18.2	57.4	1.7	100.0
4	Austria	38.70	26.73	7.6	81.7	1.3	100.0
5	Belgium	31.04	25.52	10.2	74.7	1.7	100.0
6	Brazil	4.66	7.36	1.8	16.7	2.2	87.2
7	Canada	46.66	27.13	14.4	36.2	1.5	100.0
8	Chile	20.14	9.19	4.2	34.2	2.4	95.7
9	China	2.57	4.02	2.3	11.0	1.8	78.7
10	Denmark	42.95	29.00	9.3	74.0	1.8	100.0

Since we are interested in how Internet use depends on GDP, C3 GDP is the x-variable, and C2 Internet is the y-variable.

To find the correlation :

Click: **Stat > Basic Statistics > Correlation**

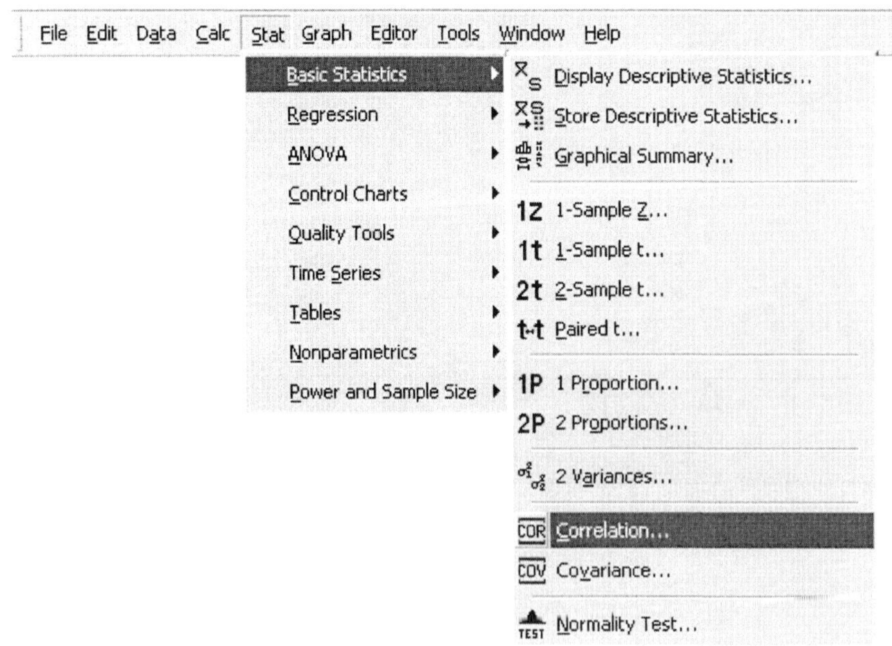

Enter: Variables: **C3 C2**
Click: **OK**

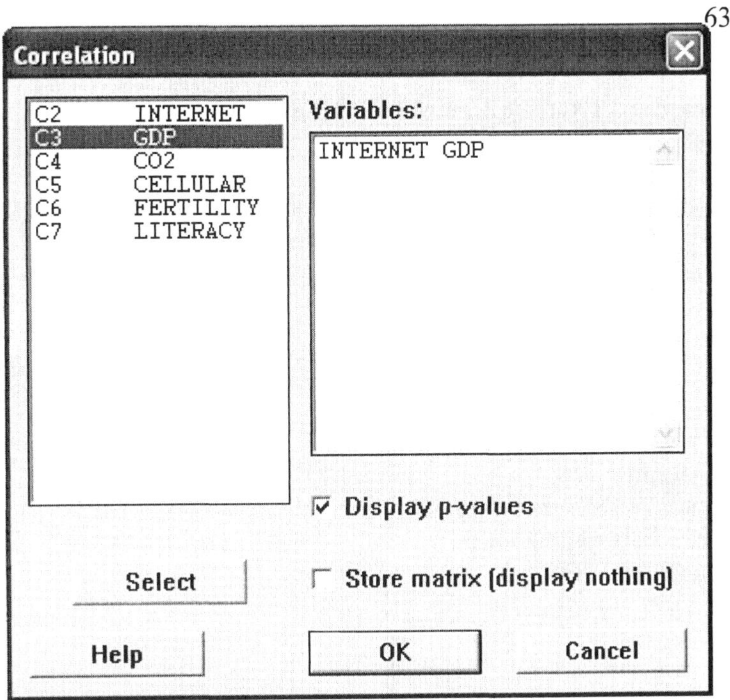

The correlation will be displayed in the Session Window.

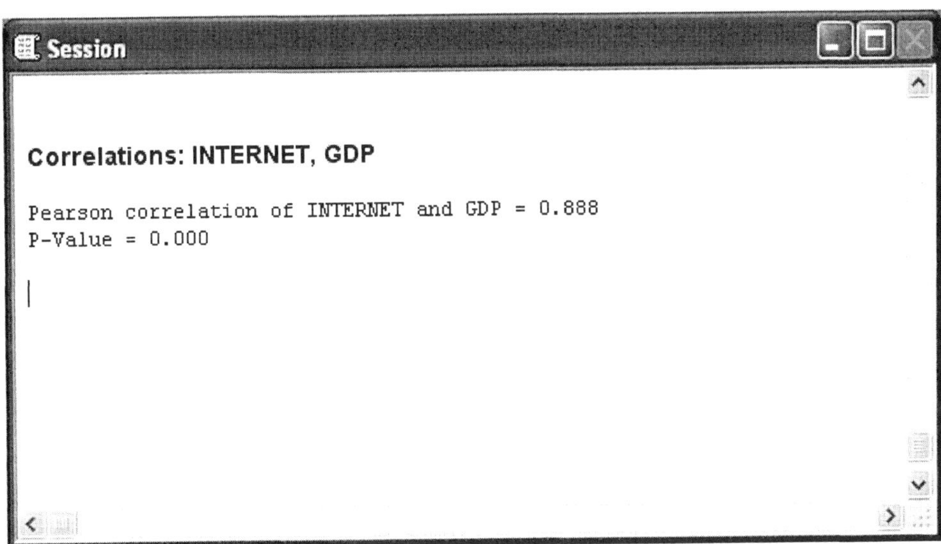

Chapter 3 Exercise 3.19 Which Mountain Bike to Buy?
Computing Correlation

Open the worksheet **mountain_bike**, which is found in the MINITAB folder of the data disk.

Since we are interested in whether and how weight affects the price, C2 weight is the x-variable, and C1 price is the y-variable.

To create a scatterplot of the data
Click: **Graph>Scatterplot > Simple > OK**
Select: Labels **enter an appropriate title**
Click: **OK OK**

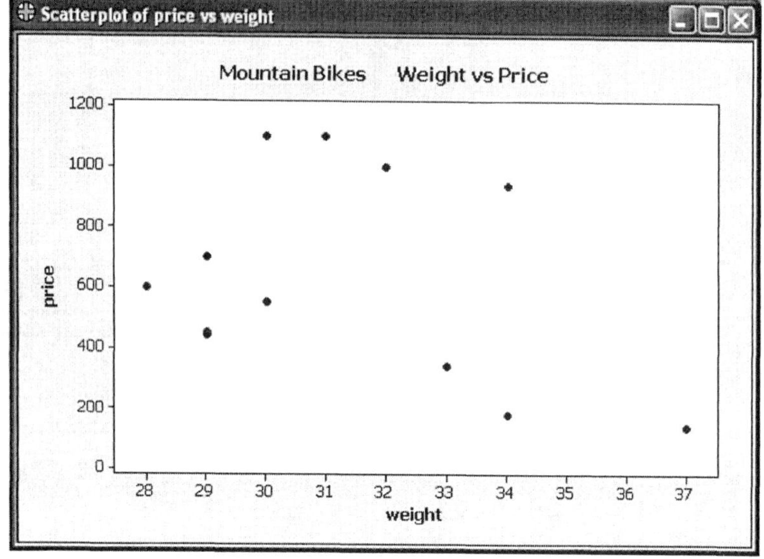

To find the correlation :
Click: **Stat > Basic Statistics > Correlation**
Enter: Variables: **C2 C1**
Click: **OK**

The correlation will be displayed in the Session Window.

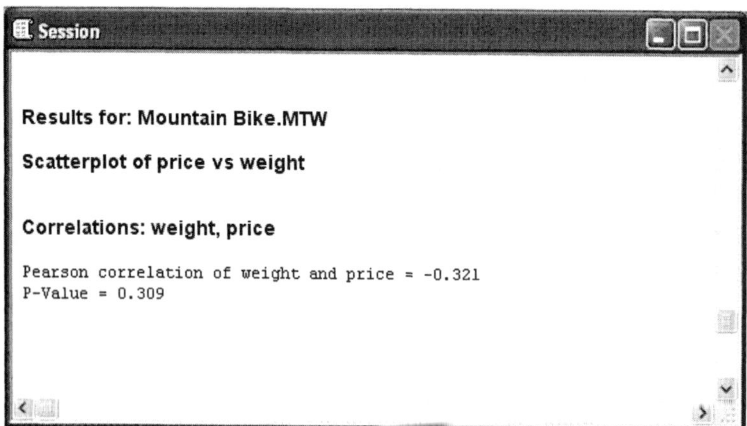

Chapter 3 Exercise 3.21 Buchanan vote
Computing Correlation

Open the worksheet **Buchanan_and_the_butterfly_ballot** which is in the MINITAB folder of the data disk.

Generating a box plot for Gore and Buchanan:

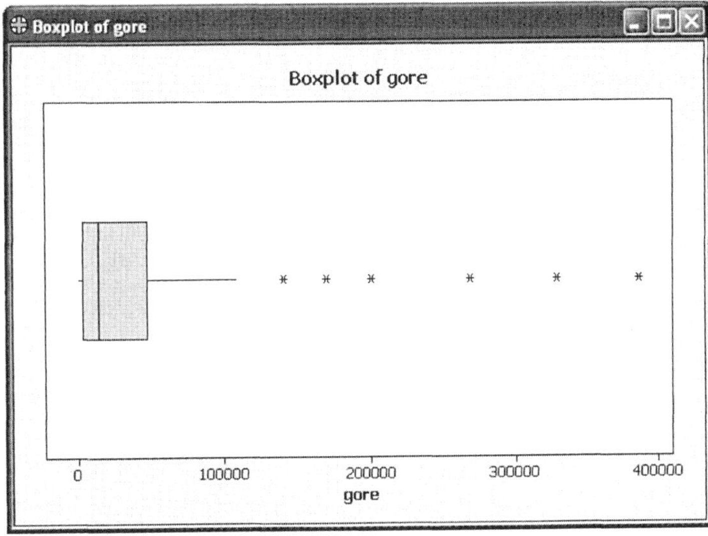

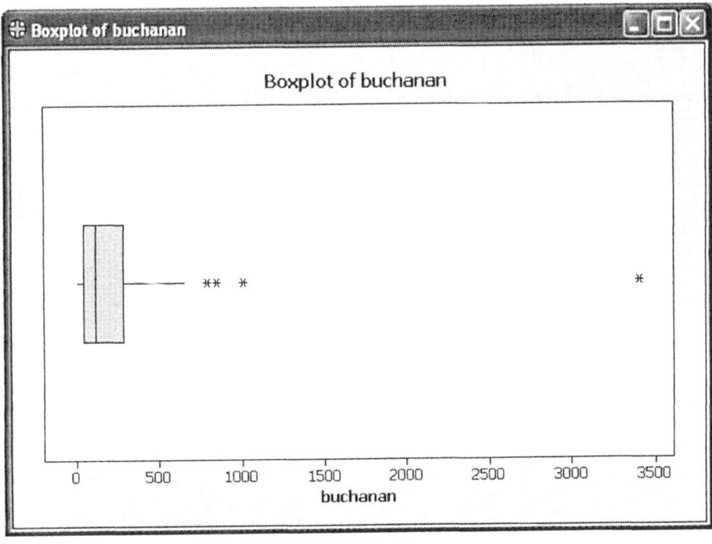

Note the different scales when comparing the two distributions.

To create a scatterplot of the data where C3 gore is the x-variable, and C5 buchanan is the y-variable

Click: **Graph > Scatterplot > Simple > OK**
Enter: Y Variables **C5**
X Variables **C3**
Select: Labels **enter an appropriate title**
Click: **OK OK**

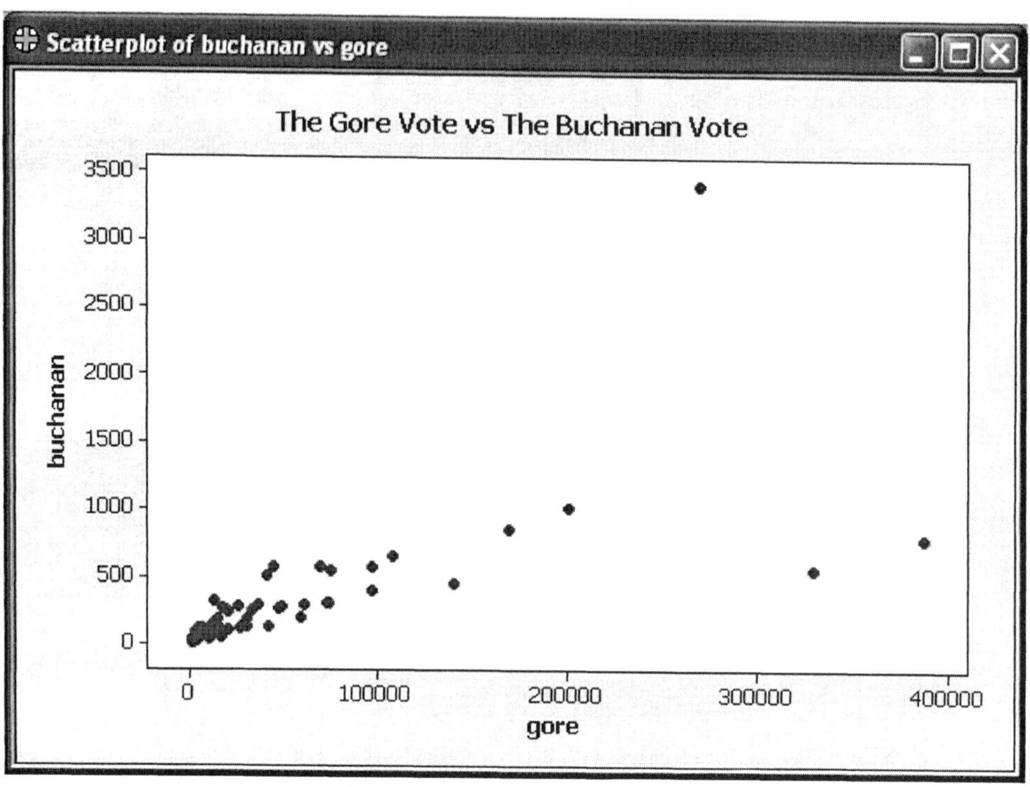

To find the correlation :

Click: **Stat > Basic Statistics > Correlation**
Enter: Variables: **C5 C3**
Click: **OK**

The correlation will be displayed in the Session Window.

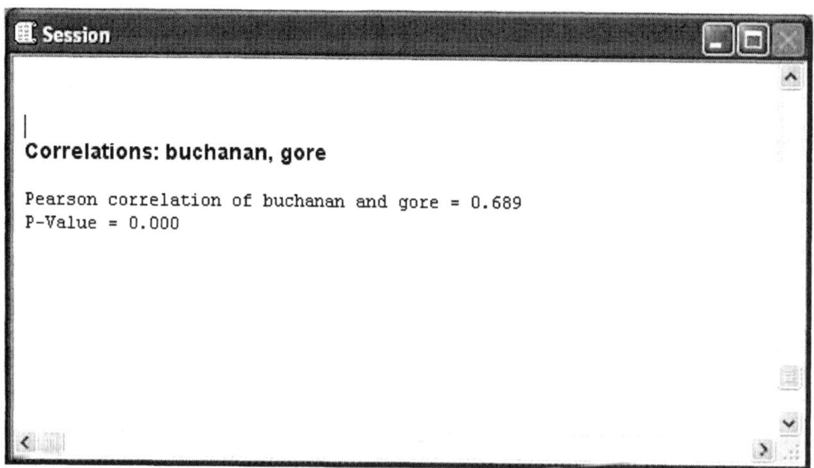

Chapter 3 Example 9 How Can We Predict Baseball Scoring Using Batting Average? Generating the Regression equation

Open the worksheet **al_team_statistics** from the MINITAB folder of the data disk.

We will first generate the scatterplot:

Choose: **Graph > Scatterplot**

Select: **Simple**

Enter: Y variables: **RUNS_AVG** X variables: **BAT_AVG**

Click: **Label:**

Title: **Predicting Scores Based on Batting Averages**

Click: **OK**

We get the following graph:

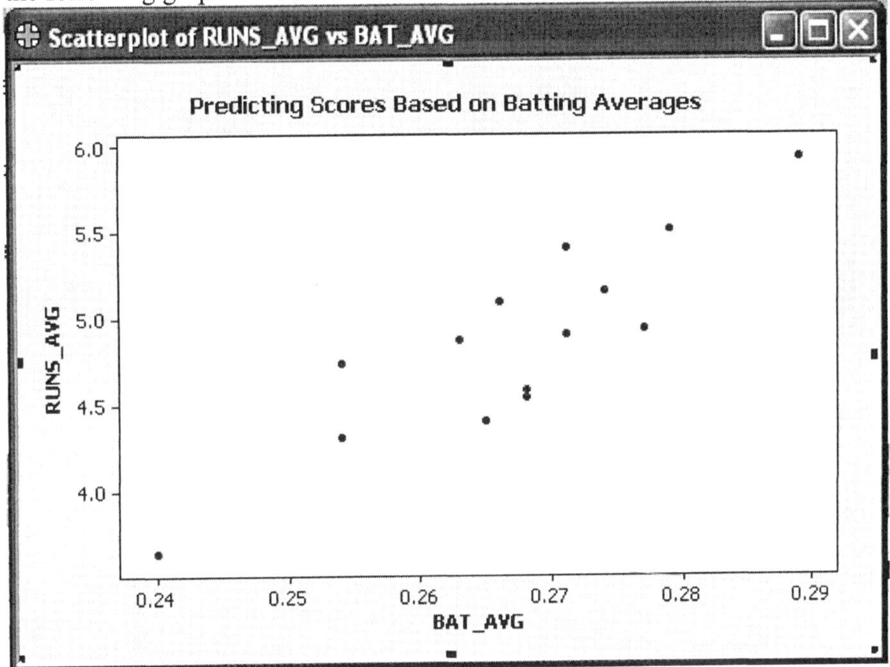

Now we will calculate the regression equation:

Choose:

Stat > Regression > Regression

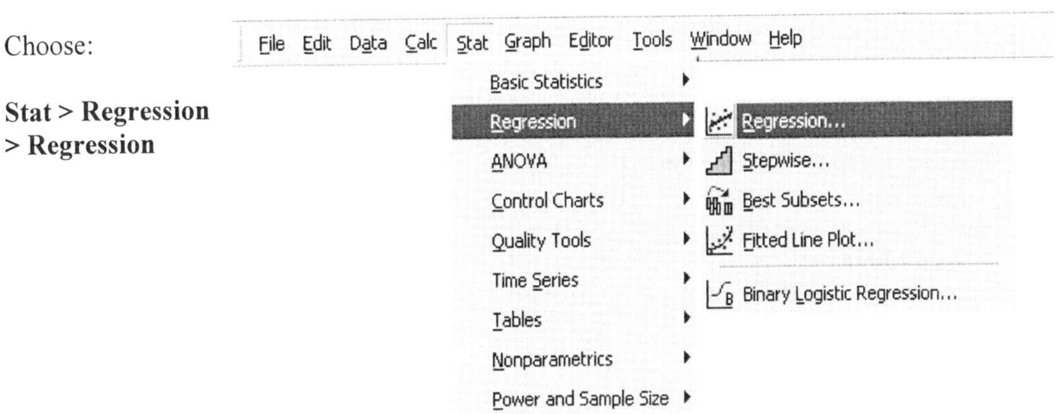

Enter:
Response: **C5**
Predictors: **C4**
Click: **OK**

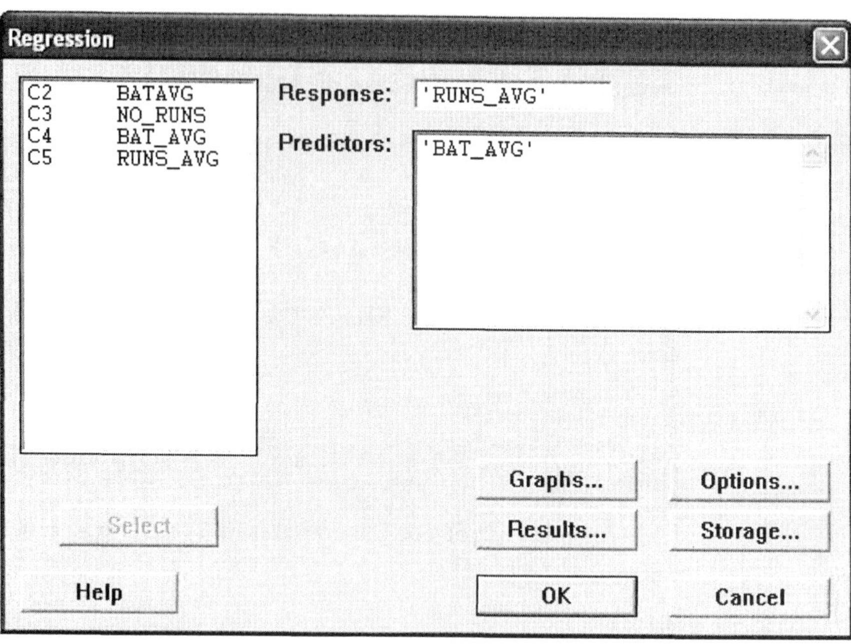

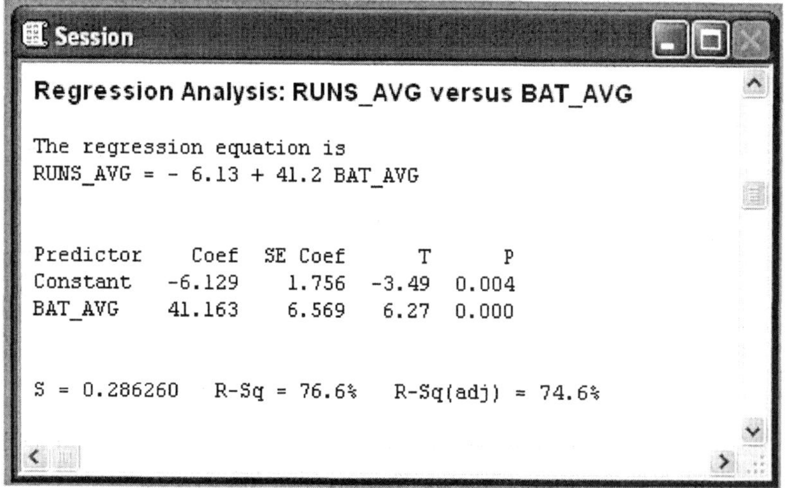

Chapter 3 Exercise 3.35 Mountain Bikes Revisited
Generating a regression equation

We first get the scatterplot:

Choose: **Graph > Scatterplot**
Select: **With Regression**
Enter: Y variables: **C1** X variables: **C2**
Click: **Label:**
Title: **Weight vs Price of Mountain Bikes**

Click: **OK**

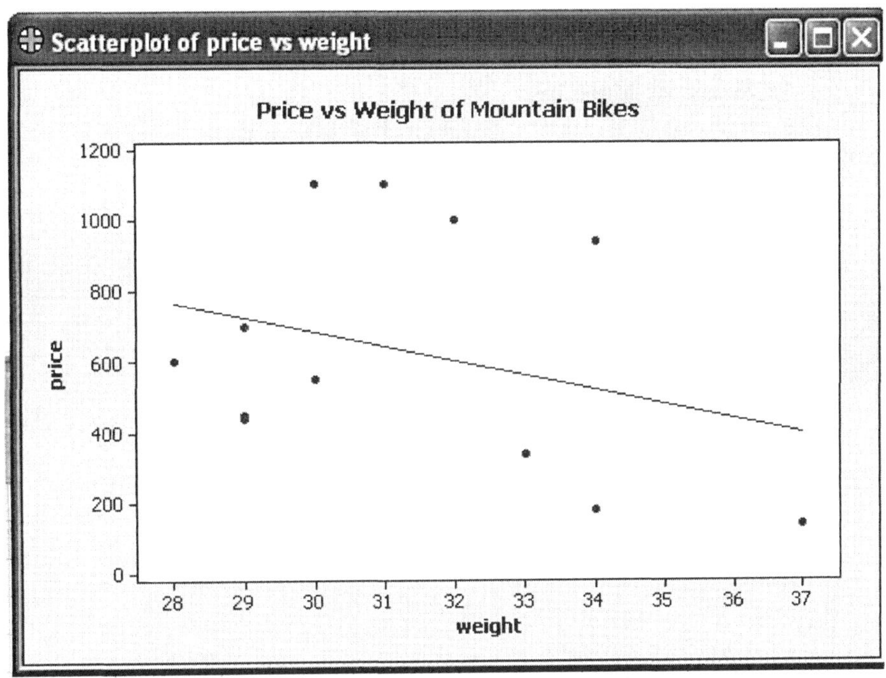

Find the regression equation:

Choose: **Stat > Regression > Regression**

Enter: Response: **price**

Predictors: **weight**

Click: **OK**

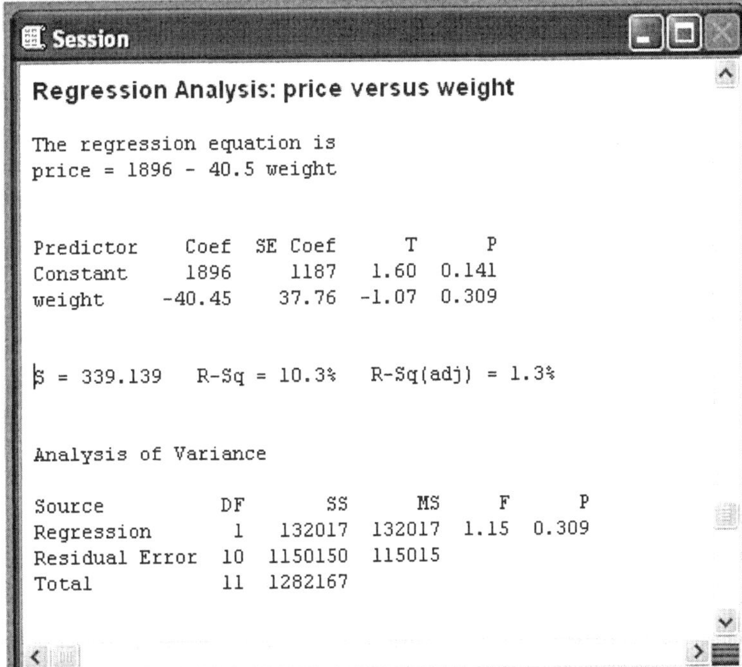

Session

Regression Analysis: price versus weight

```
The regression equation is
price = 1896 - 40.5 weight


Predictor     Coef  SE Coef      T      P
Constant      1896     1187   1.60  0.141
weight      -40.45    37.76  -1.07  0.309


S = 339.139   R-Sq = 10.3%   R-Sq(adj) = 1.3%


Analysis of Variance

Source          DF       SS      MS     F      P
Regression       1   132017  132017  1.15  0.309
Residual Error  10  1150150  115015
Total           11  1282167
```

Predicted price for a 30 lb bike:

price=1896-40.5(30)= 681

Chapter 3 Exercise 3.36: Mountain bike and suspension types
Generating Regression Equations

Choose: **Graph > Scatterplot**
Select: **with groups**

Enter: Y variables: **price**
X variables: **weight**
Enter: Categorical variables for grouping: **C3-T**

Click: **Label:**
Title: **Price vs Weight of Mountain Bikes**
Click: **OK**

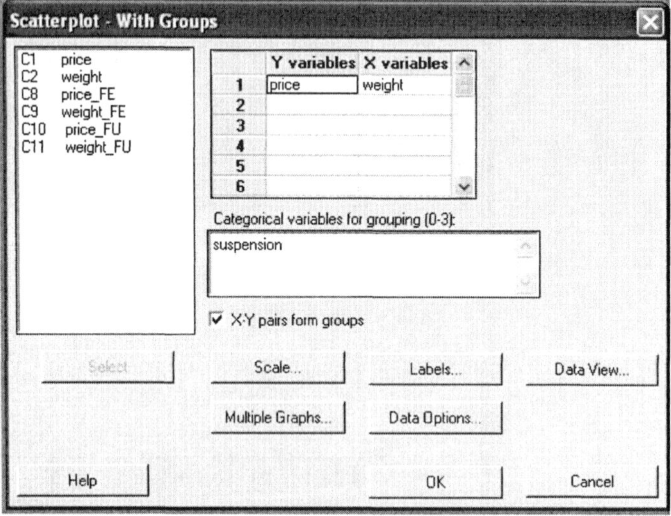

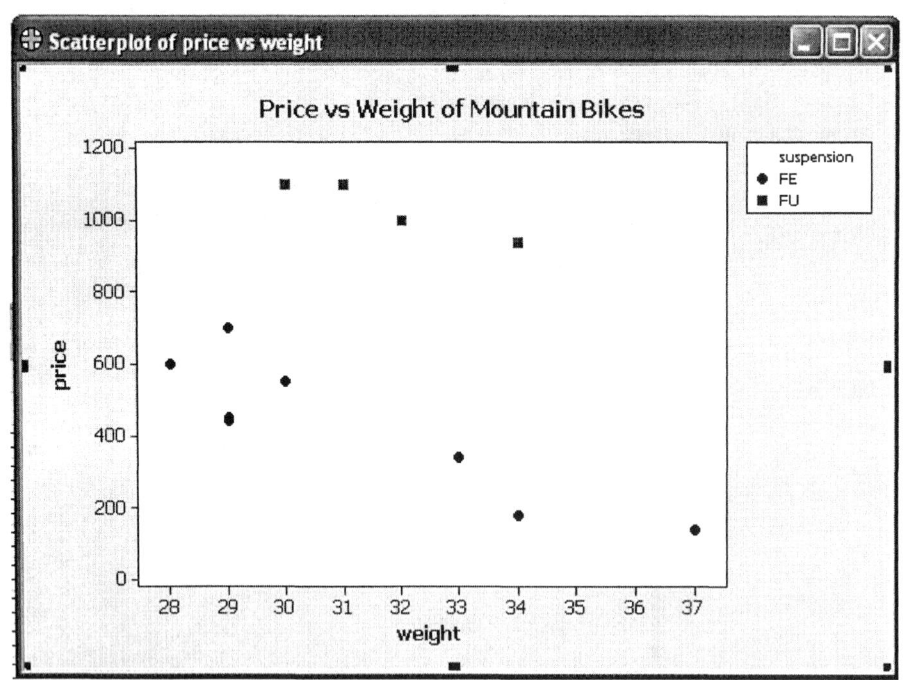

Now we do separate regression equations based on suspension type.

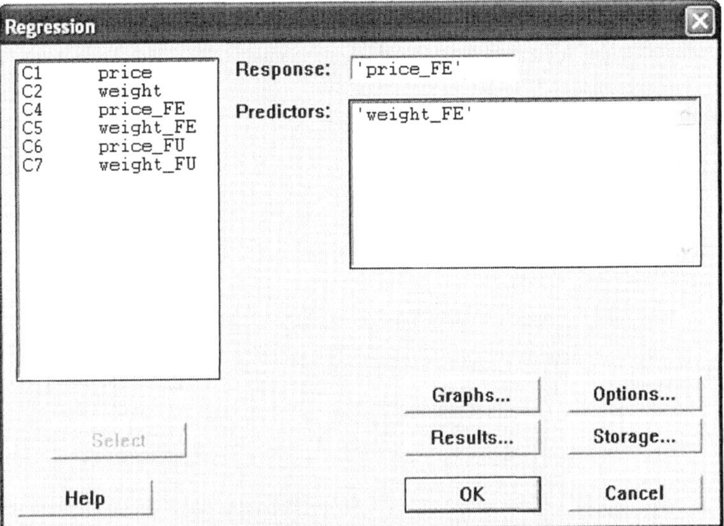

We get the following for front end suspension.

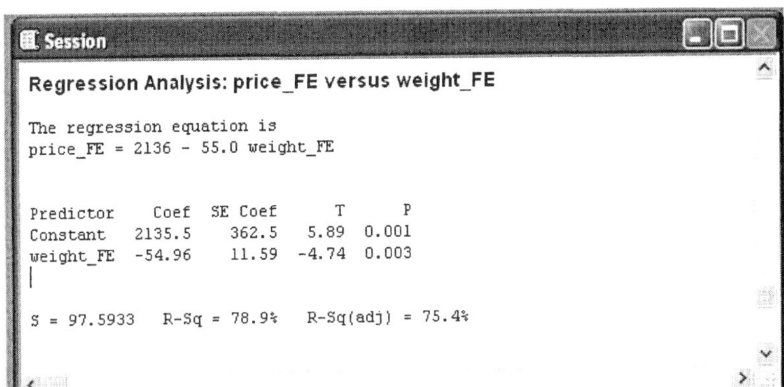

Session

Regression Analysis: price_FE versus weight_FE

```
The regression equation is
price_FE = 2136 - 55.0 weight_FE


Predictor    Coef  SE Coef      T      P
Constant   2135.5    362.5   5.89  0.001
weight_FE  -54.96    11.59  -4.74  0.003


S = 97.5933   R-Sq = 78.9%   R-Sq(adj) = 75.4%
```

Repeating the same steps for full suspension bikes we obtain:

Session

Regression Analysis: price_FU versus weight_FU

```
The regression equation is
price_FU = 2432 - 44.0 weight_FU


Predictor    Coef  SE Coef      T      P
Constant   2432.0    318.8   7.63  0.017
weight_FU  -44.00    10.03  -4.39  0.048


S = 29.6648   R-Sq = 90.6%   R-Sq(adj) = 85.9%
```

Chapter 3 Exercise 3.79 High School Graduation Rates and Health Insurance Scatterplot, Correlation, and Regression

Open the worksheet **hs_graduation_rates** from the MINITAB folder of the data disk.

Choose: **Graph > Scatterplot**
Select: **With Regression**
Enter: Y variables: **C4** X variables: **C3**
Click: **Label:**
Title: **Health Insurance vs HS Graduation Rate**
Click: **OK**

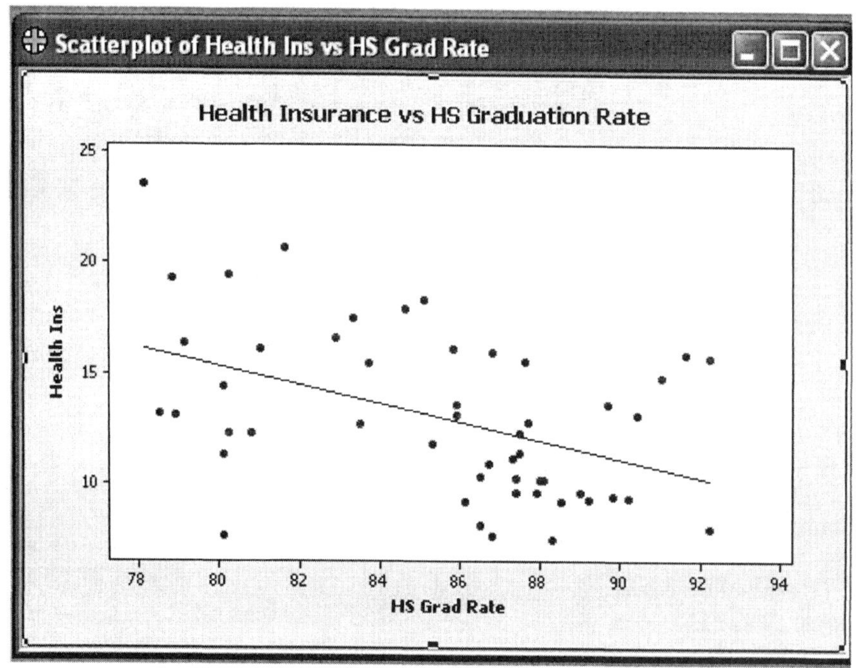

Computing correlation:

Choose: **Stat > Basic Statistics > Correlation**
Enter: Variables: **C3 C4**
Click: **OK**

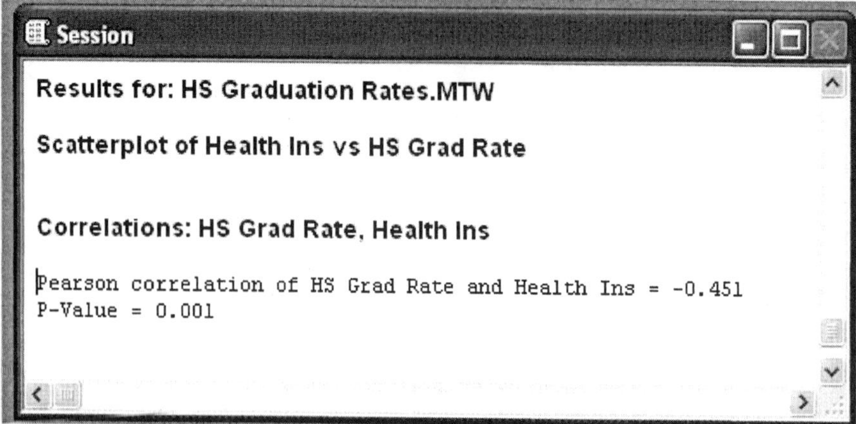

Generating the regression equation:

Choose: **Stat > Regression > Regression**
Enter: Response: **C4**
Predictors: **C3**
Click: **OK**

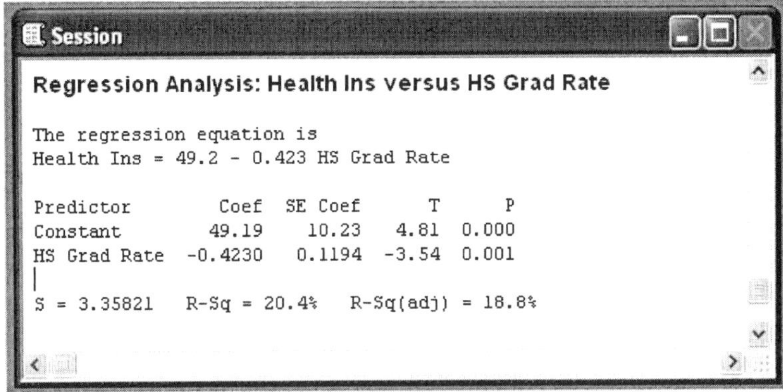

Chapter 3 Example 10 How Can We Detect An Unusual Vote Total?
Looking at residuals

Open the worksheet **Buchanan_and_the_butterfly_ballot** from the MINITAB folder of the data disk.

Choose: **Stat > Regression > Regression**
Enter: Response: **C5**
Predictors: **C2**
Click: **OK**

Click : Graphs
Select: Residual Plots / Individual Plots / **Histogram of residuals**

Click: **OK**

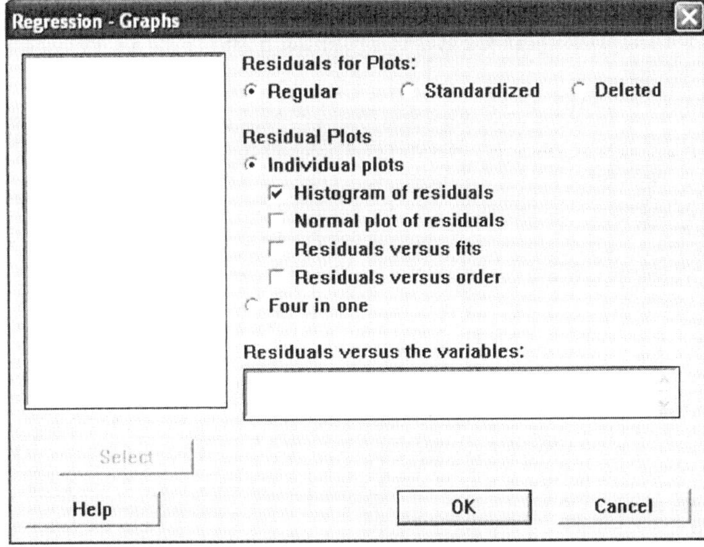

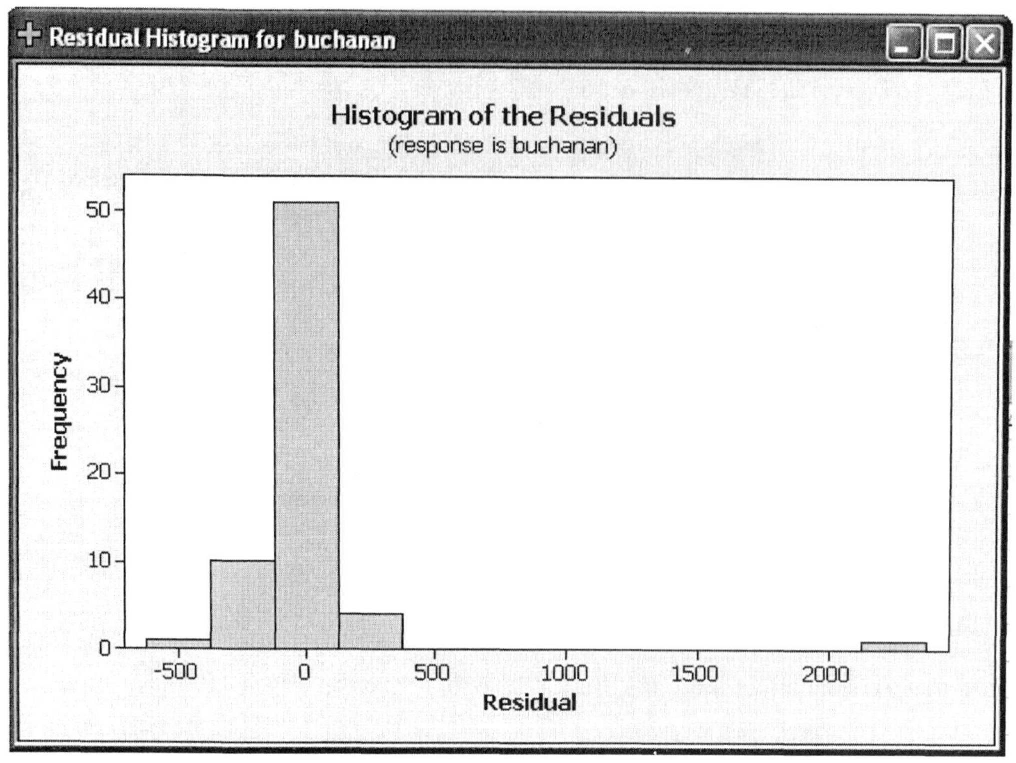

Chapter 3 Exercise 3.29 Regression between cereal sodium and sugar Residuals

Open the worksheet **CEREAL** from the MINITAB folder of the data disk.

Choose: **Graph > Scatterplot**
Select: **with regression**
Enter:
Y variables: **C2**
X variables: **C1**

Click: **Label:**
Enter: Title: **Sugar vs Sodium in Cereal**
Click: **OK**

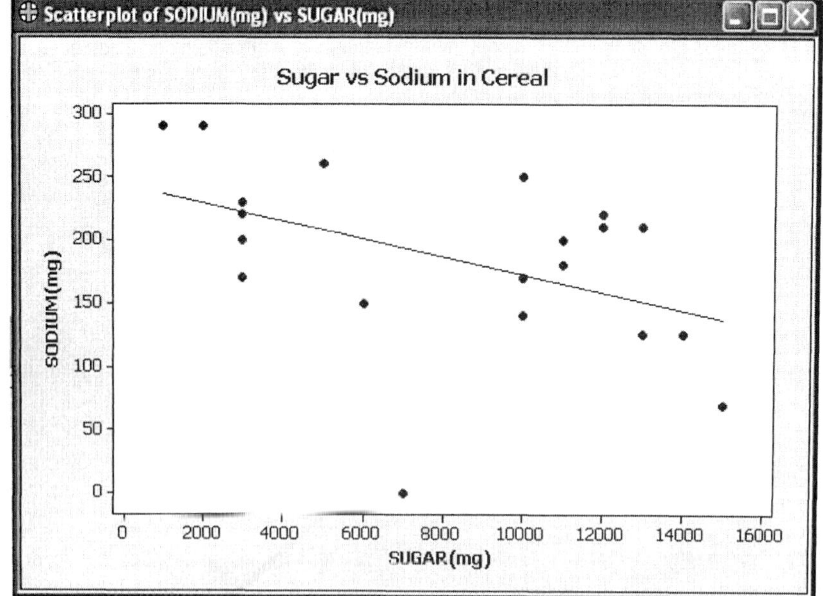

Histogram of the residuals

Choose: **Stat > Regression > Regression**
Enter: Response: **C2**
Predictors: **C7**
Click: **OK**
Click : Graphs
Select: Residual Plots / Individual Plots / **Histogram of residuals**

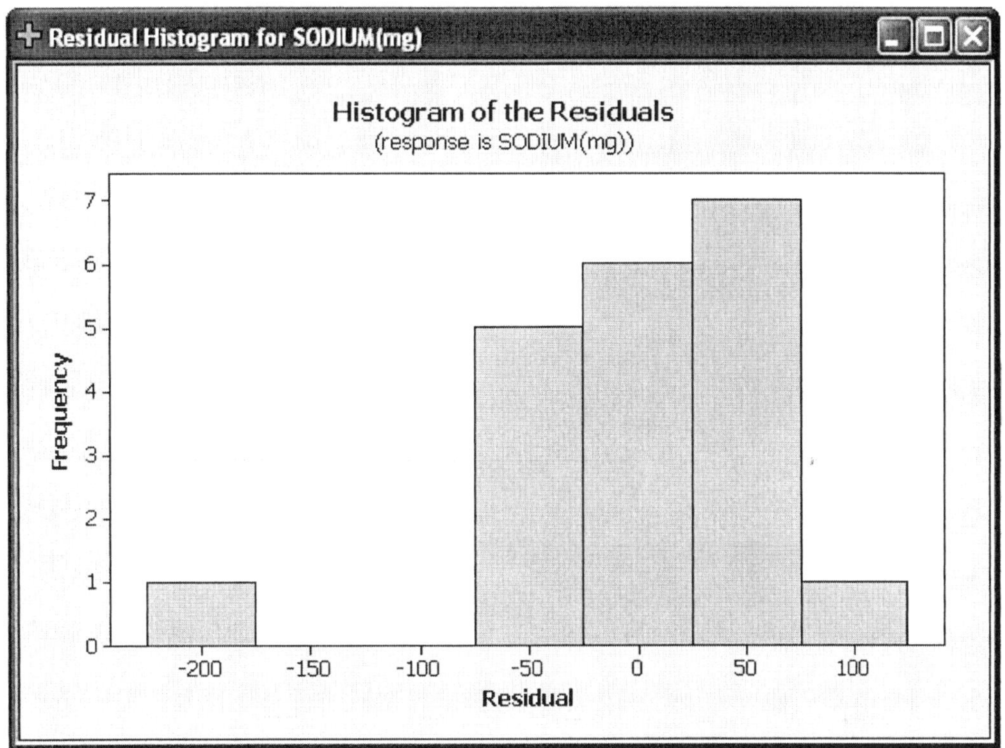

Chapter 3 Example 12 How Can We Forecast Future Global Warming?
Regression on a time series plot

Open the worksheet **central_park_yearly_temps** from the MINITAB folder of the data disk.

In order to get a regression line printed on a time series plot we are going to use the scatterplot graph with regression, and choose to connect the data points.

Choose: **Graph > Scatterplot**
Select: **With Regression**
Enter: Y variables: **C1** X variables: **C3**
Click: **Label:**
Title: **Central Park Mean Annual Temperatures**

Click Data View>Data Display
Check **symbols and connect line**
Click: **OK**

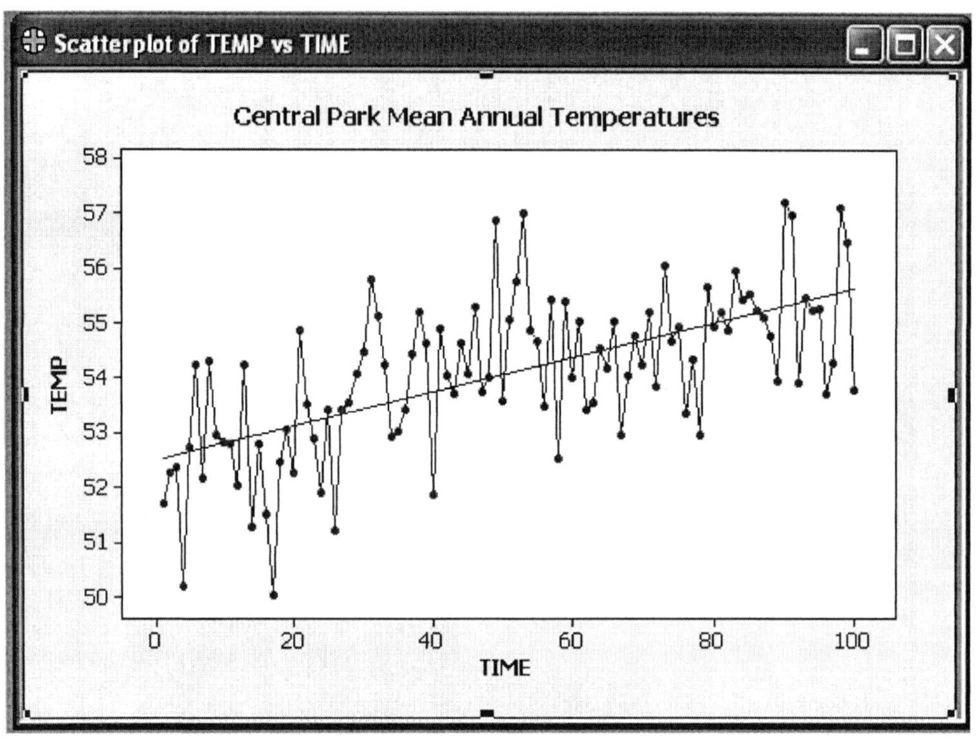

Chapter 3 Exercise 3.40 US Average Annual Temperatures
Trend Lines

Open the worksheet **us_temperatures** from the MINITAB folder of the data disk.

Choose: **Stat > Regression>Fitted Line Plot**
Enter: Response: **C2**
Predictors: **C1**
Click: **OK**

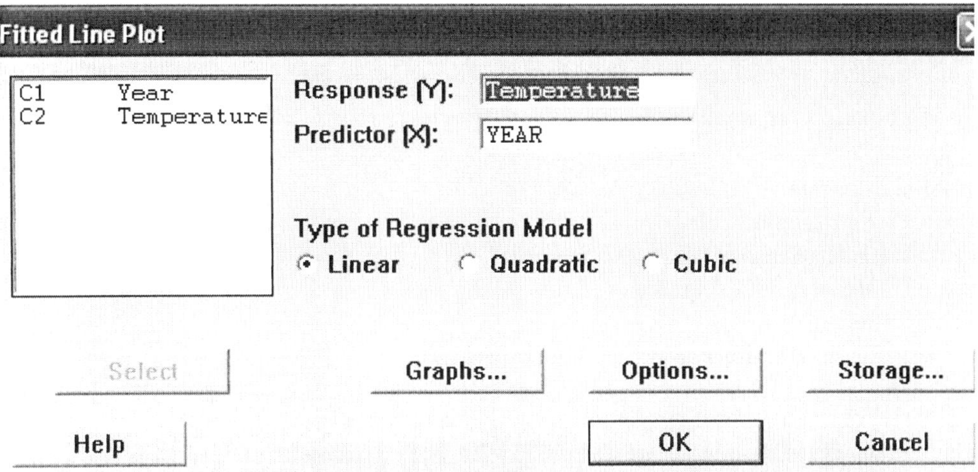

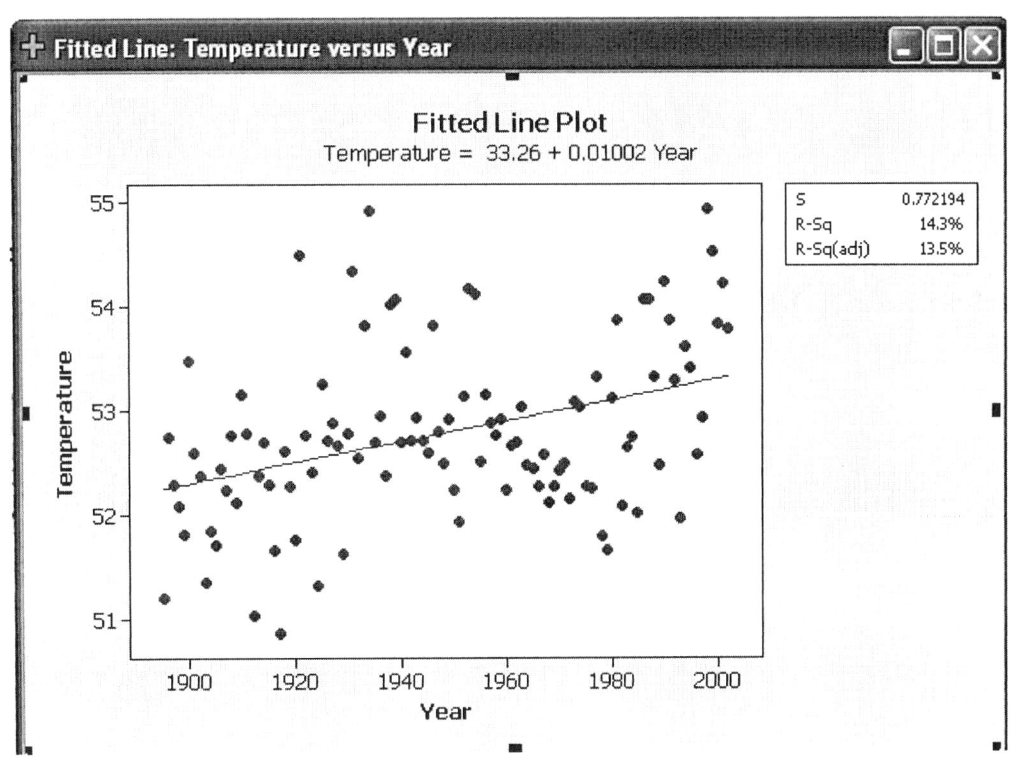

b) The equation, given at the top of the chart is: Temperature = 33.26 + 0.01002 Year. Predict the temperature for a particular year by substituting that year into the equation.

Chapter 3 Example 13 Is Higher Education Associated with Higher Murder Rates? Influential Outliers

Open the worksheet **us_statewide_crime** from the MINITAB folder of the data disk.

Scatterplot of murder rate vs college
Choose: **Graph > Scatterplot**
Select: **With Regression**
Enter: Y variables: **C3** X variables: **C6**
Click: **Label:**
Enter: Title: **Murder rate vs College education**
Click: **OK**

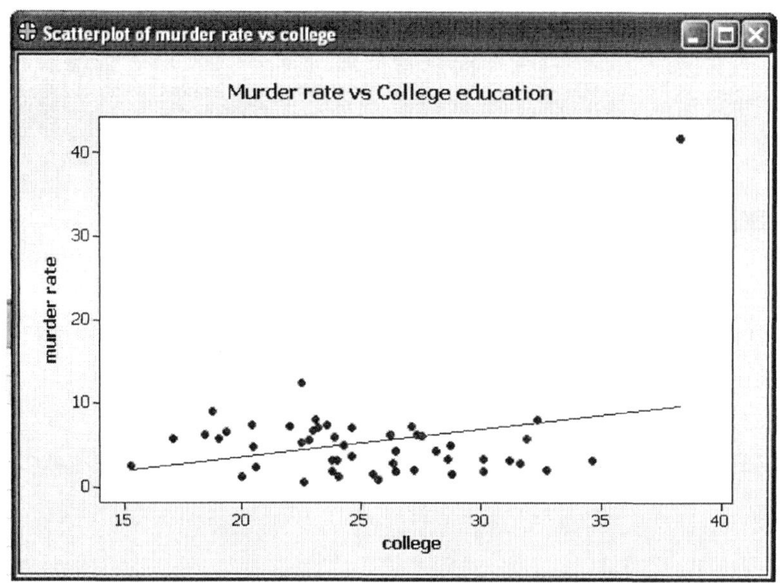

Now calculate the regression equation:

Choose: **Stat > Regression > Regression**
Enter: Response: **C3**
Predictors: **C6**
Click: **OK**

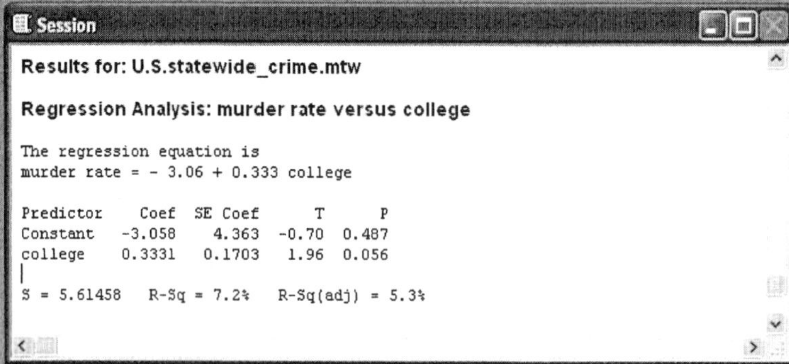

Session

Results for: U.S.statewide_crime.mtw

Regression Analysis: murder rate versus college

```
The regression equation is
murder rate = - 3.06 + 0.333 college

Predictor    Coef  SE Coef      T      P
Constant   -3.058    4.363  -0.70  0.487
college    0.3331   0.1703   1.96  0.056

S = 5.61458   R-Sq = 7.2%   R-Sq(adj) = 5.3%
```

Redoing the entire process, excluding the data for D.C. we get:

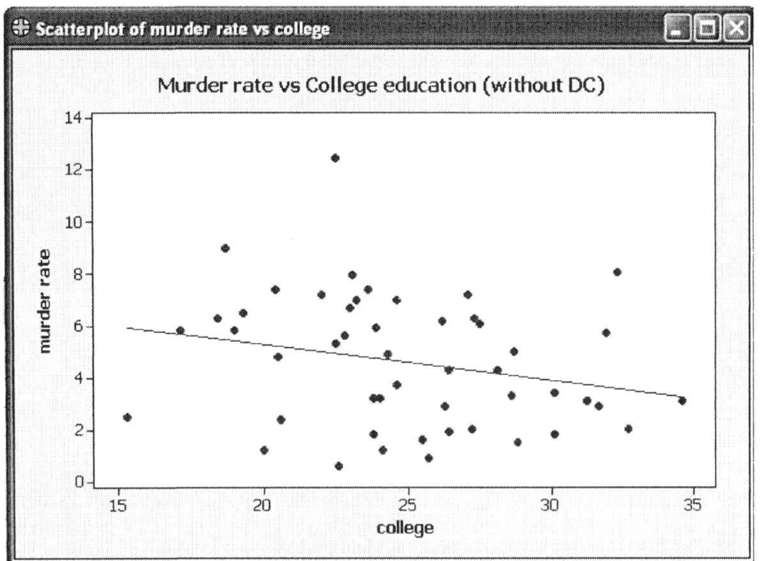

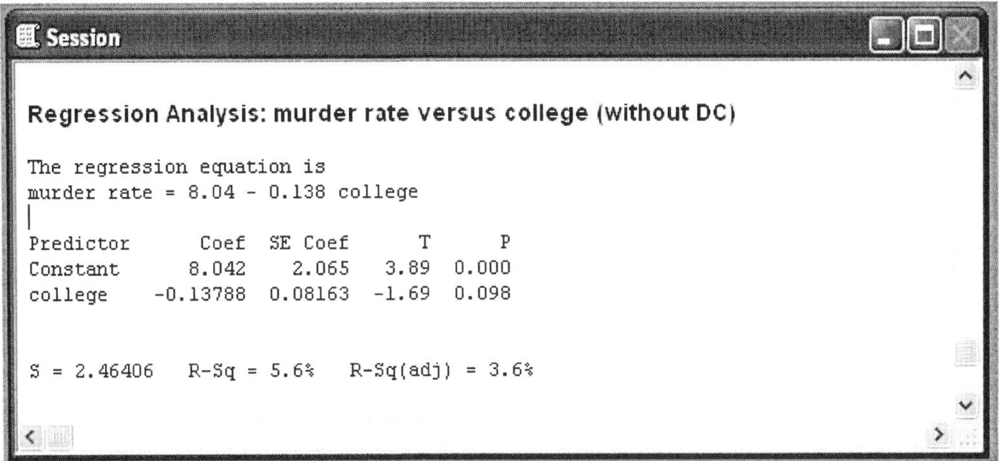
Session

Regression Analysis: murder rate versus college (without DC)

```
The regression equation is
murder rate = 8.04 - 0.138 college

Predictor       Coef  SE Coef      T      P
Constant       8.042    2.065   3.89  0.000
college     -0.13788  0.08163  -1.69  0.098


S = 2.46406   R-Sq = 5.6%   R-Sq(adj) = 3.6%
```

Chapter 3 Exercise 3.41 Murder and Education
Application of the Regression Model

Using the MINITAB results of the previous problem

a) In the equation: murder rate = - 3.06 + 0.333 college we substitute 15% and 40% we obtain 1.89 and 10.26 respectively.

b) In the equation: murder rate = 8.04 - 0.138 college we substitute 15% and 40% we obtain 5.9 and 2.4 respectively.

Chapter 3 Exercise 3.45 Regression Between Sodium and Sugar Scatterplot, Correlation, and Regression

Open the worksheet **CEREAL** from the MINITAB folder of the data disk.

a) Scatterplot:
Choose: **Stat > Regression>Fitted Line Plot**
Enter: Response: **C2** Predictor: **C5**

Note: an advantage to the fitted line plot versus the scatterplot is that the regression equation is given in the plot and is added to the scatterplot.

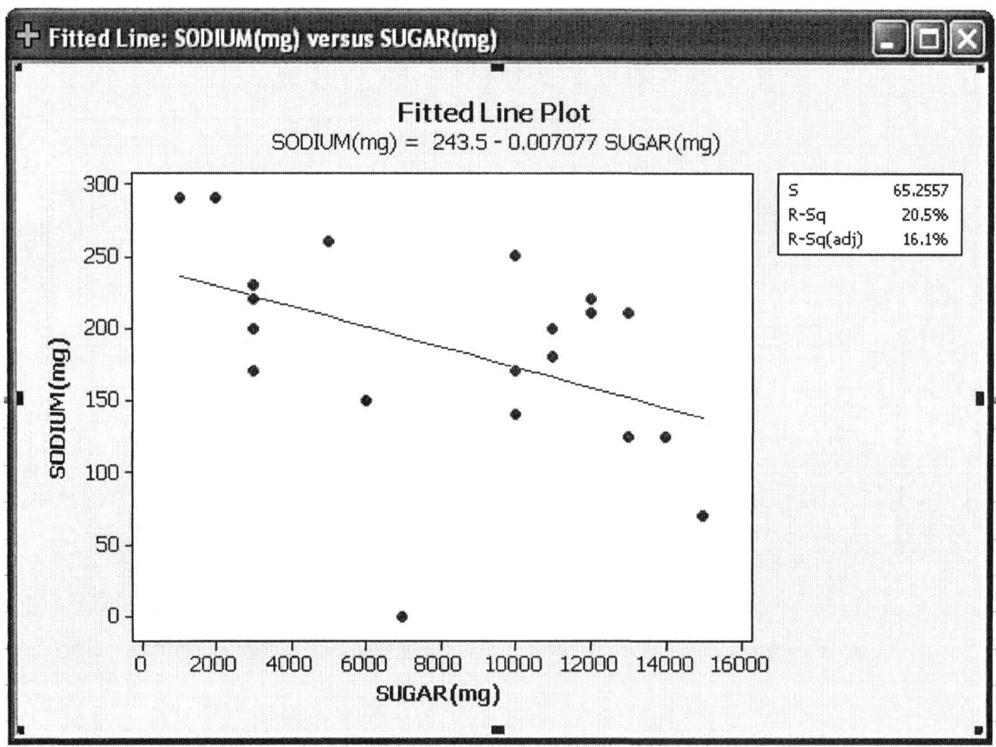

Note the outlier at (7000,0) for Frosted Mini Wheats. Repeating the regression analysis without that point we get:

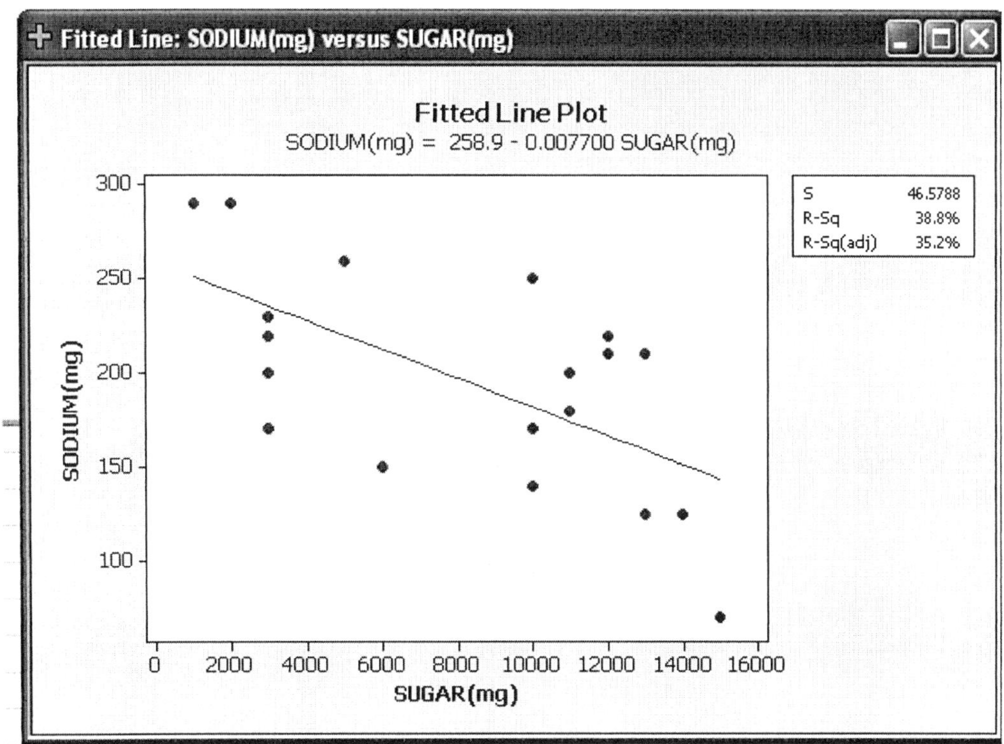

What are your conclusions?

Chapter 4 Example 5 Auditing the Accounts of a School District
Random Selection

To use random numbers within MINITAB to select 10 accounts to audit in a school district that has 60 accounts:

Choose: **Calc > Random Data > Integer**

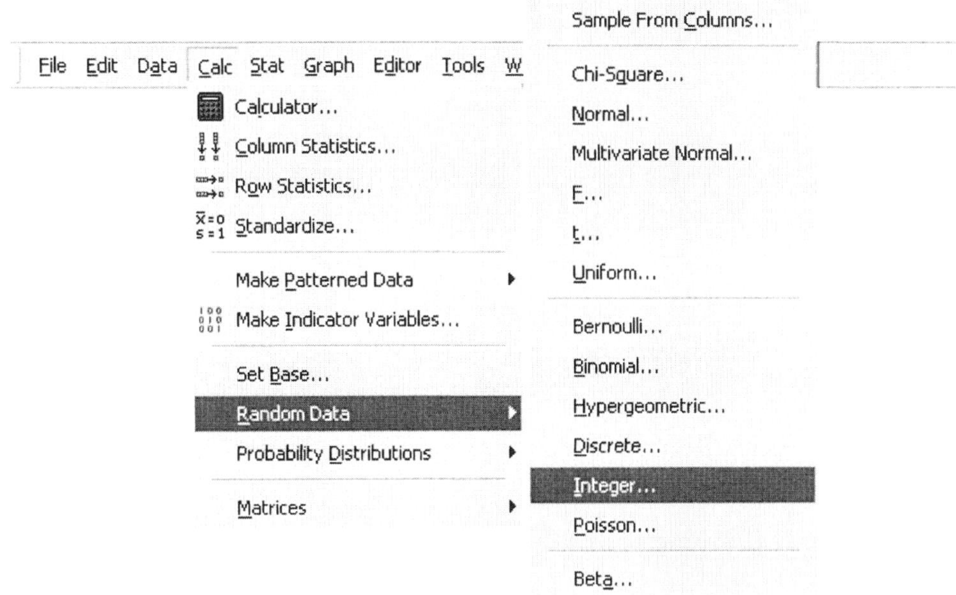

Enter: Generate **10** rows of data
Store in column(s): **C1**
Minimum value: **1**
Maximum value: **60**
Click: **OK**

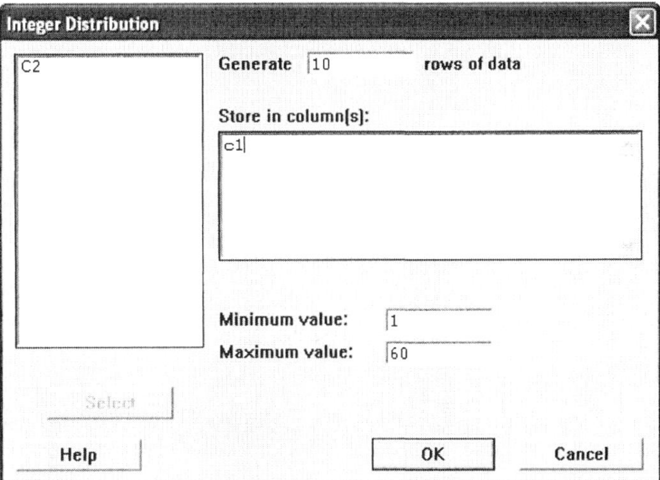

↓	C1	C2
1	18	
2	41	
3	18	
4	41	
5	51	
6	13	
7	29	
8	13	
9	35	
10	23	

There is now a random sample of 10 accounts in C1.

Note the repetition of the number 18. We will need to generate additional random numbers to actually obtain a random sample of size 10.

Chapter 4 Exercise 4.18 Auditing Accounts
Random Selection

To use random numbers within MINITAB to select 10 accounts to audit in a school district that has 60 accounts:

Choose: **Calc > Random Data > Integer**
Enter: Generate **10** rows of data
Store in column(s): **C1**
Minimum value: **1**
Maximum value: **60**
Click: **OK**

There is now a random sample of 10 accounts in C1.

Issuing the same commands again, but storing in C2, we can see that a different random sample of size 10 has been selected. Since this is a random process, the results are different every time.

↓	C1	C2
1	25	60
2	16	5
3	46	14
4	13	48
5	12	3
6	45	46
7	37	33
8	5	20
9	39	25
10	43	59

Chapter 6 Example 7 What IQ Do You Need to Get Into MENSA?
Determining Normal Probabilities

Stanford-Binet IQ scores are approximately normally distributed with $\mu = 100$ and $\sigma = 16$. To be eligible for MENSA, you must rank at the 98th percentile (i.e. you must score in the top 2%). Find the lowest IQ score that still qualifies for Mensa membership.

Choose:
Calc > Probability Distributions > Normal

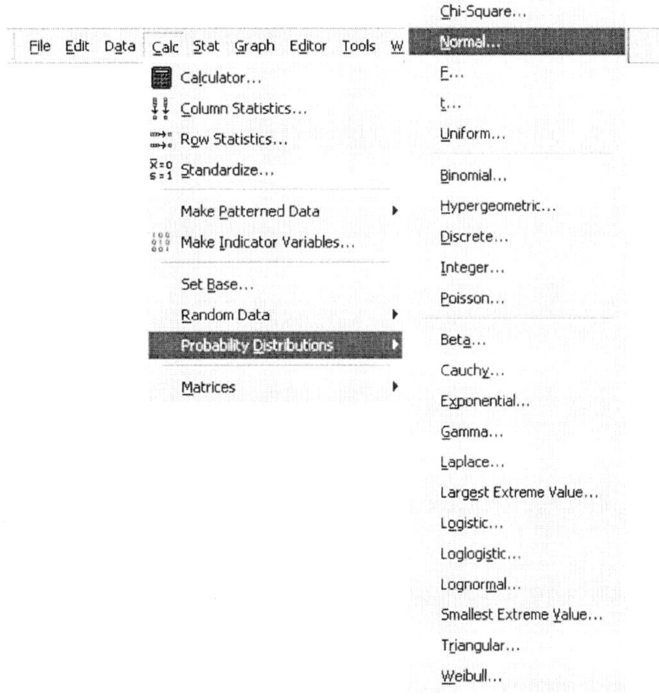

Select:
Inverse Cumulative Probability
Enter: Mean : **100**
Standard deviation: **16**
Enter: Input constant: **.98**
(area to the left of the value we wish to find)
Click: **OK**

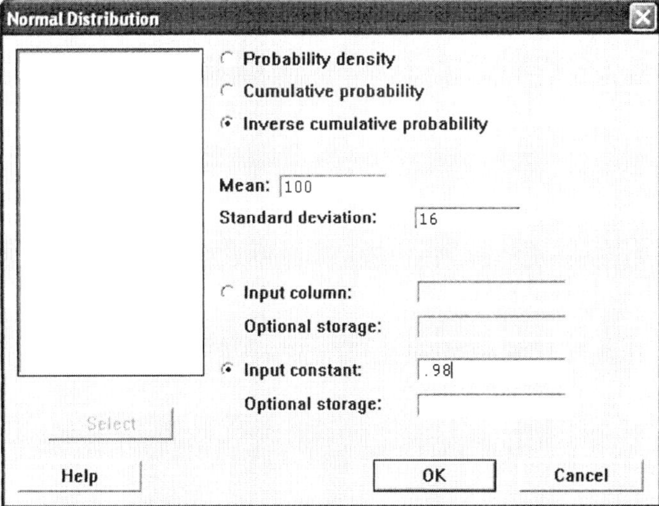

The X-value should appear in the Session window. Notice that the test score required to join Mensa is 133.

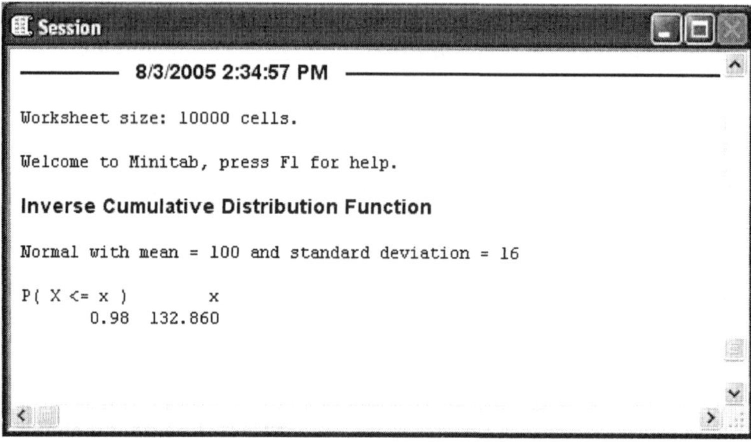
```
8/3/2005 2:34:57 PM

Worksheet size: 10000 cells.

Welcome to Minitab, press F1 for help.

Inverse Cumulative Distribution Function

Normal with mean = 100 and standard deviation = 16

P( X <= x )        x
       0.98  132.860
```

Chapter 6 Example 8 Finding Your Relative Standing on the SAT
Determining Normal Probabilities

SAT scores are approximately normally distributed with $\mu = 500$ and $\sigma = 100$. If one of your SAT scores was 650, what percentage of SAT scores were higher than yours?

Choose:
Calc > Probability Distributions > Normal
Select: **Cumulative probability**
Enter: Mean : **500**
Standard deviation: **100**
Enter: Input constant: **650**
Click: **OK**

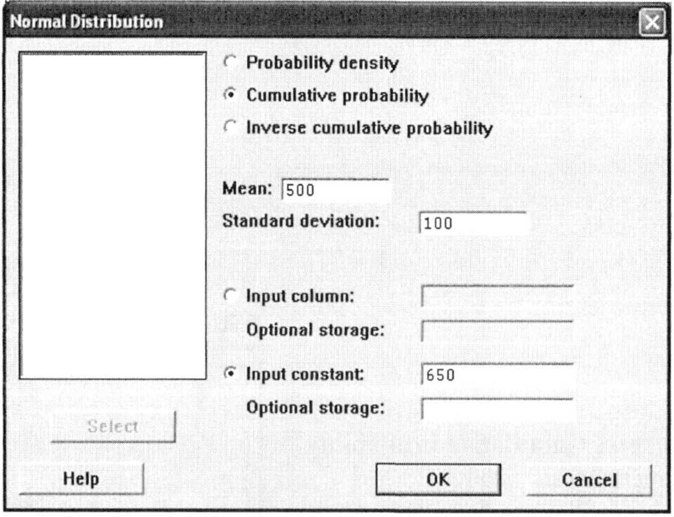

Notice that the result displayed in the Session window is the probability that X is less than 650.

Since we were asked to determine the percentage of SAT scores higher than 650 we must calculate 1 - 0.9332 = 0.0668. Only about 7% of SAT scores fall above 650.

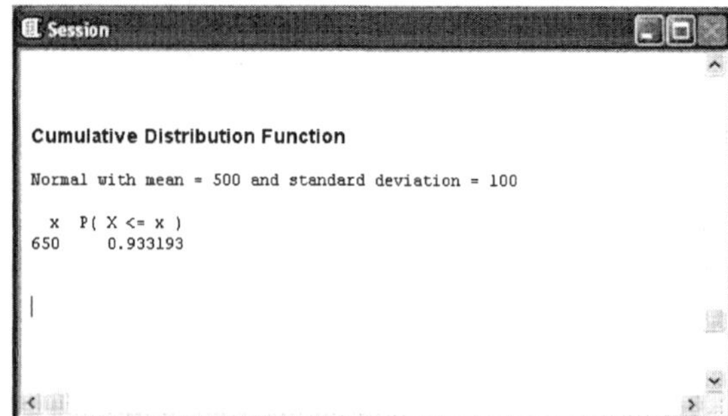
```
Cumulative Distribution Function

Normal with mean = 500 and standard deviation = 100

  x  P( X <= x )
650     0.933193
```

Chapter 6 Example 9 What Proportion of Students Get a Grade of B?
Determining Normal Probabilities

On the midterm exam, an instructor always gives a grade of B to students who score between 80 and 90. One year, the scores on the exam have an approximately normal distribution with $\mu = 83$ and $\sigma = 5$. About what proportion of students get a B?

We will have to compute two cumulative probabilities, one for X = 80 and one for X = 90, and then subtract the two probabilities.

Choose: **Calc > Probability Distributions > Normal**
Select: **Cumulative probability**
Enter: Mean : **83**
Standard deviation: **5**
Select: Input constant:
Enter: **80**
Click: **OK**

then repeat using Input constant: **90**

It follows that about 0.9192 – 0.2743 = 0.6449, or about 64%, of the exam scores were in the B range.

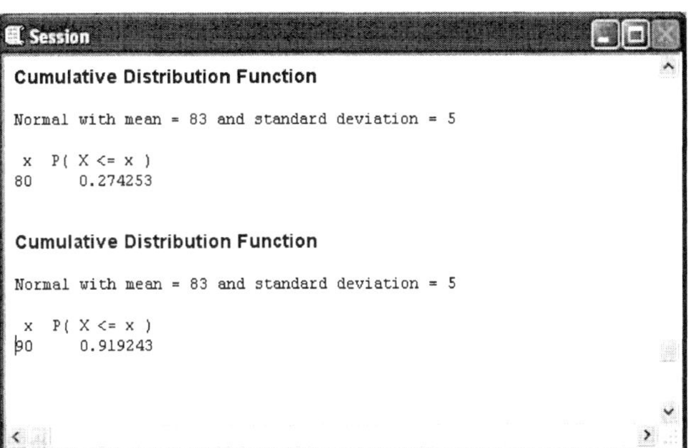

Chapter 6 Exercise 6.21 Blood pressure
Determining Normal Probabilities

In Canada, systolic blood pressure readings has are normally distributed with $\mu = 121$ and $\sigma = 16$. A reading above 140 is considered to be high blood pressure. What proportion of Canadians suffers from high blood pressure?

Choose: **Calc > Probability Distributions > Normal**

Select: **Cumulative probability**
Enter: Mean : **121**
Standard deviation: **16**

Select: Input constant:
Enter: **140**
Click: **OK**

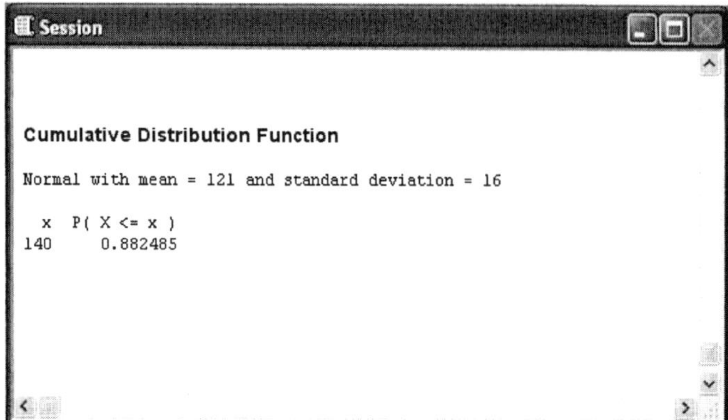

It follows that about 1 – 0.8825 = 0.1175, or about 12%, of Canadians suffer from high blood pressure.

We are also asked to determine the proportion of Canadians having systolic blood pressure in the range from 100 to 140.

We have already computed the P(X < 140) = 0.8825, so we just need to determine P(X < 100) and subtract the two results.

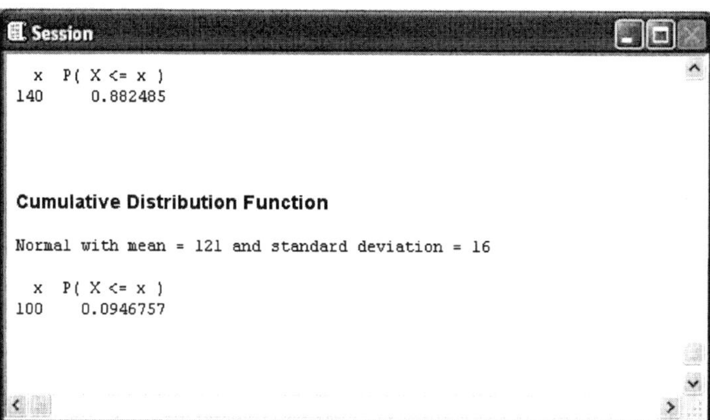

Approximately 0.8825 – 0.0947 = 0.7878, or 78% of Canadians have systolic blood pressures in the range from 100 to 140.

Chapter 6 Exercise 6.23 Mental Development Index
Determining Normal Probabilities

The Mental Development Index of the Bayley Scales of Infant Development is a standardized measure used in observing infants over time. MDI is approximately normally distributed with $\mu = 100$ and $\sigma = 16$.

What proportion of children has MDI of at least 120?

Choose: **Calc > Probability Distributions > Normal**

Select: **Cumulative probability**
Enter: Mean : **100**
Standard deviation: **16**
Select: Input constant:
Enter: **120**
Click: **OK**

It follows that about 1 – 0.8944 = 0.1056, or about 11%, of children have an MDI of at least 120.

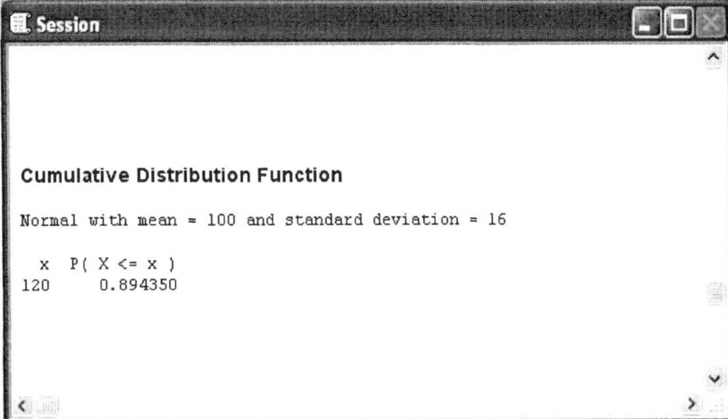

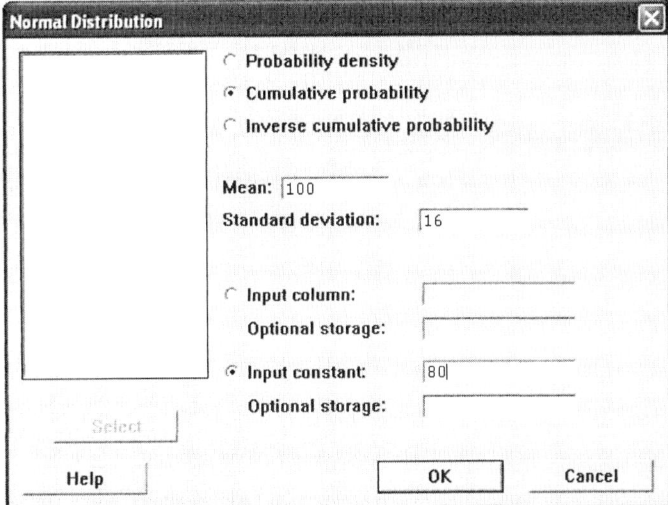

We are also asked to determine the proportion of children having an MDI of at least 80.

Approximately 1.0 – 0.1057 = 0.8943, or 89% of children have an MDI of at least 80.

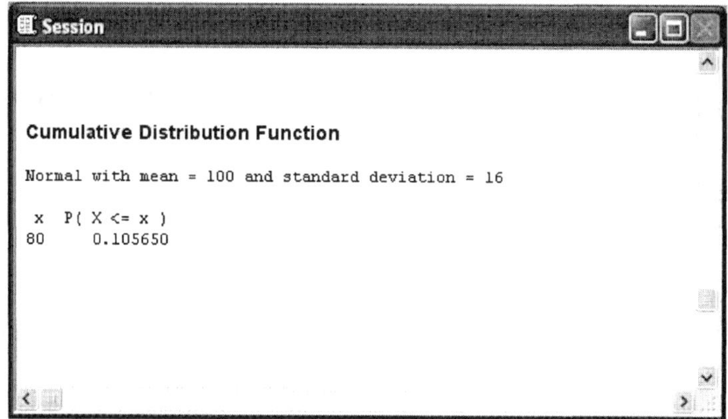

To determine the MDI score that is the 99th percentile:

Choose: **Calc > Probability Distributions > Normal**
Select: **Inverse Cumulative Probability**
Enter: Mean : **100**
Standard deviation: **16**
Select: Input constant:
Enter: **.99**
Click: **OK**

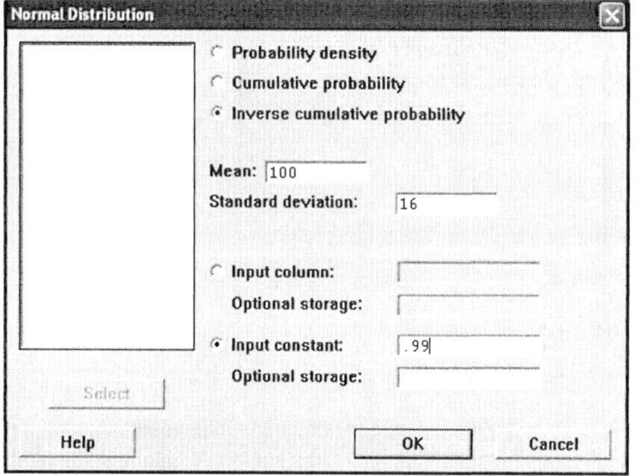

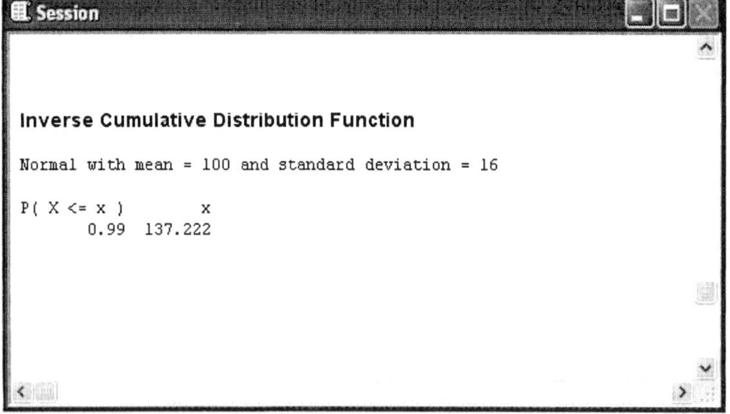

An MDI of 138 is the 99th percentile.

To determine the MDI such that only 1% of the population has an MDI below it, we will use the Input constant: **.01**

Approximately 1% of the population has an MDI less than 63.

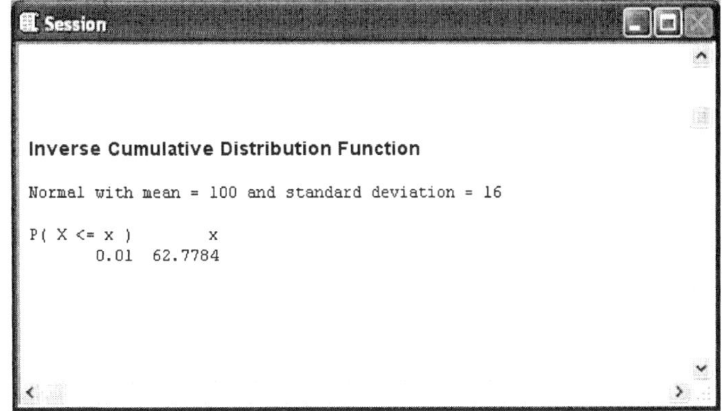

Chapter 6 Exercise 6.98 Psychomotor Development Index
Determining Normal Probabilities

Enter 97 and 103 into column 1. Then choose:

Calc>Probability Distributions>Normal>Cumulative

Enter: Mean: **100**

Standard deviation: **15**

Enter: Input column: **C1**

This gives the following result:

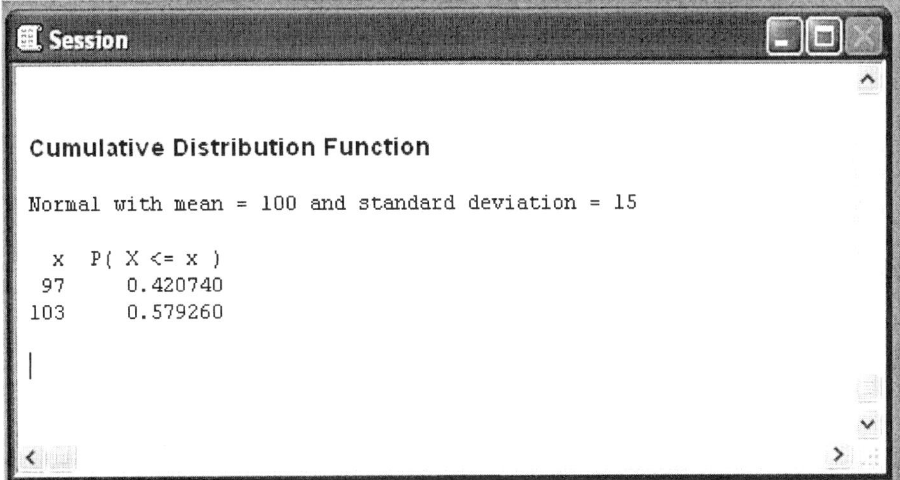

a) The probability that the PDI > 103 we can find by 1 - 0.57962.
b) The probability that the PDI is between 97 and 103 is obtained by doing the subtraction .57926 - .420740.
c) To calculate the z score for X = 90, using the calculator:

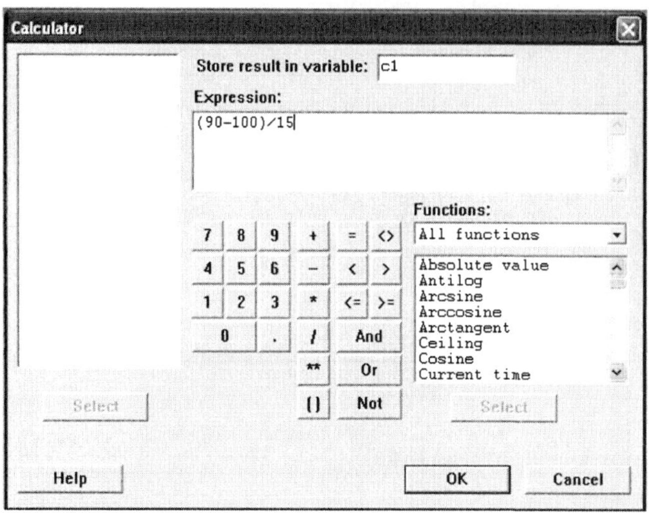

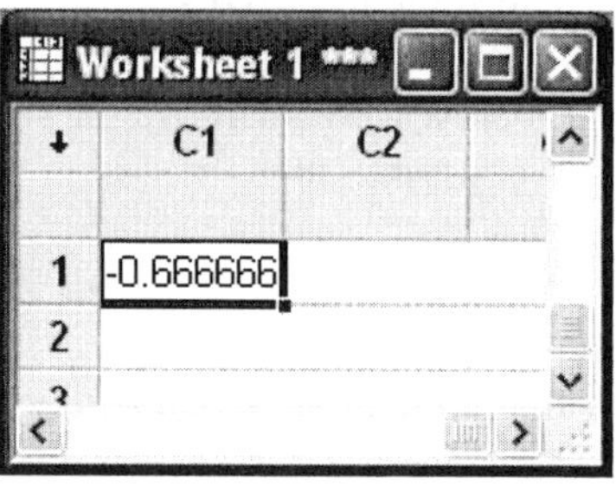

Chapter 6 Example 12 Are Women Passed Over for Managerial Training? Determining Binomial Probabilities

Let X denote the number of females selected in a random sample of ten employees (X can have any value 0, 1, 2, 3, …, 10). We can do this using the following commands:

Choose: **Calc > Make Patterned Data > Simple Set of Numbers**
Enter: Store patterned data in: **C1**
From first value: **1**
To last value: **10**
List each value **1**
List whole sequence **1**
Click: **OK**

To find the binomial probability distribution:
Choose: **Calc > Probability Distributions > Binomial**

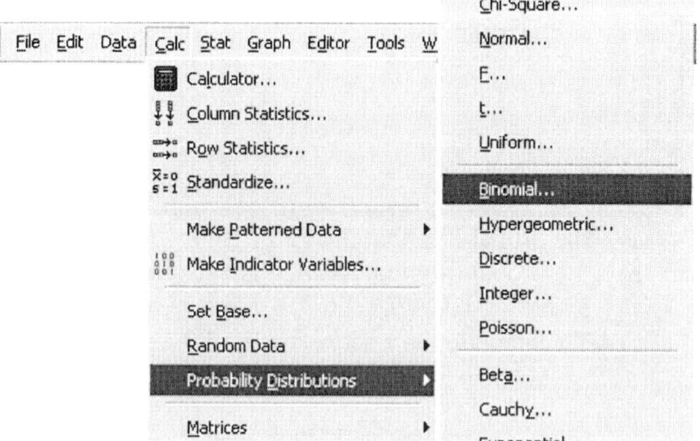

Select: **Probability**
Enter: Number of trials: **10**
Probability of success: **.5**
Input column: **C1**
Click: **OK**

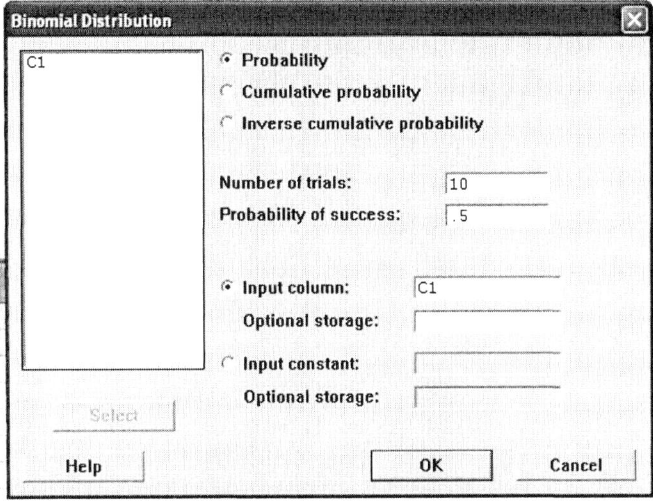

This will generate the distribution in the Session Window.

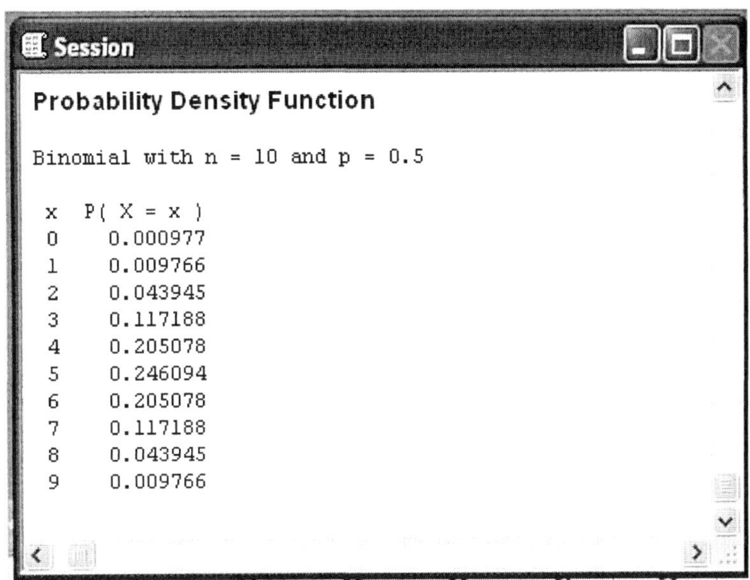

Chapter 6 Example 14 How Can We Check for Racial Profiling? Determining Binomial Probabilities

In this problem, generate numbers 0 – 262 and store in C1.

Then generate the probability distribution:
Choose: **Calc > Probability Distributions > Binomial**
Select: **Probability**
Enter: Number of trials: **262**
Probability of success: **.422**
Input column: **C1**
Click: **OK**

Worksheet 1 ***

↓	C1	C2	C3
	x	P(X=x)	
88	87	0.0006067	
89	88	0.0008808	
90	89	0.0012573	
91	90	0.0017645	
92	91	0.0024350	
93	92	0.0033044	
94	93	0.0044101	
95	94	0.0057888	
96	95	0.0074741	
97	96	0.0094927	
98	97	0.0118607	
99	98	0.0145798	
100	99	0.0176338	
101	100	0.0209854	

Here is part of the table generated. Notice where we begin to see results other than zero. What is the probability that $X \geq 207$? What conclusions can you draw from this?

Chapter 6 Exercise 6.36 Exit Poll
Mean and Standard Deviation of the Binomial Random Variable X

a) The data is binary (vote for or against), the voters were randomly selected, and each voter is separate and independent from another voter

    N = 3160   p = .5   1-p = .5

    Generate the probabilities.

First fill column 1 with numbers 0-3160.

Then generate the probability distribution:

Choose: **Calc > Probability Distributions > Binomial**
Select: **Probability**
Enter: Number of trials: **3160**
Probability of success: **.5**
Input column: **C1**
Optional storage: **C2**
Click: **OK**

b) Mean and Standard Deviation of the Binomial Distribution

Using the actual values for n, p and 1-p, the calculator within Calc>Calculator in MINITAB can be used to determine the mean and standard deviation for the binomial distribution

Recall that the mean and standard deviation for a general probability distribution is

$$\mu = \sum xP(x) \quad \text{and} \quad \sigma = \sqrt{\sum (x-u)^2 P(x)}$$

For the binomial distribution we can use the following expressions:

n * p store result in "mean"
SQRT(n*p*(1-p)) store result in "std dev"

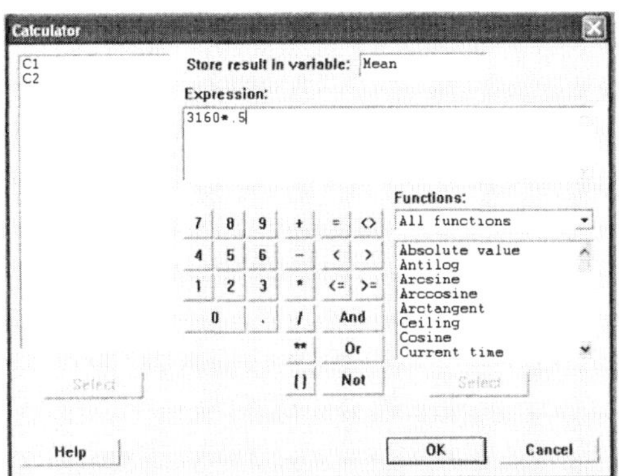

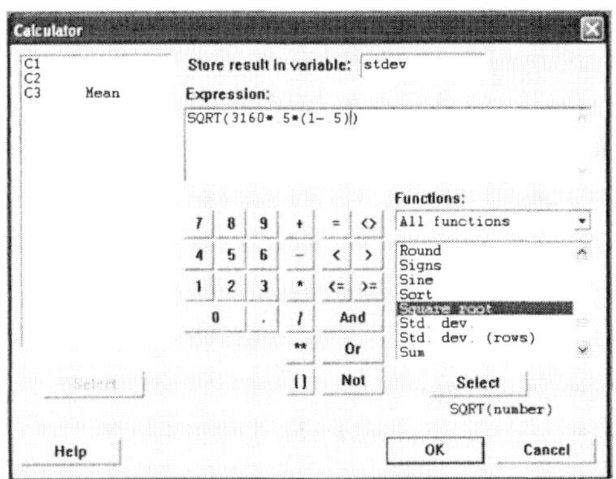

This will produce the following in the worksheet:

Worksheet 1 ***

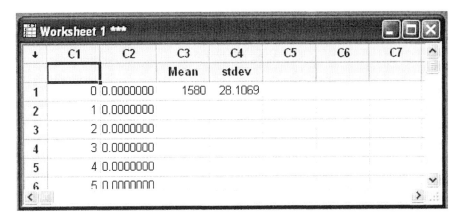

↓	C1	C2	C3	C4	C5	C6	C7
			Mean	stdev			
1	0	0.0000000	1580	28.1069			
2	1	0.0000000					
3	2	0.0000000					
4	3	0.0000000					
5	4	0.0000000					
6	5	0.0000000					

Chapter 6 Exercise 6.37 Jury Duty
Determining Binomial Probabilities

a) Check the three criteria for being binomial
b) Let n = 12, p = .4 . Enter 0 through 12 into C1, then
   Choose: **Calc > Probability Distributions > Binomial**

Select: **Probability**
Enter: Number of trials: **12**
   Probability of success: **.4**
   Input column: **C1**
   Optional storage: **C2**
Click: **OK**

Worksheet 1 ***

↓	C1	C2	C3	C4
1	0	0.002177		
2	1	0.017414		
3	2	0.063852		
4	3	0.141894		
5	4	0.212841		
6	5	0.227030		
7	6	0.176579		
8	7	0.100902		
9	8	0.042043		
10	9	0.012457		
11	10	0.002491		
12	11	0.000302		
13	12	0.000017		

Chapter 6 Example 16 Exit Poll of California Voters Revisited
The Mean and Standard Deviation of the Sampling Distribution of a Proportion

Considering the exit poll of 3160 voters, recall we had n = 3160 and p = .5
Enter patterned data into C1 and store the binomial distribution in C2.

Choose: **Calc > Make Patterned Data > Simple Set of Numbers**
   Enter: Store patterned data in: **C1**
   From first value: **1**
   To last value: **3160**
   List each value **1**
   List whole sequence **1**
  Click: **OK**

Choose: **Calc > Probability Distributions > Binomial**

Select: **Probability**
Enter: Number of trials: **3160**
Probability of success: **.5**
Input column: **C1**
Optional storage: **C2**
Click: **OK**

To calculate the mean:
Choose **Calc > Calculator**
Calculate the Mean using the formula
n * p

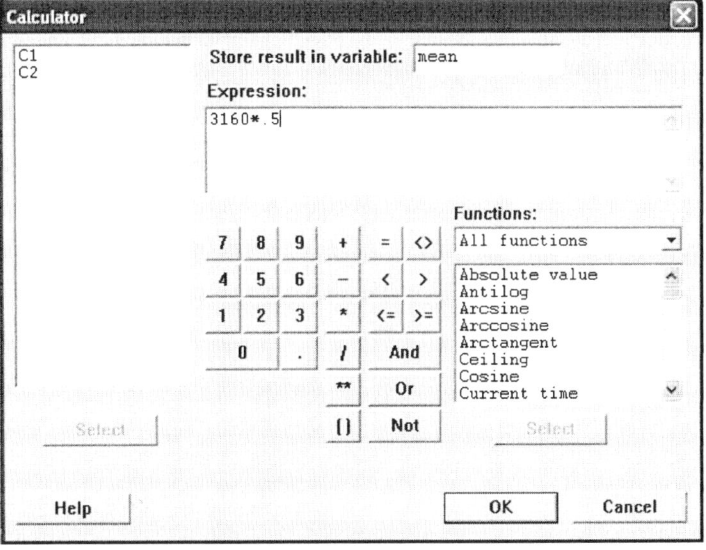

To calculate the standard deviation using Minitab's calculator

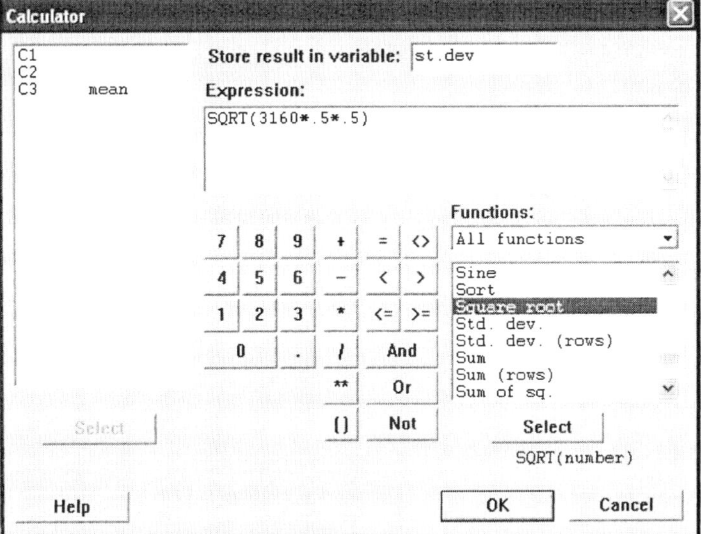

This gives the following results

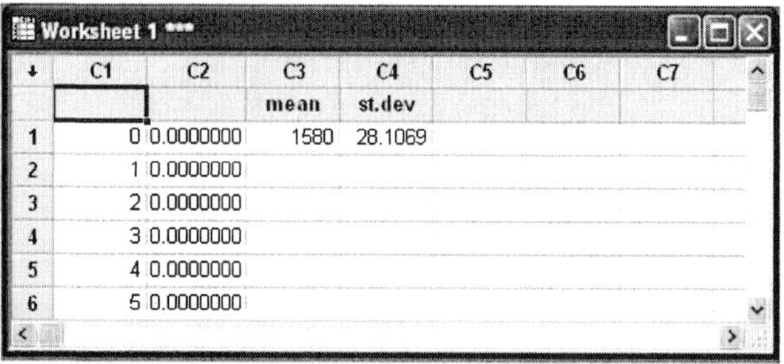

↓	C1	C2	C3	C4	C5	C6	C7
			mean	st.dev			
1	0	0.0000000	1580	28.1069			
2	1	0.0000000					
3	2	0.0000000					
4	3	0.0000000					
5	4	0.0000000					
6	5	0.0000000					

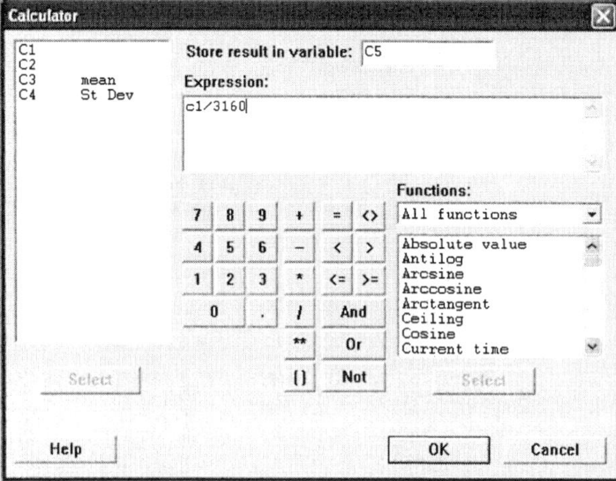

To get the proportions, divide the binomial random variable X(in C1) by 3160. Let Minitab do the work using the Calculator. Store results in C5.

Use the calculator to find the mean of the proportions we just stored in C5.

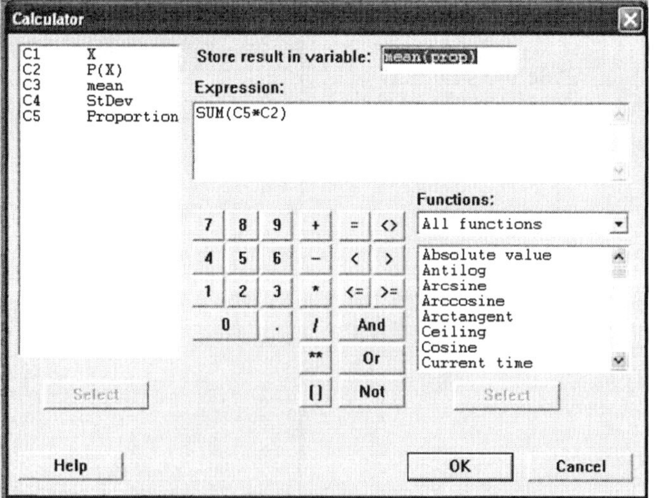

As you see in the following table, it is .5, just like the population proportion.

Worksheet 1 ***

↓	C1	C2	C3	C4	C5	C6	C7	C8	C9
	X	P(X)	mean	StDev	Proportions	mean(prop)			
1	0	0.0000000	1580	28.1069	0.00000	0.50000000			
2	1	0.0000000			0.00032				
3	2	0.0000000			0.00063				
4	3	0.0000000			0.00095				
5	4	0.0000000			0.00127				
6	5	0.0000000			0.00158				
7	6	0.0000000			0.00190				

Use the calculator to find the standard deviation of the proportions and store in C7.

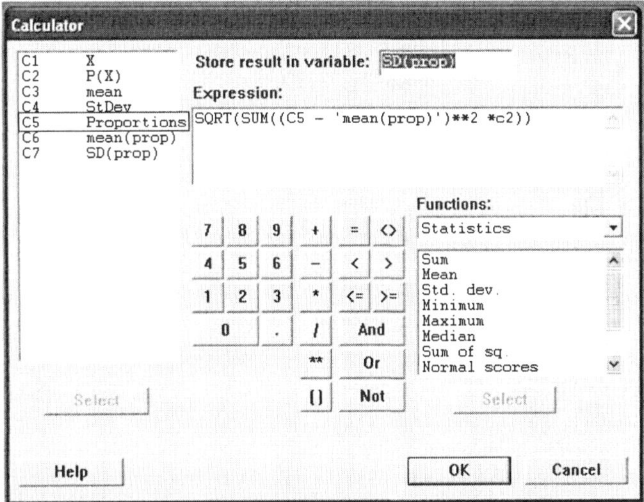

Worksheet 1 ***

↓	C1	C2	C3	C4	C5	C6	C7	C8	C9
	X	P(X)	mean	StDev	Proportions	mean(prop)	SD(prop)		
1	0	0.0000000	1580	28.1069	0.00000	0.500000	0.0088946		
2	1	0.0000000			0.00032				
3	2	0.0000000			0.00063				
4	3	0.0000000			0.00095				
5	4	0.0000000			0.00127				
6	5	0.0000000			0.00158				
7	6	0.0000000			0.00190				

We got a standard deviation of .008946 for the sampling distribution of the sample proportion.

This is the same as if we would use the formula $\sigma = \sqrt{\dfrac{p(1-p)}{n}}$

Chapter 6 Exercise 6.47 Other Scenario for Exit Poll
The Sampling Distribution of Sample Proportion

This problem repeats Example 16, but uses p = .55 instead of .5

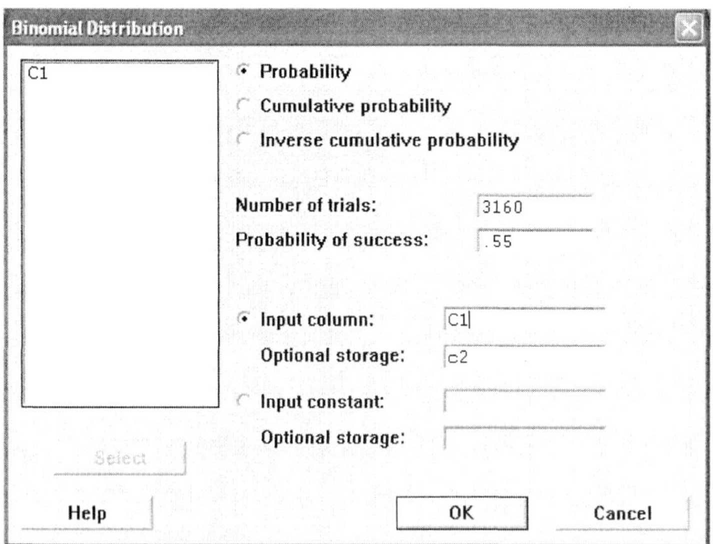

This will generate the sampling distribution for the number of voters who voted for the recall.
N = 3160 and p = .55 and store the binomial probabilities in C2.

To find the mean and standard deviation, use the **Calc> Calculator** as shown

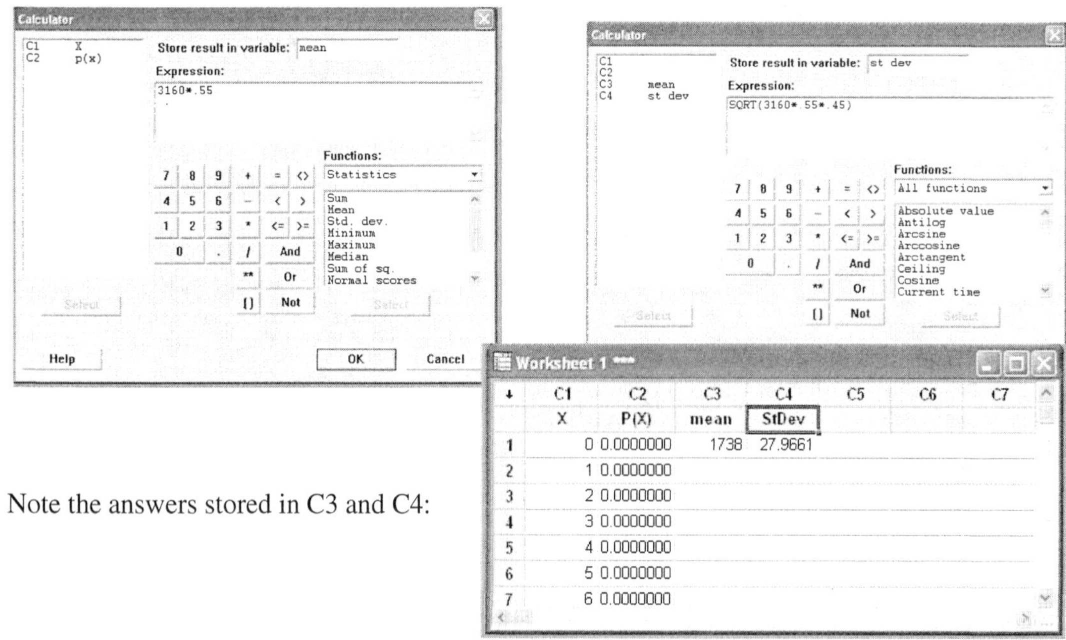

↓	C1	C2	C3	C4	C5	C6	C7
	X	P(X)	mean	StDev			
1	0	0.0000000	1738	27.9661			
2	1	0.0000000					
3	2	0.0000000					
4	3	0.0000000					
5	4	0.0000000					
6	5	0.0000000					
7	6	0.0000000					

Note the answers stored in C3 and C4:

d) Now we will find the mean and standard deviation of sampling distribution of the proportion of the people who voted for the recall. Remember we have to divide by 3160.

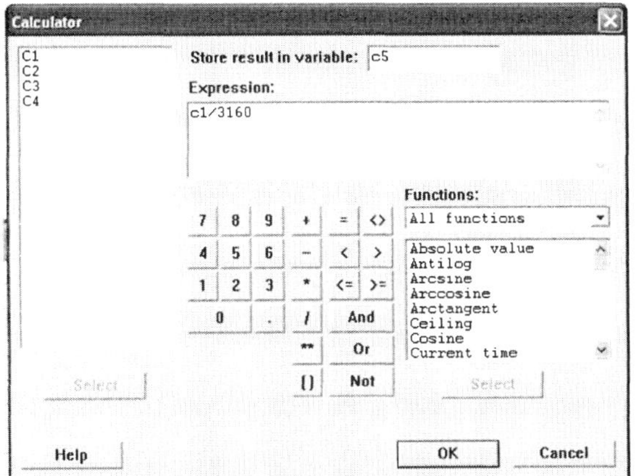

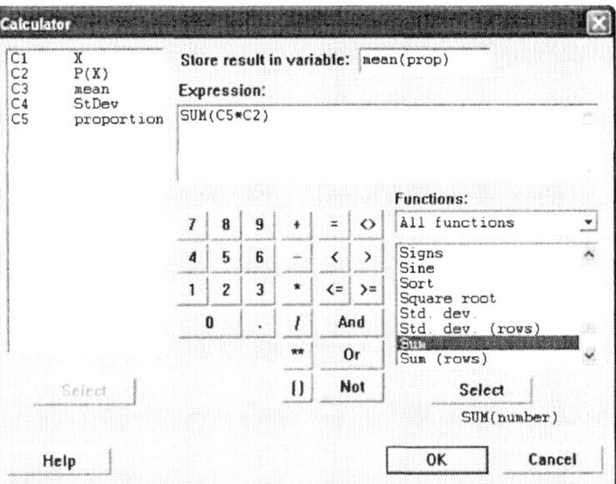

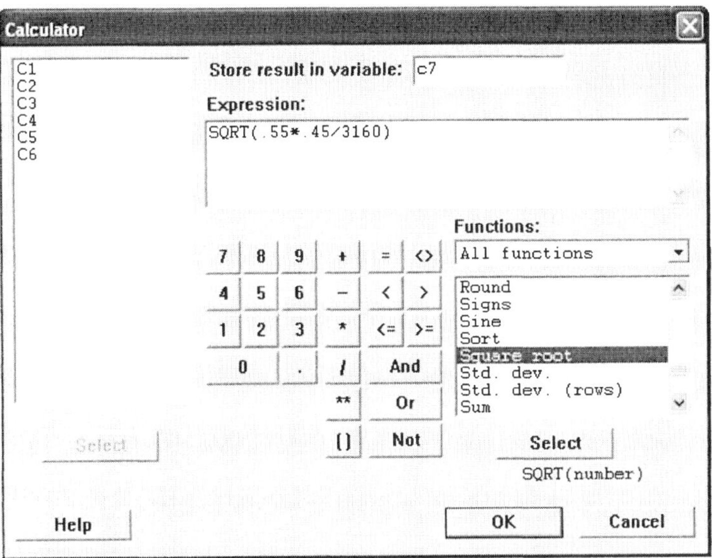

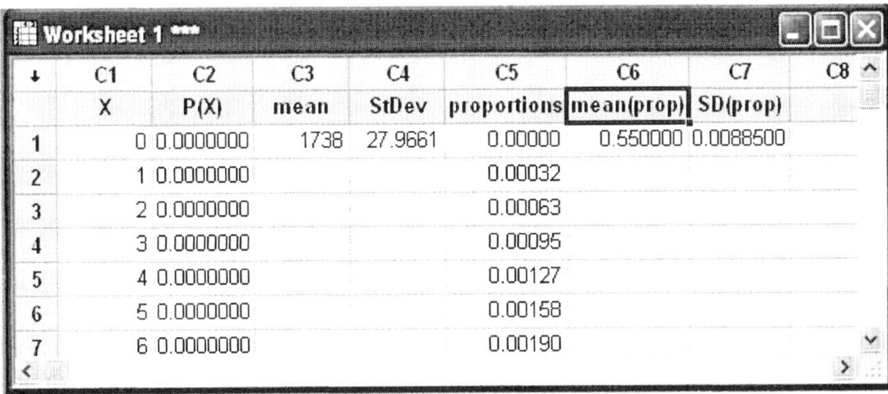
Worksheet 1 ***

↓	C1	C2	C3	C4	C5	C6	C7	C8
	X	P(X)	mean	StDev	proportions	mean(prop)	SD(prop)	
1	0	0.0000000	1738	27.9661	0.00000	0.550000	0.0088500	
2	1	0.0000000			0.00032			
3	2	0.0000000			0.00063			
4	3	0.0000000			0.00095			
5	4	0.0000000			0.00127			
6	5	0.0000000			0.00158			
7	6	0.0000000			0.00190			

Chapter 6 Class Exploration 6.124
Simulating a Sampling Distribution for a Sample Mean

Enter the data into the worksheet.
Construct a histogram.

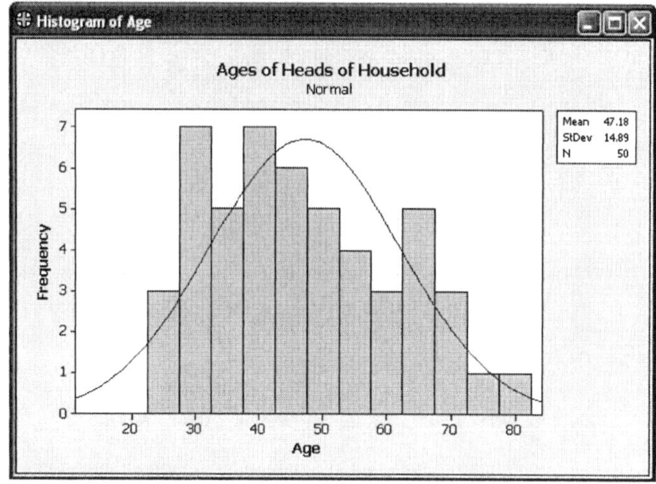

b) Lets assume a class size of 30 students. Each student can collect random samples of size 9. We will store the samples in nine columns of 30 cells each.

Calc > Random Data > Sample from Columns
Enter: Sample **30** Rows
Column **C3**
Store samples in **: enter a column**
Check box: Sample with replacement

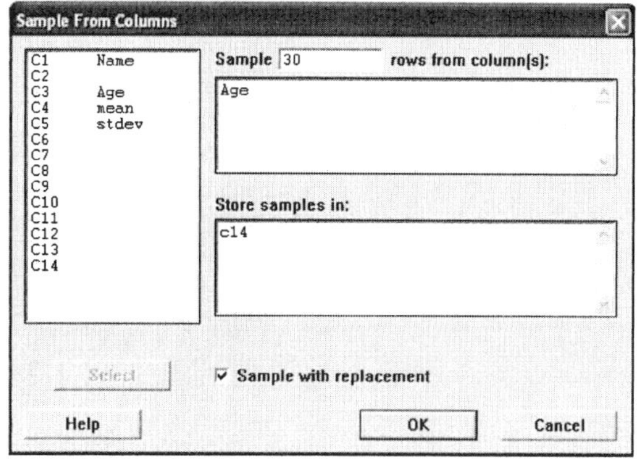

Note you will have to do this nine times, naming a new column each time.

To find the mean of each sample :
Choose: **Calc > Row Statistics**
Click **Mean**
Input variables: Enter **C6-C14**
Store in **C15**
OK

Find the mean and standard deviation of column 15. You can see these values by doing a histogram of column 15. What conclusions can you draw?

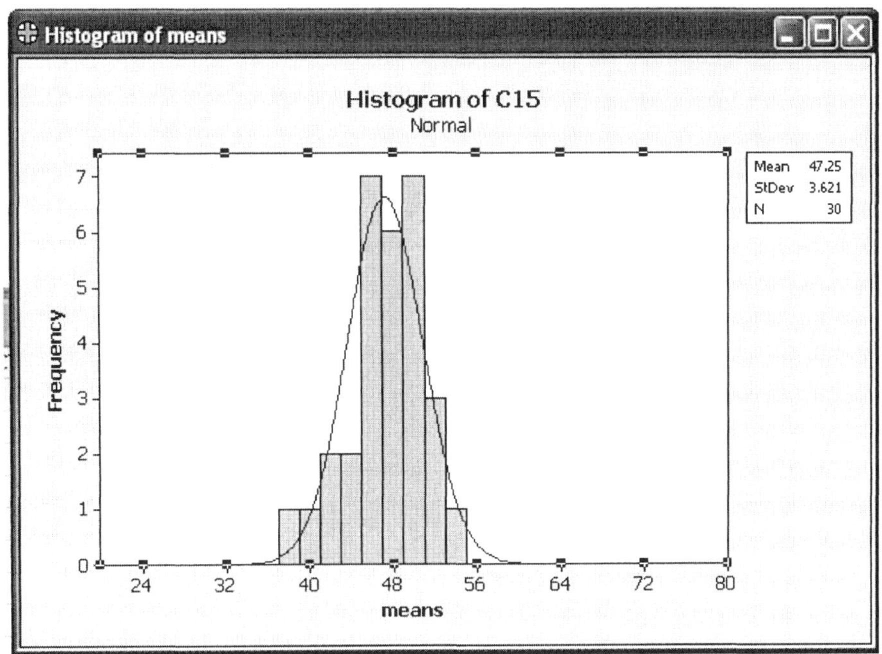

Chapter 7 Example 2 Should a Wife Sacrifice Her Career for Her Husband's? Constructing the Confidence Interval Estimate for a Population Proportion

From Chapter 6 we recall that 95% of a normal distribution falls within two standard deviations of the mean. So if we look at the (mean of the distribution) $\pm$ 1.96 (standard deviations for the distribution), we would find 95% of a normal distribution. So in this problem, if we look at 1.96(.01) we get 0.02. we can then construct the interval as the $\hat{p}$ $\pm$ 0.02, or .19 $\pm$ 0.02 which gives us (0.17,0.21).

To do this in Minitab:

Choose: **Stat > Basic Statistics > 1-Proportion**

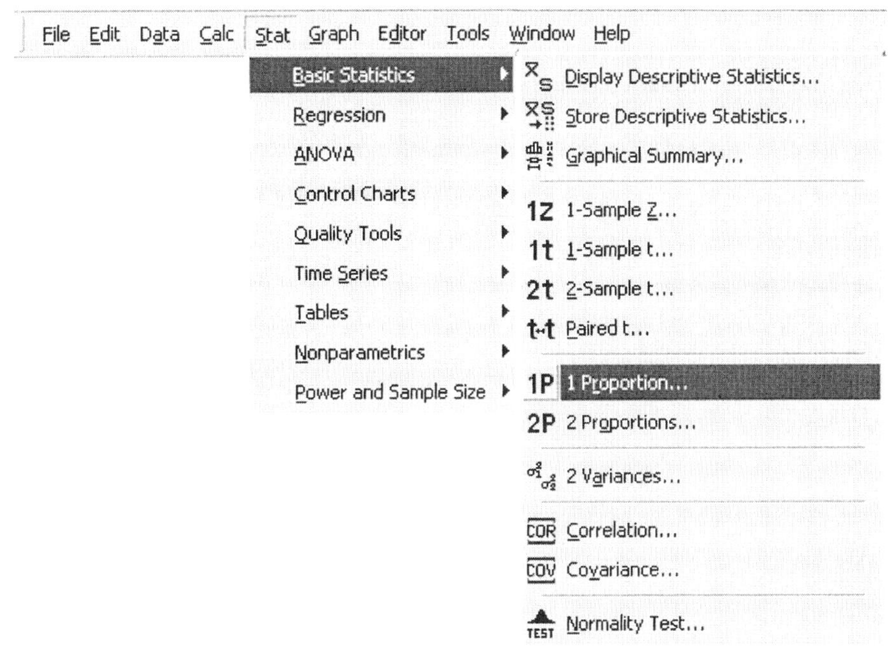

Select:
Summarized Data
Enter:
Number of trials **1823**
Number of events **346**
(Note: 19% of 1823)
Click: **OK**

Note: Minitab defaults to a 95% confidence interval.

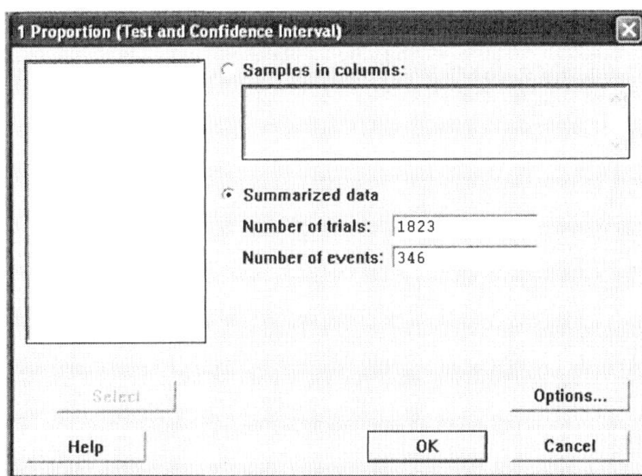

We get the following in the session window.

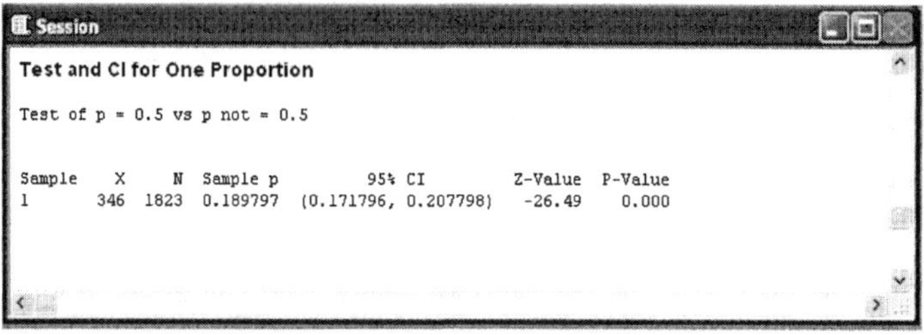

Session

Test and CI for One Proportion

Test of p = 0.5 vs p not = 0.5

Sample	X	N	Sample p	95% CI	Z-Value	P-Value
1	346	1823	0.189797	(0.171796, 0.207798)	-26.49	0.000

Given the interval in the above session window (0.17, 0.21) , and the point estimate of 0.19, do you see that the margin of error is 0.02?

Chapter 7 Exercise 7.7 Believe in heaven?
Constructing the Confidence Interval Estimate for a Proportion

In this exercise, n = 1158, and the sample proportion = 0.86

Choose: **Stat > Basic Statistics > 1-Proportion**
Select: Summarized Data
Enter: Number of trials **1156**
Number of events **996**
(Note: 86% of 1156)
Click: **OK**

1 Proportion (Test and Confidence Interval)
Samples in columns:
Summarized data
Number of trials: 1158
Number of events: 996
Select
Options...
Help
OK
Cancel

We get the following in the session window:

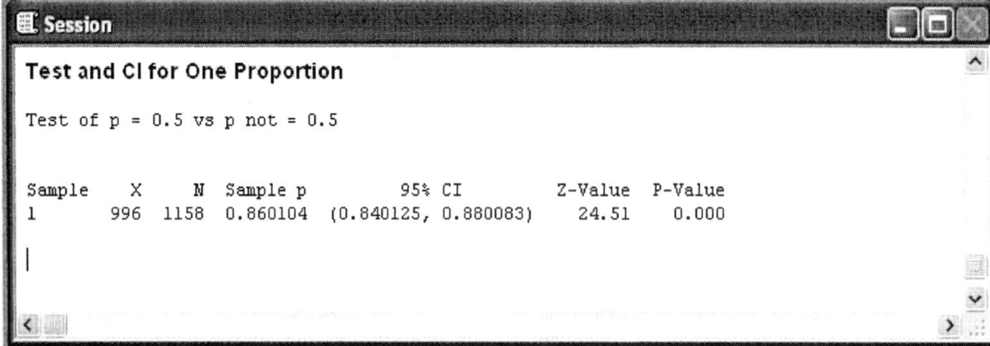

Session

Test and CI for One Proportion

Test of p = 0.5 vs p not = 0.5

Sample	X	N	Sample p	95% CI	Z-Value	P-Value
1	996	1158	0.860104	(0.840125, 0.880083)	24.51	0.000

Chapter 7 Exercise 7.21 Exit-poll Predictions
Constructing the Confidence Interval Estimate for a Proportion

In this problem we have n = 1400 voters, of whom 660 voted Democrat and 740 voted Republican. First using a 95% confidence interval:

Choose: **Stat > Basic Statistics > 1-Proportion**
Select: Summarized Data
Enter: Number of trials **1400**
Number of events **660**
Click: **OK**

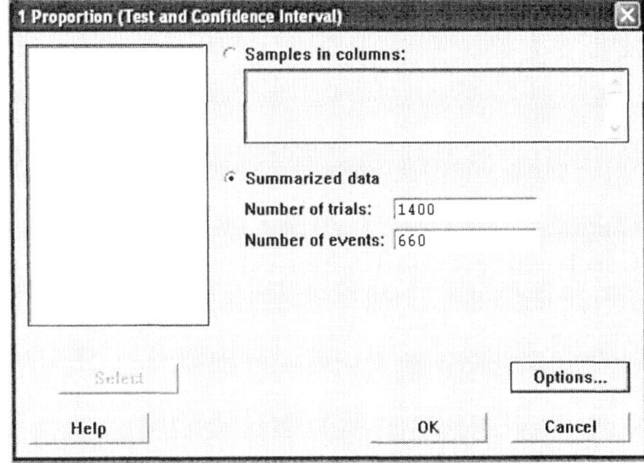

We get the following in the session window:

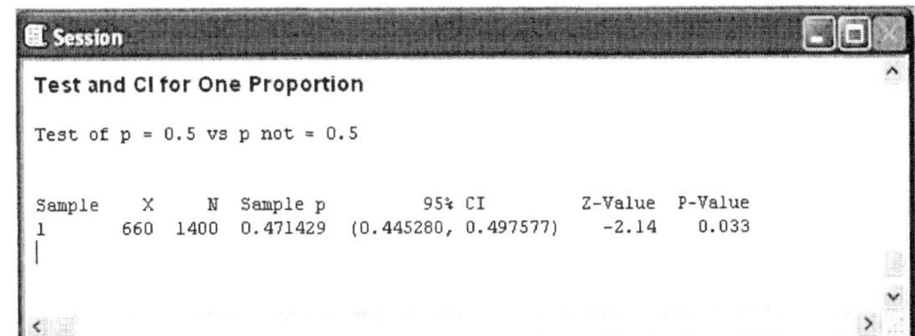

Now repeating this for a 99% confidence interval, we have to choose the Options tab to change the level.

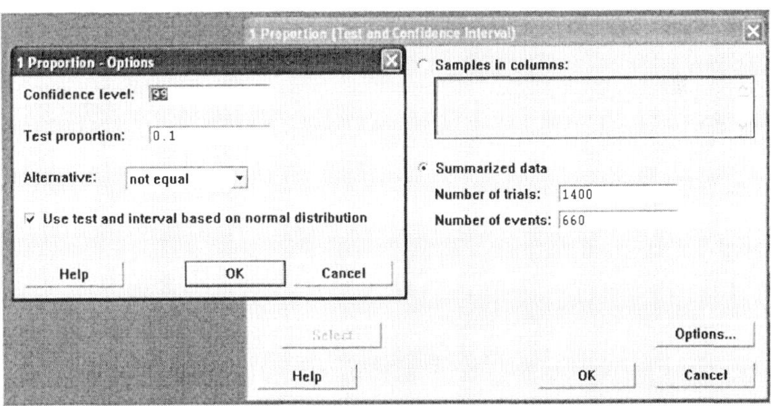

We now get these results:

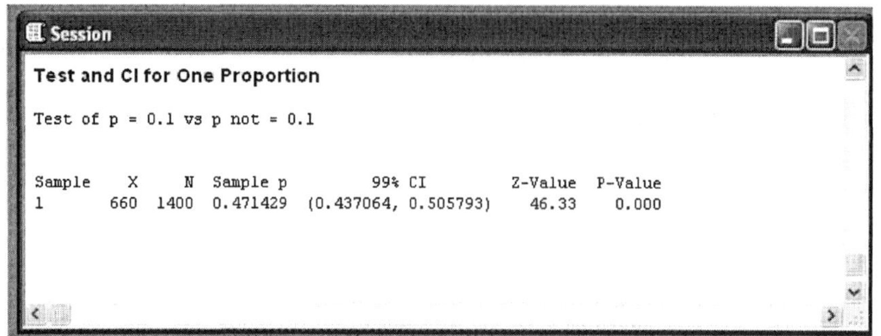
Session

Test and CI for One Proportion

Test of p = 0.1 vs p not = 0.1

Sample	X	N	Sample p	99% CI	Z-Value	P-Value
1	660	1400	0.471429	(0.437064, 0.505793)	46.33	0.000

What conclusions can you draw?

Chapter 7 Example 7 eBay Auctions of Palm Handheld Computers
Constructing a Confidence Interval Estimate for a Population Mean

Open worksheet **ebay_auctions** from the MINITAB folder of the data disk. To get the descriptive statistics

Choose **Stat> Basic Stat> Display Descriptive Statistics**
Click **Statistics** button (Choose the stats that you wish to see)
Click **OK OK**

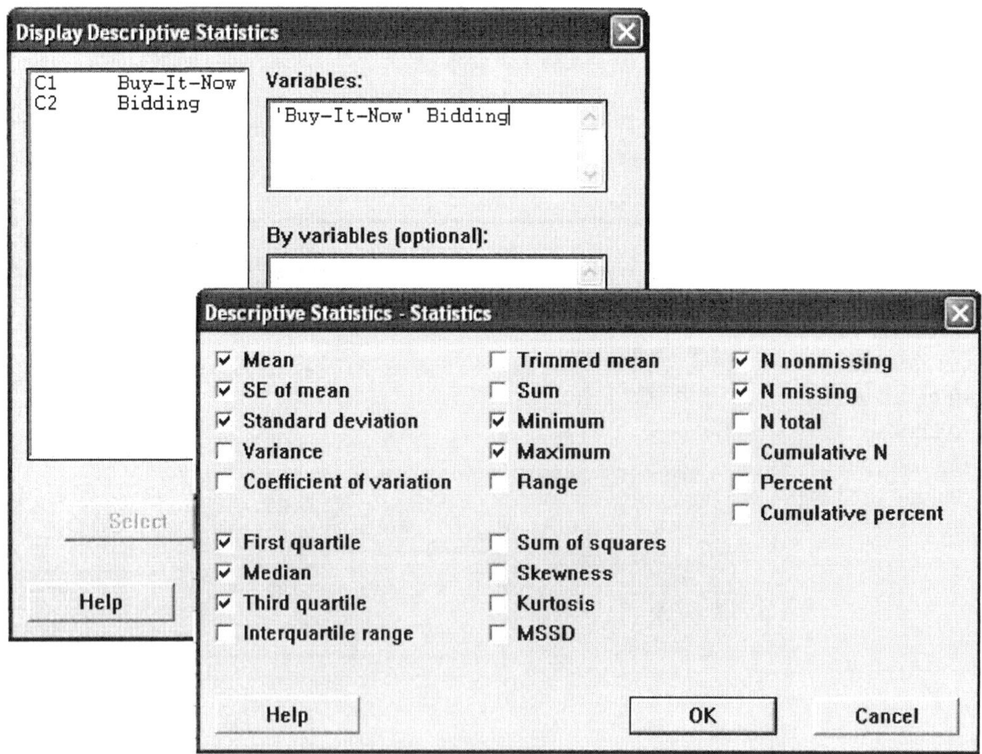

We then get the chart in the session window:

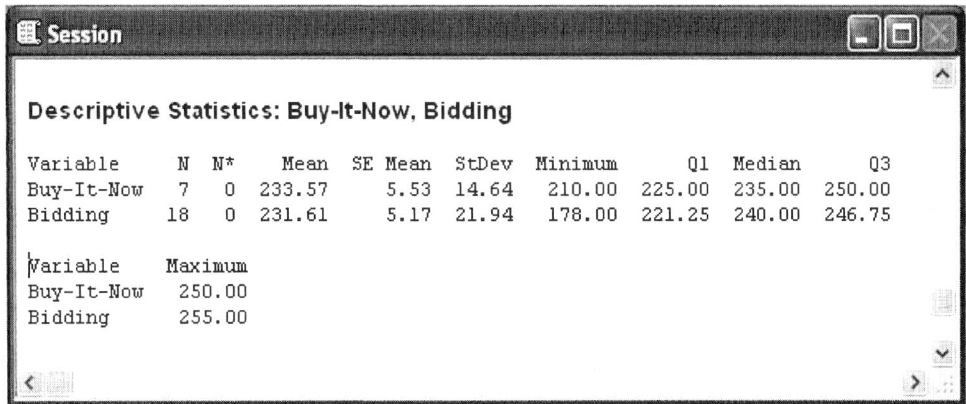

Session

Descriptive Statistics: Buy-It-Now, Bidding

Variable	N	N*	Mean	SE Mean	StDev	Minimum	Q1	Median	Q3
Buy-It-Now	7	0	233.57	5.53	14.64	210.00	225.00	235.00	250.00
Bidding	18	0	231.61	5.17	21.94	178.00	221.25	240.00	246.75

Variable	Maximum
Buy-It-Now	250.00
Bidding	255.00

We can do the dotplot for the data:
Choose: **Graph > DotPlot > Multiple y > Simple**

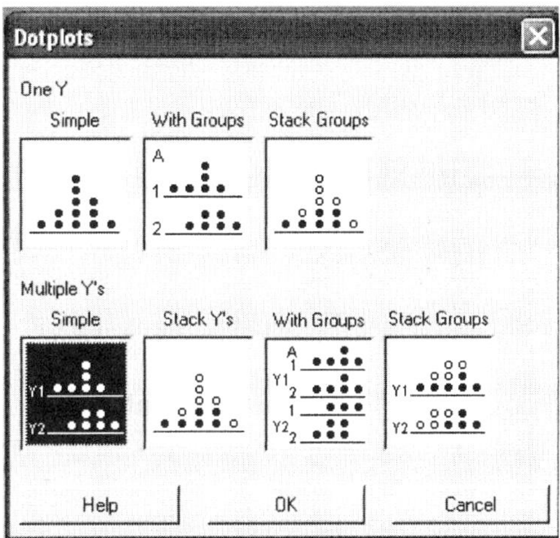

Enter Graph Variables: **C1 C2**
Enter : **appropriate title**
Click: **OK OK**

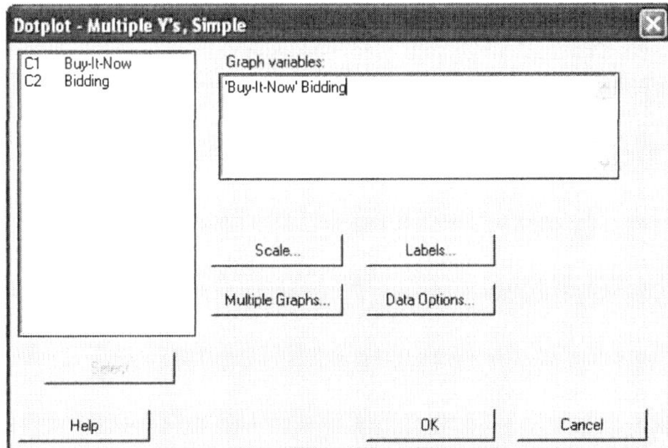

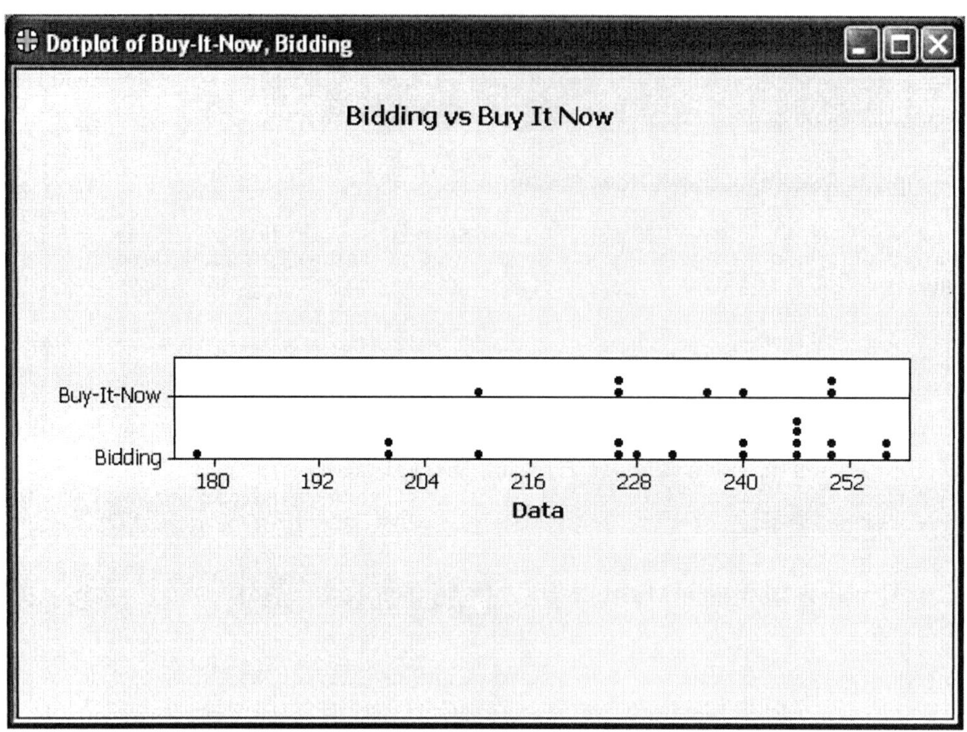

To obtain the confidence interval using the t-distribution:

Choose: **Stat>Basic Stat>1-sample t**

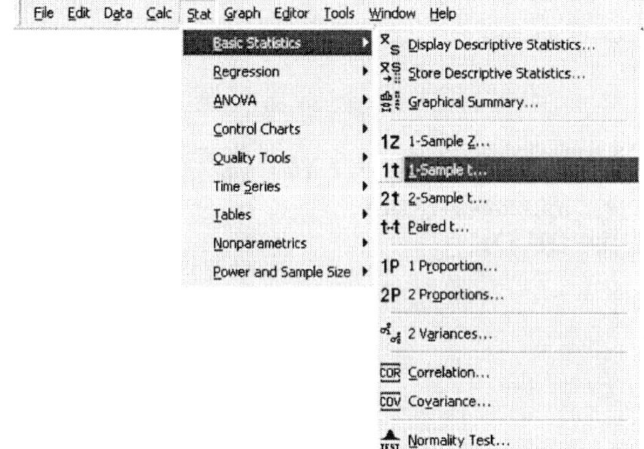

Enter: Samples in columns **C1**
Click : **OK**

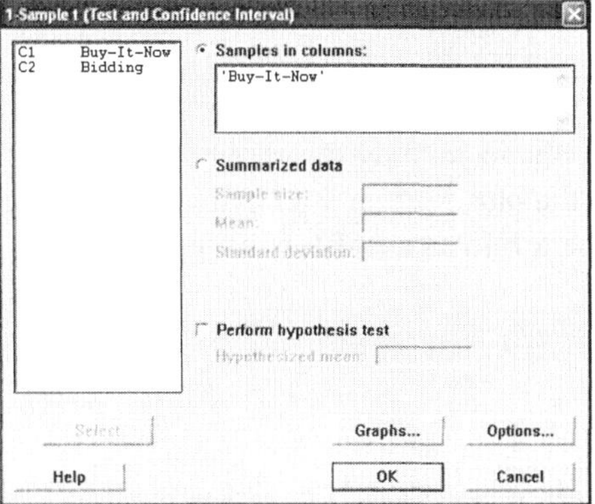

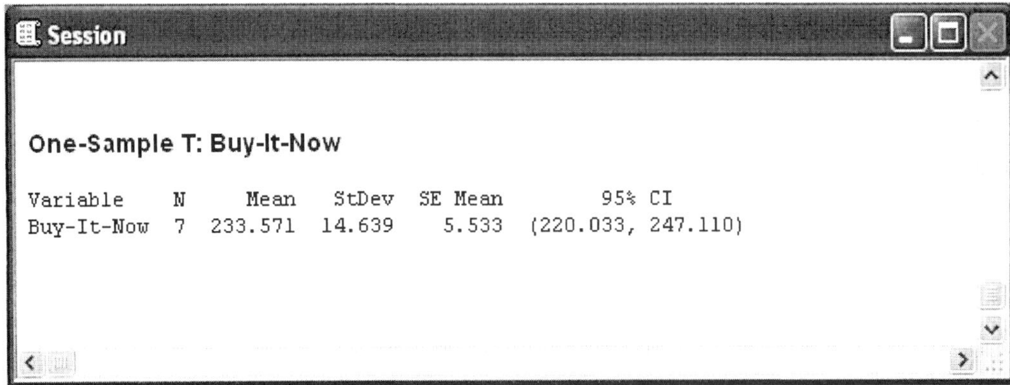
Session

One-Sample T: Buy-It-Now

Variable	N	Mean	StDev	SE Mean	95% CI
Buy-It-Now	7	233.571	14.639	5.533	(220.033, 247.110)

Using the same data, and now constructing the confidence interval for the Bidding column:
(this is exercise 7.29)

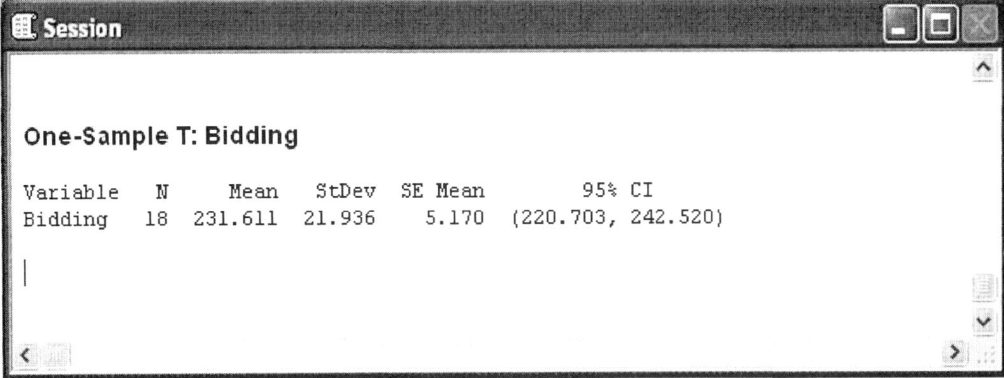
Session

One-Sample T: Bidding

Variable	N	Mean	StDev	SE Mean	95% CI
Bidding	18	231.611	21.936	5.170	(220.703, 242.520)

What conclusions can you draw?

Chapter 7 Exercise 7.51 Do You Like Tofu?
Confidence Interval Estimate of a Population Proportion with a Small Sample Size

In this problem we have a very small sample of five students. All say they like tofu. We will use the plus four method in this situation, giving us seven successes out of nine trials. Using the formulas in the text we get:

So we have N = 9, $\hat{p} = 7/9 = 0.7777$ and

$$se = \sqrt{\hat{p}(1-\hat{p})/n} = \sqrt{.7777(1-.7777)/9} = .1386.$$

The resulting interval = $\hat{p} \pm 1.96(se) = 0.7777 \pm (1.96)(0.1386)$.

This gives us the interval (0 .506, 1.05). Note that the right end of the interval is 1.05. We cannot have a value greater than 1.0 for a proportion, so we would report the interval as (0.506, 1.00).

Now we can do this using Minitab.

Choose **Stat > Basic Stat > 1 - Proportion**
Enter: Number of trials: **9**
Number of successes **7**

Click the Options tab and check the normal distribution box

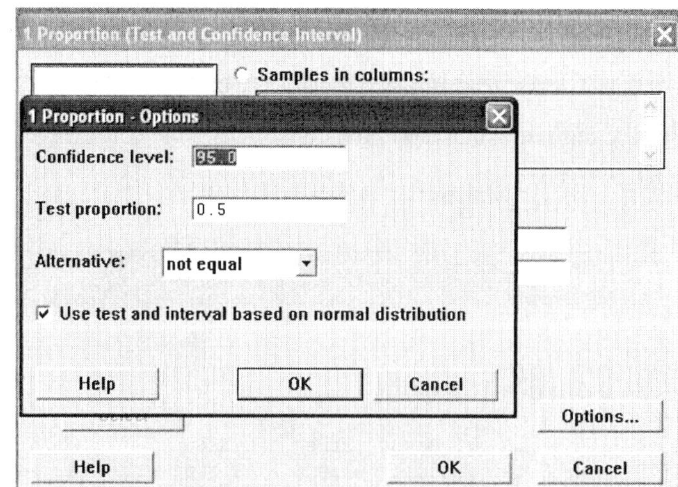

This will give you the same interval we calculated above.

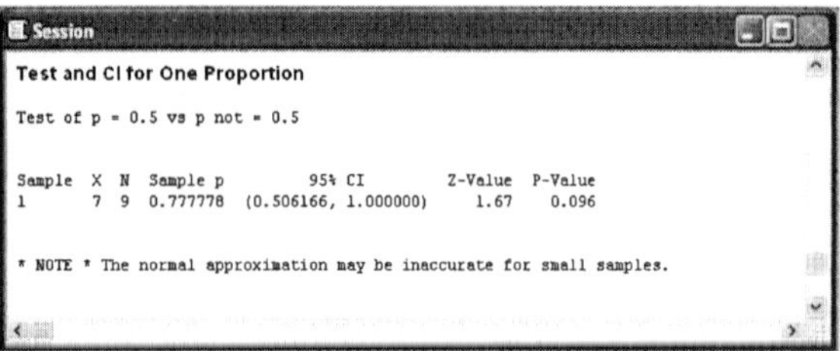

Session
Test and CI for One Proportion
Test of p = 0.5 vs p not = 0.5
Sample X N Sample p 95% CI Z-Value P-Value
1 7 9 0.777778 (0.506166, 1.000000) 1.67 0.096
* NOTE * The normal approximation may be inaccurate for small samples.

Chapter 8 Example 4 Dr Dog: Can Dogs Detect Cancer By Smell? Hypothesis Testing For a Proportion

We are asked to test H_0: $p = 1/7$ vs H_1: $p > 1/7$. The sample evidence presented is that in the total of 54 trials, the dogs made the correct selection 22 times.

Choose: **Stat > Basic Statistics > 1-Proportion**
Select: **Summarized data**
Enter: Number of trials: **54**
Number of events: **22**

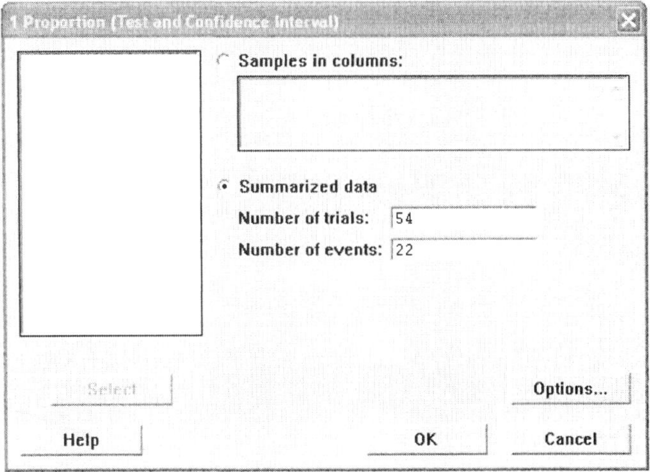

Click: **Options**
Enter: Test proportion: **.143**
Select: Alternative: **greater than**
Click: **Use test and interval based on normal distribution**
Click: **OK > OK**

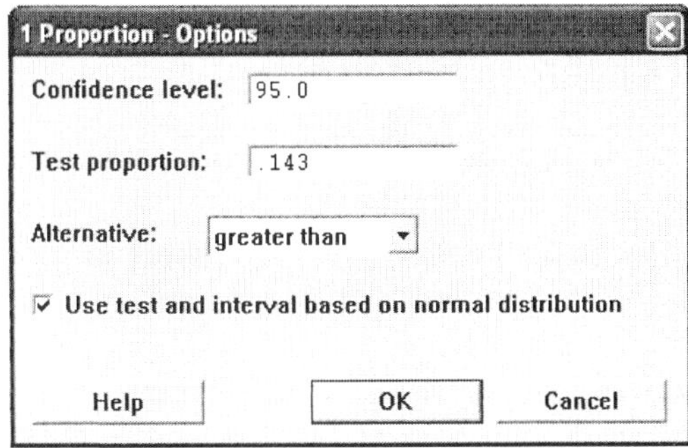

The results should be displayed in the Session window.

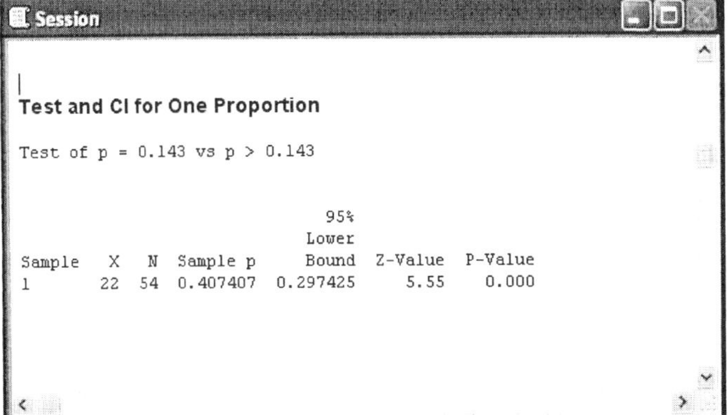

Note that the test statistic (z = 5.55) and the p-value (p = 0.000). With such a small p-value, we will reject the null hypothesis.

Chapter 8 Example 6 Can TT Practitioners Detect a Human Energy Field?
Hypothesis Testing For A Proportion

We are asked to test H_0: p = .50 vs H_1: p > .50. The sample evidence presented is that in the total of 150 trials, the TT practitioners were correct with 70 of their predictions.

Choose: **Stat > Basic Statistics > 1-Proportion**
Select: **Summarized data**
Enter: Number of trials: **150**
Number of events: **70**
Click: **Options**
Enter: Test proportion: **.50**
Select: Alternative: **greater than**
Click: **Use test and interval based on normal distribution**
Click: **OK > OK**

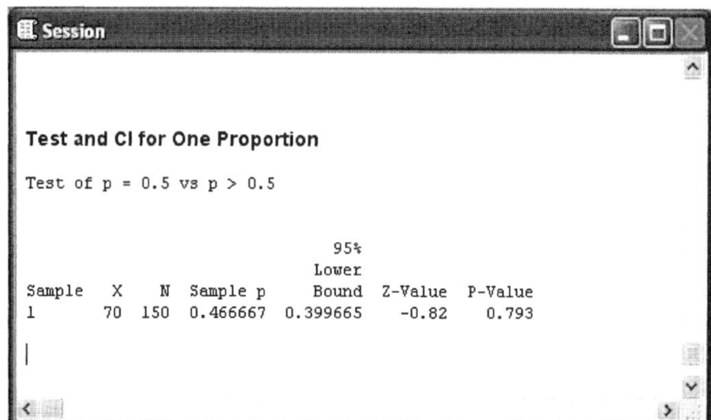
Session

Test and CI for One Proportion

Test of p = 0.5 vs p > 0.5

Sample	X	N	Sample p	95% Lower Bound	Z-Value	P-Value
1	70	150	0.466667	0.399665	-0.82	0.793

With such a large p-value (p-value = .793) the decision is Fail To Reject H_0.

Chapter 8 Exercise 8.15 Another Test of Therapeutic Touch
Hypothesis Testing For A Proportion

We are asked to test H_0: p = .50 vs H_1: p > .50. The sample evidence presented is that in the total of 130 trials, the TT practitioners were correct with 53 of their predictions.

Choose: **Stat > Basic Statistics > 1-Proportion**
Select: **Summarized data**
Enter: Number of trials: **130**
Number of events: **53**
Click: **Options**
Enter: Test proportion: **.50**
Select: Alternative: **greater than**
Click: **Use test and interval based on normal distribution**
Click: **OK > OK**

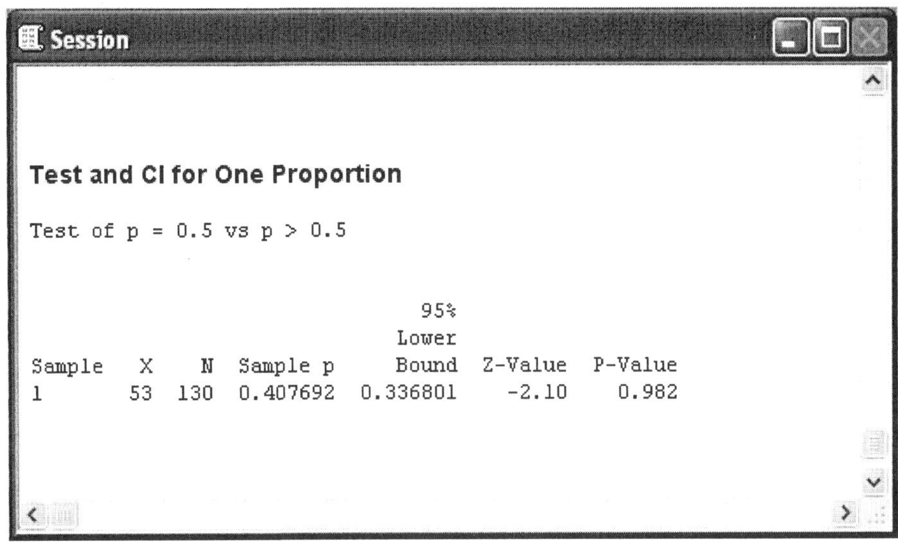

Test and CI for One Proportion

Test of p = 0.5 vs p > 0.5

Sample	X	N	Sample p	95% Lower Bound	Z-Value	P-Value
1	53	130	0.407692	0.336801	-2.10	0.982

Chapter 8 Exercise 8.17 Gender Bias in Selecting Managers
Hypothesis Testing For A Proportion

We are asked to test H_0: p = .60 vs H_1: p ≠ .60. The sample evidence presented is that in the total of 40 employees chosen for management training, 28 were male.

Choose: **Stat > Basic Statistics > 1-Proportion**
Select: **Summarized data**
Enter: Number of trials: **40**
Number of events: **28**
Click: **Options**
Enter: Test proportion: **.60**
Select: Alternative: **not equal**
Click: **Use test and interval based on normal distribution**
Click: **OK > OK**

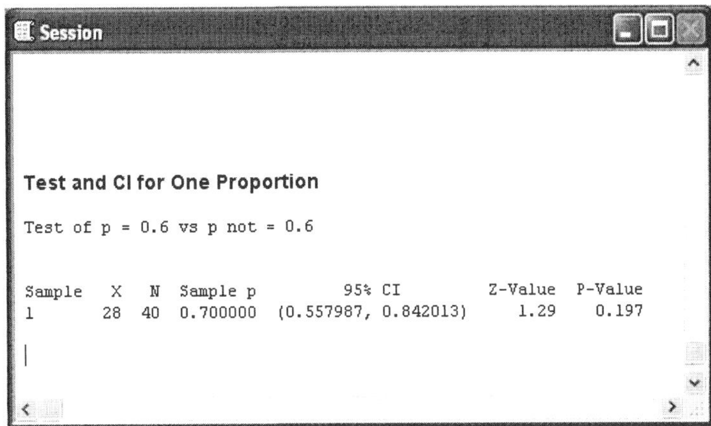

Test and CI for One Proportion

Test of p = 0.6 vs p not = 0.6

Sample	X	N	Sample p	95% CI	Z-Value	P-Value
1	28	40	0.700000	(0.557987, 0.842013)	1.29	0.197

Chapter 8 Example 7 Mean Weight Change in Anorexic Girls
Significance Test About a Mean

Open the worksheet **anorexia**, which is found in the MINITAB folder of the data disk.
We wish to perform a two-tail test on H_0: $\mu = 0$ versus H_a: $\mu \neq 0$, using the weight gains found in C7 cogchange.

Select: **Stat > Basic Statistics > 1-Sample t**

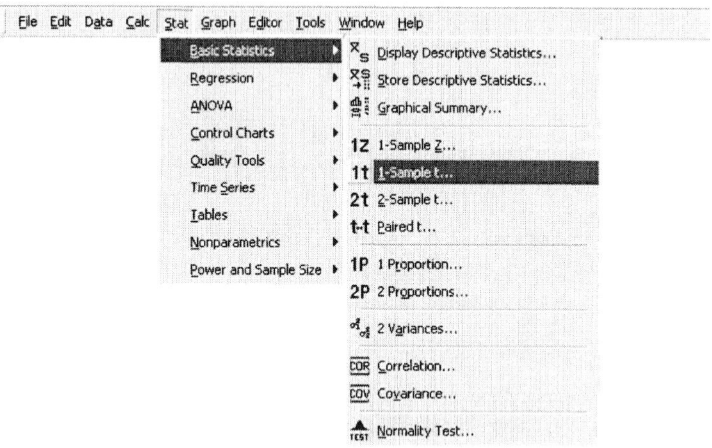

Select: **Samples in columns:**
Enter: **C7**
Test mean: **0.0**

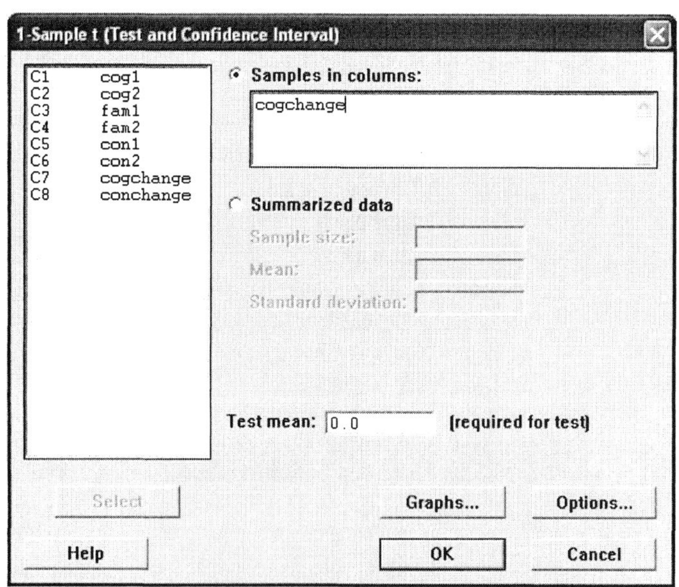

Select: **Options**
Select: Alternative: **not equal**
Click: **OK OK**

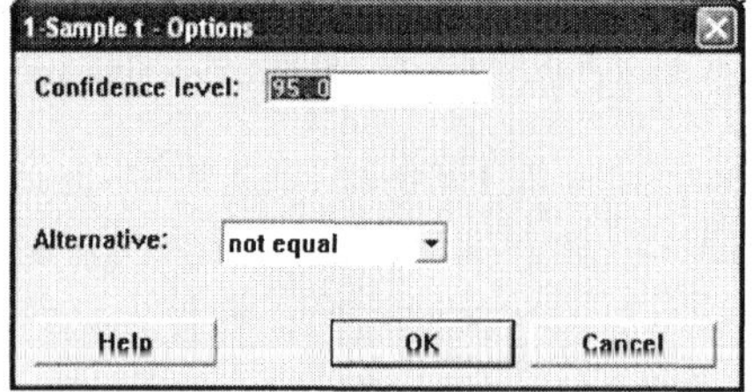

The results are displayed in the Session window.

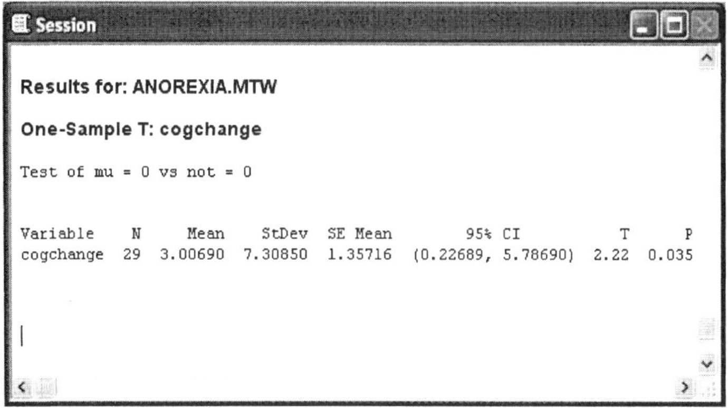

Chapter 8 Exercise 8.35 Anorexia in teenage girls
Significance Test About a Mean

Enter the data into the MINITAB Data Window.

We wish to perform a two-tail test on H_0: $\mu = 0$ versus H_a: $\mu \neq 0$, using the weight gains given.

First, we will plot the data with a box plot. There are no outliers indicated, and the data distribution is not highly skewed.

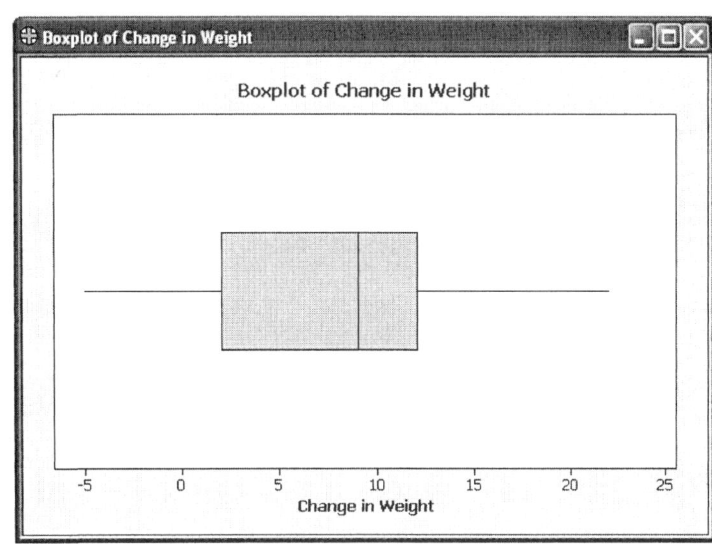

Performing the significance test about H_0: $\mu = 0$ versus H_a: $\mu \neq 0$

Select: **Stat > Basic Statistics > 1-Sample t**
Select: **Samples in columns:**
Enter: **C1**
Test mean: **0.0**

Select: **Options**
Select: Alternative: **not equal**
Click: **OK OK**

The results are displayed in the Session window.

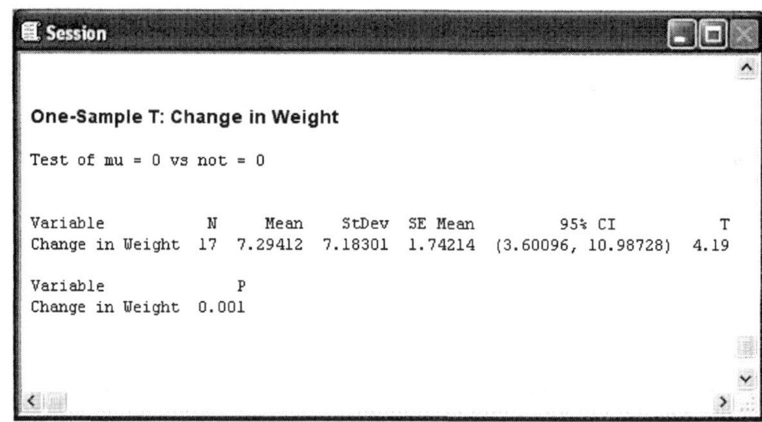

With a p-value of 0.001, the decision would be to reject H_0. There is significant evidence of a positive effect with the family therapy.

Chapter 8 Activity 2 Simulating the Performance of Significance Tests

To get a feel for the two possible errors in significance tests, we will simulate many samples from a population with a given proportion value, and perform a significance test for each sample. We will then check how often the tests make an incorrect inference.

For our first simulation, we will set the null hypothesis as H_0: p = 1/3 for a two-sided test using significance level $\alpha = 0.05$ with sample size 100.

To generate 100 samples of size 100:

Select: **Calc > Random Data > Bernoulli**

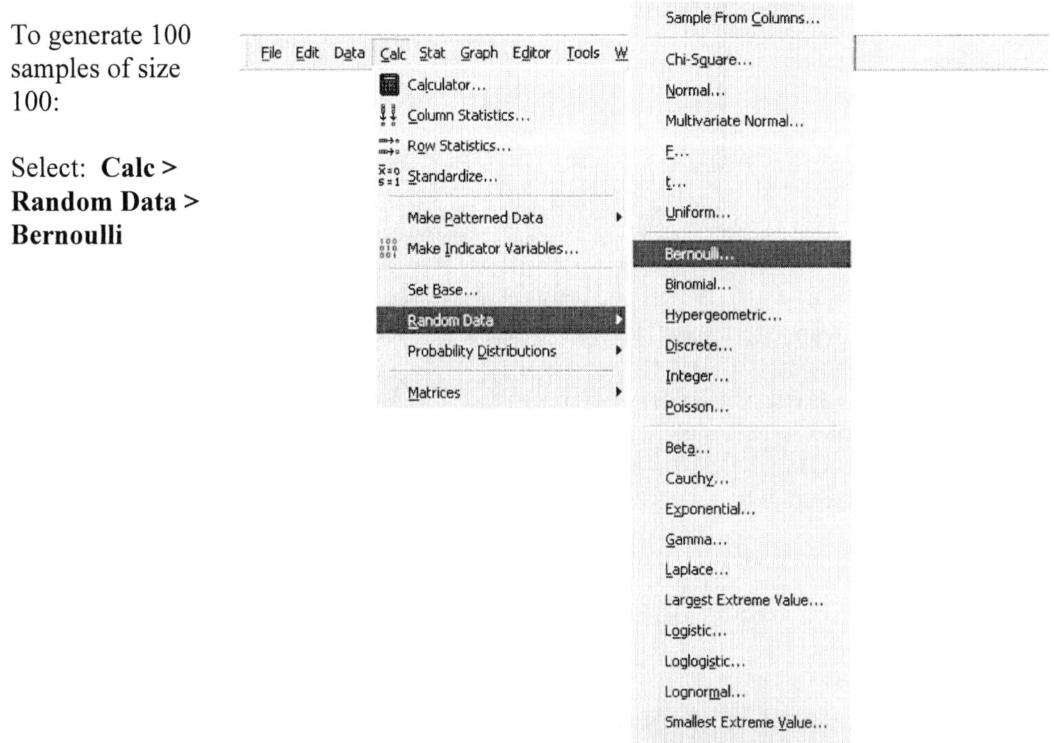

To simulate 100 samples of size 100, enter values as indicated. Click **OK**.

Bernoulli Distribution
Generate
100
rows of data
Store in column(s):
c1-c100
Probability of success:
.333333333
Select
Help
OK
Cancel

Your worksheet now contains 100 simulated samples from a binary trial where the true probability is p = 1/3.

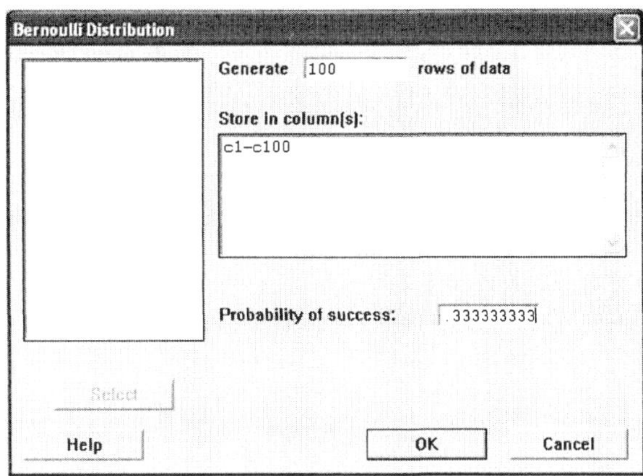

To perform the significance test on each of the 100 samples:

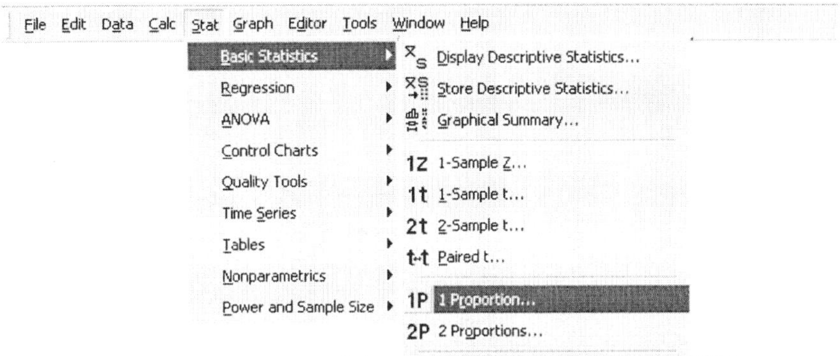

File Edit Data Calc Stat Graph Editor Tools Window Help
Basic Statistics
Regression
ANOVA
Control Charts
Quality Tools
Time Series
Tables
Nonparametrics
Power and Sample Size
Display Descriptive Statistics...
Store Descriptive Statistics...
Graphical Summary...
1-Sample Z...
1-Sample t...
2-Sample t...
Paired t...
1 Proportion...
2 Proportions...

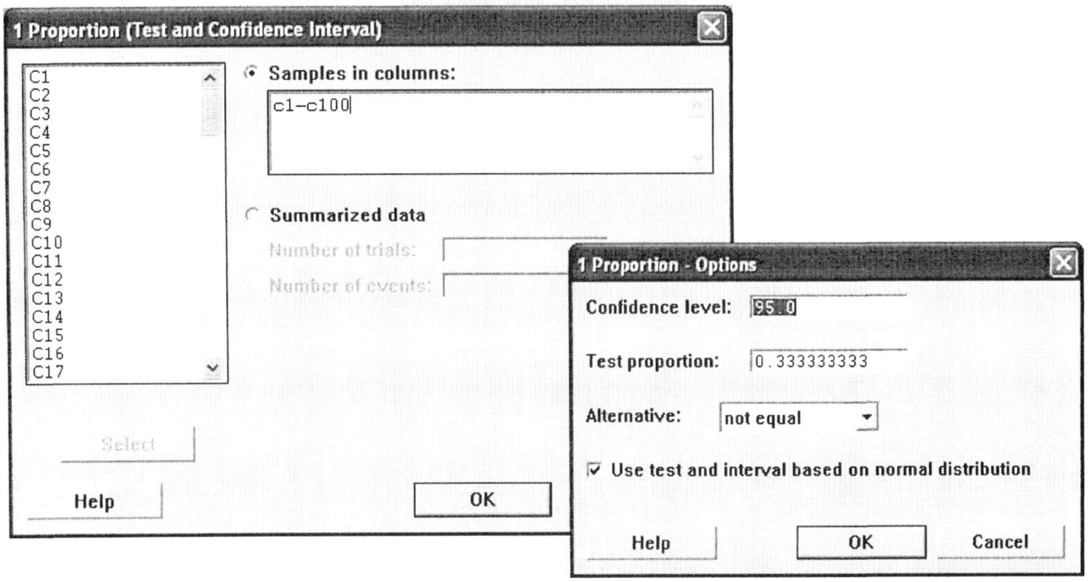

1 Proportion (Test and Confidence Interval)
C1
C2
C3
C4
C5
C6
C7
C8
C9
C10
C11
C12
C13
C14
C15
C16
C17
Samples in columns:
c1-c100
Summarized data
Number of trials:
Number of events:
Select
Help
OK
1 Proportion - Options
Confidence level:
95.0
Test proportion:
0.333333333
Alternative:
not equal
Use test and interval based on normal distribution
Help
OK
Cancel

Analyzing the results

Counting the number of samples that would result in a p-value less than 0.05 (meaning we would reject H_0, thereby making an incorrect decision), we get approximately 5/100 or 5% of our samples lead to an incorrect decision. (Your results will differ slightly, the samples selected are random.)

Session

Test and CI for One Proportion: C1, C2, C3, C4, C5, C6, C7, C8, ...

Test of p = 0.333333 vs p not = 0.333333

Event = 1

Variable	X	N	Sample p	95% CI	Z-Value	P-Value
C1	34	100	0.340000	(0.247155, 0.432845)	0.14	0.888
C2	39	100	0.390000	(0.294403, 0.485597)	1.20	0.229
C3	35	100	0.350000	(0.256516, 0.443484)	0.35	0.724
C4	34	100	0.340000	(0.247155, 0.432845)	0.14	0.888
C5	32	100	0.320000	(0.228572, 0.411428)	-0.28	0.777
C6	31	100	0.310000	(0.219353, 0.400647)	-0.49	0.621
C7	34	100	0.340000	(0.247155, 0.432845)	0.14	0.888
C8	33	100	0.330000	(0.237840, 0.422160)	-0.07	0.944
C9	37	100	0.370000	(0.275372, 0.464628)	0.78	0.437

We will now redo the simulation in a new worksheet with 100 samples based on a Bernoulli distribution with p = 0.5, so that H_0 is actually false.

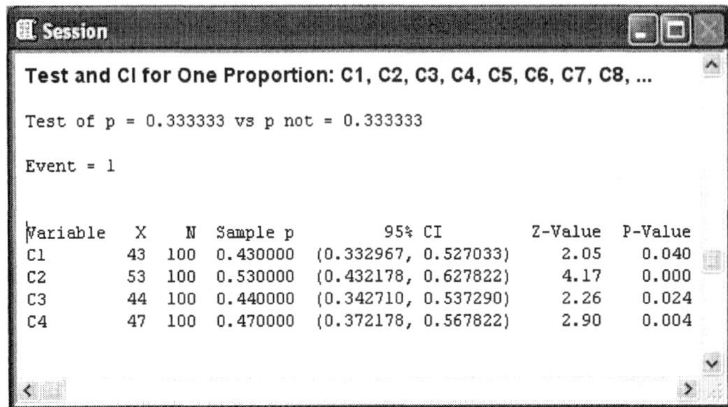

Session

Test and CI for One Proportion: C1, C2, C3, C4, C5, C6, C7, C8, ...

Test of p = 0.333333 vs p not = 0.333333

Event = 1

Variable	X	N	Sample p	95% CI	Z-Value	P-Value
C1	43	100	0.430000	(0.332967, 0.527033)	2.05	0.040
C2	53	100	0.530000	(0.432178, 0.627822)	4.17	0.000
C3	44	100	0.440000	(0.342710, 0.537290)	2.26	0.024
C4	47	100	0.470000	(0.372178, 0.567822)	2.90	0.004

Performing a significance test of H_0: p = 1/3, we will fail to reject our false Ho when the p-value is greater than alpha = 0.05. This occurs in 10 of the 100 samples. So, P(Type II) error is 0.10. Using the reasoning of example 13, we would expect P(type II) error to be 0.07.

Chapter 8 Exercise 8.55 Balancing Type I and Type II errors

Following the reasoning of Example 13, a Type II error occurs when the sample proportion falls less than z = 2.326 standard errors above the null hypothesis value of p = 1/3.

$$\hat{p} < 1/3 + 2.326\sqrt{[(1/3)(2/3)]/116} = .435$$

When p = 0.50, a Type II error has probability that $\hat{p} < 0.435$ when p = 0.50. This probability is $P(\hat{p} < 0.435) = P(z < (0.435 - 0.50)/\sqrt{[(.5)(.5)]/116}) = P(z < -1.40) = 0.0808$

To see if we get approximately the same results using simulation, we will set the null hypothesis as H_0: p = 1/3 for a one-sided test using significance level $\alpha = 0.01$ with sample size 100 (Student version of Minitab 14 will not allow more than 10,000 pieces of data, so we cannot use sample size 116.)

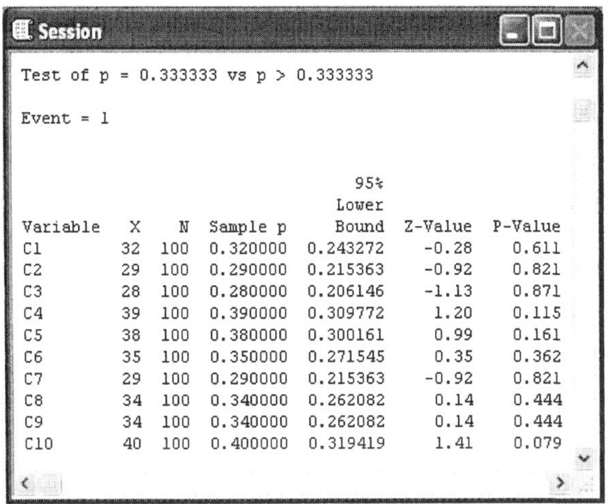

```
Test of p = 0.333333 vs p > 0.333333

Event = 1

                                    95%
                                  Lower
Variable   X    N  Sample p       Bound  Z-Value  P-Value
C1        32  100  0.320000    0.243272    -0.28    0.611
C2        29  100  0.290000    0.215363    -0.92    0.821
C3        28  100  0.280000    0.206146    -1.13    0.871
C4        39  100  0.390000    0.309772     1.20    0.115
C5        38  100  0.380000    0.300161     0.99    0.161
C6        35  100  0.350000    0.271545     0.35    0.362
C7        29  100  0.290000    0.215363    -0.92    0.821
C8        34  100  0.340000    0.262082     0.14    0.444
C9        34  100  0.340000    0.262082     0.14    0.444
C10       40  100  0.400000    0.319419     1.41    0.079
```

In only one sample (of the 100 different samples generate) we would reject a true H_0. Thus, P(Type I error) = 1/100 = 0.01 (which is α).

Once again, redoing the simulation in a new worksheet with 100 samples based on a Bernoulli distribution with p = 0.5, so that H_0 is actually false, and performing the significance test of H_0: p = 1/3 versus H_a: p > 1/3, we will fail to reject our false H_0 when the p-value is greater than $\alpha = 0.01$.

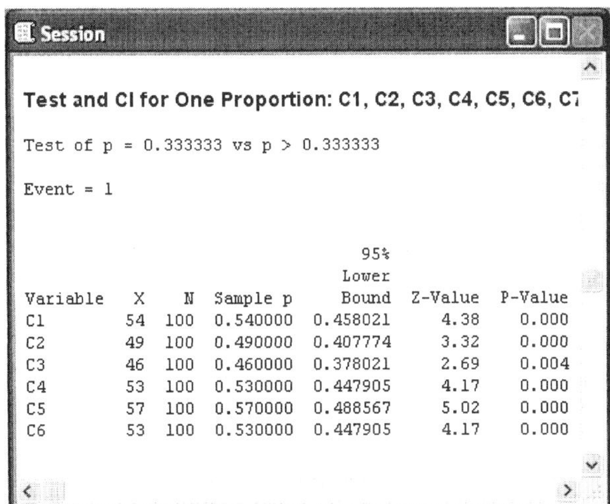

Test and CI for One Proportion: C1, C2, C3, C4, C5, C6, C7

```
Test of p = 0.333333 vs p > 0.333333

Event = 1

                                    95%
                                  Lower
Variable   X    N  Sample p       Bound  Z-Value  P-Value
C1        54  100  0.540000    0.458021     4.38    0.000
C2        49  100  0.490000    0.407774     3.32    0.000
C3        46  100  0.460000    0.378021     2.69    0.004
C4        53  100  0.530000    0.447905     4.17    0.000
C5        57  100  0.570000    0.488567     5.02    0.000
C6        53  100  0.530000    0.447905     4.17    0.000
```

This occurs in 10 of the 100 samples. So, P(Type II) error is 0.10.

(This differs from the .0808 we calculated because it is based on these particular random samples.)

Chapter 9 Example 4 - Confidence Interval Comparing Heart Attack Rates for Aspirin and Placebo - Confidence Interval for Difference Between Two Proportions

To construct a confidence interval $p_1 - p_2$, note that for the placebo group X = 189 and n = 11034, and for the aspirin group X = 104 and n = 11037.

Select: **Stat > Basic Statistics > 2 Proportions**

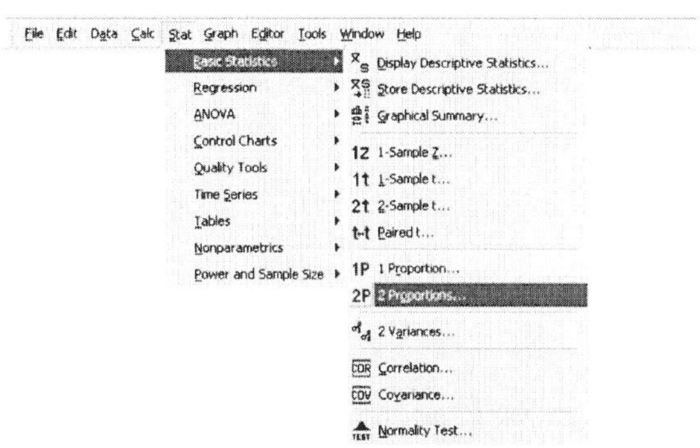

Select: **Summarized Data**
Enter: First: Trials: **11034** Events: **189**
Second: Trials: **11037** Events: **104**

Select: **Options...**
Enter: Confidence level: **95**
Click: **OK OK**

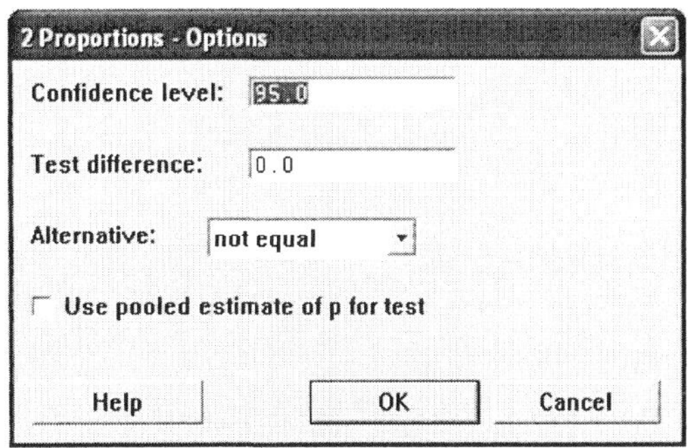

The results are displayed in the Session window.

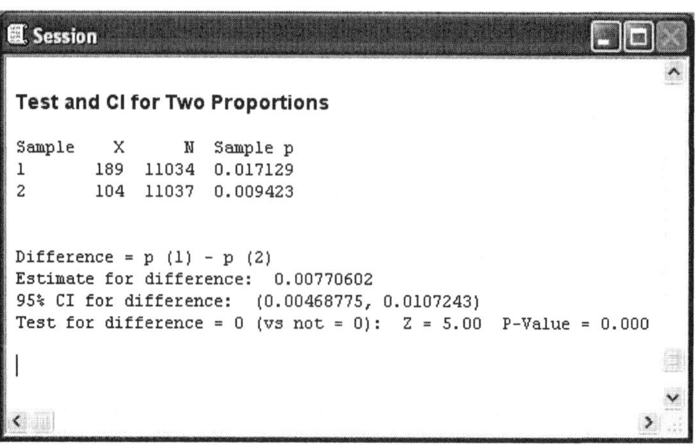

Session

Test and CI for Two Proportions

```
Sample     X      N  Sample p
1        189  11034  0.017129
2        104  11037  0.009423


Difference = p (1) - p (2)
Estimate for difference:  0.00770602
95% CI for difference:  (0.00468775, 0.0107243)
Test for difference = 0 (vs not = 0):  Z = 5.00  P-Value = 0.000
```

Chapter 9 Example 5 – Is TV Watching Associated with Aggressive Behavior? Significance Test for Difference Between Two Proportions

To perform a significance test for H_0: $p_1 - p_2 = 0$ versus H_a: $p_1 - p_2 \neq 0$ note that for the "less than 1 hour of TV per day" group X = 5 and n = 88, and for the "at least 1 hour of TV per day" group X = 154 and n = 619.

Select: **Stat > Basic Statistics > 2 Proportions**

Select: **Summarized Data**
Enter: First: Trials: **88** Events: **5**
Second: Trials: **619** Events: **154**

Select: **Options…**
Enter: Test difference: **0.0**
Select: Alternative: **not equal**
Select: **Use pooled estimate of p for test**
Click: **OK OK**

The results are displayed in the Session window.

```
Session

Test and CI for Two Proportions

Sample    X    N  Sample p
1         5   88  0.056818
2       154  619  0.248788


Difference = p (1) - p (2)
Estimate for difference:  -0.191970
95% CI for difference:  (-0.251124, -0.132816)
Test for difference = 0 (vs not = 0):  Z = -4.04  P-Value = 0.000
```

Chapter 9 Exercise 9.3 Binge Drinking
Confidence Interval for Difference Between Two Proportions

To form a 95% confidence interval for $p_1 - p_2$ note that for the 2001 group, X = 4234 (.482 * 8783) and n = 8783, and for the 1993 group, X = 5071 (.399 * 12708) and n = 12708.

Select: **Stat > Basic Statistics > 2 Proportions**

Select: **Summarized Data**
Enter: First: Trials: **8783** Events: **4234**
Second: Trials: **12708** Events: **5071**
Select: **Options…**

Enter: Confidence level: **95.0**
Click: **OK OK**

The results are displayed in the Session window.

```
Session

Test and CI for Two Proportions

Sample     X      N  Sample p
1       4234   8783  0.482068
2       5071  12708  0.399040


Difference = p (1) - p (2)
Estimate for difference:  0.0830277
95% CI for difference:  (0.0695483, 0.0965070)
Test for difference = 0 (vs not = 0):  Z = 12.08  P-Value = 0.000
```

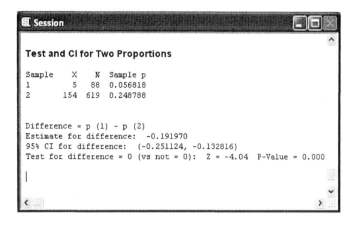

Chapter 9 Exercise 9.9 Drinking and Unplanned Sex
Significance Test for Difference Between Two Proportions

To perform a significance test for H0: $p_1 - p_2 = 0$ versus Ha: $p_1 - p_2 \neq 0$ note that for the 2001 group, X = 1871 (.213 * 8783) and n = 8783, and for the 1993 group, X = 2440 (.192 * 12708) and n = 12708.

Select: **Stat > Basic Statistics > 2 Proportions**

Select: **Summarized Data**
Enter: First: Trials: **8783** Events: **1871**
Second: Trials: **12708** Events: **2440**
Select: **Options...**

Enter: Test difference: **0.0**
Select: Alternative: **not equal**
Select: **Use pooled estimate of p for test**
Click: **OK OK**

The results are displayed in the Session window.

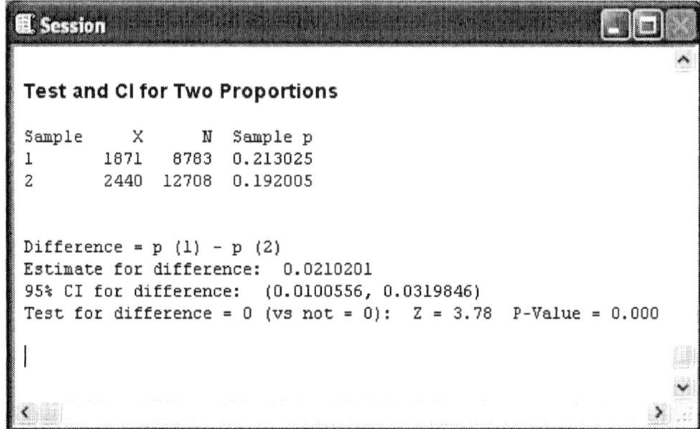

Session
Test and CI for Two Proportions
Sample X N Sample p
1 1871 8783 0.213025
2 2440 12708 0.192005
Difference = p (1) - p (2)
Estimate for difference: 0.0210201
95% CI for difference: (0.0100556, 0.0319846)
Test for difference = 0 (vs not = 0): Z = 3.78 P-Value = 0.000

Chapter 9 Example 8 Nicotine – How Much More Addicted are Smokers than Ex-Smokers?
Confidence Interval for Difference Between Population Means

To construct a confidence interval for $\mu_1 - \mu_2$, note that for the smoker group, $\bar{x} = 5.9$ and s = 3.3 for n = 75, and for the ex-smoker group, $\bar{x} = 1.0$ and s = 2.3 for n = 257.

Select: **Stat > Basic Statistics > 2-Sample t…**

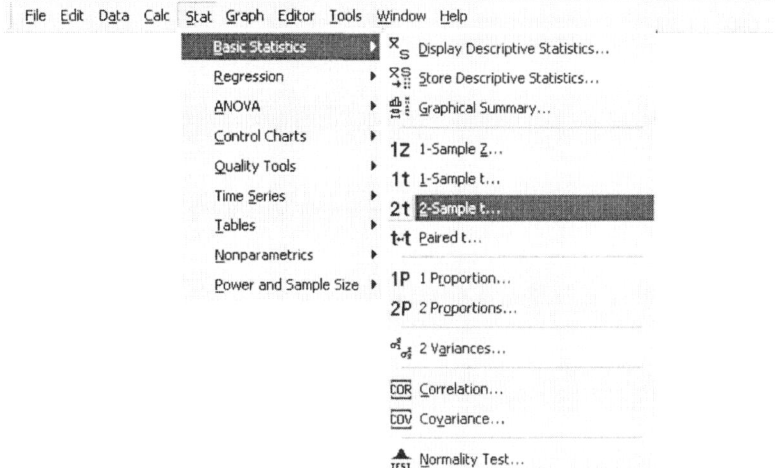

Select: **Summarized Data**

Enter:
First: Sample size:**75** Mean: **5.9** Standard deviation: **3.3**
Second: Sample size: **257** Mean: **1.0**
Standard deviation: **2.3**

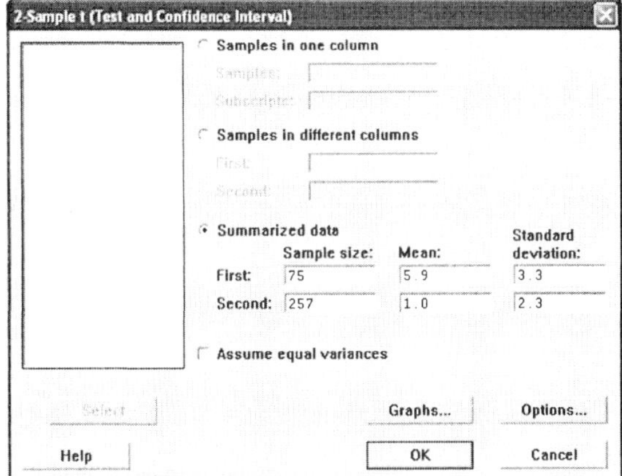

Select: **Options…**
Enter: Confidence level: **95**
Click: **OK OK**

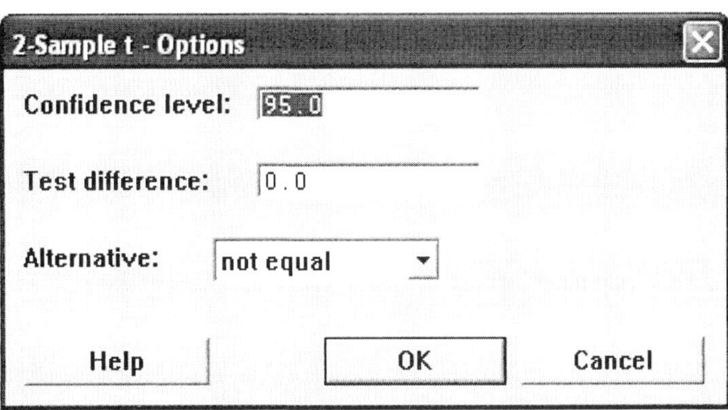

The results are displayed in the Session window.

```
Session

Two-Sample T-Test and CI

Sample    N  Mean  StDev  SE Mean
1        75  5.90   3.30     0.38
2       257  1.00   2.30     0.14


Difference = mu (1) - mu (2)
Estimate for difference:  4.90000
95% CI for difference:  (4.09167, 5.70833)
T-Test of difference = 0 (vs not =): T-Value = 12.03  P-Value = 0
```

Chapter 9 Example 9 Does Cell Phone Use While Driving Impair Reaction Times? Significance Test for Comparing Two Population Means

To perform a significance test for H_0: $\mu_1 - \mu_2 = 0$ versus H_a: $\mu_1 - \mu_2 \neq 0$ note that for the "cell phone" group, $\bar{x} = 585.2$ and s = 89.6 for n = 32, and for the "control" group, $\bar{x} = 533.7$ and s = 65.3 for n = 32.

Select: **Stat > Basic Statistics > 2-Sample t...**

Select: **Summarized Data**
Enter: First: Sample size: **32** Mean: **585.2** Standard deviation: **89.6**
Second: Sample size: **32** Mean: **533.7** Standard deviation: **65.3**

Select: **Options...**
Enter: Test difference: **0.0**
Select: Alternative: **not equal**
Click: **OK OK**

The results are displayed in the Session window.

```
Session

Two-Sample T-Test and CI

                                SE
Sample   N   Mean  StDev  Mean
1       32  585.2   89.6    16
2       32  533.7   65.3    12


Difference = mu (1) - mu (2)
Estimate for difference:  51.5000
95% CI for difference:  (12.2379, 90.7621)
T-Test of difference = 0 (vs not =): T-Value = 2.63  P-Value = 0.011  DF = 56
```

Chapter 9 Exercise 9.19 Some Smoked But Didn't Inhale
Confidence Interval for Difference Between Population Means

To construct a 95% confidence interval for $\mu_1 - \mu_2$, note that for the "Inhalers" group, $\bar{x} = 2.9$ and s = 3.6 for n = 237, and for the "Non-inhalers" group, $\bar{x} = 0.1$ and s = 0.5 for n = 95.

Select: **Stat > Basic Statistics > 2-Sample t…**

Select: **Summarized Data**
Enter: First: Sample size: **237** Mean: **2.9** Standard deviation: **3.6**
Second: Sample size: **95** Mean: **0.1** Standard deviation: **0.5**

Select: **Options…**
Enter: Confidence level: **95**
Click: **OK OK**

The results are displayed in the Session window.

```
Session

Two-Sample T-Test and CI

Sample    N   Mean  StDev  SE Mean
1       237   2.90   3.60     0.23
2        95  0.100  0.500    0.051


Difference = mu (1) - mu (2)
Estimate for difference:  2.80000
95% CI for difference:  (2.32855, 3.27145)
T-Test of difference = 0 (vs not =): T-Value = 11.70  P-Value = 0.000  DF = 257
```

Chapter 9 Exercise 9.21 Females or Males More Nicotine Dependent?
Significance Test for Comparing Two Population Means

To perform a significance test for H_0: $\mu_1 - \mu_2 = 0$ versus H_a: $\mu_1 - \mu_2 \neq 0$ note that for the "females", $\bar{x} = 2.8$ and s = 3.6 for n = 150, and for the "males", $\bar{x} = 1.6$ and s = 2.9 for n = 182.

Select: **Stat > Basic Statistics > 2-Sample t…**

Select: **Summarized Data**
Enter: First: Sample size: **150** Mean: **2.8** Standard deviation: **3.6**
Second: Sample size: **182** Mean: **1.6** Standard deviation: **2.9**

Select: **Options…**
Enter: Confidence level: **95**

Click: **OK OK**

The results are displayed in the Session window.

```
Session
Two-Sample T-Test and CI

Sample    N  Mean  StDev  SE Mean
1       150  2.80   3.60     0.29
2       182  1.60   2.90     0.21


Difference = mu (1) - mu (2)
Estimate for difference:  1.20000
95% CI for difference:  (0.48321, 1.91679)
T-Test of difference = 0 (vs not =): T-Value = 3.30  P-Value = 0.001  DF = 284
```

Chapter 9 Exercise 9.23 TV Watching and Gender
Significance Test for Comparing Two Population Means

To perform a significance test for H_0: $\mu_1 - \mu_2 = 0$ versus H_a: $\mu_1 - \mu_2 \neq 0$ note that for the "females", $\bar{x}$ = 3.06 and s = 2.12 for n = 506, and for the "males", $\bar{x}$ = 2.88 and s = 2.63 for n = 399.

Select: **Stat > Basic Statistics > 2-Sample t...**

Select: **Summarized Data**
Enter: First: Sample size: **506** Mean: **3.06** Standard deviation: **2.12**
Second: Sample size: **399** Mean: **2.88** Standard deviation: **2.63**

Select: **Options...**
Enter: Test difference: **0.0**
Alternative: **not equal**
Click: **OK OK**

The results are displayed in the Session window.

```
Session
Two-Sample T-Test and CI

Sample    N  Mean  StDev  SE Mean
1       506  3.06   2.12    0.094
2       399  2.88   2.63     0.13


Difference = mu (1) - mu (2)
Estimate for difference:  0.180000
95% CI for difference:  (-0.137866, 0.497866)
T-Test of difference = 0 (vs not =): T-Value = 1.11  P-Value = 0.267  DF = 754
```

Chapter 9 Example 10 Is Arthroscopic Surgery Better Than Placebo?
Significance Test for Comparing Two Population Means Assuming Equal Population Standard Deviations

To perform a significance test for H_0: $\mu_1 - \mu_2 = 0$ versus H_a: $\mu_1 - \mu_2 \neq 0$ note that for the "placebo" group, $\bar{x} = 51.6$ and s = 23.7 for n = 60, and for the "lavage arthroscopic" group, $\bar{x} = 53.7$ and s = 23.7 for n = 61.

Select: **Stat > Basic Statistics > 2-Sample t…**

Select: **Summarized Data**
Enter: First: Sample size: **60** Mean: **51.6** Standard deviation: **23.7**
Second: Sample size: **61** Mean: **53.7** Standard deviation: **23.7**
Select: **Assume equal variances**

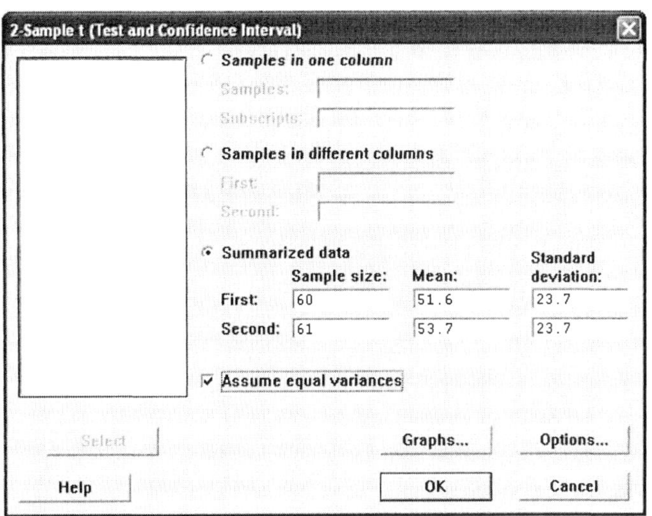

Select: **Options…**

2-Sample t - Options
Confidence level: 95.0
Test difference: 0.0
Alternative: not equal
Help OK Cancel

Enter: Test difference: **0.0**
Select: Alternative: **not equal**
Click: **OK OK**

The results are displayed in the Session window.

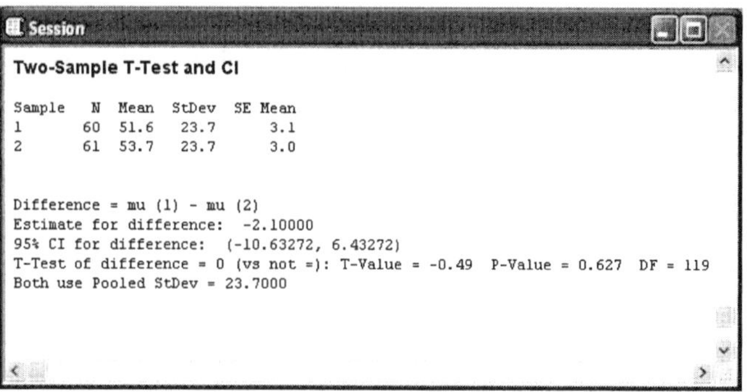

Chapter 9 Exercise 9.31 Vegetarians More Liberal?
Significance Test for Comparing Two Population Means Assuming Equal Population Standard Deviations

To perform a significance test for H_0: $\mu_1 - \mu_2 = 0$ versus H_a: $\mu_1 - \mu_2 \neq 0$ assuming equal variances note that for the "non-vegetarian" group, $\bar{x} = 3.18$ and s = 1.72 for n = 51, and for the "vegetarian" group, $\bar{x} = 2.22$ and s = .67 for n = 9.

Select: **Stat > Basic Statistics > 2-Sample t...**
Select: **Summarized Data**
Enter: First: Sample size: **51** Mean: **3.18** Standard deviation: **1.72**
Second: Sample size: **9** Mean: **2.22** Standard deviation: **.67**
Select: **Assume equal variances**
Select: **Options...**
Enter: Test difference: **0.0**
Select: Alternative: **not equal**
Click: **OK OK**

The results are displayed in the Session window.

```
Session
Two-Sample T-Test and CI

Sample   N   Mean  StDev  SE Mean
1       51   3.18   1.72     0.24
2        9  2.220  0.670     0.22


Difference = mu (1) - mu (2)
Estimate for difference:  0.960000
95% CI for difference:  (-0.209716, 2.129716)
T-Test of difference = 0 (vs not =): T-Value = 1.64  P-Value = 0.106  DF = 58
Both use Pooled StDev = 1.6162
```

Chapter 9 Example 13 Matched Pairs Analysis of Cell Phone Impact on Driver Reaction Time
Comparing Means with Matched Pairs

Enter the data from Table 9.9.
To perform a significance test for H_0: $\mu_d = 0$ versus H_a: $\mu_d \neq 0$

Select: **Stat > Basic Statistics > Paired t…**

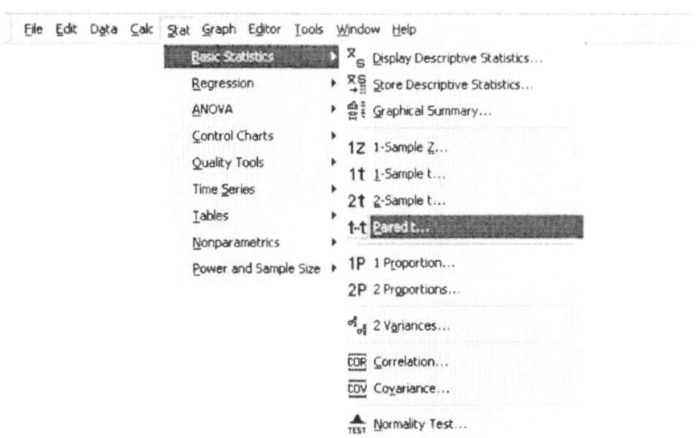

Select: **Samples in columns**
Enter: First sample: **Yes**
Second sample: **No**

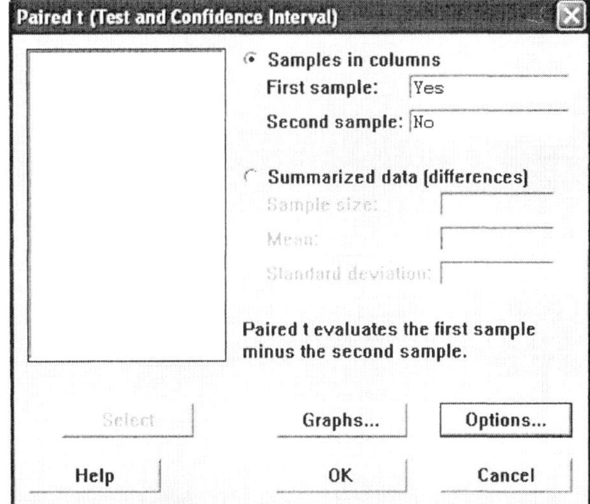

Select: **Options…**

Enter: Confidence level: **95**
Enter: Test difference: **0.0**
Select: Alternative: **not equal**
Click: **OK OK**

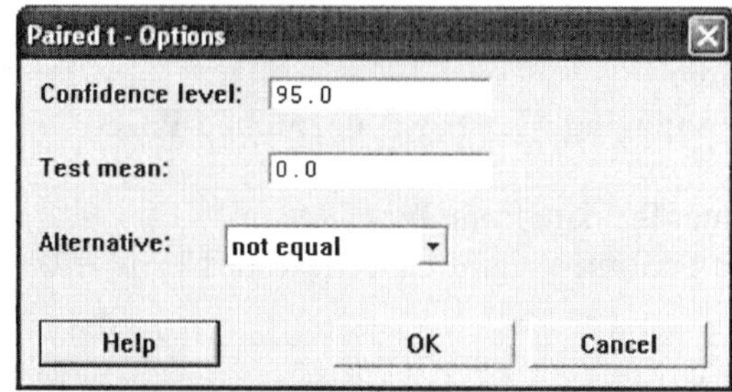

The results are displayed in the Session window.

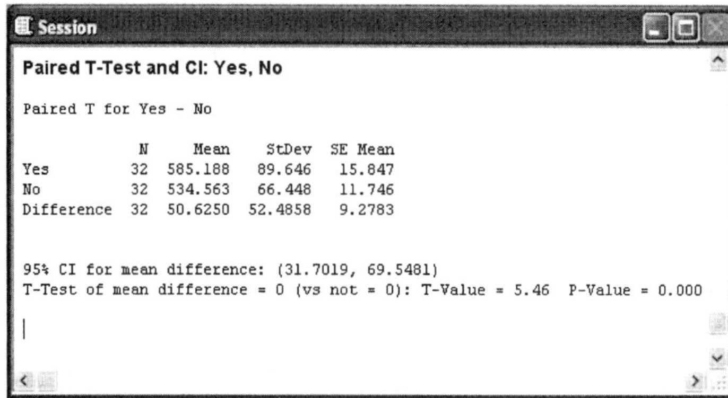
Session

Paired T-Test and CI: Yes, No

```
Paired T for Yes - No

             N     Mean    StDev  SE Mean
Yes         32  585.188   89.646   15.847
No          32  534.563   66.448   11.746
Difference  32  50.6250  52.4858   9.2783


95% CI for mean difference: (31.7019, 69.5481)
T-Test of mean difference = 0 (vs not = 0): T-Value = 5.46  P-Value = 0.000
```

Chapter 9 Exercise 9.41 MINITAB Output for Inferential Analysis: Paired t for movies - sports
Comparing Means with Matched Pairs

Enter the data into a worksheet. We wish to perform inference on "movies – sports".

To perform a significance test for H_0: $\mu_d = 0$ versus H_a: $\mu_d \neq 0$ and to generate a 95% confidence interval

Select: **Stat > Basic Statistics > Paired t…**

Select: **Samples in columns**
Enter: First sample: **Movies**
Second sample: **Sports**

Select: **Options…**

Enter: Confidence level: **95**
Enter: Test difference: **0.0**
Select: Alternative: **not equal**
Click: **OK OK**

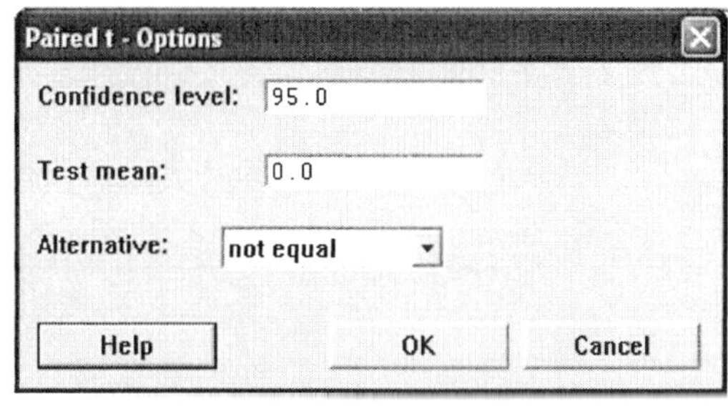

The results are displayed in the Session window.

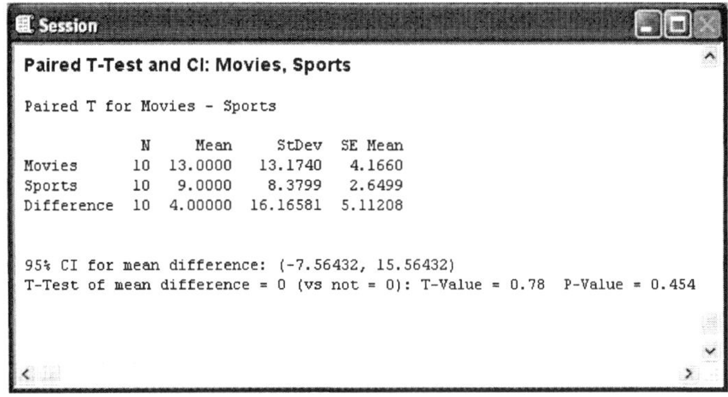

Chapter 9 Example 15 Inference Comparing Beliefs in Heaven and Hell Comparing Proportions with Dependent Samples

We will create the 95% confidence interval for population mean difference $p_1 - p_2$ using summary statistics.

First note that the sample mean of the 1120 difference scores equals

[(0)(833) + (1)(125) + (-1)(2) + (0)(160)] / 1120 = 123/1120 = 0.109821

and the sample standard deviation of the 1120 difference scores equals

$\sqrt{\{ [(0 - .109821)^2 * (833) + (1- .109821)^2 *(125)+(-1- .109821)^2 *(2) + (0- .109821)^2 *(160)] / (1120-1)\}}$
= 0.318469

Select: **Stat > Basic Statistics > Paired t…**

Select: **Summarized data (differences)**
Enter: Sample size: **1120**
Mean: **.109821**
Standard deviation: **.318469**

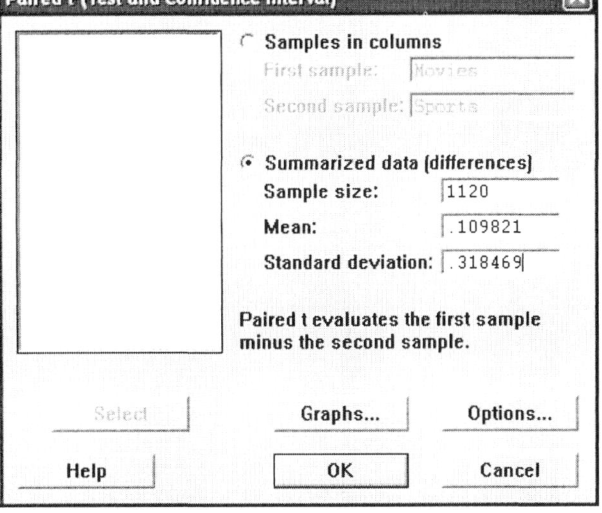

Select: **Options**

Enter: Confidence level: **95**

Click: **OK OK**

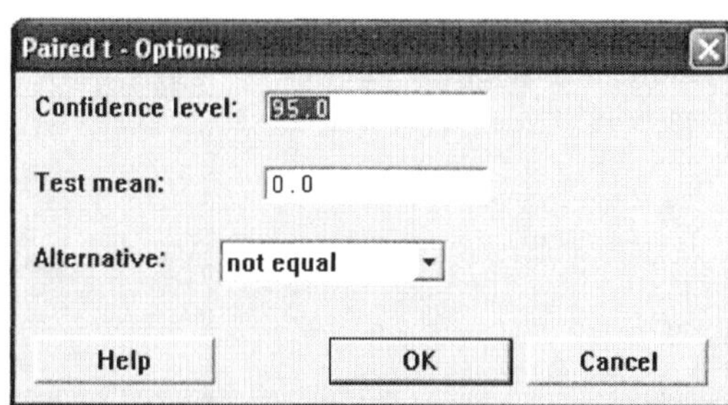

The results are displayed in the Session window.

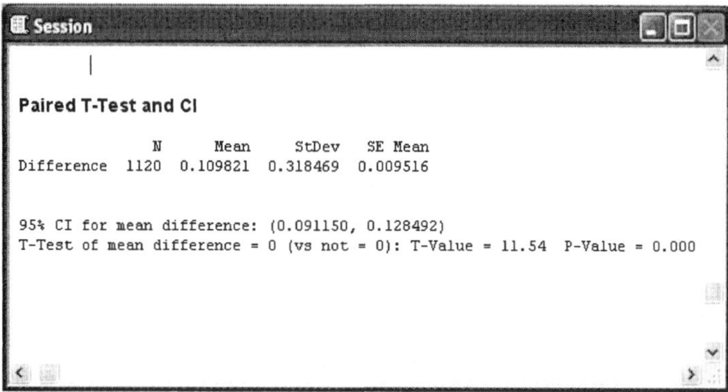

Chapter 9 Exercise 9.49 Heaven and Hell
Comparing Proportions for Dependent Samples

A point estimate for the difference between the population proportions believing in heaven and believing in hell is $\hat{p}_1 - \hat{p}_2 = .631 - .496 = .135$

Because we do not know the sample size used, but can assume dependent sampling, we'll use McNemar's test. Since McNemar's test is not a procedure included in MINITAB, we'll just use MINITAB calculator to do the computation.

Determining the z-score:

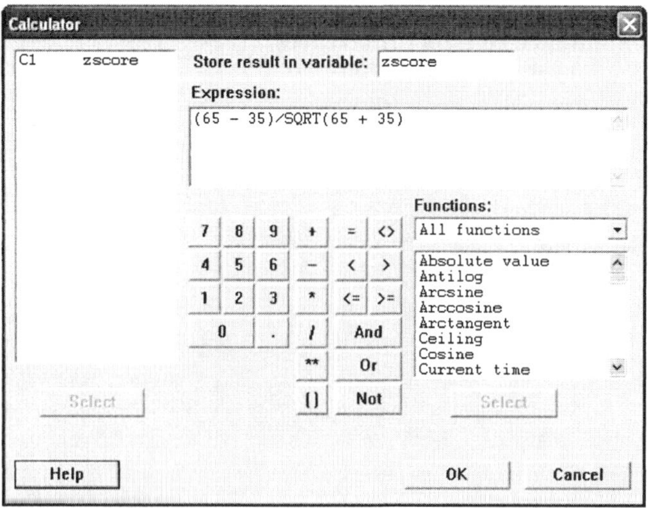

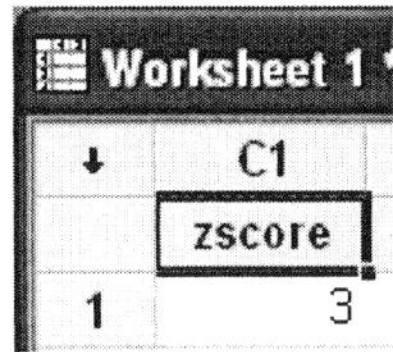

For z = 3, the two-sided p-value equals 0.0026. The sample gives extremely strong evidence that the proportion p_1 definitely believing in heaven is higher than the proportion p_2 definitely believing in hell.

Chapter 10 Exercise 10.1 Gender Gap in Politics?
Contingency Tables

Enter the data into the worksheet.

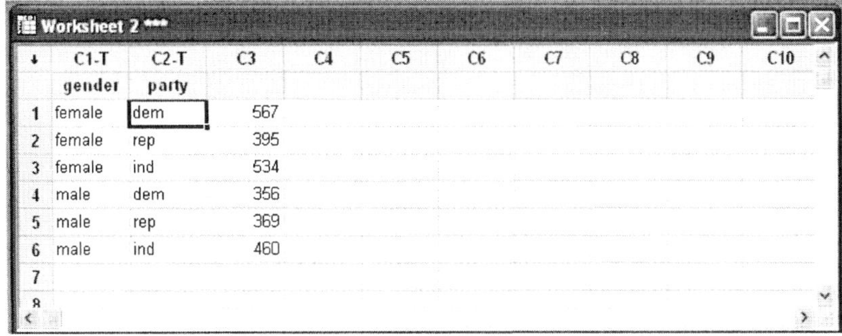

To construct the basic table choose **Stat > Tables > Cross Tabulation and Chi Square**

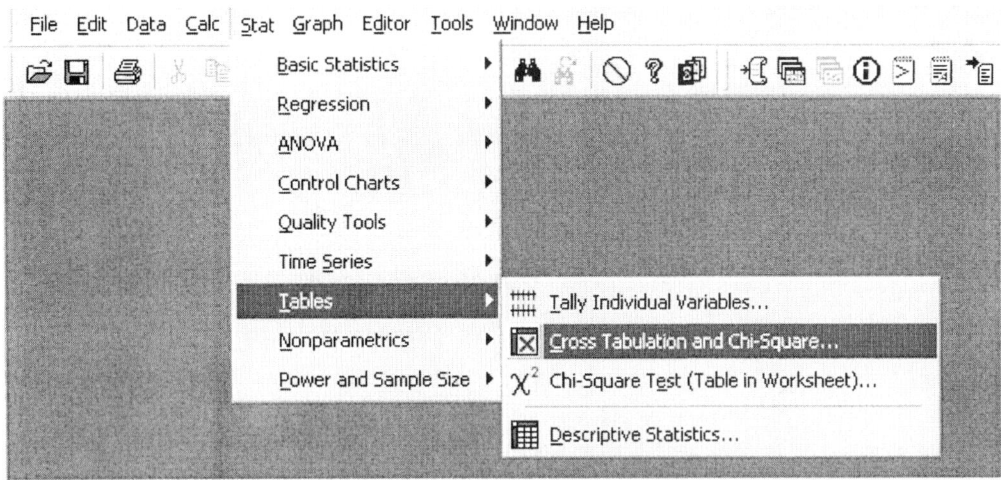

Enter the variables as shown

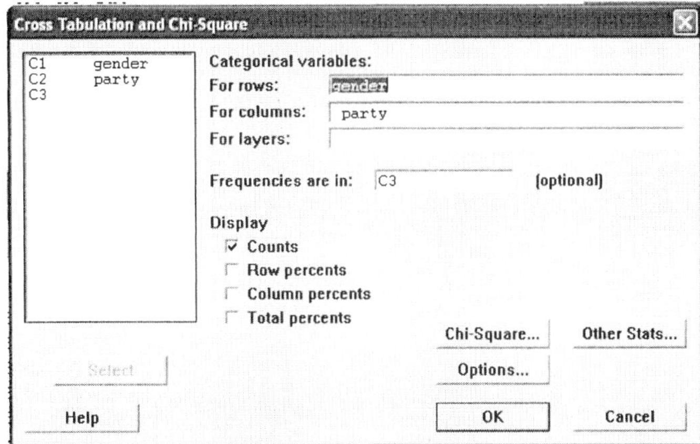

This will generate the basic table.

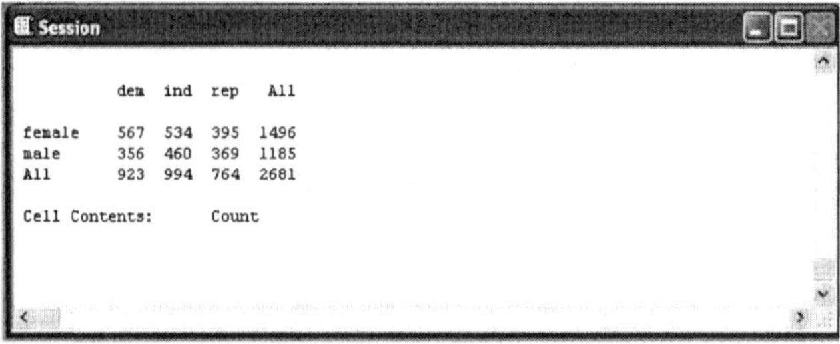

Now to show the percentages, click on Counts, Row Percents and Column Percents.

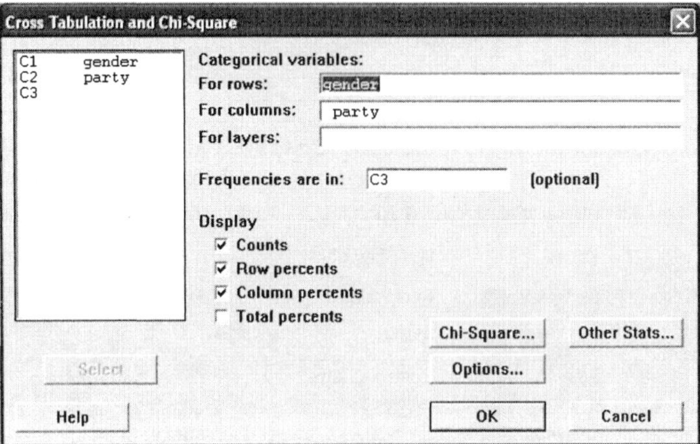

This give the following in the session window

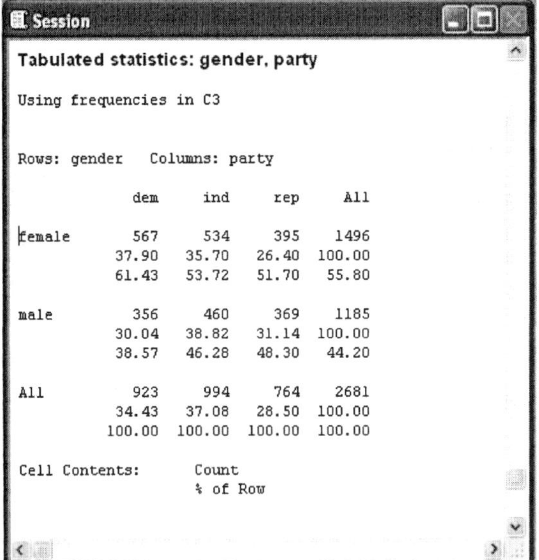

We now enter the chart into the worksheet, so that we can graph it.

Worksheet 1 ***

↓	C1-T	C2-T	C3	C4	C5-T	C6	C7	C8	C9
	gender	party			genders	dem	rep	ind	
1	female	dem	567		male	356	369	460	
2	female	rep	395		female	567	395	534	
3	female	ind	534						
4	male	dem	356						
5	male	rep	369						
6	male	ind	460						
7									

To create the bar chart
Choose: **Graph > Bar Chart > Values from a table > Two way table cluster**

Enter the values as shown.

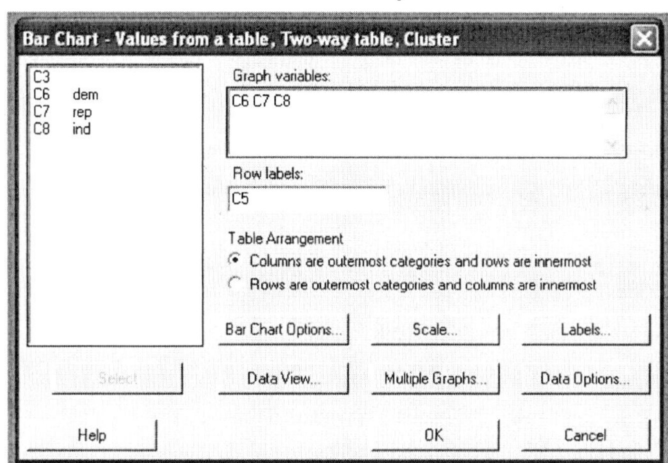

This gives you the following bar chart

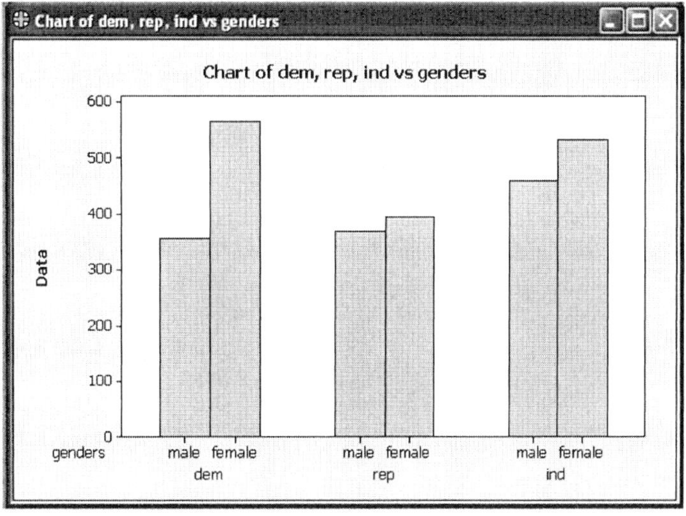

If you had chosen rows to be the outermost category, you would get the next chart.

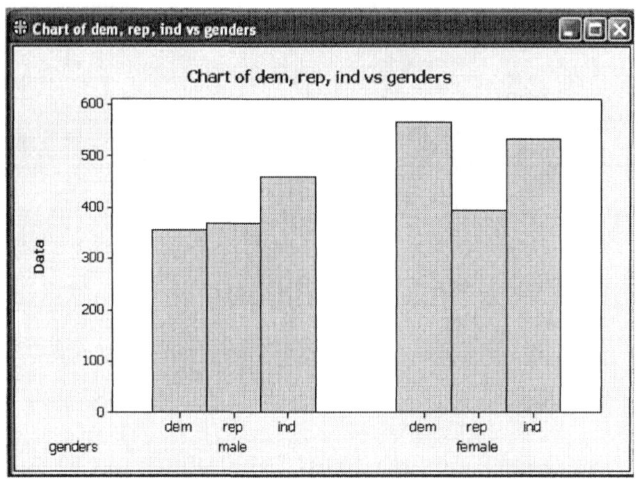

Exercise 10.5 Marital Happiness and Income Contingency Table and the Chi Squared Statistic

Following the directions in Activity1 at the end of section 10.2, you can access the data in the GSS data base for this problem. Enter the fields as shown below.

SDA Frequencies/Crosstabulation Program
Selected Study: GSS 1972-2002 Cumulative Datafile
Help: General / Recoding Variables

REQUIRED Variable names to specify

Row: FINRELA(r:1-2;3;4-5)

OPTIONAL Variable names to specify

Column: HAPMAR

Control:

Selection Filter(s): YEAR(2002) *Example: age(18-50)*

Weight: No Weight

TABLE OPTIONS

Percentaging:
☐ Column ☑ Row ☐ Total
with 1 decimal(s)

☑ Statistics with 2 decimal(s)

☐ Question text ☐ Suppress table
☑ Color coding ☐ Show Z-statistic

CHART OPTIONS

Type of chart: Stacked Bar Chart
Bar chart options:
Orientation: ◉ Vertical ○ Horizontal
Visual Effects: ◉ 2-D ○ 3-D

Show Percents: ☐ Yes
Palette: ◉ Color ○ Grayscale
Size - width: 600 height: 400

Run the Table | Clear Fields

After you have accessed the data, enter it into a Minitab worksheet as follows:

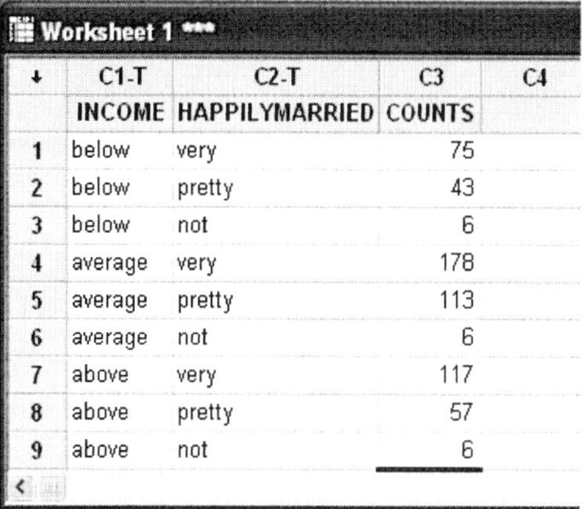

Worksheet 1 ***

↓	C1-T	C2-T	C3	C4
	INCOME	HAPPILYMARRIED	COUNTS	
1	below	very	75	
2	below	pretty	43	
3	below	not	6	
4	average	very	178	
5	average	pretty	113	
6	average	not	6	
7	above	very	117	
8	above	pretty	57	
9	above	not	6	

Choose **Stat > Tables > Cross Tabulation and Chi Square**

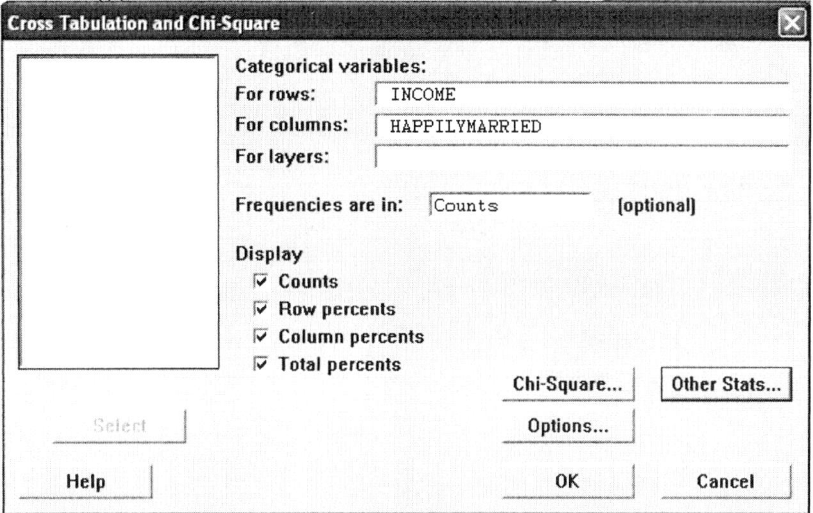

This gives us the basic contingency table.

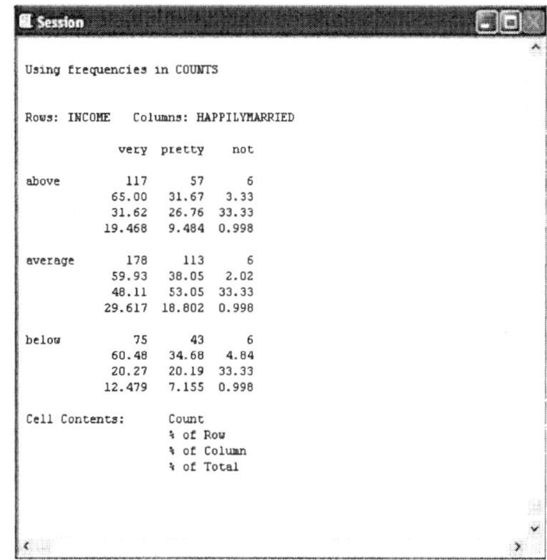

Session

Using frequencies in COUNTS

Rows: INCOME Columns: HAPPILYMARRIED

	very	pretty	not
above	117	57	6
	65.00	31.67	3.33
	31.62	26.76	33.33
	19.468	9.484	0.998
average	178	113	6
	59.93	38.05	2.02
	48.11	53.05	33.33
	29.617	18.802	0.998
below	75	43	6
	60.48	34.68	4.84
	20.27	20.19	33.33
	12.479	7.155	0.998

Cell Contents: Count
% of Row
% of Column
% of Total

To create the bar chart choose **Graph > Bar Chart > Values from a table > Two way table cluster.** Enter the values as shown.

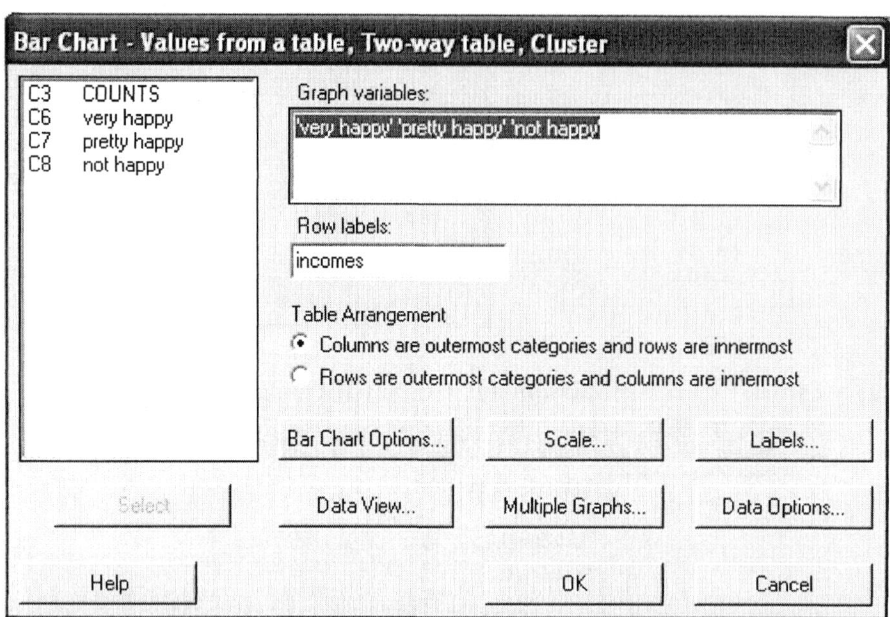

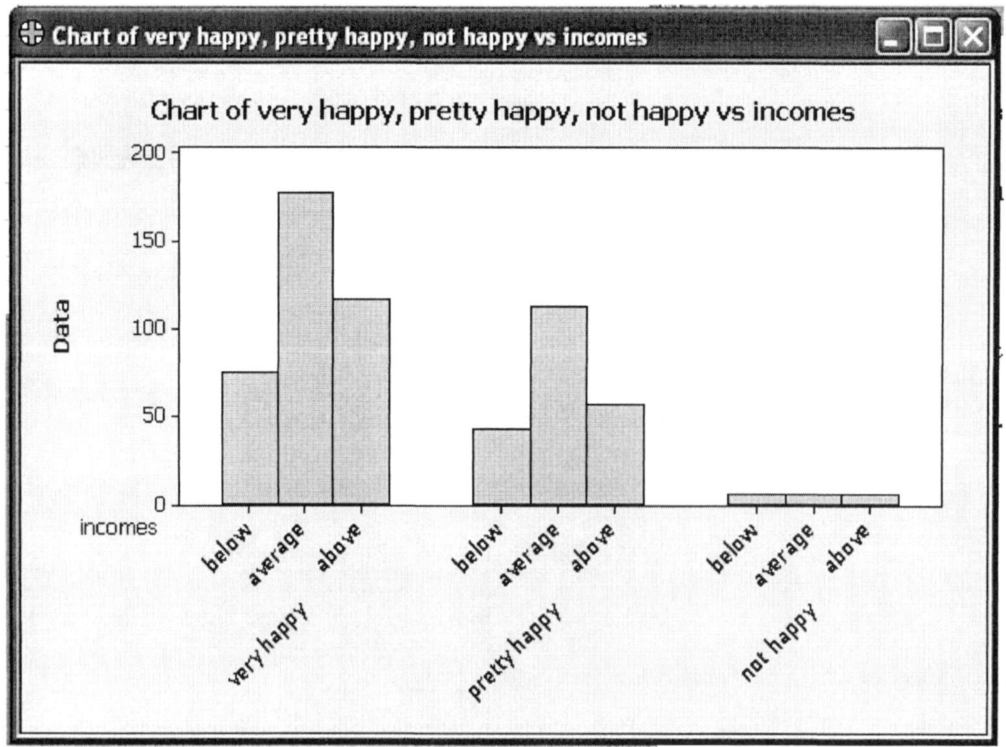

Chapter 10 Example 3: Chi Squared Test for Happiness and Family Income
Chi Squared Test of Independence

In this problem we are using the data summarized in the margin of the text, originally retrieved from the GSS. We can create a data file using it by manually entering the data.

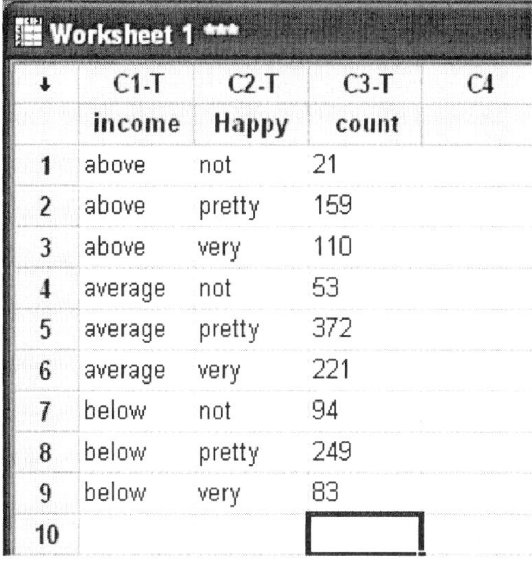
Worksheet 1 ***

↓	C1-T	C2-T	C3-T	C4
	income	Happy	count	
1	above	not	21	
2	above	pretty	159	
3	above	very	110	
4	average	not	53	
5	average	pretty	372	
6	average	very	221	
7	below	not	94	
8	below	pretty	249	
9	below	very	83	
10				

To create the table as shown in the text : Choose **Stat > Tables > Cross Tabulation and Chi Squared**

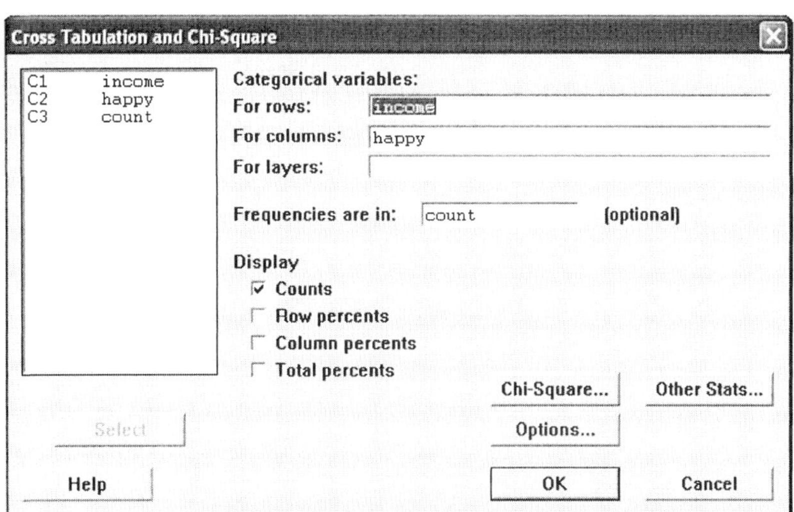

Click on the **Chi-Square** box and check the desired boxes as shown:

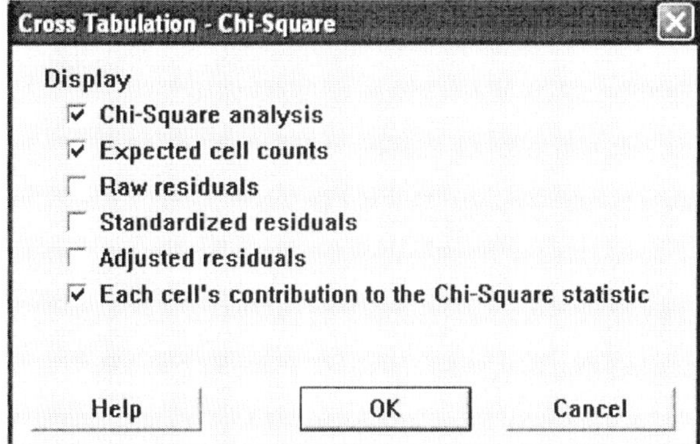

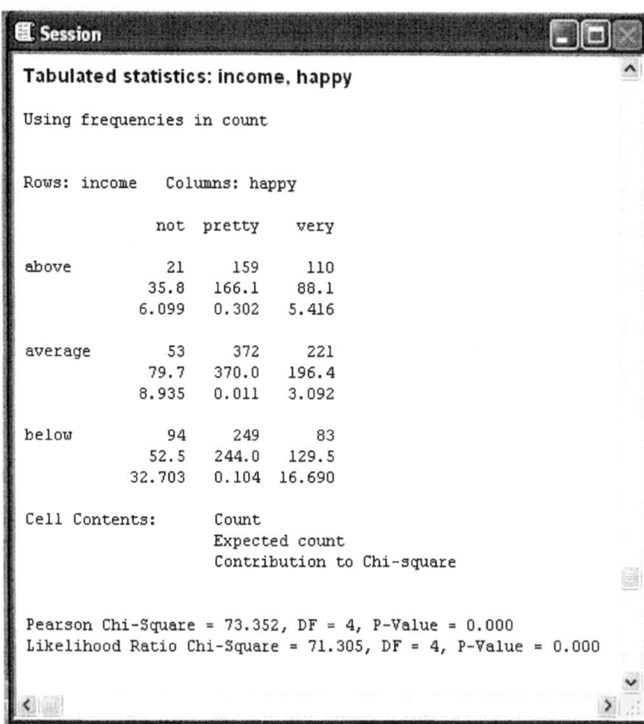

```
Session

Tabulated statistics: income, happy

Using frequencies in count


Rows: income   Columns: happy

              not  pretty    very

above          21     159     110
             35.8   166.1    88.1
            6.099   0.302   5.416

average        53     372     221
             79.7   370.0   196.4
            8.935   0.011   3.092

below          94     249      83
             52.5   244.0   129.5
           32.703   0.104  16.690

Cell Contents:      Count
                    Expected count
                    Contribution to Chi-square


Pearson Chi-Square = 73.352, DF = 4, P-Value = 0.000
Likelihood Ratio Chi-Square = 71.305, DF = 4, P-Value = 0.000
```

Chapter 10 Example 4: Are Happiness and Income Independent?
Chi Squared Test of Independence

To create the table as shown in the text :

Choose **Stat > Tables > Cross Tabulation and Chi Squared**
Click on the **Chi-Square** box and check the desired boxes

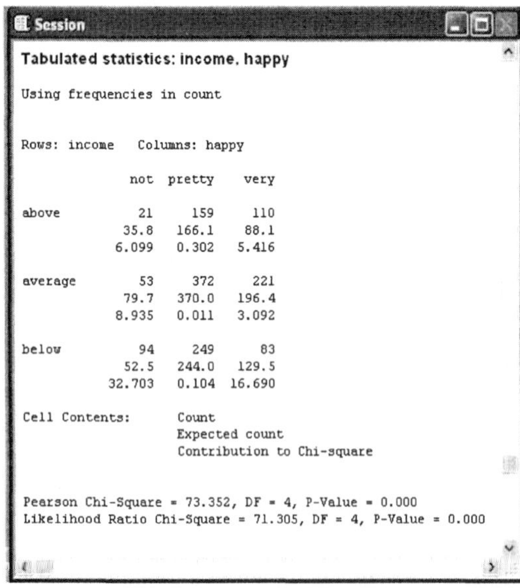


```
Session

Tabulated statistics: income, happy

Using frequencies in count


Rows: income   Columns: happy

              not  pretty    very

above          21     159     110
             35.8   166.1    88.1
            6.099   0.302   5.416

average        53     372     221
             79.7   370.0   196.4
            8.935   0.011   3.092

below          94     249      83
             52.5   244.0   129.5
           32.703   0.104  16.690

Cell Contents:      Count
                    Expected count
                    Contribution to Chi-square


Pearson Chi-Square = 73.352, DF = 4, P-Value = 0.000
Likelihood Ratio Chi-Square = 71.305, DF = 4, P-Value = 0.000
```


Notice the p-value given as P = 0.000, degrees of freedom = 4 and Chi-Square = 73.3

Since the p-value is below 0.05, we can reject the null hypothesis that happiness and family income are independent.

Chapter 10 Exercise 10.11 Life After Death and Gender Contingency Table and the Chi Squared Statistic

a) Construct a contingency table by entering the data into the worksheet.

Worksheet 1 ***

↓	C1-T	C2-T	C3	C4
	gender	belief	count	
1	male	yes	425	
2	male	no	138	
3	female	yes	550	
4	female	no	98	
5				

To create the table Choose **Stat > Tables > Cross Tabulation and Chi Squared**

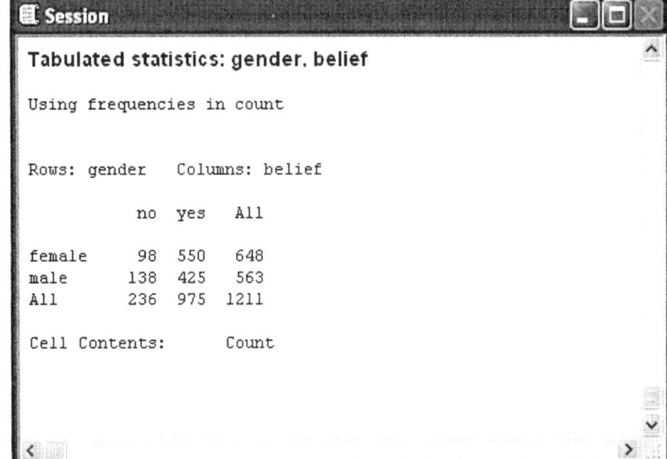

Session

Tabulated statistics: gender, belief

Using frequencies in count

Rows: gender   Columns: belief

	no	yes	All
female	98	550	648
male	138	425	563
All	236	975	1211

Cell Contents: Count

To obtain the expected counts for each cell, its contribution to the Chi-Square and the analysis, check the appropriate boxes.

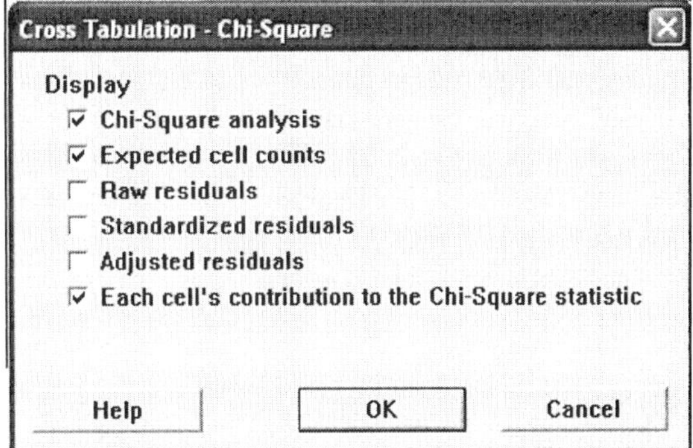

This gives the following in the session window:

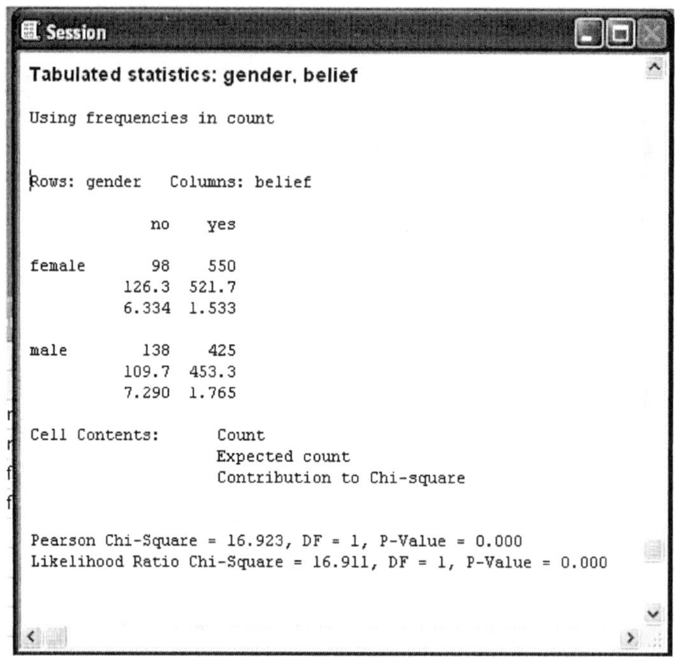
Tabulated statistics: gender, belief

Using frequencies in count

Rows: gender   Columns: belief

	no	yes
female	98	550
	126.3	521.7
	6.334	1.533
male	138	425
	109.7	453.3
	7.290	1.765

Cell Contents: Count
Expected count
Contribution to Chi-square

Pearson Chi-Square = 16.923, DF = 1, P-Value = 0.000
Likelihood Ratio Chi-Square = 16.911, DF = 1, P-Value = 0.000

Chapter 10 Exercise 10.15 Help the Environment
A Chi-Squared test

First enter the data into the worksheet.

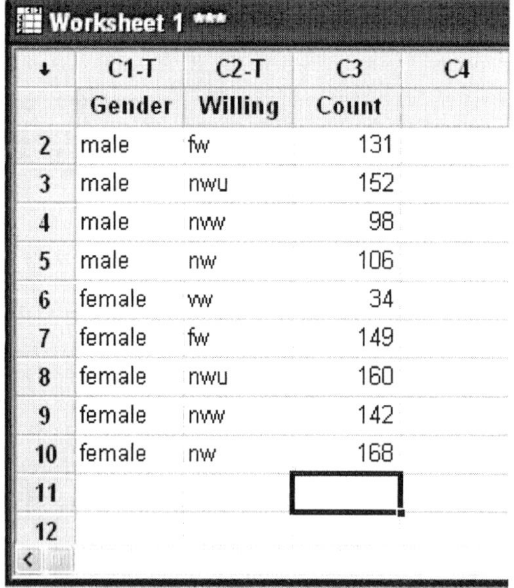
Worksheet 1 ***

↓	C1-T	C2-T	C3	C4
	Gender	Willing	Count	
2	male	fw	131	
3	male	nwu	152	
4	male	nww	98	
5	male	nw	106	
6	female	vw	34	
7	female	fw	149	
8	female	nwu	160	
9	female	nww	142	
10	female	nw	168	
11				
12				

Create the two-way table: Choose **Stat > Tables > Cross Tabulation and Chi Squared**
Click on the **Chi-Square** box and check the desired boxes

Note the degrees of freedom, the Chi Square statistic and the p-value are all given at the bottom of the chart.

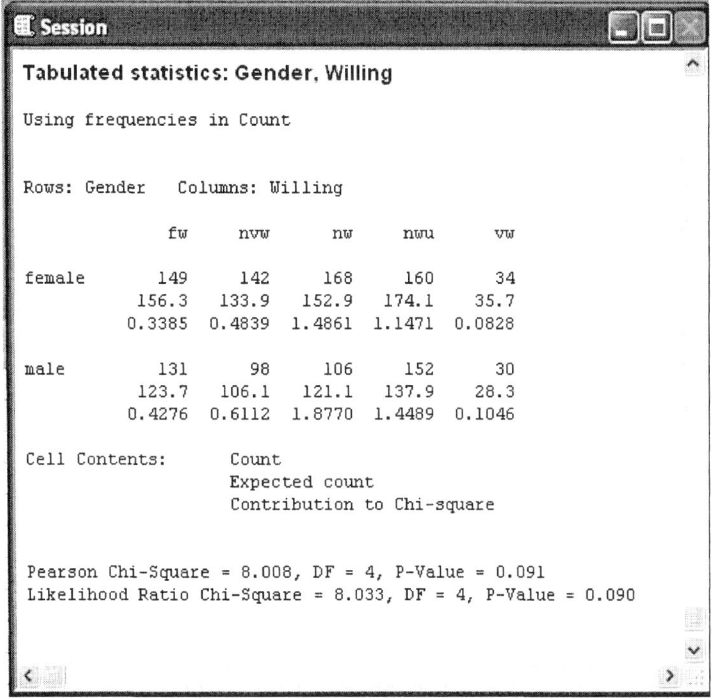
Session

Tabulated statistics: Gender, Willing

Using frequencies in Count

Rows: Gender   Columns: Willing

	fw	nvw	nw	nwu	vw
female	149	142	168	160	34
	156.3	133.9	152.9	174.1	35.7
	0.3385	0.4839	1.4861	1.1471	0.0828
male	131	98	106	152	30
	123.7	106.1	121.1	137.9	28.3
	0.4276	0.6112	1.8770	1.4489	0.1046

Cell Contents: Count
Expected count
Contribution to Chi-square

Pearson Chi-Square = 8.008, DF = 4, P-Value = 0.091
Likelihood Ratio Chi-Square = 8.033, DF = 4, P-Value = 0.090

Chapter 10 Example 9: Standardized Residuals for Religiosity and Gender

We can duplicate the chart in Example 9 by entering the data into the worksheet.

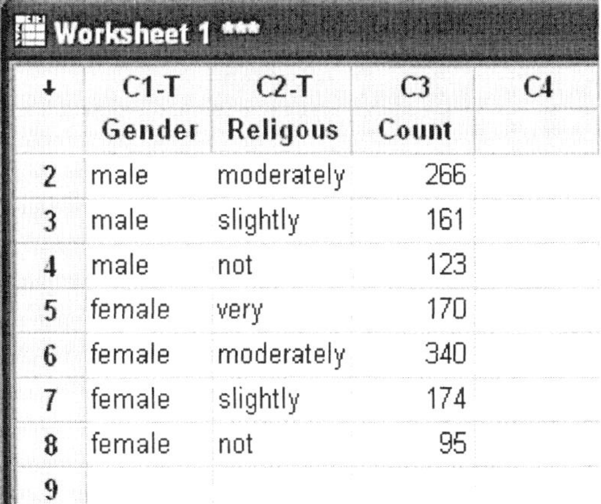
Worksheet 1 ***

↓	C1-T	C2-T	C3	C4
	Gender	Religous	Count	
2	male	moderately	266	
3	male	slightly	161	
4	male	not	123	
5	female	very	170	
6	female	moderately	340	
7	female	slightly	174	
8	female	not	95	
9				

Choose **Stat > Tables > Cross Tabulation and Chi Squared**

To show the residuals, check the adjusted residuals box.

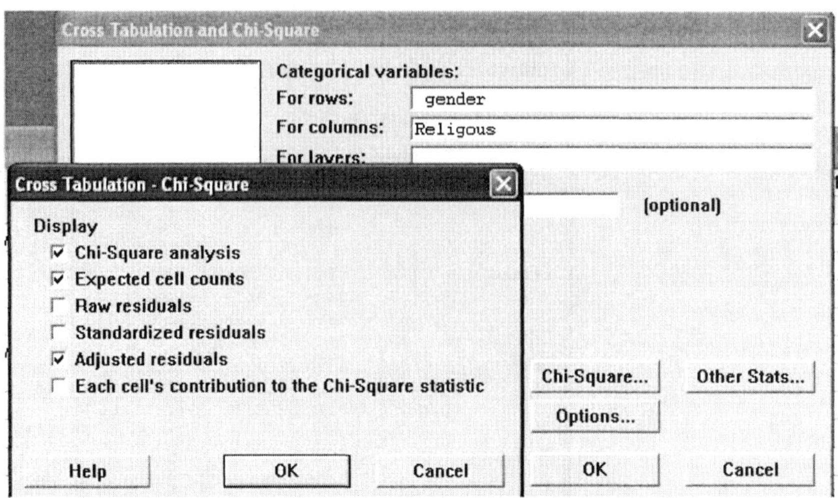

This will give you Table 10.17 in the text.

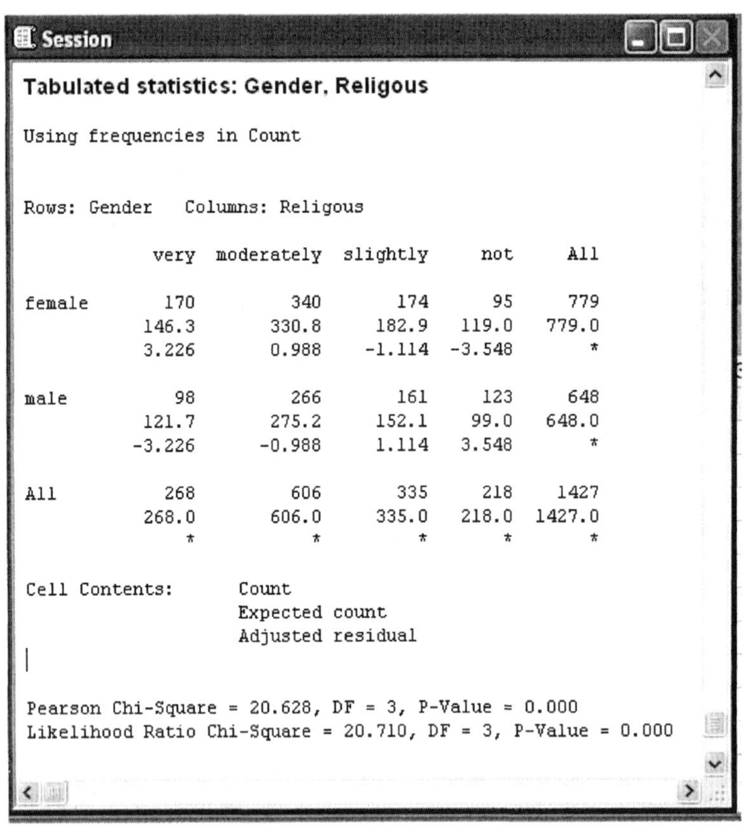

Session

Tabulated statistics: Gender, Religous

```
Using frequencies in Count


Rows: Gender   Columns: Religous

            very  moderately  slightly     not     All

female       170         340       174      95     779
           146.3       330.8     182.9   119.0   779.0
           3.226       0.988    -1.114  -3.548       *

male          98         266       161     123     648
           121.7       275.2     152.1    99.0   648.0
          -3.226      -0.988     1.114   3.548       *

All          268         606       335     218    1427
           268.0       606.0     335.0   218.0  1427.0
               *           *         *       *       *

Cell Contents:      Count
                    Expected count
                    Adjusted residual


Pearson Chi-Square = 20.628, DF = 3, P-Value = 0.000
Likelihood Ratio Chi-Square = 20.710, DF = 3, P-Value = 0.000
```

When you look at this chart, note the entry for females that are very religious. The difference between the observed and expected counts is more than three standard deviations. What conclusion can you draw? What do the negative residuals mean?

Chapter 10 Exercise 10.31 Standardized residuals for Happiness and Income

We again have this worksheet of raw data:

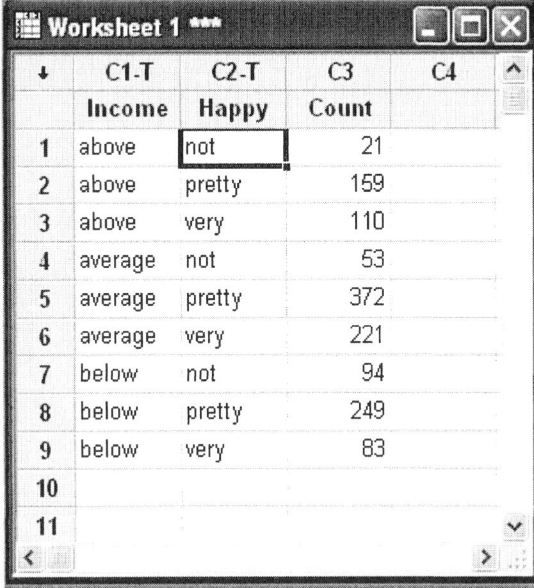

↓	C1-T Income	C2-T Happy	C3 Count	C4
1	above	not	21	
2	above	pretty	159	
3	above	very	110	
4	average	not	53	
5	average	pretty	372	
6	average	very	221	
7	below	not	94	
8	below	pretty	249	
9	below	very	83	
10				
11				

We can generate the chart by choosing

Stat > Tables > Cross Tabulation and Chi Squared

Be sure to check the appropriate boxes in the Chi Square tab.

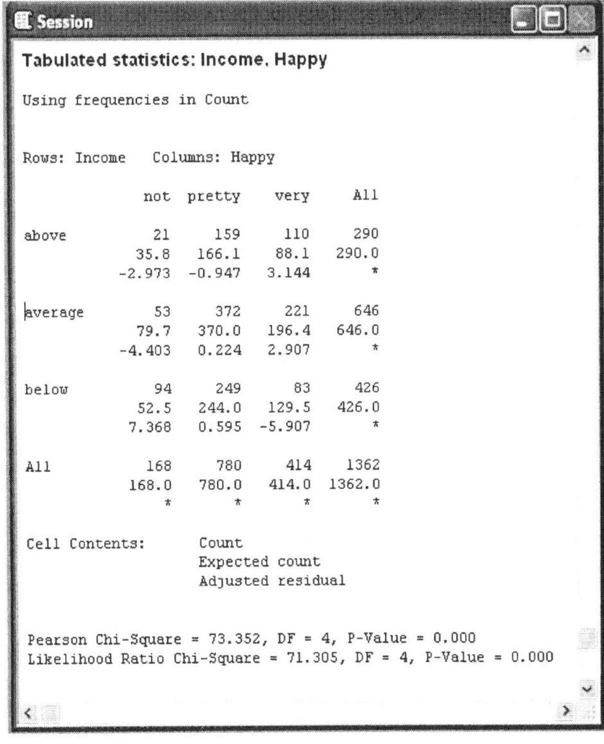

Tabulated statistics: Income, Happy

Using frequencies in Count

Rows: Income Columns: Happy

	not	pretty	very	All
above	21	159	110	290
	35.8	166.1	88.1	290.0
	-2.973	-0.947	3.144	*
average	53	372	221	646
	79.7	370.0	196.4	646.0
	-4.403	0.224	2.907	*
below	94	249	83	426
	52.5	244.0	129.5	426.0
	7.368	0.595	-5.907	*
All	168	780	414	1362
	168.0	780.0	414.0	1362.0
	*	*	*	*

Cell Contents: Count
Expected count
Adjusted residual

Pearson Chi-Square = 73.352, DF = 4, P-Value = 0.000
Likelihood Ratio Chi-Square = 71.305, DF = 4, P-Value = 0.000

Notice the high residuals in certain cells. How would you interpret these?
Notice the negative residuals in certain cells. How would you interpret these?

Chapter 10 Example 10 Tea Tastes Better with Milk Poured First? Small Sample Size – Fisher's Exact Test

We start by creating the worksheet.

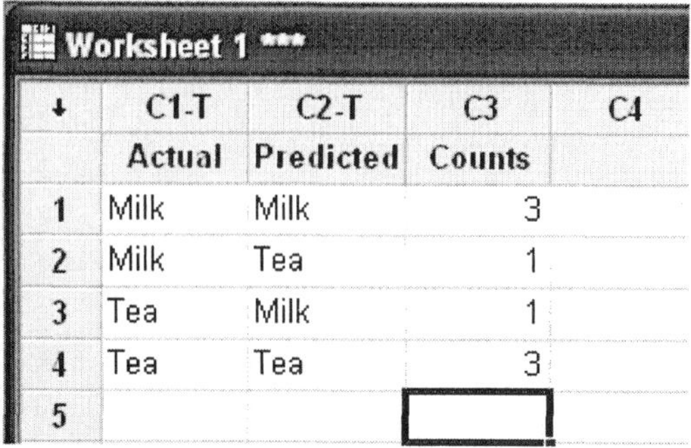

Worksheet 1 ***

↓	C1-T	C2-T	C3	C4
	Actual	Predicted	Counts	
1	Milk	Milk	3	
2	Milk	Tea	1	
3	Tea	Milk	1	
4	Tea	Tea	3	
5				

To create the table as shown in the text : Choose **Stat > Tables > Cross Tabulation and Chi Squared**

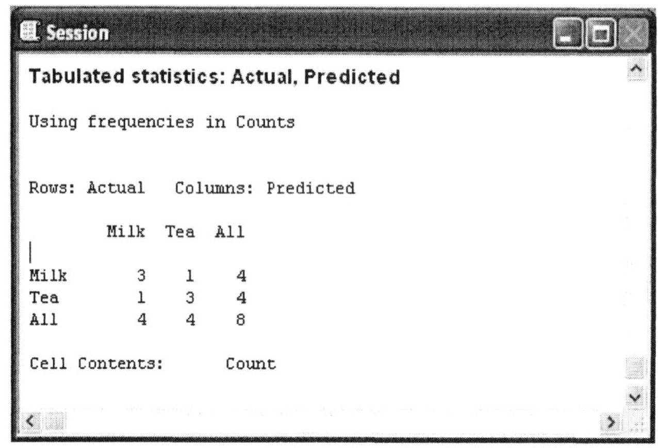

```
Session

Tabulated statistics: Actual, Predicted

Using frequencies in Counts


Rows: Actual   Columns: Predicted

       Milk  Tea  All

Milk      3    1    4
Tea       1    3    4
All       4    4    8

Cell Contents:      Count
```

In the Chi Square tab, check the Chi Square Analysis and Expected Cell Counts boxes.

In the Other Stats tab, check the box for Fishers Exact Test for 2x2 Tables.

This will give you the following chart:

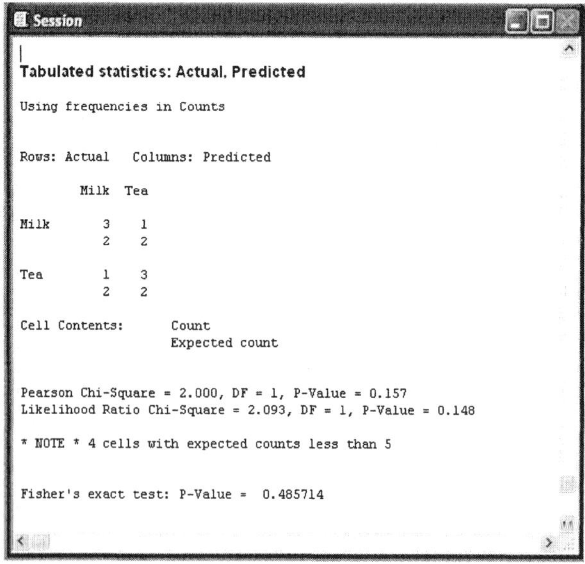

```
Session

Tabulated statistics: Actual, Predicted

Using frequencies in Counts


Rows: Actual   Columns: Predicted

       Milk  Tea

Milk      3    1
          2    2

Tea       1    3
          2    2

Cell Contents:      Count
                    Expected count


Pearson Chi-Square = 2.000, DF = 1, P-Value = 0.157
Likelihood Ratio Chi-Square = 2.093, DF = 1, P-Value = 0.148

* NOTE * 4 cells with expected counts less than 5


Fisher's exact test: P-Value =  0.485714
```

Fisher's exact test is a test based on an exact distribution rather than on the approximate chi-square distribution used for Pearson's and likelihood ratio tests. Fisher's exact test is useful when the expected cell counts are low and chi-square approximation is not very good.

Chapter 10 Exercise 10.39 Claritin and Nervousness
Small Sample Size – Fisher's Exact Test

We enter the data into the worksheet as follows:

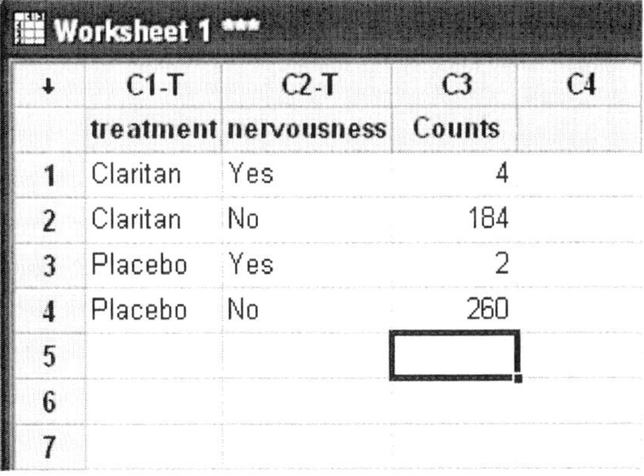
Worksheet 1 ***

↓	C1-T	C2-T	C3	C4
	treatment	nervousness	Counts	
1	Claritan	Yes	4	
2	Claritan	No	184	
3	Placebo	Yes	2	
4	Placebo	No	260	
5				
6				
7				

To create the table Choose **Stat > Tables > Cross Tabulation and Chi Squared**
In the Chi Square tab, check the Chi Square Analysis and Expected Cell Counts boxes.
In the Other Stats tab, check the box for Fishers Exact Test for 2x2 Tables.

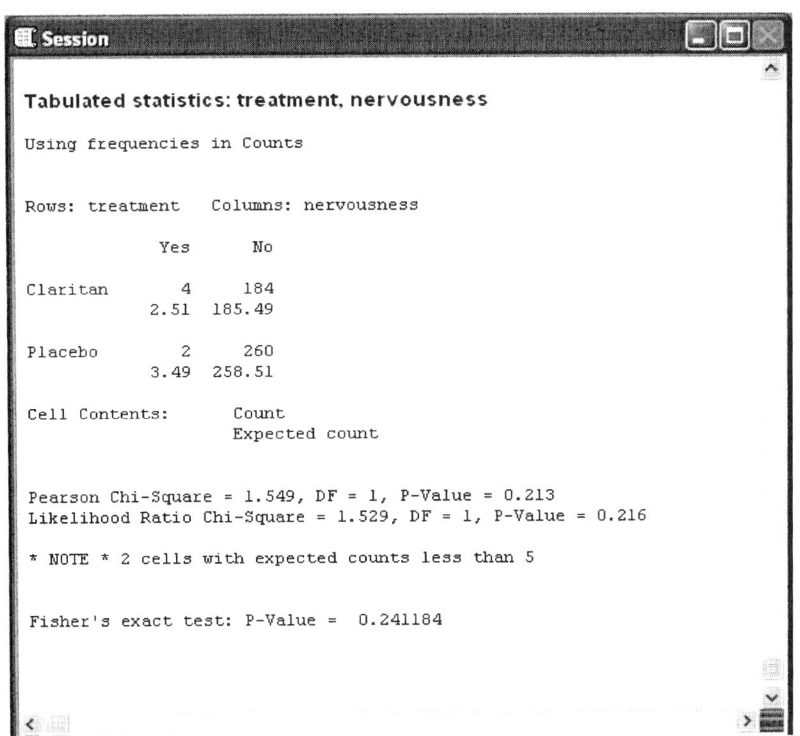
Session

```
Tabulated statistics: treatment, nervousness

Using frequencies in Counts


Rows: treatment   Columns: nervousness

             Yes      No

Claritan       4     184
            2.51  185.49

Placebo        2     260
            3.49  258.51

Cell Contents:       Count
                     Expected count


Pearson Chi-Square = 1.549, DF = 1, P-Value = 0.213
Likelihood Ratio Chi-Square = 1.529, DF = 1, P-Value = 0.216

* NOTE * 2 cells with expected counts less than 5


Fisher's exact test: P-Value =  0.241184
```

Note the p-value, the Chi Square statistic.

Chapter 11 Example 2 What Do We Learn From a Scatterplot in The Strength Study? Review of Producing a Scatterplot

Open the data file **high_school_female_athletes** from the MINITAB folder of the data disk.

To create a scatterplot of the data with C15 BP(60) as the x-variable, and C12 BP as the y-variable

Click: **Graph > Scatterplot > Simple > OK**
Enter: Y Variables **C12**
X Variables **C15**
Select: Labels **enter an appropriate title**
Click: **OK OK**

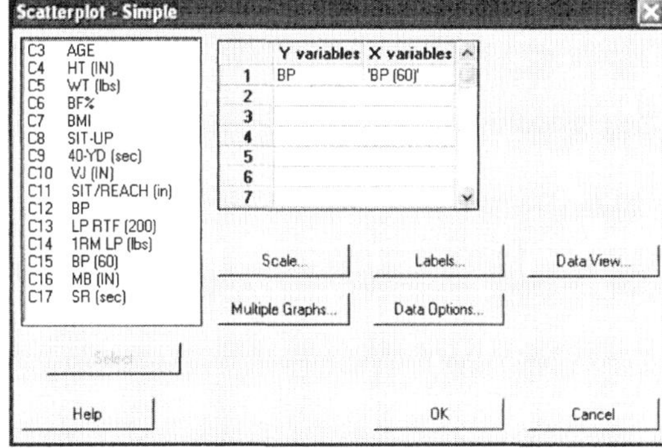

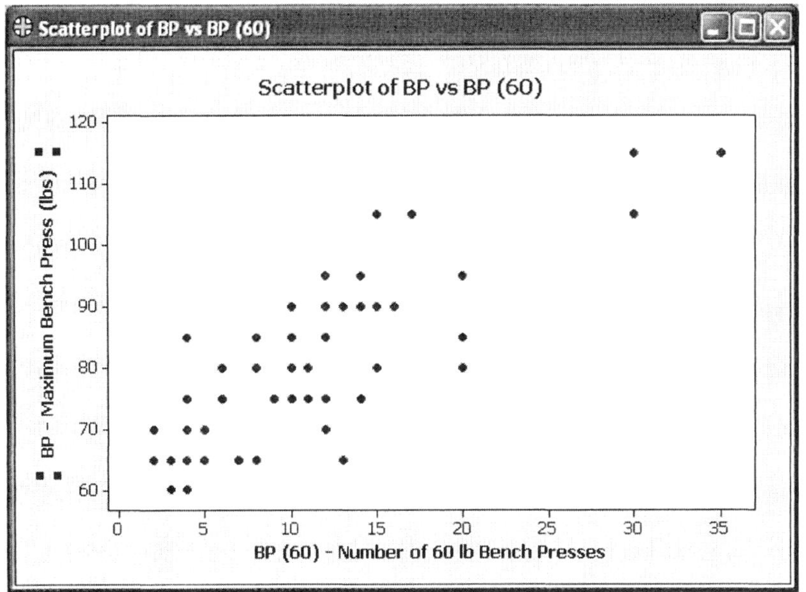

The scatterplot shows that female athletes with higher numbers of 60-pound bench presses also tended to have higher vales for the maximum bench press.

Chapter 11 Example 3 Which Regression Line Predicts Maximum Bench Press? Review of Generating a Regression Line

Open the data file **high_school_female_athletes** from the MINITAB folder of the text CD.

To generate the least squares regression line

Click: **Stat > Regression > Regression**

Enter: Response: **C12**
Predictors: **C15**
Click: **OK**

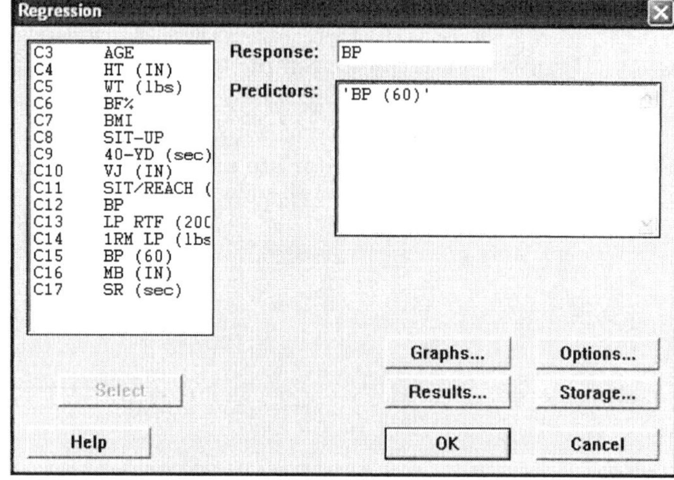

The results are displayed in the Session window.

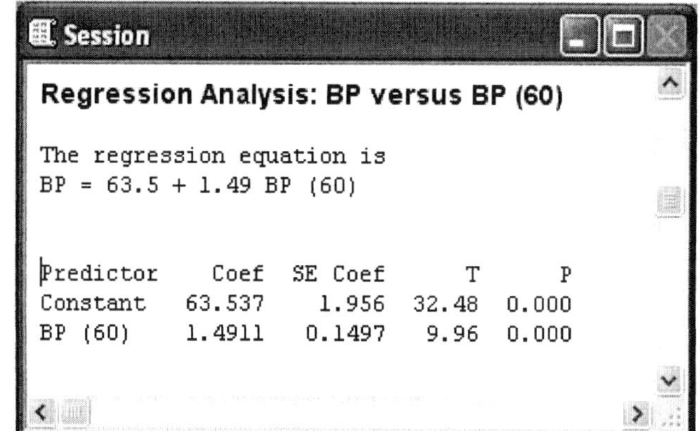

Regression Analysis: BP versus BP (60)

```
The regression equation is
BP = 63.5 + 1.49 BP (60)


Predictor     Coef  SE Coef      T      P
Constant    63.537    1.956  32.48  0.000
BP (60)     1.4911   0.1497   9.96  0.000
```

To create a scatterplot of the data with the regression line added:

Click: **Graph > Scatterplot > With Regression > OK**
Enter: Y Variables **C12**
X Variables **C15**
Select: Labels **enter an appropriate title**
Click: **OK OK**

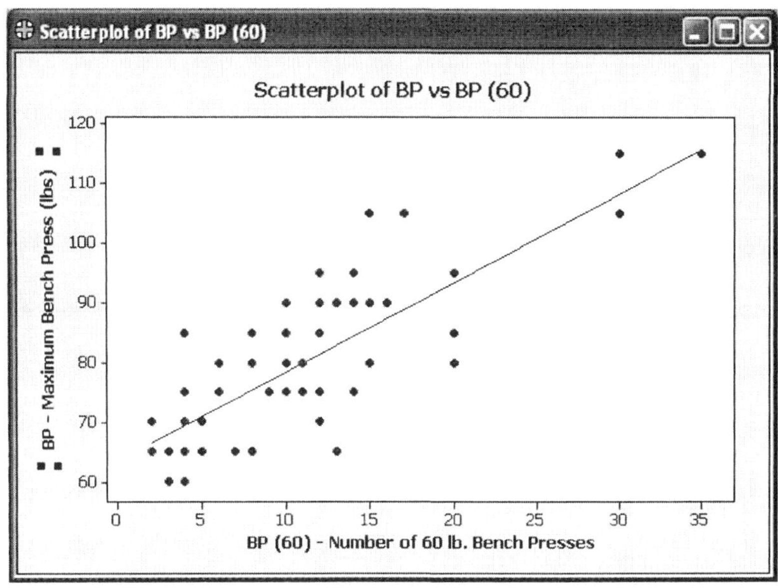

Chapter Exercise 11.1 Car Mileage and Weight
Generating a Regression Line

Open the data file **car_weight_and_mileage** from the MINITAB folder of the text CD.

To generate the least squares regression line
Click: **Stat > Regression > Regression**

Enter: Response: **mileage**
Predictors: **weight**
Click: **OK**

The results are displayed in the Session window.

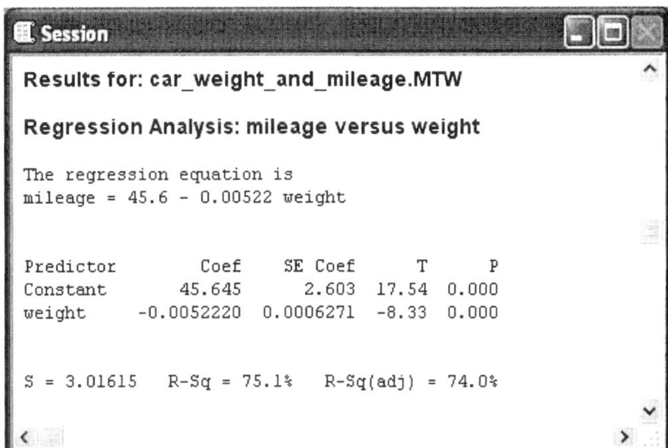

Session

Results for: car_weight_and_mileage.MTW

Regression Analysis: mileage versus weight

```
The regression equation is
mileage = 45.6 - 0.00522 weight


Predictor        Coef    SE Coef      T      P
Constant       45.645      2.603  17.54  0.000
weight     -0.0052220  0.0006271  -8.33  0.000


S = 3.01615   R-Sq = 75.1%   R-Sq(adj) = 74.0%
```

Chapter Exercise 11.7 Predicting College GPA
Generating a Regression Line

Open the data file **georgia_student_survey** from the MINITAB folder of the text CD. We are interested in predicting college GPA based on high school GPA.

explanatory: high school GPA
response: college GPA

To generate the scatterplot:

Select: **Graph > Scatterplot > Simple > OK**

Enter: Y variables: **CGPA** X variables: **HSGPA**
Click: **OK**

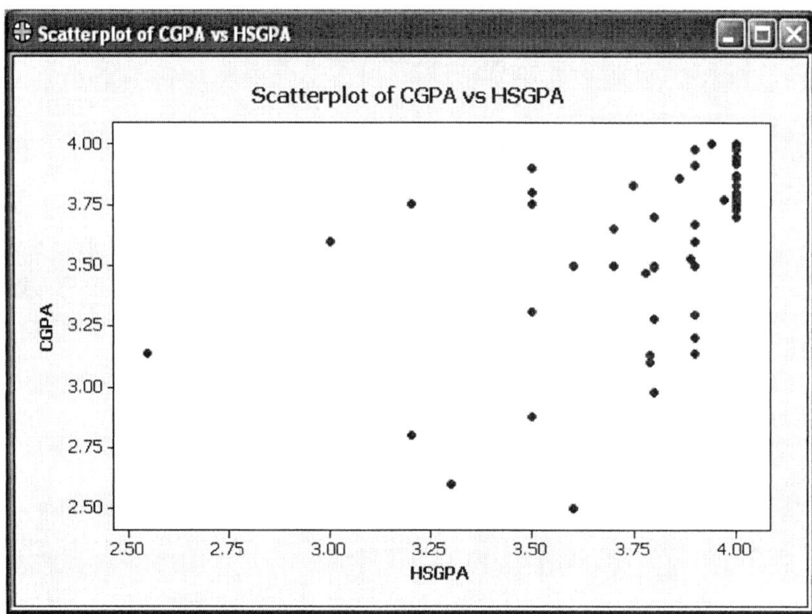

To generate the least squares regression line

Click: **Stat > Regression > Regression**

Enter: Response: **CGPA**
Predictors: **HSGPA**
Click: **OK**

The results are displayed in the Session window.

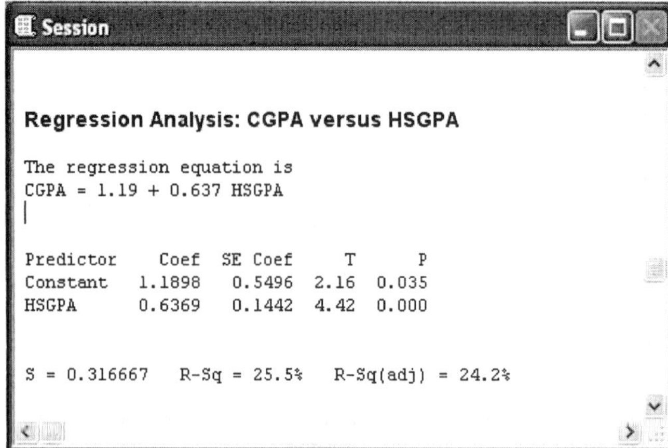

Chapter 11 Example 6 What's the Correlation for Predicting Strength?
Review of Determining Correlation

Open the data file **high_school_female_athletes** from the MINITAB folder of the text CD. We have already identified C12 BP as the response variable and C15 BP(60) as the explanatory variable.

To determine the correlation using MINITAB:

Select: **Stat > Basic Statistics > Correlation**
Enter: Variables: **C15 C12**
Click: **OK**

The results are displayed in the Session window.

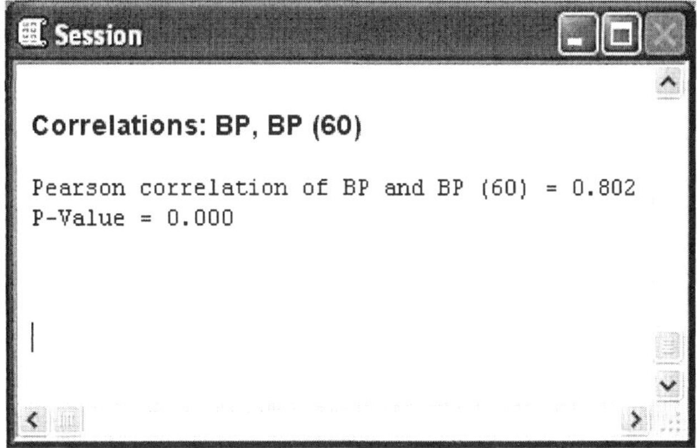

Chapter 11 Example 9 What Does r^2 Tell Us In The Strength Study?
Determining r^2

Open the data file **high_school_female_athletes** from the MINITAB folder of the text CD. We have already identified C12 BP as the response variable and C15 BP(60) as the explanatory variable.

To determine the value of r^2 using MINITAB:

Select: **Stat > Regression > Regression**
Enter: Variables: **C15 C12**
Click: **OK**

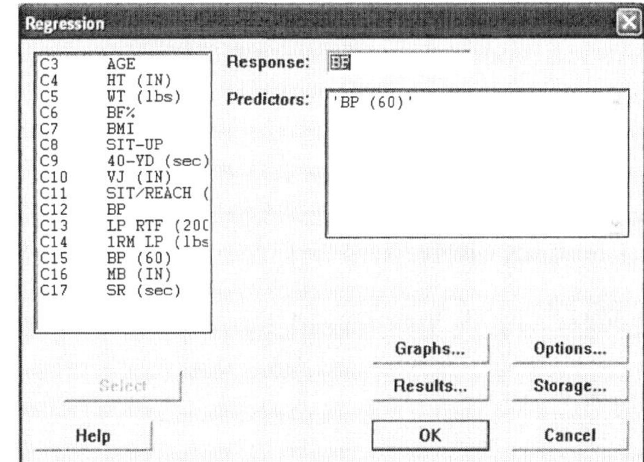

Select: **Results**

Select: **Regression equation, table of coefficients, s, R-squared, and basic analysis of variance**

Click: **OK OK**

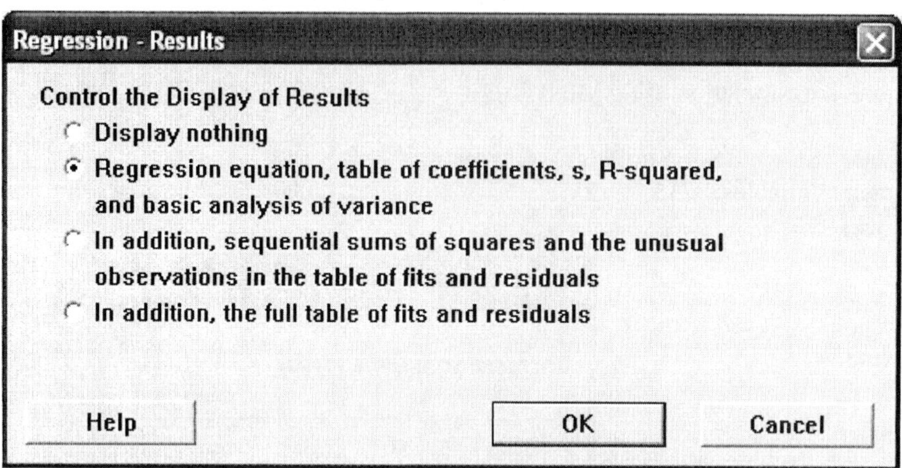

The results are displayed in the Session window.

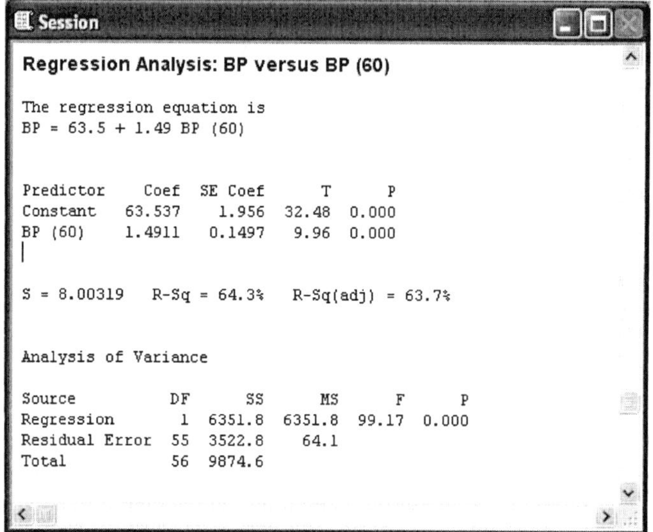

Chapter 11 Exercise 11.13 Sit-ups and the 40-yard dash
Determining the Value of r^2

Open the data file **high_school_female_athletes** from the MINITAB folder of the text CD. For this exercise we have C8 SIT-UP as the explanatory variable and C9 40-YD (sec) as the response variable.

To determine the value of r^2 using MINITAB:

Select: **Stat > Regression > Regression**
Enter: Variables: **C8 C9**

Select: **Results**
Select: **Regression equation, table of coefficients, s, R-squared, and basic analysis of variance**
Click: **OK OK**

The results are displayed in the Session window.

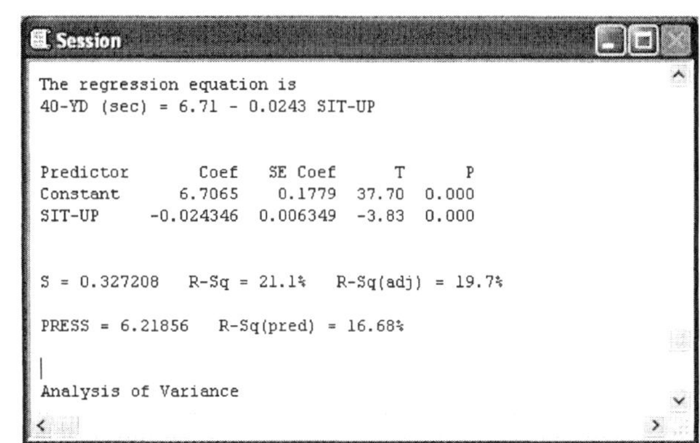

```
The regression equation is
40-YD (sec) = 6.71 - 0.0243 SIT-UP

Predictor       Coef    SE Coef      T      P
Constant      6.7065     0.1779  37.70  0.000
SIT-UP     -0.024346   0.006349  -3.83  0.000

S = 0.327208   R-Sq = 21.1%   R-Sq(adj) = 19.7%

PRESS = 6.21856   R-Sq(pred) = 16.68%

Analysis of Variance
```

Note the regression equation is $\hat{y} = 6.71 - 0.0243x$. Using this equation we can predict time in the 40-yard dash for any subject who can do a given number of sit-ups.

Note the displayed value of R-Sq = 21.1%

Chapter 11 Exercise 11.15 Student ideology
Determining the Value of r^2

Open the data file **fla_student_survey** from the MINITAB folder of the text CD. For this exercise we have C10 newspapers as the explanatory variable and C14 political_ideology as the response variable.

Doing a complete analysis of the association of these two variables, we start with the scatterplot:
Select: **Graph > Scatterplot > Simple > OK**

Enter: Y variables: **C14** X variables: **C10**
Click: **OK**

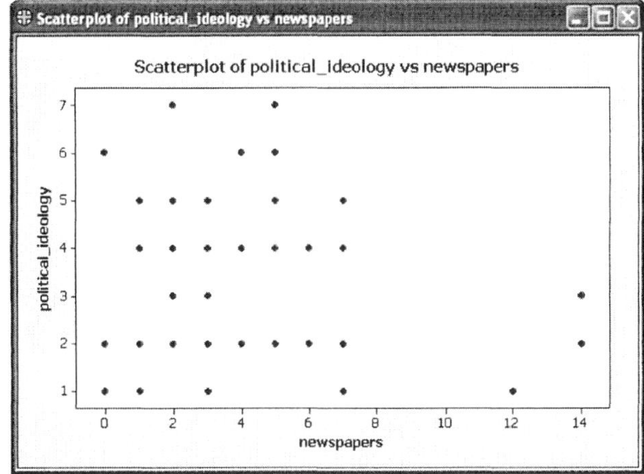

From this scatterplot, we can see that there is very little linear association.

Determining the correlation:

Select: **Stat > Basic Statistics > Correlation**
Enter: **C10 C14**
Click: **OK**

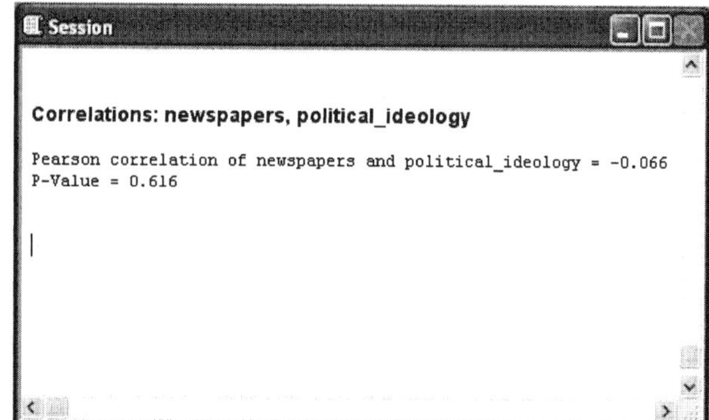

```
Correlations: newspapers, political_ideology

Pearson correlation of newspapers and political_ideology = -0.066
P-Value = 0.616
```

Again, based on the value of r = -0.066, we see very little evidence of a linear association between these two variables.

Determining the value of r^2:

Select: **Stat > Regression > Regression**
Enter: Response: **C14**
Predictors: **C10**

Select: **Results**
Select: **Regression equation, table of coefficients, s, R-squared, and basic analysis of variance**
Click: **OK OK**

The results are displayed in the Session window.

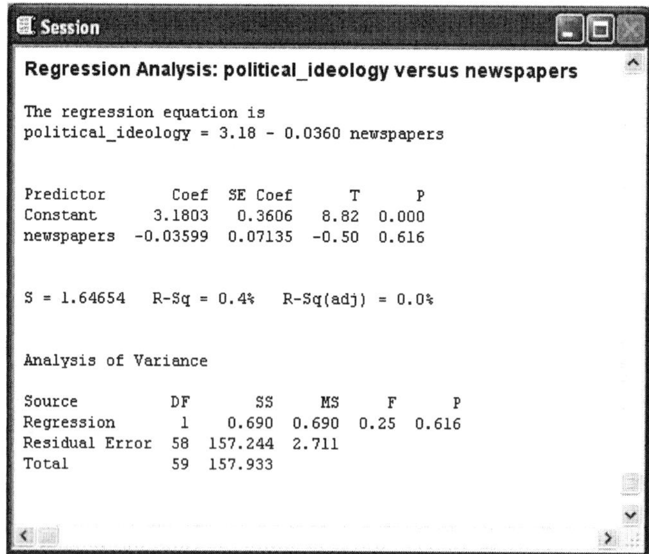

```
Regression Analysis: political_ideology versus newspapers

The regression equation is
political_ideology = 3.18 - 0.0360 newspapers


Predictor       Coef   SE Coef      T      P
Constant      3.1803    0.3606   8.82  0.000
newspapers  -0.03599   0.07135  -0.50  0.616


S = 1.64654   R-Sq = 0.4%   R-Sq(adj) = 0.0%


Analysis of Variance

Source          DF       SS     MS     F      P
Regression       1    0.690  0.690  0.25  0.616
Residual Error  58  157.244  2.711
Total           59  157.933
```

Note the displayed value of R-Sq = .4%

Chapter 11 Example 11 Is Strength Associated with 60-pound Bench Presses? A Significance Test of Independence

Open the worksheet **high_school_female_student_athletes,** which is found in the MINITAB folder.
The number of bench presses before fatigue, C15 BP(60), is the x-variable, and maximum bench press, C12 BP, is the y-variable. The regression output includes the t test statistic and the p-value.

Select: **Stat > Regression > Regression**
Enter: Response: **C12**
Predictors: **C15**
Click: **OK**

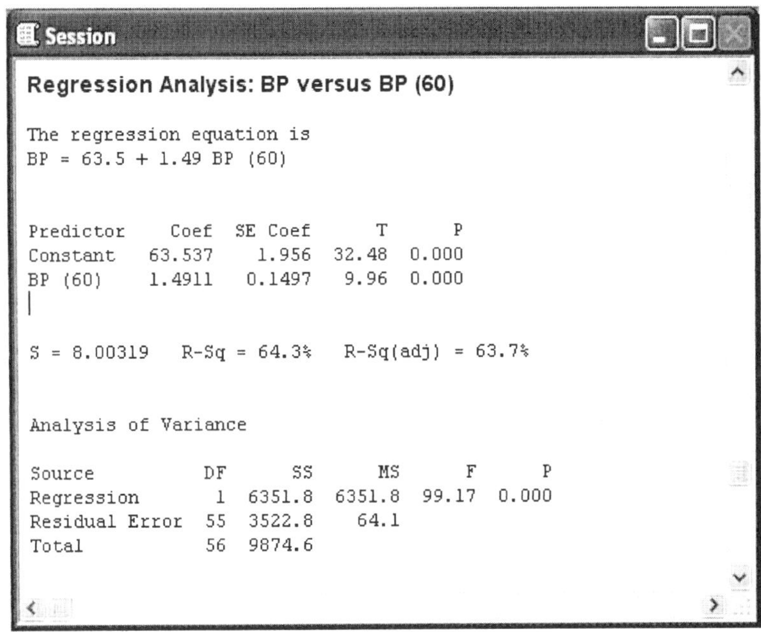

```
Session

Regression Analysis: BP versus BP (60)

The regression equation is
BP = 63.5 + 1.49 BP (60)


Predictor     Coef  SE Coef      T      P
Constant    63.537    1.956  32.48  0.000
BP (60)     1.4911   0.1497   9.96  0.000


S = 8.00319   R-Sq = 64.3%   R-Sq(adj) = 63.7%


Analysis of Variance

Source          DF      SS      MS      F      P
Regression       1  6351.8  6351.8  99.17  0.000
Residual Error  55  3522.8    64.1
Total           56  9874.6
```

Notice the *t* test statistic ($t = 9.96$) appears under the column heading "T" in the row for the predictor BP (60), and the p-value ($p = 0.000$) for the two-sided alternative H_a: $\beta \neq 0$ appears under the heading "P" in the row for the predictor BP (60).

Chapter 11 Exercise 11.29 Predicting house prices Significance Test of Independence and a Confidence Interval for Slope

Open the worksheet **house_selling_prices**, which is found in the MINITAB folder.
The size of the house is in C8 size, and selling price is in C7 price.

Select: **Stat > Regression > Regression**
Enter: Response: **price**
Predictors: **size**
Click: **OK**

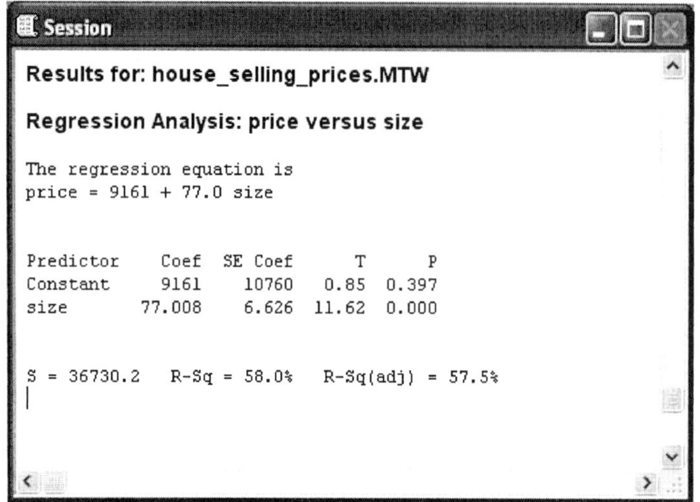

Session
Results for: house_selling_prices.MTW
Regression Analysis: price versus size
The regression equation is
price = 9161 + 77.0 size
Predictor Coef SE Coef T P
Constant 9161 10760 0.85 0.397
size 77.008 6.626 11.62 0.000
S = 36730.2 R-Sq = 58.0% R-Sq(adj) = 57.5%

Notice the t test statistic (t = 11.62), and the p-value (p = 0.000) for the two-sided alternative H_a: $\beta \neq 0$. Since p-value = 0.000 is less than a 0.05 significance level, there is evidence that these two variables are not independent, and that the sample association between these two variables is not just random variation.

To compute a 95% confidence interval for the population slope, note that the $t_{.025}$ value for $df = n - 2$ = 98 is t = 1.9845, and that b = 77.008, and se = 6.626. (MINITAB does not include a t-interval estimate for a population slope.)

$$b \pm t_{.025}(se) = 77.008 \text{ +/- } 1.9845\ (6.626)$$
$$= 77.008 \text{ +/- } 13.149$$
$$(63.859, 90.157)$$

This interval does not support the builder's claim that selling price increases $100, on the average, for every extra square foot. (The interval does not contain 100.)

Chapter 11 Exercise 11.35 Advertising and Sales
A Significance Test of Independence

Enter the data into the MINITAB Data Window. Enter the x's into C1 and name the column Advertising. Enter the y's into C2 and name the column Sales.

Select: **Stat > Regression > Regression**
Enter: Response: **Sales**
Predictors: **Advertising**

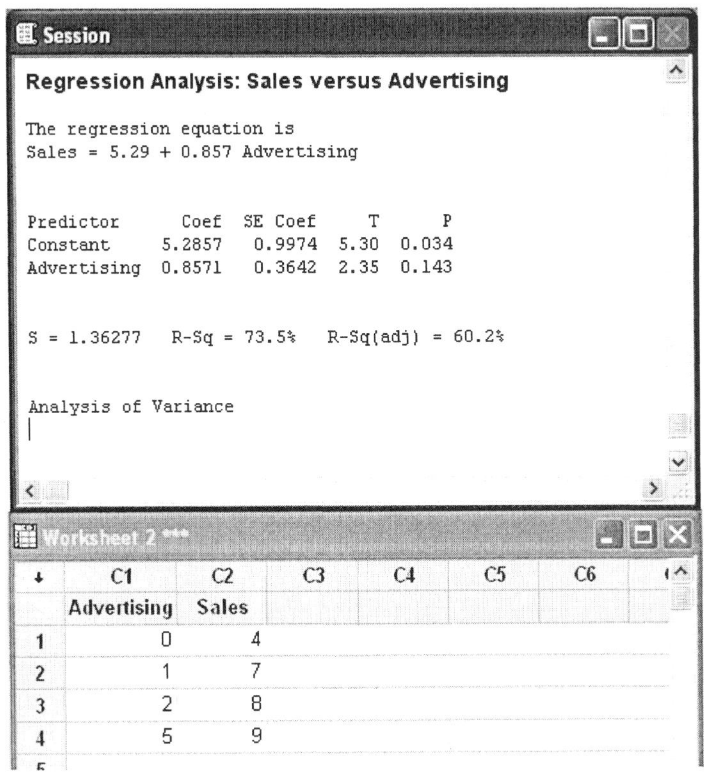

Notice the t test statistic (t = 2.35), and the p-value (p = 0.143) for the two-sided alternative H_a: $\beta \neq 0$. Since p-value = 0.143 is not less than significance level 0.05, there is not evidence that these two variables are independent. A word of caution though, were the assumptions validated? This is a very small sample, so results are suspect.

Chapter 11 Example 13 Detecting an Underachieving College Student
How Data Vary Around the Regression Line

Open the worksheet **georgia_student_survey**, found in the MINITAB folder. High school GPA is found in column C8 and college GPA is in column C9.

As part of the regression analysis, MINITAB highlights observations that have standardized residuals with absolute value larger than 2. To generate this table of "unusual observations":

Select: **Stat > Regression > Regression**
Enter: Response: **CGPA**
Predictors: **HSGPA**

Select: **Results**
Select: **In addition, sequential sums of squares and the unusual observations in the table of fits and residuals**
Click: **OK OK**

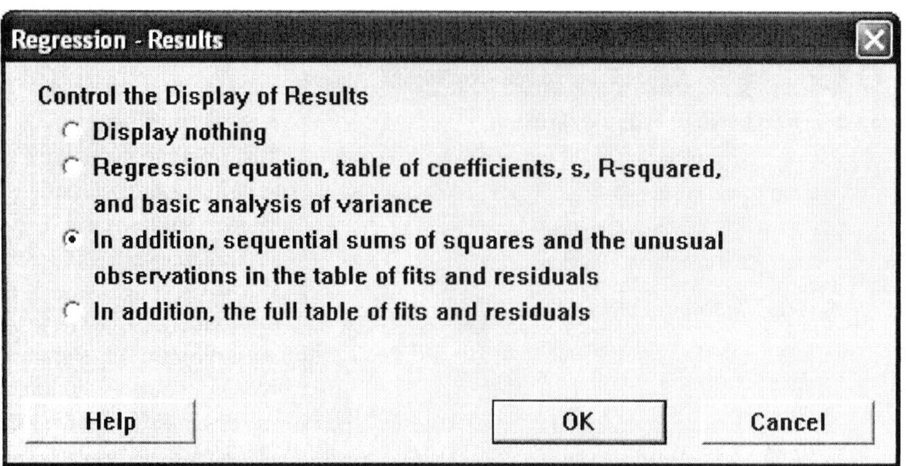
Regression - Results
Control the Display of Results
Display nothing
Regression equation, table of coefficients, s, R-squared, and basic analysis of variance
In addition, sequential sums of squares and the unusual observations in the table of fits and residuals
In addition, the full table of fits and residuals
Help
OK
Cancel

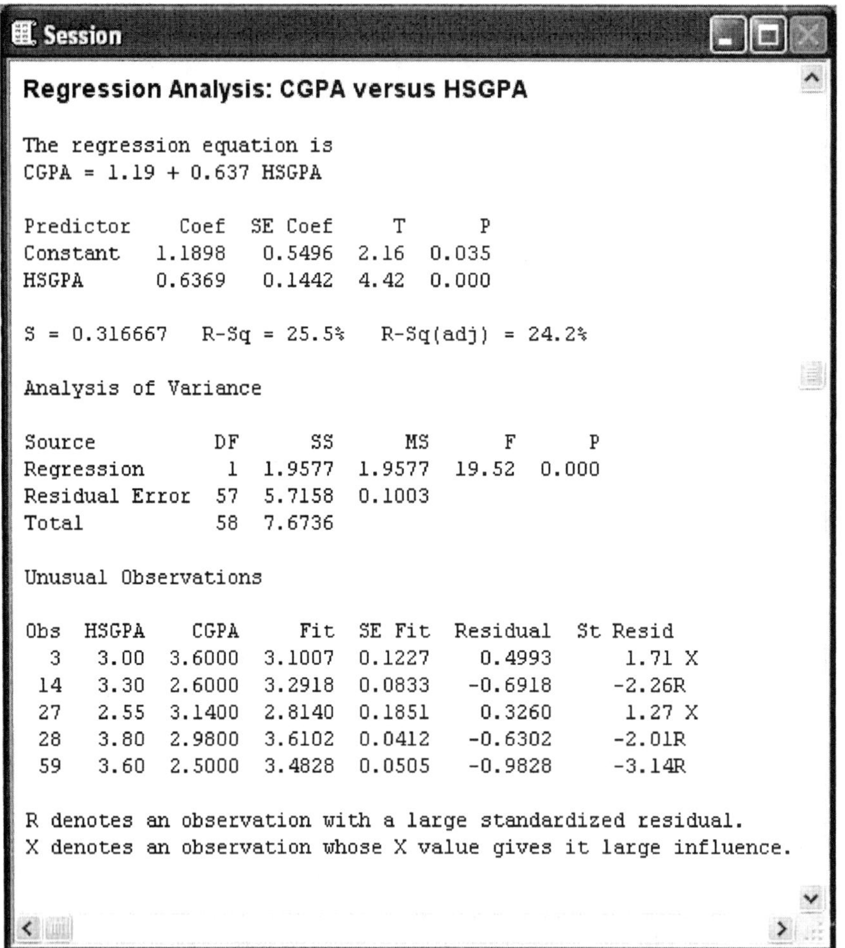
Session
Regression Analysis: CGPA versus HSGPA
The regression equation is
CGPA = 1.19 + 0.637 HSGPA
Predictor Coef SE Coef T P
Constant 1.1898 0.5496 2.16 0.035
HSGPA 0.6369 0.1442 4.42 0.000
S = 0.316667 R-Sq = 25.5% R-Sq(adj) = 24.2%
Analysis of Variance
Source DF SS MS F P
Regression 1 1.9577 1.9577 19.52 0.000
Residual Error 57 5.7158 0.1003
Total 58 7.6736
Unusual Observations
Obs HSGPA CGPA Fit SE Fit Residual St Resid
3 3.00 3.6000 3.1007 0.1227 0.4993 1.71 X
14 3.30 2.6000 3.2918 0.0833 -0.6918 -2.26R
27 2.55 3.1400 2.8140 0.1851 0.3260 1.27 X
28 3.80 2.9800 3.6102 0.0412 -0.6302 -2.01R
59 3.60 2.5000 3.4828 0.0505 -0.9828 -3.14R
R denotes an observation with a large standardized residual.
X denotes an observation whose X value gives it large influence.

Chapter 11 Example 16 Predicting Maximum Bench Press and Estimating Its Mean Confidence Interval for Population Mean of y and Prediction Interval for a Single y-value

Open the worksheet **high_school_female_athletes**, found in the MINITAB folder.

Proceed with the regular regression command:

Select: **Stat > Regression > Regression**
Enter: Response: **BP**
Predictors: **BP (60)**

Now, to find both the confidence interval and the prediction interval:

Select: **Options**
(Fit intercept is selected by default)
Enter: Prediction intervals for new observations: **11**
Enter: Confidence level: **95**
Select: **Confidence limits**
Select: **Prediction limits**
Click: **OK OK**

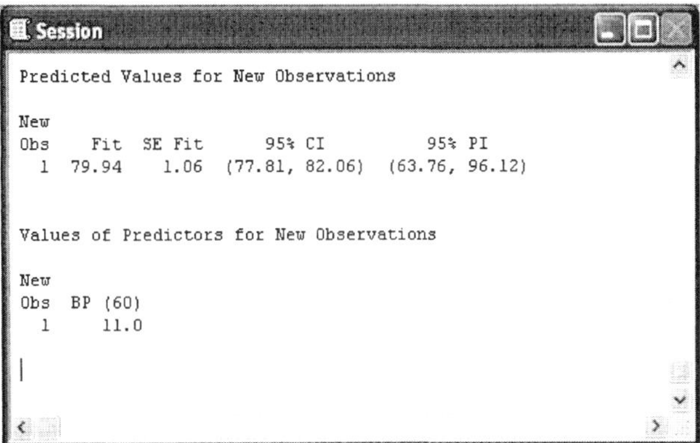

```
Predicted Values for New Observations

New
Obs    Fit  SE Fit       95% CI           95% PI
  1  79.94    1.06  (77.81, 82.06)  (63.76, 96.12)


Values of Predictors for New Observations

New
Obs  BP (60)
  1     11.0
```

Chapter 11 Exercise 11.37 Poor predicted strengths How Data Vary Around the Regression Line

Open the worksheet **high_school_female_athletes**, found in the MINITAB folder.

Select: **Stat > Regression > Regression**
Enter: Response: **BP**
Predictors: **BP (60)**

Select: **Results**
Select: **In addition, sequential sums of squares and the unusual observations in the table of fits and residuals**
Click: **OK OK**

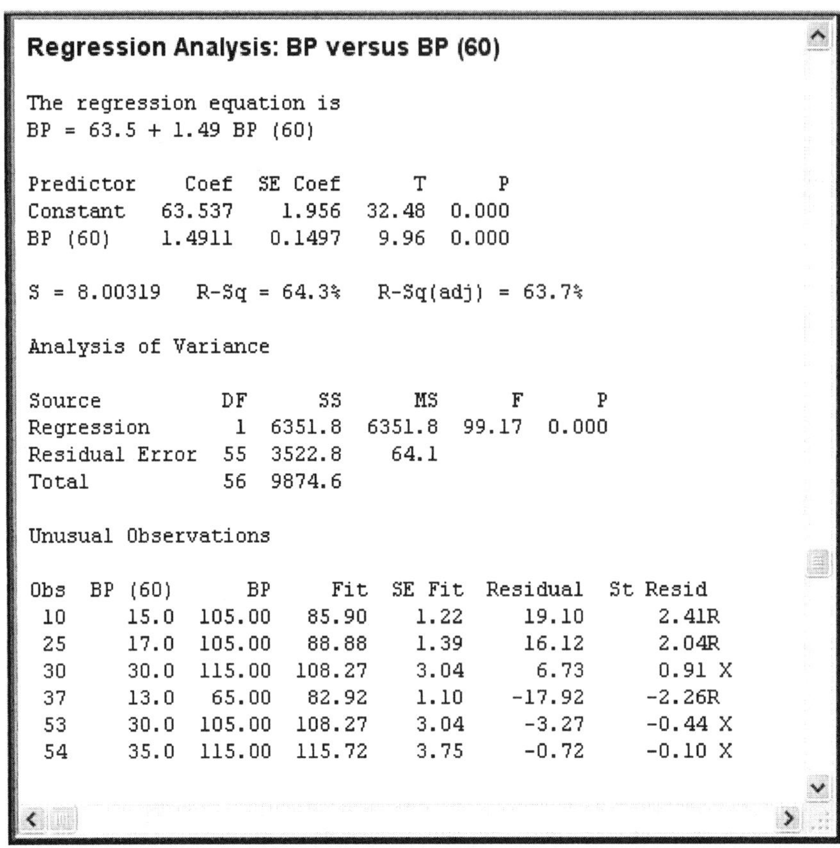

Regression Analysis: BP versus BP (60)

```
The regression equation is
BP = 63.5 + 1.49 BP (60)

Predictor     Coef  SE Coef      T      P
Constant    63.537    1.956  32.48  0.000
BP (60)     1.4911   0.1497   9.96  0.000

S = 8.00319   R-Sq = 64.3%   R-Sq(adj) = 63.7%

Analysis of Variance

Source          DF      SS      MS      F      P
Regression       1  6351.8  6351.8  99.17  0.000
Residual Error  55  3522.8    64.1
Total           56  9874.6

Unusual Observations

Obs  BP (60)      BP     Fit  SE Fit  Residual  St Resid
 10     15.0  105.00   85.90    1.22     19.10     2.41R
 25     17.0  105.00   88.88    1.39     16.12     2.04R
 30     30.0  115.00  108.27    3.04      6.73     0.91 X
 37     13.0   65.00   82.92    1.10    -17.92    -2.26R
 53     30.0  105.00  108.27    3.04     -3.27    -0.44 X
 54     35.0  115.00  115.72    3.75     -0.72    -0.10 X
```

Chapter 11 Exercise 11.43 ANOVA Table for Leg Press

Open the worksheet **high_school_female_student_athletes,** which is found in the MINITAB folder.
The number of leg presses before fatigue, C13 LP (200), is the x-variable, and maximum leg press, C14 LP , is the y-variable

Proceed with the regular regression command:
Select: **Stat > Regression > Regression**
Enter: Response: **LP**
Predictors: **LP (200)**

Now, to find to generate the ANOVA table

Select: **Results**

Select: **Regression equation, table of coefficients, s, R-squared, and basic analysis of variance**

Click: **OK OK**

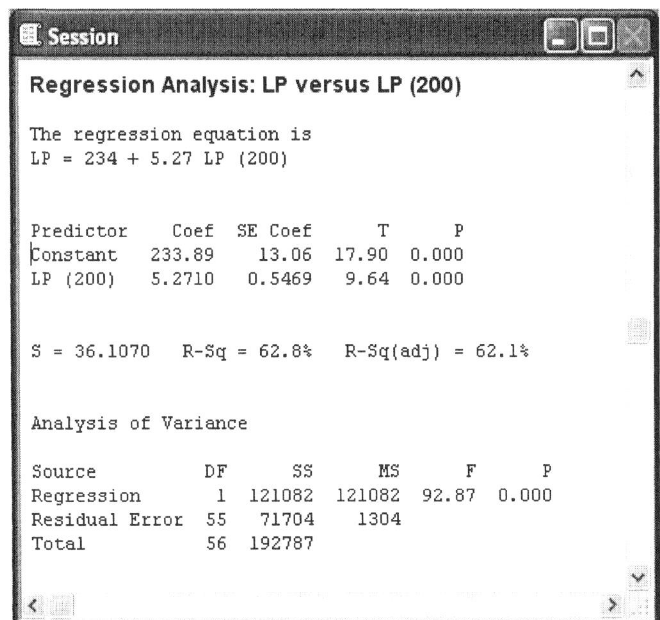

Session

Regression Analysis: LP versus LP (200)

The regression equation is
LP = 234 + 5.27 LP (200)

Predictor	Coef	SE Coef	T	P
Constant	233.89	13.06	17.90	0.000
LP (200)	5.2710	0.5469	9.64	0.000

S = 36.1070   R-Sq = 62.8%   R-Sq(adj) = 62.1%

Analysis of Variance

Source	DF	SS	MS	F	P
Regression	1	121082	121082	92.87	0.000
Residual Error	55	71704	1304		
Total	56	192787			

From the ANOVA table, we can use SS Residual Error = 71704 to determine the residual standard deviation of y-values. Note that $df = 57 - 2 = 55$.

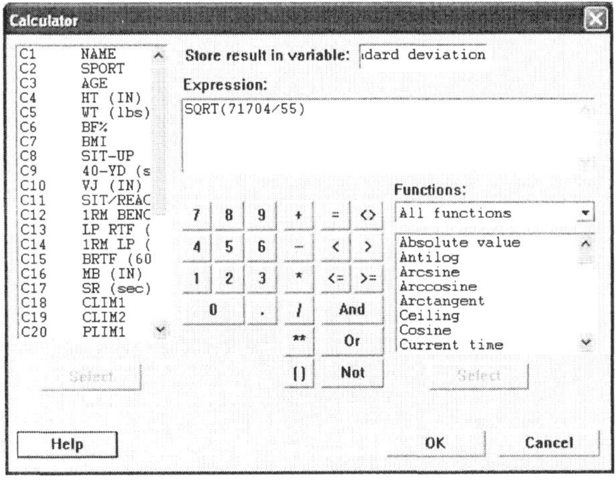

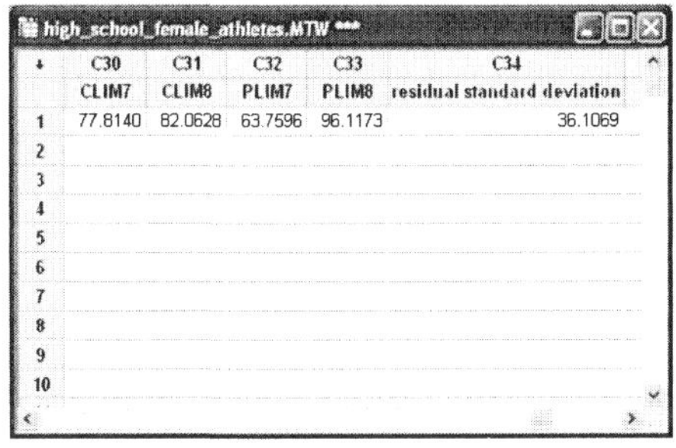

high_school_female_athletes.MTW ***

	C30	C31	C32	C33	C34
	CLIM7	CLIM8	PLIM7	PLIM8	residual standard deviation
1	77.8140	82.0628	63.7596	96.1173	36.1069
2					
3					
4					
5					
6					
7					
8					
9					
10					

At any fixed value x of number of 200-pound leg presses, we estimate that the maximum leg press values have a standard deviation of 36.1 pounds.

For female athletes with $x = 22$, we would estimate the mean maximum leg press to be `LP = 234 + 5.27 (22)` = 349.94 pounds and the variability of their maximum leg press values to be 36.1 pounds.

If y-values are approximately normal, then 95% of the y-values would fall in the interval approximately $\hat{y} \pm 2s = 349.94 \pm 2(36.1) = (277.74, 422.14)$

Using MINITAB to generate the prediction interval, notice we get similar results.

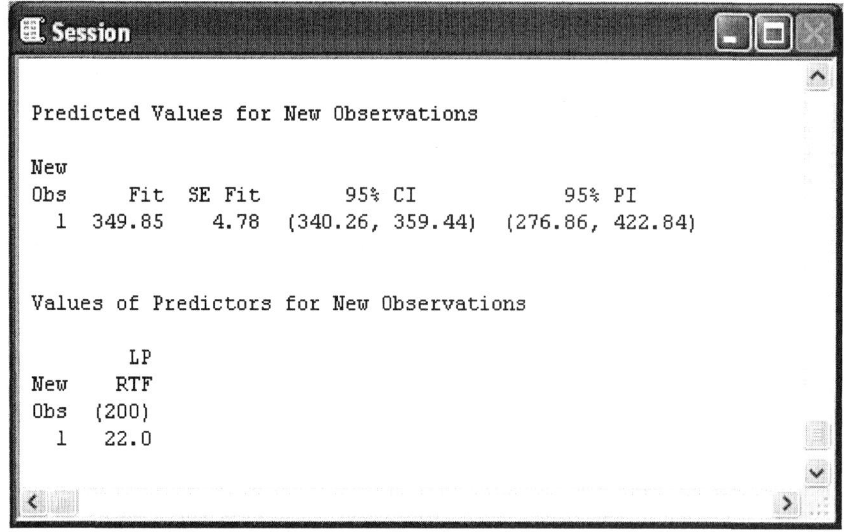

```
Session

Predicted Values for New Observations

New
Obs     Fit  SE Fit        95% CI              95% PI
  1  349.85    4.78  (340.26, 359.44)  (276.86, 422.84)


Values of Predictors for New Observations

         LP
New     RTF
Obs   (200)
  1    22.0
```

Chapter 11 Example 18 Explosion in Number of People Using the Internet Exponential Regression

Enter data into the Data Window, using C1 for number of years since 1995, and C2 for number of people (in millions).

internet_use.MTW ***

↓	C1	C2	C3	C4	C5
	No. Years Since 1995	Number People			
1	0	16			
2	1	36			
3	2	76			
4	3	147			
5	4	195			
6	5	369			
7	6	513			
8					
9					
10					

Producing a scatterplot, we see that the number of people increases over time, and the amount of increase from one year to the next seems itself to increase over time.

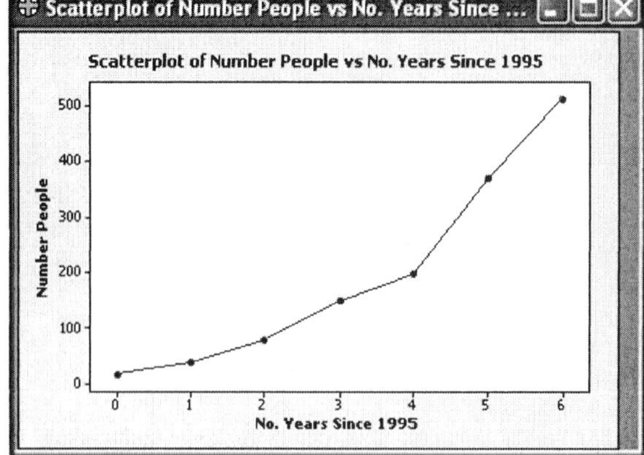

Generating the logarithm of the Number People:

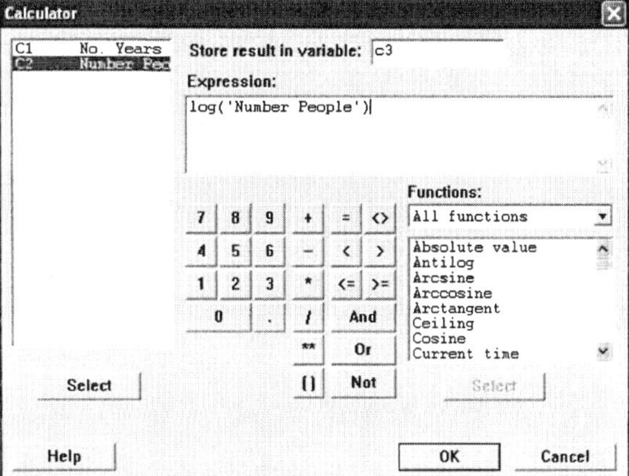

And then producing a scatterplot of C1 versus the logarithms:

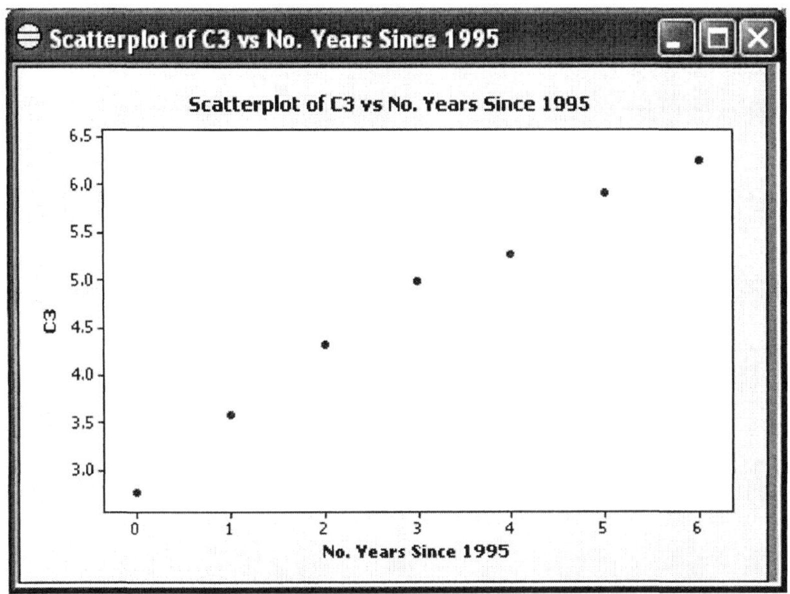

Determining the correlation between the logarithms and number of years since 1995

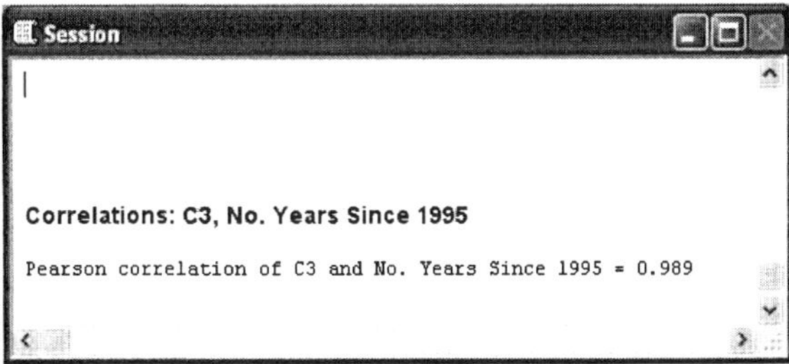

This value of 0.99 suggests that growth in Internet users over this time period was approximately exponential.

To generate the exponential regression model:

Select: **Stat > Time Series > Trend Analysis**

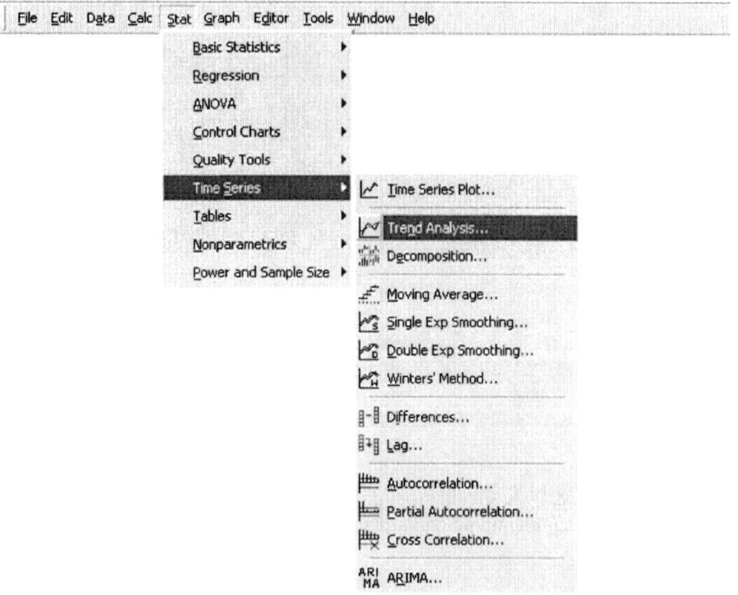

Enter: Variable: **Number People**
Select: **Exponential growth**
Click: **OK**

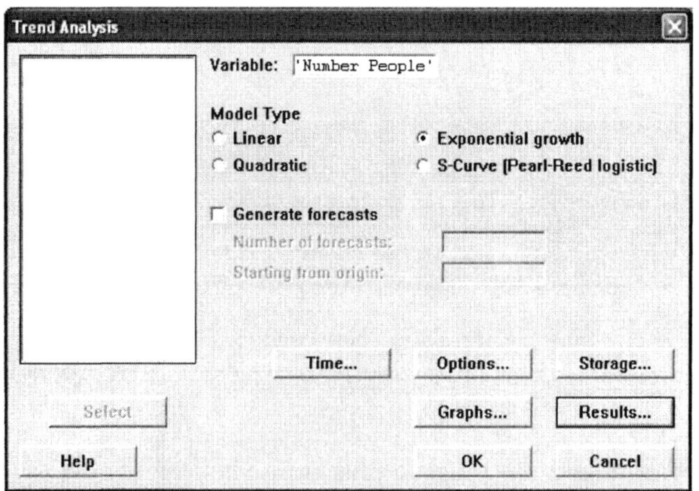

MINITAB produces a trend analysis plot, which includes the exponential regression model,
Yt = 11.5083 * (1.77079**t)

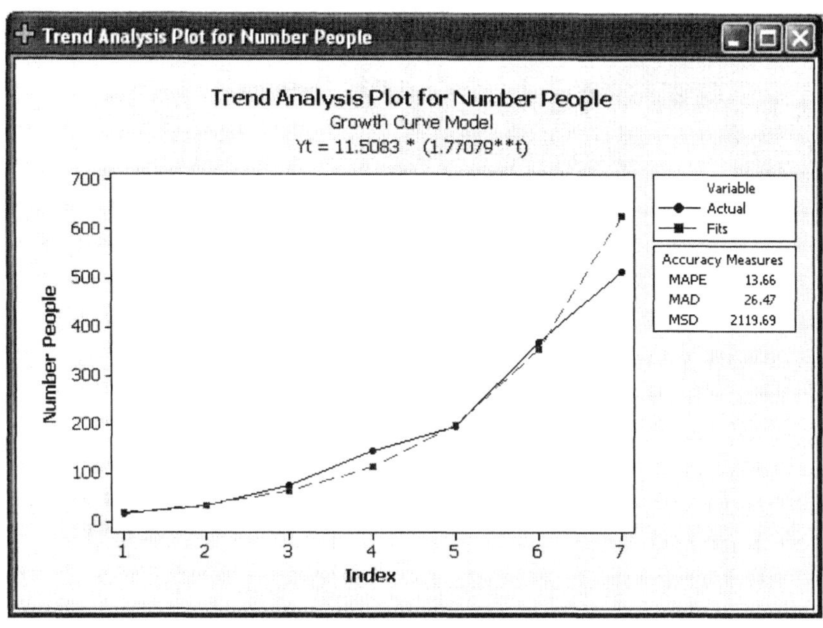

Note that Index = 1 means 1995, Index = 2 means 1996, and so on.

For the year 2000, Index = 6, and the predicted $\hat{y} = 11.5083 * 1.77079^6 = 354.8$ million.

Chapter 11 Exercise 11.55 Leaf Litter Decay
Exponential Regression

Worksheet 1 ***

↓	C1	C2	C3
	weeks	weight	
1	0	75.0	1.87506
2	1	60.9	1.78462
3	2	51.8	1.71433
4	3	45.2	1.65514
5	4	34.7	1.54033
6	5	34.6	1.53908
7	6	26.2	1.41830
8	7	20.4	1.30963
9	8	14.0	1.14613
10	9	12.3	1.08991
11	10	*	*
12	11	8.2	0.91381
13	12	*	*
14	13	*	*
15	14	*	*
16	15	3.1	0.49136
17	16	*	*
18	17	*	*
19	18	*	*
20	19	*	*
21	20	1.4	0.14613

Enter data into the Data Window, using C1 for number of weeks, and C2 for the weight.

Note the inclusion of weeks 10, 12, 13, 14, 16, 17, 18, and 19, even though we do not have weights for these particular weeks. The time series trend analysis assumes that the variable values (weight) are given at time intervals having an increment of one.

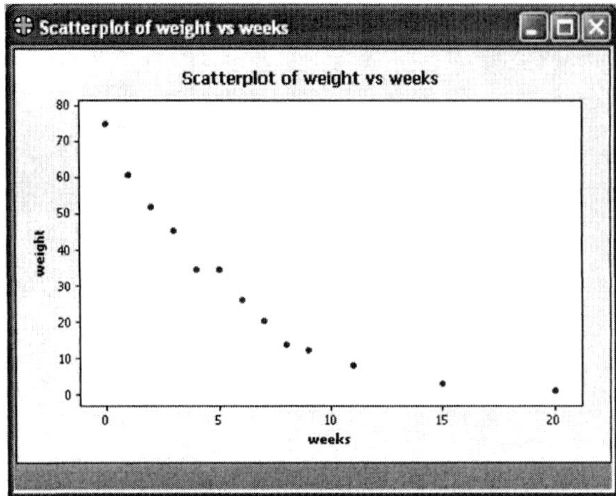

Producing a scatterplot, we see that the weight of the organic mass in the bag decreases over time, and the amount of decrease from one week to the next seems itself to decrease over time. Thus, a straight-line model is inappropriate.

Generating the logarithm of the weight and then producing a scatterplot of C1 versus the logarithms:

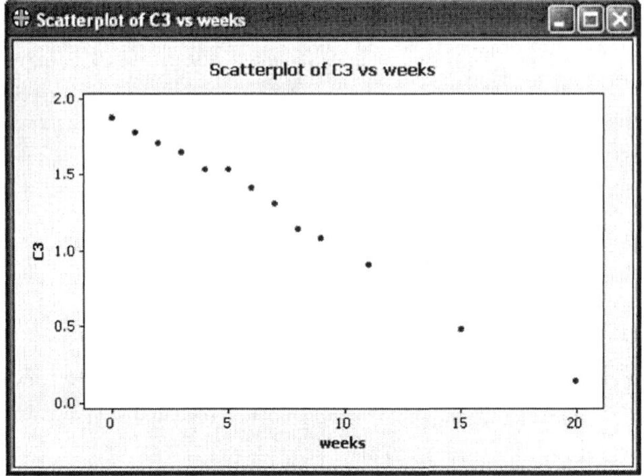

Determining the correlation between the logarithms and number of weeks

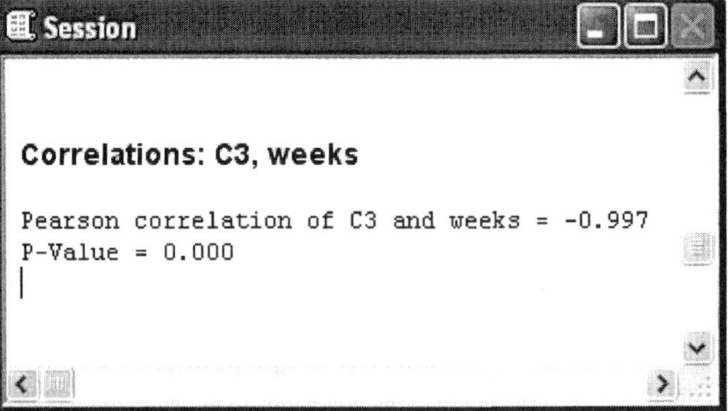

This value of -0.997 suggests that the decay of the organic mass over this time period was approximately exponential.

To generate the exponential regression model:

Select: **Stat > Time Series > Trend Analysis**

Enter: Variable: **weight**
Select: **Exponential growth**
Click: **OK**

MINITAB produces a trend analysis plot, which includes the exponential regression model,
Yt = 99.1144 * (0.813443**t)

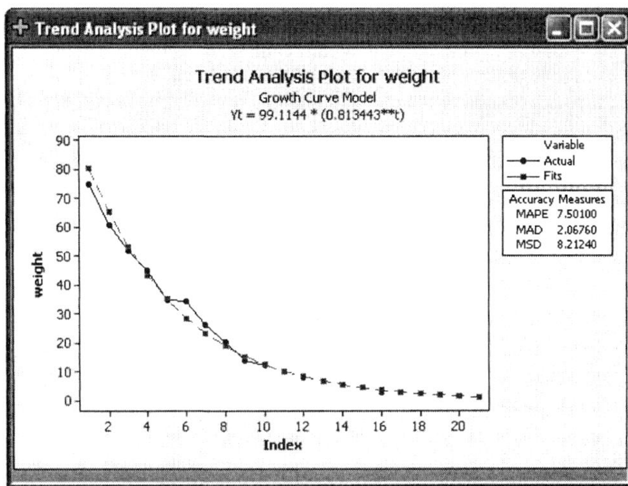

Note that Index = 1 means "initially", Index = 2 means "after one week", and so on.

For after the 20th week Index = 21, and the predicted $\hat{y} = 99.1144 * 0.813443^{21} = 1.30$ kg.

Chapter 12 Example 2 Predicting Selling Price Using House and Lot Sizes Multiple Regression and Plotting the Relationships

Open the worksheet **house_selling_prices** , which is found in the MINITAB folder. The price is found in C7, house size is found in C8, and lot size is found in C9.

Select: **Stat > Regression > Regression**
Enter: Response: **price**
Predictors: **size lot**
Select: **Results**
Select: **Regression equation, table of coefficients, s, R-squared, and basic analysis of variance**
Click: **OK OK**

The results are displayed in the Session Window.

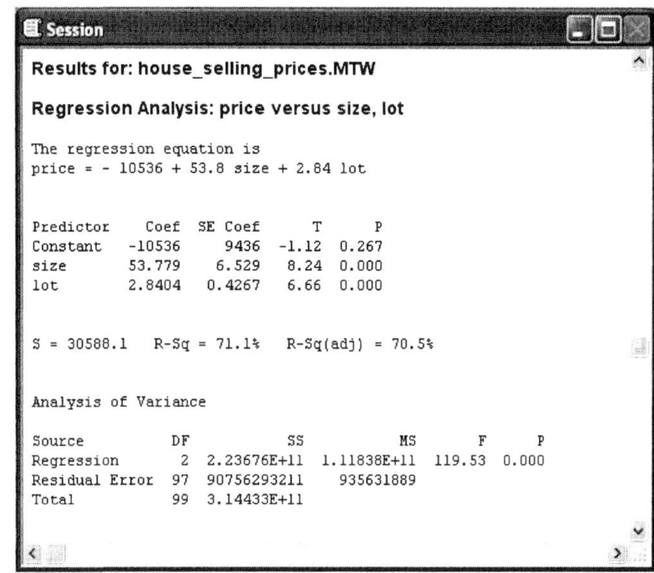

To use MINITAB to generate a scatterplot matrix

Select: **Graph > Matrix plot**

Select: **Matrix of plots / Simple > OK > OK**

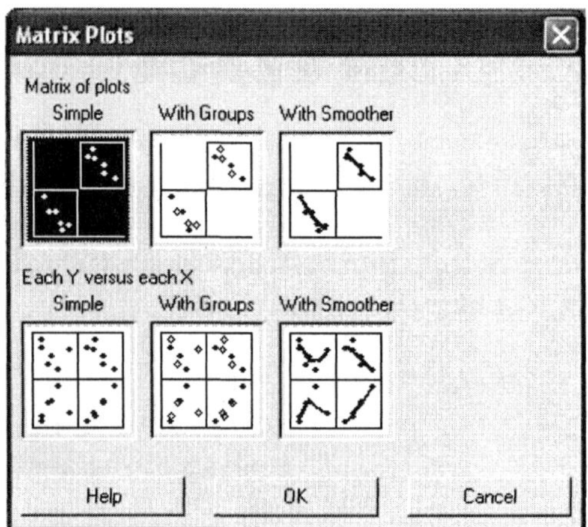

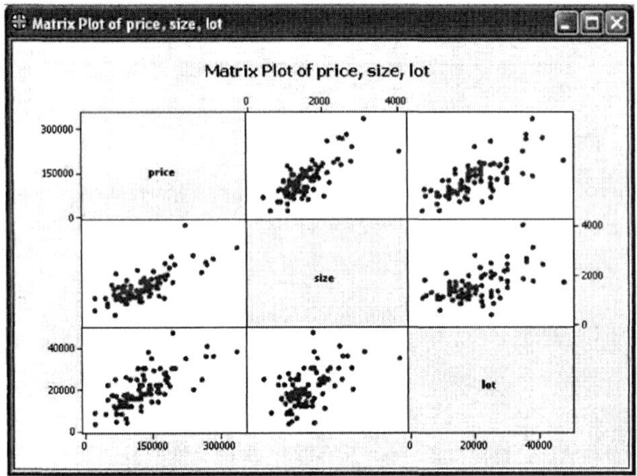

Since we are truly interested in the association between price and house size, and price and lot size, another option is

Select: **Graph > Matrix of plots**
Select: **Each Y versus each X, Simple**
Click: **OK**

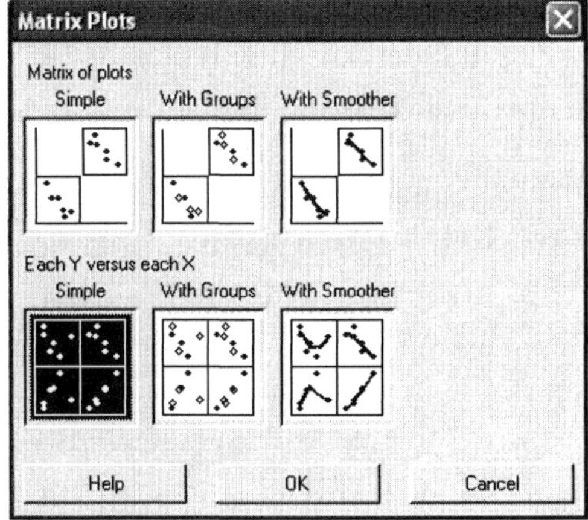

Enter: Y variables: **price**
X variables: **size lot**
Click: **OK**

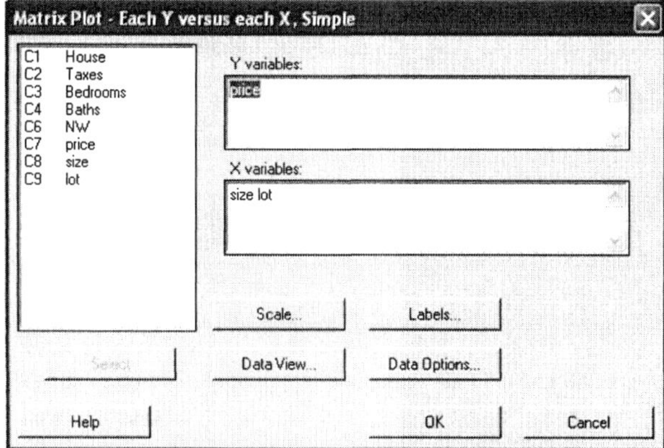

These commands produce the following graph window.

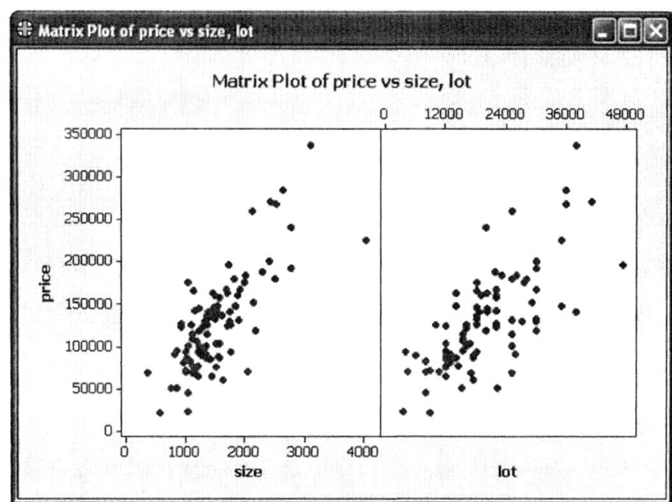

Chapter 12 Exercise 12.5 Does more education cause more crime?
Multiple Regression

Open the worksheet **fla_crime**, which is found in the MINITAB folder. The crime rate is found in C2, education is found in C3, and urbanization is found in C4.

Select: **Stat > Regression > Regression**
Enter: Response: **crime**
Predictors: **education urbanization**
Select: **Results**
Select: **Regression equation, table of coefficients, s, R-squared, and basic analysis of variance**
Click: **OK OK**

The results are displayed in the Session Window.

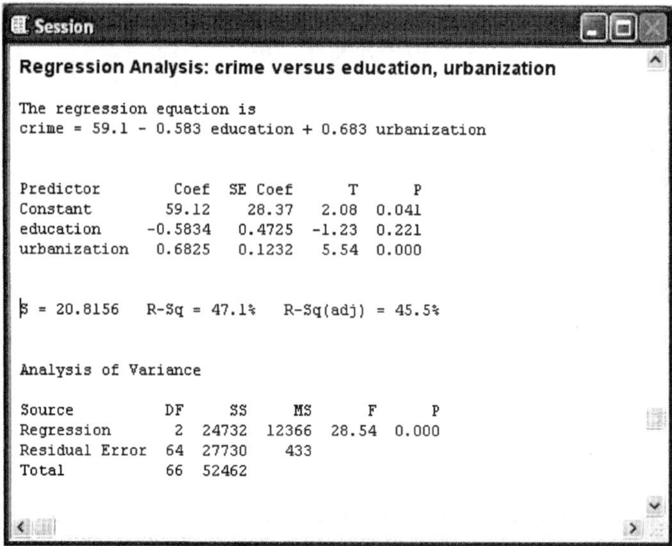

Session

Regression Analysis: crime versus education, urbanization

```
The regression equation is
crime = 59.1 - 0.583 education + 0.683 urbanization


Predictor        Coef  SE Coef      T      P
Constant        59.12    28.37   2.08  0.041
education     -0.5834   0.4725  -1.23  0.221
urbanization   0.6825   0.1232   5.54  0.000


S = 20.8156   R-Sq = 47.1%   R-Sq(adj) = 45.5%


Analysis of Variance

Source          DF     SS     MS      F      P
Regression       2  24732  12366  28.54  0.000
Residual Error  64  27730    433
Total           66  52462
```

To generate a scatterplot matrix

Select: **Graph > Matrix plot**

Select: **Matrix of plots / Simple > OK > OK**

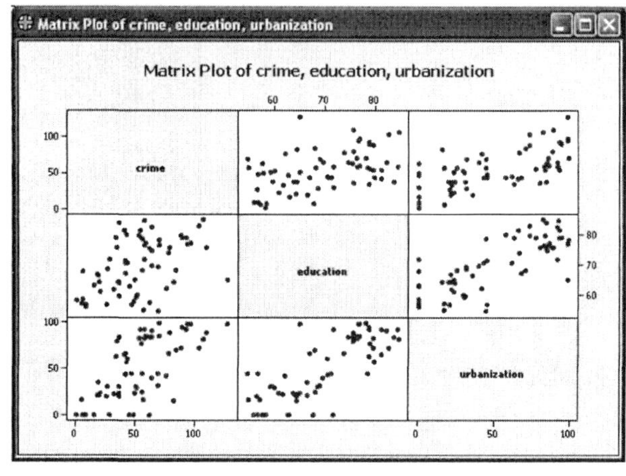

Since we are truly interested in the association between crime rate and education, and crime rate and urbanization, another option is

Select: **Graph > Matrix of plots**
Select: **Each Y versus each X, Simple**
Click: **OK**

Enter: Y variables: **crime**
X variables: **education urbanization**
Click: **OK**

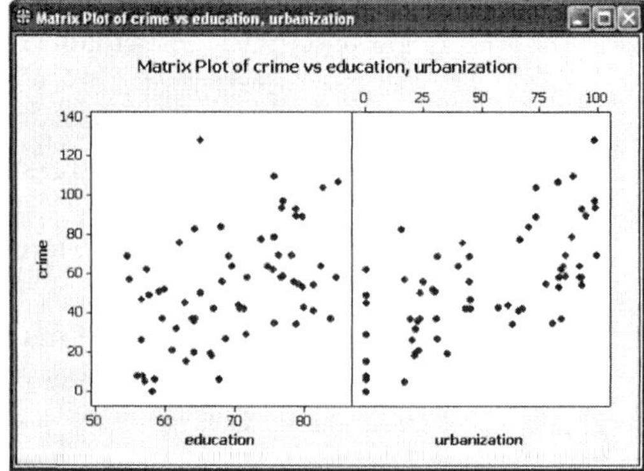

Using the regression equation generated by MINITAB,

```
crime = 59.1 - 0.583 education + 0.683 urbanization
```

we can predict crime rates for a county that has 0% in an urban environment by substituting 0 for "urbanization" and using the resulting equation crime = 59.1 – 0.583 education . With education set equal to 70% we determine the crime rate to be 59.1 – 0.583(70) = 18.29. For an 80% high school graduation rate, the crime rate is predicted to be 59.1 – 0.583(80) = 12.46.

Chapter 12 Example 3 How Well Can We Predict House Selling Prices? ANOVA and R

Open the worksheet **house_selling_prices** , which is found in the MINITAB folder. The price is found in C7, house size is found in C8, and lot size is found in C9. The regression command can be tailored to include the value of R-squared. We have done this numerous times.

Select: **Stat > Regression > Regression**
Enter: Response: **price**
Predictors: **size lot**
Select: **Results**
Select: **Regression equation, table of coefficients, s, R-squared, and basic analysis of variance**
Click: **OK OK**

The results are displayed in the Session Window.

Note the value of R-squared is 71.1%. The multiple correlation between selling price and the two explanatory variables is $R = \sqrt{R^2} = \sqrt{.711} = 0.84$.

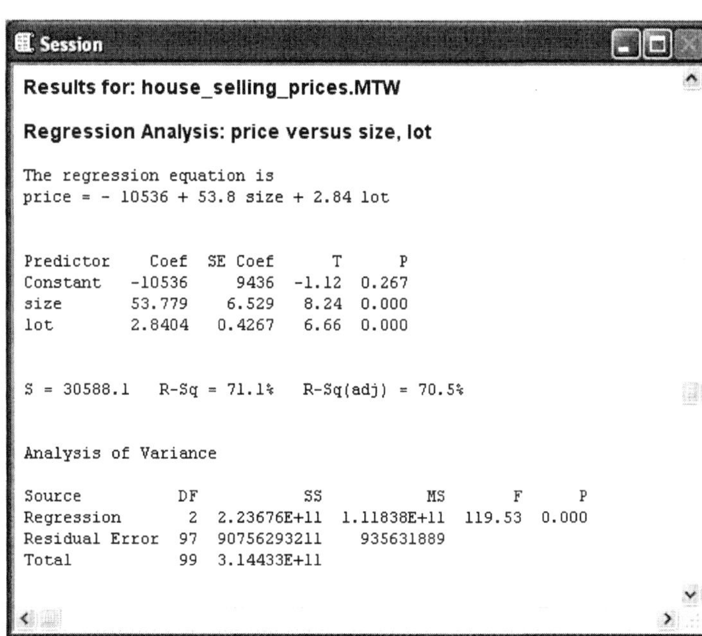
```
Session

Results for: house_selling_prices.MTW

Regression Analysis: price versus size, lot

The regression equation is
price = - 10536 + 53.8 size + 2.84 lot


Predictor     Coef  SE Coef      T      P
Constant    -10536     9436  -1.12  0.267
size        53.779    6.529   8.24  0.000
lot         2.8404   0.4267   6.66  0.000


S = 30588.1   R-Sq = 71.1%   R-Sq(adj) = 70.5%


Analysis of Variance

Source          DF           SS           MS       F      P
Regression       2  2.23676E+11  1.11838E+11  119.53  0.000
Residual Error  97  90756293211    935631889
Total           99  3.14433E+11
```

Chapter 12 Example 4 What Helps Predict a Female Athlete's Weight?
Significance Test and Confidence Interval about a Multiple Regression Parameter β

Open the worksheet **college_athletes**, which is found in the MINITAB folder. The total body weight (TBW) is found in C1, height (HGT) in inches is found in C2, the percent of body fat (%BF) is found in C3, and age (AGE) is found in C11.

Select: **Stat > Regression > Regression**
Enter: Response: **TBW**
Predictors: **HGT %BF AGE**
Select: **Results**
Select: **Regression equation, table of coefficients, s, R-squared, and basic analysis of variance**
Click: **OK OK**

The results are displayed in the Session Window.

Note the value of R-squared is 66.9%. The predictive power is good.

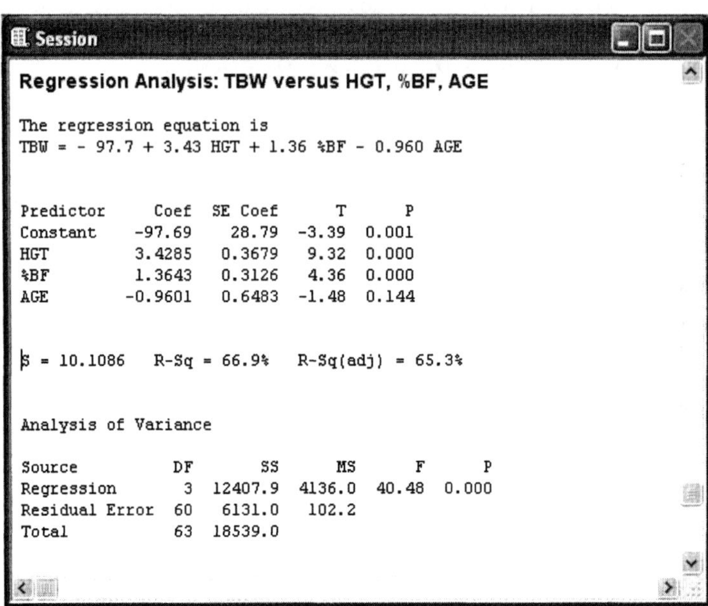
Session

Regression Analysis: TBW versus HGT, %BF, AGE

```
The regression equation is
TBW = - 97.7 + 3.43 HGT + 1.36 %BF - 0.960 AGE


Predictor     Coef  SE Coef      T      P
Constant    -97.69    28.79  -3.39  0.001
HGT         3.4285   0.3679   9.32  0.000
%BF         1.3643   0.3126   4.36  0.000
AGE        -0.9601   0.6483  -1.48  0.144


S = 10.1086   R-Sq = 66.9%   R-Sq(adj) = 65.3%


Analysis of Variance

Source          DF       SS      MS      F      P
Regression       3  12407.9  4136.0  40.48  0.000
Residual Error  60   6131.0   102.2
Total           63  18539.0
```

To test whether age helps us to predict weight, if we already know height and percent body fat, we perform a significance test on $H_0 : \beta_3 = 0$ versus $H_a: \beta_3 \neq 0$. The *t* test statistic is reported in the table as -1.48 and the p-value is .144. This p-value does not give much evidence against the null hypothesis. Age does not significantly predict weight, if we already know height and percentage of body fat.

Chapter 12 Example 5 What's Plausible for the Effect of Age on Weight? Confidence Interval about a Multiple Regression Parameter β

Open the worksheet **college_athletes**, which is found in the MINITAB folder. The total body weight (TBW) is found in C1, height (HGT) in inches is found in C2, the percent of body fat (%BF) is found in C3, and age (AGE) is found in C11.

Select: **Stat > Regression > Regression**
Enter: Response: **TBW**
Predictors: **HGT %BF AGE**
Select: **Results**
Select: **Regression equation, table of coefficients, s, R-squared, and basic analysis of variance**
Click: **OK OK**

The results are displayed in the Session Window.

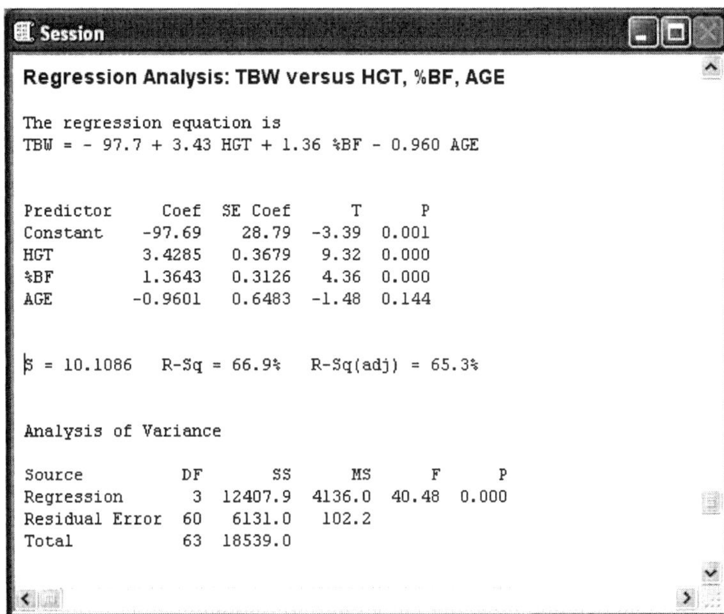

```
Regression Analysis: TBW versus HGT, %BF, AGE

The regression equation is
TBW = - 97.7 + 3.43 HGT + 1.36 %BF - 0.960 AGE


Predictor      Coef  SE Coef      T      P
Constant     -97.69    28.79  -3.39  0.001
HGT          3.4285   0.3679   9.32  0.000
%BF          1.3643   0.3126   4.36  0.000
AGE         -0.9601   0.6483  -1.48  0.144


S = 10.1086   R-Sq = 66.9%   R-Sq(adj) = 65.3%


Analysis of Variance

Source          DF       SS      MS      F      P
Regression       3  12407.9  4136.0  40.48  0.000
Residual Error  60   6131.0   102.2
Total           63  18539.0
```

To compute a 95% confidence interval for the population slope β_3, note that the $t_{.025}$ value for $df = n - 2 = 60$ is $t = 2.00$, and that the estimate of β_3 is -0.9601, and $se = 0.6483$. (MINITAB does not include a t-interval estimate for a population slope.)

$$b \pm t_{.025}(se) = -0.9601 \pm 2.00\,(0.6483) = -0.9601 \pm 1.2966$$

$$(-2.26,\ 0.34)$$

At fixed values of height and percent body fat, we infer that the population mean weight changes very little (and may not change at all, since this interval includes 0), making these results consistent with what we found in example 4.

Chapter 12 Example 7 The F Test For Predictors of Athletes' Weight
The F Test that All Multiple Regression Parameters β = 0

Open the worksheet **college_athletes**, which is found in the MINITAB folder. The total body weight (TBW) is found in C1, height (HGT) in inches is found in C2, the percent of body fat (%BF) is found in C3, and age (AGE) is found in C11.

Select: **Stat > Regression > Regression**
Enter: Response: **TBW**
Predictors: **HGT %BF AGE**
Select: **Results**
Select: **Regression equation, table of coefficients, s, R-squared, and basic analysis of variance**
Click: **OK OK**

The results are displayed in the Session Window.

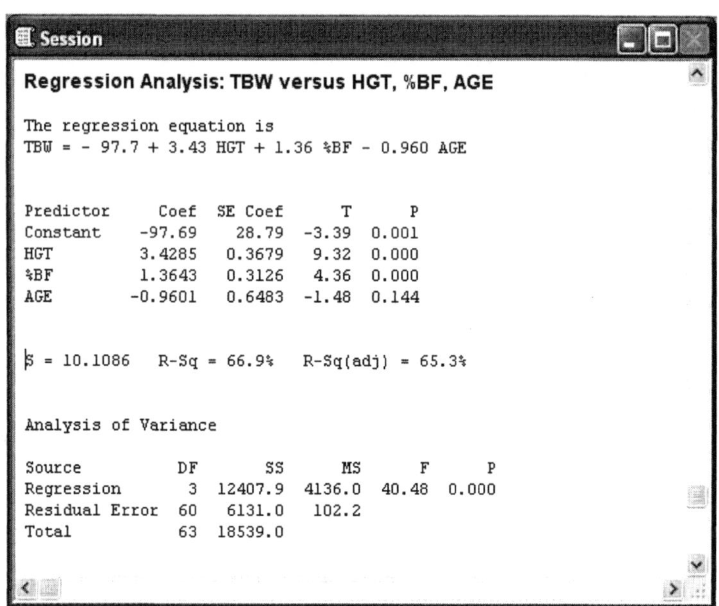

Session

Regression Analysis: TBW versus HGT, %BF, AGE

The regression equation is
TBW = - 97.7 + 3.43 HGT + 1.36 %BF - 0.960 AGE

Predictor	Coef	SE Coef	T	P
Constant	-97.69	28.79	-3.39	0.001
HGT	3.4285	0.3679	9.32	0.000
%BF	1.3643	0.3126	4.36	0.000
AGE	-0.9601	0.6483	-1.48	0.144

S = 10.1086   R-Sq = 66.9%   R-Sq(adj) = 65.3%

Analysis of Variance

Source	DF	SS	MS	F	P
Regression	3	12407.9	4136.0	40.48	0.000
Residual Error	60	6131.0	102.2		
Total	63	18539.0			

The ANOVA table for the multiple regression output includes the value of the F test statistic (F = 40.48) and the p-value (p = 0.000) for testing H_0: $\beta_1 = \beta_2 = \beta_3 = 0$. We can reject H_0 and conclude that at least one predictor has an effect on weight.

Chapter 12 Exercise 12.20 Study time help GPA?
Hypothesis Test and Confidence Interval About β

Open the worksheet **georgia student survey**, which is found in the MINITAB folder. Performing a multiple regression for college GPA based on high school GPA and study time:

Select: **Stat > Regression > Regression**
Enter: Response: **CGPA**
Predictors: **HSGPA Studytime**
Select: **Results**
Select: **Regression equation, table of coefficients, s, R-squared, and basic analysis of variance**
Click: **OK OK**

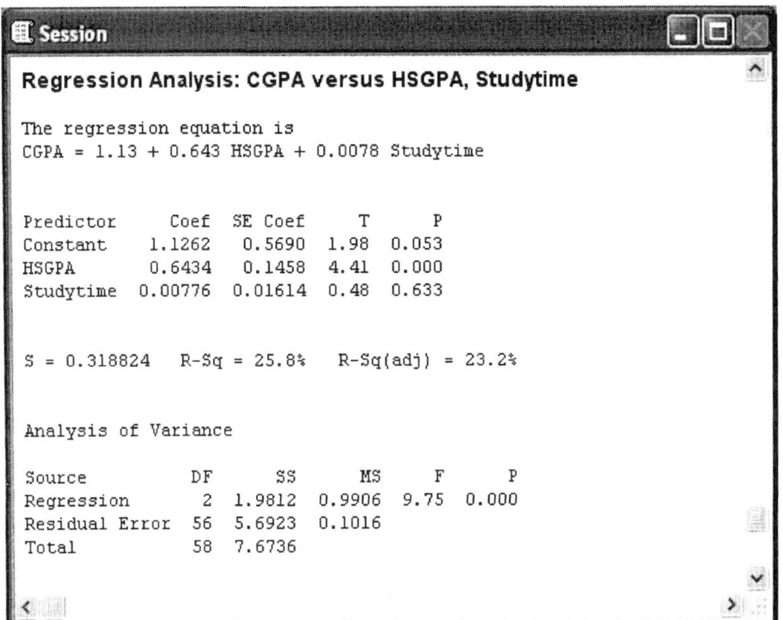

Session

Regression Analysis: CGPA versus HSGPA, Studytime

```
The regression equation is
CGPA = 1.13 + 0.643 HSGPA + 0.0078 Studytime


Predictor      Coef  SE Coef     T      P
Constant     1.1262   0.5690  1.98  0.053
HSGPA        0.6434   0.1458  4.41  0.000
Studytime   0.00776  0.01614  0.48  0.633


S = 0.318824   R-Sq = 25.8%   R-Sq(adj) = 23.2%


Analysis of Variance

Source          DF      SS      MS     F      P
Regression       2  1.9812  0.9906  9.75  0.000
Residual Error  56  5.6923  0.1016
Total           58  7.6736
```

To test whether study time helps us to college GPA, if we already know high school GPA, we perform a significance test on $H_0 : \beta_2 = 0$ versus H_a: $\beta_2 \neq 0$. The *t* test statistic is reported in the table as 0.48 and the p-value is 0.633. This p-value does not give evidence against the null hypothesis. Study time does not significantly predict college GPA if we already know high school GPA.

To compute a 95% confidence interval for the population slope β_2, note that the $t_{.025}$ value for $df = n - 2 = 57$ is $t =$ 2.0025, and that the estimate of β_2 is 0.00776, and se = 0.01614. (MINITAB does not include a t-interval estimate for a population slope.)

$$b \pm t_{.025}\,(se) = 0.00776 \pm 2.0025\,(0.01614)$$
$$= 0.00776 \text{ +/- } 0.0323$$
$$(-0.0247,\ 0.0399)$$

At a fixed value of high school GPA, we infer that the population mean college GPA changes very little (and may not change at all, since this interval includes 0), making these results consistent with the significance test we have already completed.

Chapter 12 Exercise 12.29 More predictors for house price
The F Test that All Multiple Regression Parameters β = 0

Open the worksheet **house_selling_prices**, which is found in the MINITAB folder. Performing multiple regression for predicting house selling price using size of home, lot size, and real estate tax:

Select: **Stat > Regression > Regression**
Enter: Response: **price**
Predictors: **size lot Taxes**
Select: **Results**
Select: **Regression equation, table of coefficients, s, R-squared, and basic analysis of variance**
Click: **OK OK**

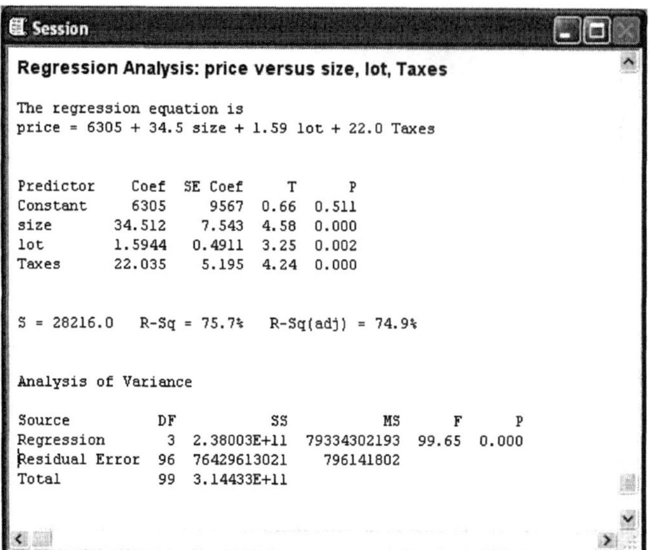

Session
Regression Analysis: price versus size, lot, Taxes
The regression equation is
price = 6305 + 34.5 size + 1.59 lot + 22.0 Taxes
Predictor Coef SE Coef T P
Constant 6305 9567 0.66 0.511
size 34.512 7.543 4.58 0.000
lot 1.5944 0.4911 3.25 0.002
Taxes 22.035 5.195 4.24 0.000
S = 28216.0 R-Sq = 75.7% R-Sq(adj) = 74.9%
Analysis of Variance
Source DF SS MS F P
Regression 3 2.38003E+11 79334302193 99.65 0.000
Residual Error 96 76429613021 796141802
Total 99 3.14433E+11

The ANOVA table for the multiple regression output includes the value of the F test statistic (F = 99.65) and the p-value (p = 0.000) for testing H_0: $\beta_1 = \beta_2 = \beta_3 = 0$. We can reject H_0 and conclude that at least one predictor has an effect on price.

The p-values of 0.000, 0.002, 0.000 for *t*-tests for each of the explanatory variables, respectively, indicate that each predictor (house size, lot size, and taxes) does significantly predict price, if we already know the values of the other two variables.

Chapter 12 Example 9 Another Residual Plot for House Selling Prices
Plots of Residuals Against Explanatory Variables

Open the worksheet **house_selling_prices** , which is found in the MINITAB folder. The price is found in C7, house size is found in C8, and lot size is found in C9.

To produce a histogram of standardized residuals for the multiple regression model predicting selling price by the house size and the lot size:

Select: **Stat > Regression > Regression**
Enter: Response: **price**
Predictors: **size lot**
Select: **Graphs**

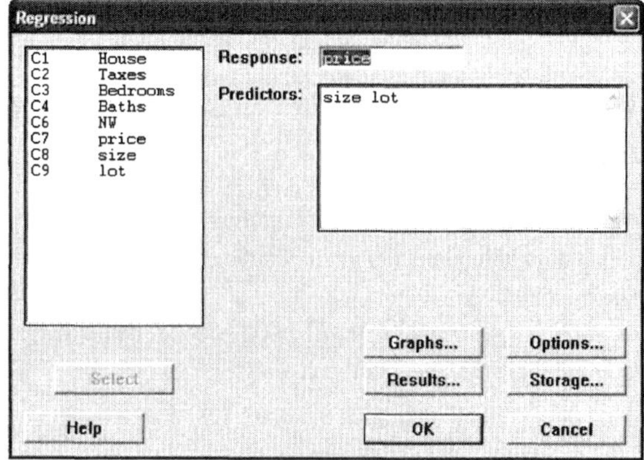

Select: Residuals for Plots: **Standardized**
Residual Plot: **Histogram of residuals**
Click: **OK OK**

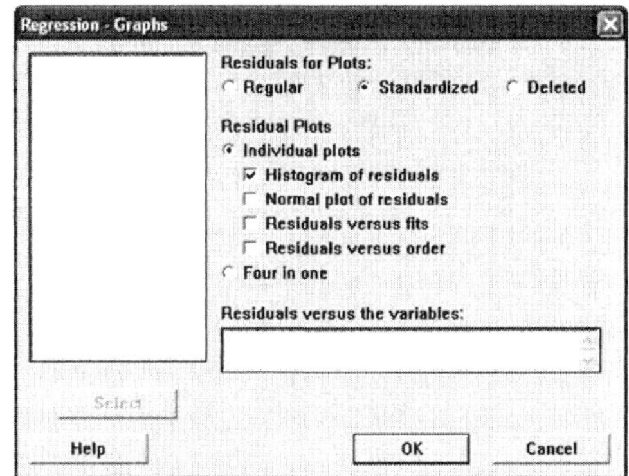

The histogram is displayed in a Graph Window.

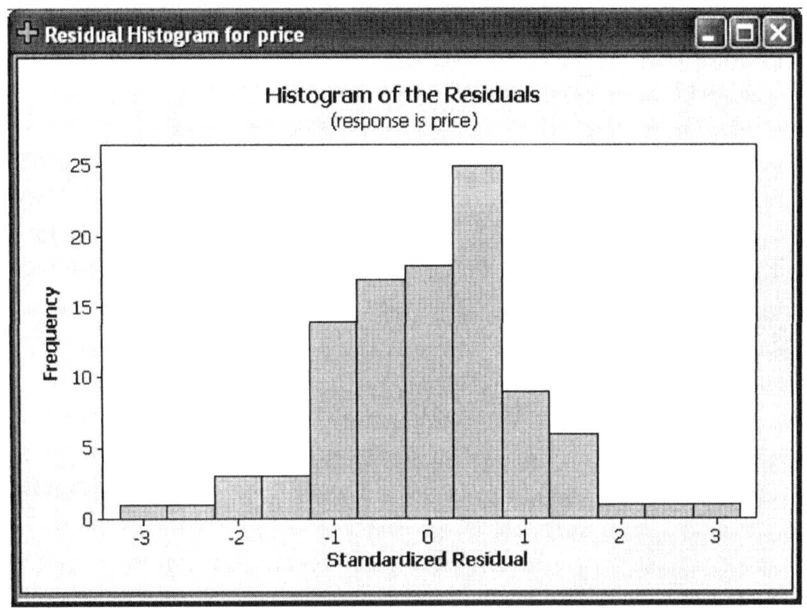

To plot the residuals against the explanatory variable house size:

Select: **Stat > Regression > Regression**
Enter: Response: **price**
Predictors: **size lot**
Select: **Graphs**
Select: Residuals for Plots: **Standardized**
Enter: Residuals versus the variables: **size**
Click: **OK OK**

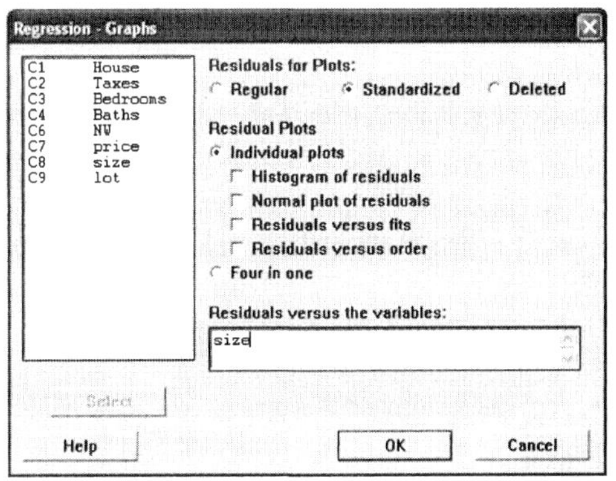

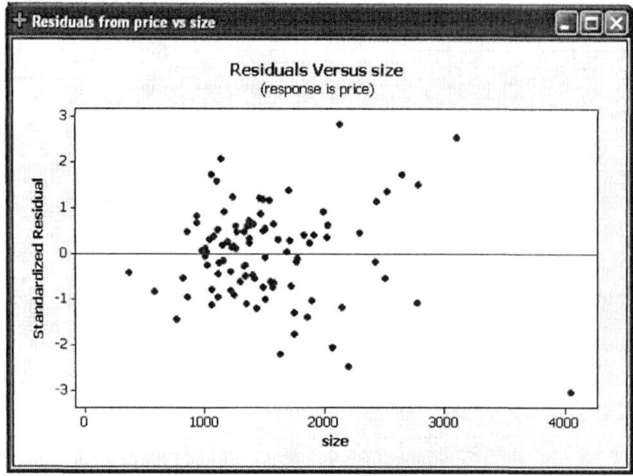

Chapter 12 Example 11 Comparing Winning High Jumps For Men and Women Including Categorical Predictors in Regression

Open the worksheet **high_jump** , which is found in the MINITAB folder. The following screen shot indicates the organization of the data file.

high_jump.MTW ***

↓	C1	C2	C3	C4
	Men_Meters	Year_Men	Women_Meters	Year_Women
4	1.905	1908	*	*
5	1.930	1912	*	*
6	1.935	1920	*	*
7	1.980	1924	*	*
8	1.940	1928	1.590	1928
9	1.970	1932	1.657	1932
10	2.030	1936	1.600	1936
11	1.980	1948	1.680	1948
12	2.040	1952	1.670	1952
13	2.120	1956	1.760	1956

Note that 1928 is the first year that women participated in the high jump.

We need to modify the data file so that each winning high jump (whether for male or female) is listed in one column and add columns to indicate x_1 = number of years since 1928 and x_2 = gender (1 = male, 0 = female). Since we're only interested in years starting at 1928 we will not use the data for the men for prior years.

high_jump.MTW ***

↓	C1	C2	C3	C4	C5	C6	C7
	Men_Meters	Year_Men	Women_Meters	Year_Women	Meters	x2	x1
1	1.94	1928	1.590	1928	1.940	1	0
2	1.97	1932	1.657	1932	1.970	1	4
3	2.03	1936	1.600	1936	2.030	1	8
4	1.98	1948	1.680	1948	1.980	1	20
5	2.04	1952	1.670	1952	2.040	1	24
6	2.12	1956	1.760	1956	2.120	1	28
7	2.16	1960	1.850	1960	2.160	1	32
8	2.16	1964	1.900	1964	2.160	1	36
9	2.24	1968	1.820	1968	2.240	1	40
10	2.23	1972	1.920	1972	2.230	1	44

Performing multiple regression for predicting winning height (in meters) as a function of x1 = number of years since 1928 and x2 = gender (1 = male, 0 = female)

Select: **Stat > Regression > Regression**
Enter: Response: **Meters**
Predictors: **x1 x2**
Click: **OK**

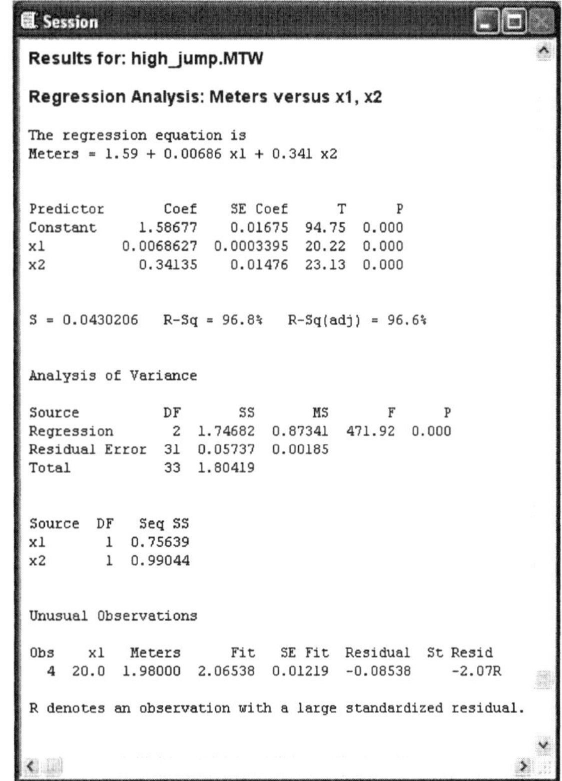

```
Session

Results for: high_jump.MTW

Regression Analysis: Meters versus x1, x2

The regression equation is
Meters = 1.59 + 0.00686 x1 + 0.341 x2


Predictor        Coef    SE Coef      T      P
Constant      1.58677    0.01675  94.75  0.000
x1          0.0068627  0.0003395  20.22  0.000
x2            0.34135    0.01476  23.13  0.000


S = 0.0430206   R-Sq = 96.8%   R-Sq(adj) = 96.6%


Analysis of Variance

Source          DF       SS       MS       F      P
Regression       2  1.74682  0.87341  471.92  0.000
Residual Error  31  0.05737  0.00185
Total           33  1.80419


Source  DF   Seq SS
x1       1  0.75639
x2       1  0.99044


Unusual Observations

Obs    x1   Meters      Fit   SE Fit  Residual  St Resid
  4  20.0  1.98000  2.06538  0.01219  -0.08538    -2.07R

R denotes an observation with a large standardized residual.
```

To produce a scatterplot that includes the categorical distinction:
Select: **Graph > Scatterplot > With Groups > OK**
Select: **With Groups**

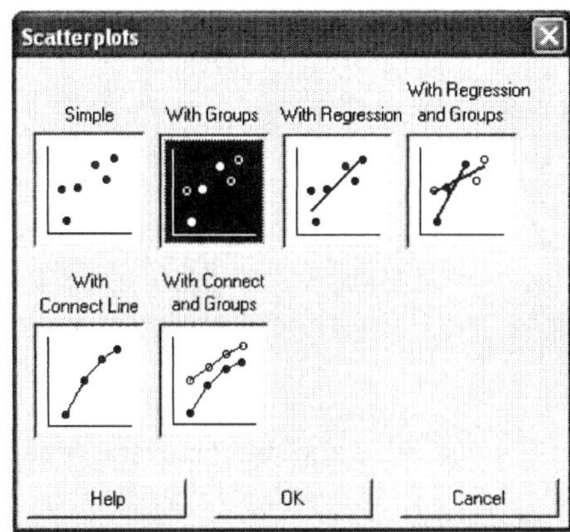

Enter the appropriate variables and select **X-Y pairs form groups > OK**

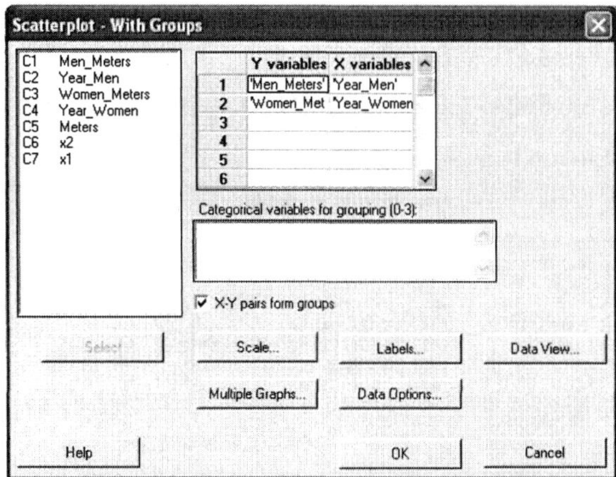

Note that the scatterplot was generated from the modified data file, so it does not include the men's winning high jumps prior to 1928.

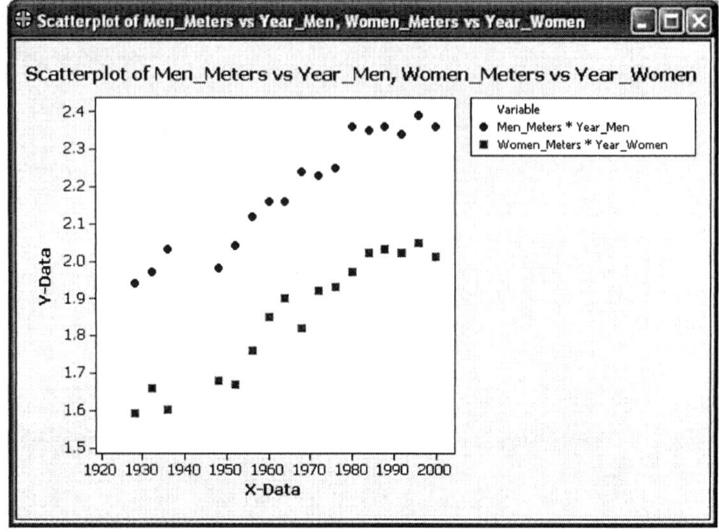

Chapter 12 Exercise 12.45 Houses, tax, and NW
Including Categorical Predictors in Regression

Open the worksheet **house_selling_prices** , which is found in the MINITAB folder. The following screen shot indicates the organization of the data file.

The data file includes C6 NW where 1 = NW and 0 = other.

house_selling_prices.MTW ***

↓	C1	C2	C3	C4	C5-T	C6	C7	C8	C9
	House	Taxes	Bedrooms	Baths	Quadrant	NW	price	size	lot
1	1	1360	3	2.0	NW	1	145000	1240	18000
2	2	1050	1	1.0	NW	1	68000	370	25000
3	3	1010	3	1.5	NW	1	115000	1130	25000
4	4	830	3	2.0	SW	0	69000	1120	17000
5	5	2150	3	2.0	NW	1	163000	1710	14000
6	6	1230	3	2.0	NW	1	69900	1010	8000
7	7	150	2	2.0	NW	1	50000	860	15300
8	8	1470	3	2.0	NW	1	137000	1420	18000
9	9	1050	3	2.0	NW	1	121300	1270	16000
10	10	320	3	2.0	NW	1	70000	1160	8000

Performing multiple regression for predicting price as a function of real estate tax and whether the home is in the NW region:

Select: **Stat > Regression > Regression**
Enter: Response: **price**
Predictors: **Taxes NW**
Click: **OK**

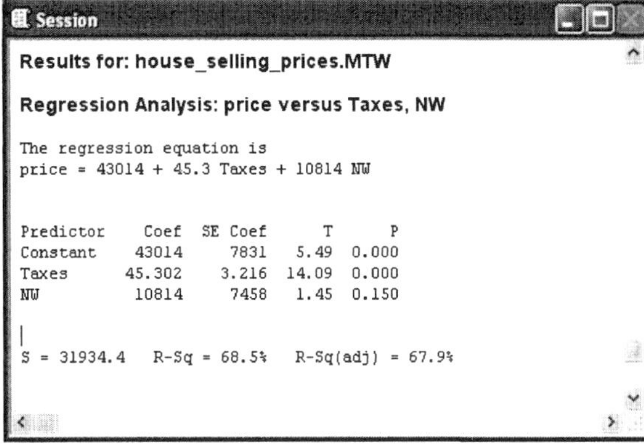
Session

Results for: house_selling_prices.MTW

Regression Analysis: price versus Taxes, NW

The regression equation is
price = 43014 + 45.3 Taxes + 10814 NW

Predictor	Coef	SE Coef	T	P
Constant	43014	7831	5.49	0.000
Taxes	45.302	3.216	14.09	0.000
NW	10814	7458	1.45	0.150

S = 31934.4 R-Sq = 68.5% R-Sq(adj) = 67.9%

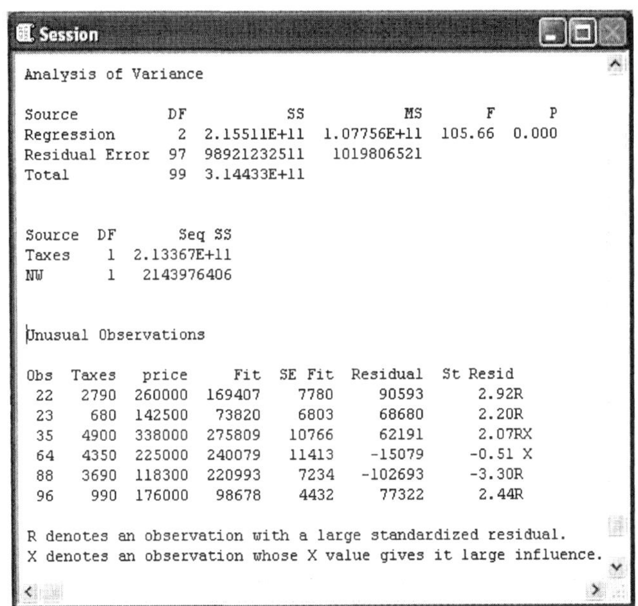
Session

Analysis of Variance

Source	DF	SS	MS	F	P
Regression	2	2.15511E+11	1.07756E+11	105.66	0.000
Residual Error	97	98921232511	1019806521		
Total	99	3.14433E+11			

Source	DF	Seq SS
Taxes	1	2.13367E+11
NW	1	2143976406

Unusual Observations

Obs	Taxes	price	Fit	SE Fit	Residual	St Resid
22	2790	260000	169407	7780	90593	2.92R
23	680	142500	73820	6803	68680	2.20R
35	4900	338000	275809	10766	62191	2.07RX
64	4350	225000	240079	11413	-15079	-0.51 X
88	3690	118300	220993	7234	-102693	-3.30R
96	990	176000	98678	4432	77322	2.44R

R denotes an observation with a large standardized residual.
X denotes an observation whose X value gives it large influence.

From the coefficient for NW we see that the selling price increases by $10,814 just because the house is located in the NW region.

Chapter 12 Example 12 Annual Income and Having a Travel Credit Card Logistic Regression

Open the data file **credit_card_and_income** from the MINITAB folder of the text CD.

Choose: **Stat > Regression > Binary Logistic Regression**

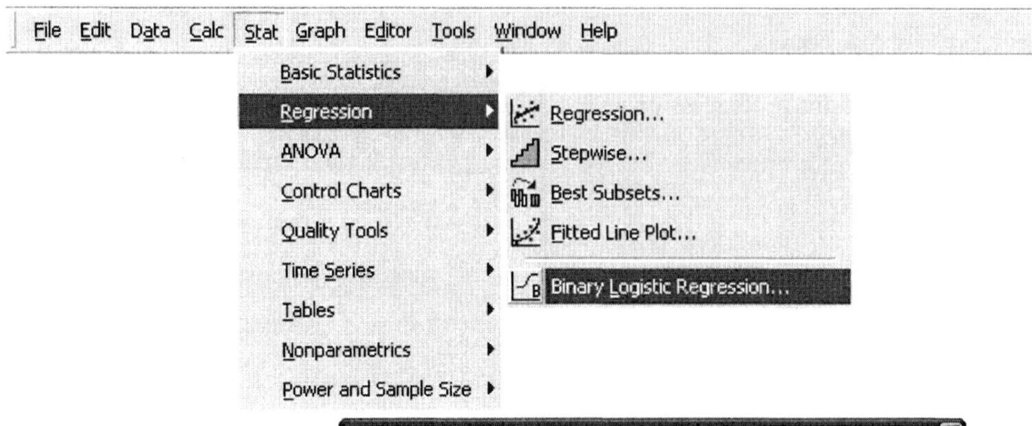

Enter: Response: **y**
Model: **income**
Click: **OK**

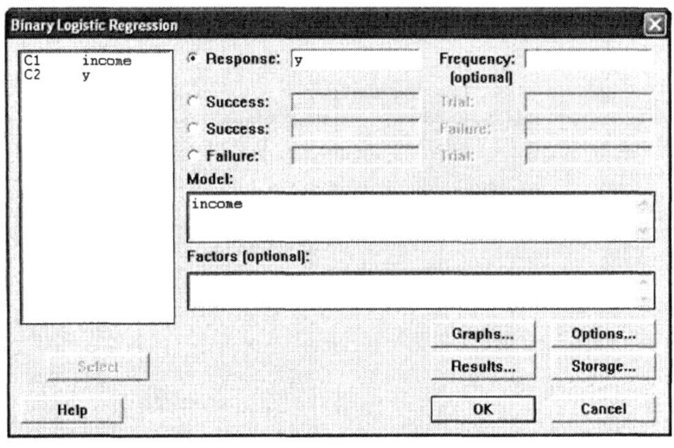

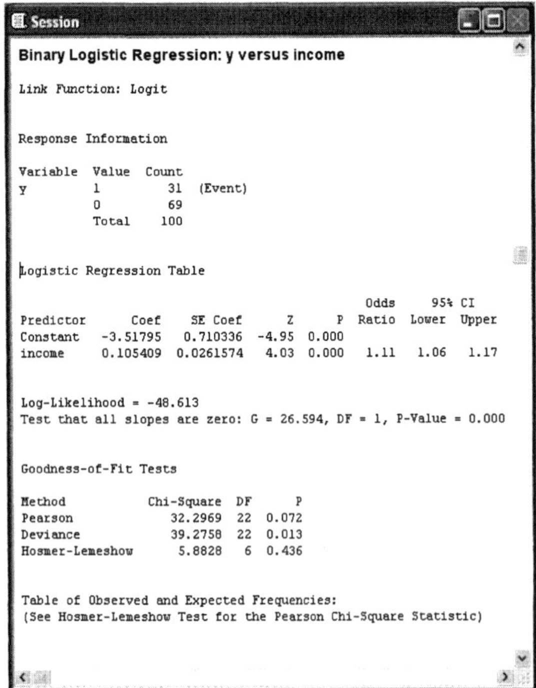

Quite a bit of information is displayed in the Session Window.

The values we need are α and β. They are found under the Coef heading.
$\alpha = -3.51795$ and $\beta = 0.105409$

So, the equation for estimated probability $\hat{p}$ of possessing a travel credit card is

$$\hat{p} = \frac{e^{-3.52 + .105x}}{1 + e^{-3.52 + .105x}}$$

Chapter 12 Example 14 Estimating Proportion of Students Who Have Used Marijuana Logistic Regression

For this problem we will need to enter the data. The response variable "marijuana use" will be coded 1 = yes and 0 = no. Likewise we will code the indicator variables, alcohol use, and cigarette use, 1 = yes and 0 = no. We also need to include a column with the corresponding frequencies.

Worksheet 3 ***

↓	C1 Alcohol Use	C2 Cigarette Use	C3 Marijuana Use	C4 Count
1	1	1	1	911
2	1	1	0	538
3	1	0	1	44
4	1	0	0	456
5	0	1	1	3
6	0	1	0	43
7	0	0	1	2
8	0	0	0	279
9				
10				

Choose: **Stat > Regression > Binary Logistic Regression**

Enter: Response: **Marijuana Use**
Model: **Alcohol Use** **Cigarette Use**
Click: **OK**

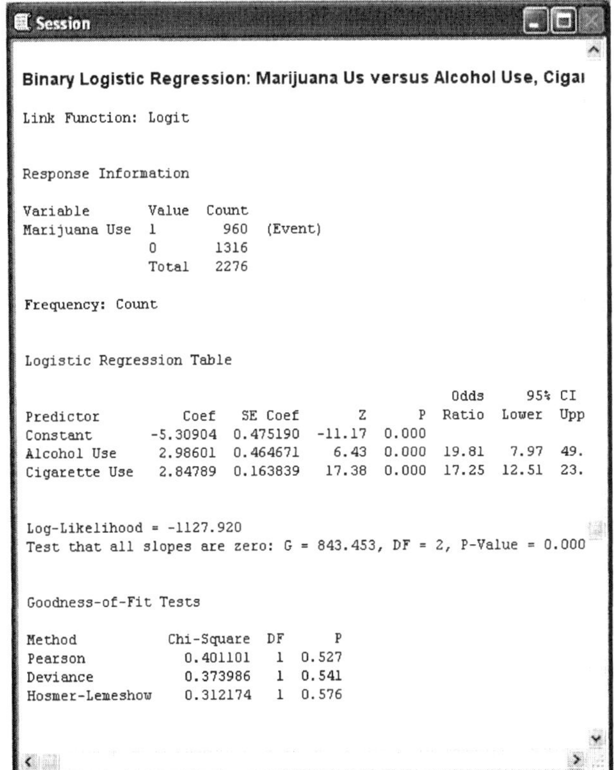

Session

Binary Logistic Regression: Marijuana Us versus Alcohol Use, Cigai

```
Link Function: Logit


Response Information

Variable       Value  Count
Marijuana Use  1        960  (Event)
               0       1316
               Total   2276

Frequency: Count


Logistic Regression Table

                                                     Odds     95% CI
Predictor          Coef   SE Coef       Z      P   Ratio   Lower  Upp
Constant       -5.30904  0.475190  -11.17  0.000
Alcohol Use     2.98601  0.464671    6.43  0.000   19.81    7.97  49.
Cigarette Use   2.84789  0.163839   17.38  0.000   17.25   12.51  23.


Log-Likelihood = -1127.920
Test that all slopes are zero: G = 843.453, DF = 2, P-Value = 0.000


Goodness-of-Fit Tests

Method           Chi-Square  DF      P
Pearson            0.401101   1  0.527
Deviance           0.373986   1  0.541
Hosmer-Lemeshow    0.312174   1  0.576
```

Quite a bit of information is displayed in the Session Window.

The values we need are α and β_1 and β_2. They are found under the Coef heading.
$\alpha = -5.30904$ and $\beta1 = 2.98601$ and $\beta_2 = 2.84789$

So, the equation for estimated probability $\hat{p}$ of using marijuana is

$$\hat{p} = \frac{e^{-5.31 + 2.99x_1 + 2.85x_2}}{1 + e^{-5.31 + 2.99x_1 + 2.85x_2}}$$

Chapter 12 Exercise 12.57 Death Penalty and Race
Logistic Regression

Enter the data into the worksheet as indicated.

Worksheet 4 ***

↓	C1	C2	C3	C4
	d	v	y	count
1	1	1	1	53
2	1	1	0	414
3	1	0	1	0
4	1	0	0	16
5	0	1	1	11
6	0	1	0	37
7	0	0	1	4
8	0	0	0	139
9				
10				

Choose: **Stat > Regression > Binary Logistic Regression**

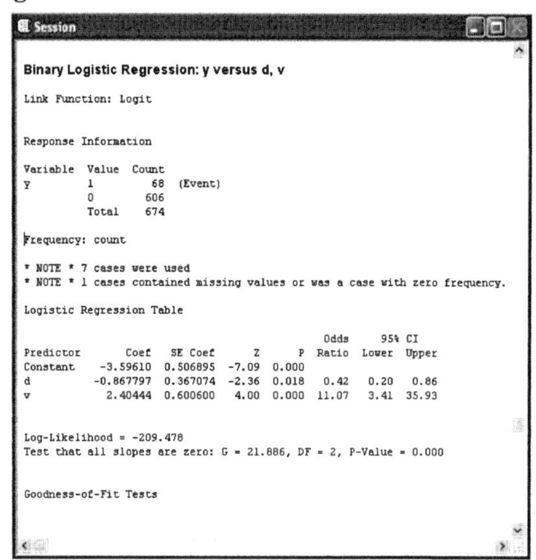
Session

Binary Logistic Regression: y versus d, v

```
Link Function: Logit


Response Information

Variable  Value  Count
y         1         68  (Event)
          0        606
          Total    674

Frequency: count

* NOTE * 7 cases were used
* NOTE * 1 cases contained missing values or was a case with zero frequency.

Logistic Regression Table

                                                  Odds     95% CI
Predictor       Coef   SE Coef      Z      P  Ratio  Lower  Upper
Constant    -3.59610  0.506895  -7.09  0.000
d          -0.867797  0.367074  -2.36  0.018   0.42   0.20   0.86
v            2.40444  0.600600   4.00  0.000  11.07   3.41  35.93


Log-Likelihood = -209.478
Test that all slopes are zero: G = 21.886, DF = 2, P-Value = 0.000


Goodness-of-Fit Tests
```

Enter: Response: **y**
Model: **d v**
Click: **OK**

The results are displayed in the Session Window.

The values we need are α and β_1 and β_2. They are found under the Coef heading.
$\alpha = -3.59610$ and $\beta_1 = -0.867797$ and $\beta_2 = 2.40444$

So, the equation for estimated probability $\hat{p}$ of death penalty verdict is

$$\hat{p} = \frac{e^{-3.596 - 0.868d + 2.404v}}{1 + e^{-3.596 - 0.868d + 2.404v}}$$

Chapter 13 Example 3: Customers' Telephone Holding Time ANOVA

We first must enter the data into the spreadsheet. The data is found in Example 2 of the previous section.

Worksheet 1 ***

↓	C1	C2	C3	C4	C5
	Advertisement	Muzak	Classical		
1	5	0	13		
2	1	1	9		
3	11	4	8		
4	2	6	15		
5	8	3	7		
6					
7					

If we do a dotplot on this data some interesting things can be shown.

Choose: Graph > Dotplot> Multiple y's Simple
Enter the variables as shown:

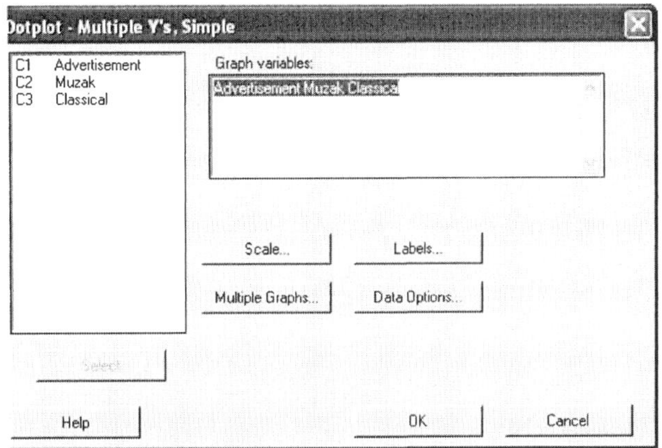

Notice where the plots overlap and where they do not.

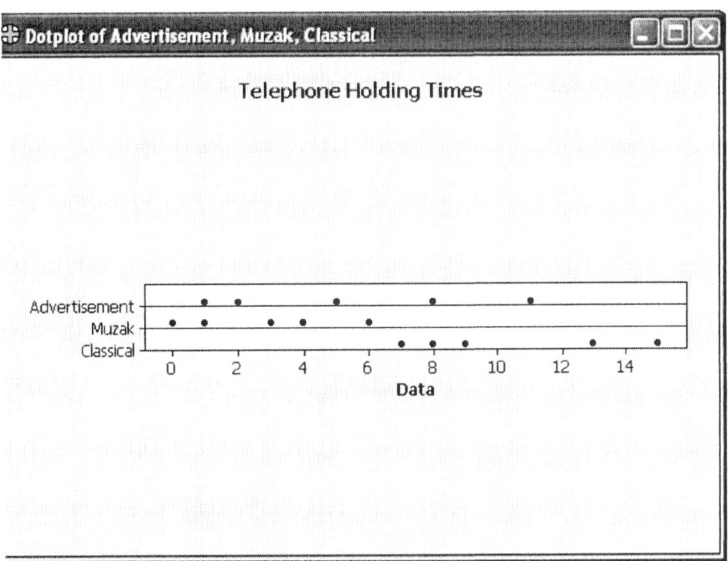

To do the ANOVA analysis

Choose: **Stat > ANOVA >Oneway [Unstacked]**

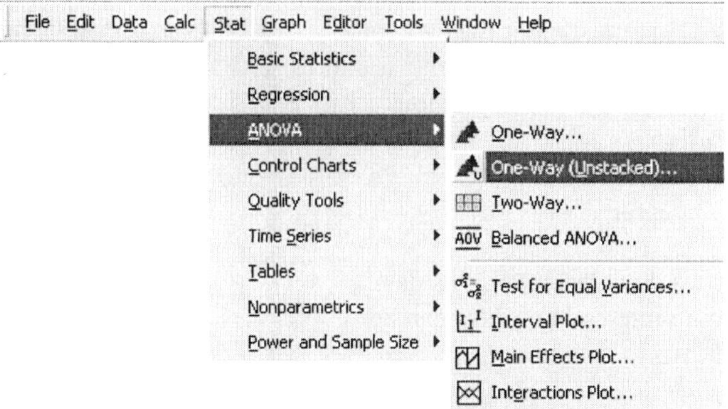

Enter the variables:

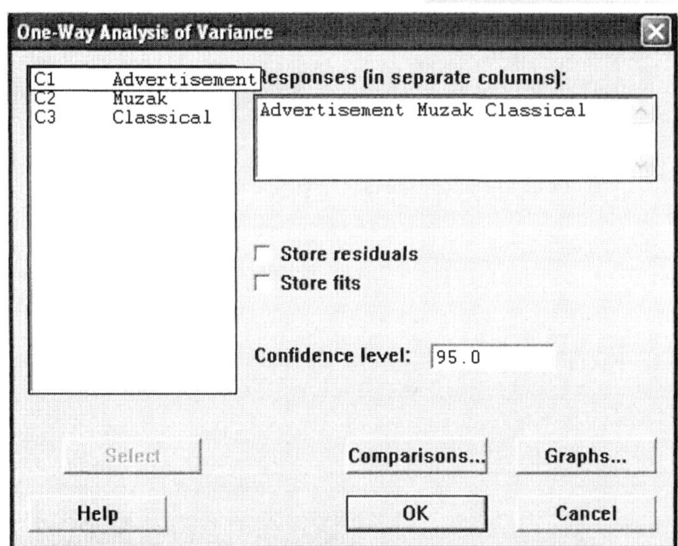

This gives you the following output in the Session window.

Notice the chart lists the degrees of freedom, the F statistic and the p-value.

```
Session

One-way ANOVA: Advertisement, Muzak, Classical

Source  DF     SS    MS     F      P
Factor   2  149.2  74.6  6.43  0.013
Error   12  139.2  11.6
Total   14  288.4

S = 3.406   R-Sq = 51.73%   R-Sq(adj) = 43.69%

                                   Individual 95% CIs For Mean Based on
                                   Pooled StDev
Level          N    Mean  StDev  --+---------+---------+---------+-------
Advertisement  5   5.400  4.159        (--------*-------)
Muzak          5   2.800  2.387  (-------*-------)
Classical      5  10.400  3.435                    (-------*-------)
                                 --+---------+---------+---------+-------
                                 0.0       4.0       8.0      12.0

Pooled StDev = 3.406
```

Chapter 13 Exercise 13.3 What's the best way to learn French?
ANOVA analysis

We are given data of quiz scores for three groups that are students studying French:
Group I: never studied a foreign language but have good English skills
Group II: never studied a foreign language but have poor English skills
Group III: Studied at least one other foreign language.

We enter the data as follows:

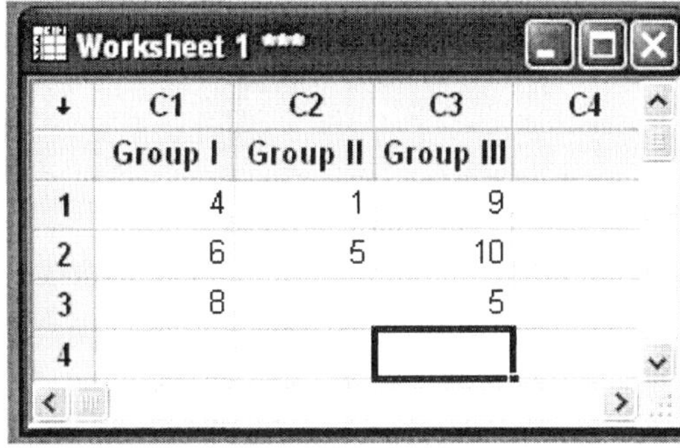

Worksheet 1 ***

↓	C1	C2	C3	C4
	Group I	Group II	Group III	
1	4	1	9	
2	6	5	10	
3	8		5	
4				

Step 1: Assumptions: independent random samples, normal population distributions with equal standard deviations.

Step 2: Hypothesis: H_0: $\mu_1 = \mu_2 = \mu_3$
H_a: at least 2 of the population means are unequal

We then give the following commands to do the ANOVA analysis to find F and p:
Choose: **Stat > ANOVA >Oneway [Unstacked]**

Enter the variables as shown:

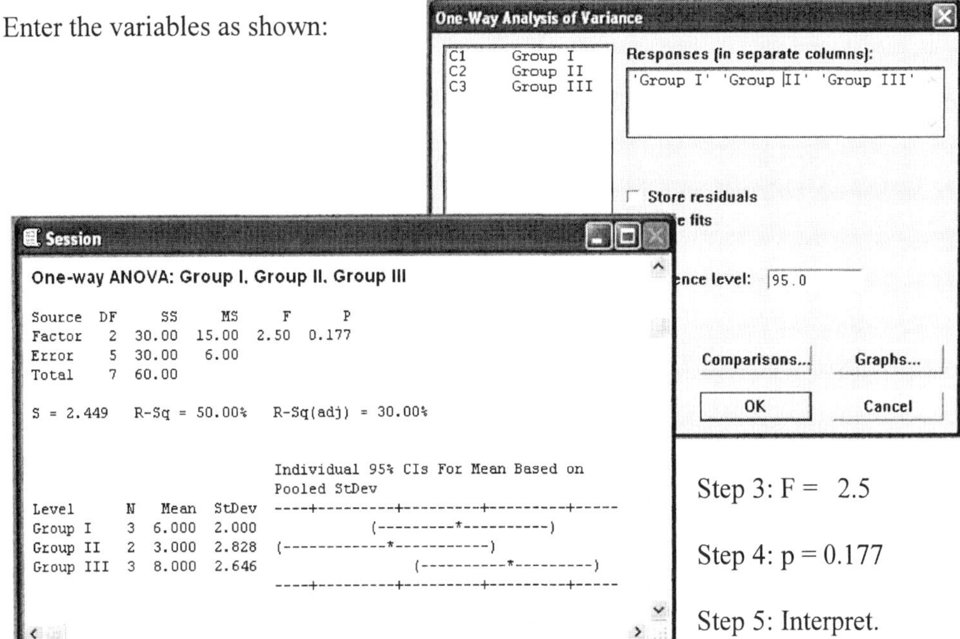

One-Way Analysis of Variance
C1 Group I
C2 Group II
C3 Group III
Responses (in separate columns):
'Group I' 'Group II' 'Group III'
Store residuals
e fits
ence level: 95.0
Comparisons...
Graphs...
OK
Cancel
Session
One-way ANOVA: Group I, Group II, Group III
Source DF SS MS F P
Factor 2 30.00 15.00 2.50 0.177
Error 5 30.00 6.00
Total 7 60.00
S = 2.449 R-Sq = 50.00% R-Sq(adj) = 30.00%
Individual 95% CIs For Mean Based on Pooled StDev
Level N Mean StDev
Group I 3 6.000 2.000
Group II 2 3.000 2.828
Group III 3 8.000 2.646

Step 3: F = 2.5

Step 4: p = 0.177

Step 5: Interpret.

Chapter 13 Exercise 13.15 Telephone Holding Time Comparisons
The ANOVA F Test Tukey method

We first must enter the data into the spreadsheet. The data is found in Example 2.

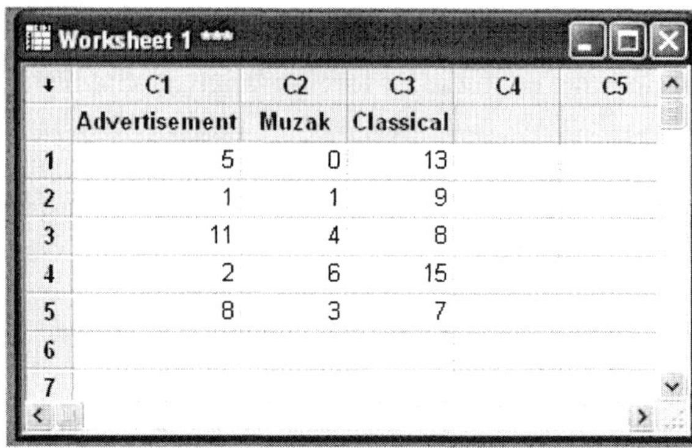

↓	C1	C2	C3	C4	C5
	Advertisement	Muzak	Classical		
1	5	0	13		
2	1	1	9		
3	11	4	8		
4	2	6	15		
5	8	3	7		
6					
7					

To do the ANOVA analysis

Choose: **Stat > ANOVA >Oneway [Unstacked]**

Enter the variables:

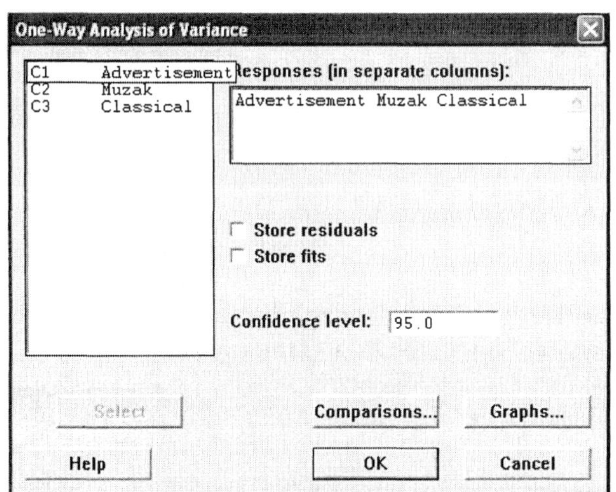

Select: **Comparisons**

Select: **Tukey's, family error rate**

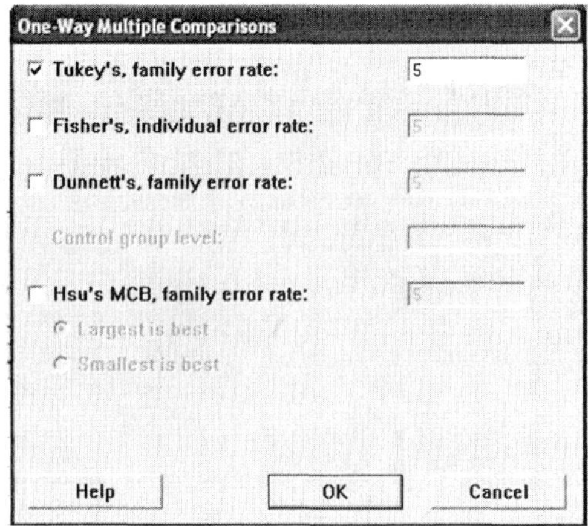

In the session window you will see the results.

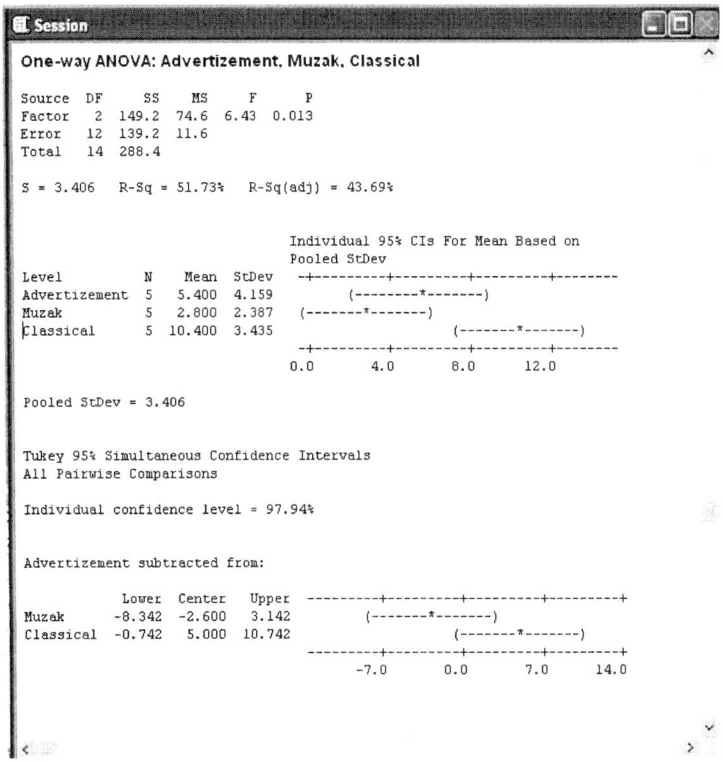

Chapter 13 Exercise 21 French ANOVA
Comparing Means using Fisher and Tukey methods

We enter the data as shown:

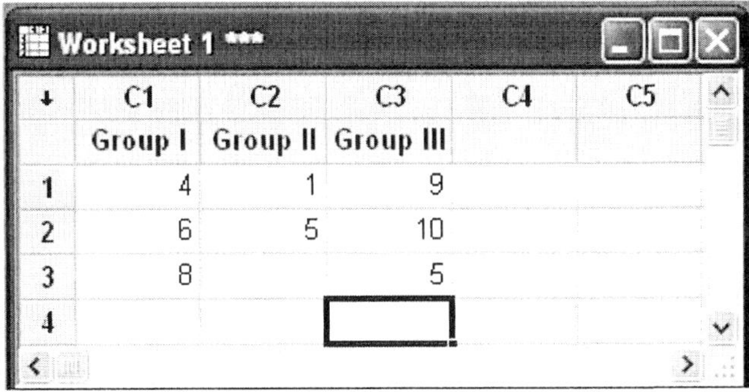

↓	C1	C2	C3	C4	C5
	Group I	Group II	Group III		
1	4	1	9		
2	6	5	10		
3	8		5		
4					

We can do the ANOVA as before, checking the Fisher box under the Comparisons screen
Choose: **Stat > ANOVA >Oneway [Unstacked**

```
Session
One-way ANOVA: Group I, Group II, Group III
Source  DF     SS     MS     F      P
Factor   2  30.00  15.00  2.50  0.177
Error    5  30.00   6.00
Total    7  60.00
S = 2.449   R-Sq = 50.00%   R-Sq(adj) = 30.00%

                            Individual 95% CIs For Mean Based on
                            Pooled StDev
Level      N   Mean  StDev  ----+---------+---------+---------+-----
Group I    3  6.000  2.000              (---------*----------)
Group II   2  3.000  2.828  (------------*-----------)
Group III  3  8.000  2.646                  (----------*---------)
                            ----+---------+---------+---------+-----
                              0.0       3.5       7.0      10.5
Pooled StDev = 2.449
Fisher 95% Individual Confidence Intervals
All Pairwise Comparisons
Simultaneous confidence level = 89.32%
Group I subtracted from:

            Lower  Center  Upper  --------+---------+---------+---------+-
Group II   -8.748  -3.000  2.748     (---------*---------)
Group III  -3.141   2.000  7.141              (-------*--------)
                                  --------+---------+---------+---------+-
                                       -6.0       0.0       6.0      12.0
Group II subtracted from:
Lower  Center  Upper  --------+---------+---------+---------+-
Group III  -0.748   5.000  10.748                (--------*---------)
                                  --------+---------+---------+---------+-
                                       -6.0       0.0       6.0      12.0
```

Now doing it again checking the Tukey box we get the following results:

Make your comparisons between the two results.

```
Session
Source  DF     SS     MS     F      P
Factor   2  30.00  15.00  2.50  0.177
Error    5  30.00   6.00
Total    7  60.00
S = 2.449   R-Sq = 50.00%   R-Sq(adj) = 30.00%
                              Individual 95% CIs For Mean Based on
                            Pooled StDev
Level      N   Mean  StDev  ----+---------+---------+---------+-----
Group I    3  6.000  2.000              (---------*----------)
Group II   2  3.000  2.828  (------------*-----------)
Group III  3  8.000  2.646                  (----------*---------)
                            ----+---------+---------+---------+-----
                              0.0       3.5       7.0      10.5
Pooled StDev = 2.449
Tukey 95% Simultaneous Confidence Intervals
All Pairwise Comparisons
Individual confidence level = 97.74%
Group I subtracted from:

             Lower  Center  Upper  --------+---------+---------+---------+-
Group II   -10.273  -3.000  4.273     (----------*---------)
Group III   -4.505   2.000  8.505             (--------*--------)
                                   --------+---------+---------+---------+-
                                        -7.0       0.0       7.0      14.0
Group II subtracted from:
Lower  Center  Upper  --------+---------+---------+---------+-
Group III  -2.273   5.000  12.273             (---------*----------)
                                   --------+---------+---------+---------+-
                                        -7.0       0.0       7.0      14.0
```

Chapter 13 Example 9: Testing the Main Effects for Corn Yield Two Way ANOVA

We first enter the data as shown:

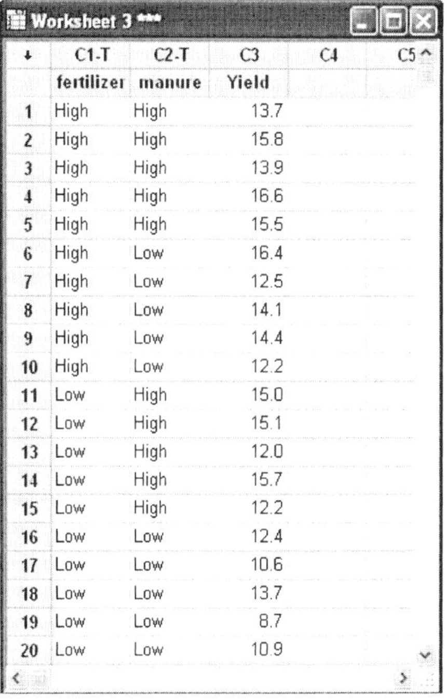

Worksheet 3 ***

↓	C1-T fertilizer	C2-T manure	C3 Yield	C4	C5
1	High	High	13.7		
2	High	High	15.8		
3	High	High	13.9		
4	High	High	16.6		
5	High	High	15.5		
6	High	Low	16.4		
7	High	Low	12.5		
8	High	Low	14.1		
9	High	Low	14.4		
10	High	Low	12.2		
11	Low	High	15.0		
12	Low	High	15.1		
13	Low	High	12.0		
14	Low	High	15.7		
15	Low	High	12.2		
16	Low	Low	12.4		
17	Low	Low	10.6		
18	Low	Low	13.7		
19	Low	Low	8.7		
20	Low	Low	10.9		

The two factors we are considering are fertilizer levels (High, Low) and Manure levels (High, Low). We are given five sample yields for each level, for a total of twenty data points.

To perform a two-way ANOVA , Choose **Stat > ANOVA > Two Way**

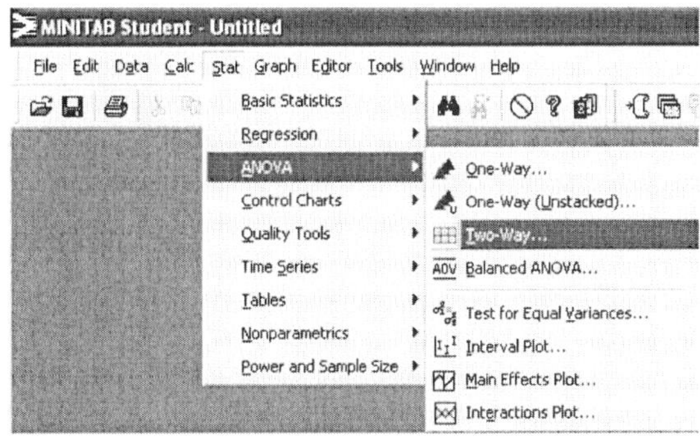

Enter the response, Row Factor, Column Factor as indicated.
Check the Fit Additive model box.

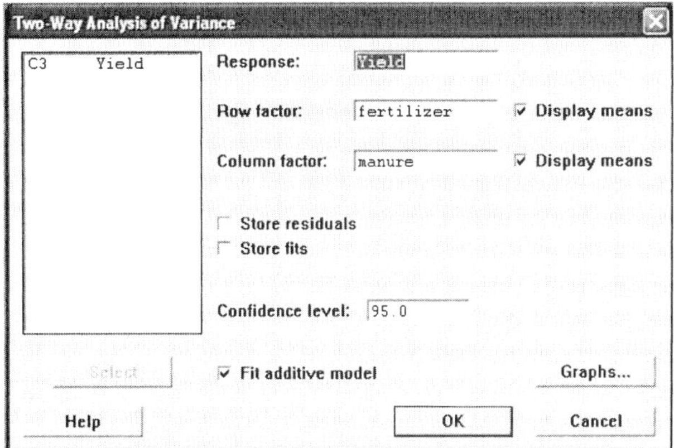

The results of the ANOVA table will be in the session window.

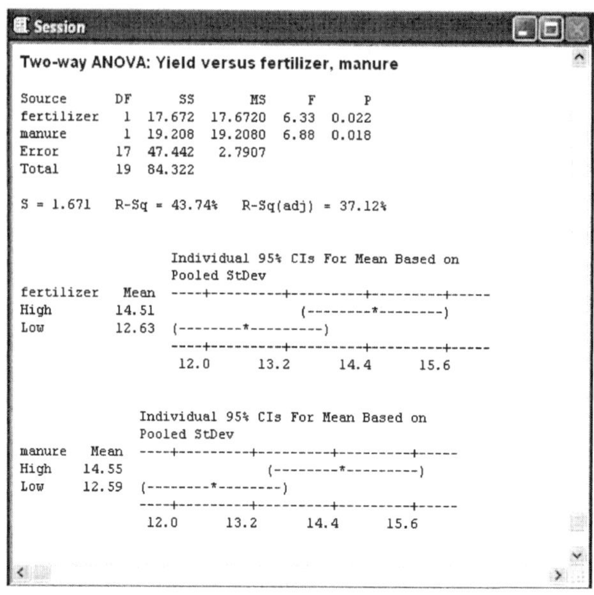

Session

Two-way ANOVA: Yield versus fertilizer, manure

```
Source      DF      SS       MS     F      P
fertilizer   1  17.672  17.6720  6.33  0.022
manure       1  19.208  19.2080  6.88  0.018
Error       17  47.442   2.7907
Total       19  84.322

S = 1.671   R-Sq = 43.74%   R-Sq(adj) = 37.12%


                       Individual 95% CIs For Mean Based on
                       Pooled StDev
fertilizer   Mean  ----+---------+---------+---------+-----
High        14.51                (--------*--------)
Low         12.63  (--------*---------)
                   ----+---------+---------+---------+-----
                     12.0      13.2      14.4      15.6


                  Individual 95% CIs For Mean Based on
                  Pooled StDev
manure   Mean  ----+---------+---------+---------+-----
High    14.55                (--------*---------)
Low     12.59  (--------*--------)
               ----+---------+---------+---------+-----
                 12.0      13.2      14.4      15.6
```

Note that this example is assuming no interaction. To see if this is a valid assumption, redo the ANOVA table to allow for interaction in assessing the effects of fertilizer level and manure level in the mean corn yield. We can do this simply by unchecking **the Fit Additive model box**. This tests the hypothesis H_0: There is no interaction. This results in the following chart:

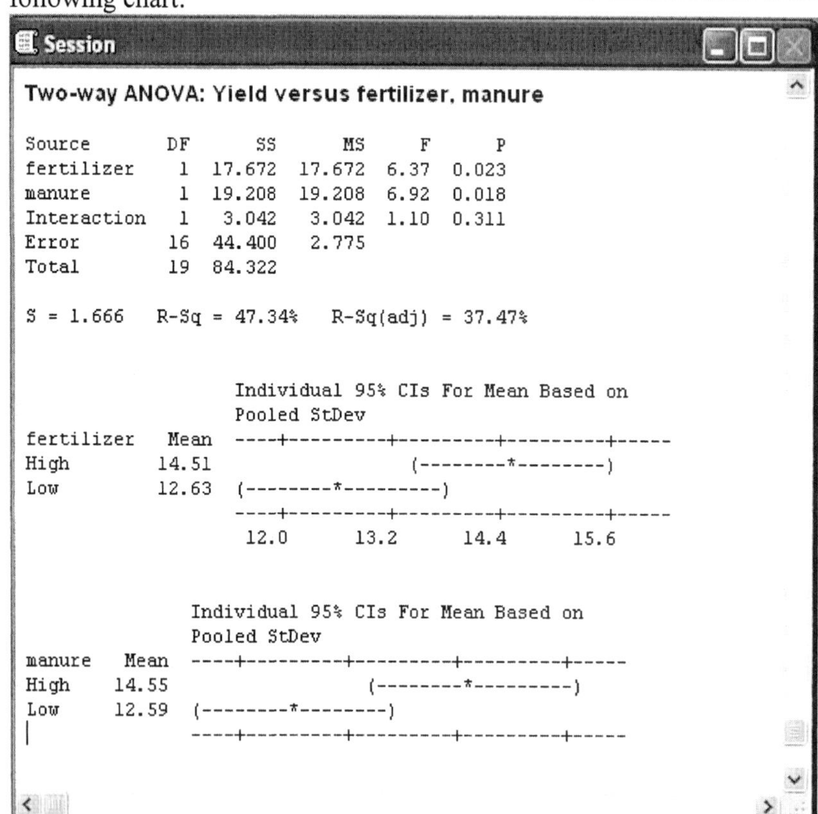

Session

Two-way ANOVA: Yield versus fertilizer, manure

```
Source       DF      SS      MS     F      P
fertilizer    1  17.672  17.672  6.37  0.023
manure        1  19.208  19.208  6.92  0.018
Interaction   1   3.042   3.042  1.10  0.311
Error        16  44.400   2.775
Total        19  84.322

S = 1.666   R-Sq = 47.34%   R-Sq(adj) = 37.47%


                       Individual 95% CIs For Mean Based on
                       Pooled StDev
fertilizer   Mean  ----+---------+---------+---------+-----
High        14.51                (--------*--------)
Low         12.63  (--------*---------)
                   ----+---------+---------+---------+-----
                     12.0      13.2      14.4      15.6


                  Individual 95% CIs For Mean Based on
                  Pooled StDev
manure   Mean  ----+---------+---------+---------+-----
High    14.55                (--------*---------)
Low     12.59  (--------*--------)
               ----+---------+---------+---------+-----
```

Based on the test statistic F = 1.10, and the reported p-value = 0.31, there is not much evidence of interaction so we would not reject H_0 .

In doing a Two Way ANOVA in Minitab Student, you must have the same number of observations in each cell of the contingency table. In this example we had five observations in each of the four cells of the table.

Chapter 13 Example 10 Regression Modeling to Estimate and Compare Mean Corn Yields

To do the regression analysis, we must use numeric data to represent High (1) and Low (0). The worksheet now looks like this:

Worksheet 3 ***

↓	C1-T	C2-T	C3	C4	C5
	fertilizer	manure	Yield	F	M
1	High	High	13.7	1	1
2	High	High	15.8	1	1
3	High	High	13.9	1	1
4	High	High	16.6	1	1
5	High	High	15.5	1	1
6	High	Low	16.4	1	0
7	High	Low	12.5	1	0
8	High	Low	14.1	1	0
9	High	Low	14.4	1	0
10	High	Low	12.2	1	0
11	Low	High	15.0	0	1
12	Low	High	15.1	0	1
13	Low	High	12.0	0	1
14	Low	High	15.7	0	1
15	Low	High	12.2	0	1
16	Low	Low	12.4	0	0
17	Low	Low	10.6	0	0
18	Low	Low	13.7	0	0
19	Low	Low	8.7	0	0
20	Low	Low	10.9	0	0

To perform the regression
Choose: **Stat > Regression > Regression**

Enter Response: **Yield**
Predictors: **F M**
OK

The chart will appear in the session window.

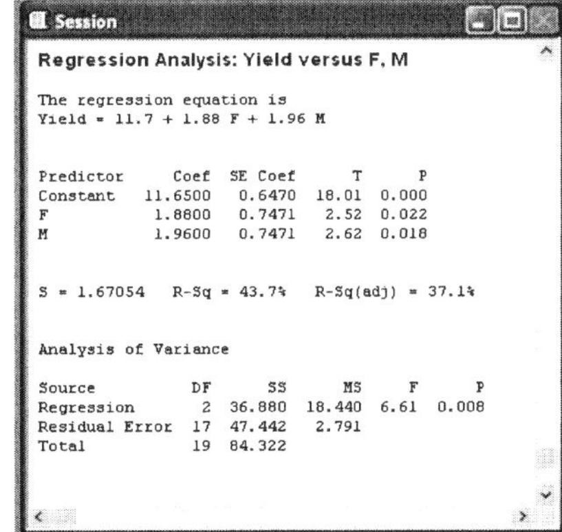

Session

Regression Analysis: Yield versus F, M

```
The regression equation is
Yield = 11.7 + 1.88 F + 1.96 M


Predictor      Coef  SE Coef      T      P
Constant    11.6500   0.6470  18.01  0.000
F            1.8800   0.7471   2.52  0.022
M            1.9600   0.7471   2.62  0.018


S = 1.67054   R-Sq = 43.7%   R-Sq(adj) = 37.1%


Analysis of Variance

Source          DF      SS      MS     F      P
Regression       2  36.880  18.440  6.61  0.008
Residual Error  17  47.442   2.791
Total           19  84.322
```

Chapter 13 Exercise 13.33 Diet and Weight Gain
Two Way ANOVA, with and without interaction.

Enter the data into the worksheet either by hand or opening the file **protein_and_weight_gain** worksheet from the MINITAB folder. A portion is shown here. The file is 60 rows by 3 columns. Note that there are ten observations in each cell, so we have a balanced design.

protein_and_weight_gain.M...

↓	C1-T	C2-T	C3
	source	level	weight_gain
1	beef	high	73
2	beef	high	102
3	beef	high	118
4	beef	high	104
5	beef	high	81
6	beef	high	107
7	beef	high	100
8	beef	high	87
9	beef	high	117
10	beef	high	111

a) Conduct an ANOVA without interaction.
Choose **Stat > ANOVA > Two Way …** and enter the variables as shown

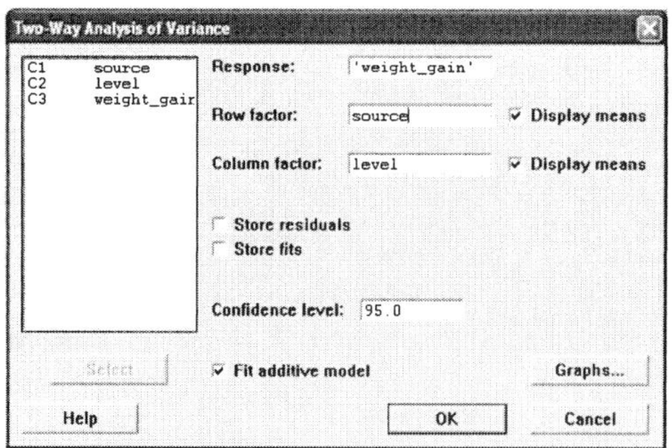

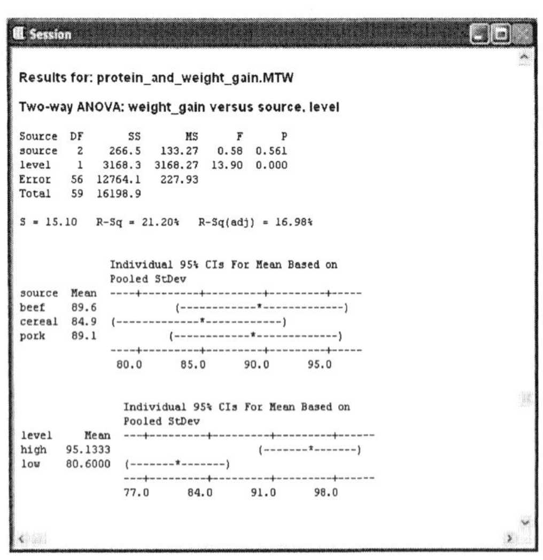

Session

Results for: protein_and_weight_gain.MTW

Two-way ANOVA: weight_gain versus source, level

```
Source  DF       SS       MS      F      P
source   2    266.5   133.27   0.58  0.561
level    1   3168.3  3168.27  13.90  0.000
Error   56  12764.1   227.93
Total   59  16198.9

S = 15.10   R-Sq = 21.20%   R-Sq(adj) = 16.98%


              Individual 95% CIs For Mean Based on
              Pooled StDev
source  Mean  ----+---------+---------+---------+-----
beef    89.6            (------------*------------)
cereal  84.9  (------------*------------)
pork    89.1           (------------*------------)
              ----+---------+---------+---------+-----
               80.0      85.0      90.0      95.0


                Individual 95% CIs For Mean Based on
                Pooled StDev
level     Mean  ---+---------+---------+---------+------
high   95.1333                        (-------*-------)
low    80.6000  (-------*-------)
                ---+---------+---------+---------+------
                 77.0      84.0      91.0      98.0
```

The results will be in the session window.

b) Conduct an ANOVA with interaction.
Choose **Stat > ANOVA > Two Way …** and enter the variables as appropriate.
This time uncheck the box for Fit Additive Model. Again the results will be in the session window.

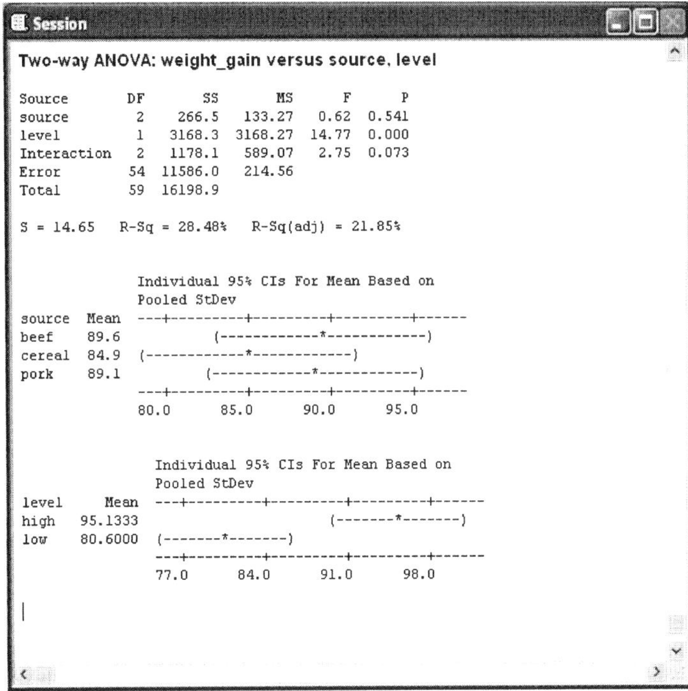

```
Session

Two-way ANOVA: weight_gain versus source, level

Source       DF       SS       MS      F      P
source        2    266.5   133.27   0.62  0.541
level         1   3168.3  3168.27  14.77  0.000
Interaction   2   1178.1   589.07   2.75  0.073
Error        54  11586.0   214.56
Total        59  16198.9

S = 14.65   R-Sq = 28.48%   R-Sq(adj) = 21.85%

              Individual 95% CIs For Mean Based on
              Pooled StDev
source  Mean  ---+---------+---------+---------+------
beef    89.6           (------------*------------)
cereal  84.9  (------------*------------)
pork    89.1          (------------*------------)
              ---+---------+---------+---------+------
              80.0      85.0      90.0      95.0


                 Individual 95% CIs For Mean Based on
                 Pooled StDev
level     Mean   ---+---------+---------+---------+------
high   95.1333                        (-------*-------)
low    80.6000   (-------*-------)
                 ---+---------+---------+---------+------
                 77.0      84.0      91.0      98.0
```

Chapter 14 Example 4: Estimating the Difference between Median Reaction Times Two sample Wilcoxon Test (Mann-Whitney)

You can perform a two-sample Wilcoxon rank sum test of the equality of two population medians, and calculate the corresponding point estimate and confidence interval. MINITAB calls this the Mann-Whitney test.

The hypotheses are: H_0: $\eta_1 = \eta_2$ versus H_1: $\eta_1 \neq \eta_2$, where η is the population median.

Open the worksheet entitled **cell_phones** from the MINITAB folder of the data disk.

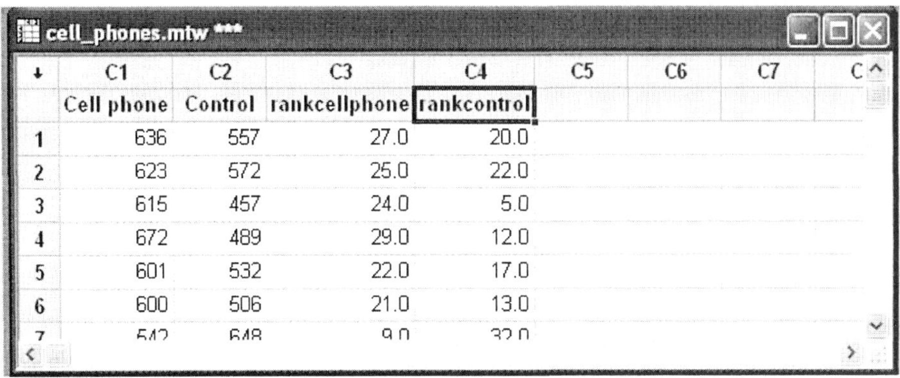

↓	C1	C2	C3	C4	C5	C6	C7
	Cell phone	Control	rankcellphone	rankcontrol			
1	636	557	27.0	20.0			
2	623	572	25.0	22.0			
3	615	457	24.0	5.0			
4	672	489	29.0	12.0			
5	601	532	22.0	17.0			
6	600	506	21.0	13.0			

To perform the test Choose **Stat > Nonparametrics > Mann- Whitney**

Enter the variables as shown:

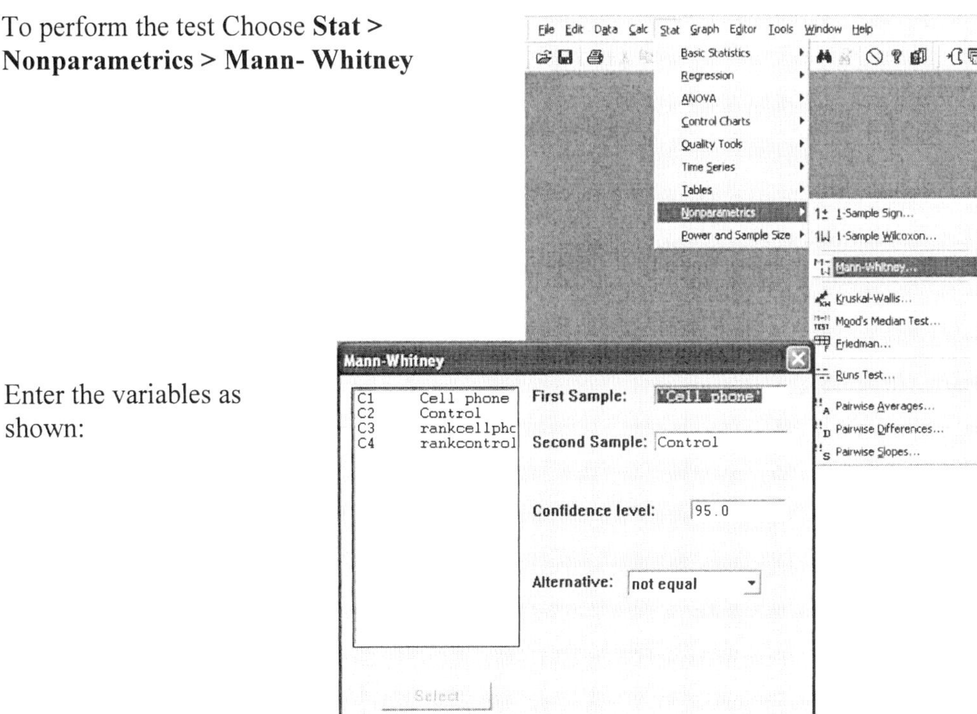

The results are in the session window.

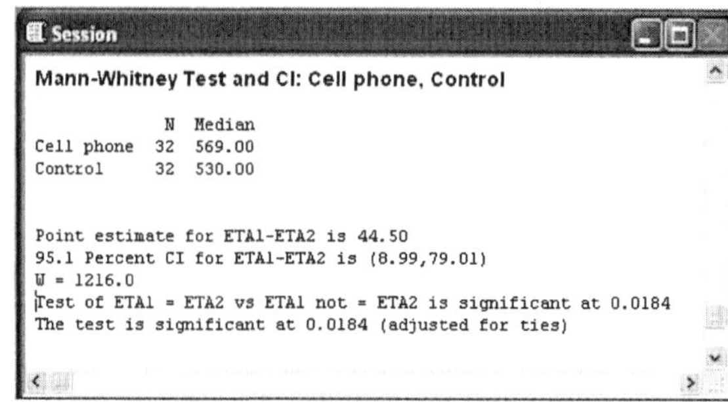

Chapter 14 Exercise 14.5 Estimating Hypnosis Effect
Two sample Wilcoxon Test (Mann-Whitney)

Enter the data into the worksheet:

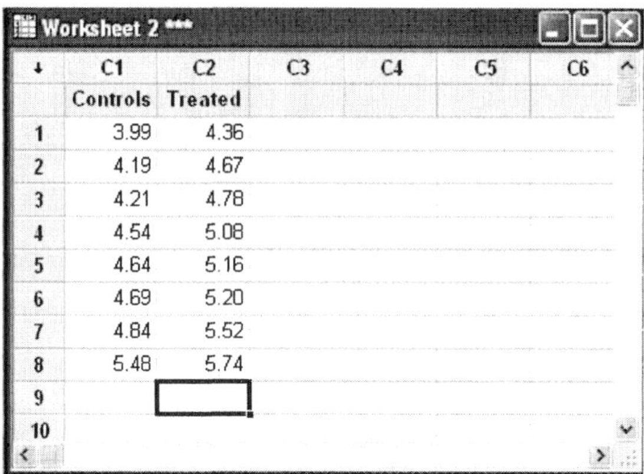

	C1 Controls	C2 Treated
1	3.99	4.36
2	4.19	4.67
3	4.21	4.78
4	4.54	5.08
5	4.64	5.16
6	4.69	5.20
7	4.84	5.52
8	5.48	5.74

To perform the test Choose **Stat > Nonparametrics > Mann- Whitney**

Enter the columns, leaving the default confidence level and alternatives.

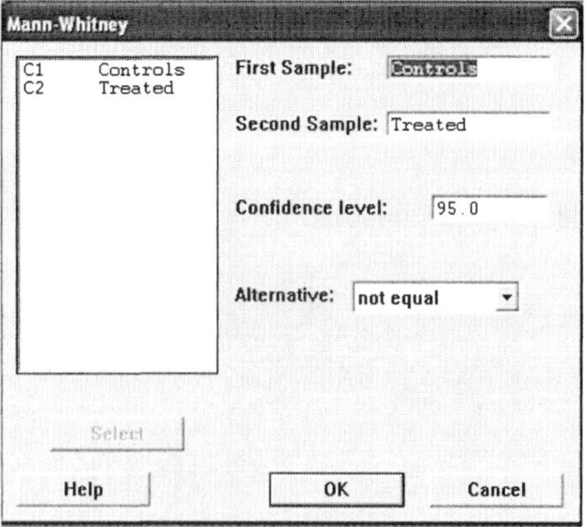

The results appear in the Session window.

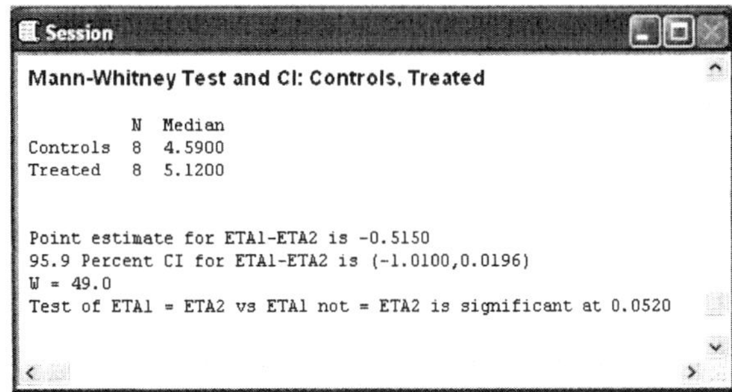

Exercise 14.7 Teenage Anorexia
Two – Way Wilcoxon (Mann-Whitney)

Open the worksheet entitled **anorexia** from the MINITAB folder of the data disk.

To perform the test: Choose **Stat > Nonparametrics > Mann-Whitney**
And choose the columns appropriately:

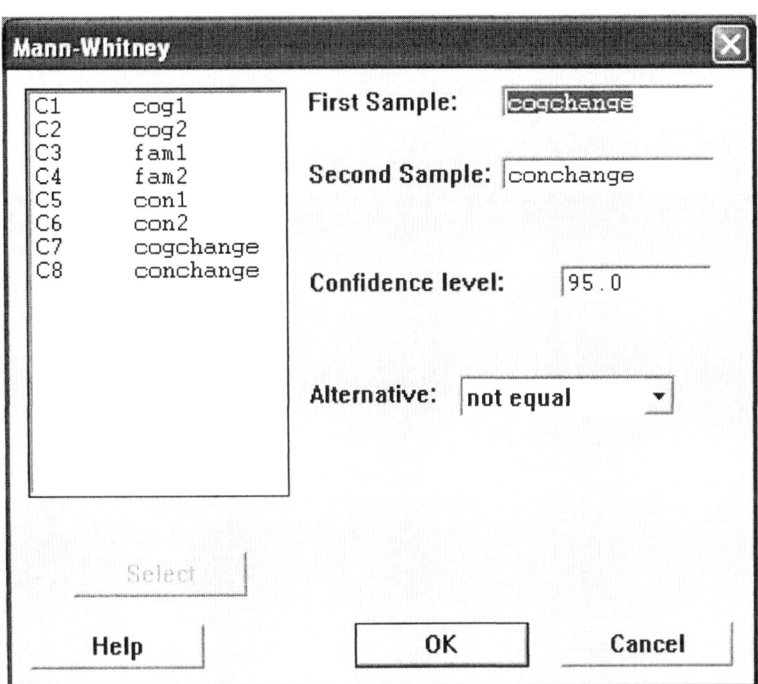

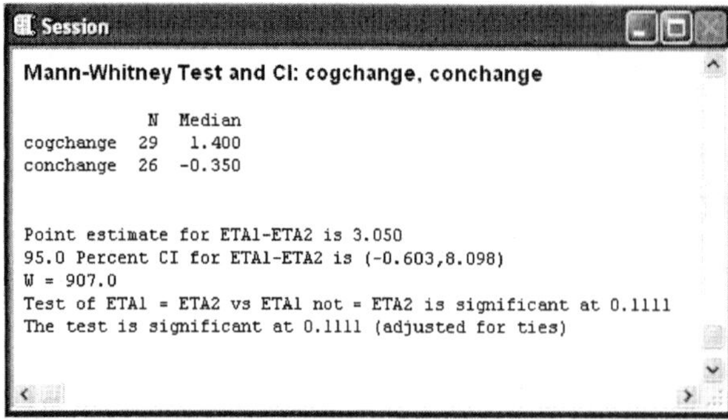

```
Session
Mann-Whitney Test and CI: cogchange, conchange

            N  Median
cogchange  29   1.400
conchange  26  -0.350


Point estimate for ETA1-ETA2 is 3.050
95.0 Percent CI for ETA1-ETA2 is (-0.603,8.098)
W = 907.0
Test of ETA1 = ETA2 vs ETA1 not = ETA2 is significant at 0.1111
The test is significant at 0.1111 (adjusted for ties)
```

Note the point estimate = 3.050 and the confidence interval is (-0.603, 8.098).

The hypotheses are: H_0: $\eta_1 = \eta_2$ versus H_1: $\eta_1 \neq \eta_2$, where η is the population median. The test statistic W = 907 has a p-value of 0.111 when adjusted for ties. What conclusions can you draw?

Chapter 14 Example 5 Does Heavy Dating Affect College GPA?
Kruskal-Wallis Test

The Kruskal-Wallis hypotheses are:
H_0: the population medians are all equal versus H_1: the medians are not all equal

First enter the GPA data in column 1. Then you can rank the data by choosing **Data > Rank** and entering the ranks into Column 2. Enter the dating factors into column 3.

Worksheet 4 ***

↓	C1	C2	C3-T	C4	C5
	GPA	rank	dating		
1	1.75	1.0	Rare		
2	2.00	2.0	Occasional		
3	2.40	3.0	Regular		
4	2.95	4.0	Regular		
5	3.15	5.0	Rare		
6	3.20	6.0	Occasional		
7	3.40	7.0	Regular		
8	3.44	8.0	Occasional		
9	3.50	9.5	Rare		
10	3.50	9.5	Occasional		
11	3.60	11.0	Occasional		
12	3.67	12.0	Regular		
13	3.68	13.0	Rare		
14	3.70	14.0	Regular		
15	3.71	15.0	Occasional		
16	3.80	16.0	Occasional		
17	4.00	17.0	Regular		

To perform the test choose **Stats> Nonparametric> Kruskal-Wallis** and enter the response and Factor as shown:

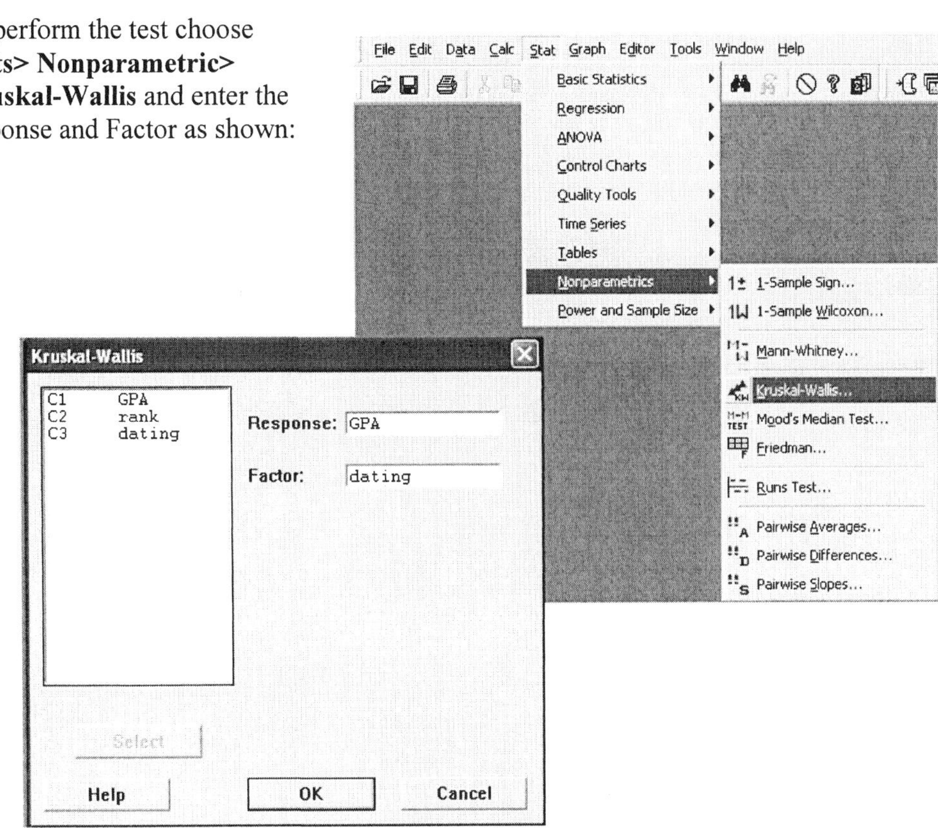

The results are shown in the session window.

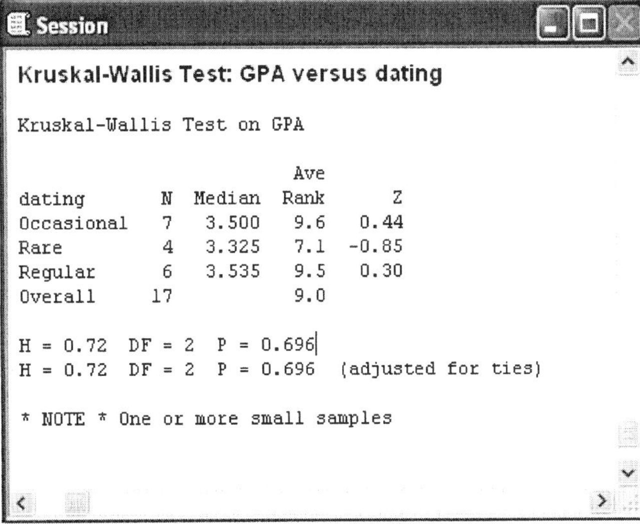

Chapter 14 Example 6 Spend More Time Browsing the Internet or Watching TV? Redone with The Wilcoxin Signed Ranks Test

Open the worksheet for the **georgia_student_survey** from the MINITAB folder of the data disk. This is a large data set, but we are only interested in two columns: "Watching TV" and "Browse Internet". We want to look at the paired differences between these two columns. To create the column of paired differences choose **Calc> Calculator** and enter the dialog boxes as shown. A section of the worksheet is also shown.

To perform the test choose

Stat > Nonparametrics > One-Sample Wilcoxon

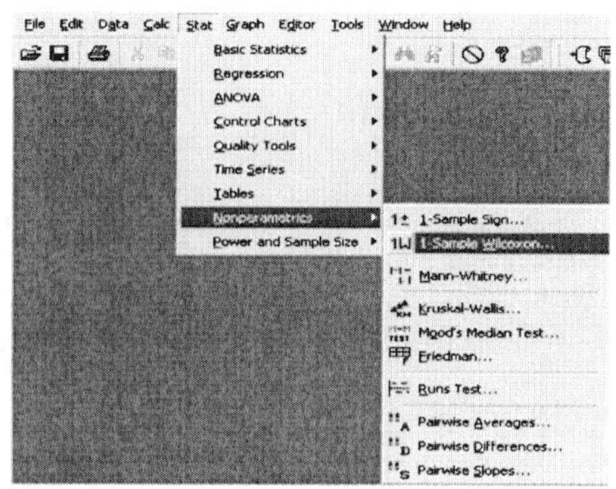

Enter the dialog boxes as shown.

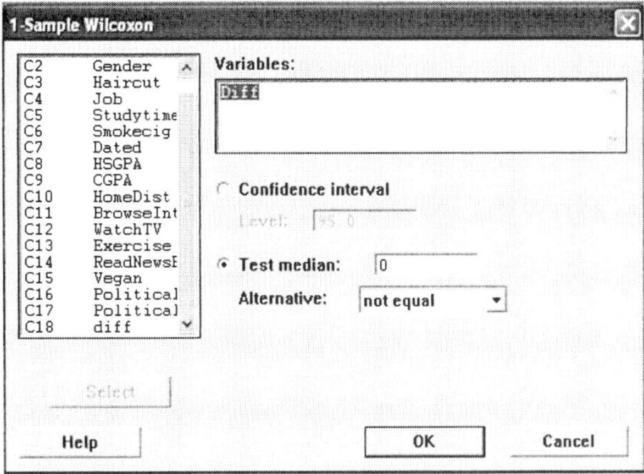

The results as in the session window.

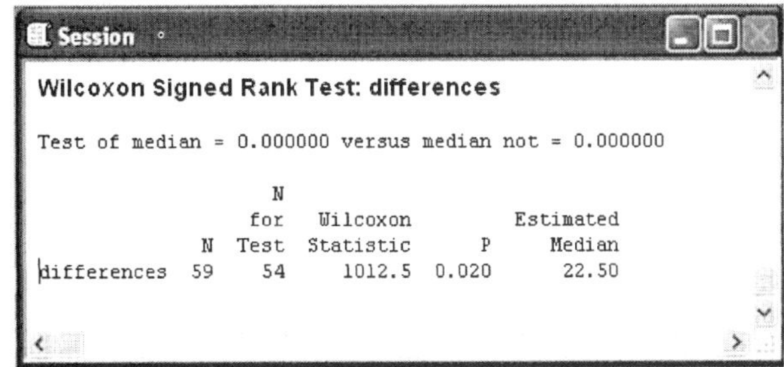

Chapter 14 Exercise 14.8 How Long do You Tolerate Being Put on Hold? Kruskal Wallis Test

The Kruskal-Wallis hypotheses are
H_0: the population medians are all equal versus H_1: the medians are not all equal

First enter the GPA data in column 1.

You can rank the data by choosing **Data > Rank** and entering the ranks into Column 2. Enter the message types into column 3.

Worksheet 6 ***

↓	C1	C2	C3-T	C4
	Holding Time	Rank	message Type	
1	0	1.0	Muzak	
2	1	2.5	Muzak	
3	4	6.0	Muzak	
4	6	8.0	Muzak	
5	3	5.0	Muzak	
6	5	7.0	Advertisement	
7	1	2.5	Advertisement	
8	11	13.0	Advertisement	
9	2	4.0	Advertisement	
10	8	10.5	Advertisement	
11	13	14.0	Classical	
12	9	12.0	Classical	
13	8	10.5	Classical	
14	15	15.0	Classical	
15	7	9.0	Classical	
16				

To perform the test choose

Stats> Nonparametric> Kruskal-Wallis

and enter the response and Factor as shown

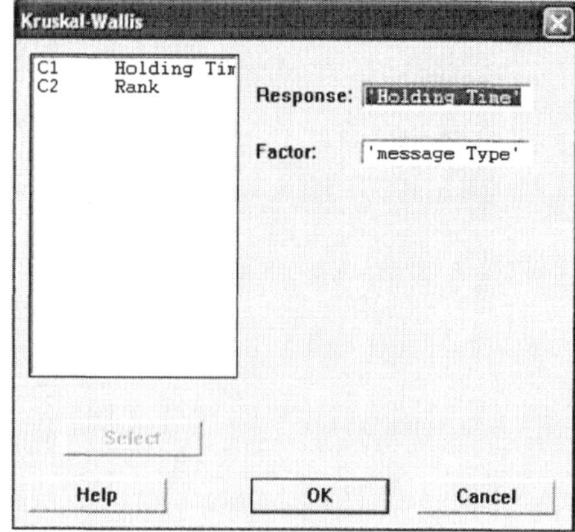

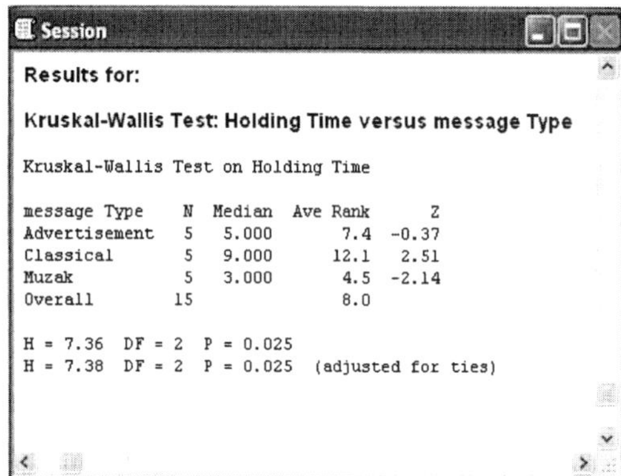

Session

Results for:

Kruskal-Wallis Test: Holding Time versus message Type

```
Kruskal-Wallis Test on Holding Time

message Type    N  Median  Ave Rank      Z
Advertisement   5   5.000       7.4  -0.37
Classical       5   9.000      12.1   2.51
Muzak           5   3.000       4.5  -2.14
Overall        15               8.0

H = 7.36  DF = 2  P = 0.025
H = 7.38  DF = 2  P = 0.025  (adjusted for ties)
```

The sample medians for the three treatments were calculated 5.0,9.0,3.0. The test statistic (H) had a p-value of 0.0125 both unadjusted and adjusted for ties, indicating that the null hypothesis can be rejected at levels higher than 0.025 in favor of the alternative hypothesis of at least one difference among the treatment groups.

Chapter 14 Exercise 14.16 More on cell phones
Wilcoxon Signed Rank Test

Using the data in Example 12 from Chapter 9, create the worksheet. A small segment of the worksheet is shown here:

REACTION_TIME_CELL_PHONE_USE.MTW ***

↓	C1	C2	C3	C4	C5
	Student	No	Yes	Difference	
1	1	604	636	32	
2	2	556	623	67	
3	3	540	615	75	
4	4	522	672	150	
5	5	459	601	142	
6	6	544	600	56	
7	7	513	542	29	

To perform the test on this data choose

Stat> Nonparametrics > 1 - Sample Wilcoxon… and enter the dialog boxes as shown:

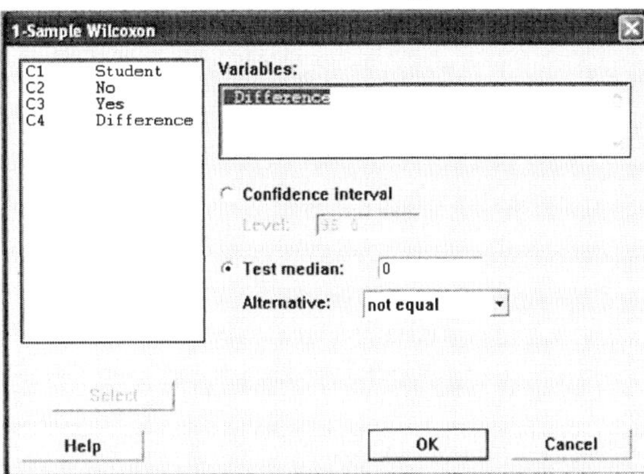

The results are shown in the session window.

Note the hypothesis in the box. It is testing whether the population median of difference scores is 0.

The Wilcoxon test statistic is the sum of ranks for the positive differences. Note the P-Value and the Estimated Median.

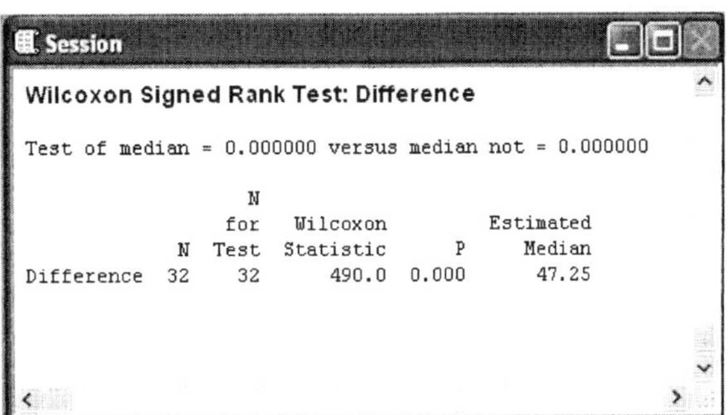

	N	N for Test	Wilcoxon Statistic	P	Estimated Median
Difference	32	32	490.0	0.000	47.25

TI MANUAL

KIRK ANDERSON
Grand Valley State University

STATISTICS

THE ART AND SCIENCE OF LEARNING FROM DATA

Agresti • Franklin

Table of Contents

Chapter One: Introduction to the TI-83 Plus

The chapters of this manual correspond to the Agresti/Franklin text, all except for this one. This chapter covers some basic information to get you started using your calculator. Subsequent chapters will use examples and exercises from the Agresti/Franklin text to show various statistical methods on the calculator.

What calculator do I need?
Note that many of the graphing calculators manufactured by Texas Instruments are quite similar. For our purposes here, the TI-83, TI-83 Plus, TI-83 Plus Silver Edition, TI-84 Plus, and TI-84 Plus Silver Edition are all equivalent calculators. The keypads (and therefore keystroke sequences) are identical, with one exception. The TI-83 has no APPS key; only FINANCE. This is switched with the MATRIX key. Owners of the regular TI-83 will have no problems using this manual. "Plus" and "Silver Edition" denote more memory, faster processors and special applications already built-in. The CellSheet app, for example, is useful for statistics, but will not be covered here. To keep things simple, the calculator will be referred to simply as the TI-83 in this manual. The term "TI-83" will be used to denote any and all of the models mentioned above. All screen shots are taken from a TI-83 Plus model.

Other models
Statistics applications are available for other TI models such as the TI-86 and TI-89. The `Infstat` program is quite simple to transfer from one TI-86 to another, if you happen to know someone that has it. The Stats/List Editor flash application for the TI-89 is very impressive; once loaded, the TI-89 will be able to perform more statistical functions than the TI-83 series. Try downloading it from education.ti.com/us/product/apps/statsle.html.

Don't forget the "real" manual
This document is merely a supplement to the Agresti/Franklin text. The manual that came with your calculator should be kept handy as well. If you have misplaced yours, free downloads are available from education.ti.com/us/global/guides.html.

The Cursor
By default, the cursor will appear as a blinking solid square. Typing in this mode will allow you to overwrite any characters to the right of the cursor. You may insert characters without overwriting by pressing INS (2nd DEL). In this mode, the cursor will appear as an underline. Before applying a secondary function, an up-arrow will be visible in the cursor square. Likewise, the letter A will appear in the cursor block during ALPHA mode.

The Keypad
Hopefully you are somewhat comfortable with your calculator already. Primary key functions are printed on the buttons themselves: digits 0-9, arithmetic operators, log and trig functions, etc. Some generally useful buttons will be discussed here. The caret symbol (^) button is used for raising numbers or expressions to an exponential power. Parentheses (above 8 and 9) are very useful for complicated arithmetic – they help you to keep the order of operations correct.

Commas (above 7) are often needed – to separate arguments for a stat function, for example. The MATH button produces a menu with a few functions of use to the statistics student. CLEAR can be used to exit a menu screen, clear the current line, or the whole screen. The STAT menu contains most of what we will need. Notice the EDIT, CALC, and TESTS sub-menus. The DEL button works like the delete key on a computer keyboard – it deletes characters to the right of the cursor. The arrow keypad is needed to direct our way through menus, navigate graphs, and other uses. STO> can be used to store information in a list, matrix or single-value variable.

Secondary Functions
Above (and to the left) of most keys you will find a secondary function – in yellow on the TI-83 series, blue on the TI-84's. Certainly you have used the 2nd function already – you need to in order to shut the calculator off! Some of the most useful secondary functions for us are 2nd 1-6, to call up the data lists L1-L6, and DISTR (2nd VARS), the menu of probability distributions.

ALPHA Functions
The ALPHA key allows us to type letters – notice that the letters of the alphabet appear above (and to the right) of each key, in teal on the TI-83's and light green on the TI-84 models. The five buttons on the top of the keypad are used for graphs, including plots of data. It is also used to SOLVE equations (see chapter 7).

Home Screen
This is the default screen, where calculations are performed. Pressing QUIT (2nd MODE) or CLEAR from other menus and screens will get you back to the home screen.

Miscellaneous Tips and "Tricks"
Here are a few time- and effort-saving "tricks" that are quite convenient. First of all, students should realize that the arithmetic operator "minus", for subtracting a value or expression from another, is found on the right side of the keypad, above + and below ×. When you need to designate a value or expression as negative, you must use the negative sign key (-), found to the left of ENTER. The two keys are not interchangeable. ENTRY (2nd ENTER) can be used to call up previous expressions when using the home screen. This can save you many keystrokes. ANS (2nd (-)) can be used to call up the most recently calculated value. This not only saves keystrokes – it also minimizes rounding error. Lastly, don't assume that your screen is hard to read due to low battery power. The lightness and darkness of the screen is controlled by 2nd (up arrow) and 2nd (down arrow). Up is darker, down is lighter.

Downloading Data
Mostly, it is assumed you will type in the (small) data sets. For the larger data sets, you may be able to download them to your calculator. This is assuming two things. Firstly, that the data are provided by the authors. Nearly 40 data files are supplied with the textbook. Second, you must have the appropriate cable and software to connect your computer to your calculator. The "TI Connectivity Kit" includes a cable that connects your calculator to your Windows PC via the USB port, and the "TI Connect" software. This software is necessary for the data transfer, regardless of which of the following two methods you choose.

Downloading Data using the CellSheet application

Owners of "Silver Edition" versions of the TI-83 plus or TI-84 plus have a spreadsheet application automatically loaded on their calculators. Press APPS, then choose 0:CelSheet. Downloading data to the CellSheet is easy to do. See EXAMPLE 12 in chapter 3 for details. The CellSheet application will run on any TI-83 plus or TI-84 plus, but must be purchased from Texas Instruments. More information (including the guidebook for the application) can be found at http://education.ti.com/us/product/apps/cellsheet.html.

Downloading Data from Excel files

If you do not have the CellSheet application (or even if you do), you can download data from the provided Excel files. Some editing of the Excel file is necessary first, but this is a relatively painless way to get large data sets into your calculator without having to type them in. See EXAMPLE 1 in chapter 11 of this manual for details.

Chapter Two: Exploring Data with Graphs and Numerical Summaries

Entering Data

Data can be stored in lists. For example, suppose we wish to store the numbers 12, 16, and 35 in the first default list, L1. We could enter the numbers, separated by commas and enclosed in braces, and use STO> to put them in L1.

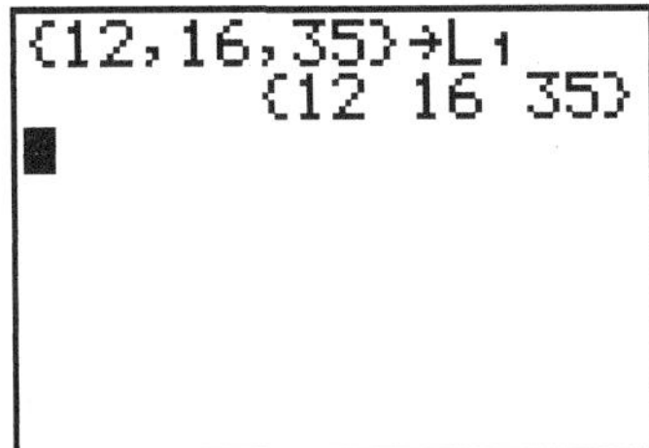

An easier way to enter a list of data is to use the data editor. Pressing STAT EDIT 1:Edit... produces a spreadsheet-like data entry and viewing device. You should see, by default, lists L1-L3. If you scroll to the right, you'll see L4-L6.

If this is not what you see – if any of the lists are missing, for example – you can reset to the default like so: STAT EDIT 5:SetUpEditor. Note that you can scroll down to 5, or just press 5. Press ENTER in the home screen. Don't worry; no data will be lost.

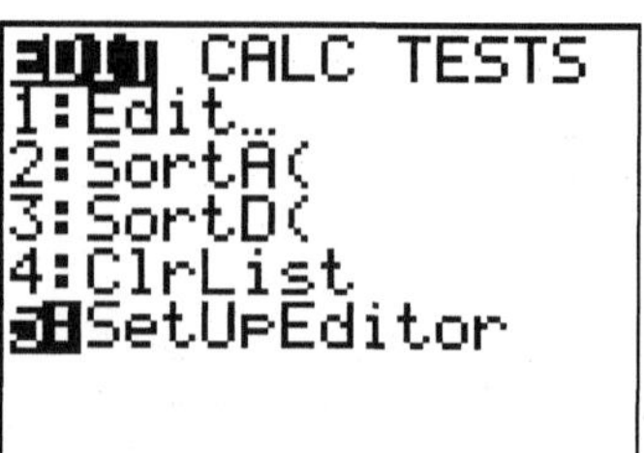

If you need to clear old data from a list, one way is the following: move the cursor up to highlight the list name, press CLEAR, then ENTER. You can clear the data from all lists by using MEM (2nd +), ClrAllLists, then ENTER.

Pie Charts and Bar Graphs

Pie charts are not produced by the TI-83, unless you have the CellSheet application. Bar graphs can be created using the histogram option in STAT PLOT.

EXAMPLE 3: How much electricity comes from renewable energy resources? (pp. 30-32 of Agresti/Franklin)

Let's first get a bar graph of the U.S. percentages, in the order that they appear in table 2.2. We will enter the categories coal, hydropower, etc. as numbers 1-6 in L1, then the percentages in L2.

L1	L2	L3 2
1	51	------
2	6	
3	16	
4	21	
5	3	
6	3	

L2(7) =

Next press STAT PLOT (2nd Y=). You will need to have one of the three plots turned on, and the other two plots turned off. Sometimes it is quickest to turn all three of the plots off by using the PlotsOff function, then turn on the plot you want (usually Plot1).

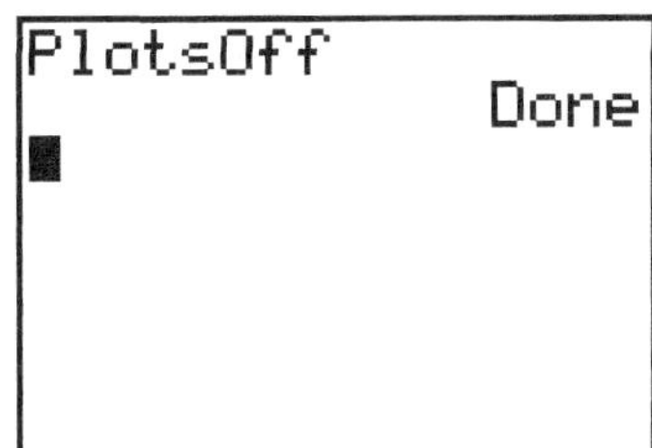

Go back to STAT PLOT. Choose Plot1. Turn it on by choosing On. Change the type to the histogram icon (first row, third column of plot icons). Enter L1 for the Xlist, and L2 after Freq. Realize that we are "tricking" the calculator to produce a bar graph. We are simply plotting the six categories, and specifying how high each bar will be. Admittedly, software such as SPSS is a better bet for this type of graph.

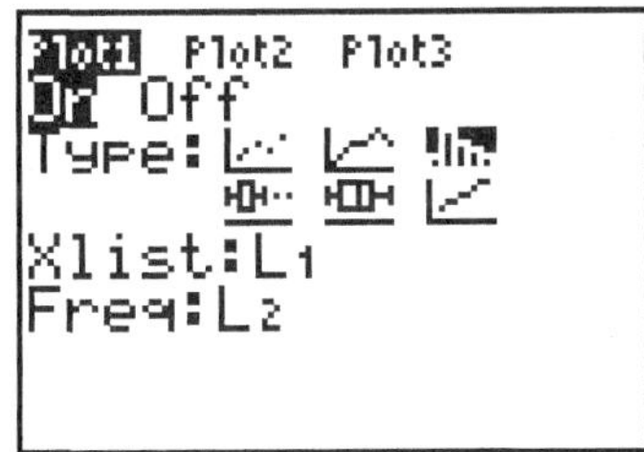

We must specify certain aspects of the graph "window" in order for it to appear as intended. Press WINDOW. Horizontally, we need room for six categories, so enter Xmin=1 and Xmax=7. Xscl (X scale, the interval width) should be set to 1 for this example. Vertically, we need room to see the bottom and top of the graph. Start a bit lower than zero with Ymin=-10 (be sure to use the negative sign (-), not the minus sign –). End up a bit higher than the largest percentage (51%) by entering Ymax=55. Setting the Y scale to 10 just gives tick marks every 10 percent.

Xres is not used for bar graphs. Make sure there are no equations in the Y= screen, since those will be plotted along with the bar graph. Now we are ready to press GRAPH.

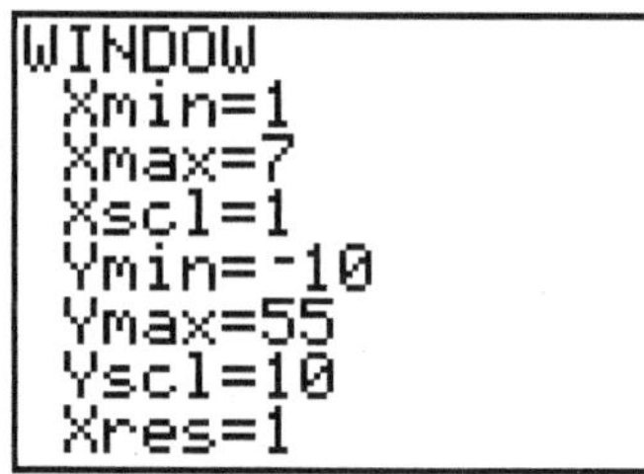

Plot1 Plot2 Plot3
\Y1=
\Y2=
\Y3=
\Y4=
\Y5=
\Y6=
\Y7=

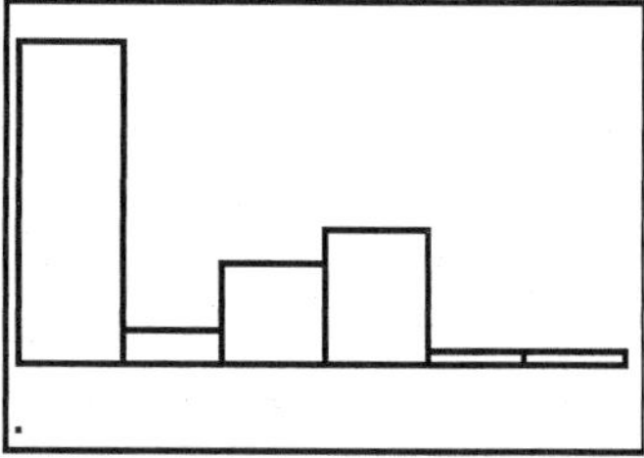

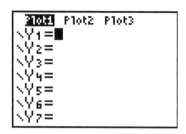

When viewing the graph, pressing TRACE will allow you to get more information. For example, the image below shows that the first bar is category 1, since the histogram "interval" goes from one to, but not including, two. Also, the "frequency" (actually percentage) is given as 51. Scroll over to the right. You will be able to verify all of the frequencies in table 2.2.

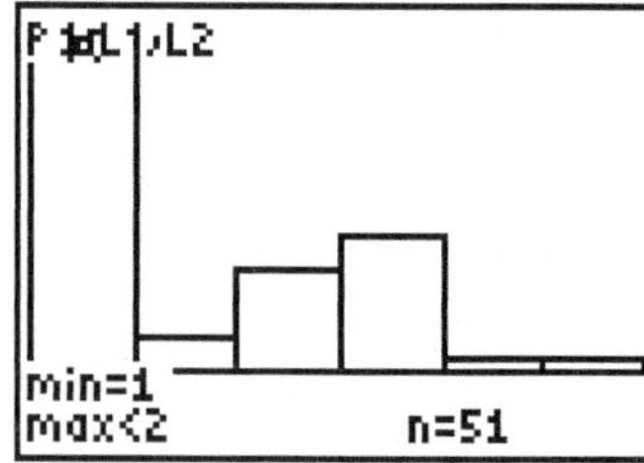

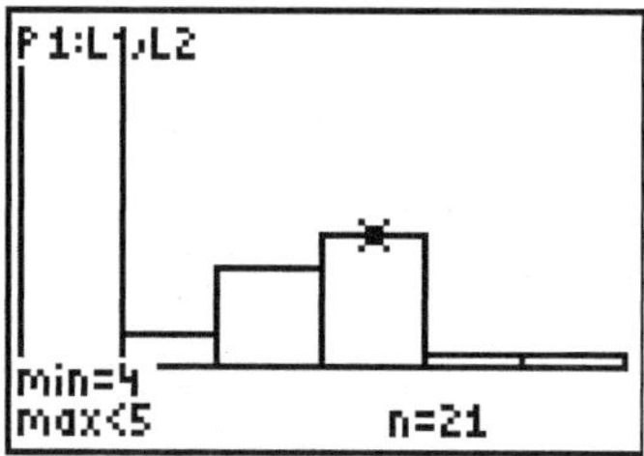

Next, let's plot the same bar graph, this type as a Pareto chart. This simply means that the categories should be ordered from the largest to smallest percentages (tallest to shortest bar). We should leave L1 as it is, since those numbers were there only to construct the bar graph, and sort L2 from highest to lowest. Instead of re-entering the percentages, let's instead use the sortD (sort descending) function. Press STAT. in the EDIT menu, choose sortD. Provide L2 as the argument for the function. Next go back to the data editor to verify the sorted list.

```
EDIT CALC TESTS
1:Edit…
2:SortA(
3:SortD(
4:ClrList
5:SetUpEditor
```

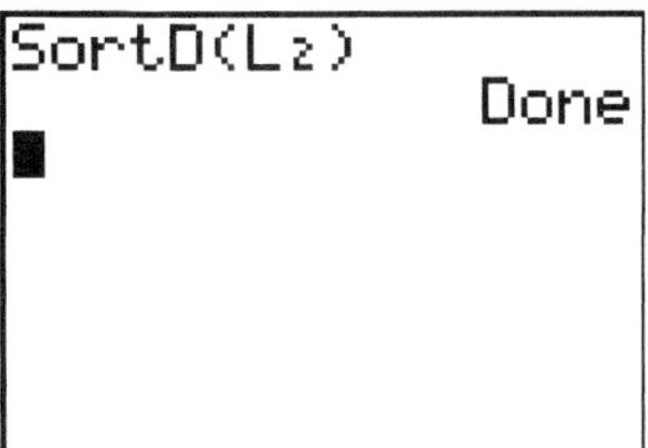


```
L1      L2      L3     2
1       51      ------
2       21
3       16
4       6
5       3
6       3
------  ------
L2(1)=51
```

All options selected in STAT PLOT and WINDOW are still exactly what we want; we simply need to press GRAPH to see our new graph. Again, TRACE allows us to see more information.

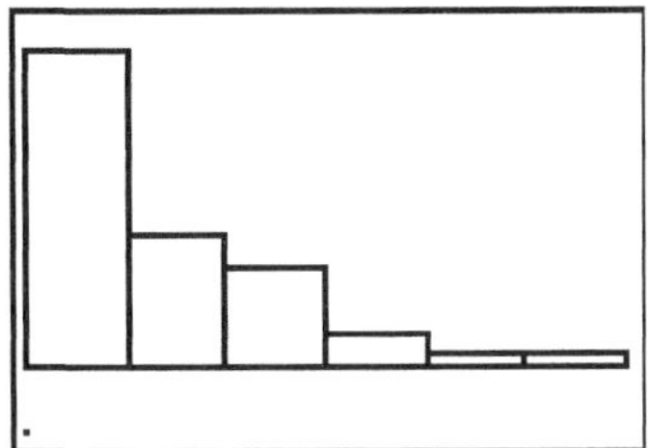

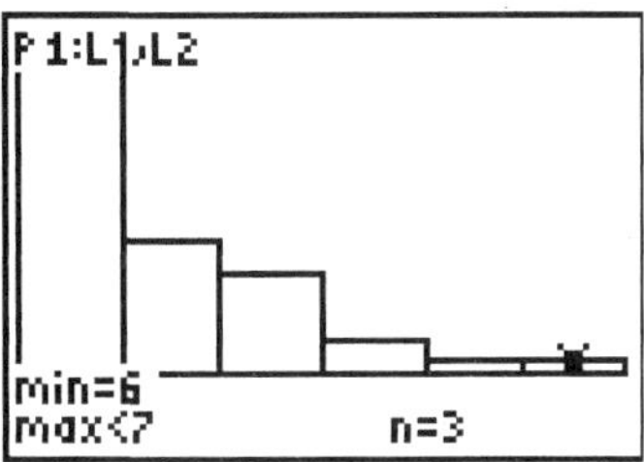

Dotplots and Stem-and-Leaf Plots

We now move from categorical to quantitative data. Neither dotplots or stem-and-leaf plots are available with the TI-83, but histograms are.

Histograms

EXAMPLE 7: Exploring the health value of cereals (pp. 38-39 of Agresti/Franklin)

To create a histogram of the cereal sodium values, enter (or download) the 20 numbers into L1.

```
L1      L2      L3     1
0       ------  ------
210
260
125
220
290
L1(7)=210
```

```
L1      L2      L3     1
170
70
230
200
290
180
L1(21) =
```

Back in STAT PLOT, make sure that Plot1 is turned on and the other two plots are off. Change Freq to 1, which is the default. This simply means that each of the 20 sodium values in L1 occurs once.

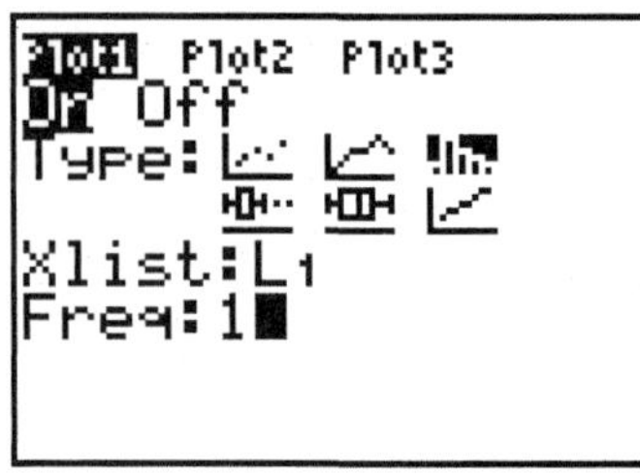

At this point, we are ready to GRAPH. But most likely, nothing will show up on screen. Get used to this – the window dimensions from the last graph rarely work very well for the next graph, unless you're plotting the same data values. We had to manually specify the window dimensions for the bar graph, but here's some good news: when creating most graphs using STAT PLOT, we can use the ZoomStat function to automatically "fix" the window dimensions to fit our data. Press ZOOM, then choose 9:ZoomStat.

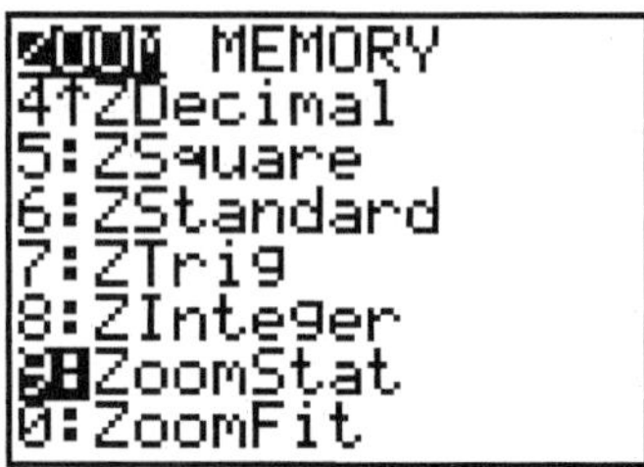

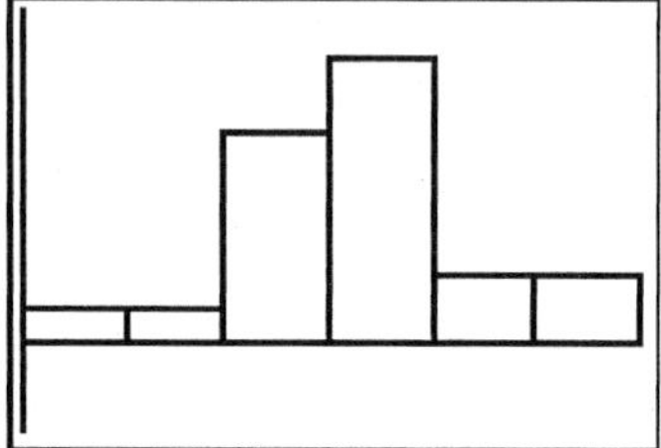

Viola! Now we can view our histogram. Notice that it doesn't look exactly like the histogram on p. 39 of Agresti/Franklin. This is to be expected. For any data set, an infinite amount of slightly different histograms can be created. The choice of interval width and number of intervals is endless. Too few (or too many) intervals can result in a histogram that doesn't illustrate the data very well. The good news is that the default histogram created via ZoomStat will always be adequate.

However, it is possible to adjust the window dimensions to match figure 2.6. The entries below will do the trick. Once you've entered the window dimensions below, press GRAPH (not ZoomStat). Notice that the intervals are 40 mg wide, and they range from 0 to 320 mg. The maximum frequency is 7, and some room is left for the bottom of the graph.

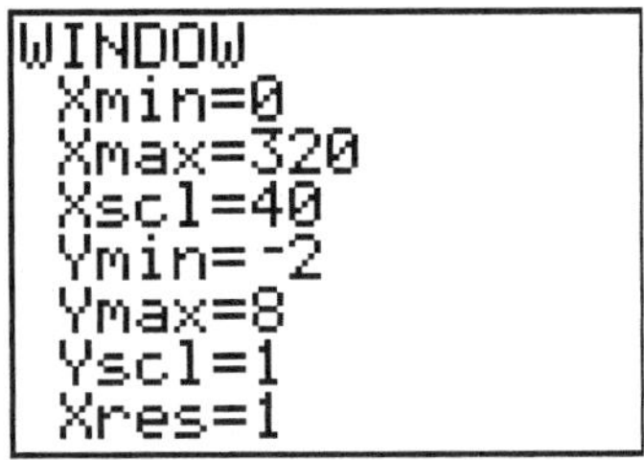

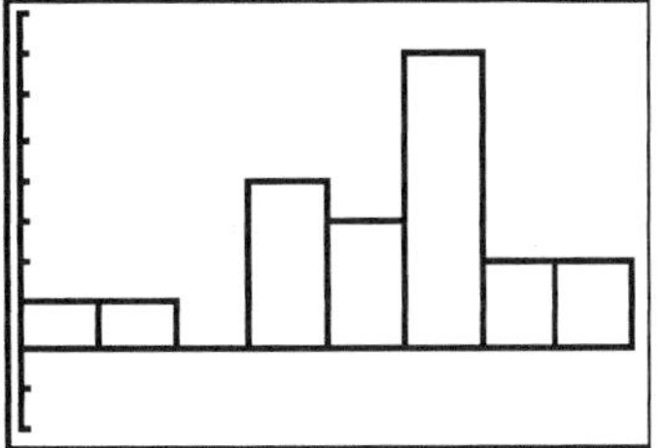

The TRACE function is used below to show that no sodium values fell between 80 and 120 mg.

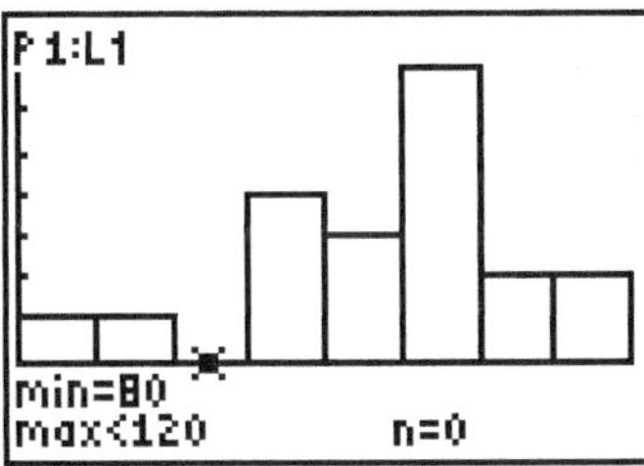

Exercise 2.114: Controlling asthma (p. 85 of Agresti/Franklin)
Enter the formoterol values into L1, and the salbutamol values into L2.

L1	L2	L3 2
320	290	
330	365	
250	210	
380	350	
340	260	
220	90	

L2(14) =

We can look at histograms for each medication separately. Turn all plots off except for Plot1. Remember to use ZoomStat. Verify that there are no PEF values between 258 and 296 for the F group.

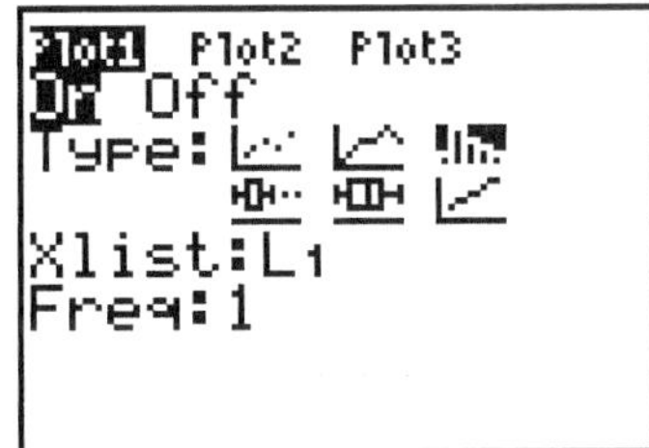

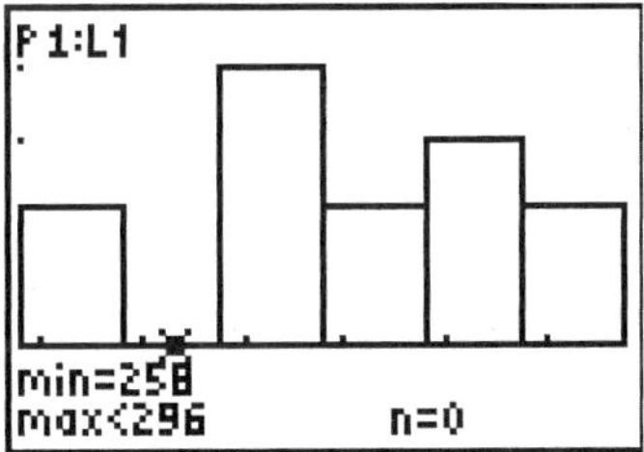

Next, change the Xlist to L2 to plot the S group. Press GRAPH. Notice what happens here. If we don't use ZoomStat again, the S group is plotted on the same scale (which is good), but doesn't show all the data (which is not good). Where's the last PEF value of 90? It's there, but at the moment, it's off-screen.

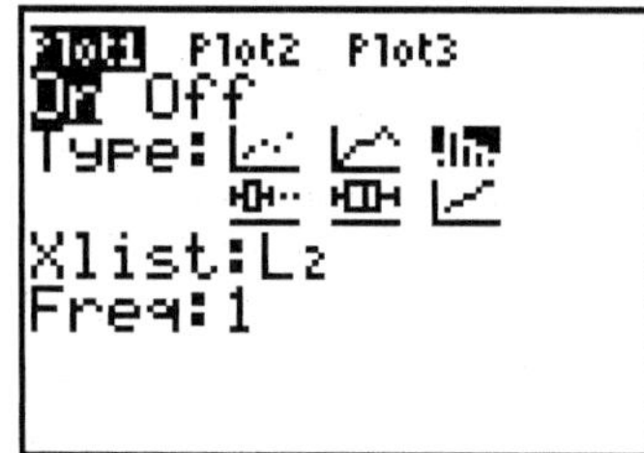

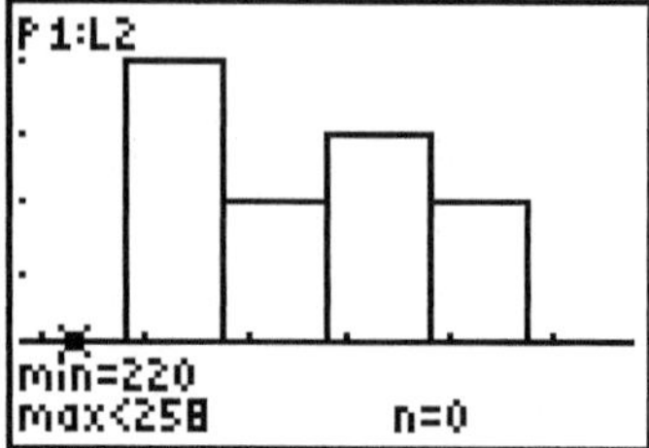

Maybe a better idea is to use ZoomStat on the S group, since it has a wider range. After using ZoomStat on the S group (L2), change the Xlist to L1 (F group) and press GRAPH. Then we can view both groups of PEF values on the same scale.

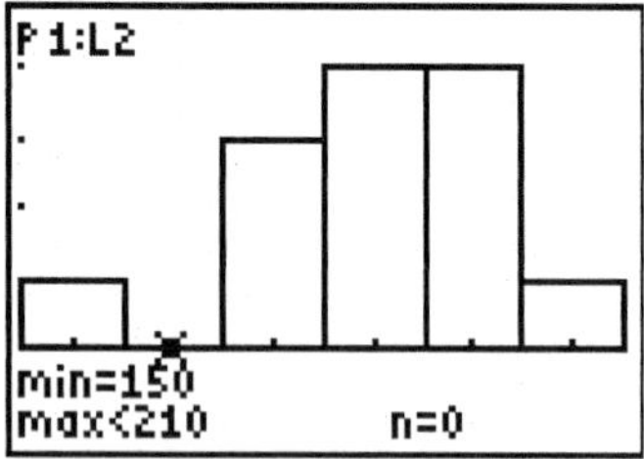

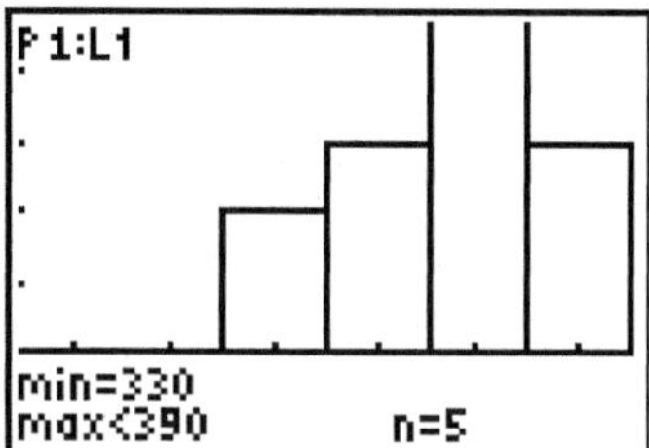

Now the window dimensions are such that the tallest bar is too tall for the screen. This could be fixed, but may not be worth the effort – after all, with paired data such as this, we will be interested in the *differences* between the two measurements for each individual in the study. See section 9.3 for many examples involving paired data. For now, let's anticipate this. Calculate the (A – F) difference for each child, and create a histogram of the differences. Instead of calculating all 13 differences manually, we can ask the calculator to do it for us. In the screen below, notice that the cursor is poised to enter the first row of data for L3. Move up to the top of the list, so that the entire list can be defined. Since we want A – F, simply type L1 – L2.

L1	L2	L3 3
310	270	
385	370	
400	310	
310	260	
410	380	
370	300	
410	390	

L3(1)=

L1	L2	L3 3
310	270	------
385	370	
400	310	
310	260	
410	380	
370	300	
410	390	

L3 =

L1	L2	L3
310	270	------
385	370	
400	310	
310	260	
410	380	
370	300	
410	390	

L3 =L1−L2

L1	L2	L3
310	270	40
385	370	15
400	310	90
310	260	50
410	380	30
370	300	70
410	390	20

L3(1)=40

L1	L2	L3
320	290	30
330	365	-35
250	210	40
380	350	30
340	260	80
220	90	130
------	------	

L3(14) =

If you scroll down, you'll see that the formoterol values are higher for 12 of the 13 children. Let's create a histogram of *these* values, the differences. Don't forget to use ZoomStat.

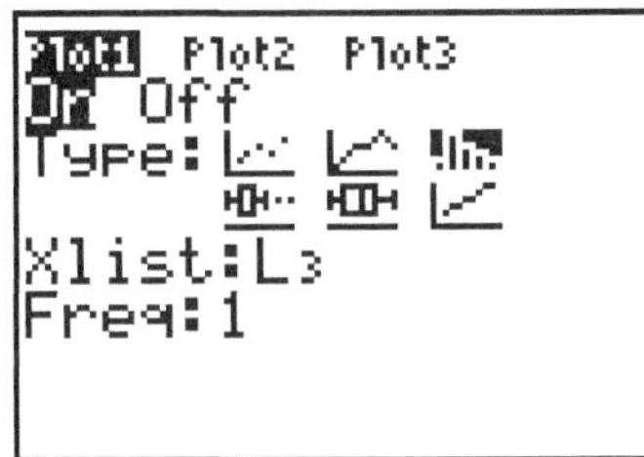

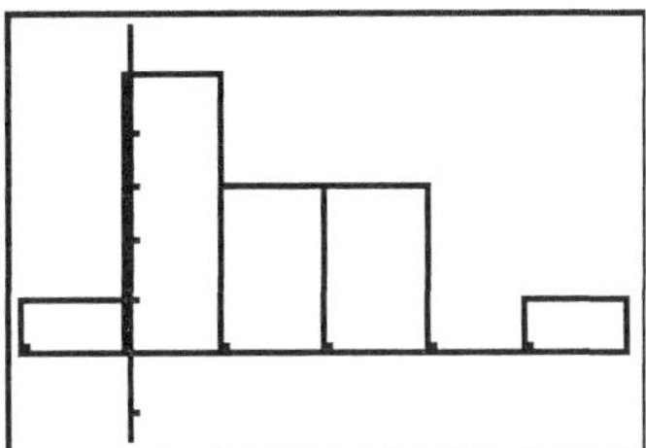

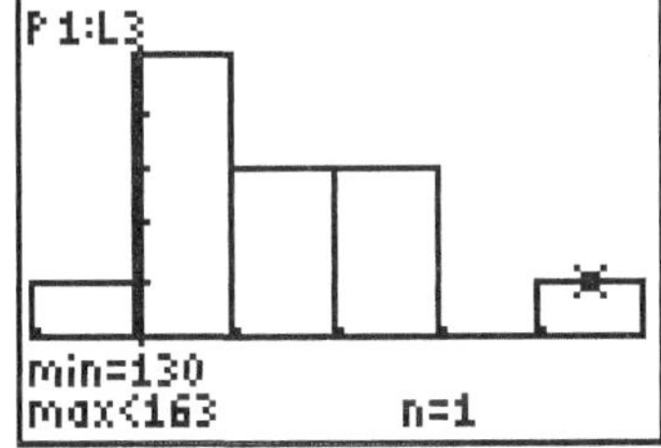

Notice the vertical axis; it's at zero. We only observed one negative difference. The largest difference was positive, 220 – 90 = 130 for child #13.

Time Plots

Exercise 2.26: Is whooping cough close to being eradicated? (p. 47 of Agresti/Franklin)
To create a plot of incidence rates for whooping cough from 1925 to 1970, we will use the xyLine option in STAT PLOT. The X data, time, will go in L1. For this example, we could get away with simply entering a sequence from 1 to 10 to represent the ten years of data given, but

instead let's enter the years as given on p. 47. This way, the TRACE option will give us better information, and the plot would be more accurate if, say, we didn't have data at equally spaced intervals like we do here. The xyLine option is the second plot icon available in STAT PLOT. The Mark is your choice – aesthetic purposes only. Don't forget ZoomStat. TRACE can be used to scroll through the years and verify, for example, that the rate was 38.2 per 100,000 in 1955.

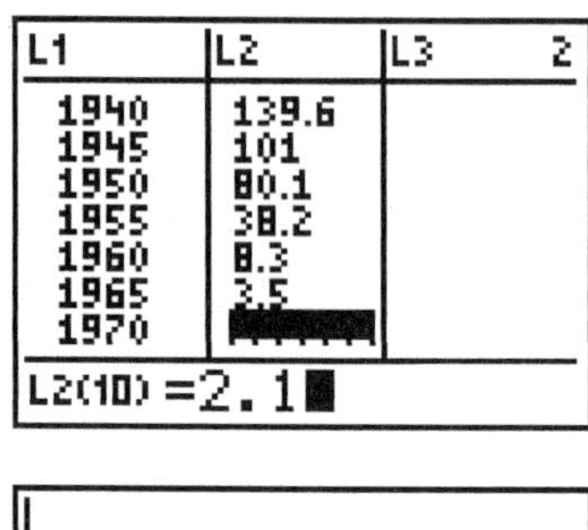

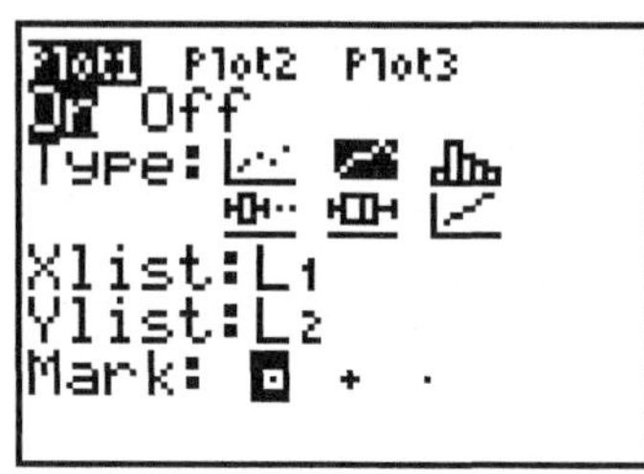

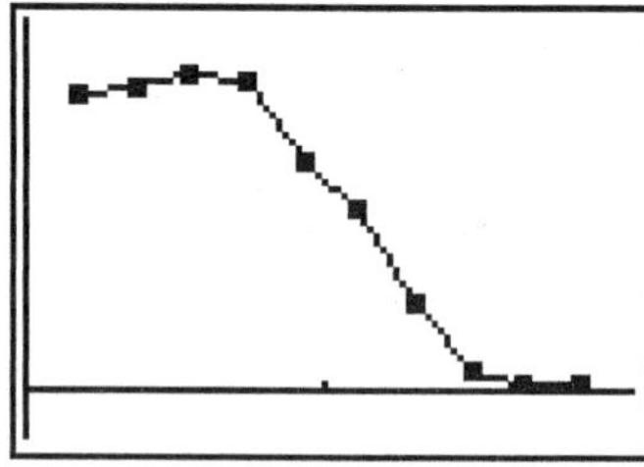

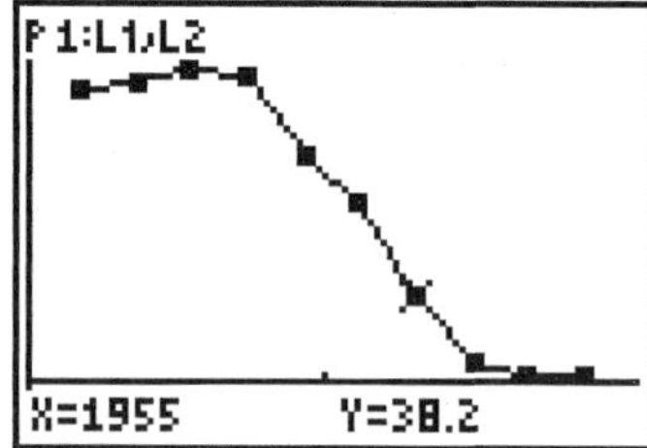

Descriptive Statistics: Finding the Mean, Median, Standard Deviation, and Quartiles

EXAMPLE 10: What's the center of the cereal sodium data? (pp. 48-49 of Agresti/Franklin)

The 20 sodium levels are given on p. 48. Notice that the values are given in order, from smallest to largest. This helps considerably when finding the median by hand. In general, we can't expect data to be presented to us in order. The cereal data first appears in table 2.3 on p. 34. Enter the sodium values as they appear there, then apply the sortA function to do an *ascending* sort. Remember that the sort functions are in the STAT EDIT menu.

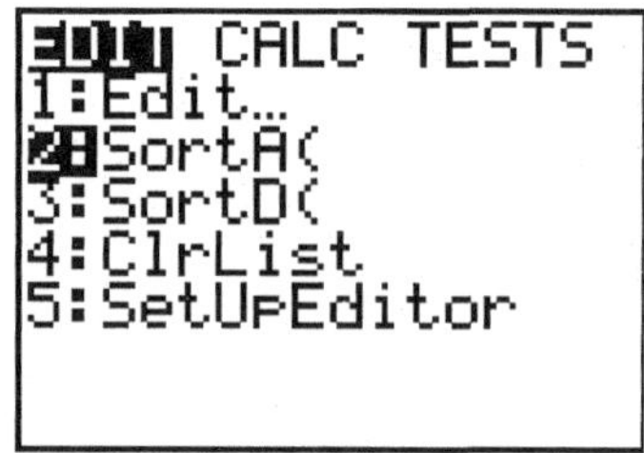

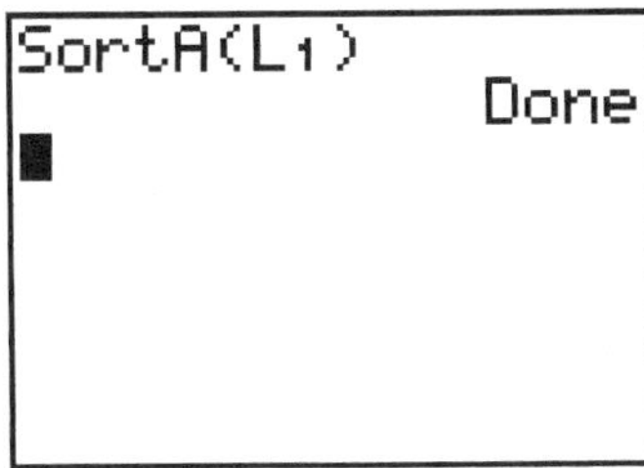

Of course, this is not necessary when we allow the calculator to do the manual labor for us. However, note that the median is the average of the 11th and 12th ordered values, as shown at the top of p. 49.

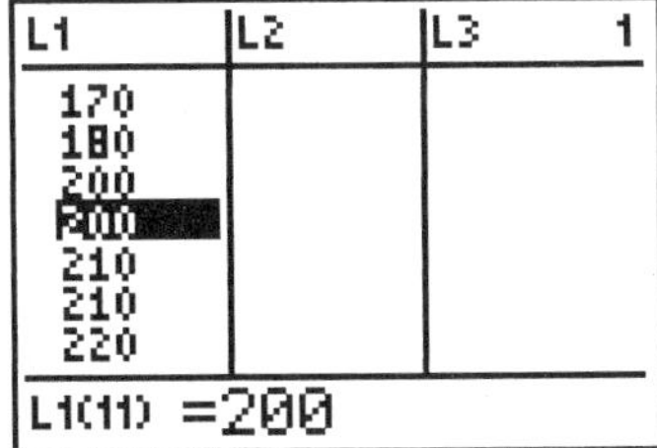

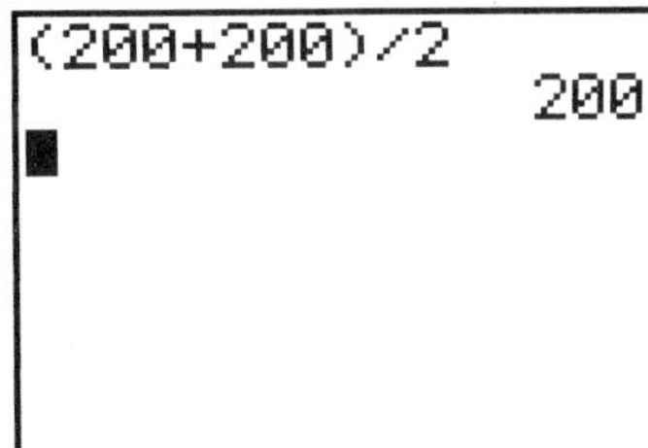

Mean and median functions are available from the LIST menu. We haven't considered this menu yet; let's do so now. Press LIST (2nd STAT). NAMES allows us to access the list names. The screen below shows the default lists L1-L6, plus another that was named AGES2 by the user. The OPS menu provides various list operations, including the sort functions we used from the STAT EDIT menu before. MATH contains various functions, including mean and median.

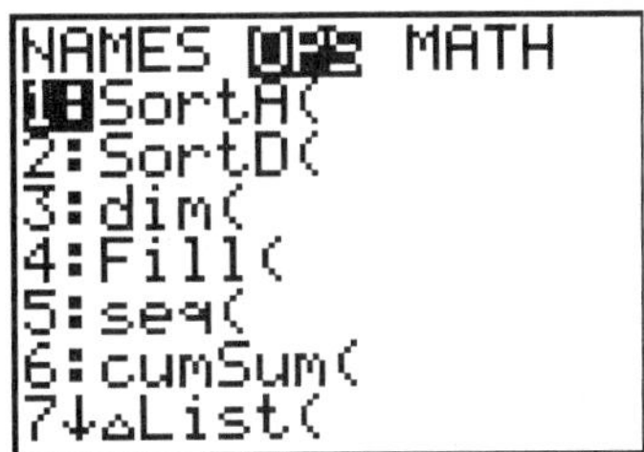

Verify that the mean and median of the sodium data are 185.5 and 200, respectively.

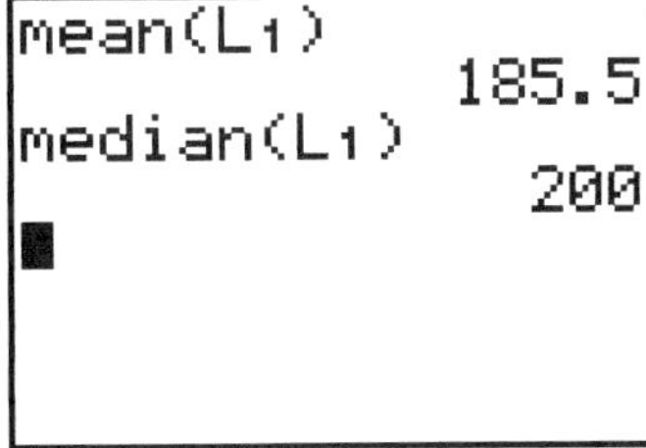

An easier way to get the mean, median, and other descriptive statistics is to use the 1-Var Stats function, which is found in the STAT CALC menu. By default, the statistics are calculated on L1, but it is good practice to always specify which list you want the stats for.

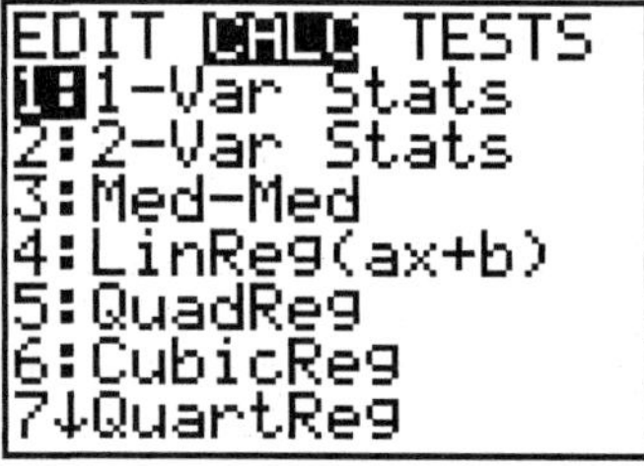

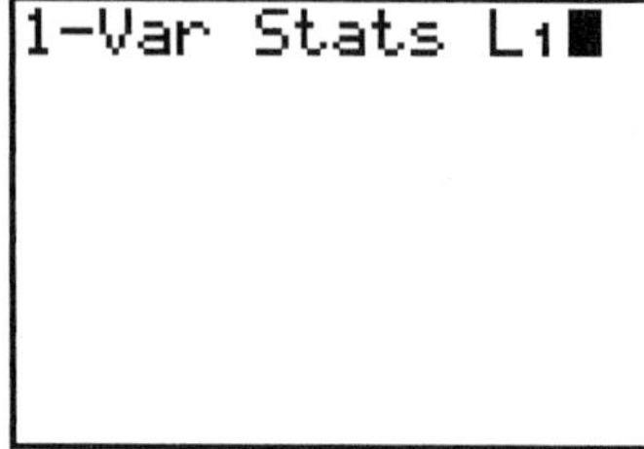

Two screens' worth of statistics are produced – scroll down to read it all.

```
1-Var Stats
 x̄=185.5
 Σx=3710
 Σx²=784650
 Sx=71.24642189
 σx=69.44242219
↓n=20
```

```
1-Var Stats
↑n=20
 minX=0
 Q1=145
 Med=200
 Q3=225
 maxX=290
```

The first screen gives the mean, sum, sum of squared values, sample standard deviation, population standard deviation (not very useful since we typically work with samples of data), and sample size. The second screen gives the five number summary: minimum, 1st quartile, median, 3rd quartile, and maximum.

Exercise 2.29: More on CO_2 emissions (p. 55 of Agresti/Franklin)

Enter the seven emissions values into L1 (or any list you'd like). Use 1-Var Stats to find the mean and median.

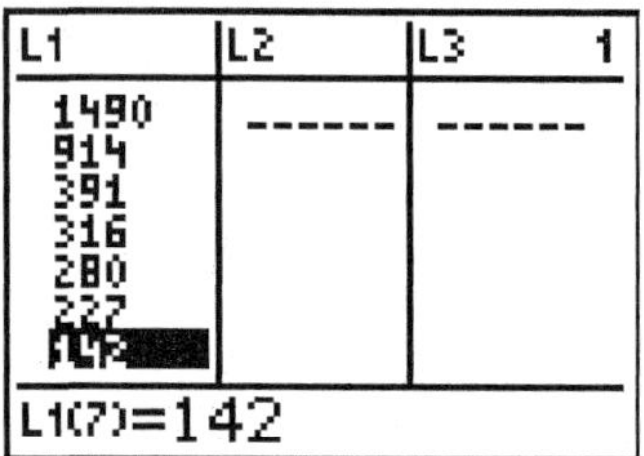

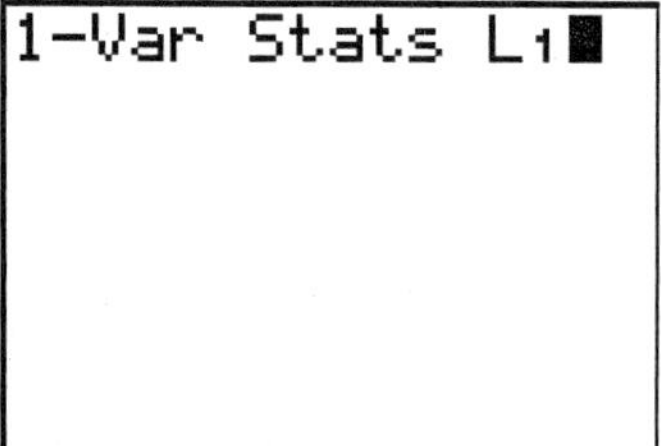

```
1-Var Stats
 x̄=537.1428571
 Σx=3760
 Σx²=3458326
 Sx=489.6714642
 σx=453.3476838
↓n=7
```

```
1-Var Stats
↑n=7
 minX=142
 Q1=227
 Med=316
 Q3=914
 maxX=1490
```

EXAMPLE 13: Comparing women's and men's ideal number of children (pp. 58-59 of Agresti/Franklin)

Put the men's responses in L1, and the women's responses in L2. To just find the standard deviations, we could use LIST MATH again, this time choosing function 7.

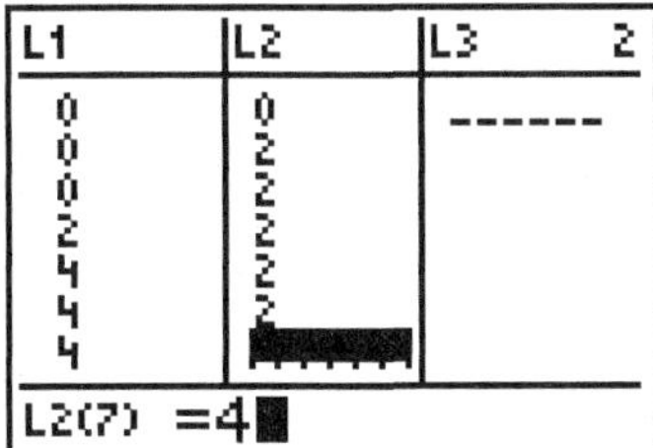


```
NAMES OPS MATH
1:min(
2:max(
3:mean(
4:median(
5:sum(
6:prod(
7↓stdDev(
```

```
stdDev(L1)
                2
stdDev(L2)
      1.154700538
```

The 1-Var Stats results for the men appear below.

```
1-Var Stats
 x̄=2
 Σx=14
 Σx²=52
 Sx=2
 σx=1.8516402
↓n=7
```

```
1-Var Stats
↑n=7
 minX=0
 Q1=0
 Med=2
 Q3=4
 maxX=4
```

The 1-Var Stats results for the women appear below. Note that, for either gender, the range is max – min = 4 – 0 = 4. The standard deviation provides a much better idea of the difference in variability between men and women.

```
1-Var Stats
 x̄=2
 Σx=14
 Σx²=36
 Sx=1.154700538
 σx=1.069044968
↓n=7
```

```
1-Var Stats
↑n=7
 minX=0
 Q1=2
 Med=2
 Q3=2
 maxX=4
```

Exercise 2.43: Sick leave (p. 63 of Agresti/Franklin)
Enter the sick leave data into L1. Use the 1-Var Stats function. Range = max – min = 6 – 0 = 6, and s = 2.38.

```
L1      L2      L3     1
0       ------  ------
0
4
0
0
0
6
L1 ={0,0,4,0,0,0...
```

```
1-Var Stats
 x̄=1.25
 Σx=10
 Σx²=52
 Sx=2.375469878
 σx=2.222048604
↓n=8
```

```
1-Var Stats
↑n=8
 minX=0
 Q1=0
 Med=0
 Q3=2
 maxX=6
```

Re-code the "6" as 60. Apply the 1-Var-Stats function to L1 again. Now range = 60, and s = 21.1.

```
L1      L2      L3     1
0
4
0
0
0
6
0
L1(7)=60
```

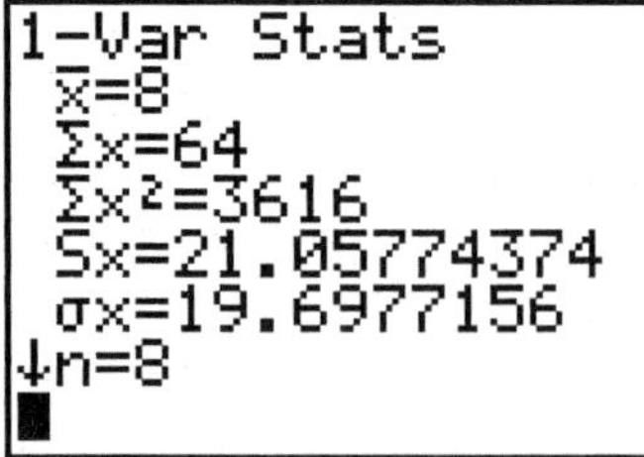

```
1-Var Stats
↑n=8
 minX=0
 Q1=0
 Med=0
 Q3=2
 maxX=60
```

EXAMPLE 16: What are the quartiles for the cereal sodium data? (p. 66 of Agresti/Franklin)
Refer back to example 10. We applied the 1-Var Stats function to the sodium values then, and got $Q_1 = 145$, and $Q_3 = 225$.

Exercise 2.58: European unemployment (p. 72 of Agresti/Franklin)
Enter the 15 unemployment rates into L1.

```
L1      L2      L3     1
8.3     ------  ------
6
9.2
9.3
11.2
9.5

L1(7)=6.7
```

```
1-Var Stats
 x̄=7.053333333
 Σx=105.8
 Σx²=823.08
 Sx=2.342729381
 σx=2.263291605
↓n=15
```

```
1-Var Stats
↑n=15
 minX=3.9
 Q1=4.6
 Med=6.7
 Q3=9.2
 maxX=11.2
```

Median = 6.7, $Q_1 = 4.6$, $Q_3 = 9.2$, and mean = 7.05.

Creating Boxplots

EXAMPLE 17: Boxplot for the breakfast cereal sodium data (p. 69 of Agresti/Franklin)
Enter the sodium data, most recently displayed on p. 66, into L1. In the STAT PLOT menu, make sure plots 2 and 3 are turned off. Also delete any functions that may be in the Y= screen. Turn Plot1 on, and select the modified boxplot icon. This version of the boxplot draws special attention to outlying values by displaying them as one of three marks (this is purely aesthetic and is up to you to choose). Don't forget to use ZoomStat.

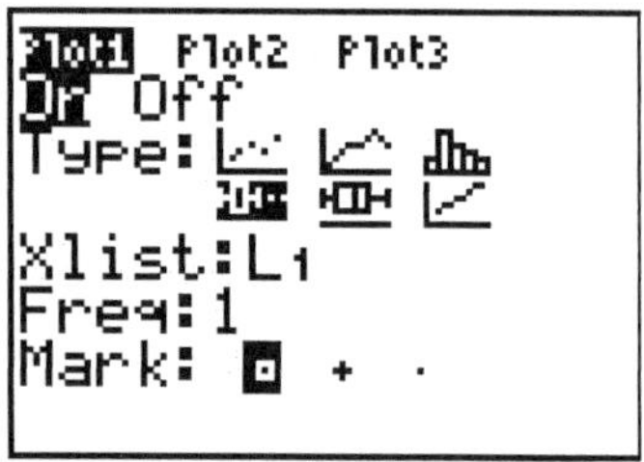

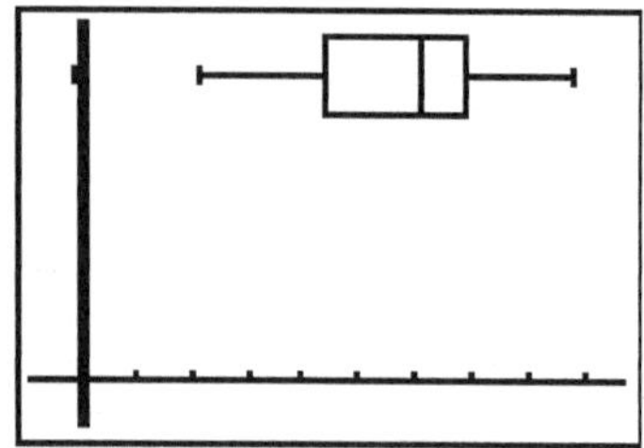

There is one outlier. It is easy to miss, because it is the zero sodium value for Frosted Mini Wheats. It is partially obscured by the vertical axis.

Note: The TI-83 only displays one mark type for outliers. There is nothing to distinguish between "potential" and "extreme" outliers, like there is with most statistical software. SPSS, for example, uses a circle for potential outliers and an asterisk for extreme outliers.

It is also worth noting that the calculator provides a "simple" boxplot option. This is the icon that appears to the right of the modified boxplot icon. By way of exploring this, let's take advantage of the situation and display two boxplots on the screen at the same time. For the most part, you will always want just one of the three plots turned on. But for boxplots, we are allowed to have three at a time, and sometimes this is very useful for comparison purposes. Leave Plot1 as is. Turn Plot2 on, choose the simple boxplot icon, and specify the same list of data, the sodium values in L1. Now press GRAPH. No need for ZoomStat; the window dimensions are set for the sodium data.

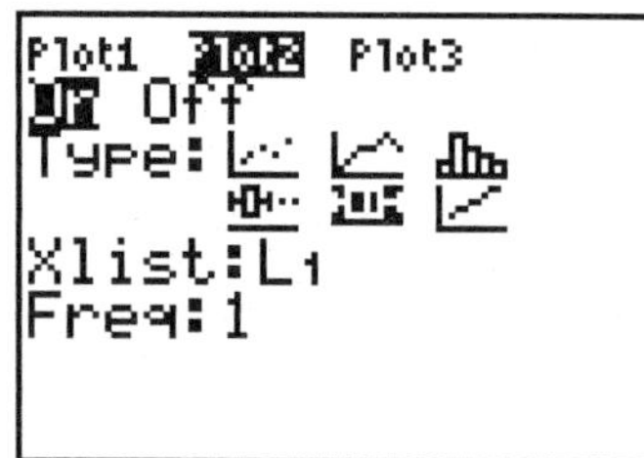

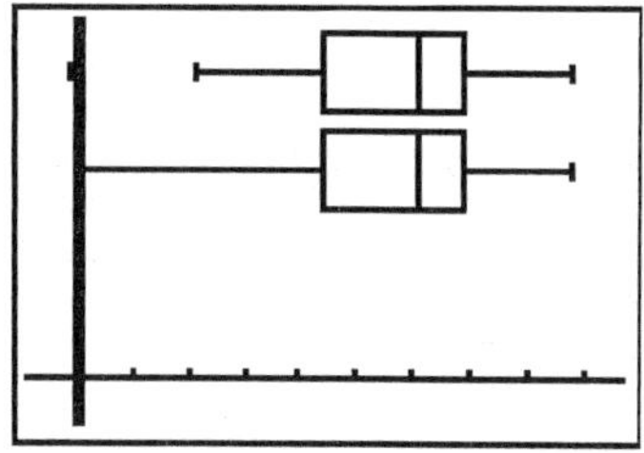

The lower boxplot, Plot2, is the simple boxplot. Notice that the left tail extends all the way to the minimum value of zero.

Exercise 2.67: Public transportation (p. 73 of Agresti/Franklin)
Enter the ten miles-per-day values into L1. Even with ZoomStat, it appears that we have only half of our boxplot! Actually, it's all there. Use the TRACE function to learn more.

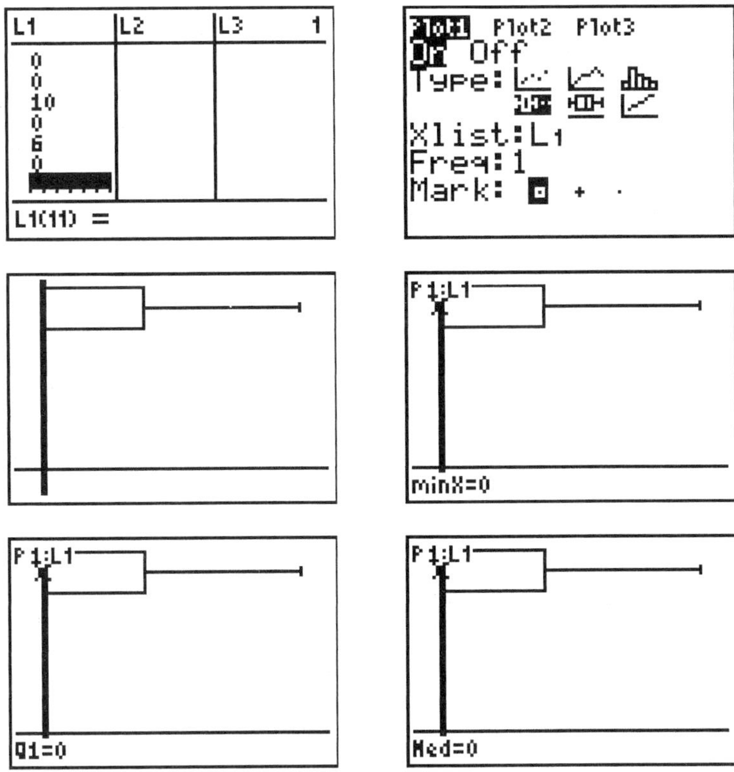

For this sample of data, the minimum, 1st quartile, and median are all equal to zero! Notice that seven of the ten values are zero. This is more than half, so the above boxplot is to be expected.

Exercise 2.69: European Union unemployment rates (p. 74 of Agresti/Franklin)
Consider again the unemployment rates of exercise 2.58. Note that there are no outliers in this data set. For example, on the high side, $Q_3 + 1.5\ IQR = 9.2 + 1.5\ (9.2 - 4.6) = 16.1$. Since the maximum rate is 11.2 (Spain), none of the data values qualify for outlier status.

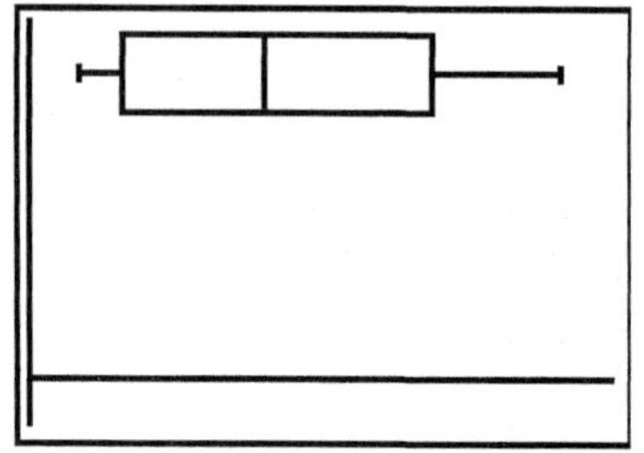

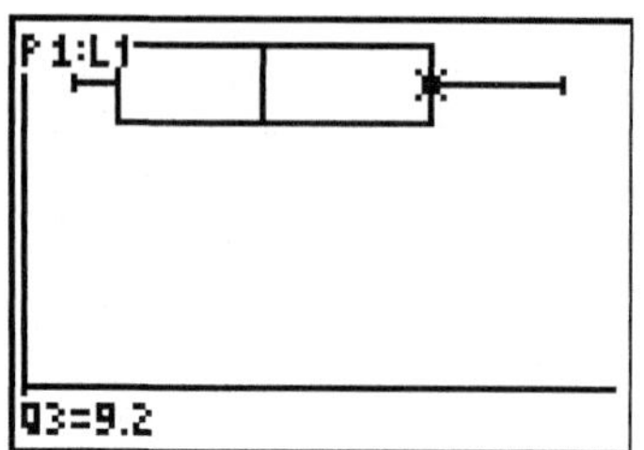

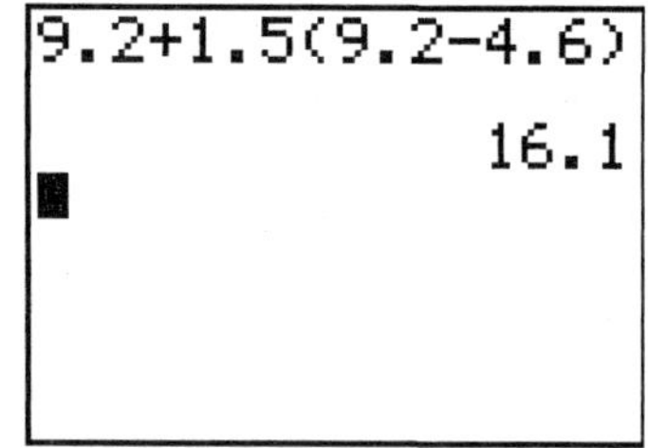

Chapter Three: Association: Contingency, Correlation, and Regression

EXAMPLE 3: How can we compare pesticide residues for the food types graphically? (pp. 93-94 of Agresti/Franklin)
This will be quite similar to the bar graph we created for example 3 of chapter 2. Again, we are "tricking" the calculator to give us a bar graph by supplying the percentages and using the histogram option under STAT PLOT. Our goal is to create a clustered bar graph like figure 3.2 on p. 94 of Agresti/Franklin. First enter the categories as numbers in L1. We'll need the numbers 1;5 to cover the two food types (twice), plus a "spacer" between the two levels of pesticide status. Put the percentages from table 3.2 (p. 93) in L2.

L1	L2	L3 2
1	27	------
2	77	
3	0	
4	73	
5	23	

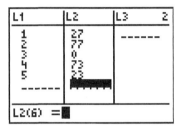

L2(6) =

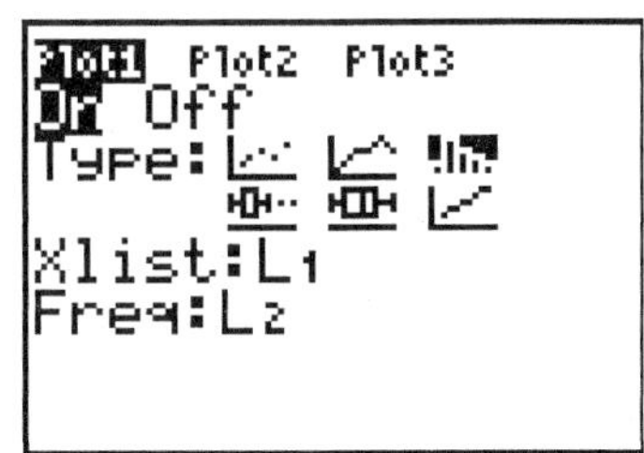

So "1" represents (no pesticide, conventional), "2" represents (no pesticide, organic), "3" inserts a space, "4" represents (pesticide, conventional), and "5" represents (pesticide, organic). Note that a zero is entered on the third line. Next press STAT PLOT. Making sure that Plots 2 and 3 are turned off, choose Plot1. Select the histogram icon, specify L1 for Xlist, and L2 for Freq. Unfortunately, ZoomStat doesn't help us here – we need to specify the WINDOW dimensions in order to produce the intended graph.

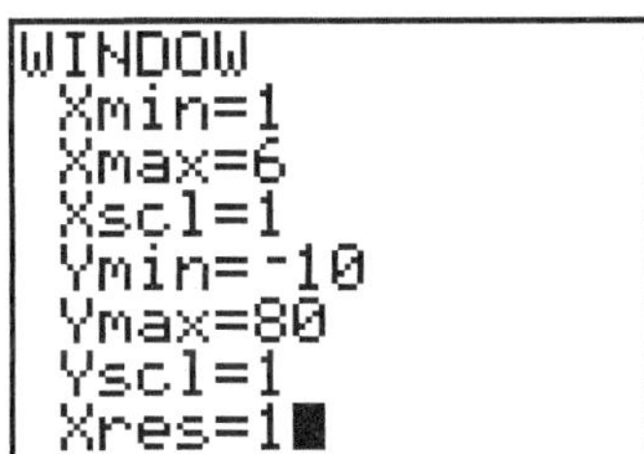

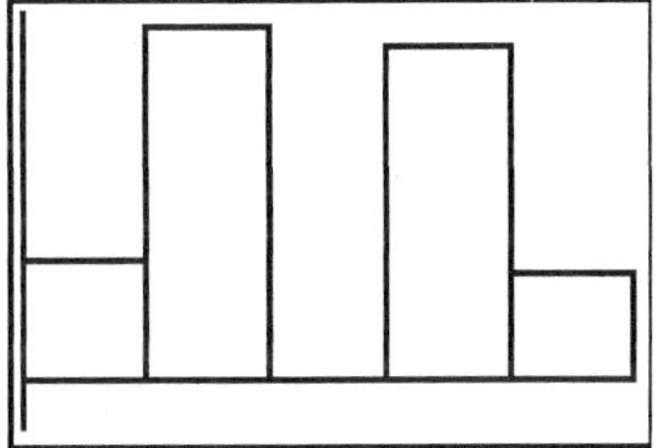

Give an X range to accommodate the five categories; Xscl should be one – everything here is similar to example 3 in chapter 2. TRACE tells us that, for example, 73% of the conventional foods had pesticide residue.

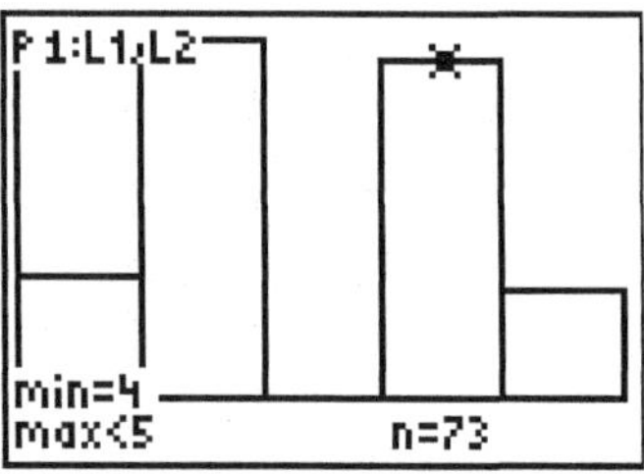

EXAMPLE 5: Constructing a scatterplot for internet use and GDP (pp. 166-161 of Agresti/Franklin)

The goal here is to create a graph to illustrate what relationship, if any, exists between internet use (in % for each country) and GDP (per capita, in thousands of US dollars). The data appear in table 3.4 on p. 98 of Agresti/Franklin. Enter or download the INTERNET and GDP columns into lists L1 and L2.

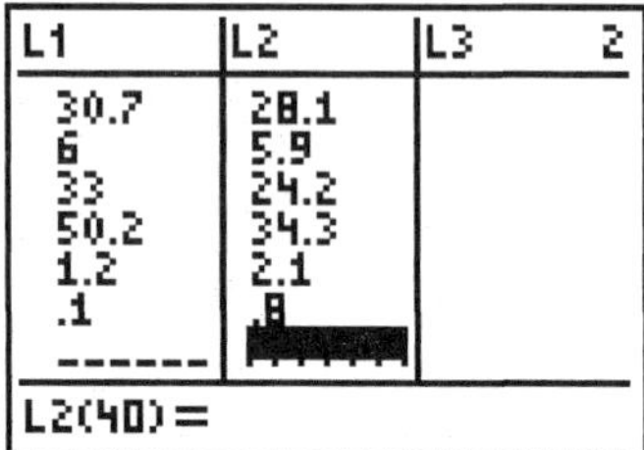

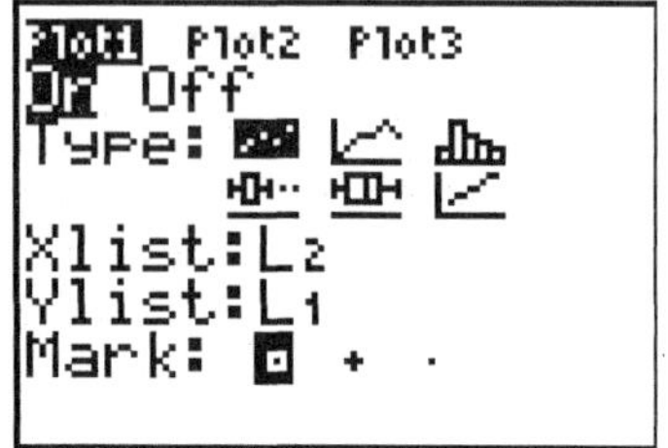

Refer to p. 100 of Agresti/Franklin. Since internet use is the response variable, we will label it Y, and plot it on the vertical axis. Since GDP is the explanatory variable, we will label it X, and plot it on the horizontal axis. We'll need to remember that we placed the X data in L2, and the Y data in L1. Press STAT PLOT. Make sure that Plot1 is the only plot turned on. The first plot icon given is the scatterplot. Note that we are given three options for how the points should appear. This can be useful later – for now, it doesn't matter which you choose. Before pressing GRAPH, realize that the window dimensions won't be set for the new data. You may wish to go straight to ZoomStat.

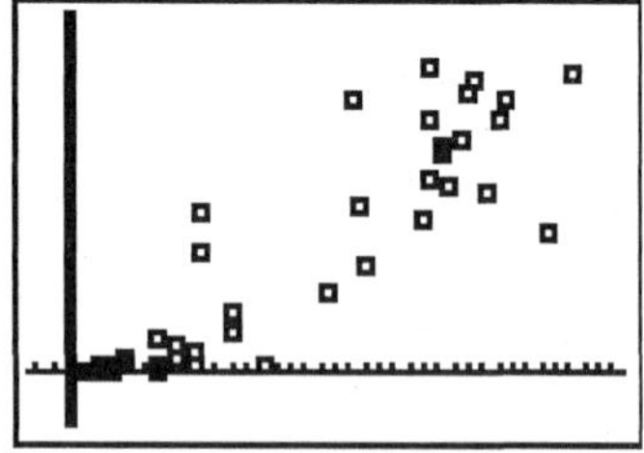

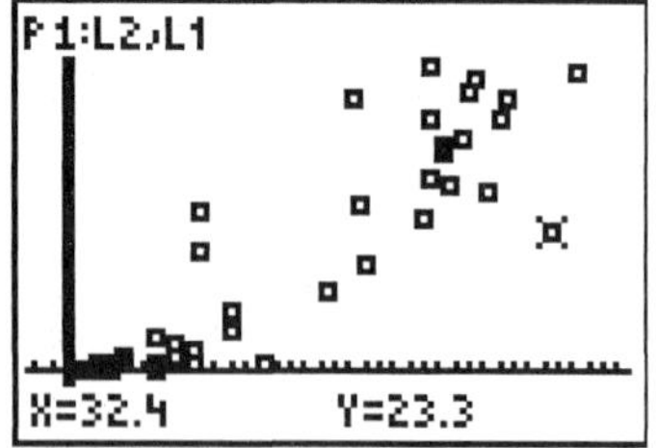

The TRACE function allows us to scroll through the points. Pressing the right arrow moves you down the original list, so eventually you can highlight (32.4, 23.3), which is the (GDP, internet use %) for Ireland.

EXAMPLE 7: The correlation between internet use and GDP (pp. 165-167 of Agresti/Franklin)

To get the correlation coefficient between two quantitative variables, we must use LinReg, the linear regression function. There is no "corr" function in the catalog, and the 2;Var Stats function in the STAT CALC menu doesn't produce the correlation coefficient. So we may be getting ahead of ourselves a bit, but let's ask the calculator for the linear equation for predicting internet use based on GDP. From the STAT CALC menu, choose 8: LinReg(a+bx). This model matches up with the simple linear regression equation given at the bottom of p. 109 of Agresti/Franklin. Without any arguments, the calculator will assume that the X data is in L1, and the Y data is in L2. This is not the case for our example, which forces us to get into the good habit of always specifying the lists explicitly.

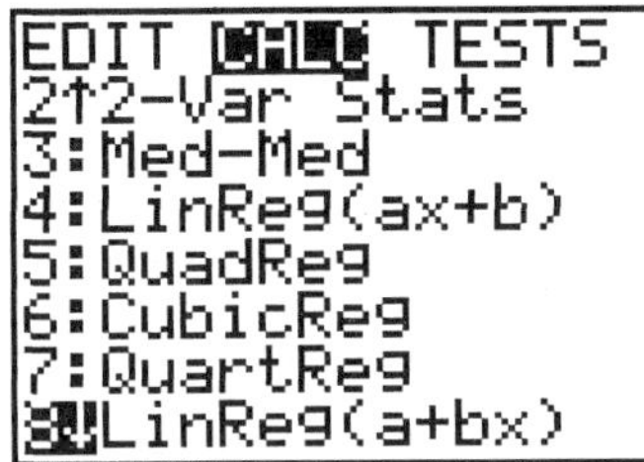

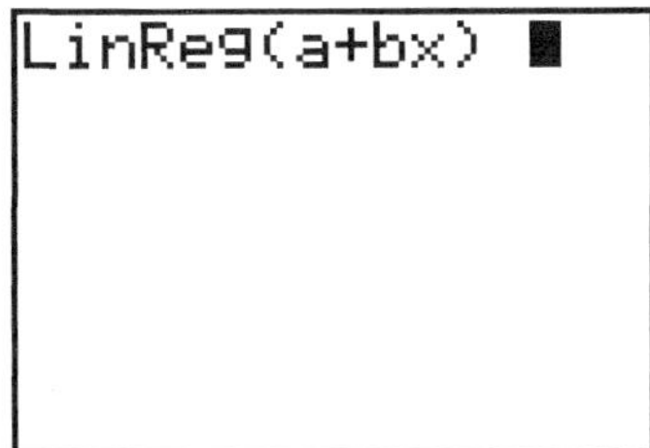

So instead of simply pressing ENTER after the LinReg(a+bx) function, clearly state where the X and Y data are. The format is LinReg(a+bx) Xlistname, Ylistname.

```
LinReg(a+bx) L2,
L1
```

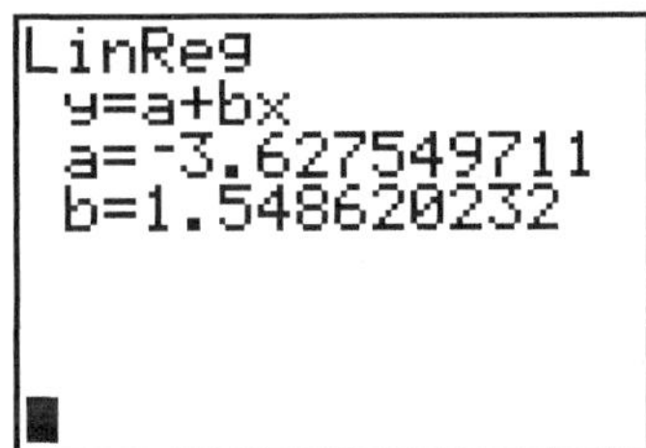

By default, the calculator will not produce **r**, the correlation coefficient. We need to turn this option on. Press CATALOG (2nd 0). Note that the "alpha" symbol is visible in the upper right of the screen. This means that you can easily jump to certain letters in the catalog list. We are looking for the DiagnosticOn function, so press D (x^{-1}). Now you can scroll down to the desired function.

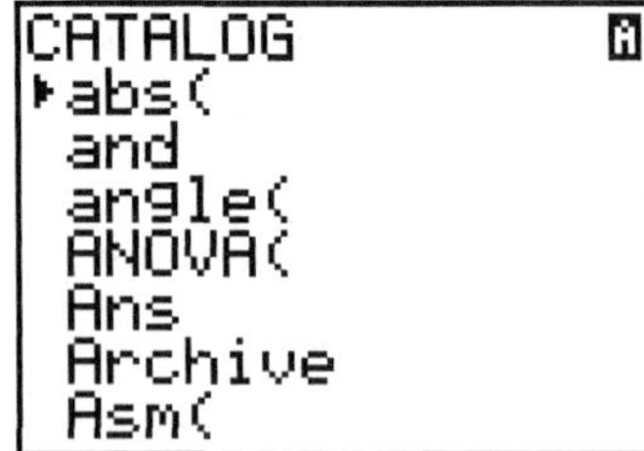


```
CATALOG
 Degree
 DelVar
 DependAsk
 DependAuto
 det(
 DiagnosticOff
▸DiagnosticOn
```

Once it is echoed back to the home screen, press enter. Now we will get the correlation coefficient as part of the LinReg output.

```
DiagnosticOn
                Done
```

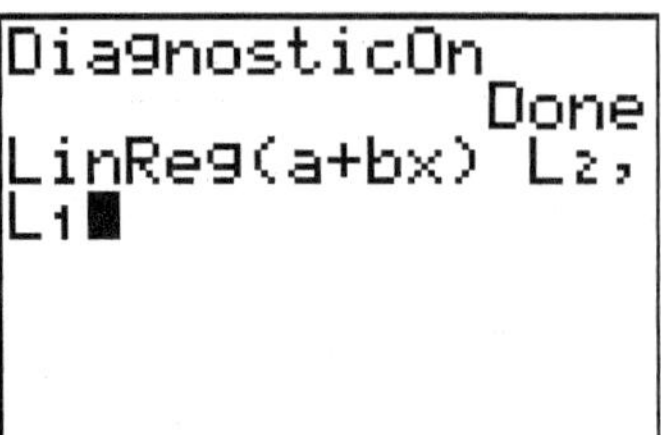


```
LinReg
 y=a+bx
 a=-3.627549711
 b=1.548620232
 r²=.7895959318
 r=.8885921065
```

The correlation coefficient between internet use and GDP is r = .889.

Exercise 3.19 (p. 169 of Agresti/Franklin)
Enter the Price data in L1, and the Weight data in L2. If we wish to ultimately predict price based on weight, then Y=price and X=weight. The scatterplot should have weight on the X axis and price on the Y axis. Don't forget ZoomStat.

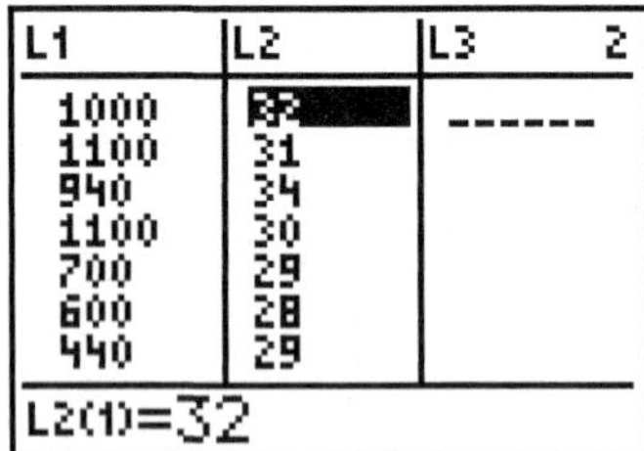

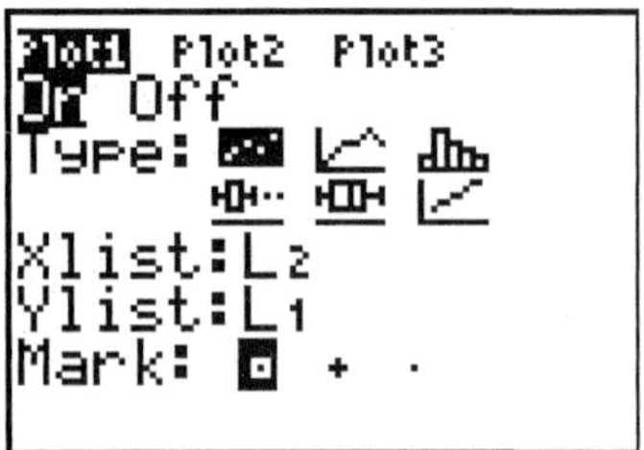

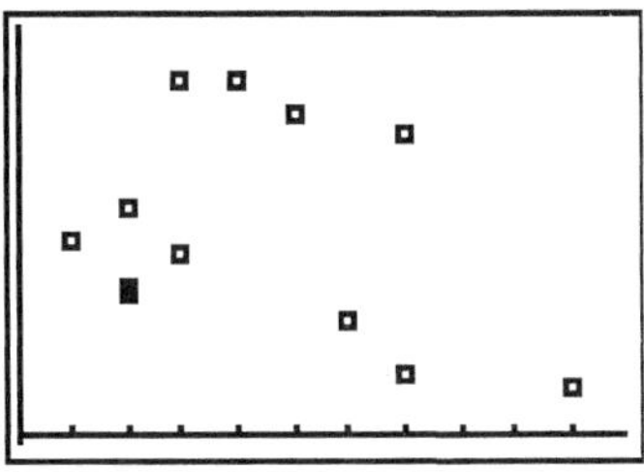

The TRACE function allows us to explore the “solid square”, actually two points that are nearly on top of each other.

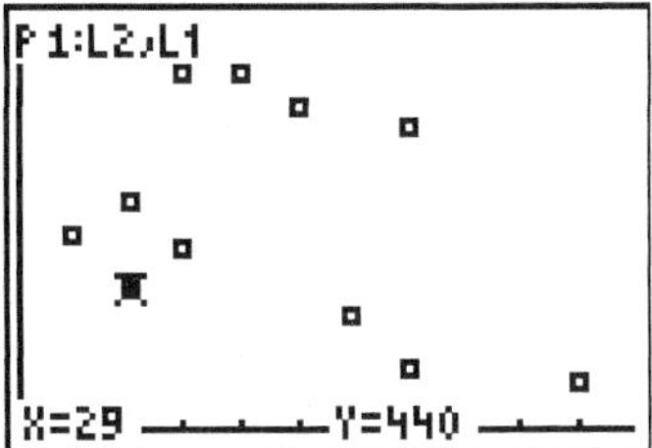

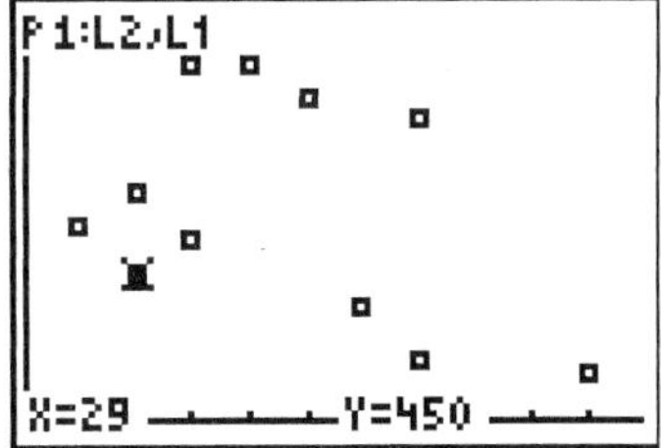

The Haro Escape A7.1 and the Giant Yukon SE weigh the same (29 lbs), and cost nearly the same ($440 and $450, respectively). Next let’s get the correlation coefficient between price and weight. We should expect a negative number, based on the scatterplot.

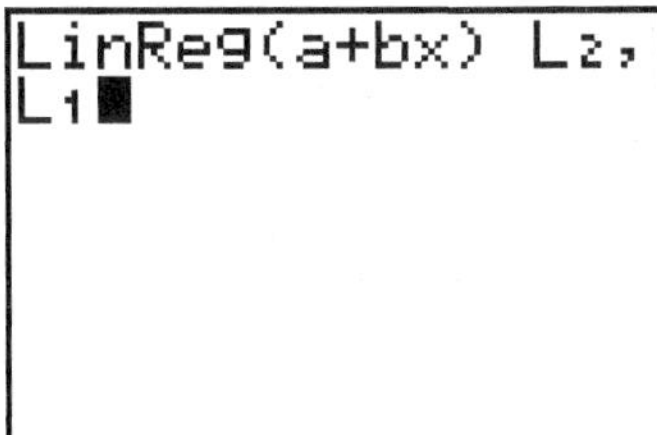


```
LinReg
 y=a+bx
 a=1895.909091
 b=-40.45454545
 r²=.1029637333
 r=-.3208796243
```

Remember that the X data is in L2, the Y data is in L1, and the calculator expects the arguments to be supplied in the order (Xlist, Ylist) for the LinReg(a+bx) function. The correlation coefficient is r = – .32.

Next, let's consider the other variable that is given on p. 109, Type of suspension. We can create a scatterplot of price vs. weight like the one above, with different symbols for FU=full suspension and FE=front end suspension bikes. To accomplish this, we need to enter the FU pairs of data (the first four bikes) in say, L1 and L2, then enter the FE pairs of data (the last eight bikes) in, say, L3 and L4. Then we can use Plot1 and Plot2, both turned on at the same time, and using different symbols for the points.

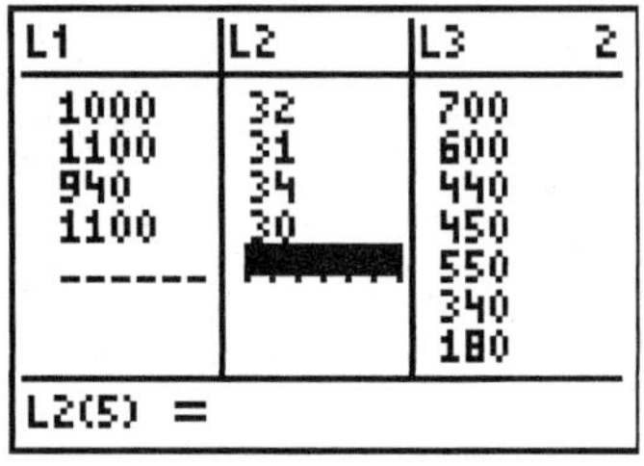

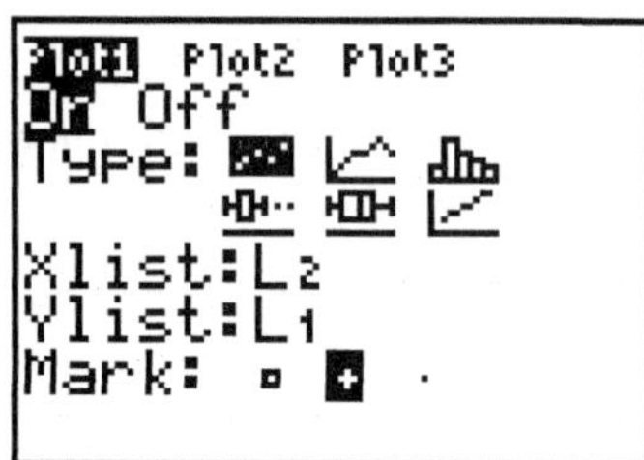

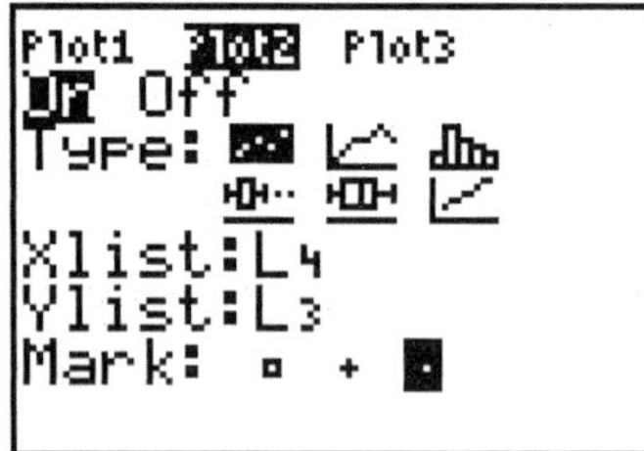

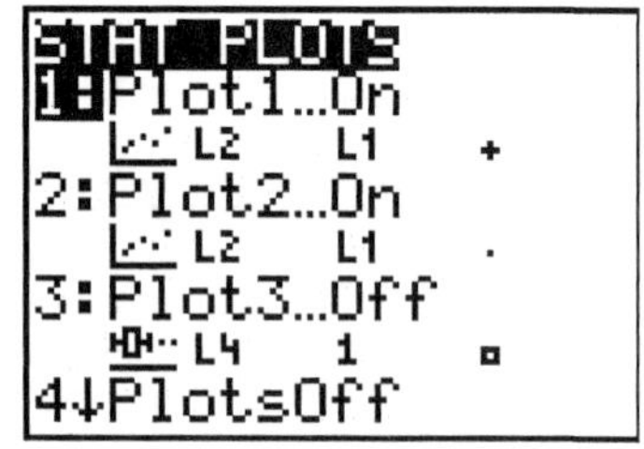

Once you have the data entered and the plots specified, press GRAPH. The FU and FE bikes are distinguished from each other quite dramatically. Note that when using the TRACE function, you must use the up/down arrows to move between Plot1 and Plot2.

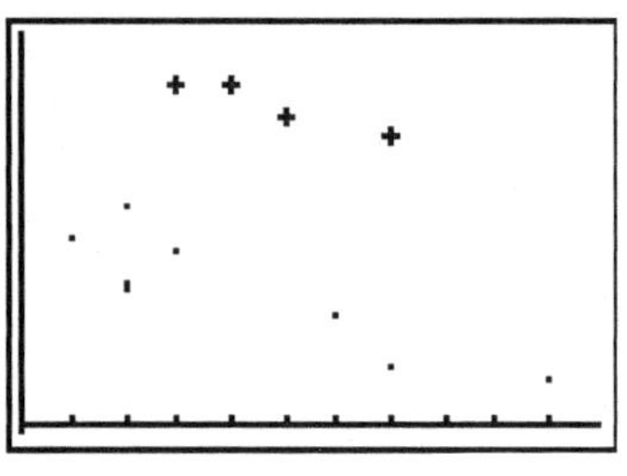

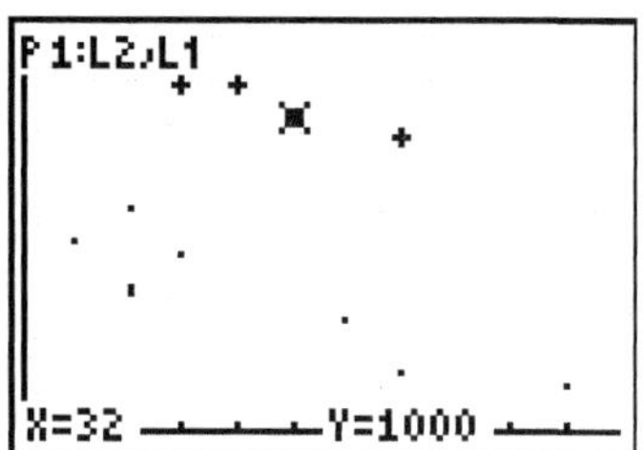

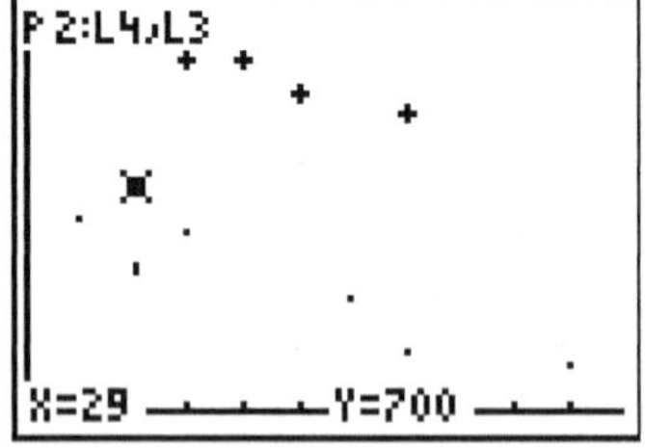

You may wish to explore exercises 3.35 and 3.36 before deleting the mountain bike data from your calculator.

EXAMPLE 9: How can we predict baseball scoring using batting average? (pp. 112-114 of Agresti/Franklin)

Enter the Batting Averages in L1, and the Team Scoring data in L2. Let's get the scatterplot, correlation coefficient, and linear regression equation for predicting team scoring from batting average. Note that if we wish to predict team scoring from batting average, this makes X=batting average and Y=team scoring.

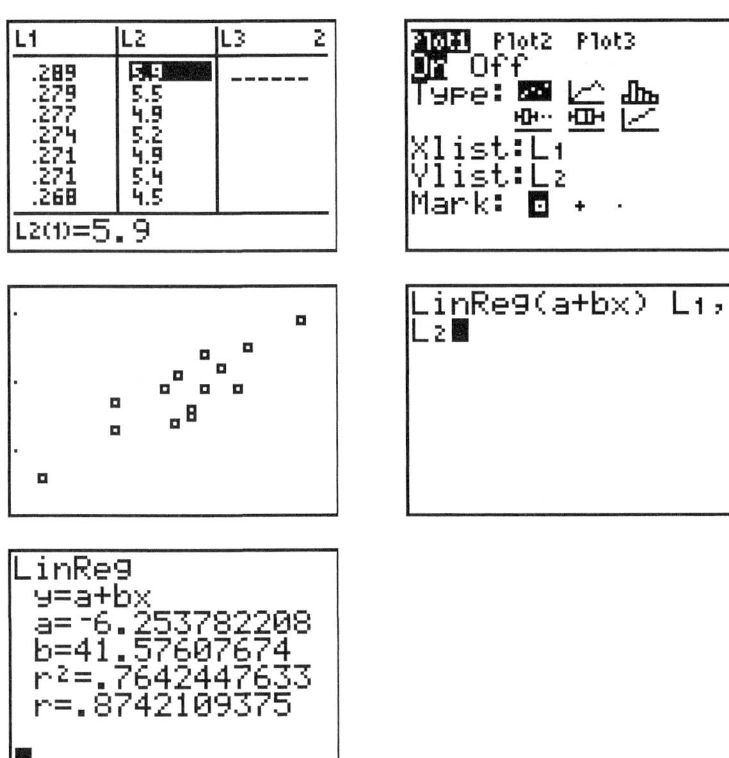

We can add the regression line to the scatterplot in one of two ways. Now that we know that the linear equation is $\hat{y} = -6.1 + 41.2x$, we can enter this in the Y= screen. For "X", use the [X,T,θ,n] button. Next, press GRAPH again.

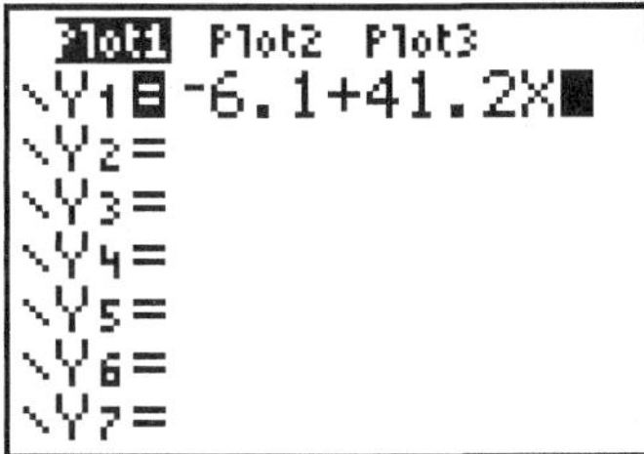

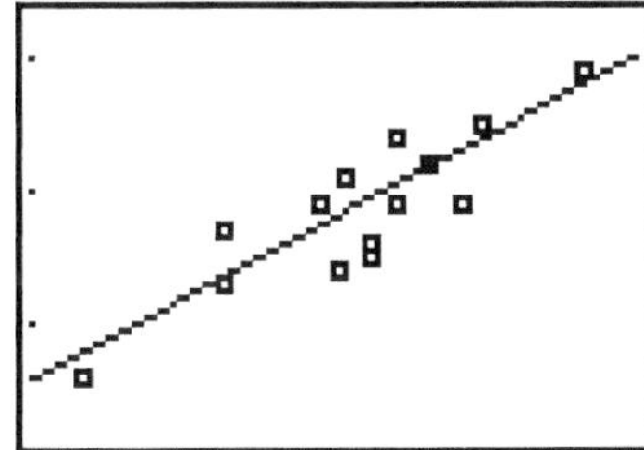

Another option is the following. Supply a third argument to the LinReg function. If you specify Y1, the regression equation will be stored there.

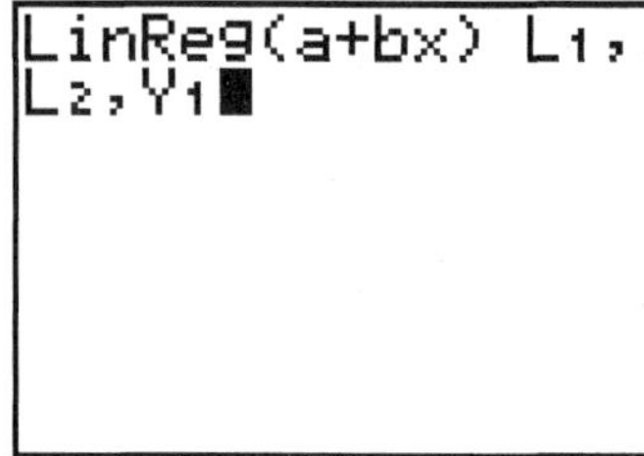

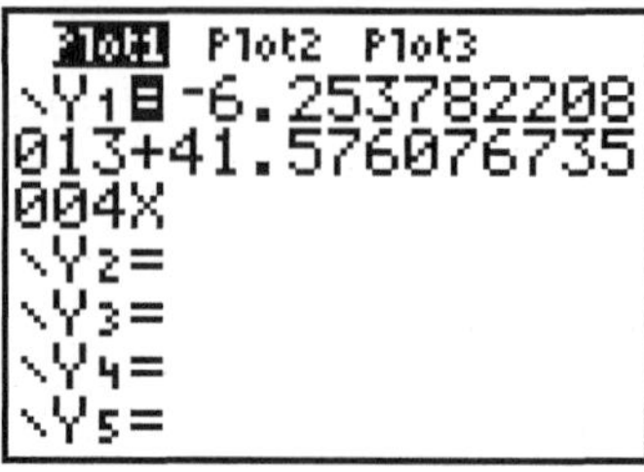

Note that you cannot type in alpha "Y", then the number 1. You must call up Y1 from the Y; VARS menu. After choosing the LinReg function and supplying the L1 and L2 lists, press the VARS button, right arrow to Y;VARS, and choose 1:Function. Pressing ENTER chooses Y1, and echoes this choice back to the home screen.

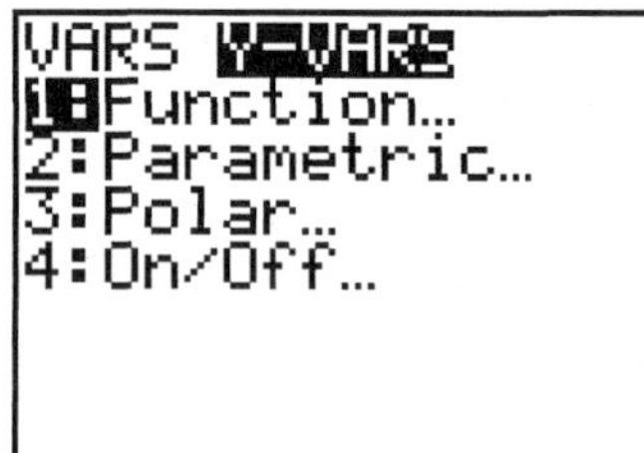

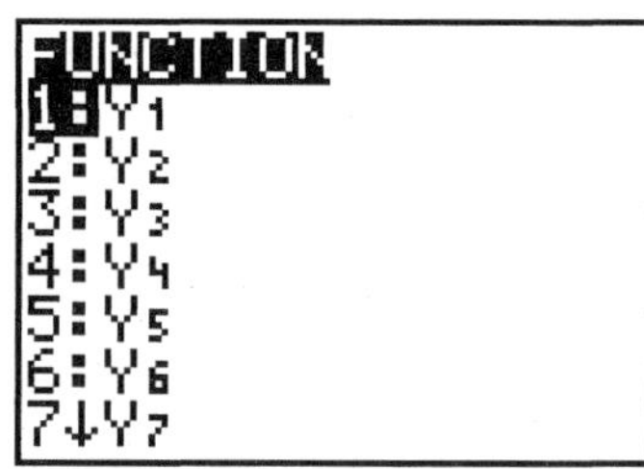

Press the TRACE button. Next press the down arrow. Now we can move along the regression line by using the left/right arrow keys.

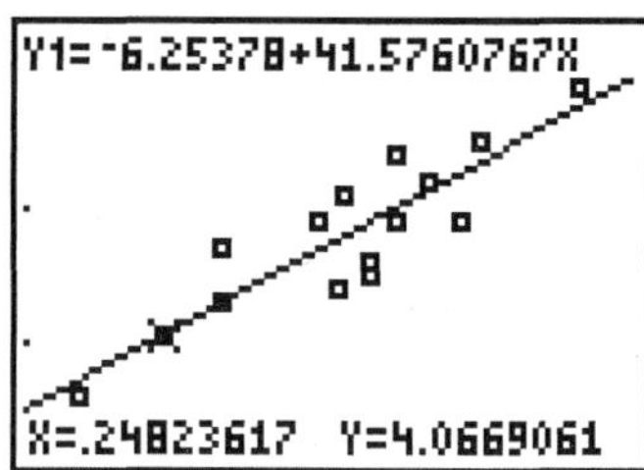

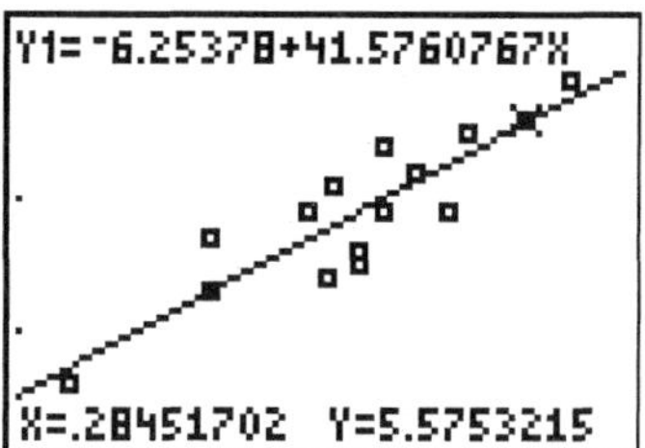

Note that for a team with a batting average of .285, scoring is predicted to be 5.575 runs per game.

Exercise 3.35 (p. 122 of Agresti/Franklin)
Enter the mountain bike data into L1 (Y=price) and L2 (X=weight) again if need be. Request the regression equation to be put into Y1.

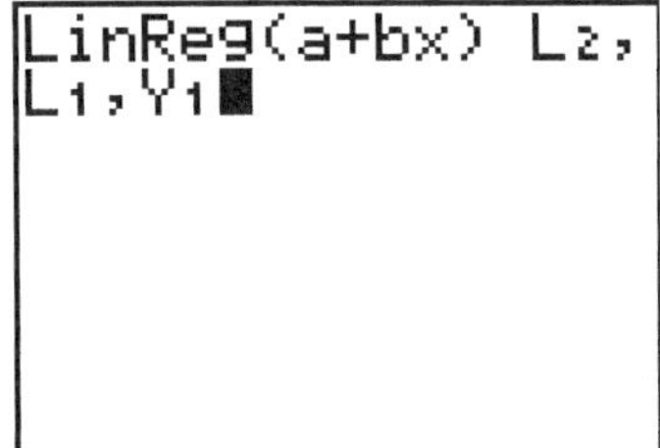


```
LinReg
 y=a+bx
 a=1895.909091
 b=-40.45454545
 r²=.1029637333
 r=-.3208796243
```

The linear equation to predict price based on weight is $\hat{y} = 1895.9 - 40.45x$. This line will superimposed on the plot automatically.

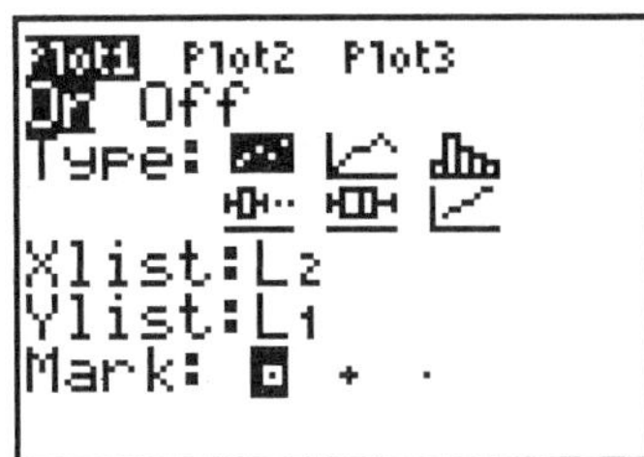

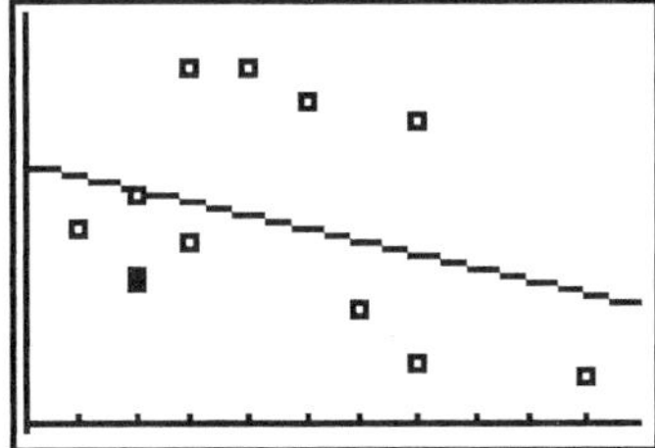

Press TRACE, then the down arrow. Now enter the number 30. We can specify X values, and the calculator will provide the predicted Y value.

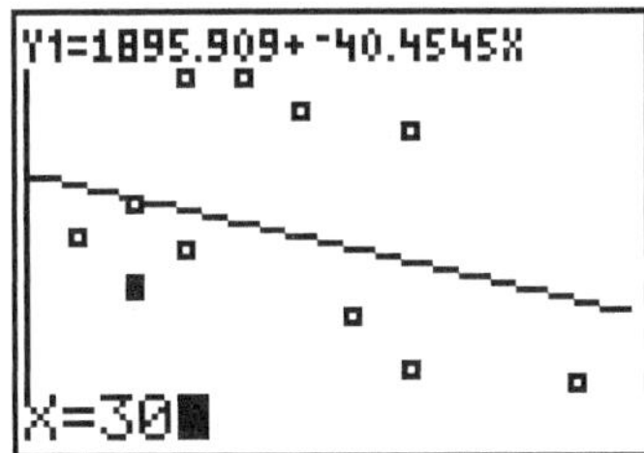

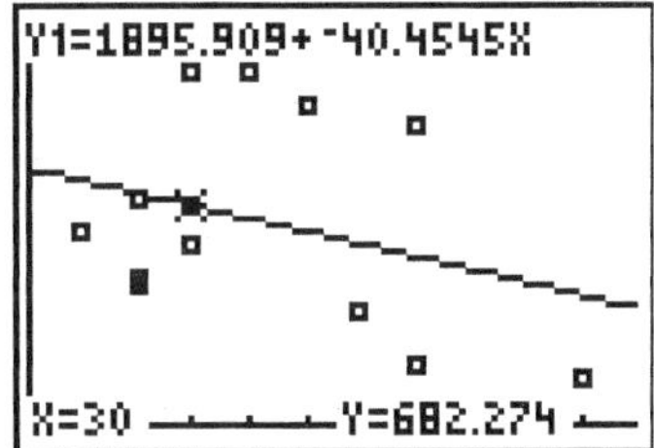

For a bike that weighs 30 pounds, we predict the price to be $\hat{y} = 1895.9 - 40.45(30) = \$682.27$. Another way to get this result is to supply 30 as an argument to the Y1 equation. Press VARS, scroll right to Y;VARS, and select 1:Function. Now supply 30 in parentheses.

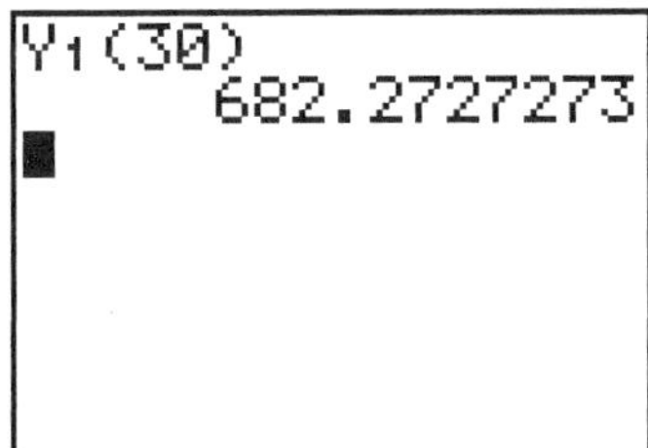

Exercise 3.36 (p. 122 of Agresti/Franklin)

Refer to exercise 3.19 above. With the FU bike data in lists L1 and L2, and the FE bike data in lists L3 and L4, we can get the correlation coefficients and regression equations for each type of bike separately. Below, we see the results for the FU bikes.

```
LinReg
 y=a+bx
 a=2432
 b=-44
 r²=.9058823529
 r=-.9517785209
```

The FE bike data analysis appears below. Note that Y2 is specified for the FE bike regression equation.

```
LinReg(a+bx) L4,
L3,Y2
```

```
LinReg
 y=a+bx
 a=2135.502646
 b=-54.95590829
 r²=.7892819424
 r=-.888415411
```

Press the Y= button to view both equations. Using the same plotting method as in exercise 3.19 above, we can plot the bike data with different symbols for FU and FE bikes, plus both of the regression lines.

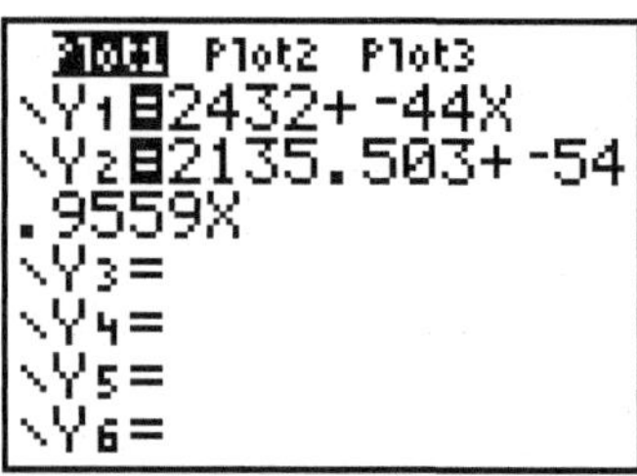

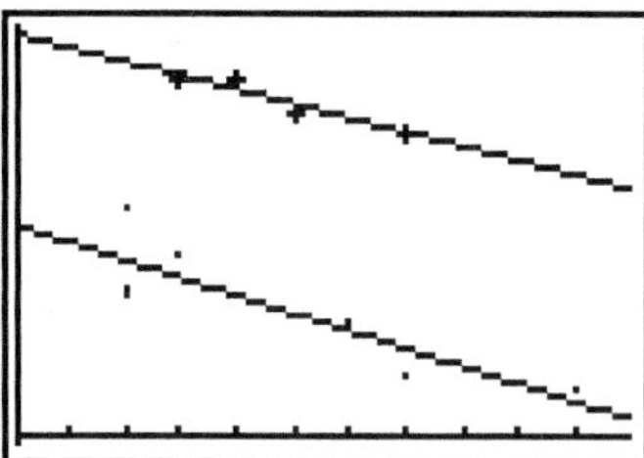

Exercise 3.29 (p. 121 of Agresti/Franklin)
The sodium and sugar content data for 20 breakfast cereals appears in table 2.3 on p. 34 of Agresti/Franklin. Enter the sodium data into L1 and the sugar data into L2. Let's get the scatterplot and regression line as it appears on p. 121. Realize that X=sugar (L2) and Y=sodium (L1).

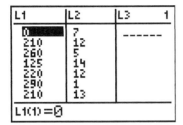

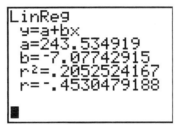

L1	L2	L3 1
0	7	------
210	12	
260	5	
125	14	
220	12	
290	1	
210	13	

L1(1) =0

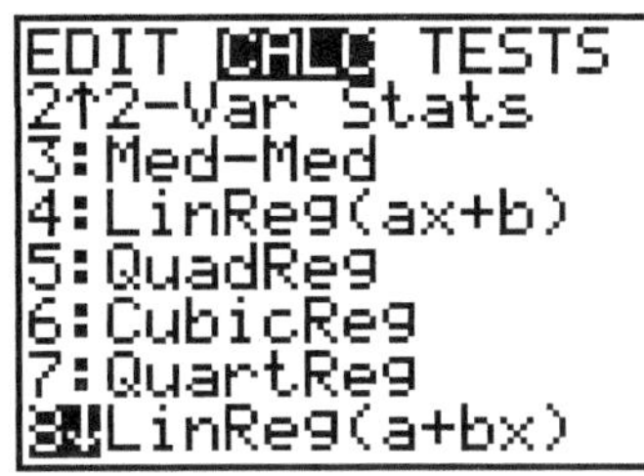

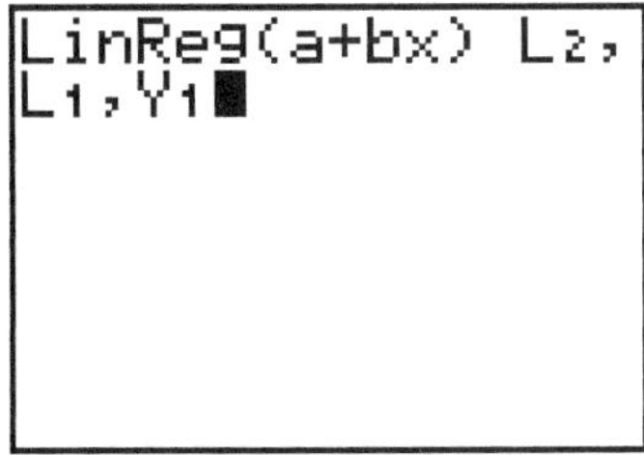


```
LinReg
 y=a+bx
 a=243.534919
 b=-7.07742915
 r²=.2052524167
 r=-.4530479188
```

Make sure that Y1 is the only equation in the Y= menu. Otherwise, other lines that have nothing to do with this problem will be plotted!

```
Plot1 Plot2 Plot3
\Y1=243.53491902
834+-7.077429149
7976X
\Y2=
\Y3=
\Y4=
\Y5=
```

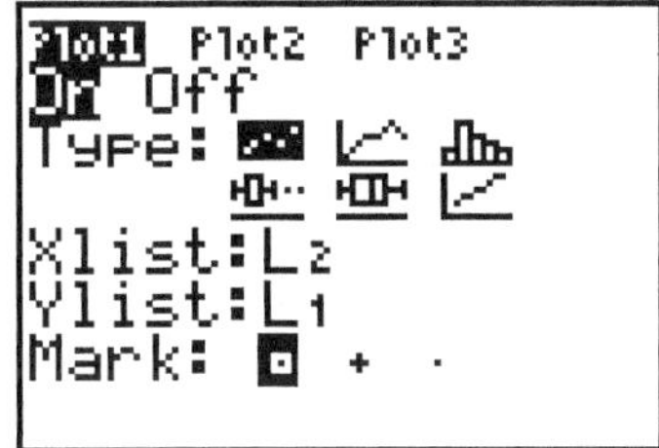

Don't forget ZoomStat to fit the graph into the window optimally. Refer to the original data in table 2.3 on p. 34 of Agresti/Franklin. Note that Frosted Mini Wheats have no sodium. The TRACE function locates this point right away, since it is the first pair of data.

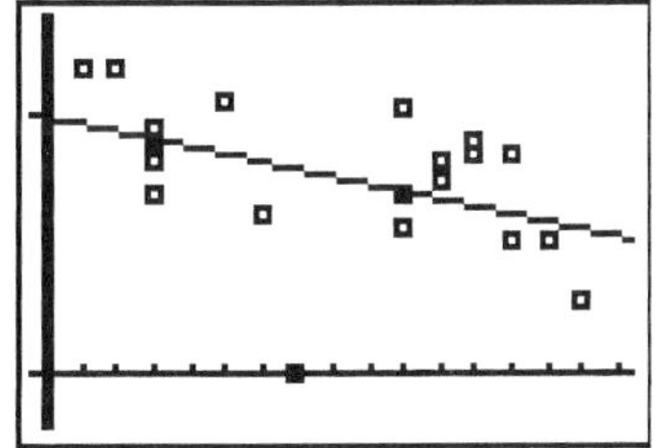

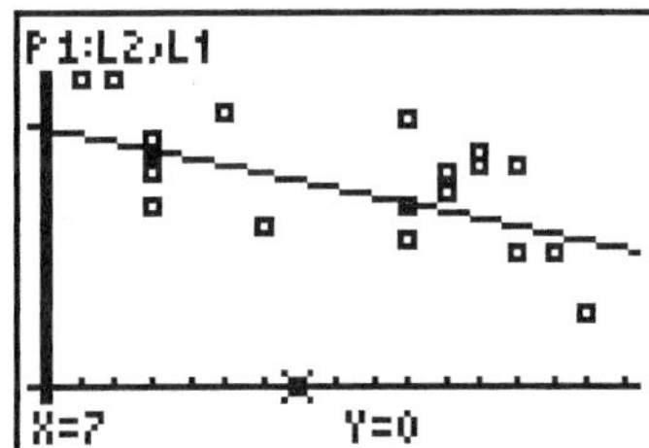

The vertical distance from (7,0) and the regression line is the residual for the Frosted Mini Wheats data point. While using TRACE, press the down arrow. Now type in "7". Recall that this will allow us to predict Y (sodium) for X (sugar) equal to 7 mg.

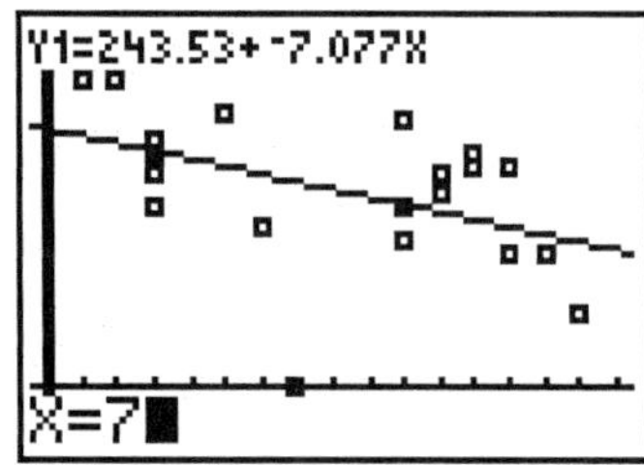

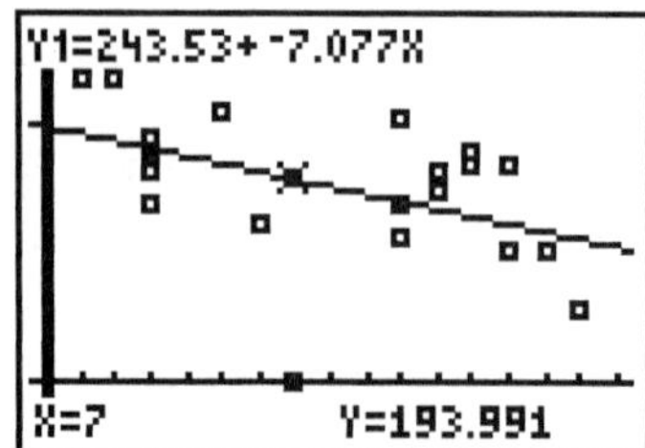

Since the residual is the difference between the observed Y value and the predicted Y value, or $y - \hat{y}$, we have 0 – 193.991 = – 193.991. The Frosted Mini Wheats point is *below* the regression line, 193.991 mg of sodium *less* than predicted.

EXAMPLE 12: How can we forecast future global warming? (pp. 123-124 of Agresti/Franklin)

To get the data for this problem into your calculator, you have three options.

1. Type it in. Since the data are not listed in the textbook, you would have to open up one of the provided data files in order to type it in. Not very efficient!
2. Download it from the "central_park_yearly_temps" Excel file. For details on how to get data from Excel into your calculator, see EXAMPLE 1 in chapter 11 of this manual. Those instructions can be modified to fit this example.
3. Download it from the CENTRALP.8xv TI data file. To do this, you must have the CellSheet application. This application comes loaded on the Silver Edition versions of the TI;83 plus and TI;84 plus. Owners of non;Silver Edition versions of the TI;83 plus or TI;84 plus may download the CellSheet application from Texas Intruments for a fee. See http://education.ti.com/us/product/apps/cellsheet.html.

Downloading data into CellSheet

We will use this example to illustrate how to download data into the CellSheet application. It is assumed that you have a Windows PC, the TI Connect software, and cable to connect your calculator to the USB port of your computer.

1. Connect your calculator to your computer and open the TI Connect software.
2. Open Device Explorer and select your calculator.
3. Drag and drop the CENTRALP.8xv file onto the Device Explorer window.
4. After the file transfers to your calculator, press APPS. Select 0:CelSheet.

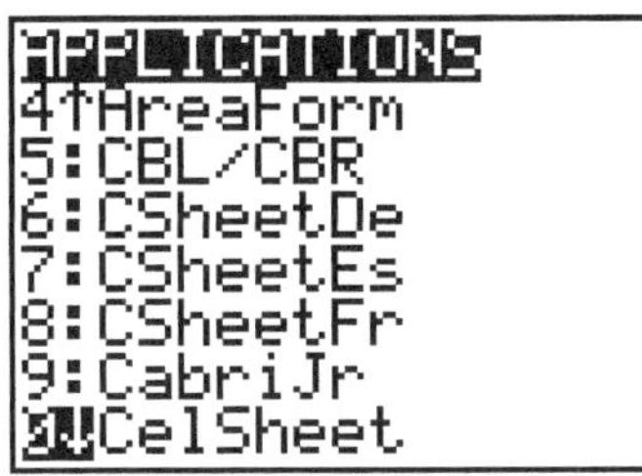

5. Press any key to continue. Note that F5 (ALPHA GRAPH) is the Menu key. Actually, you need only to press the GRAPH key to access the CellSheet menu.

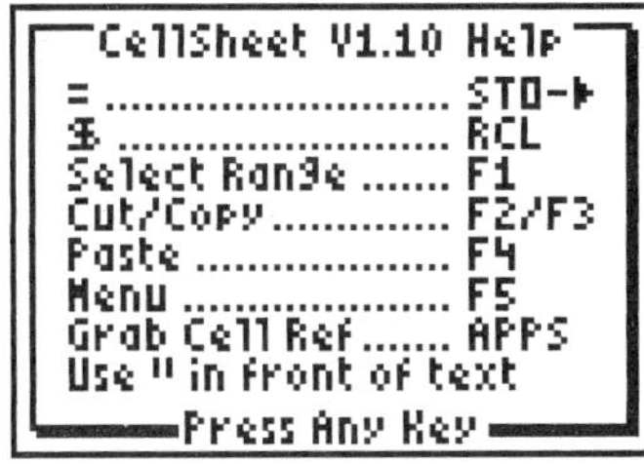

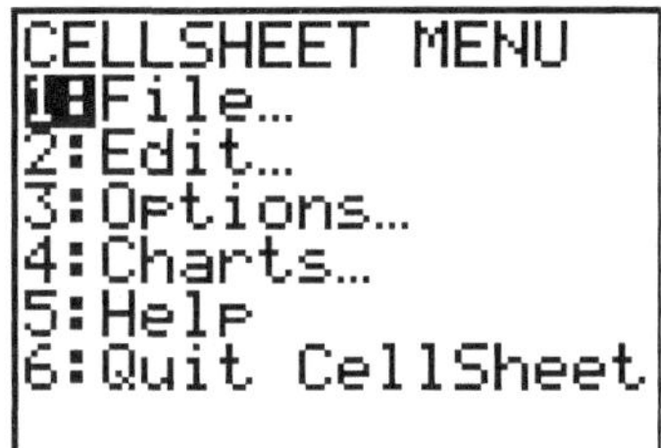

6. Once in CellSheet, press GRAPH, which produces the MENU screen. Choose File, then Open. Choose CENTRALP.

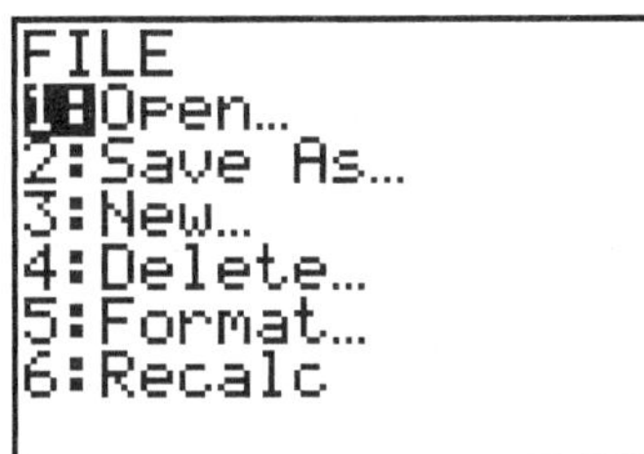

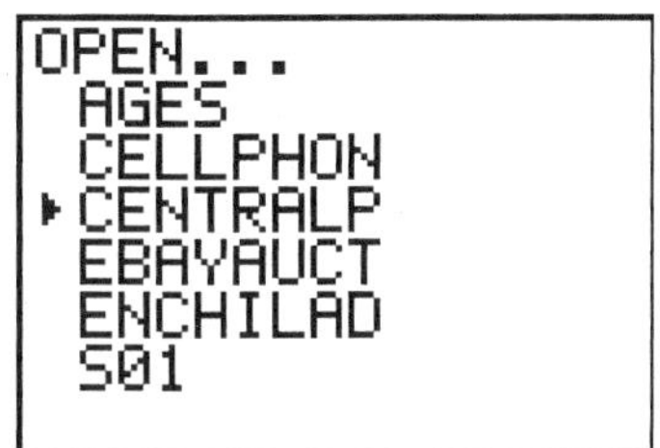

CENT	A	B	C
1	TEMP	YEAR	TIME
2	51.73	1901	1
3	52.27	1902	2
4	52.37	1903	3
5	50.19	1904	4
6	52.75	1905	5

A1: "TEMP Menu

CELLSHEET MENU
1:File...
2:Edit...
3:Options...
4:Charts...
5:Help
6:Quit CellSheet

A few statistical options are available here in CellSheet. Press GRAPH again to access the menu. Choose 3:Options.

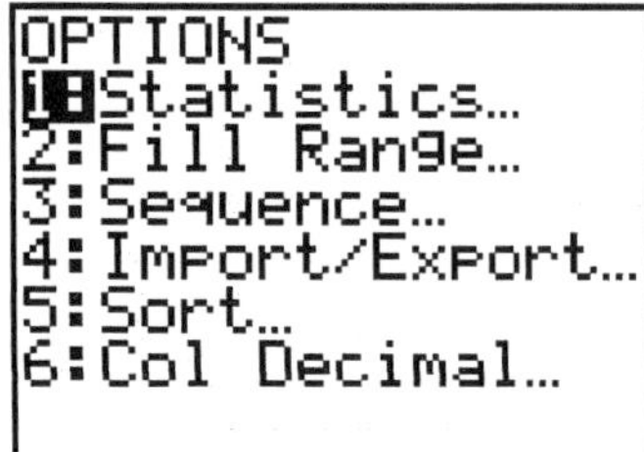

We *could* produce the linear regression output asked for in this example, including the plot. However, there are more options available to us if we transfer the data from CellSheet into our old friend, the STAT Editor.

Transferring data from CellSheet into STAT Editor

1. Press QUIT (2nd MODE) to get back to the CellSheet menu. Choose 3:Options.
2. Choose 4:Import/Export.

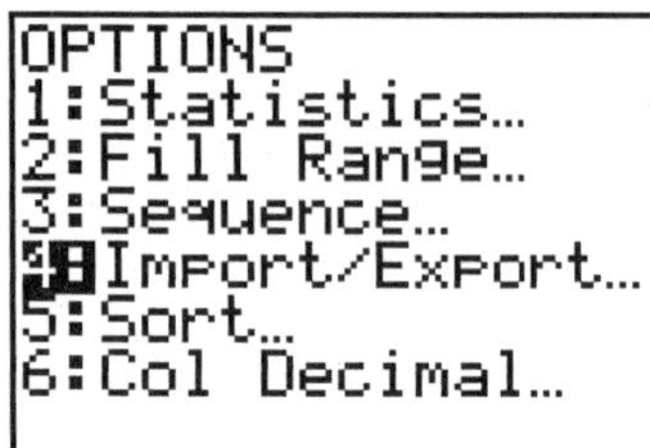

2. Choose 2: Export List. The temperature data resides in cells a2 to a101, since the name "TEMP" is in a1, and there are 100 years of data. Export this list first. You'll have to type in the info as you see it below using the ALPHA key as needed. With the cursor blinking on Enter, press the ENTER key.

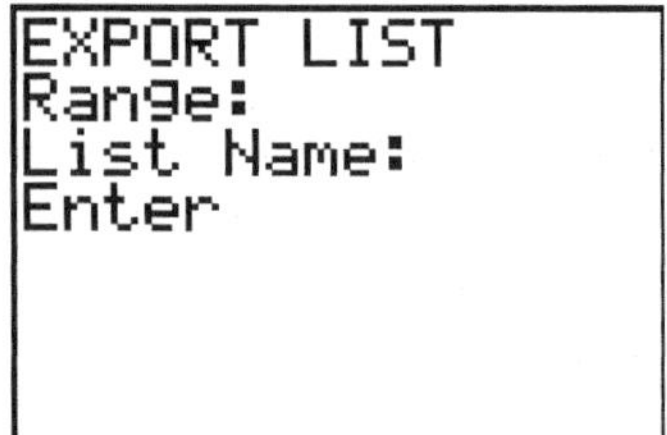

3. Repeat for YEAR and TIME. Ranges will be B2:B101 and C2:C101, respectively.
4. Exit CellSheet by pressing QUIT (2nd MODE). Access the STAT Editor by pressing STAT, then 1:Edit. Clear out any unwanted data. Move the cursor up to the L1 header. From here, press LIST (2nd STAT). Scroll down until you find TEMP. Choose it.

L1	L2	L3 1
------	------	------

L1 =

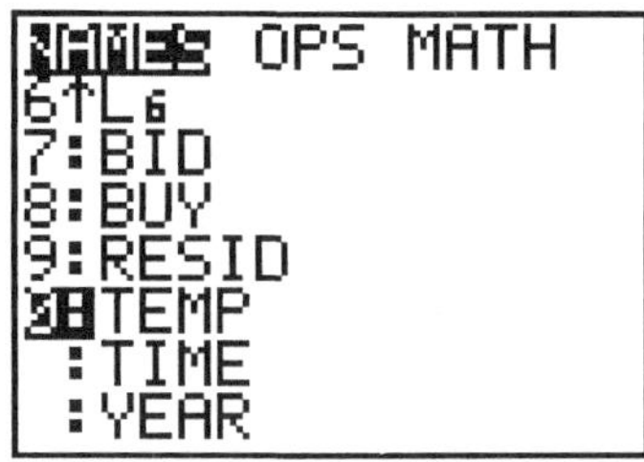

5. Note that you are about to place the TEMP data into L1. Press ENTER.

L1	L2	L3 1
------	------	------

L1 =LTEMP

L1	L2	L3 1
51.73	------	------
52.27		
52.37		
50.19		
52.75		
54.25		
52.19		

L1(1)=51.73

6. Repeat for YEAR (L2) and TIME (L3).

L1	L2	L3 3
51.73	1901	1
52.27	1902	2
52.37	1903	3
50.19	1904	4
52.75	1905	5
54.25	1906	6
52.19	1907	7

L3(1)=1

Now we have the Central Park mean annual temperature data for the years 1901 to 2000 in lists L1;L3. Finally, we can address the questions posed in EXAMPLE 12.

First, let's plot the data. Press STAT PLOT. Making sure Plot2 and Plot3 are turned Off, choose Plot1 and turn it On. Select the second plot icon, which gives a time series plot. It will "connect;the;dots" over the 100;year period. Suppl y L3 (or L2) for the Xlist. Either one will produce the same graph. Supply L1 for the Ylist. Choose whichever Mark you think will look the best. Press ZOOM, then 9: ZoomStat.

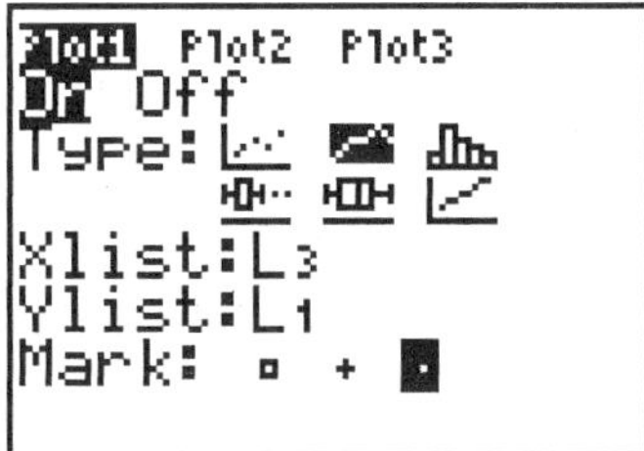

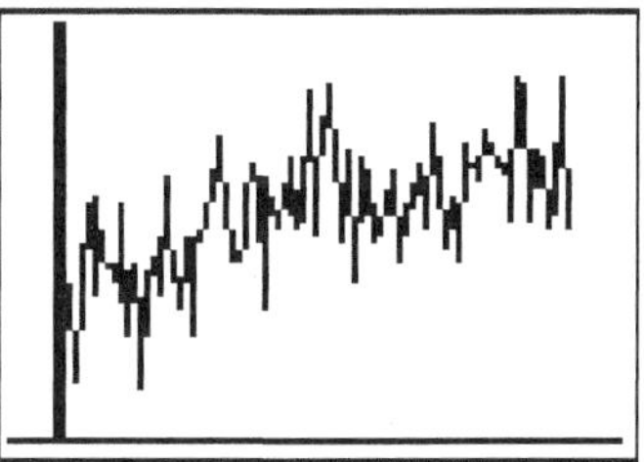

The TRACE function allows us to cycle through the years. For example, we can see that the average temperature was 53.43 in 1927.

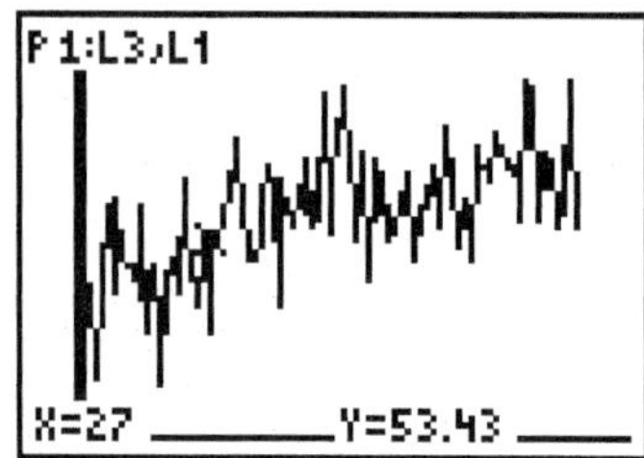

What does a regression line tell us about the trend over the twentieth century? The response variable is Y=TEMP (L1), and let's use TIME (L3) for the X variable. Press STAT CALC, then choose 8:LinReg(a+bx). Supply the arguments as shown below. Remember that you must supply the Xlist first, then the Ylist. An optional third argument can be given to store the regression equation. Recall that you must press VARS, Y;VARS and 1:Function in order to supply the Y1 argument. Typing in "Y" and "1" will give an error (don't try it).

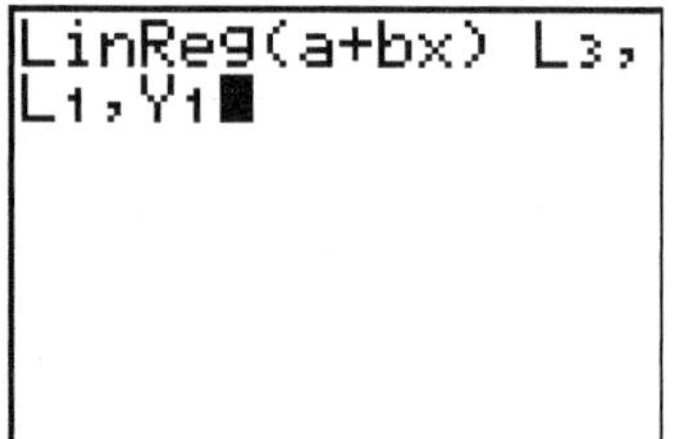

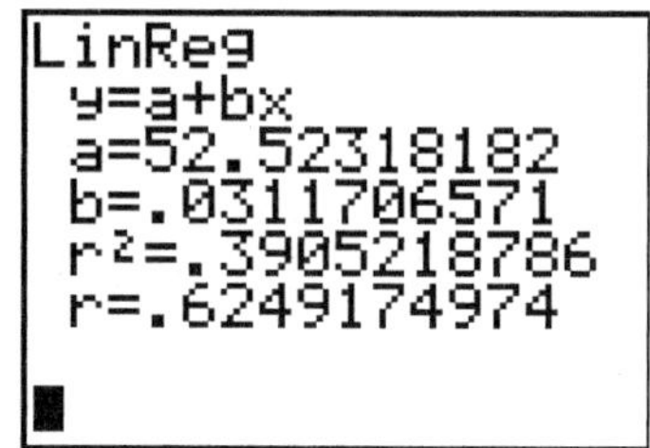

The trend line is $\hat{y} = 52.5 + 0.031x$. Pressing GRAPH again will produce the time plot, along with the linear trend line.

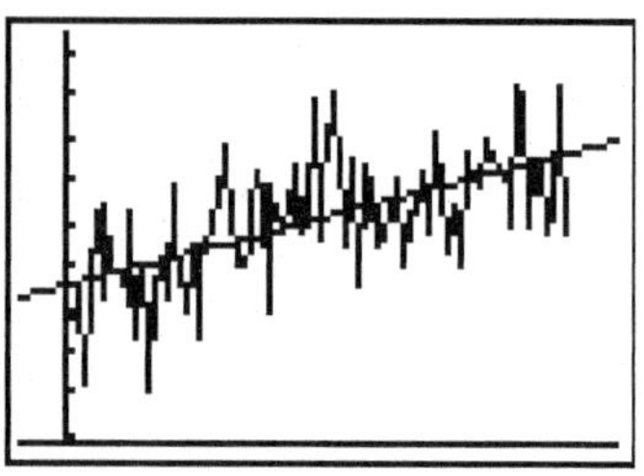

Exercise 3.45 (p. 137 of Agresti/Franklin)

Recall that when using sugar content to predict sodium content for the cereal data, we saw a negative linear pattern in the scatterplot. The linear equation was estimated as $\hat{y} = 243.5 - 7.08x$. Remember also that the point attributed to Frosted Mini Wheats stood out from the majority of the data. How influential is this point? Let's find out by deleting it from the data lists and re;running the regression analysis. In the STAT Editor, use the DEL key to remove the first row of data, the Y=0 and X=7 pair.

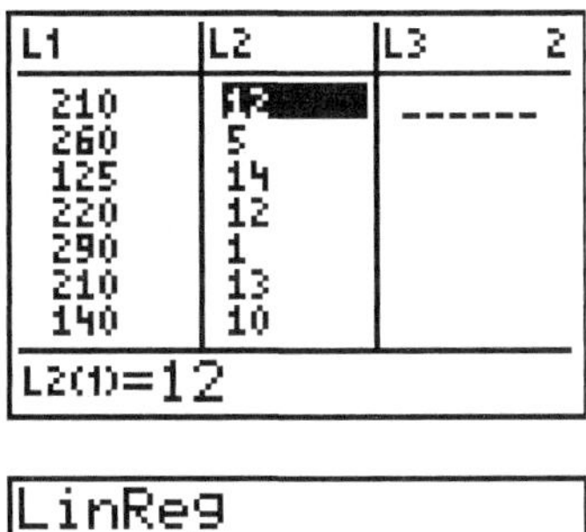

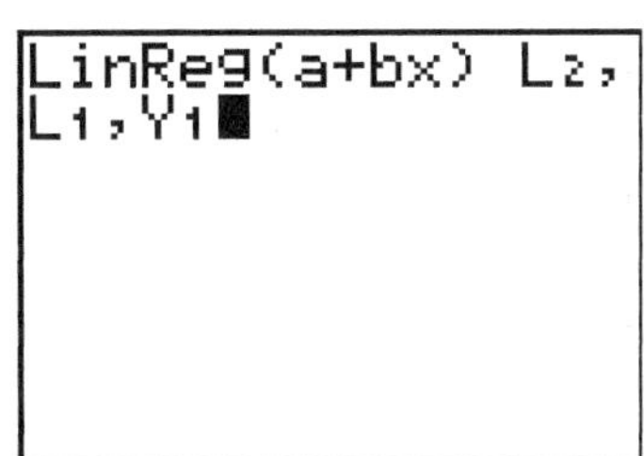

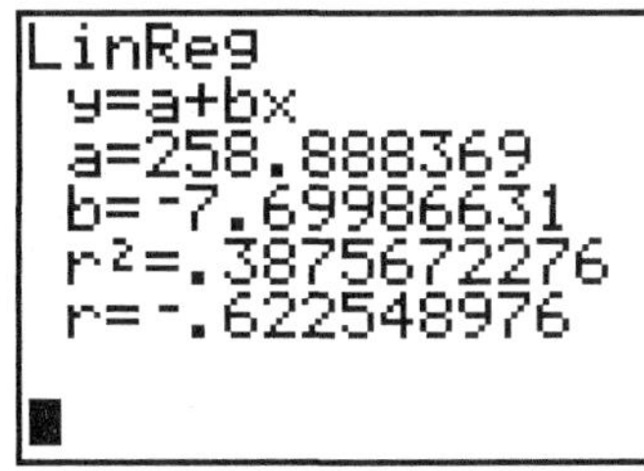

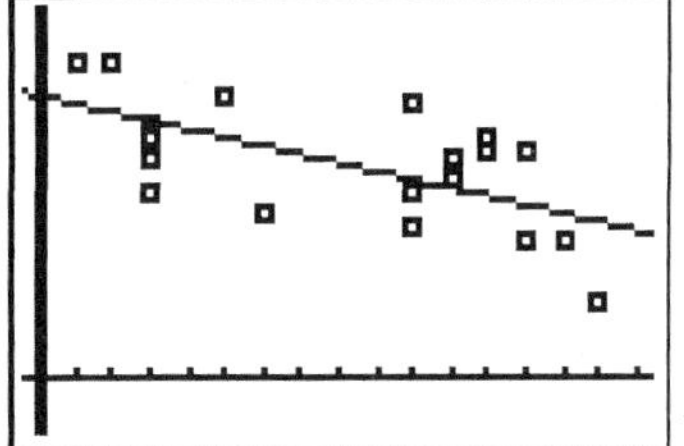

Now the regression equation is $\hat{y} = 258.9 - 7.70x$. Deleting the (7,0) point didn't change the equation very much. (Contrast this with the murder rate in D.C example in figure 3.20 on p. 128 of Agresti/Franklin.) The Frosted Mini Wheats point wasn't very influential on the regression analysis. This is because the X=7 value is neither low nor high relative to the rest of the sugar data.

Chapter Four: Gathering Data

EXAMPLE 5: Auditing the accounts of a school district (pp. 158-159 of Agresti/Franklin)
In this example, the authors use a table of random numbers to generate a simple random sample of size 10, where the numbers are between 1 and 60. This is possibly the best feature of the TI-83 – no need for tables! We will use the randInt function, which can be found in the MATH PRB menu. Press MATH, then scroll to PRB. Choose 5:randInt. This function randomly generates an integer (or a collection of several integers) in a certain range. So we need to supply a lower and upper bound for these integers. In this example, the lower bound is 1, and the upper bound is 60. We can also give a third argument for how many integers we would like.

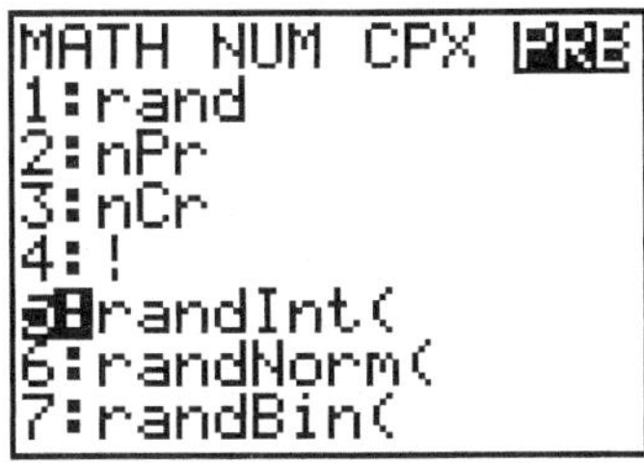

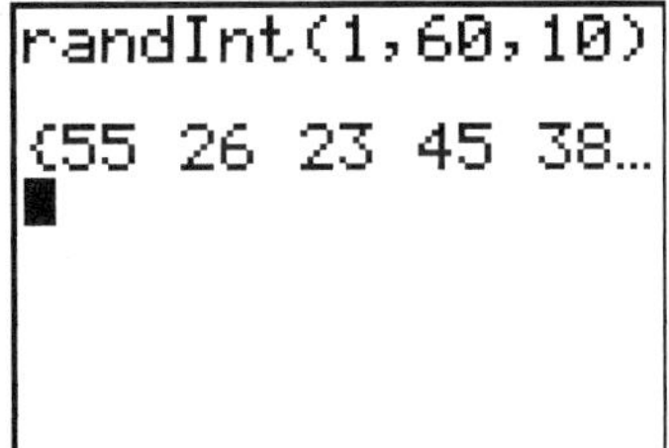

NOTE: You will get different results on your calculator! Above, you can see that 10 integers were requested. The first five are visible: 55, 26, 23, 45, and 38. Scroll right to see the rest. We can't select account 33 and 38 twice, so at this point we only have selected eight accounts. Let's use the randInt function to generate integers one-at-a-time until we have a total of ten unique account numbers from 1 to 60.

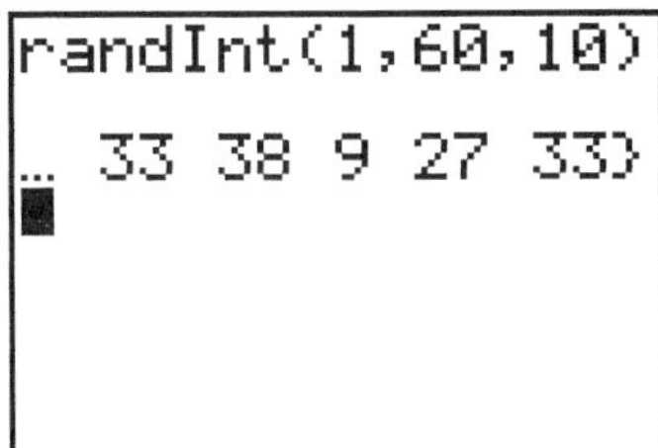

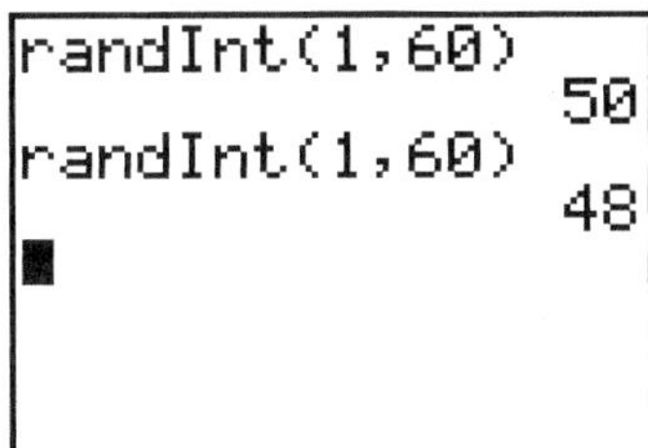

Twice more did the trick; now we have our simple random sample. We should select accounts 55, 26, 23, 45, 38, 33, 9, 27, 50 and 48.

Exercise 4.15: Sample students (p. 165 of Agresti/Franklin)
To randomly sample 3 students from 50, we will use a lower integer bound of 1 (same as last time), and an upper integer bound of 50. Remember that you will get different integers from your calculator!

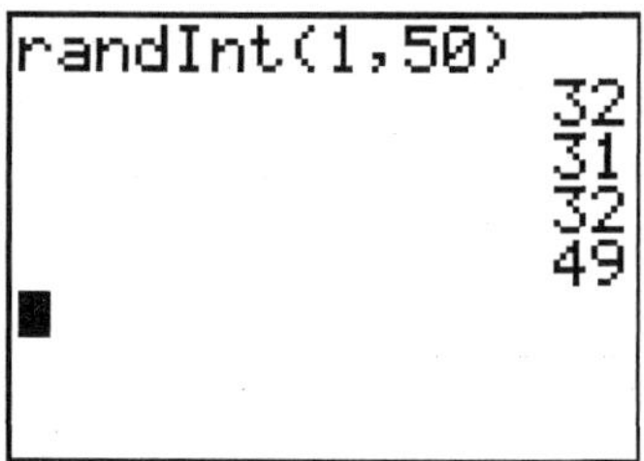

The method above is maybe the best way to select a simple random sample using the TI-83. The randInt function was used, along with the appropriate lower and upper integer bounds. After generating the first integer, you can just press ENTER again, and another integer will be generated. You can see above that since 32 appeared a second, ENTER was pressed four times in order to get three unique numbers. The random sample of three students from the class is: student #32, 31, and 49.

Chapter Five: Probability in Our Daily Lives

EXAMPLE 2: Is a die fair? (pp. 194-196 of Agresti/Franklin)
We can simulate the die rolling experiment with the TI-83. We'll need to use the randBin (random binomial) function. Without discussing binomial experiments in detail here, think of the die rolling example like so. With a fair die, the chance of rolling a "six" is one-out-of-six, or 1/6. We'll use the randBin function with three arguments: 1, 1/6, and 100. This will simulate the outcome of the die as follows: Roll a die once. If a "six" comes up, record as "1", otherwise "0". Repeat for a total of 100 rolls. In this way, we can easily count up how many sixes were rolled. We can create a table similar to table 5.1 on p. 195 of Agresti/Franklin. To start, let's put the sequence 1,>,…,100 in L1. This would be tedious to type in manually, so instead let's use the seq function. Find this function by pressing LIST (>nd STAT), then right arrow to OPS. Many types of sequences can be generated with this function. The arguments are seq(expression, variable, begin, end, increment). We do not need to apply any sort of "expression" to the "variable", and since the increment is one by default, we can leave this off. The letter A is typically used for the variable, so it will appear twice here. (Please don't worry about this minor detail! It's just an easy way to avoid typing in all the numbers from 1 to 100.) Remember to use the ALPHA key in order to type the letter A (ALPHA MATH).

```
NAMES OPS MATH
1:SortA(
2:SortD(
3:dim(
4:Fill(
5:seq(
6:cumSum(
7↓ΔList(
```

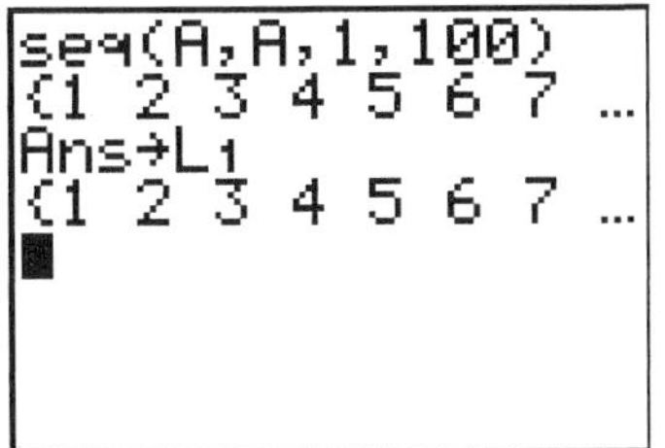

Use the STO> key to store the sequence in L1. Next, let's simulate the 100 rolls of the die. Instead of "no" and "yes" like you see in table 5.1, we will generate 0's and 1's to denote whether or not a six was rolled. Press MATH, then arrow over to PRB. Choose 7:randBin.

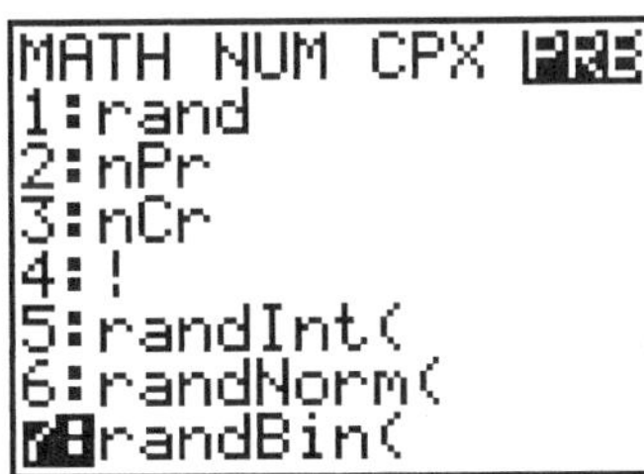

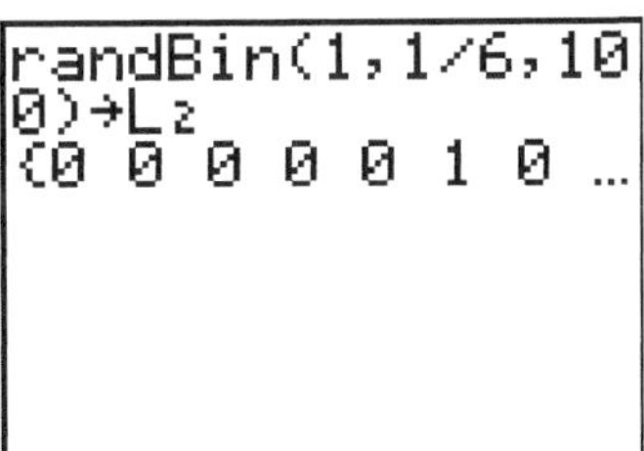

Note that the first five rolls were not sixes. The sixth roll produced a six, and the seventh did not. We have such results for 100 rolls. Also note that the output can be placed directly into L> using the STO> key. And remember that **your** calculator will produce different results! Indeed, the above output differs from the simulation in table 5.1.

If you like, you can use the STAT Editor to scroll through the output.

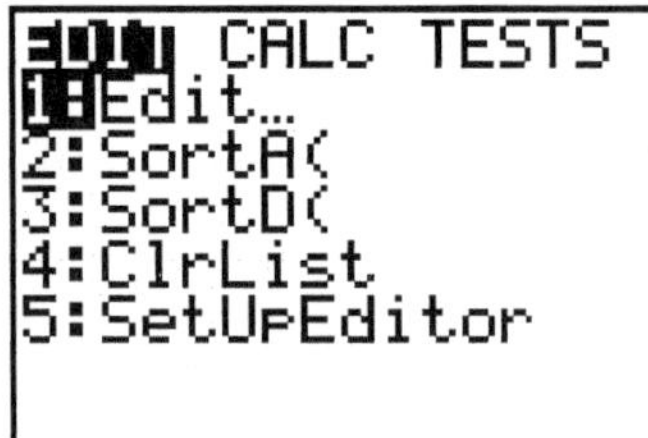

L1	L2	L3 1
10	0	
11	0	
12	0	
13	1	
14	1	
15	0	
16	0	

L1(16) =16

Above, we can see that rolls 13 and 14 produced a "six". Next, we need to add up those sixes as we go down the list. This can be accomplished with the cumSum function. Press LIST, then arrow right to OPS. Choose 6:cumSum. A running total of L> can be placed into L3.

```
NAMES OPS MATH
1:SortA(
2:SortD(
3:dim(
4:Fill(
5:seq(
6:cumSum(
7↓ΔList(
```

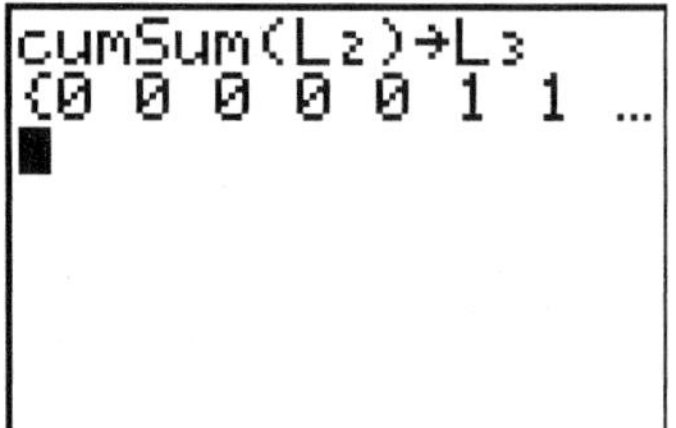

L1	L2	L3 3
3	0	0
4	0	0
5	0	0
6	1	1
7	0	1
8	0	1
9	1	2

L3(9) =2

In the above simulation, two sixes were observed by the time we got to roll #9. Next, we can obtain the cumulative proportion of sixes, like we see in table 5.1. For this, we just need to divide each cumulative frequency by the number of rolls at that stage, which is L3 divided by L1. Place this result into L4.

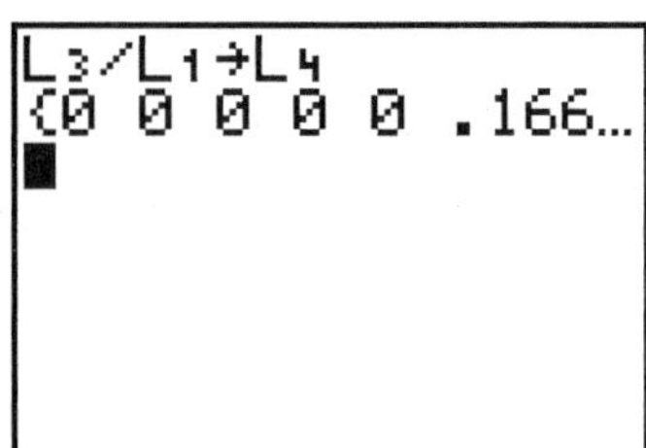

L2	L3	L4 4
0	0	0
0	0	0
0	0	0
1	1	.16667
0	1	.14286
0	1	.125
1	2	.22222

L4(9) =.22222222...

Now we are ready to illustrate the long run behavior of rolling a die. We can produce a graph similar to figure 5.1 on p. 196 of Agresti/Franklin. We need to place the trial number, L1, on the X axis, and the cumulative proportion, L4, on the Y axis. Press STAT PLOT. Turn Plot1 On, and make sure Plots > and 3 are Off. First, let's create a scatterplot. It will look better with the smallest plotting symbol. Don't forget to press ZoomStat.

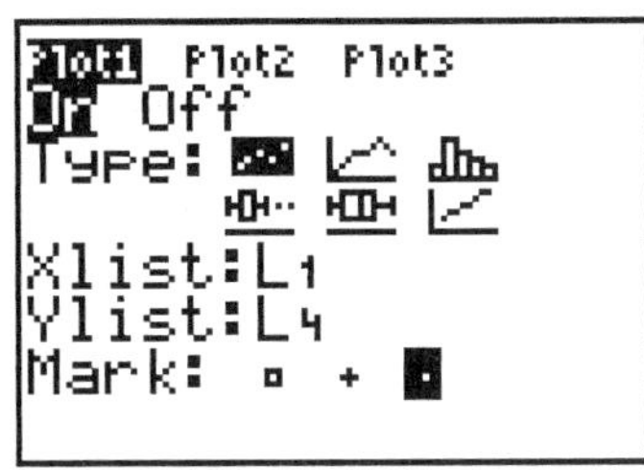

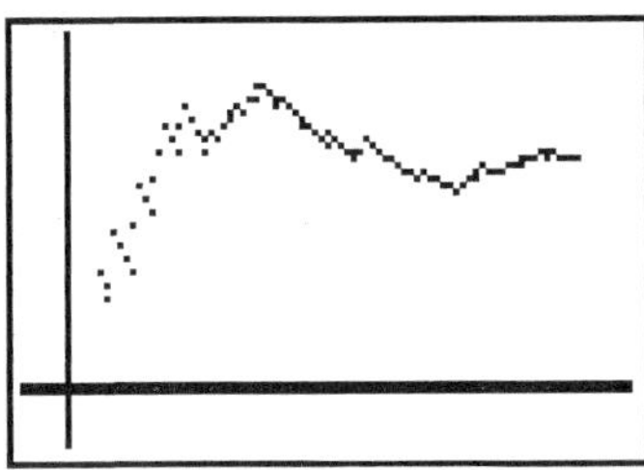

The plot icon next to scatterplot is xyLine. This will "connect-the-dots" and produce a plot more like we see in figure 5.1.

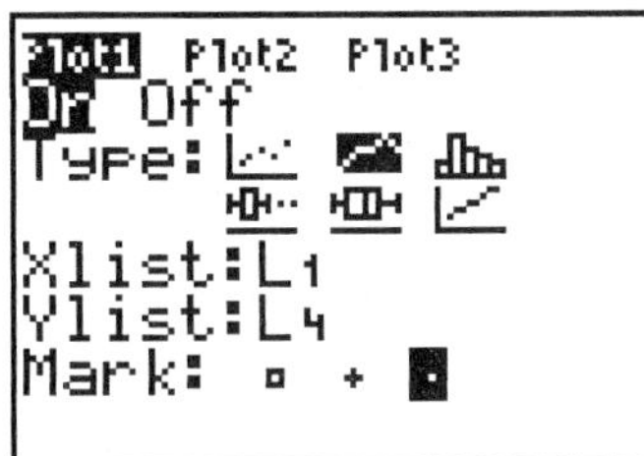

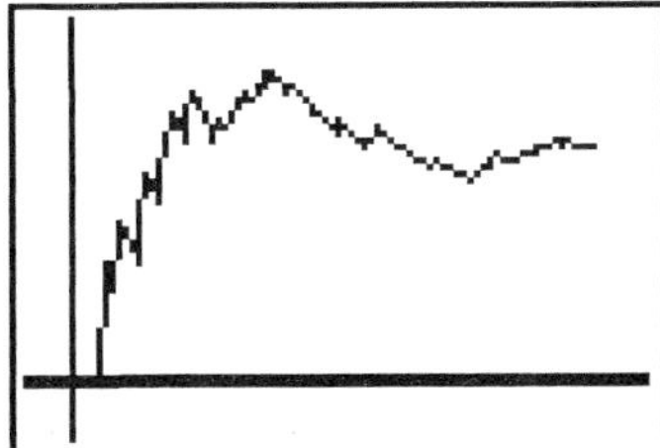

Note how the above simulation differs from figure 5.1. There, sixes occurred quite a lot at first, then leveled out at >3%, somewhat higher than 1/6 = 17%. Above, we see that sixes didn't come up very much at first, but eventually we saw 3> out of the 100 rolls. You can explore this using the TRACE function.

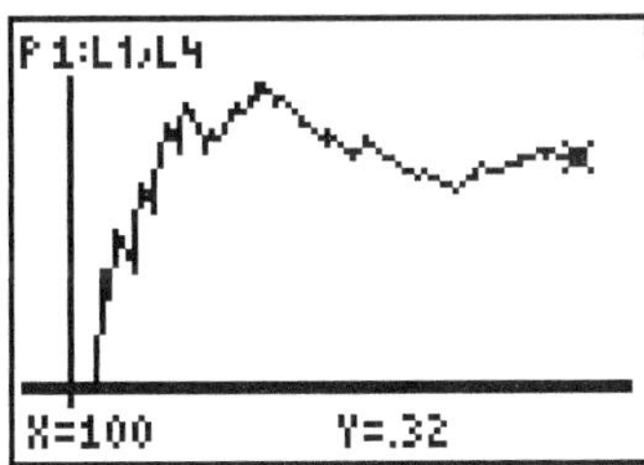

It would also be worthwhile to scroll down the lists in the STAT Editor to see the final results in the simulation. Here we see nearly twice the expected number of 17 sixes in 100 rolls. What happened on your calculator when you tried it?

L1	L2	L3 1
94	1	31
95	0	31
96	0	31
97	0	31
98	0	31
99	1	32
100	0	32

L1(100) =100

L2	L3	L4 4
1	31	.32979
0	31	.32632
0	31	.32292
0	31	.31959
0	31	.31633
1	32	.32323
0	32	.32

L4(100) =.32

Chapter Six: Probability Distributions

The Agresti/Franklin textbook will show you how to use statistical tables to find probabilities and percentiles for certain probability distributions. Depending on the application, the probabilities may be called "p-values", and the percentiles might be called "critical values". Knowing how to use tables to find these "magic numbers" is a useful skill, but why not use your calculator instead, if you have the choice?

Many useful functions for certain probability distributions can be found under the distributions (DISTR) menu. Press 2nd VARS. Scroll down to view them all. Seven probability distributions are represented: Normal, Student's t, Chi-square, F, Binomial, Poisson, and Geometric.

```
DISTR DRAW
1:normalpdf(
2:normalcdf(
3:invNorm(
4:tpdf(
5:tcdf(
6:χ²pdf(
7↓χ²cdf(
```

```
DISTR DRAW
9↑Fcdf(
0:binompdf(
A:binomcdf(
B:poissonpdf(
C:poissoncdf(
D:geometpdf(
E:geometcdf(
```

Note that the first three functions have the word (or partial word) "normal" or "norm" in them. The first one, normalpdf, refers to the probability density function (pdf) for the normal distribution. If you supply a z-score, normalpdf will tell you the height of the normal curve at that point. This is not of much use to us – statistics courses with more of an emphasis on probability (and a calculus prerequisite) might solve some interesting problems with it.

The second function, **normalcdf**, refers to the cumulative distribution function (cdf) of the normal distribution. This function returns **probabilities**; the accumulated area under the normal curve between two z-scores. We'll use this function in the example after next. You'll use this function quite a bit.

The name of the third function, **invNorm**, is short for "inverse normal". This function returns **percentiles**. For example, if you supply a value between 0 and 1, say .70, the invNorm function will return the 70th percentile of that particular normal distribution. You will always have the choice of working with the standard normal distribution (where the mean is zero and the standard deviation is one) or specifying the mean and standard deviation for some other normal distribution. We'll use the invNorm function in the next example.

Probabilities and Percentiles from the Normal Distribution: Section 6.2 of Agresti/Franklin

EXAMPLE 7: What IQ do you need to get into MENSA? (pp. 261-262 of Agresti/Franklin)

We are told that the IQ scores are normally distributed with mean 100 and standard deviation 16. What is the z-score corresponding to the 98th percentile of the standard normal distribution? Choose the invNorm function from the DISTR menu. Supply just one argument, the desired percentile.

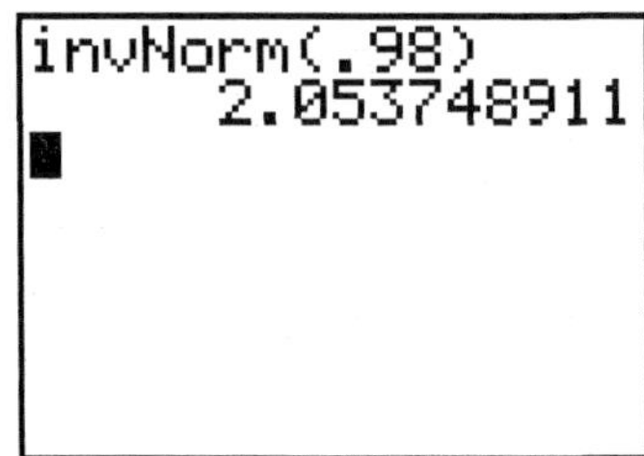

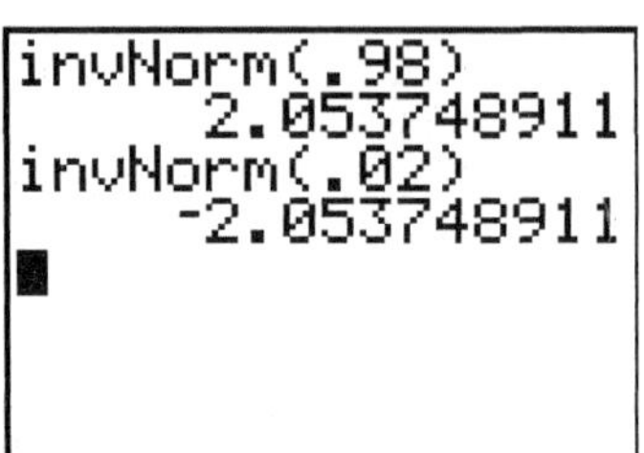

Note what happens if we supply the tail area (see figure 6.9 on p. 261 of Agresti/Franklin) of 2% instead. By the symmetry of the normal curve, we get the same z-score, just on the other side of $\mu = 0$. This is the 2nd percentile. Next we can convert the 98th percentile z-score of 2.05 back into an IQ score. See the conversion calculation on p. 262 of Agresti/Franklin.

```
invNorm(.98)
           2.053748911
Ans*16+100
           132.8599826
```

If you like, you can skip a step by specifying the mean and standard deviation of the IQ scores in the invNorm function. The arguments follow the form invNorm(area, mean, stdev) where mean=0 and stdev=1 is assumed if the last two arguments are not supplied.

```
invNorm(.98,0,1)

           2.053748911
invNorm(.98,100,
16)
           132.8599826
```

EXAMPLE 8: Finding your relative standing on the SAT (pp. 262-263 of Agresti/Franklin)

We are working with normally distributed SAT scores with mean 500 and standard deviation 100. Since 650 is one-and-a-half standard deviations *above* the mean, we get a z-score of (*positive*) 1.5.

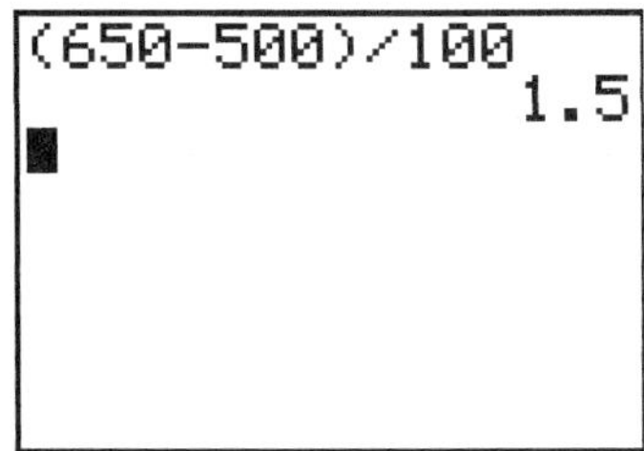

To find the percentage of SAT scores higher than 650, we will use the normalcdf function. Note that this is equivalent to finding the percentage of z-scores above 1.5 (see p. 343 of Agresti/Franklin). The normalcdf function takes at least two arguments: a lower and upper bound. The area (or probability) under the normal curve between the lower and upper bounds is returned. Here, we desire to know the area under the curve for z = 1.5 or greater. The lower bound is 1.5, and the upper bound is (theoretically) infinity. An easy habit to get into is to use 10^{99} (a one with 99 zeroes after it – a very large number!), which is the largest number the calculator will accept. You can type 10^99 (the carat key is below the CLEAR key) or 1E99 (press "EE", which is 2nd comma, for the exponentiation).

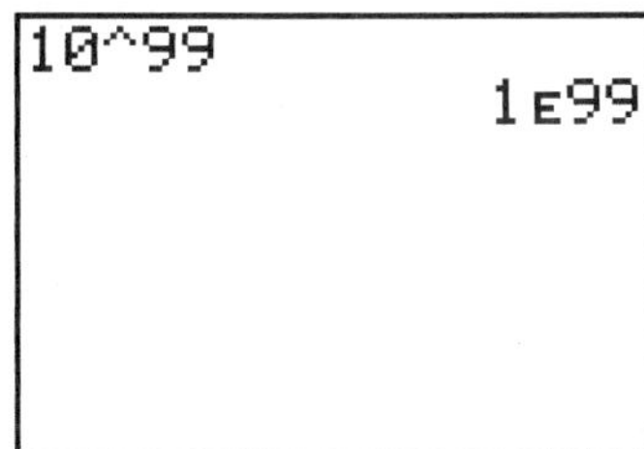

Below we see that the amount of tail area to the right of z = 1.5 on the standard normal curve is .0668. Less than 7% of the area is to the right of 1.5 for a normal distribution with mean zero and standard deviation one. Note that we obtained this probability directly – there was no need to subtract .9332 from 1 (see p. 263) as with Table A in Agresti/Franklin.

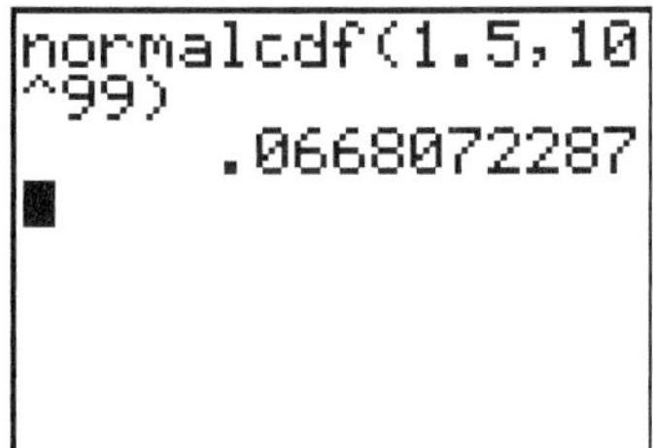

We can work directly with the distribution of SAT scores if we like. That is, when working with normalcdf, we are allowed to specify μ=500 and σ=100 and skip the z-score step. Supply a third and fourth argument to the normalcdf function. Now we are asking for the area under the curve to the right of 650 for a normal curve with μ=500 and σ=100.

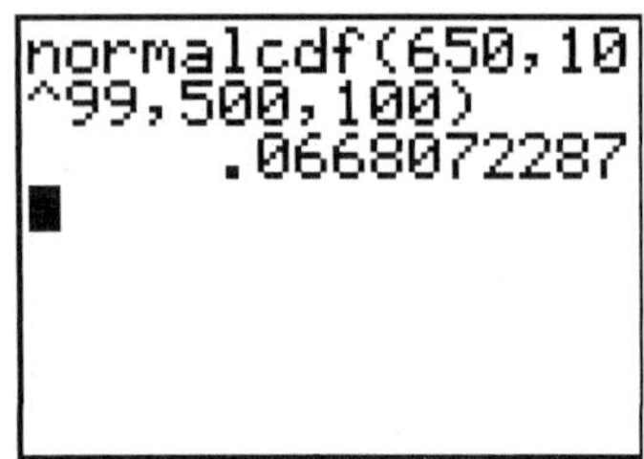

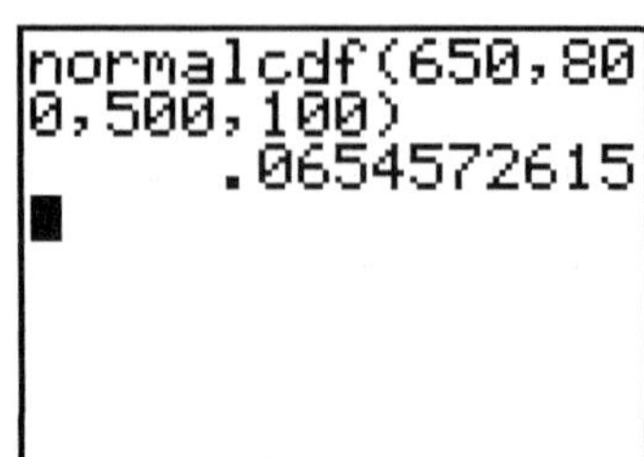

Of course, one cannot achieve a score of infinity (or 10^{99} for that matter) on the SAT, so we might rather supply the upper bound of 800. Either way, we see that about 6 or 7% of SAT scores fall above 650. Remember that these SAT scores are *approximately* normally distributed!

Another useful function here is ShadeNorm. Press DISTR (2nd VARS) again, but this time, arrow right to DRAW. Choose ShadeNorm. The arguments are the same as normalcdf. Be sure to turn off all Stat Plots and Y= functions.

If you don't see an image similar to figure 6.10 (p. 263 of Agresti/Franklin), you'll have to fix the WINDOW settings. ZoomStat won't help here, unfortunately. Use your knowledge of the standard normal distribution and the empirical rule (illustrated again in figure 6.5 on p. 258 of Agresti/Franklin). Since nearly all of the observations fall within three standard deviations of the mean, we should specify the X axis to go from –3 to 3. Setting the vertical dimension of the window to go from –0.1 to 0.4 allows us to view the curve clearly. Xscl=1 puts a space of one between tick marks. Yscl and Xres do not affect the graph.

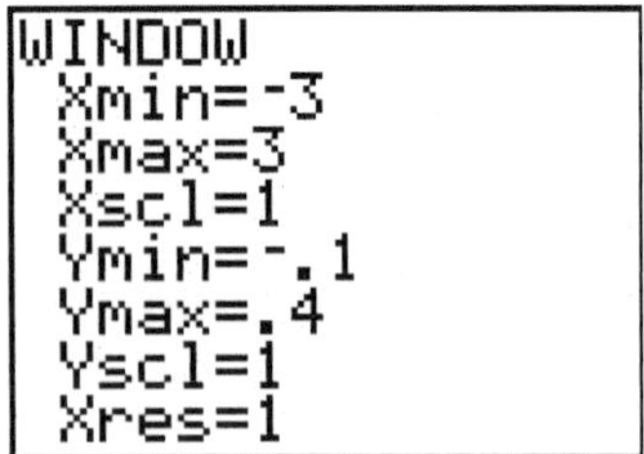

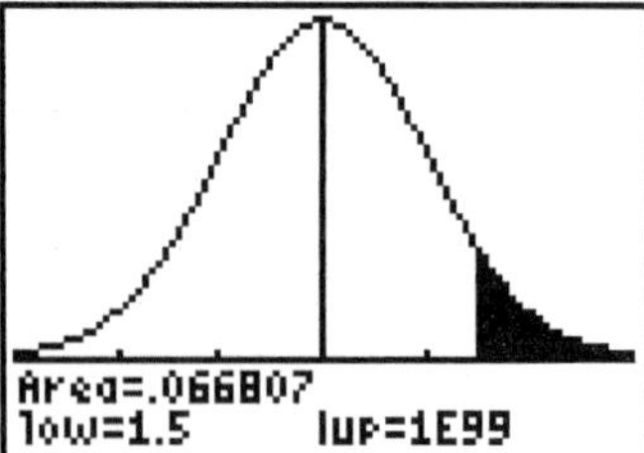

Before creating your next graph, you may need to clear the current one. Use the ClrDraw function under the DRAW menu. Press 2nd PRGM to access the DRAW menu.

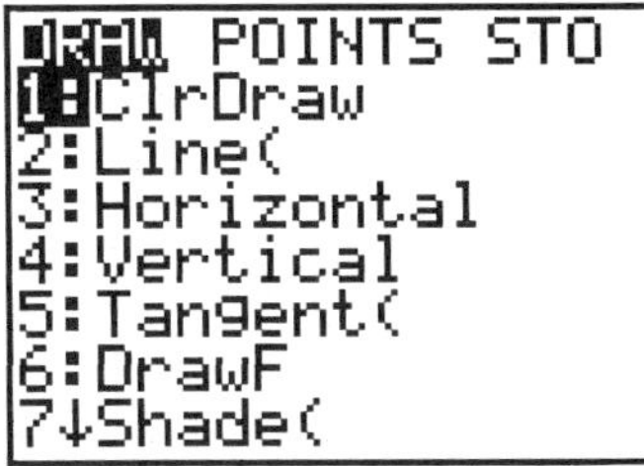

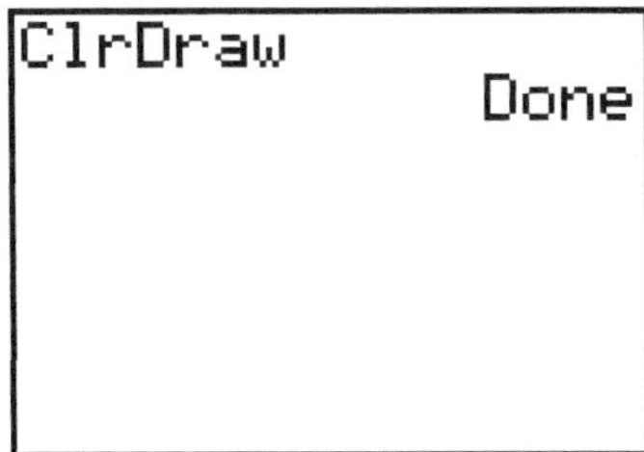

It is possible to use the SAT scores distribution (normal with μ=500 and σ=100) with ShadeNorm, but we'll have to re-specify the WINDOW dimensions. See below. The X scale is easy – we know that the SAT scores range from 200 to 800. The Y scale is a little trickier. For Ymax, since 0.4 works so well for the standard normal distribution, we can divide 0.4 by σ, in this case 100. So use 0.4/100 = 0.004 for Ymax. For Ymin, we can always use zero, but the graph looks better if a little room is provided underneath. Play around with it until it looks good.

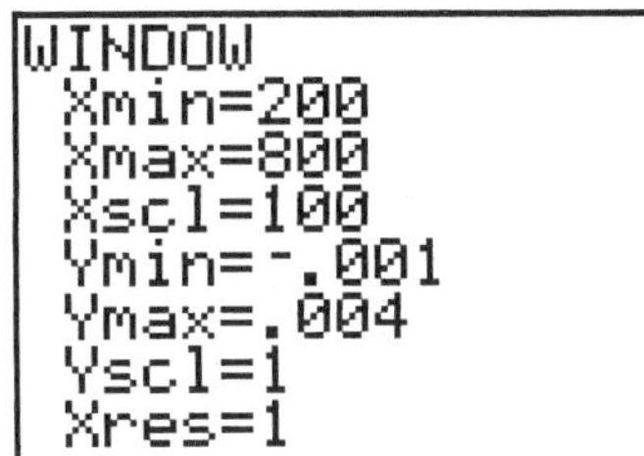

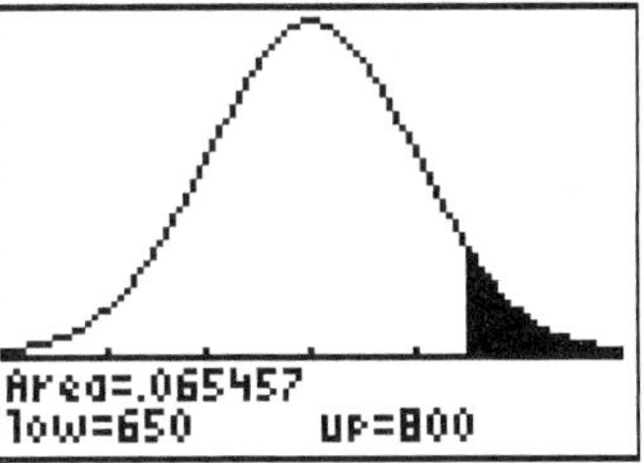

EXAMPLE 9: What proportion of students get a grade of B? (pp. 263-264 of Agresti/Franklin)

Now we are working with a normal distribution that has μ = 83 and σ = 5. What is the proportion of students that score between 80 and 90? Let's first find the z-scores (see calculations on p. 344).

```
(80-83)/5
                  -.6
(90-83)/5
                  1.4
```

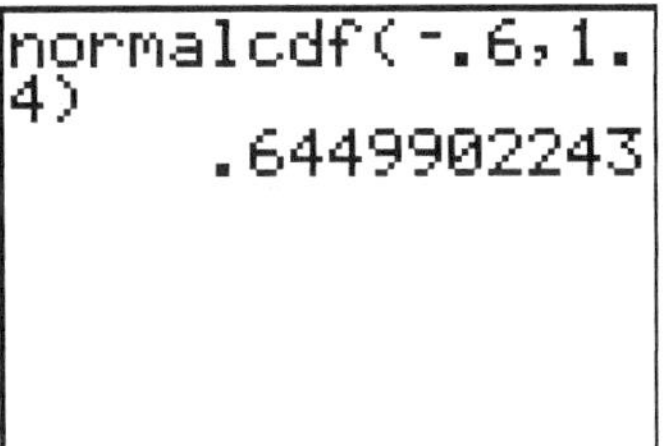

The area under the Z curve between –0.6 and 1.4 is about 65%. Note that we didn't need to use 10^{99} since this is not a "tail probability". Also note that we did not have to find two probabilities and subtract one from the other as shown on p. 264. This is only necessary when using the normal table in Agresti/Franklin (Table A).

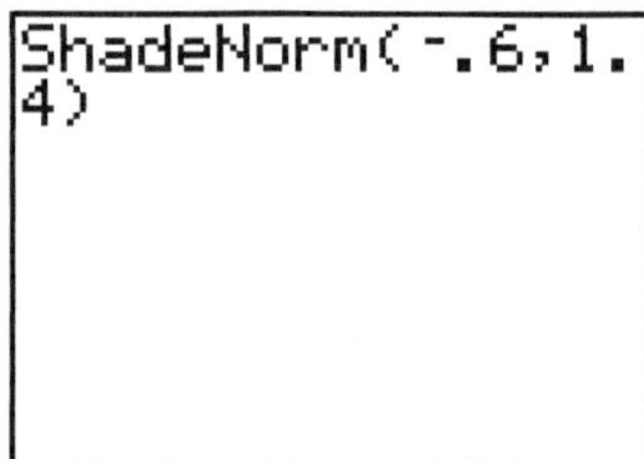

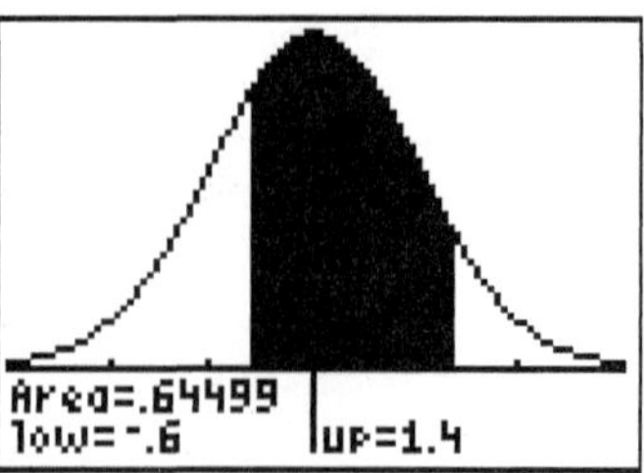

ShadeNorm can be used to get a picture of the distribution. Remember that you may have to use ClrDraw and reset the WINDOW settings. See above for details.

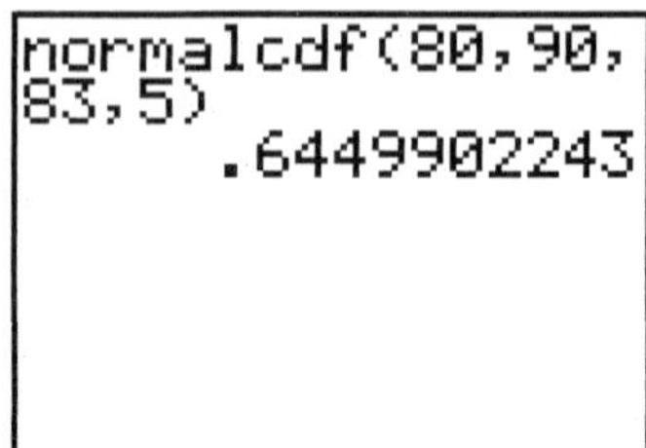

The proportion of students that score in the B range can be obtained without calculating the z-scores first, as long as you specify $\mu = 83$ and $\sigma = 5$ in the normalcdf command.

Exercise 6.21: Blood pressure (p. 267 of Agresti/Franklin)
The systolic blood pressure readings are normal with $\mu = 121$ and $\sigma = 16$.

a. Convert 140 to a z-score.

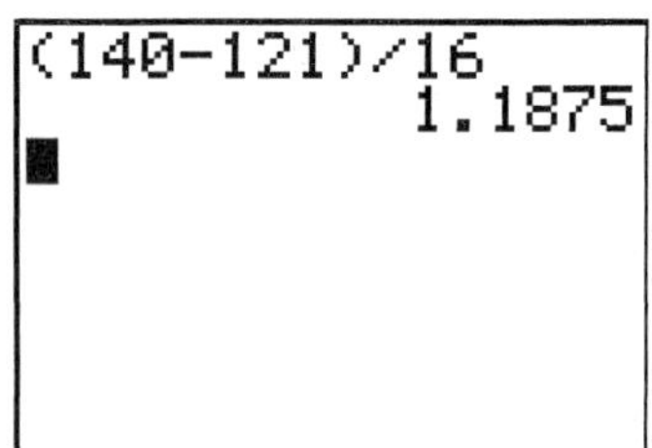

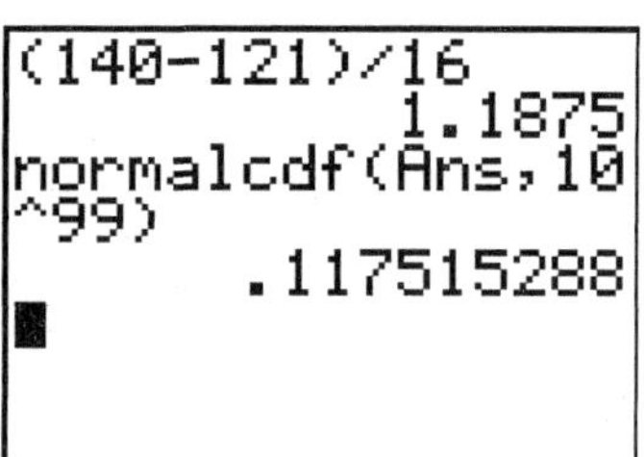

b. The proportion of blood pressure readings above 140 is equal to the area to the right of 1.1875 on the z-curve. Recall that you can type ANS (2nd (-)) to call up the most recently calculated value. ShadeNorm is shown below.

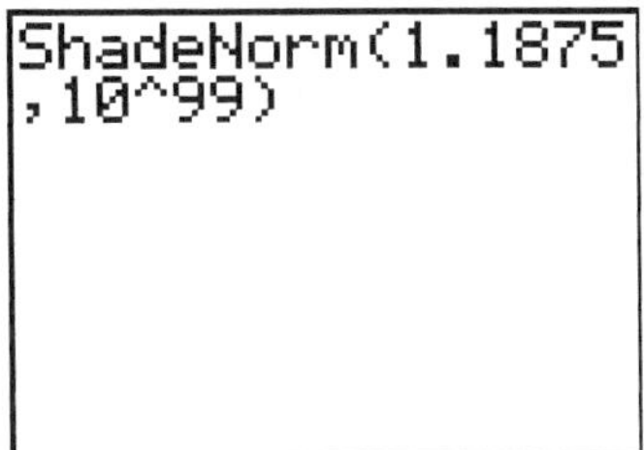

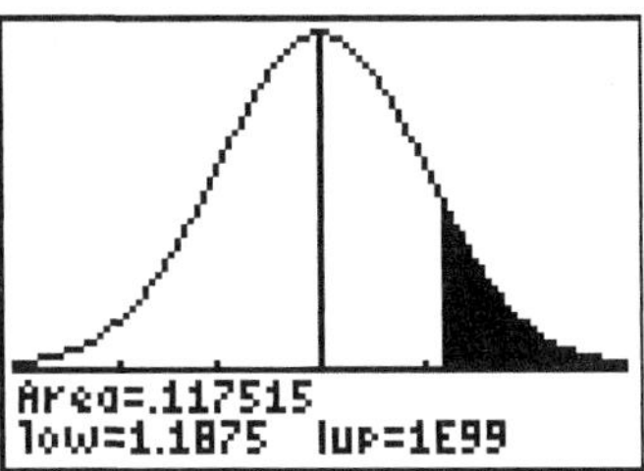

c. Recall that the z-score for 140 is 1.1875. The z-score for a blood pressure reading of 100 is −1.3125 (shown below). We need the area under the z-curve between −1.3125 and 1.1875.

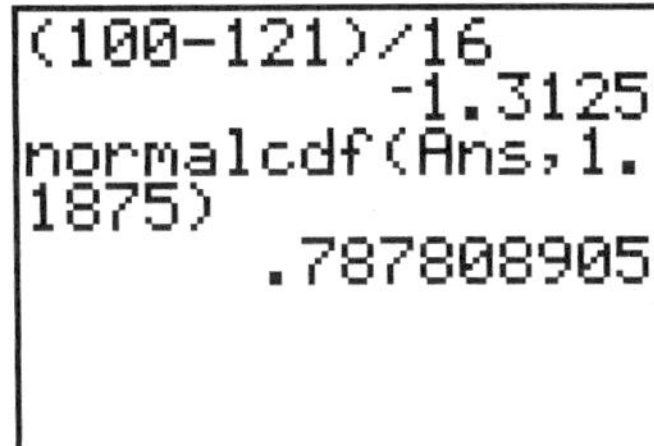

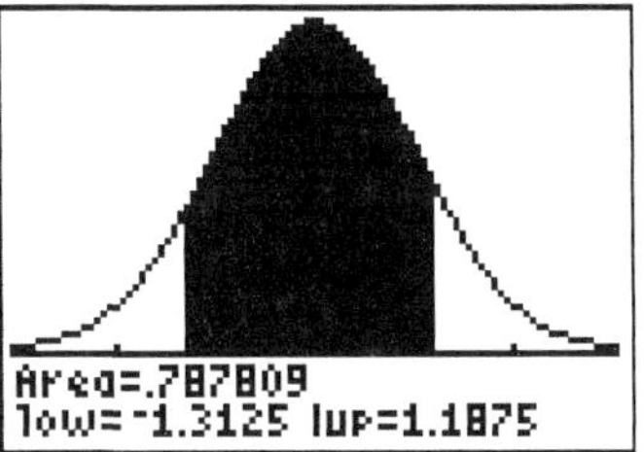

Don't forget to use ClrDraw before using ShadeNorm. If you like using ShadeNorm, you are better off always working with z-scores. That way, you will not have to adjust the WINDOW dimensions. Then, to check your work, you can use the normalcdf command specifying the mean and standard deviation of the original distribution.

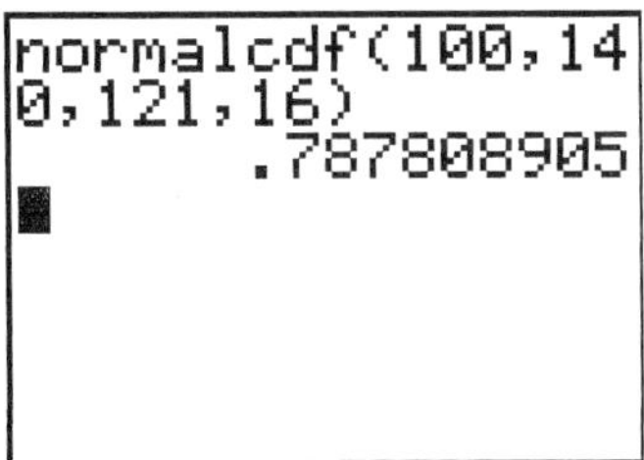

Exercise 6.23: MDI (pp. 267-268 of Agresti/Franklin)

The MDI scores are normal with $\mu = 100$ and $\sigma = 16$.

a. Note that 120 and 80 are 1¼ standard deviations above and below, respectively, the mean. "At least 120" is 120 or more, so we need the area to the right of 120. Similarly for "at least 80".

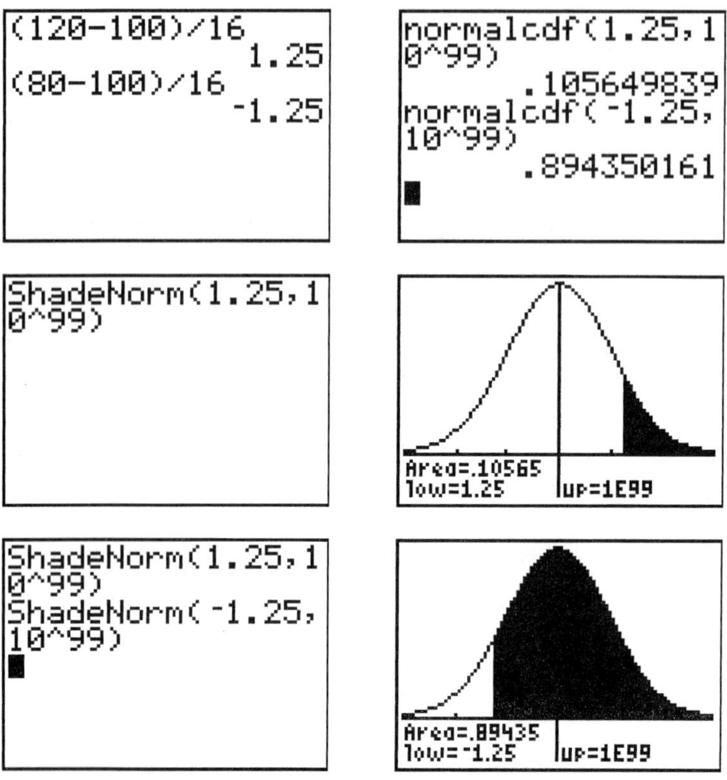

b. The 99th percentile can be found using the invNorm command. Below, the 99th percentile of z-scores is obtained first and then converted to an MDI score.

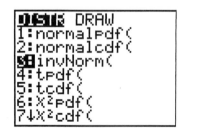

```
DISTR DRAW
1:normalpdf(
2:normalcdf(
3:invNorm(
4:tpdf(
5:tcdf(
6:X²pdf(
7↓X²cdf(
```

```
invNorm(.99)
          2.326347877
Ans*16+100
          137.221566
```

Recall that you can specify the mean and standard deviation in the InvNorm command to skip the z-score step and get the 99th percentile of MDI scores directly.

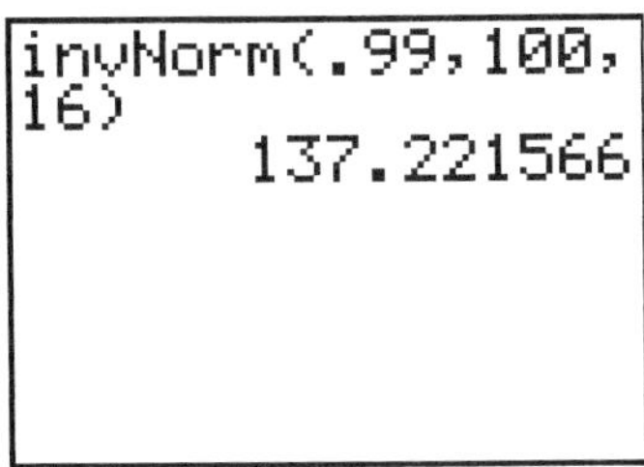

c. The MDI score such that only 1% are below it is the 1st percentile. This is on the opposite of distribution from the 99th percentile. You should not be surprised that the z-score is just the negative version of the one we obtained in part b.

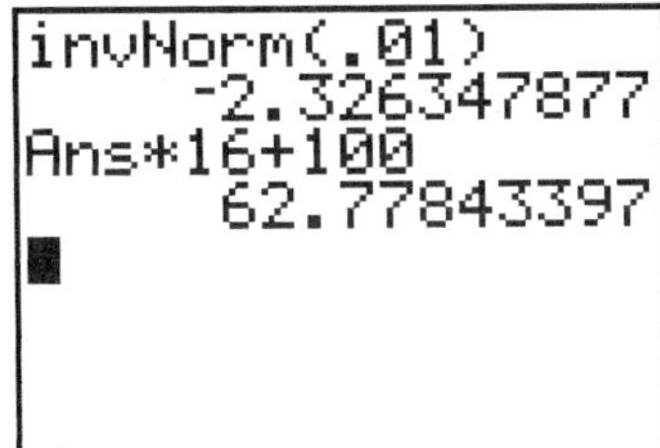

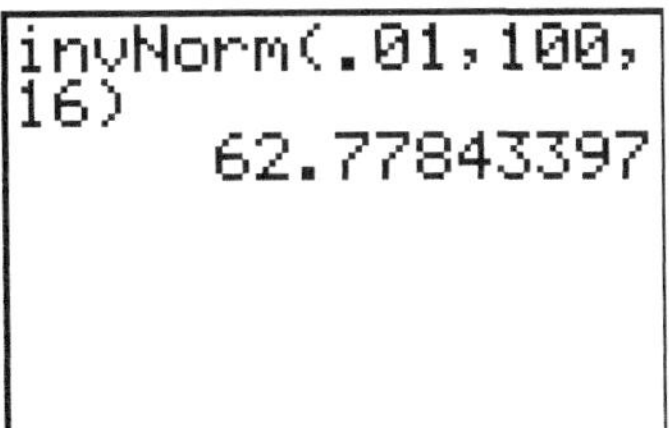

The Binomial Distribution: Section 6.3 of Agresti/Franklin

The normal distribution is a *continuous* probability distribution, a smooth curve. We were interested in the proportion of the distribution that fell in a certain interval. *Discrete* phenomena may be explained by the binomial distribution, under certain conditions (see p. 269 of Agresti/Franklin). Return to the DISTR (2nd VARS) menu. This time, scroll down (or up, which is quicker) until you see the binompdf and binomcdf functions. Both of these commands will be useful for us.

```
DISTR DRAW
9↑Fcdf(
0:binompdf(
A:binomcdf(
B:poissonpdf(
C:poissoncdf(
D:geometpdf(
E:geometcdf(
```

EXAMPLE 12: Are women passed over for managerial training? (pp. 271-272 of Agresti/Franklin)

Note that the conditions for a binomial experiment are satisfied. If X is the number of females selected, then X is binomial with n = 10 and p = .50. X can take any value from 0, 1, 2, …, 10. The probability that zero females are chosen (see calculation on p. 272) is shown below. Use the binom**pdf** function, since this gives the probability that X is equal to a specific value. The arguments are (n, p, x), where x=0 for the problem at hand.

```
binompdf(10,.5,0
)
         9.765625E-4
```

This is a small probability, so the calculator resorted to scientific notation. The probability that no females are chosen is .000977, or about .001.

Let's use this example to further explore the binomial distribution. If we only supply two arguments to binompdf, we get the distribution of probabilities for the entire range of X values (see the margin of p. 272 of Agresti/Franklin). It's hard to see; you must scroll right or left, and it's not clear what X values each probability corresponds to.

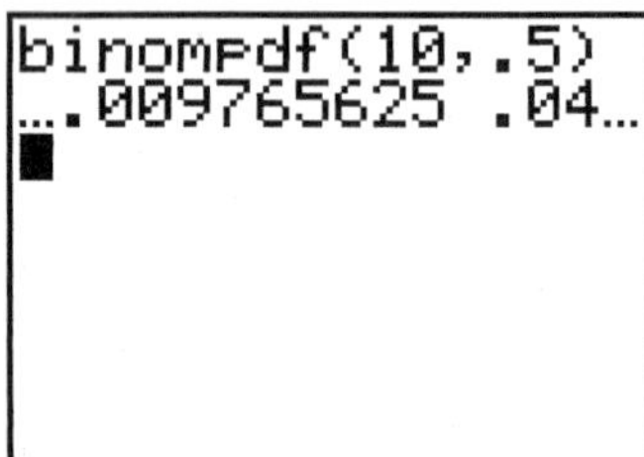

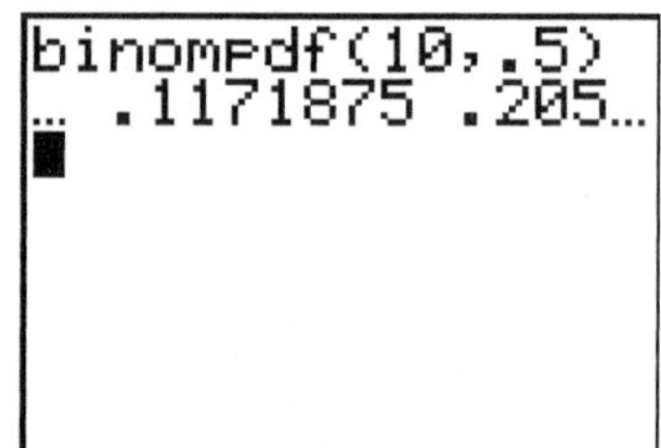

Let's use the STAT Editor to overcome this. Press STAT, then Edit, and clear out any data that may be there (remember how useful the ClrAllLists function can be – press 2nd MEM). Once in the STAT Editor, put the range of possible X values in L1. Typing in 0, 1, …, 10 isn't very hard, but remember the seq function from chapter 5 if you'd like to make the calculator generate the sequence for you. It's under the LIST OPS (2nd STAT) menu.

```
L1      L2      L3      1
0       ------  ------
1
2
3
4
5

L1(7)=6
```

```
NAMES OPS MATH
1:SortA(
2:SortD(
3:dim(
4:Fill(
5:seq(
6:cumSum(
7↓ΔList(
```

Recall that the STO> key can be used to store data in lists.

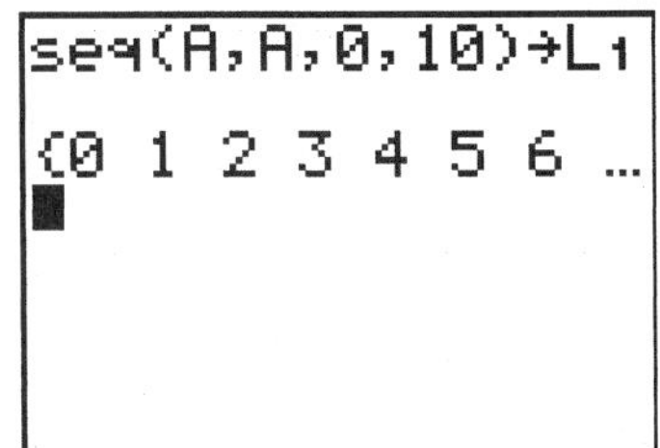

L1	L2	L3
0	------	------
1		
2		
3		
4		
5		
6		

L2 =

Next, let's use the binompdf function again with only the two arguments, n=10 and p=.5. If we store these probabilities in L2, then we can use the STAT Editor to easily see the probability of selecting no women, one woman, two women, and so on, up to the probability of selecting all ten women. Note that, for example, the probability of selecting exactly six women is .205.

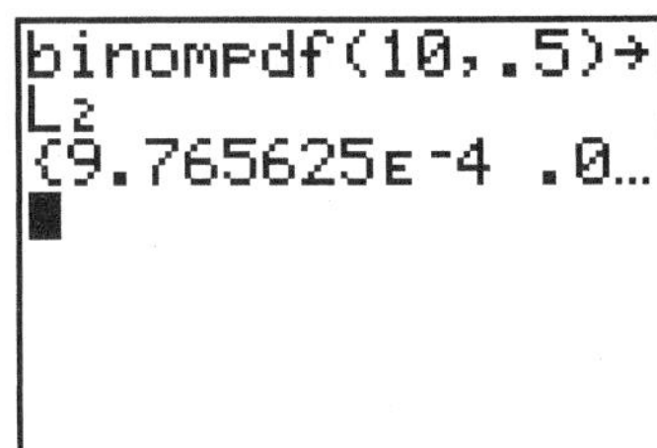

L1	L2	L3
0	9.8E-4	------
1	.00977	
2	.04395	
3	.11719	
4	.20508	
5	.24609	
6	.20508	

L2(7) =.205078124…

Scroll down to verify that this table matches the one in the margin of p. 272 of Agresti/Franklin. We could also make a graph of this probability distribution, using the histogram in STAT PLOT. Specify L1 as the Xlist, and L2 for Freq.

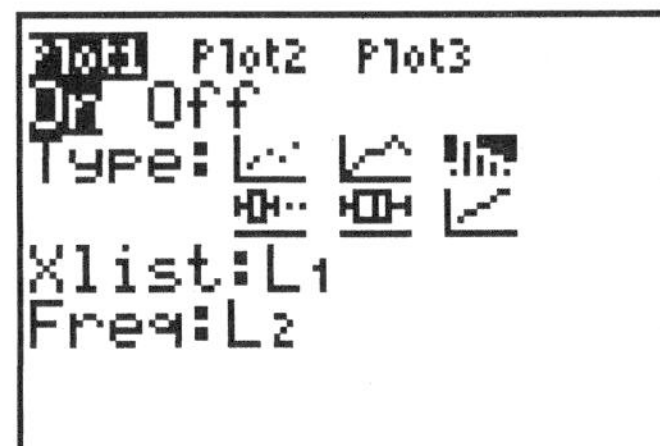

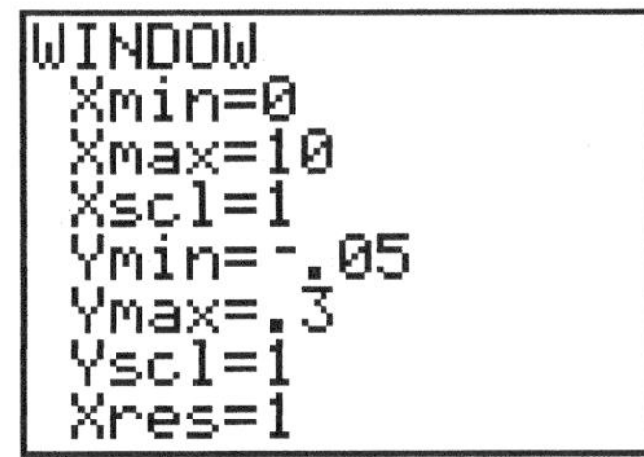

You will most likely need to adjust the WINDOW dimensions. See above.

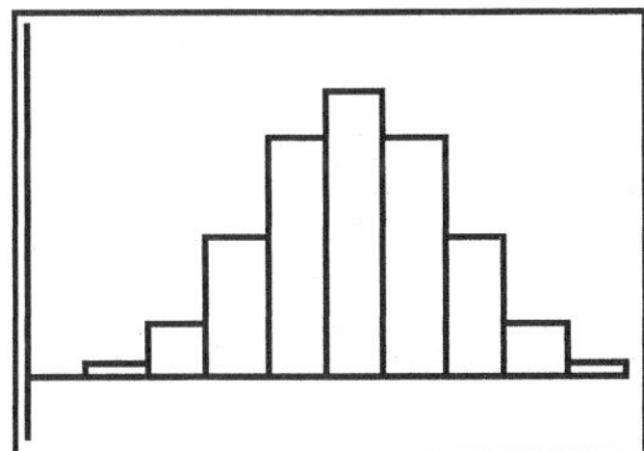

The binomcdf command is also of use here. Instead of returning the probability of selecting *exactly* 6 women, for example, it gives the probability of selecting 6 women *or less*. This can save calculation time, since we don't have to add up the individual probabilities for X = 0, 1, 2, 3, 4, 5, and 6. Let's go back to the STAT Editor and place the cumulative binomial probabilities in L3.

```
L1     L2       L3     3
0      9.8E-4   ------
1      .00977
2      .04395
3      .11719
4      .20508
5      .24609
6      .20508
L3 =
```

```
L1     L2       L3     3
0      9.8E-4   ------
1      .00977
2      .04395
3      .11719
4      .20508
5      .24609
6      .20508
L3 =binomcdf(10,
```

This can be done while in the data editor. Scroll up so that the cursor is on the L3 header, not in the first row of the spreadsheet. Now go to the DISTR menu (without QUITing first) and choose binomcdf. This allows you to define L3 all at once. We won't need the STO> key. Specify n=10 and p=.5.

```
L1     L2       L3       3
0      9.8E-4   9.8E-4
1      .00977   .01074
2      .04395   .05469
3      .11719   .17187
4      .20508   .37695
5      .24609   .62305
6      .20508   .82812
L3(7) =.828124999...
```

```
L1     L2       L3       3
5      .24609   .62305
6      .20508   .82812
7      .11719   .94531
8      .04395   .98926
9      .00977   .99902
10     9.8E-4   1
------ ------
L3(12) =
```

Verify that the individual probabilities of selecting 0, 1, …, 6 women add up to .82812. Note also that we have accumulated all 100% of the probability once we get to X=10!

Exercise 6.33: NBA shooting (pp. 276-277 of Agresti/Franklin)

Read the first sentence of the problem. This tells us that p=.9. Read part a. Now we know that n=10. For part c(i), we must find the probability that X=10. We can use binompdf here. See below (left).

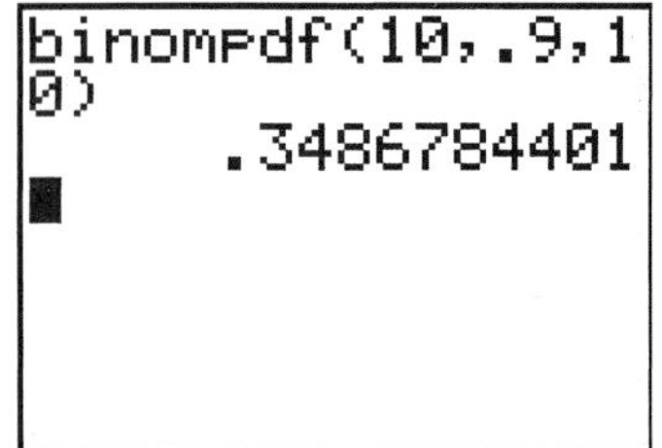

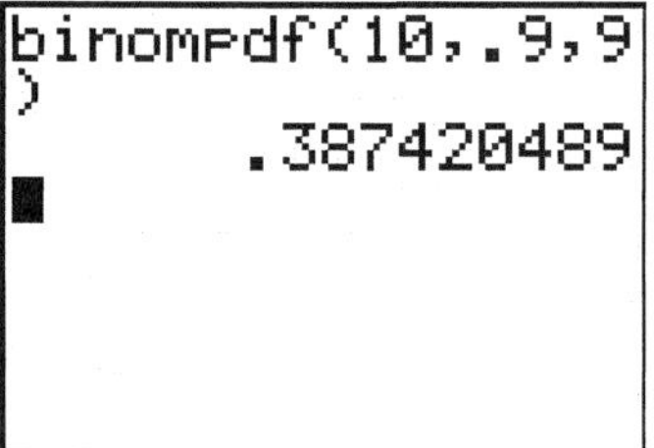

For part c(ii), we need the probability that X=9. See above (right).

Note that both probabilities involved making exactly a certain number of free throws. What about the probability of making at least eight? Or at most five? Creating a table of binomial probabilities would really pay off if we were faced with such questions.

If you need to, clear the data from the data editor using ClrAllLists. Press 2nd MEM to access the MEMORY menu.

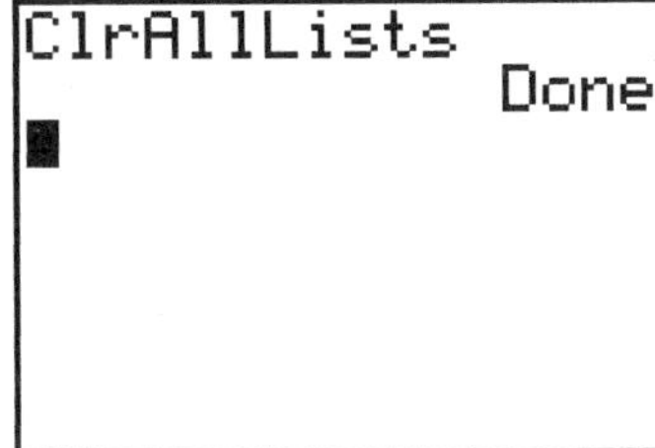

Press STAT to get into the Editor. Scroll up so that your cursor is ready to define L1.

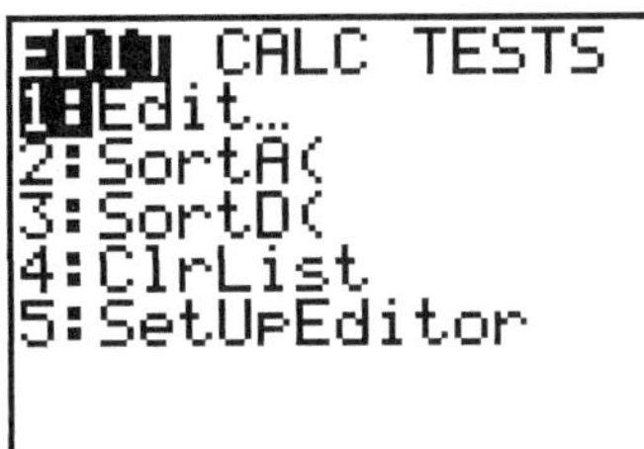

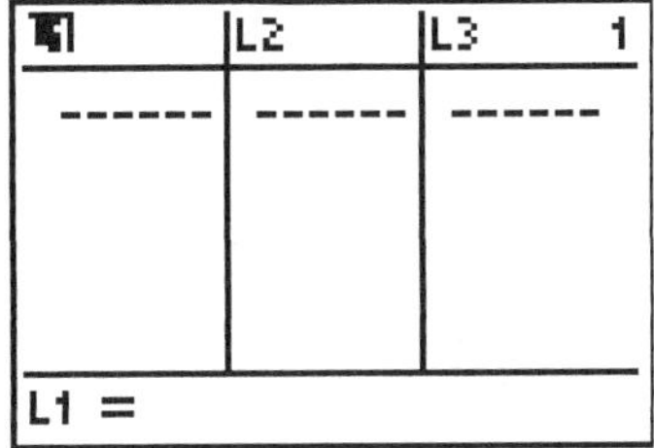

From here, press 2nd STAT to access the LIST menu. Scroll right to get the OPS menu. Choose seq.

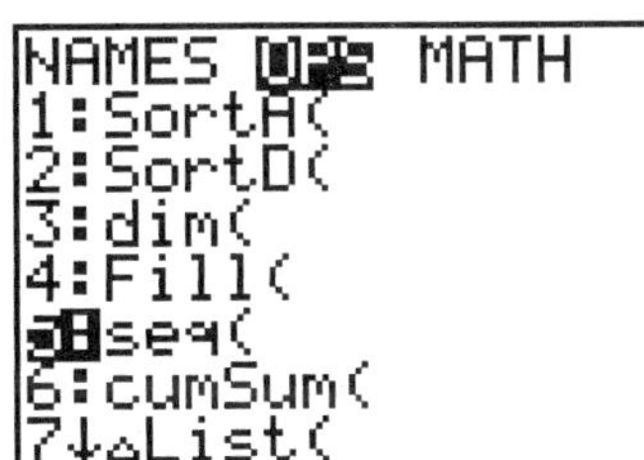

L1 =seq(A,A,0,10)

Recall that we enter "A" for the first two arguments since we simply want the sequence from 0 to 10. If you'd rather just type the digits from 0 to 10 into L1, by all means do that.

Next, move the cursor so that it is on the L2 header. From here, access the DISTR menu by typing 2nd VARS. Choose binompdf.

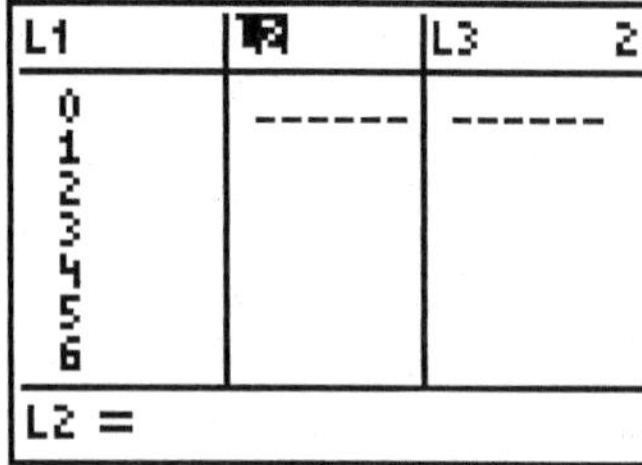

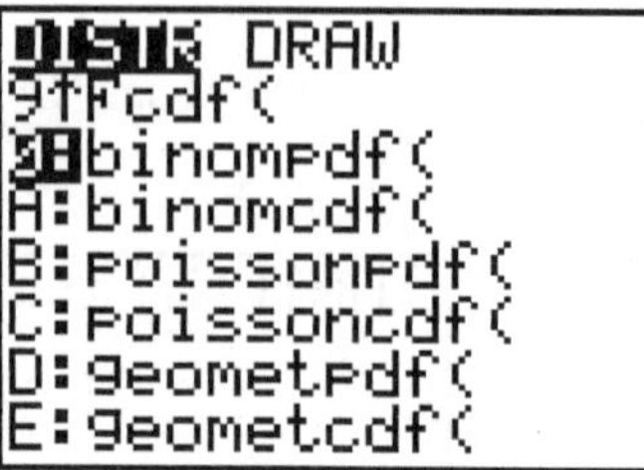

Supply two arguments only, n=10 and p=.9.

```
L1    L2      L3     2
0     ------  ------
1
2
3
4
5
6
L2 =...ompdf(10,.9
```

```
L1    L2      L3     2
0     1E-10   ------
1     9E-9
2     3.6E-7
3     8.7E-6
4     1.4E-4
5     .00149
6     .01116
L2(1)=1E-10
```

Next, move the cursor to the L3 header. Choose binomcdf from the DISTR menu.

```
L1    L2      L3     3
0     1E-10   ------
1     9E-9
2     3.6E-7
3     8.7E-6
4     1.4E-4
5     .00149
6     .01116
L3 =
```

```
DISTR DRAW
9↑Fcdf(
0:binompdf(
A:binomcdf(
B:poissonpdf(
C:poissoncdf(
D:geometpdf(
E:geometcdf(
```

Supply the same two arguments as before.

```
L1    L2      L3     3
0     1E-10   ------
1     9E-9
2     3.6E-7
3     8.7E-6
4     1.4E-4
5     .00149
6     .01116
L3 =...omcdf(10,.9
```

```
L1    L2      L3     3
0     1E-10   1E-10
1     9E-9    9.1E-9
2     3.6E-7  3.7E-7
3     8.7E-6  9.1E-6
4     1.4E-4  1.5E-4
5     .00149  .00163
6     .01116  .0128
L3(1)=1E-10
```

Now we have the possible values for X in L1, the probability that X is *exactly* equal to any of these values in L2, and the probability that X is *at most* any of these values in L3.

Let's answer the questions posed above. What is the probability of making at least eight free throws out of ten? There are two ways to answer this. If we use the exact probabilities in L2, we must add up the probabilities for X=8, 9, and 10. This is .19371 + .38742 + .34868 = .92981.

L1	L2	L3
5	.00149	.00163
6	.01116	.0128
7	.0574	.07019
8	.19371	.2639
9	.38742	.65132
10	.34868	1

L2(12) =

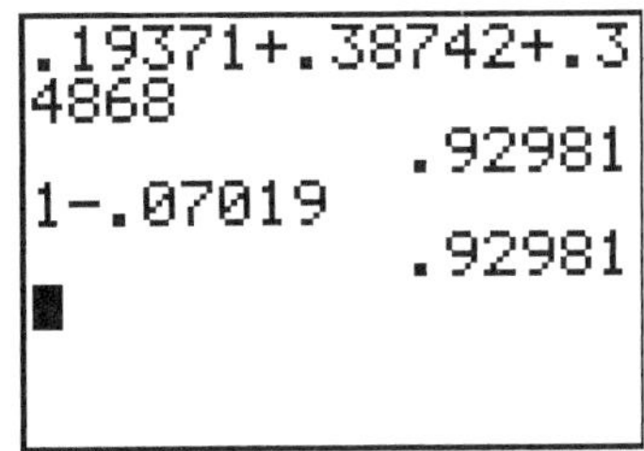

Or, we can use the cumulative probabilities in L3. The probability of making *at most* seven free throws is .07019. The probabilities of "at least eight" and "at most seven" must sum to one, since that covers the entire range of possibilities for X. Thus, we can simply subtract .07019 from one to get the probability of making at least eight free throws: 1 – .07019 = .92981.

What is the probability of making at most five free throws out of ten? Again, we can add exact probabilities from L2. Now we need them for X=0, 1, 2, 3, and 4. It is much easier, however, to use the cumulative probabilities in L3. The probability of making at most five free throws is .00163.

The Sampling Distribution of the Mean: Section 6.5 of Agresti/Franklin

Exercise 6.58: Restaurant profit? (p. 297 of Agresti/Franklin)

We may have a skewed distribution, but we know that the expense per customer data distribution has μ=\$8.20 and σ=\$3.00. We are working with the mean of a random sample of size n=100, so the sampling distribution of the mean should be normal. The sampling distribution will have mean μ=\$8.20 and standard error $\sigma/\sqrt{n} = 3/\sqrt{100} = 0.3$. To find the probability that the sample mean is less than \$8.95, we need to standardize into a z-score appropriately. Using the above standard error, we obtain z = 2.5. The probability to the left of z = 2.5 is .9938. See the below TI-83 screen (on the left).

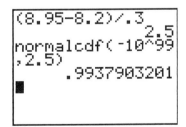

```
(8.95-8.2)/.3
                2.5
normalcdf(-10^99
,2.5)
       .9937903201
```

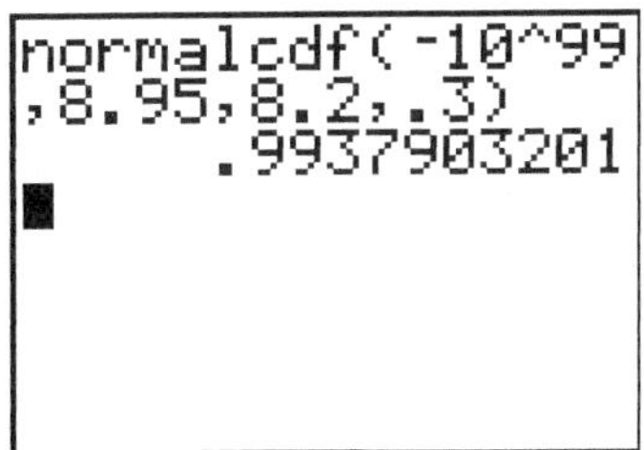

The above screen on the right shows how to skip the z-score step and get the same probability.

Chapter Seven: Statistical Inference – Confidence Intervals

This chapter explores methods of interval estimation. Two of the menus on the TI-83 will be useful to us here: The DISTR menu, and the STAT TESTS menu.

Point and Interval Estimates: Section 7.1 of Agresti/Franklin

EXAMPLE 2: Should a wife sacrifice her career for her husband's? (p. 320 of Agresti/Franklin)
Here we are estimating a proportion. We have a sample proportion of 19%, which has a standard error of 1%. To construct an interval in which the population proportion falls with 95% confidence, we multiply the standard error by ±1.96. This number is the percentile of the standard normal distribution such that the middle 95% of the distribution falls within −1.96 and 1.96. In other words, −1.96 is the 2.5th percentile, and 1.96 is the 97.5th percentile. See the margin of p. 415 for an illustration. Recall that the invNorm command, found in the DISTR menu, gives normal percentiles.

```
DISTR DRAW
1:normalpdf(
2:normalcdf(
3:invNorm(
4:tpdf(
5:tcdf(
6:χ²pdf(
7↓χ²cdf(
```

```
invNorm(.025)
     -1.959963986
invNorm(.975)
      1.959963986
```

Now that we know where the 1.96 comes from, we can construct the interval.

```
.19-1.96*.01
            .1704
.19+1.96*.01
            .2096
```

Confidence Intervals for p: Section 7.2 of Agresti/Franklin

EXAMPLE 3: Would you pay higher prices to protect the environment? (p. 324 of Agresti/Franklin)

The sample proportion and standard error of $\hat{p}$ are calculated below. Remember that you can call up the most recently calculated value via ANS (2nd (-)). This saves time typing, and results in less rounding error. Recall from example 2 above that we must multiply the standard error by 1.96 to achieve 95% confidence.

```
518/1154
      .4488734835
√(Ans*(1-Ans)/11
54)
      .0146414714
Ans*1.96
       .028697284
```

This problem can be made a little easier if we use the 1-PropZInt function. Press STAT, then move right to TESTS. Scroll down to A: 1-PropZInt.

```
EDIT CALC TESTS
6↑2-PropZTest...
7:ZInterval...
8:TInterval...
9:2-SampZInt...
0:2-SampTInt...
A:1-PropZInt...
B↓2-PropZInt...
```

```
1-PropZInt
 x:518
 n:1154
 C-Level:.95
 Calculate
```

The 1-PropZInt (one proportion Z interval) function takes three inputs. First we enter x, the number of "successes", which is the number of people willing to pay higher prices to protect the environment in this example. Then we supply n, the sample size, and the C (confidence) level. After supplying this information, move the cursor down to Calculate and press ENTER.

```
1-PropZInt
 (.42018,.47757)
 p̂=.4488734835
 n=1154
```

Note that $\hat{p}$ is displayed along with the confidence interval. The above method shouldn't be used in place of doing it the "long way", rather, you should work out all the details like on p. 421, then use 1-PropZInt to check your work.

Exercise 7.18: z-score and confidence level (p. 332 of Agresti/Franklin)
We know how to use the invNorm function to find 1.96, the normal percentile needed for 95% confidence intervals. See the margin of p. 327 for an illustration of 99% confidence. If we mark off the middle 99% of the distribution, that leaves only ½%, or .005, in each tail. These are the two percentiles we need.

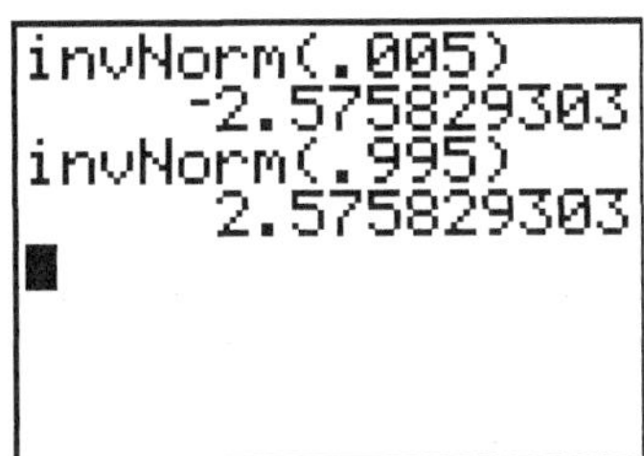

Of course, we only need to find one of the above percentiles. By the symmetry of the normal curve, we know to multiply by ±2.58.

Part (a) asks for 90% confidence. Draw a normal curve and mark off the middle 90%. How much area is left over for the two tails? (Answer: 5% in each!) We'll need the 5th and 95th percentiles of the standard normal distribution. See below left. Z = ±1.65.

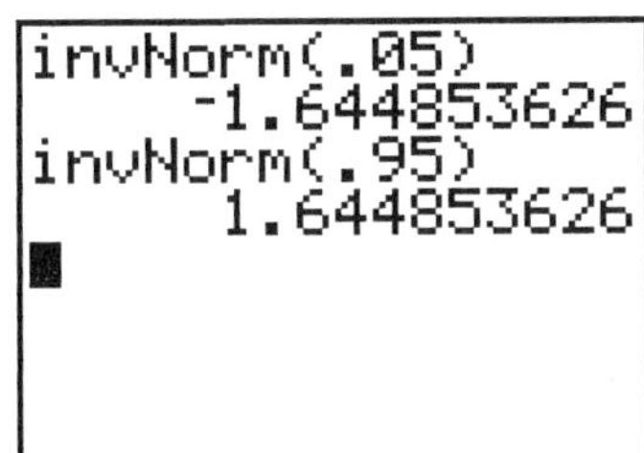

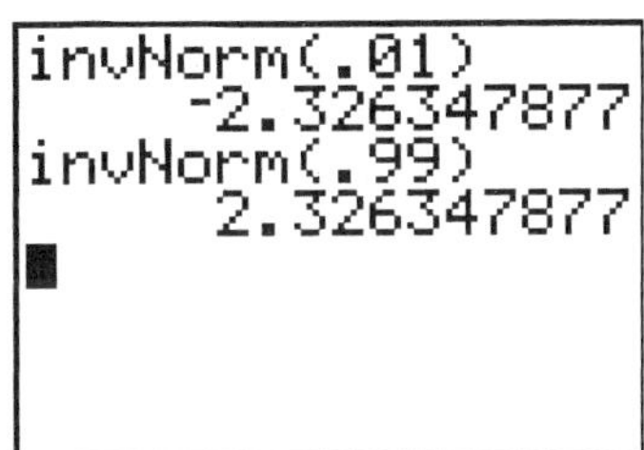

Part (b) specifies 98% confidence. With 98% in the middle, that leaves 2% left over, to be split evenly between the two tails. Draw a curve to help you visualize it. We need the 1st and/or 99th percentiles of the standard normal distribution. See above right. Z = ±2.33.

For part (c), there's only 1−.999 = .001 area left to be split in half. Each tail has only .0005, or .05% of the area. To have that much confidence we must multiply the standard error by ±3.29.

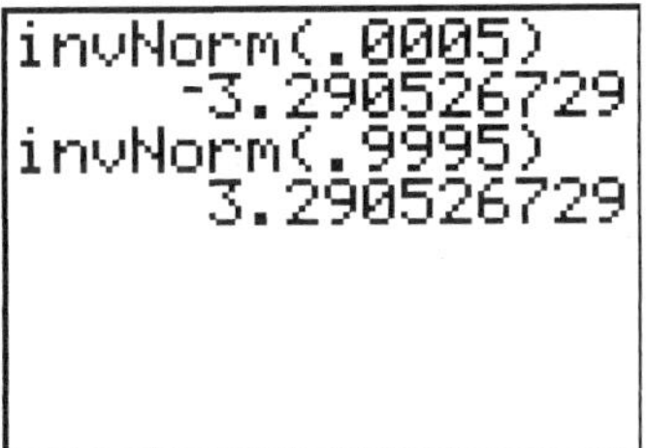

Exercise 7.21: Exit poll predictions (p. 333 of Agresti/Franklin)
We have the choice to either estimate the proportion of Democratic votes, or the proportion of Republican votes. Note that 660 + 740 = 1400, so it's just one or the other. Let's estimate the proportion of Republican votes. The sample proportion is about 53%, with a standard error of .013. Part (a) asks for a 95% confidence interval.

```
740/1400
           .5285714286
√(Ans*(1-Ans)/14
00)
            .013341227
```

```
invNorm(.975)
           1.959963986
.529-1.96*.013
                .50352
.529+1.96*.013
                .55448
```

So we can be 95% confident that the proportion of Republican votes falls between 50.3% and 55.4%. Close, but it looks like we have a majority.

Part (b) asks for a 99% confidence interval. Recall from exercise 7.18 that z = 2.58 for this level of confidence. This larger multiplier makes the interval wider. We have more confidence, but a less precise estimate.

```
invNorm(.995)
           2.575829303
.529-2.58*.013
                .49546
.529+2.58*.013
                .56254
```

We can no longer predict a Republican majority. With 99% confidence, we can only say that the true proportion of Republican votes falls between 49.5% and 56.3%.

Remember the 1-PropZInt function under the STAT TESTS menu. Let's revisit part (a).

```
1-PropZInt
 x:740
 n:1400
 C-Level:.95
 Calculate
```

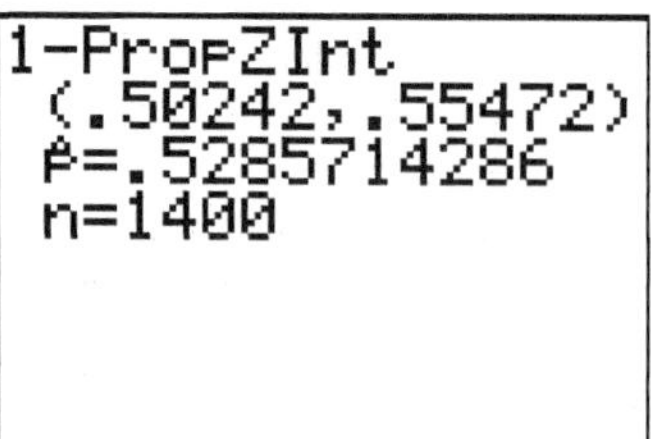

The above answer is slightly more accurate than before, since we rounded off some of the decimal places.

The interval in part (b) is shown below.

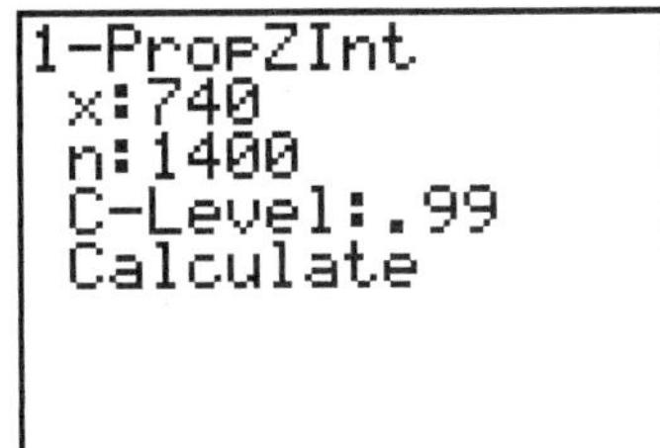

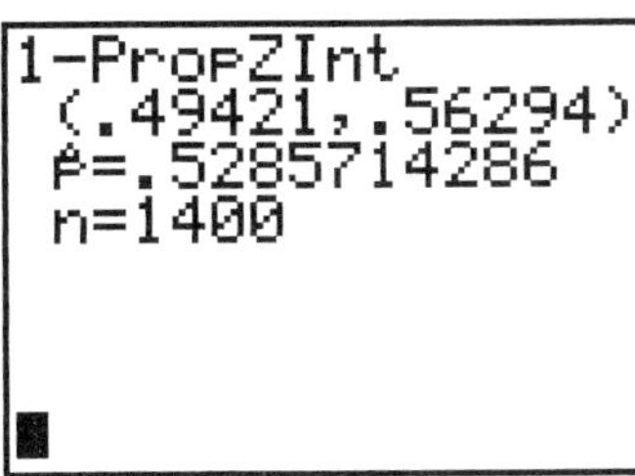

Confidence Intervals for the Mean: Section 7.3 of Agresti/Franklin

EXAMPLE 7: eBay auctions of palm handheld computers (pp. 338-340 of Agresti/Franklin)

We have a sample of n=7 final prices with which to estimate the true average. Let's enter these data into L1 in the data editor. Clear out any unwanted data first.

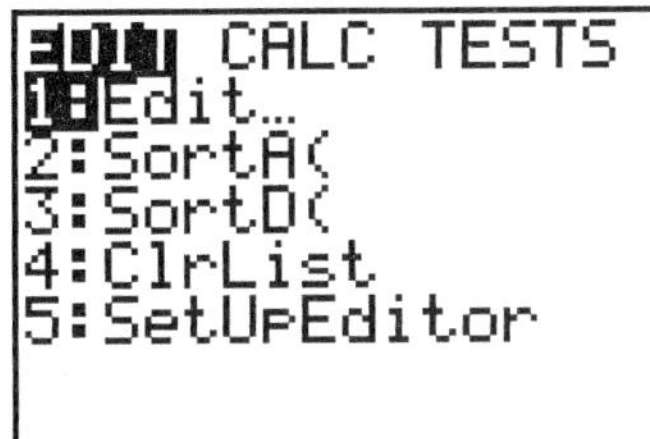

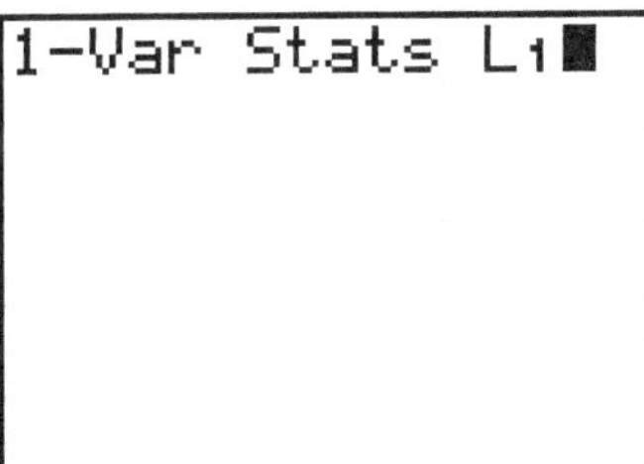

Take a look at the confidence interval formula in the summary box on p. 337. We are going to need the mean and standard deviation of the sample. Let's use 1-Var Stats from the STAT CALC menu. Press STAT, move right to CALC, then ENTER to choose 1-Var Stats. Press (2nd 1) to specify L1.

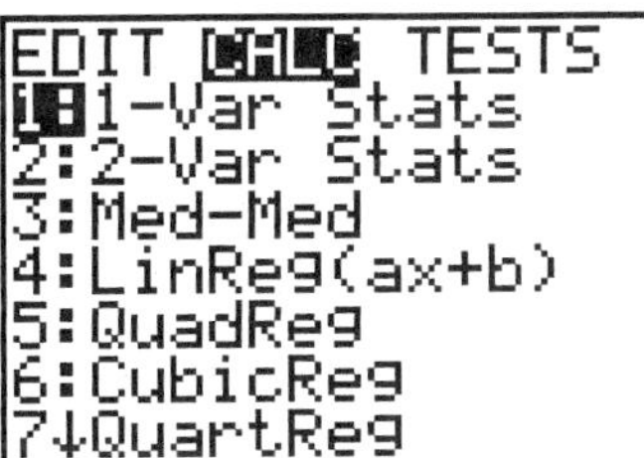

```
1-Var Stats
 x̄=233.5714286
 Σx=1635
 Σx²=383175
 Sx=14.63850109
 σx=13.55261854
↓n=7
```

We see that $\bar{x}$=233.57 and s = 14.64. Thus we have a standard error equal to $s/\sqrt{n} = 14.64/\sqrt{7} = 5.53$. Now all we need is the t critical value for 95% confidence and n – 1 = 6 degrees of freedom. For this, consider the DISTR menu (press 2nd VARS).

```
DISTR DRAW
1:normalpdf(
2:normalcdf(
3:invNorm(
4:tpdf(
5:tcdf(
6:χ²pdf(
7↓χ²cdf(
```

Recall that normalcdf gives probabilities for the normal distribution. The arguments are normalcdf(lower, upper), which returns the area under the standard normal curve between the lower and upper bound. tcdf works the same way, with a third argument for the degrees of freedom. So tcdf(lower, upper, df) gives the area under the t curve between the lower and upper bound. For the current problem, df = n – 1 = 6.

Note that there is an invNorm function in order to get percentiles of the normal distribution, but there is no inverse t function! We can get around this, if we use the EQUATION SOLVER along with the tcdf function. Consider the following relationship:

tcdf(lower, upper, df) = area or tcdf(L, U, D) = A

If we enter the above equation into the SOLVER, we can supply three of the four arguments, and solve for the fourth. Draw a symmetric bell curve to represent a t distribution. Mark off the middle 95% to represent the 95% confidence level. Note that each tail contains .025 of the area (A). Consider the left tail. For the left tail, the lower bound (L) is negative infinity. This makes the upper bound (U) the t critical value of interest.

For the problem at hand, we can set A = .025, D = 6, L = -10^{99}, and solve for U = $t_{.025}$. Press MATH, then scroll down to 0:Solver. Press ENTER.

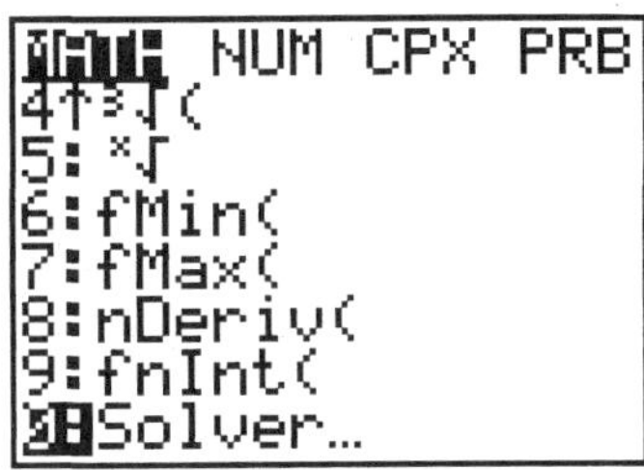

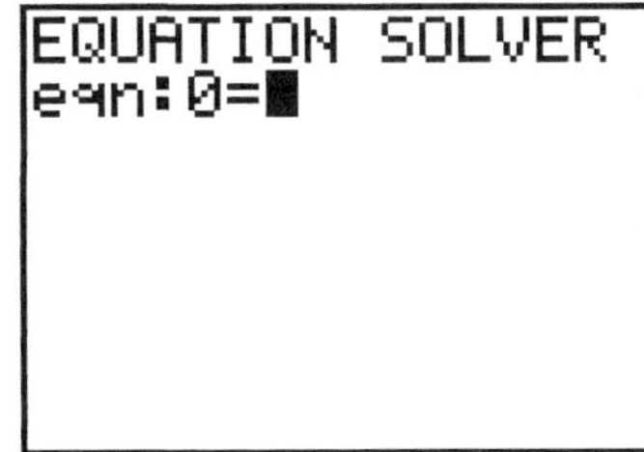

Clear out any equation that may there. You may need to arrow UP once to get to the above (right) screen. Note that "0=" is hard-coded on the left hand side of the equation. Instead of entering tcdf(L, U, D) = A directly, we must enter 0 = tcdf(L, U, D) – A.

While in the EQUATION SOLVER, press 2nd VARS to access the DISTR menu. Choose 5:tcdf. This will be echoed back into the SOLVER.

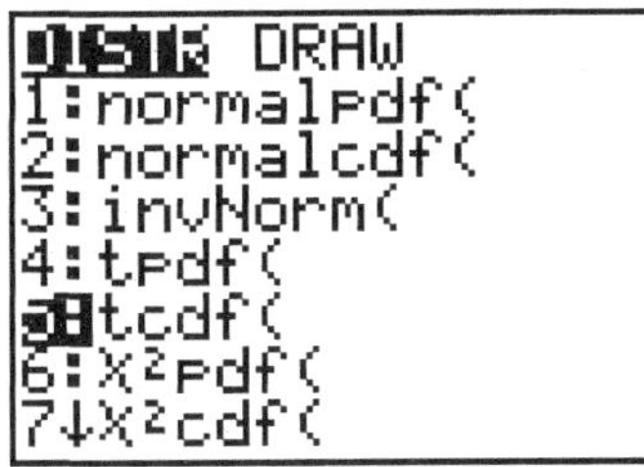

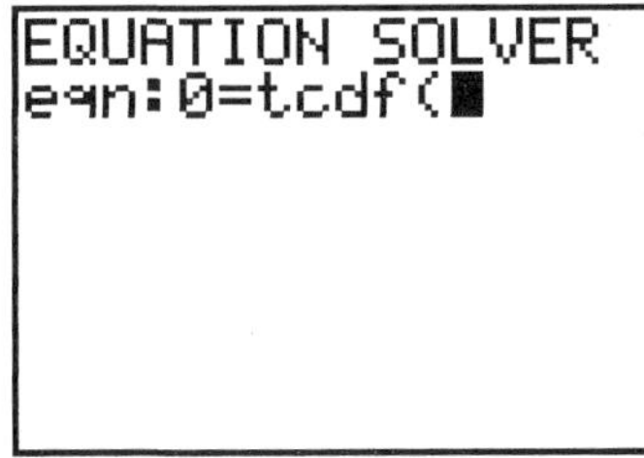

Next, we need to enter four letters. L is ALPHA ")", U is ALPHA "5", D is ALPHA "x $^{-1}$", and A is ALPHA MATH.

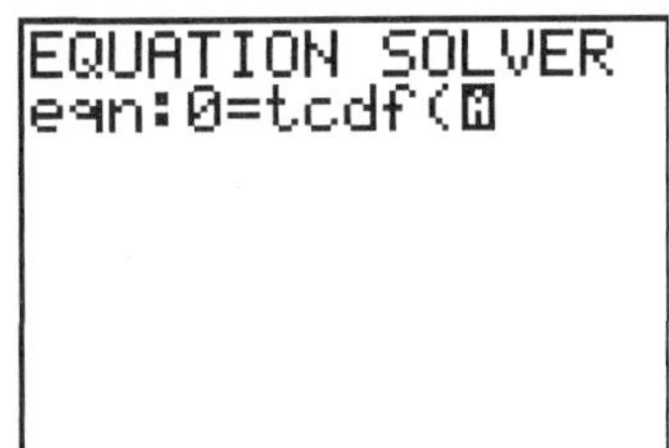

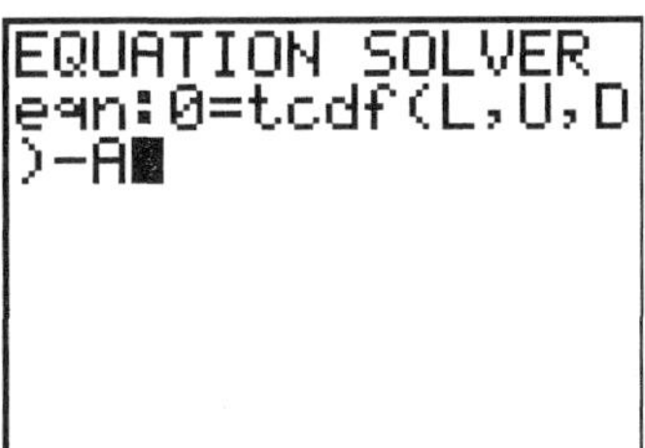

Press ENTER. Now we are taken to another screen, where we can enter all known information and solve for the unknown. For this problem, we are focusing on the left tail of a t distribution with 6 degrees of freedom, so the lower bound is negative infinity, and D is 6. A is .025 since we desire 95% confidence. Once we solve for the upper bound, we'll know the value of the t-score. To solve for U, position the cursor on the U= line. We must have a number here; zero will do. Press SOLVE (ALPHA ENTER). Be patient, it takes a few seconds to compute.

```
tcdf(L,U,D)-A=0
  L=0
  U=0
  D=0
  A=0
  bound={-1E99,1...
```

```
tcdf(L,U,D)-A=0
  L=-1E99
  U=00
  D=6
  A=.025
  bound={-1E99,1...
```

```
tcdf(L,U,D)-A=0
  L=-1E99
▪ U=-2.446911846...
  D=6
  A=.025
  bound={-1E99,1...
▪ left-rt=0
```

Now we see that $t_{.025} = 2.447$. Note that U came back negative from the Solver – this is because we used the left tail of the distribution. No matter, we know by the symmetry of the curve that the middle 95% of the t curve with df = 6 falls between –2.447 and 2.447. The absolute value of the t-score is what we'll plug into the formula.

That was a lot of work; you may prefer using Table B in the back of Agresti/Franklin! Actually, that *was* kind of a pain, but from now on t critical values will be easy to find with this method. All we'll have to do is return to the SOLVER, and adjust the L,U,D,A values. We can always leave L equal to -10^{99}, and always solve for U, the t-score. The only adjustment we'll have to make is to D, depending on the sample size, and A, depending on the confidence level.

To finish the problem, plug the mean, standard error and t-score into the confidence interval formula in the summary box on p. 337.

```
233.57-2.447*5.5
3
            220.03809
233.57+2.447*5.5
3
            247.10191
```

For an easier method, press STAT, then move right to TESTS. Choose TInterval.

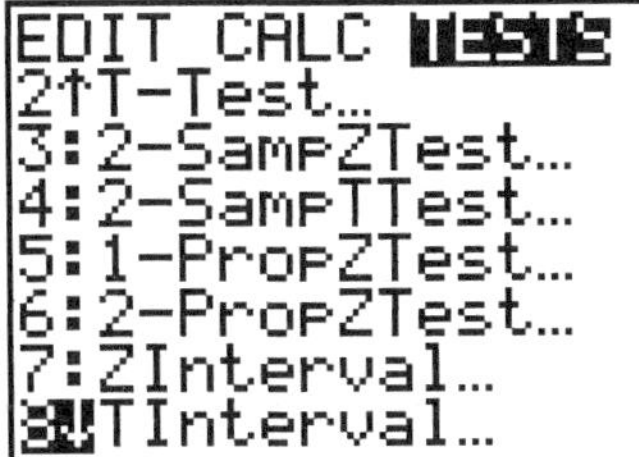


```
TInterval
 Inpt:Data Stats
 List:L1
 Freq:1
 C-Level:.95
 Calculate
```

Note that there are two "inputs". Since we have the eBay data entered in L1, choose Data. Specify L1 for the List. By default, the Freq: value will be 1. This is what we want, since we merely have 7 data observations in L1. Specify a 95% confidence level, and press ENTER with the cursor on Calculate.

```
TInterval
 (220.03,247.11)
 x̄=233.5714286
 Sx=14.63850109
 n=7
```

What a time saver! Of course, you gained a more detailed understanding of the problem doing it "the long way".

On page 340 of Agresti/Franklin, the seven "buy-it-now" eBay auction prices are used to obtain a 99% confidence interval for the population mean price. To get the new t critical value, press MATH, then choose 0:Solver to access the EQUATION SOLVER. It looks exactly how we left it! The only adjustment we need to make is to A. Now, with the middle 99% of the t curve marked off, that leaves only 1% to be split evenly between the two tails. Set A equal to .005. There is no need to delete the previous t-score value – just position the cursor on the "U=" line and press SOLVE (2nd ENTER).

```
tcdf(L,U,D)-A=0
 L=-1E99
 U=2.446911846...
 D=6
 A=.005
 bound={-1E99,1...
```

```
tcdf(L,U,D)-A=0
 L=-1E99
▪U=-3.707428020...
 D=6
 A=.005
 bound={-1E99,1...
▪left-rt=0
```

The new t-score is 3.707. See p. 340 of Agresti/Franklin for the calculation of the interval.

To obtain the interval using the TInterval function, access the STAT TESTS menu. Change the confidence level to 99%. Cursor down to Calculate and press ENTER.

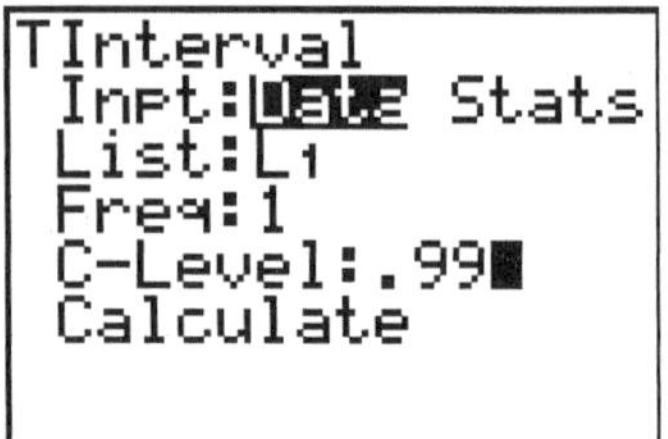

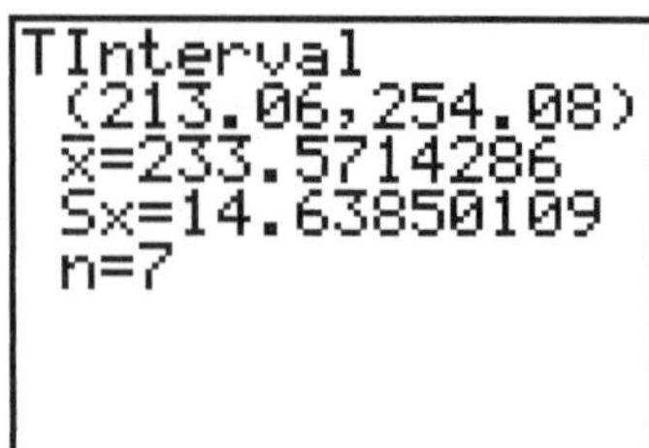

Exercise 7.28: Anorexia in teenage girls (p. 343 of Agresti/Franklin)
Enter the 17 weight changes into L1.

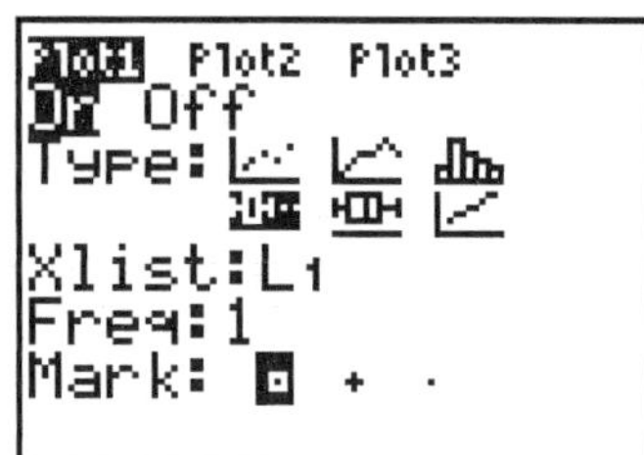

(a) Let's illustrate the data using either a histogram or boxplot. Or both, if we like. Request a modified boxplot for Plot1, in order to check for outliers. Ask for a histogram for Plot2, so we can get a better idea of the shape of the data.

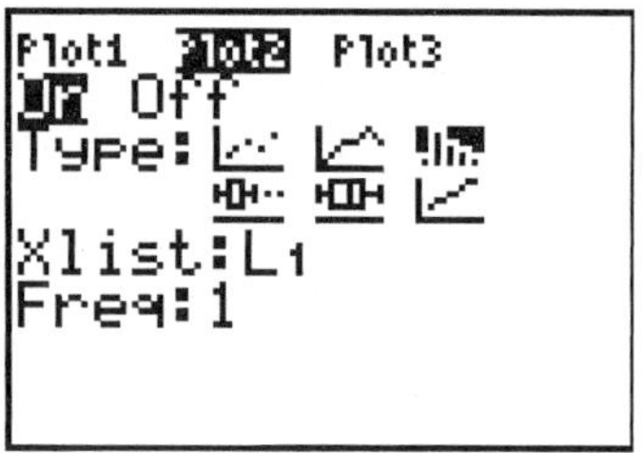

Plot1 Plot2 Plot3
On Off
Type:
Xlist:L1
Freq:1

Don't forget ZoomStat, the 9th option in the ZOOM menu.

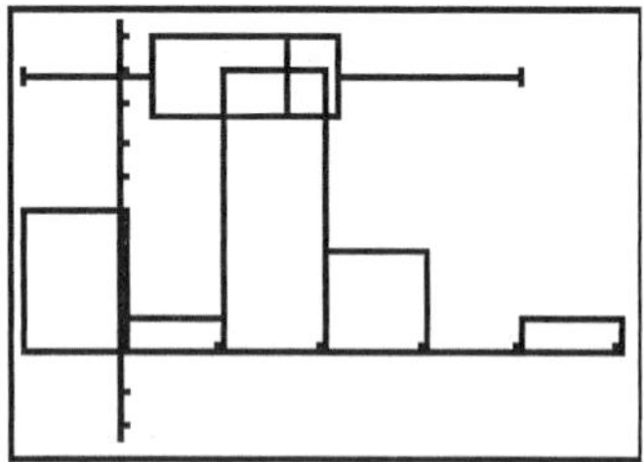

The data are slightly right-skewed, but the t methods should work well since no outliers are present.

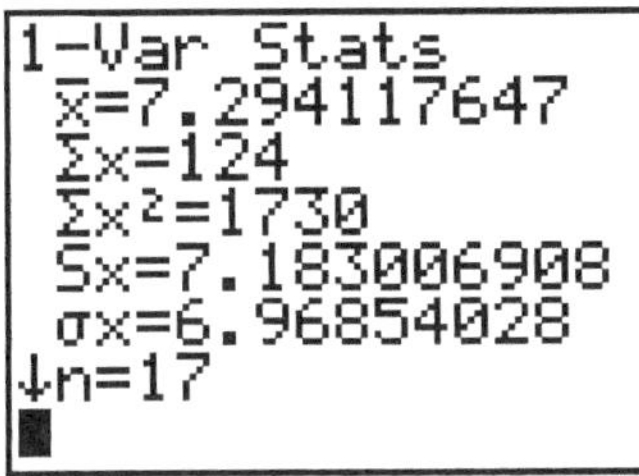

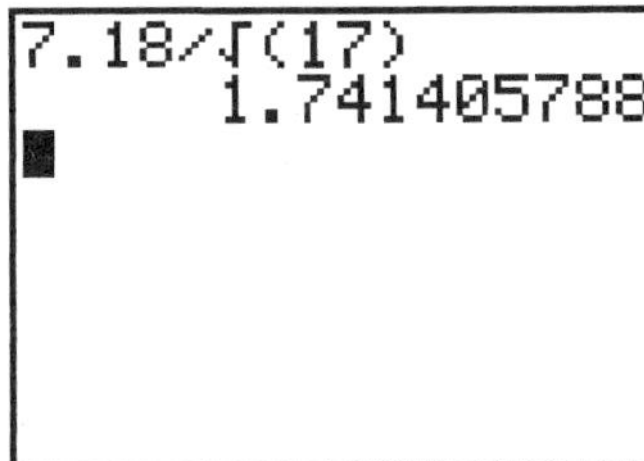

(b) Use the 1-Var Stats function to verify the sample mean and standard deviation.
(c) See above (right).
(d) Let's return to the EQUATION SOLVER to get the t critical value for df = 16 and 95% confidence. Press MATH, then press 0. Leave L = -10^{99}. Change D to 16, and A to .025, since that is the amount of (left) tail area. Leave whatever the last value was for U, and press SOLVE (ALPHA ENTER). t = 2.120.

```
tcdf(L,U,D)-A=0
  L=-1E99
  U=3.707428020…
  D=16
  A=.025
  bound={-1E99,1…
```

```
tcdf(L,U,D)-A=0
  L=-1E99
▪ U=-2.119905285…
  D=16
  A=.025
  bound={-1E99,1…
▪ left-rt=1E-14
```

(e) We can get the confidence interval estimate using the TInterval function from STAT TESTS. Specify Data, L1, 1, and .95.

```
TInterval
 Inpt:Data Stats
 List:L1
 Freq:1
 C-Level:.95
 Calculate
```

```
TInterval
 (3.601,10.987)
 x̄=7.294117647
 Sx=7.183006908
 n=17
```

Exercise 7.35: Political views (p. 344 of Agresti/Franklin)

(a) We can get the confidence interval using the statistics provided: $\bar{x} = 4.12$ and $s = 1.39$. Press STAT, TESTS, then 8 (for TInterval). This time, specify Stats instead of Data. Enter the values for the mean, standard deviation, sample size, and confidence level. Calculate.

```
TInterval
 Inpt:Data Stats
 x̄:4.12
 Sx:1.39
 n:1331
 C-Level:.95
 Calculate
```

```
TInterval
 (4.0453,4.1947)
 x̄=4.12
 Sx=1.39
 n=1331
```

(c)(i) If we increase the confidence level to 99%, we will obtain a wider confidence interval. We can answer this by thinking it through: we lose precision if we increase the confidence level. The t critical value will be a larger number, which makes the interval wider. Let's see just how much wider by using TInterval again. Change C-Level to .99 and leave everything else the same.

```
TInterval
 Inpt:Data Stats
 x̄:4.12
 Sx:1.39
 n:1331
 C-Level:.99
 Calculate
```

```
TInterval
 (4.0217,4.2183)
 x̄=4.12
 Sx=1.39
 n=1331
```

(c)(ii) Reset the confidence level back to 95%. If we had a sample size of merely 500, this would also make the original confidence interval wider. Consider the formula in the summary box on p. 337 of Agresti/Franklin. The smaller n is, the larger the standard error is. Let's see how wide the confidence interval is with n=500. Keep all other arguments the same.

```
TInterval
 Inpt:Data Stats
 x̄:4.12
 Sx:1.39
 n:500
 C-Level:.95
 Calculate
```

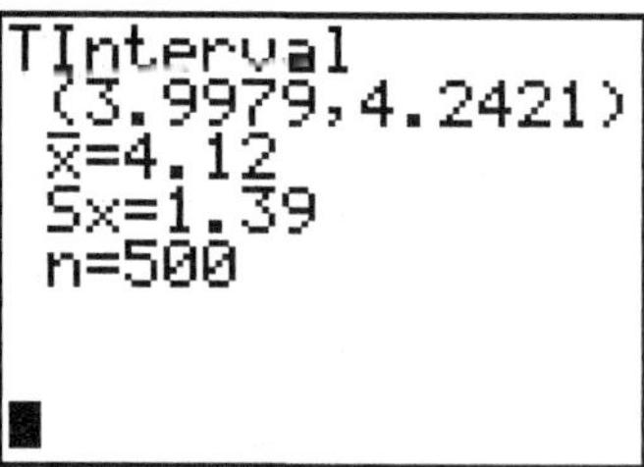

Chapter Eight: Statistical Inference – Significance Tests about Hypotheses

This chapter explores another type of statistical inference: hypothesis tests. Just like with confidence intervals, we'll need the DISTR and STAT TESTS menus.

Significance Tests about Proportions: Section 8.2 of Agresti/Franklin
In this section, we consider hypothesis tests about one proportion. This method makes use of the normal distribution, so we'll use the normalcdf function from the DISTR menu. There is also a very useful function under STAT TESTS called 1-PropZTest.

EXAMPLE 4: Dr. Dog: Can dogs detect cancer by smell? (pp. 377-378 of Agresti/Franklin)
The sample proportion and standard error calculations on pp. 377-378 are verified below (left).

```
22/54
          .4074074074
√((1/7)*(6/7)/54
)
          .0476190476
```

```
√((1/7*(6/7)/54)

          .0476190476
(22/54-1/7)/Ans
          5.555555556
```

The test statistic calculation on p. 378 is verified above. Be careful when rounding your calculations – the results can vary quite a bit when working with proportions. For our purposes here, $z = 5.6$ is accurate enough. The p-value for this test is illustrated on p. 378. We can use the normalcdf function to get this small p-value. Note that $z = 5.6$ is too large to be shown on Table A, the normal table in the back of Agresti/Franklin. Press 2nd VARS to access the DISTR menu.

```
DISTR DRAW
1:normalpdf(
2:normalcdf(
3:invNorm(
4:tpdf(
5:tcdf(
6:X²pdf(
7↓X²cdf(
```

```
normalcdf(5.6,10
^99)
   1.074621722E-8
Ans*2
   2.149243444E-8
```

Each tail probability is only .00000001, so the p-value = .00000002. P-values this small are usually reported as "p-value < .0001". We have very convincing evidence to reject H_0.

There is a function available for tests such as this. Press STAT, then move right to TESTS.

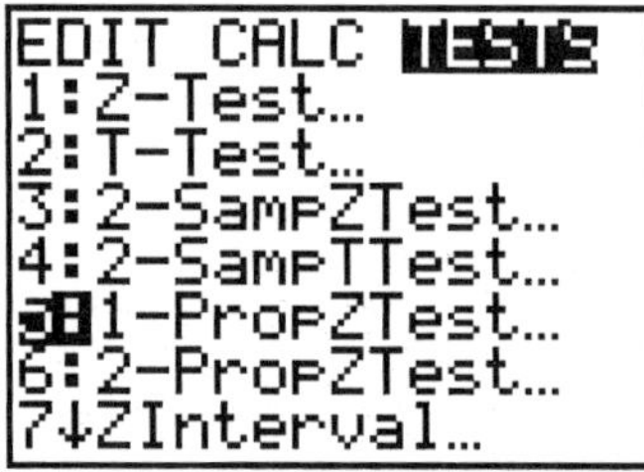

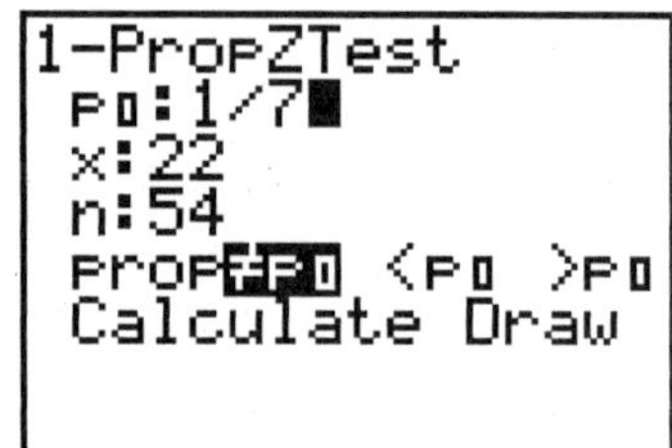

We can enter the null value of p as 1/7. The calculator will convert it to a decimal, which is OK. Remember that $\hat{p} = \frac{x}{n}$, where x is the number of successes (correct selections by the dog), and n is the sample size. Choose the $\neq p_0$ option for a two-tailed test. Calculate. The results list the alternative hypothesis, test statistic (z), PV(p), sample proportion($\hat{p}$), and sample size(n).

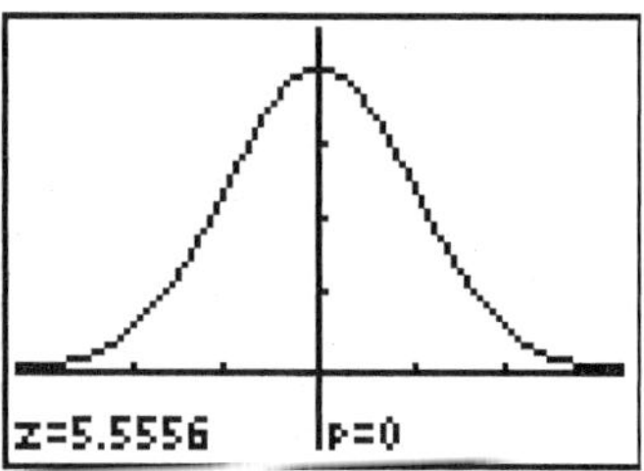

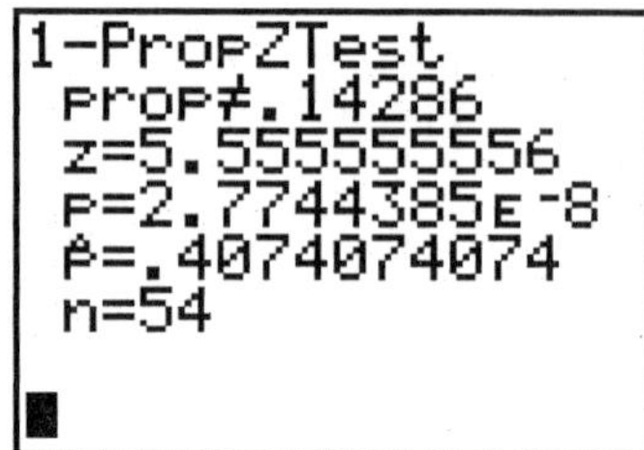

We can also choose Draw. However, our test statistic is too large to show up on the screen!

Exercise 8.19: Garlic to repel ticks (p. 387 of Agresti/Franklin)
Here we wish to test H_0: p = ½ vs. H_a: p = ½ . Note that we have a sample size of n = 37 + 29 = 66. Let's call garlic being more effective as the "success". Thus we have $\hat{p} = \frac{37}{66} = .56$. The test statistic calculation is shown below (left). We can use normalcdf (from the DISTR menu) to get the p-value. We must double the tail probability for a two-tailed test.

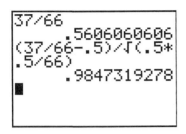

```
37/66
          .5606060606
(37/66-.5)/√(.5*
.5/66)
          .9847319278
```

```
.5/66)
          .9847319278
normalcdf(Ans,10
^99)
          .1623778823
Ans*2
          .3247557646
```

This is a large p-value, consistent with random variation. We do not have sufficient evidence of a real difference between garlic and placebo.

Next, let's press STAT, then TESTS, to access the 1-PropZTest function. We can double check our work above.

```
EDIT CALC TESTS
1:Z-Test...
2:T-Test...
3:2-SampZTest...
4:2-SampTTest...
5:1-PropZTest...
6:2-PropZTest...
7↓ZInterval...
```

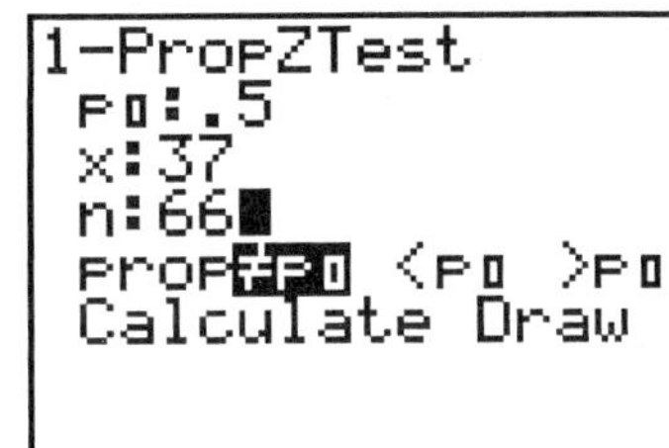

Pressing ENTER with the cursor on Calculate will verify our previous results. Pressing ENTER with the cursor on Draw will give a picture of the large, two-sided p-value.

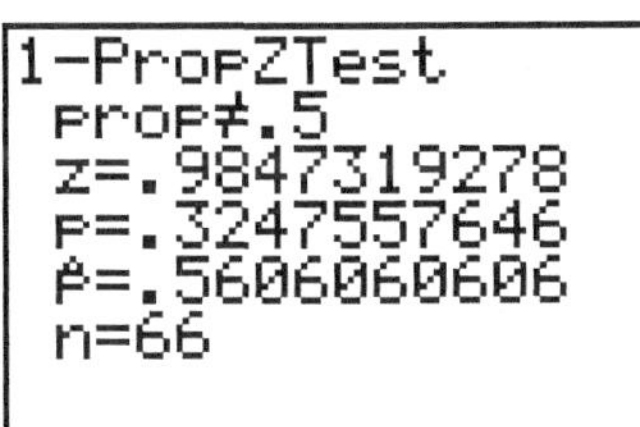

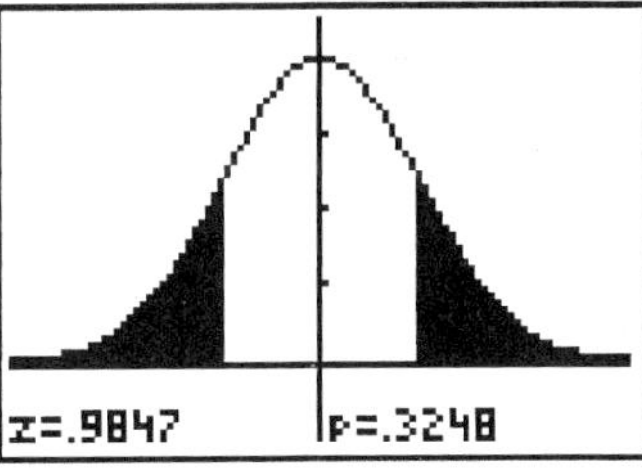

Exercise 8.14: Another test of astrology (p. 386 of Agresti/Franklin)
Here we wish to test H_0: p = 1/3 vs. H_a: p > 1/3. The sample proportion and test statistic calculation are shown below (left). Note that we do not need to double the tail probability, since this is a one-tailed test.

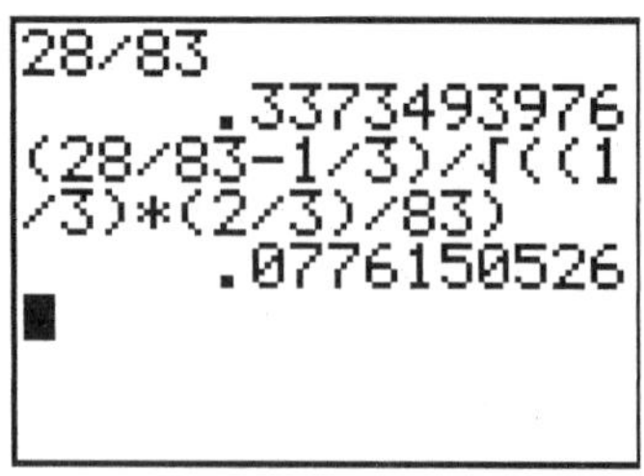

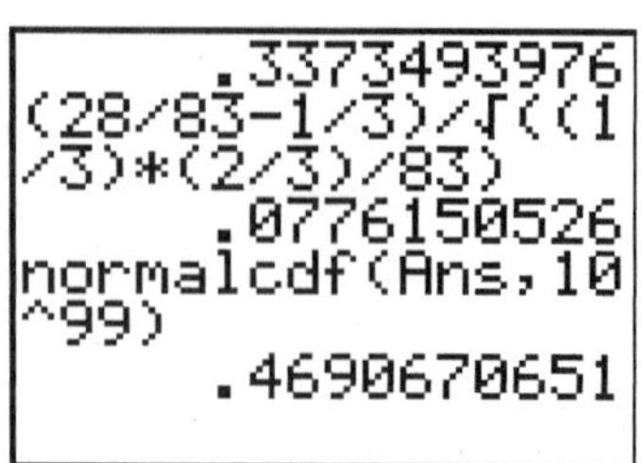

This is a large p-value, consistent with random guessing. We do not have sufficient evidence that people are more likely to select their "correct" horoscope.

Let's double-check our work using 1-PropZTest. Note that we are choosing the "right-tail" alternative.

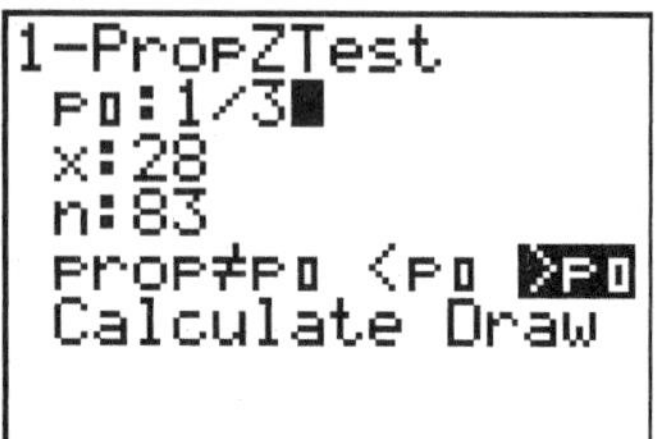

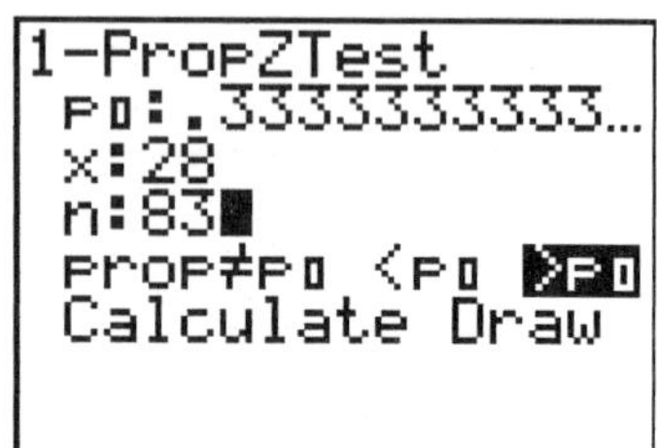

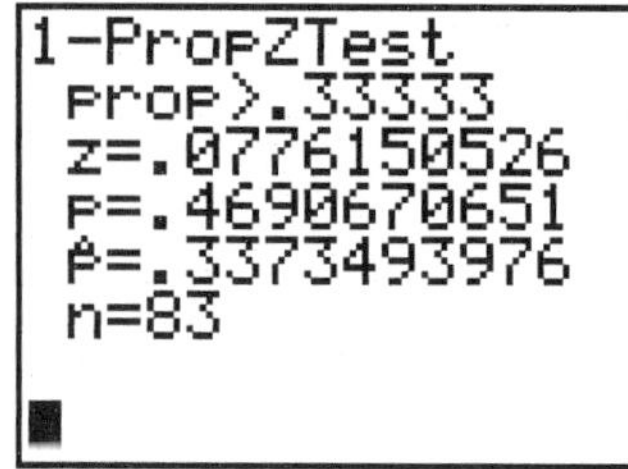

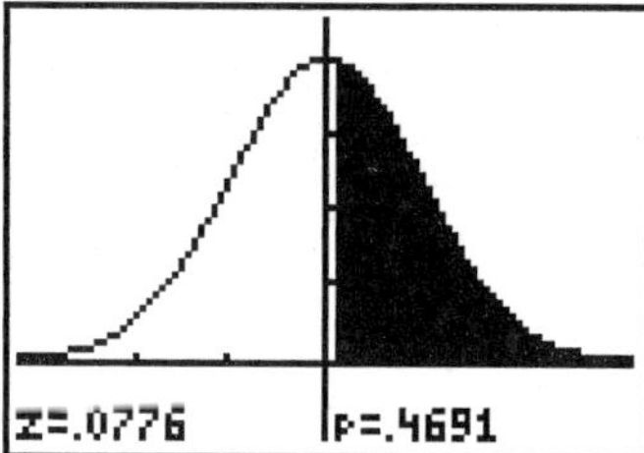

"Calculate" verifies our work, and "Draw" shows how close the p-value is to .5. This is a very small test statistic.

Significance Tests about Means: Section 8.3 of Agresti/Franklin
In this section, we consider hypothesis tests about one mean. This method makes use of the t distribution, so we'll use the tcdf function from the DISTR menu. There is also a very useful function under STAT TESTS called T-Test.

EXAMPLE 8: Does a low-carbohydrate diet work? (pp. 392-393 of Agresti/Franklin)
For this problem, we are given the summary statistics. The variable is weight change in kg, and we have n = 41, $\bar{x} = -9.7$, and s = 3.4. We will initially assume (see the null hypothesis) that the sample of weight changes came from a population with mean zero (no change), and try to show (see the alternative hypothesis) that population mean is less than zero (in other words, a significant weight *loss*). The t test statistic calculation shown on p. 393 is verified below.

```
3.4/√(41)
         .5309907904
-9.7/Ans
           -18.2677368
```

Note that this is a very large (negative) test statistic! The p-value is the area to the left of –18.3, which is tiny (see illustration p. 515). To obtain this p-value, we can use the tcdf function. This function works just like the normalcdf function, except it returns probabilities from the t distribution. We need to specify the number of degrees of freedom. The arguments are tcdf(lower, upper, df). For this problem, we want the area under the t curve to the left of –18.3, so the lower bound is negative infinity, and the upper bound is –18.3. The degrees of freedom value is df = n – 1 = 41 – 1 = 40. The p-value is so small, the calculator simply reports it as zero. We have overwhelming evidence to reject H_0 and conclude that the diet was effective for weight loss.

```
DISTR DRAW
1:normalpdf(
2:normalcdf(
3:invNorm(
4:tpdf(
5:tcdf(
6:χ²pdf(
7↓χ²cdf(
```

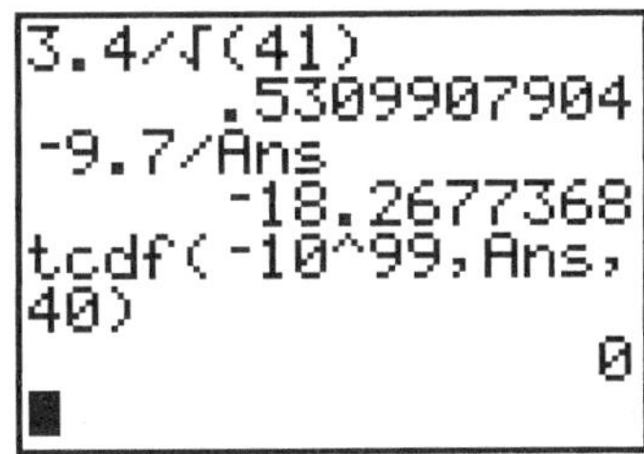

Next, let's check our work using the T-Test function. Press STAT, then cursor right to TESTS. Choose 2:T-Test. Specify Stats, since we don't have the entire sample of data to work with. Enter the null mean value of zero, $\bar{x} = -9.7$, s = 3.4, n = 41, and the direction of the alternative hypothesis. Now Calculate.

```
T-Test
 Inpt:Data Stats
 μ0:0
 x̄:-9.7
 Sx:3.4
 n:41
 μ:≠μ0 <μ0 >μ0
 Calculate Draw
```

"Calculate" gives the screen shown below left. The test statistic and p-value are verified. "Draw" gives the screen shown below right. Since the test statistic is so far away from zero, the p-value shading doesn't appear on the screen.

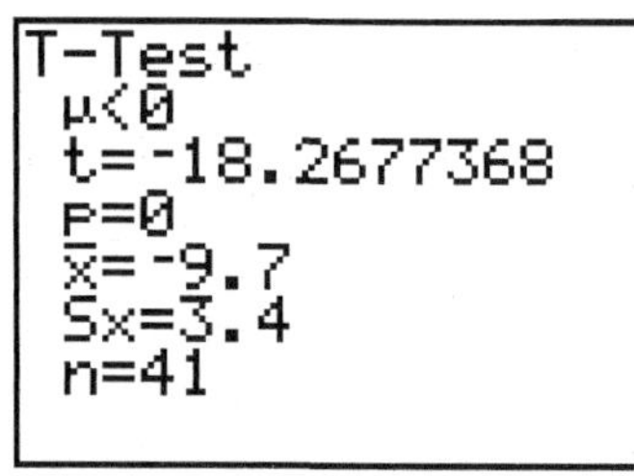

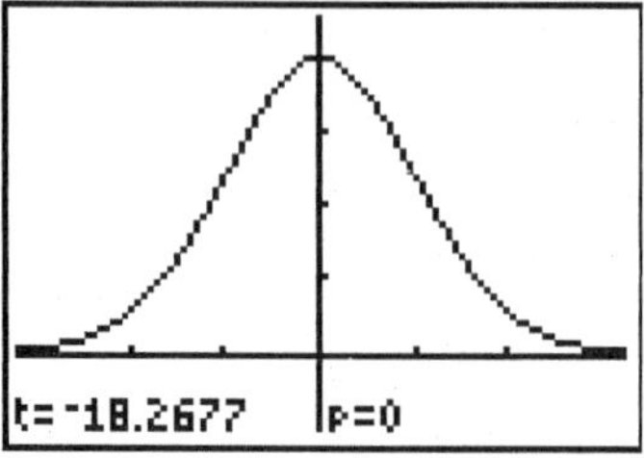

Exercise 8.26: Practice mechanics of a t-test (p. 398 of Agresti/Franklin)
Note that n = 20. We are given the test statistic and are asked for the p-value under a variety of alternative hypotheses. We can use tcdf to find these p-values instead of Table B in the appendix of Agresti/Franklin. The calculator holds a considerable advantage over the table. With the table, we must be satisfied with finding a range that our p-value falls in. So for example, we may determine from Table B that .01 < p-value < .05 for a certain problem. This does the trick, because we know that H_0 is rejected at a significance level of 5%. However, if we use the calculator, we can simply report the p-value.

(a) A two-sided alternative. Find the probability to the right of t = 2.40 on a t curve with 19 degrees of freedom (df = n – 1 = 20 – 1 = 19) and then double it. P-value = .0268.

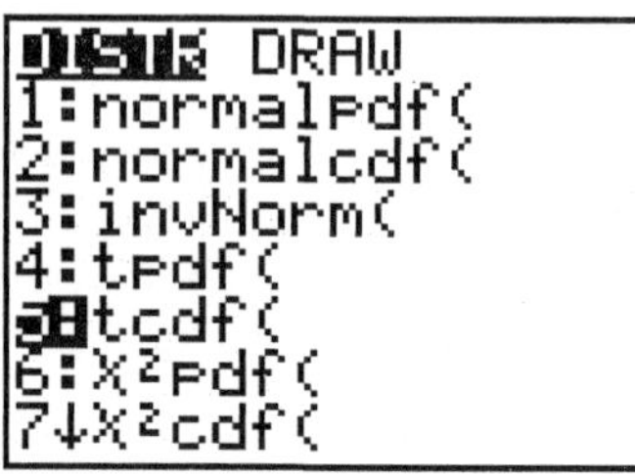

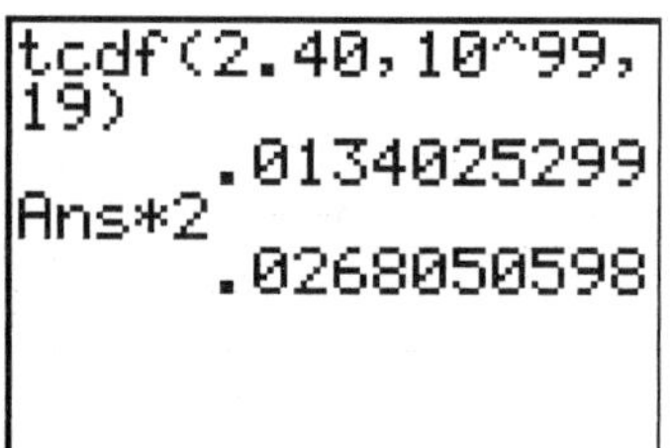

(b) A one-sided (specifically right-sided) alternative. If we are trying to show that μ > 100, it would make sense to get a positive test statistic. This simply means that $\bar{x}$ is larger than 100. The p-value will be a right-tail probability. This is the same as the probability computed above, but this time we do not double it. P-value = .0134. See below left.

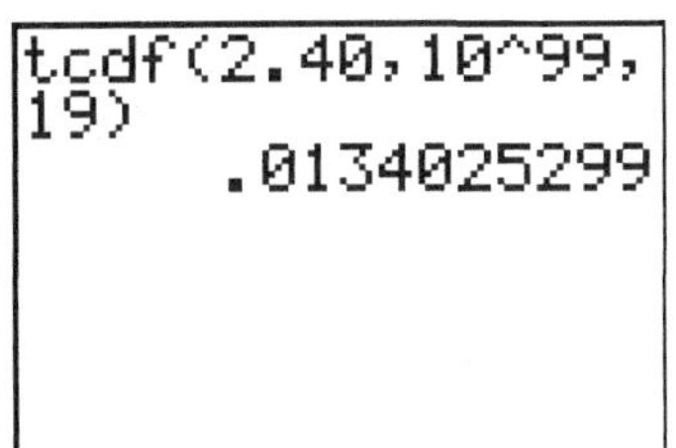

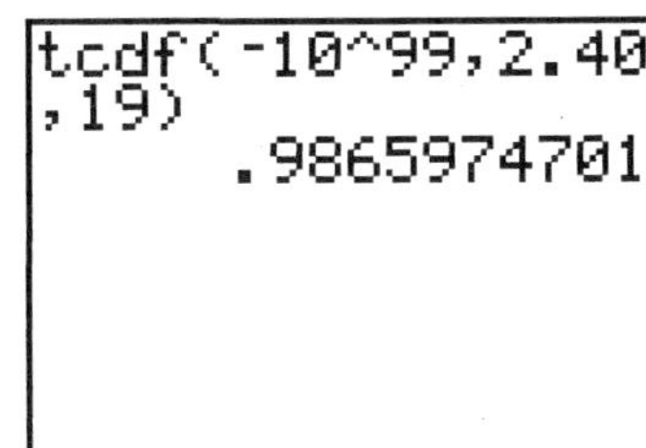

(c) Another one-sided alternative. This time it's a left "tail" probability. Before we calculate this p-value, think about it and draw a picture. If we are trying to show $\mu < 100$, you would most likely expect to get a sample mean less than 100, and consequently a negative test statistic. Whenever hypothesis tests work out in this logical way, we get a p-value that is truly a tail probability. For this problem, our test statistic is positive, but we have H_a: $\mu < 100$. This means that the p-value will be greater than 50%! We can hardly expect to reject H_0. See example 6 on pp. 500-501 for an example of this. Okay, let's find the p-value for this problem. We need the area to the left of 2.40 on a t curve with df = 19, so the lower bound is negative infinity, and the upper bound is 2.40. The p-value = .9866.

Exercise 8.29: Lake pollution (p. 398 of Agresti/Franklin)
We have a very small sample; n = 4! Let's enter the data into L1. Press STAT and choose Edit. Clear out any unwanted data first.

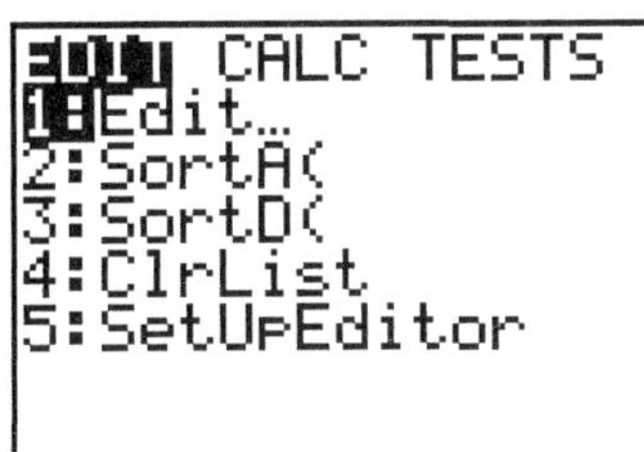

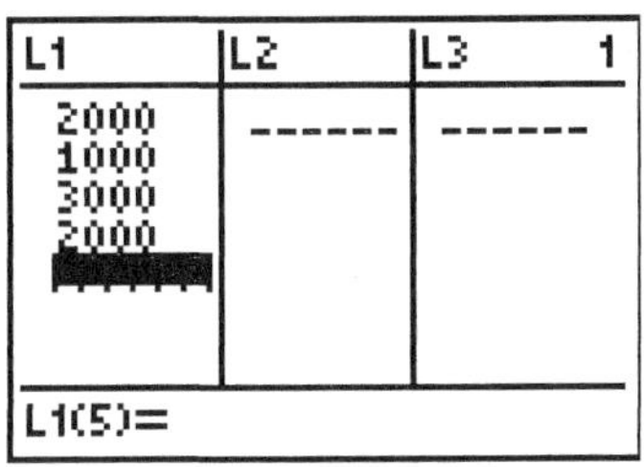

(a) The summary statistics and standard error are verified below. Press STAT, then CALC. When using the 1-Var Stats function, don't forget to indicate that the data are entered in L1.

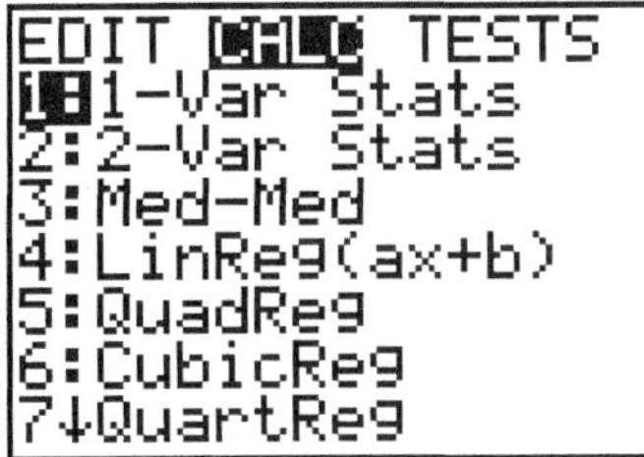

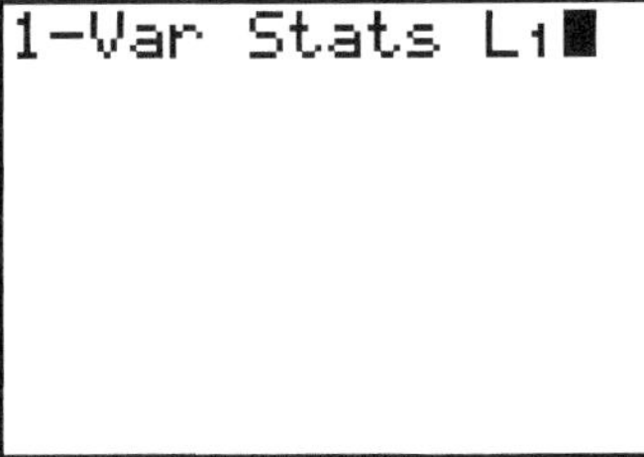

```
1-Var Stats
 x̄=2000
 Σx=8000
 Σx²=18000000
 Sx=816.4965809
 σx=707.1067812
↓n=4
```

```
816.5/√(4)
              408.25
```

(b) The test statistic calculation is verified below.

```
816.5/√(4)
              408.25
(2000-1000)/Ans
         2.449479486
■
```

(c) Use tcdf to find the p-value. Press 2nd VARS to access the DISTR menu. Since H_a: $\mu >$ 1000, we need the area under the t curve with 3 degrees of freedom (df = n – 1 = 4 – 1 = 3) from 2.45 to positive infinity. P-value = .0459.

```
DISTR DRAW
1:normalpdf(
2:normalcdf(
3:invNorm(
4:tpdf(
5:tcdf(
6:χ²pdf(
7↓χ²cdf(
```

```
816.5/√(4)
              408.25
(2000-1000)/Ans
         2.449479486
tcdf(Ans,10^99,3
)
        .0458609752
■
```

Since p-value = .0459, we reject H_0 at the 5% significance level. Let's double-check our work using the T-Test function. Press STAT, then cursor right to TESTS. Choose T-Test. Specify Data this time, since we have it stored in L1. There's no sense in specifying Stats, and then entering in the summary statistics we calculated. Choose Data, then enter the null value of the population mean (μ = 1000), supply the list where the data are stored (L1), leave Freq:1 as is (since each data value simply represents one observation), and specify a right-sided alternative.

```
EDIT CALC TESTS
1:Z-Test...
2:T-Test...
3:2-SampZTest...
4:2-SampTTest...
5:1-PropZTest...
6:2-PropZTest...
7↓ZInterval...
```

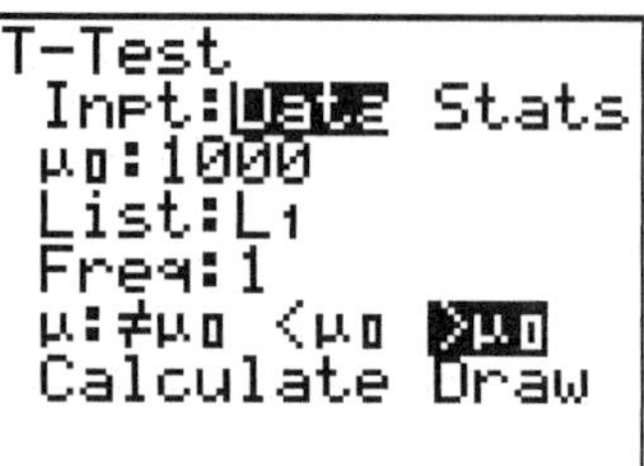

"Calculate" produces the results shown below left. "Draw" gives the picture shown below right.

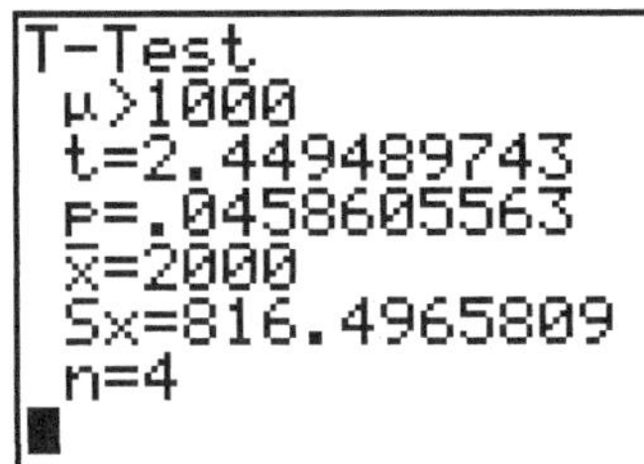

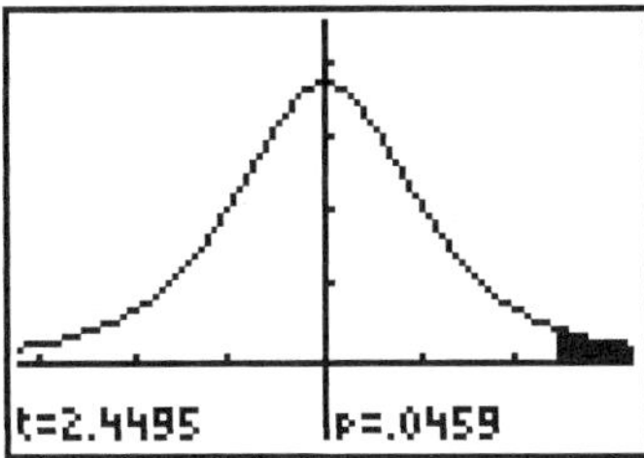

Exercise 8.32: Too little or too much wine? (p. 399 of Agresti/Franklin)
Let's enter the data into L1 and get the sample statistics.

```
L1      L2      L3     1
745     ------  ------
740
750
745

L1(5)=
```

```
1-Var Stats
 x̄=745
 Σx=2980
 Σx²=2220150
 Sx=4.082482905
 σx=3.535533906
↓n=4
```

We will be testing against a two-tailed alternative. The test statistic calculation is shown below left, and the p-value is obtained below right.

```
4.08/√(4)
                  2.04
(745-752)/Ans
          -3.431372549
```

```
(745-752)/Ans
          -3.431372549
tcdf(-10^99,Ans,
3)
          .0207482113
Ans*2
          .0414964226
```

Note that the test statistic is negative, since the sample mean is smaller than the null value of the population mean. So we can obtain the p-value by first computing the tail probability to the left of the test statistic, then doubling it. P-value = .0415, which is significant at $\alpha = .05$.

Now let's check our work with the T-Test function. Press STAT, then arrow right to TESTS. Specify Data, the null mean value of 752, L1 if you entered the data there, Freq should be left "1" by default, and a two-sided alternative.

```
EDIT CALC TESTS
1:Z-Test...
2:T-Test...
3:2-SampZTest...
4:2-SampTTest...
5:1-PropZTest...
6:2-PropZTest...
7↓ZInterval...
```

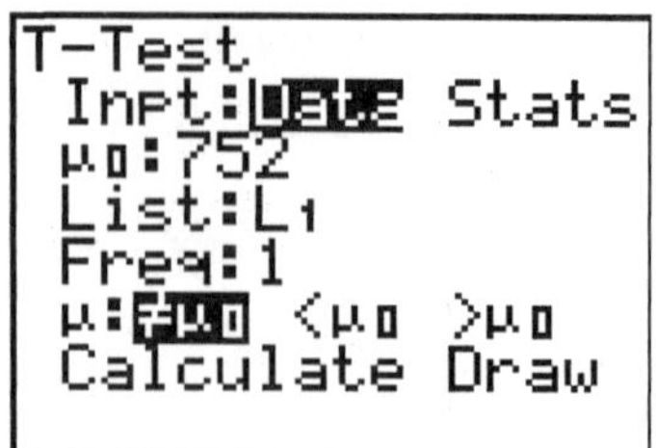

"Calculate" verifies our previous results (see below left). "Draw" produces the picture below right. The test statistic is just large enough to be off-screen.

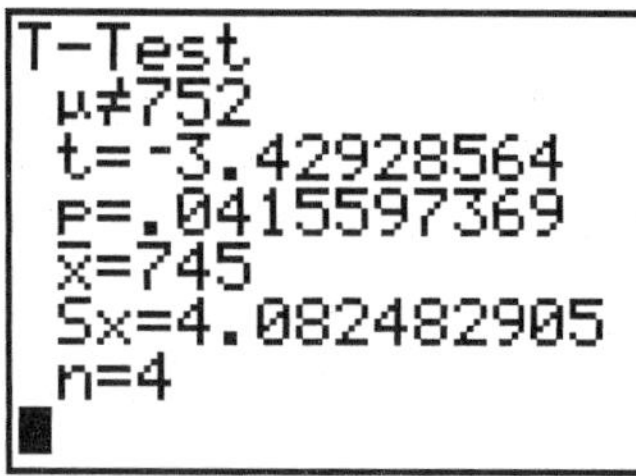

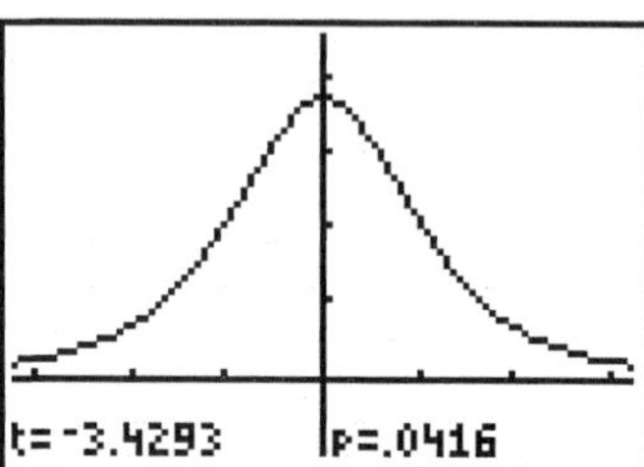

Chapter Nine: Comparing Two Groups

Confidence Intervals and Significance Tests for Two Proportions: Section 9.1 of Agresti/Franklin

In this section, we will find the STAT TESTS functions 2-PropZInt and 2-PropZTest useful, in addition to normalcdf and invNorm from the DISTR menu.

EXAMPLE 4: Confidence interval comparing heart attack rates for aspirin and placebo (pp. 431-432 of Agresti/Franklin)

This is a continuation of a previous example. Let's review the basic information given. Call the doctors that took the placebo "group 1". Call the doctors that took aspirin "group 2". Of those on placebo, 189 had a heart attack. Of those on aspirin, 104 had a heart attack. The two sample proportions are calculated below left. The standard error calculation (see p. 429) is shown below right.

```
189/11034
      .0171288744
104/11037
      .0094228504
■
```

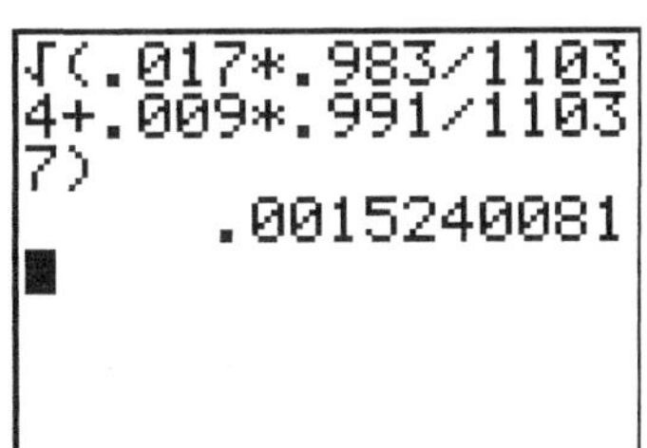

We can replicate the interval calculations shown on p. 431 as well. Remember that critical values from the normal distribution can be found using invNorm. Press 2nd VARS to access the DISTR menu. By now, you may have memorized 1.96 as the normal percentile for 95% confidence. However, recall that with the middle 95% of the distribution marked off, this leaves 5% of the area to be split evenly between the two tails. We need either the 2.5th or 97.5th percentile of the standard normal distribution. See below left. The confidence interval bounds are computed below right.

```
invNorm(.025)
     -1.959963986
invNorm(.975)
      1.959963986
■
```

```
.017-.009
              .008
.008-1.96*.0015
            .00506
.008+1.96*.0015
            .01094
```

Let's check our above work by using the 2-PropZInt function. Press STAT, then cursor right to TESTS. Choose B:2-PropZInt. Recall that group 1, the placebo group, had 189 heart attacks out of 11034 doctors. Recall that group 2, the aspirin group, had 104 heart attacks out of 11037 doctors. Specify 95% confidence, and press ENTER with the cursor on Calculate.

```
EDIT CALC TESTS
0↑2-SampTInt...
A:1-PropZInt...
B:2-PropZInt...
C:χ²-Test...
D:2-SampFTest...
E:LinRegTTest...
F:ANOVA(
```

```
2-PropZInt
 x1:189
 n1:11034
 x2:104
 n2:11037
 C-Level:.95
 Calculate
```

```
2-PropZInt
 (.00469,.01072)
 p̂1=.0171288744
 p̂2=.0094228504
 n1=11034
 n2=11037
```

The results above are a bit more precise; remember that we did some rounding along the way when we first did the problem. Rounding is OK – this is introductory statistics, not a numerical analysis course – but you should realize that any rounding you do can produce different answers. This is especially true with categorical data.

Exercise 9.3: Binge drinking (p. 438 of Agresti/Franklin)

(a) The difference in proportions is computed below left. Don't let the negative sign confuse you. We see an increase in the proportion of binge drinkers from 39.9% in 1993 to 48.2% in 2001. The difference reflects the 8.3% increase.

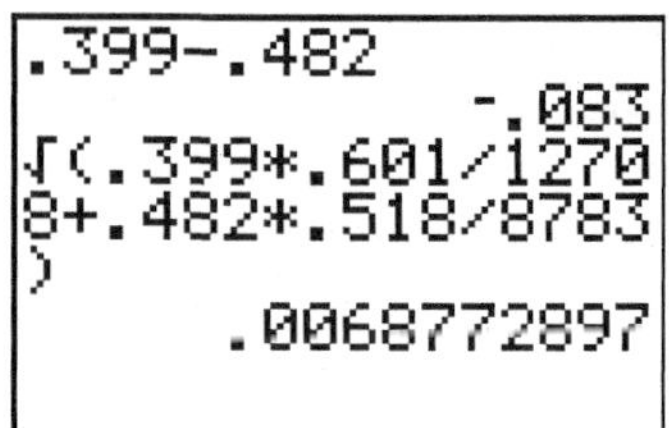

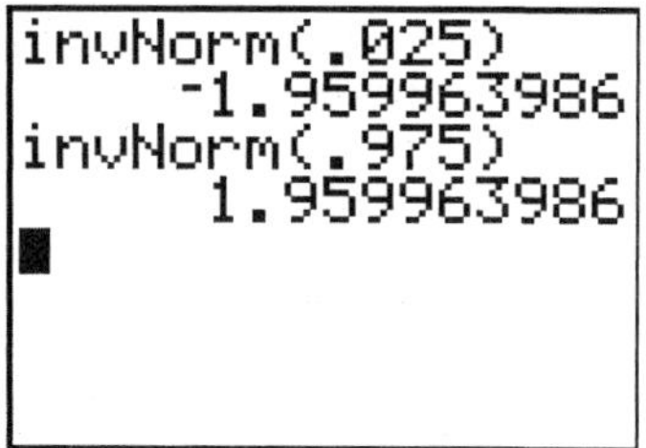

(b) The standard error is computed above left.

(c) Recall that we can use invNorm to get normal percentiles. For 95% confidence, we need either the 2.5th or 97.5th percentile. See above right. The interval is calculated below left. Realize that we can arbitrarily decide which group is "1" or "2", so our initial calculation of the difference in proportion could have been .482 – .399 = .083. Either way, we have an 8.3% increase between 1993 and 2001. The interval is calculated again using this positive difference below right. The standard error does not change.

```
-.083-1.96*.0069
           -.096524
-.083+1.96*.0069
           -.069476
```

```
.083-1.96*.0069
           .069476
.083+1.96*.0069
           .096524
```

The intervals (–.097, –.069) and (.069, .097) are equivalent. Either one tells us that we can be 95% confident that there was an increase in binge drinking between 6.9% and 9.7% from 1993 to 2001.

(d) Assumptions: We most definitely have large enough samples here. If a "success" is a binge drinker, we have (.399)12708 = 5070 successes in the 1993 sample, and 12708 – 5070 = 7638 failures. Similarly, you can check that the 2001 sample has far more than the required 10 successes and 10 failures.

EXAMPLE 5: Is TV watching associated with aggressive behavior? (pp. 434-436 of Agresti/Franklin)

Group 1 will be those who watched less than one hour of TV a day, and group 2 will be those who watched at least one hour a day. The sample proportion for group 1, group 2, and the pooled sample proportion, respectively, are calculated below left. The standard error and test statistic are computed below right.

```
5/88
        .0568181818
154/465
        .3311827957
(5+154)/(88+619)
         .224893918
```

```
√(.225*.775*(1/8
8+1/619))
        .0475735058
(.057-.249)/.047
6
       -4.033613445
```

Recall that normalcdf will give normal probabilities. We need the area to the left of –4.0, and to the right of 4.0, since this is a two-tailed test. By the symmetry of the curve, we can double the left tail area. The p-value is .00006.

```
normalcdf(-10^99
, -4)
  3.168603459E-5
Ans*2
  6.337206918E-5
```

Now let's try the 2-PropZTest function to check our work above. Press STAT, then arrow right to TESTS. Choose 6:2-PropZTest. Specify 5 aggressive acts out of 88 teenagers in group 1, and 154 aggressive acts out of 619 teenagers in group 2. Indicate that the test is two-tailed by choosing $p_1 \neq p_2$. With the cursor blinking on Calculate, press ENTER.

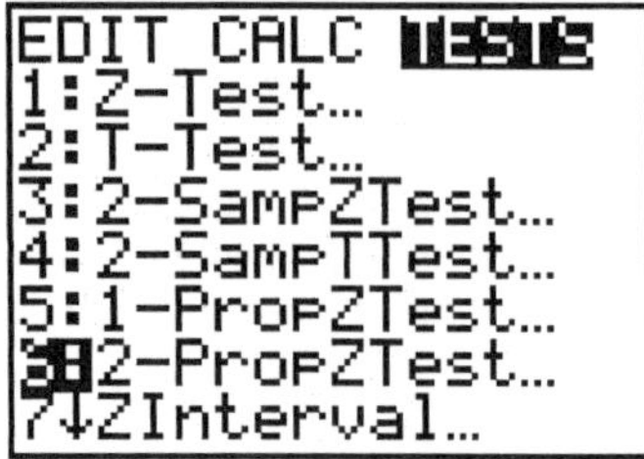

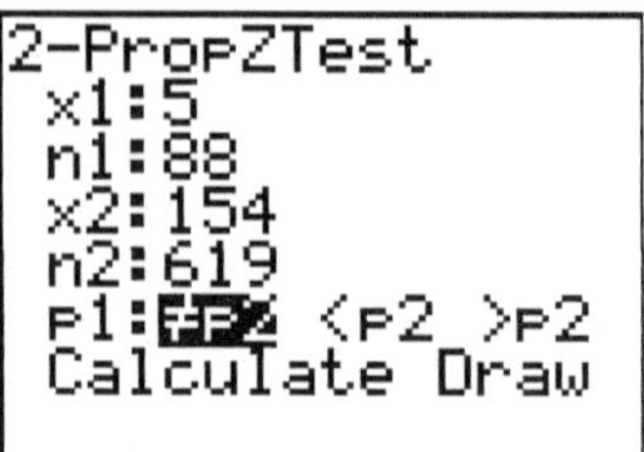

More results are produced than can fit on one screen. Below left we see the test statistic, p-value (slightly more precise than what we calculated before), sample proportions, and pooled sample proportion. Scrolling further down (see below right) just confirms the two sample sizes.

```
2-PropZTest
 p1≠p2
 z=-4.035908513
 p=5.4418244E-5
 p̂1=.0568181818
 p̂2=.2487883683
↓p̂=.224893918
```

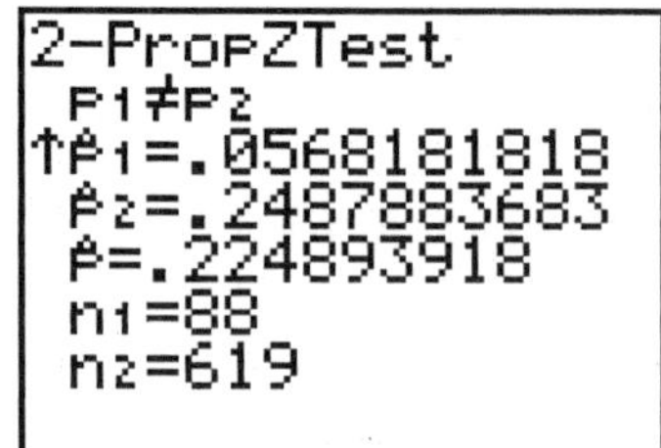

The "Draw" option produces the below picture. The test statistic is too large to show up on screen.

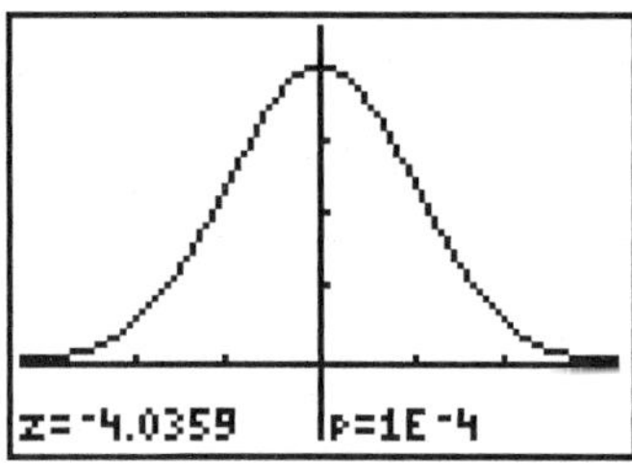

Exercise 9.9: Drinking and unplanned sex (p. 439 of Agresti/Franklin)
Let's use the 2-PropZTest function to solve this problem. Recall that we need the number of "successes" in each group. Refer to exercise 9.3 for the sample sizes. For group 1 (1993), we have $n_1 = 12708$. For group 2 (2001), we have $n_2 = 8783$. Therefore we have .192(12708) = 2440 "successes" in 1993, and .213(8783) = 1871 "successes" in 2001. Under STAT TESTS, choose 2-PropZTest. Supply the previously mentioned counts and sample sizes, and choose a two-sided alternative since we are only looking for a "change" (not specifically an *increase*, for

example). With the cursor blinking on "Calculate", pressing ENTER produces the screen below right.

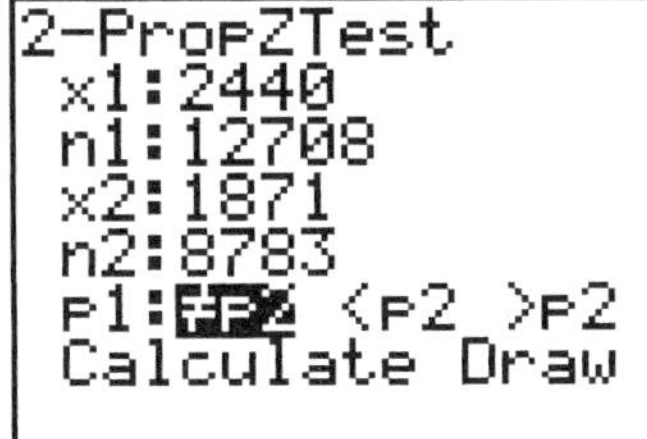

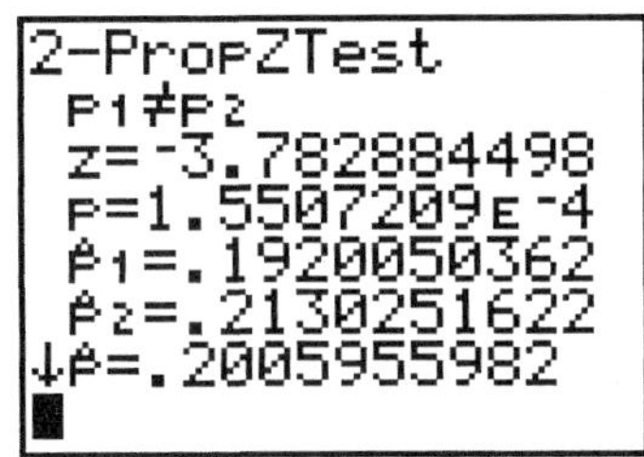

The test statistic is z = – 3.78, and the p-value is .0002. We have evidence of a statistically significant change. Do you think the large sample sizes have anything to do with it? With the cursor blinking on "Draw", we get the image below. The test statistic is too large to show up on screen.

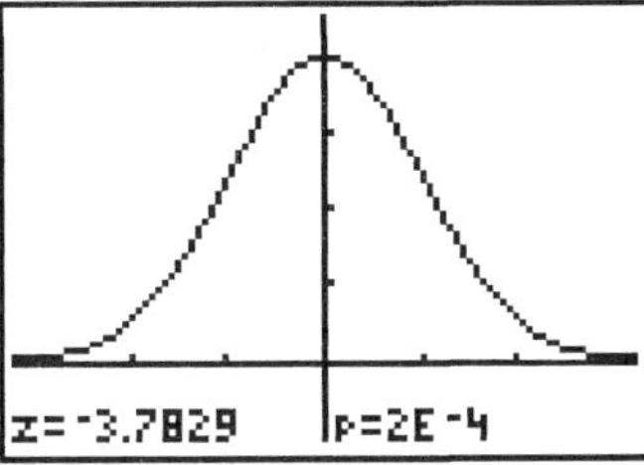

Confidence Intervals and Significance Tests for Two Means: Section 9.2 of Agresti/Franklin

In this section, we will find the STAT TESTS functions 2-SampTInt and 2-SampTTest useful, in addition to tcdf from the DISTR menu.

EXAMPLE 8: Nicotine – How much more addicted are smokers than ex-smokers? (pp. 443-444 of Agresti/Franklin)

Call the smokers group 1 and the ex-smokers group 2. The summary statistics are given under "Picture the Scenario" on p. 443. The number of degrees of freedom for this problem can be difficult to obtain; note that we are told that df = 95 under "Think it through", part b. From this, we can use the TI-83 to get the t critical value. Recall from chapter 7 that we must use the EQUATION SOLVER to get t-scores. Press MATH, then choose 0:Solver. Set the lower bound to negative infinity and with the cursor blinking on the "U=" line, press SOLVE (ALPHA ENTER). The t-score for 95% confidence is 1.985.

```
MATH NUM CPX PRB
4↑³√(
5:ˣ√
6:fMin(
7:fMax(
8:nDeriv(
9:fnInt(
0:Solver...
```

```
tcdf(L,U,D)-A=0
 L=-1E99
▪U=-1.985250955...
 D=95
 A=.025
 bound={-1E99,1...
▪left-rt=1E-14
```

See p. 444 of Agresti/Franklin for calculation of the interval using the sample statistics and t-score. Let's use the 2-SampTInt function to corroborate those results. Press STAT, then arrow to TESTS. Choose 0:2-SampTInt. Choose "Stats", since we don't have the original data stored in lists. Specify the sample mean, standard deviation, and sample size for the two groups.

```
EDIT CALC TESTS
4↑2-SampTTest...
5:1-PropZTest...
6:2-PropZTest...
7:ZInterval...
8:TInterval...
9:2-SampZInt...
0:2-SampTInt...
```

```
2-SampTInt
 Inpt:Data Stats
 x̄1:5.9
 Sx1:3.3
 n1:75
 x̄2:1
 Sx2:2.3
↓n2:257
```

Continuing, we need to specify 95% confidence, and choose the non-pooled method. Note that the standard error formula on p. 443 does not "pool" the two sample standard deviations together. Sometimes, a "pooled" method is used, but not here. With the cursor blinking on "Calculate", press ENTER.

```
2-SampTInt
↑n1:75
 x̄2:1
 Sx2:2.3
 n2:257
 C-Level:.95
 Pooled:No Yes
 Calculate
```

```
2-SampTInt
 (4.0918,5.7082)
 df=95.91055203
 x̄1=5.9
 x̄2=1
 Sx1=3.3
↓Sx2=2.3
```

The confidence interval matches the one shown on p. 444. Note that the number of df is actually 95.9! This is a detail we won't spend any time worrying about. Scrolling down further just verifies the sample sizes.

```
2-SampTInt
 (4.0918,5.7082)
↑x̄2=1
 Sx1=3.3
 Sx2=2.3
 n1=75
 n2=257
```

Exercise 9.16: eBay auctions (pp. 449-450 of Agresti/Franklin)
For this problem, we are given the "raw" data, so to speak. Let's enter group 1 (buy-it-now) into L1, and group 2 (bidding only) into L2. These data can be downloaded (see chapter 1 for downloading options and locations of examples). Choose Edit under STAT EDIT.

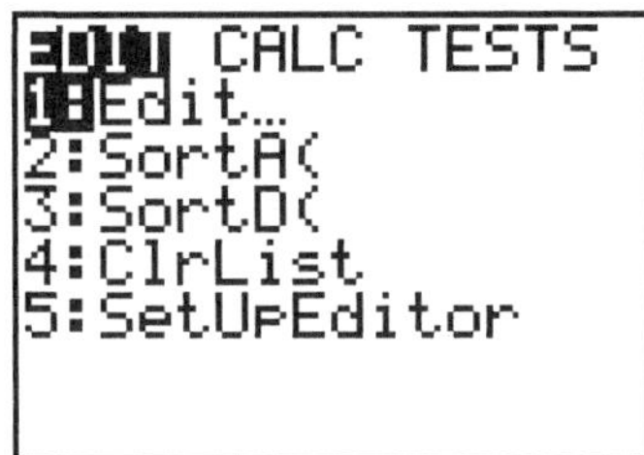

L1	L2	L3 2
250	240	
210	228	
------	255	
	232	
	246	
	210	

L2(12) =178

Before constructing the confidence interval, let's take a look at our data. The best graphical method of comparing quantitative data among two or more groups is the side-by-side boxplot. We can look at a boxplot of the buy-it-now prices and a boxplot of the bidding-only prices on the same graph. Go to STAT PLOTS (2nd Y=). Let's use Plot1 for group 1 and Plot2 for group 2. Choose Plot1 and turn it "On". Specify a modified boxplot for the group 1 data in L1.

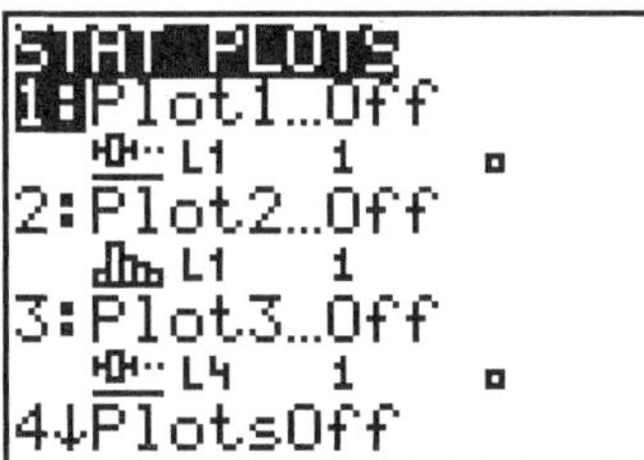

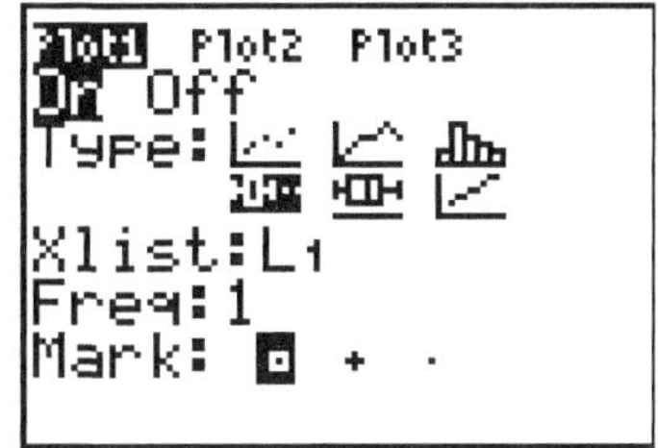

Note that from the Plot1 screen, you can cursor up and over to Plot2. In the Plot2 screen, turn it "On", specify a modified boxplot, and supply L2 for the Xlist. Now GRAPH. Don't forget that the WINDOW dimensions likely won't set optimally for the data at hand, so use ZoomStat under the ZOOM menu.

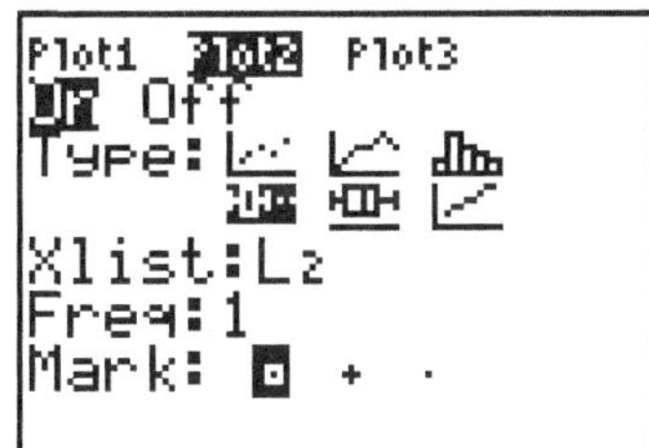

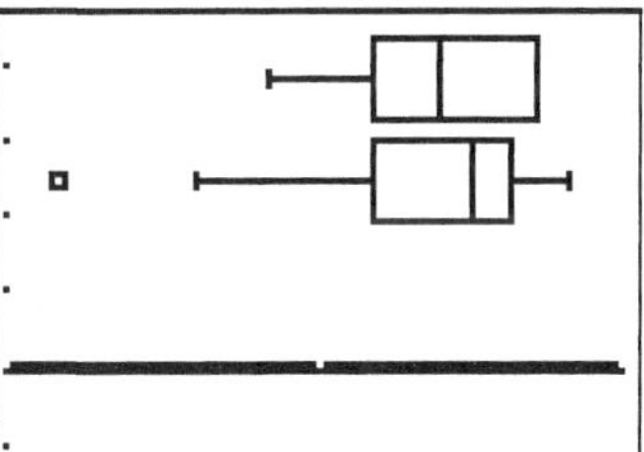

Note the outlier in the bidding-only group.

The TRACE function (shown below) will tell you that the outlier is the minimum of group 2, a price of merely \$178.

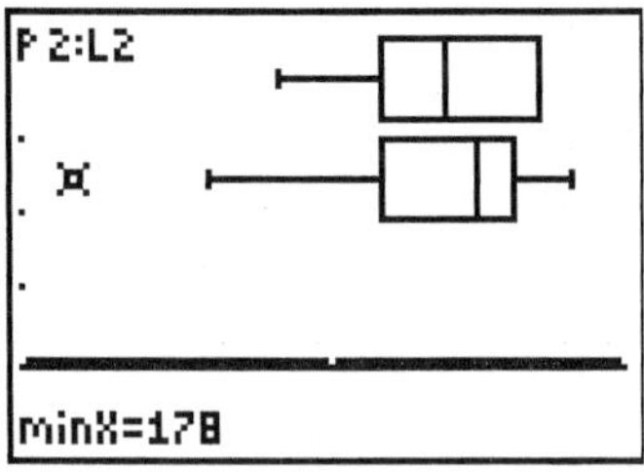

Note also the amount of overlap. Do the two groups seem to differ much in terms of final prices? Do you expect the confidence interval for the difference in means to include zero?

From STAT TESTS, choose 2-SampTInt. Choose "Data", and specify lists L1 and L2. Leave Freq1 and Freq2 equal to "1", the default, since each price in the data set represents one auction. We want 95% confidence, and do not want to pool the standard deviations together.

```
EDIT CALC TESTS
4↑2-SampTTest...
5:1-PropZTest...
6:2-PropZTest...
7:ZInterval...
8:TInterval...
9:2-SampZInt...
0↓2-SampTInt...
```

With the cursor blinking on "Calculate", press ENTER.

```
2-SampTInt
↑List1:L1
 List2:L2
 Freq1:1
 Freq2:1
 C-Level:.95
 Pooled:No Yes
 Calculate
```

```
2-SampTInt
 (-14.05,17.967)
 df=16.58973553
 x̄1=233.5714286
 x̄2=231.6111111
 Sx1=14.6385011
↓Sx2=21.9361075
```

Most of the results are shown above right. Scrolling down shows the two sample sizes (see next page).

```
2-SampTInt
 (-14.05,17.967)
↑x̄2=231.6111111
 Sx1=14.6385011
 Sx2=21.9361075
 n1=7
 n2=18
```

EXAMPLE 9: Does cell-phone use while driving impair reaction times? (pp. 446-448 of Agresti/Franklin)

We are not given the data for this example, so we are not able to replicate the boxplots shown in figure 9.6 with the calculator. See exercise 9.16 for an example of this. We are given the following sample statistics. For group 1 (cell phone), we have $\bar{x}_1 = 585.2$, $s_1 = 89.6$, and $n_1 = 32$. For group 2 (control), we have $\bar{x}_2 = 533.7$, $s_2 = 65.3$, and $n_2 = 32$. We are told that we have 56 degrees of freedom for this problem. The standard error and test statistic calculation shown on p. 447 are verified below left. For the p-value, we can use the tcdf function. We need the area to the right of 2.63 on a t curve with df=56. This is a two-tailed test, so double that tail area. P-value = .0111.

```
√(89.6²/32+65.3²
/32)
        19.59930643
(585.2-533.7)/An
s
        2.627644003
```

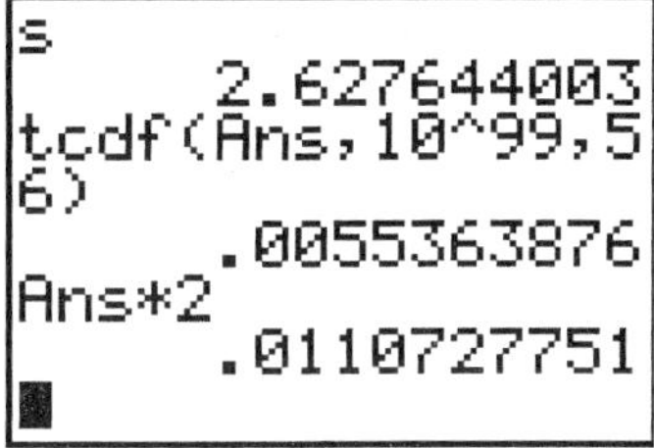

Let's use the 2-SampTTest function to check our work. This calculator function is quite valuable for problems like these because it does not require knowledge of the df. That is part of the computation that the TI-83 does for us. Press STAT, then arrow right to TESTS. Choose 4:2-SampTTest. Specify Stats. Enter the sample means, standard deviations, and sizes.

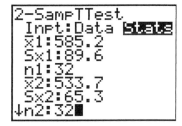

```
EDIT CALC TESTS
1:Z-Test...
2:T-Test...
3:2-SampZTest...
4:2-SampTTest...
5:1-PropZTest...
6:2-PropZTest...
7↓ZInterval...
```

```
2-SampTTest
 Inpt:Data Stats
 x̄1:585.2
 Sx1:89.6
 n1:32
 x̄2:533.7
 Sx2:65.3
↓n2:32
```

Indicate a two-tailed test, no pooling, then "Calculate". Below right we see the test statistic, p-value, and "real" df.

```
2-SampTTest
↑n1:32
 x̄2:533.7
 Sx2:65.3
 n2:32
 μ1:≠μ2 <μ2 >μ2
 Pooled:No Yes
 Calculate Draw
```

```
2-SampTTest
 μ1≠μ2
 t=2.627644003
 p=.0110413561
 df=56.68480423
 x̄1=585.2
↓x̄2=533.7
```

Scrolling down further just confirms our sample statistics. If you go back to STAT TESTS and choose "Draw" from 2-SampTTest, you'll see the picture below right. The test statistic and two-tailed p-value just barely show up on-screen.

```
2-SampTTest
 μ1≠μ2
↑x̄2=533.7
 Sx1=89.6
 Sx2=65.3
 n1=32
 n2=32
```

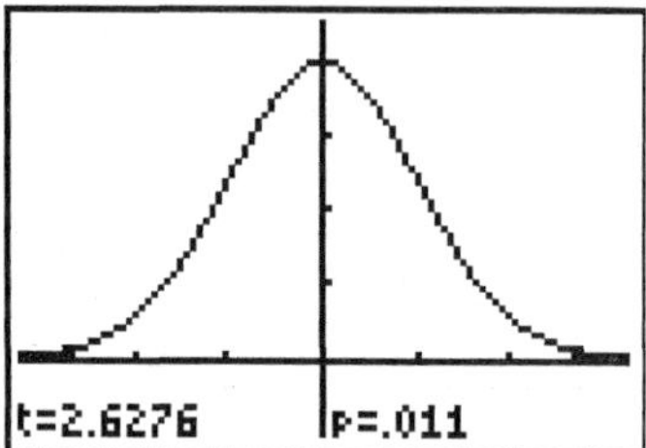

Exercise 9.21: Females or males more nicotine dependent? (p. 450 of Agresti/Franklin)
For the females, we have $\bar{x}_1 = 2.8$, $s_1 = 3.6$, and $n_1 = 150$. For the males, we have $\bar{x}_2 = 1.6$, $s_2 = 2.9$, and $n_2 = 182$. Press STAT, then arrow right to TESTS. Choose 2-SampTTest. Specify Stats. Enter the sample means, standard deviations, and sizes.

```
EDIT CALC TESTS
1:Z-Test...
2:T-Test...
3:2-SampZTest...
4:2-SampTTest...
5:1-PropZTest...
6:2-PropZTest...
7↓ZInterval...
```

```
2-SampTTest
 Inpt:Data Stats
 x̄1:2.8
 Sx1:3.6
 n1:150
 x̄2:1.6
 Sx2:2.9
↓n2:182
```

Specify a two-tailed test, no pooling, then "Calculate". Below right we see the test statistic, p-value, and "real" df.

```
2-SampTTest
↑n1:150
 x̄2:1.6
 Sx2:2.9
 n2:182
 μ1:≠μ2 <μ2 >μ2
 Pooled:No Yes
 Calculate Draw
```

```
2-SampTTest
 μ1≠μ2
 t=3.295300983
 p=.0011079094
 df=284.1007536
 x̄1=2.8
↓x̄2=1.6
```

With the cursor blinking on "Draw", we get the picture below. Note that the test statistic is too large to show up on screen.

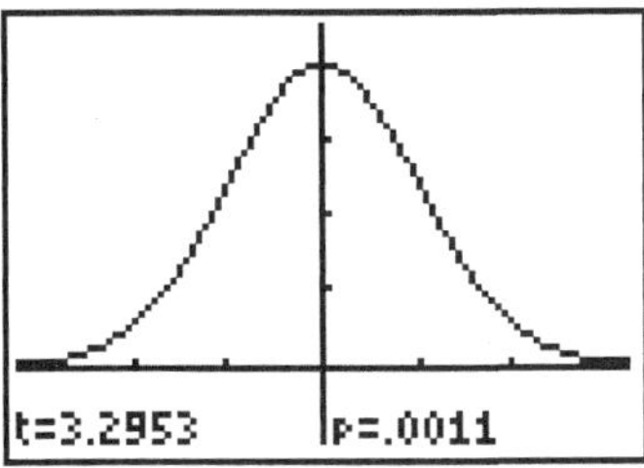

Exercise 9.24: Student survey (p. 451 of Agresti/Franklin)
Type or download the data into your calculator. Note that we have $n_1 = 31$ females and $n_2 = 29$ males.

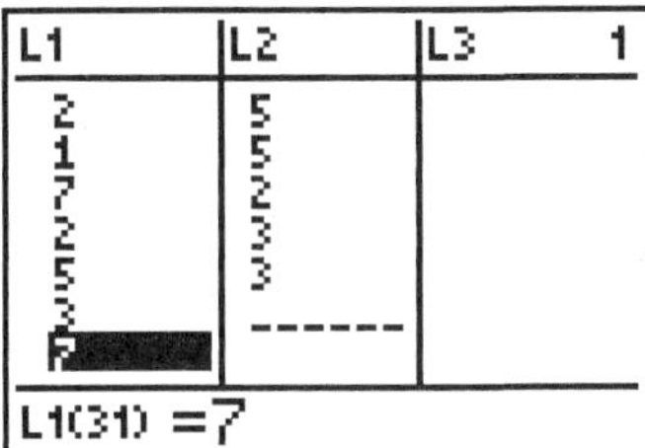

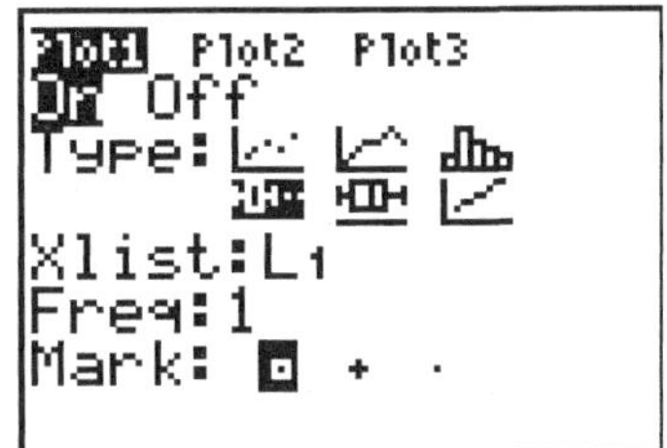

(a) Let's produce boxplots to compare the two groups, similar to what we did for exercise 9.16. Define Plot1 to be a modified boxplot for the females group, and Plot2 to be a modified boxplot for the males group. Make sure they are both On and Plot3 is turned Off. Don't forget ZoomStat (ZOOM, then option 9) to fix the WINDOW dimensions.

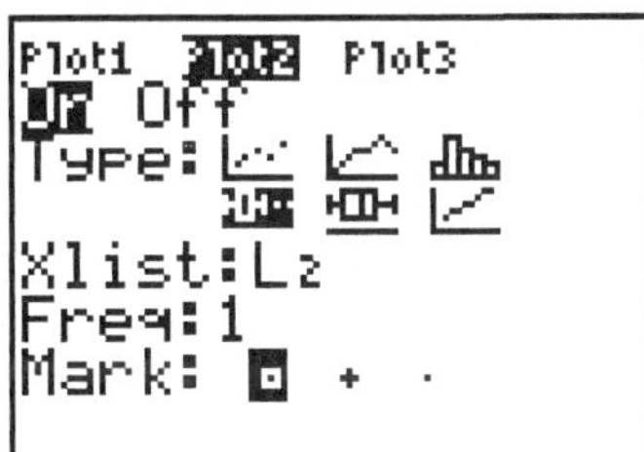

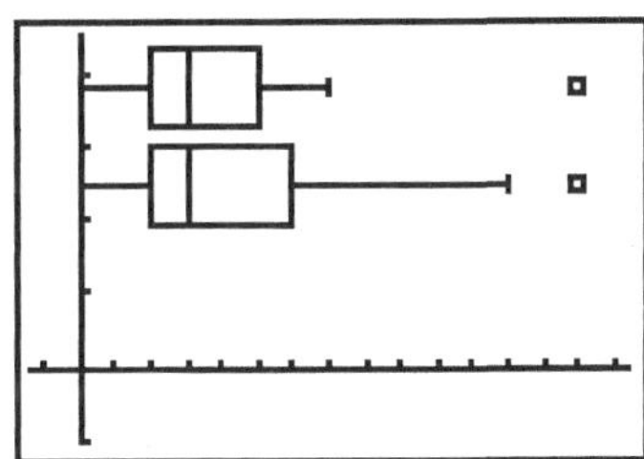

(b) From STAT TESTS, choose 2-SampTInt to get the desired confidence interval.

```
EDIT CALC TESTS
4↑2-SampTTest...
5:1-PropZTest...
6:2-PropZTest...
7:ZInterval...
8:TInterval...
9:2-SampZInt...
0↓2-SampTInt...
```

```
2-SampTInt
 Inpt:Data Stats
 List1:L1
 List2:L2
 Freq1:1
 Freq2:1
 C-Level:.95
↓Pooled:No Yes
```

Choose Data, and specify the list names where the data are stored. Freq1 and Freq2 should be left as "1". We need 95% confidence, and are not pooling the standard deviations together. Place the cursor on "Calculate" and press ENTER.

```
2-SampTInt
↑List1:L1
 List2:L2
 Freq1:1
 Freq2:1
 C-Level:.95
 Pooled:No Yes
 Calculate
```

```
2-SampTInt
 (-2.069,1.0997)
 df=56.72674577
 x̄1=3.774193548
 x̄2=4.25862069
 Sx1=2.92927383
↓Sx2=3.180916
```

We now have the desired interval. Scrolling down just confirms the sample sizes.

```
2-SampTInt
 (-2.069,1.0997)
↑x̄2=4.25862069
 Sx1=2.92927383
 Sx2=3.180916
 n1=31
 n2=29
```

(c) From STAT TESTS, choose 2-SampTTest. Again, choose Data, specify the appropriate lists, and keep Freq1 and Freq2 set to 1. We are asked to perform "a significance test comparing the population means", so no direction is implied. Choose a two-tailed test. We want the "unpooled" procedure.

```
EDIT CALC TESTS
1:Z-Test...
2:T-Test...
3:2-SampZTest...
4:2-SampTTest...
5:1-PropZTest...
6:2-PropZTest...
7↓ZInterval...
```

```
2-SampTTest
 Inpt:Data Stats
 List1:L1
 List2:L2
 Freq1:1
 Freq2:1
 μ1:≠μ2 <μ2 >μ2
↓Pooled:No Yes
```

"Calculate" produces the results shown below right. We can see the test statistic, p-value, degrees of freedom, and sample means.

```
2-SampTTest
↑List1:L1
 List2:L2
 Freq1:1
 Freq2:1
 μ1:≠μ2 <μ2 >μ2
 Pooled:No Yes
 Calculate Draw
```

```
2-SampTTest
 μ1≠μ2
 t=-.6124147026
 p=.5427106508
 df=56.72674577
 x̄1=3.774193548
↓x̄2=4.25862069
```

Scrolling down further on the results screen shows the sample standard deviations and sizes (see below left). Selecting "Draw" instead produces the image below right. Don't forget to turn the boxplots Off!

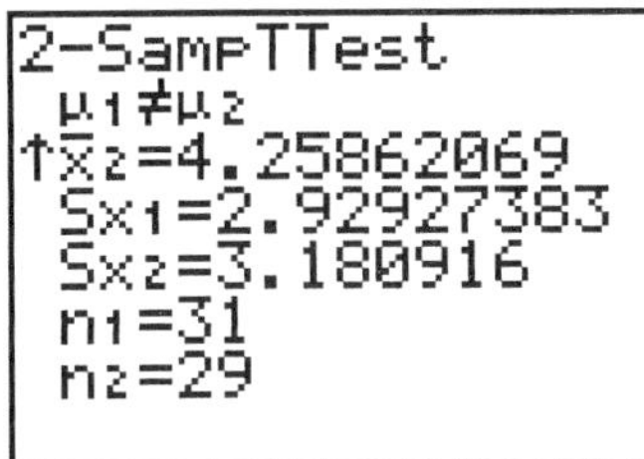

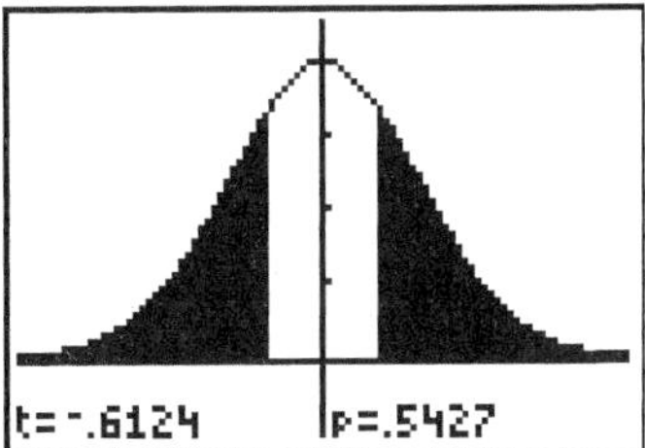

The Pooled t-test: Section 9.3 of Agresti/Franklin

EXAMPLE 10: Is arthroscopic surgery better than placebo? (pp. 452-454 of Agresti/Franklin)
We are comparing group 1 (placebo) with group 2 (lavage arthroscopic surgery). We have $n_1 = 60$, $\bar{x}_1 = 51.6$, $s_1 = 23.7$ and $n_2 = 61$, $\bar{x}_2 = 53.7$, $s_2 = 23.7$. Under STAT TESTS, choose 2-SampTInt to obtain the confidence interval shown on p. 453. Indicate Stats, since we only have the summary statistics to work with, and not the data samples. Enter the statistics for each group.

```
EDIT CALC TESTS
4↑2-SampTTest...
5:1-PropZTest...
6:2-PropZTest...
7:ZInterval...
8:TInterval...
9:2-SampZInt...
0↓2-SampTInt...
```

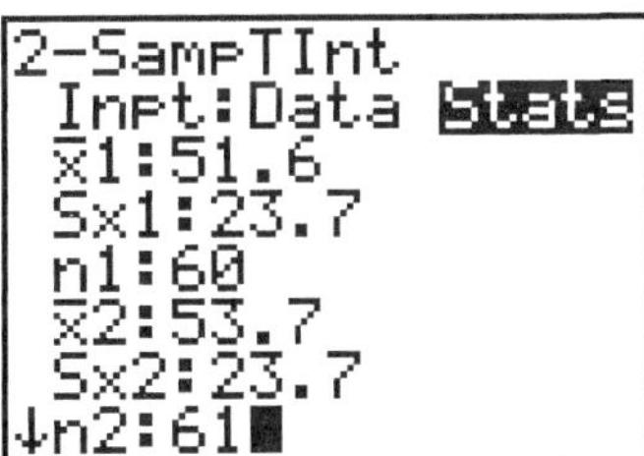

Specify 95% a confidence level, and this time request the Pooled procedure. Calculate. The interval is (–10.63, 6.43).

```
2-SampTInt
↑n1:60
 x̄2:53.7
 Sx2:23.7
 n2:61
 C-Level:.95
 Pooled:No Yes
 Calculate
```

```
2-SampTInt
 (-10.63,6.4327)
 df=119
 x̄1=51.6
 x̄2=53.7
 Sx1=23.7
↓Sx2=23.7
```

Note that df = $n_1 + n_2 - 2 = 60 + 61 - 2 = 119$. Also note (see below) that the pooled standard deviation is 23.7.

```
2-SampTInt
 (-10.63,6.4327)
↑Sx1=23.7
 Sx2=23.7
 Sxp=23.7
 n1=60
 n2=61
```

To confirm the hypothesis test results on p. 453, choose 2-SampTTest from the STAT TESTS menu. Choose Stats. Notice that the means, standard deviations and sample sizes are already ready to go!

```
EDIT CALC TESTS
1:Z-Test...
2:T-Test...
3:2-SampZTest...
4:2-SampTTest...
5:1-PropZTest...
6:2-PropZTest...
7↓ZInterval...
```

```
2-SampTTest
 Inpt:Data Stats
 x̄1:51.6
 Sx1:23.7
 n1:60
 x̄2:53.7
 Sx2:23.7
↓n2:61
```

Indicate a two-tailed test and pooled procedure. Calculate. The test statistic is t = –0.487, and p-value = .6269.

```
2-SampTTest
↑n1:60
 x̄2:53.7
 Sx2:23.7
 n2:61
 μ1:≠μ2 <μ2 >μ2
 Pooled:No Yes
 Calculate Draw
```

```
2-SampTTest
 μ1≠μ2
 t=-.4873251288
 p=.6269244633
 df=119
 x̄1=51.6
↓x̄2=53.7
```

The "Draw" option produces the image below. This is a large p-value!

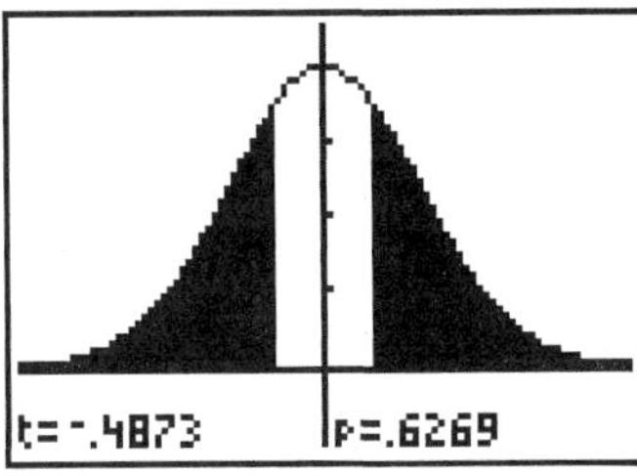

Exercise 9.13: Vegetarians more liberal? (pp. 457-458 of Agresti/Franklin)
Call group 1 the non-vegetarians and group 2 the vegetarians. We have $n_1 = 51$, $\bar{x}_1 = 3.18$, $s_1 = 1.72$ and $n_2 = 9$, $\bar{x}_2 = 2.22$, $s_2 = 0.67$. Let's verify the results given on p. 458. First, we will use a pooled procedure (assumes equal population standard deviations). The confidence interval can be obtained by pressing STAT TESTS, then 2-SampTInt.

```
2-SampTInt
 Inpt:Data Stats
 x̄1:3.18
 Sx1:1.72
 n1:51
 x̄2:2.22
 Sx2:.67
↓n2:9
```

```
2-SampTInt
↑n1:51
 x̄2:2.22
 Sx2:.67
 n2:9
 C-Level:.95
 Pooled:No Yes
 Calculate
```

Enter the Stats, choose 95% confidence, Yes on Pooled, and Calculate. The confidence interval is (–0.21, 2.13), which indicates no difference between groups. Note that the pooled standard deviation is 1.62. This value is much closer to s_1, since n_1 is a much larger sample. Still, we may question the wisdom of assuming equal standard deviations for this problem.

```
2-SampTInt
 (-.2097,2.1297)
 df=58
 x̄1=3.18
 x̄2=2.22
 Sx1=1.72
↓Sx2=.67
```

```
2-SampTInt
 (-.2097,2.1297)
↑Sx1=1.72
 Sx2=.67
 Sxp=1.61624938
 n1=51
 n2=9
```

Next, let's confirm the significance test results. Press STAT TESTS, then 2-SampTTest. You will not have to re-enter the statistics. Specify a two-tailed test and the pooled procedure. Calculate.

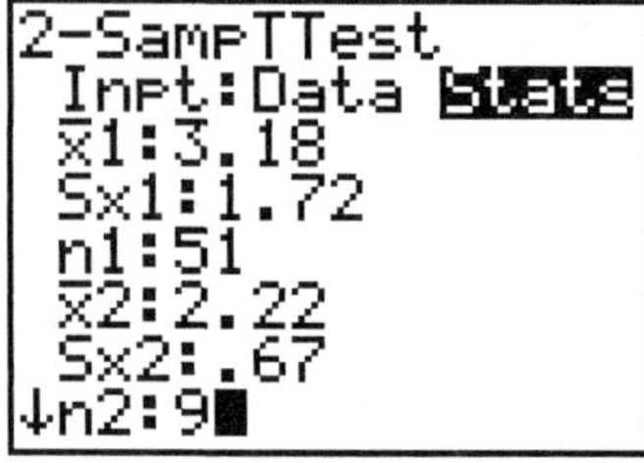


```
2-SampTTest
 μ1≠μ2
 t=1.642833607
 p=.1058285135
 df=58
 x̄1=3.18
↓x̄2=2.22
```

Draw produces the below image.

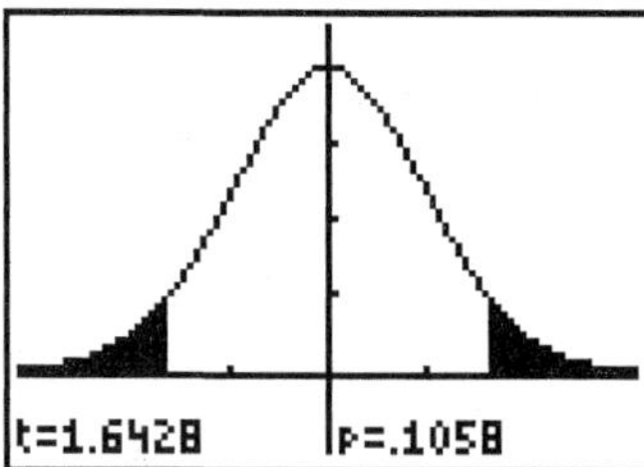

Maybe we shouldn't pool the standard deviations. Let's re-run the confidence interval and significance test, this time choosing the unpooled procedure, which does not assume equal population standard deviations. Press STAT TESTS again, and choose 2-SampTInt. The only thing you'll need to change is Pooled from Yes to No.

```
2-SampTInt
↑n1:51
 x̄2:2.22
 Sx2:.67
 n2:9
 C-Level:.95
 Pooled:No Yes
 Calculate
```

```
2-SampTInt
 (.2899,1.6301)
 df=30.76965352
 x̄1=3.18
 x̄2=2.22
 Sx1=1.72
↓Sx2=.67
```

Now we have the interval (0.29, 1.63), which indicates a nonzero difference between the two groups. Press STAT TESTS and choose 2-SampTTest to confirm the significance test results. This time you won't have to change a thing! Just press ENTER with your cursor on Calculate.

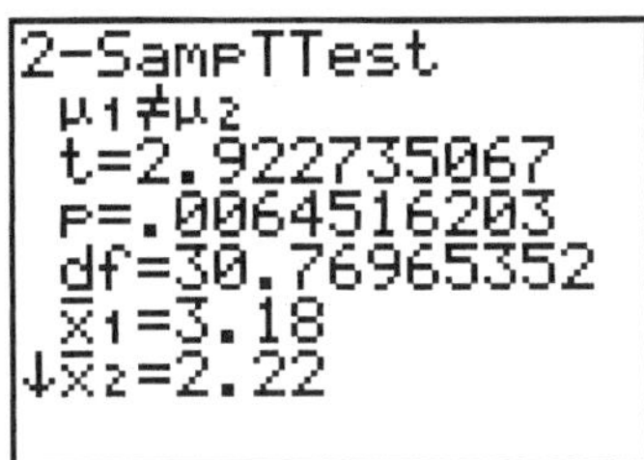

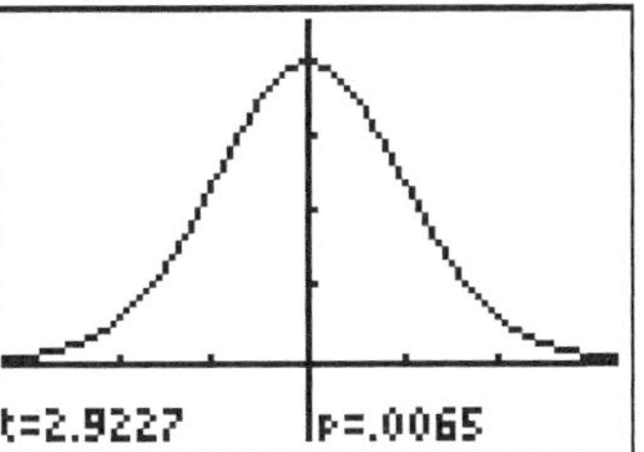

Now we have a significant result. Note that the test statistic is now too large (and subsequently the p-value too small) to show up on screen.

Matched Pairs Designs: Section 9.4 of Agresti/Franklin

EXAMPLE 12: Matched pairs designs for cell phones and driving study (pp. 459-460 of Agresti/Franklin)

Type or download the data into your calculator. Assuming you have the "No" reaction times in L1 and the "Yes" reaction times in L2, we can have the calculator compute the differences for us and store them in L3. To be consistent with table 9.9, let's compute diff = yes – no. Define L3 to be L2 – L1.

L1	L2	L3
604	636	------
556	623	
540	615	
522	672	
459	601	
544	600	
513	542	

L3 =L2−L1

L1	L2	L3
604	636	32
556	623	67
540	615	75
522	672	150
459	601	142
544	600	56
513	542	29

L3(1)=32

Let's replicate the comparison boxplots given on p. 460 of Agresti/Franklin. Press STAT PLOT. Define Plot1 to be a modified boxplot for L1, and Plot2 to be a modified boxplot for L2. Make sure that Plots 1 and 2 are turned On, and Plot3 is turned Off.

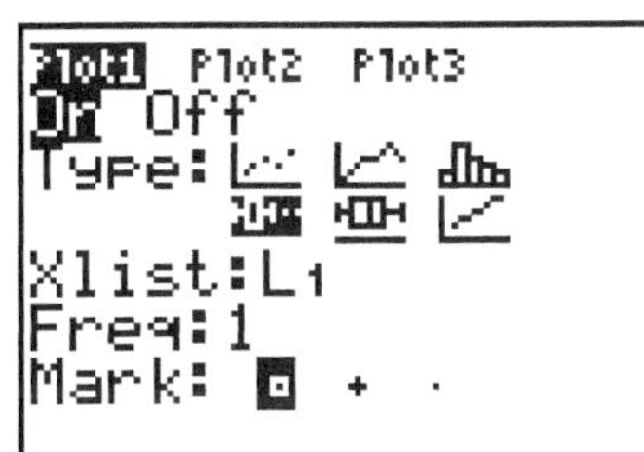

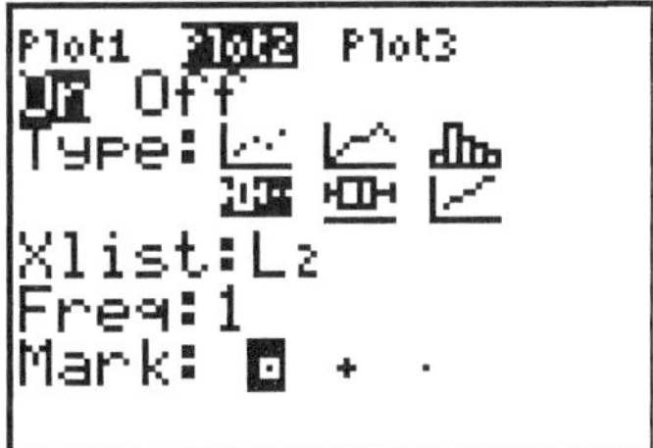

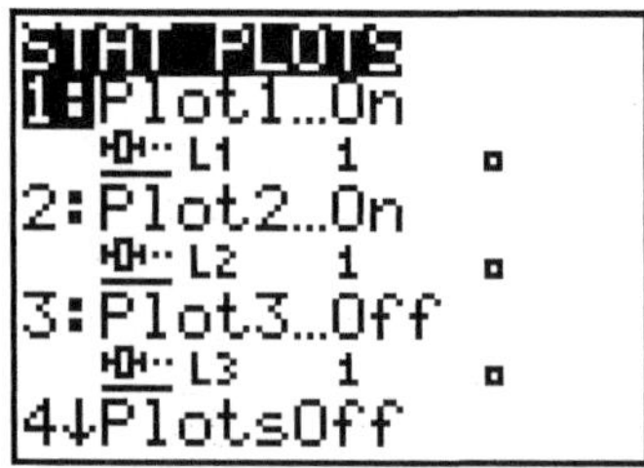

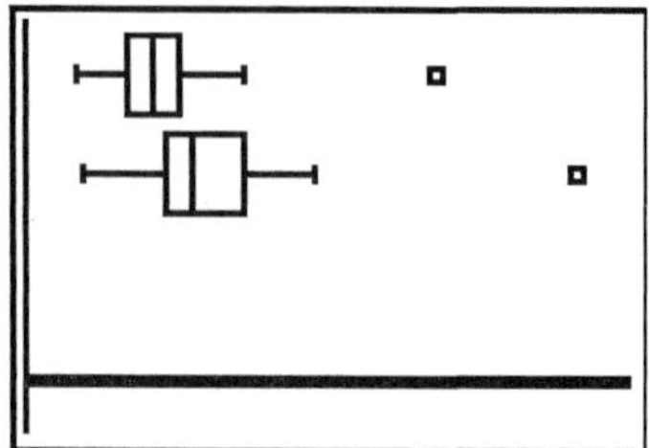

Realize that this graph doesn't really get at the heart of the matter – what are the differences for the 32 people in the study? Two outliers are visible in the above graph. Using the TRACE function, can you determine which students they belong to?

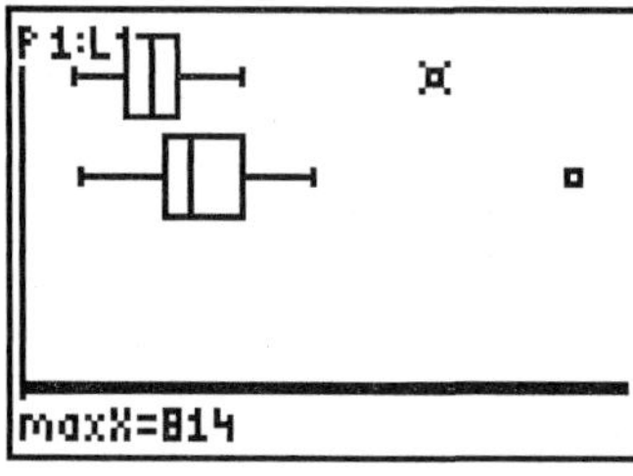

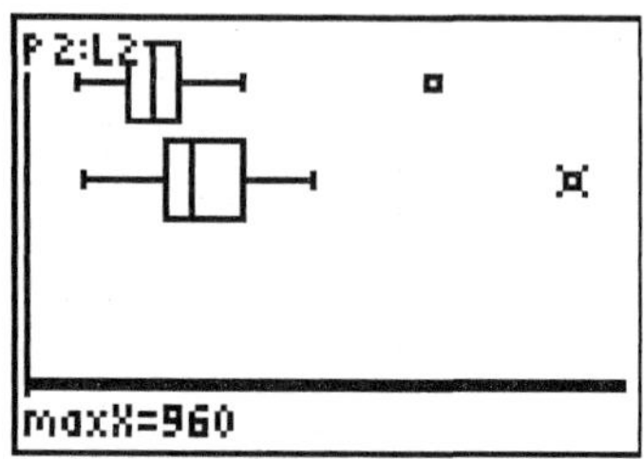

This is the same student, #28, who seems to be slow to react regardless of whether or not he or she was using a cell phone. The differences between cell phone "No" and "Yes" for each student are really what we're interested in here. Note also that the average reaction time for "No" and the average reaction time for "Yes" are not the best summary statistics to consider here. We should work exclusively with the differences for this example. Continue to EXAMPLE 13.

EXAMPLE 13: Matched pairs analysis of cell phone impact on driver reaction time (pp. 462-464 of Agresti/Franklin)

Assuming you just worked EXAMPLE 12, turn off Plots 1 and 2 and define Plot3 to be a modified boxplot for the differences. You'll need to use ZoomStat in order to view the new graph. The screen below right replicates figure 9.9 on p. 462 of Agresti/Franklin.

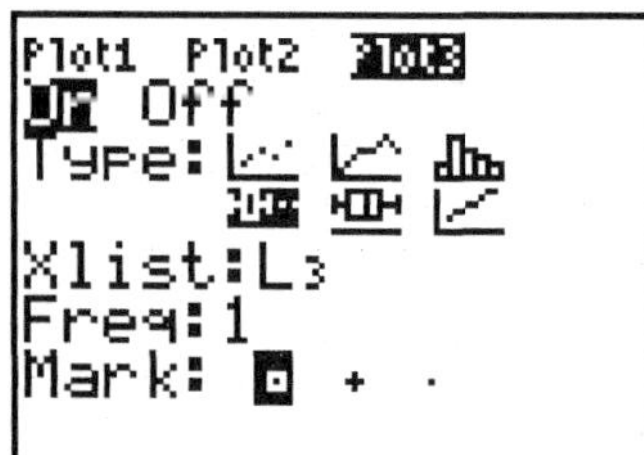

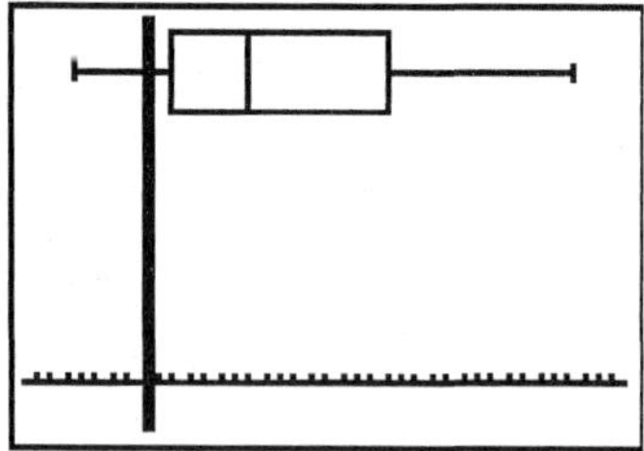

Let's find the mean and standard deviation of the differences using 1-Var Stats. Press STAT CALC. Choose 1-Var Stats, and be sure to specify L3, if you have stored the differences there.

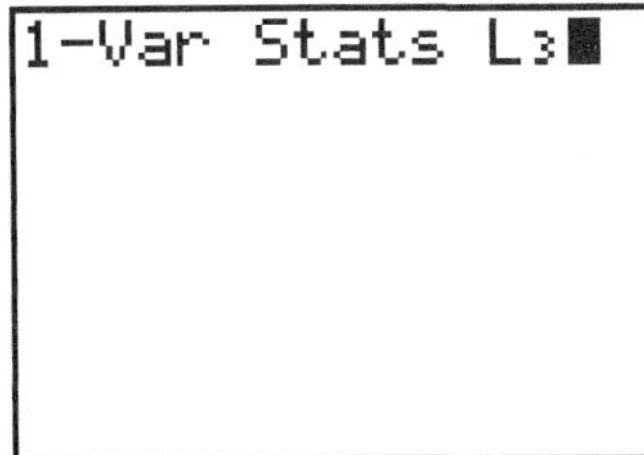


```
1-Var Stats
 x̄=50.625
 Σx=1620
 Σx²=167410
 Sx=52.48578917
 σx=51.65918965
↓n=32
```

We can see that $\bar{x}_d = 50.6$ and $s_d = 52.5$.

Consider the Questions to Explore on p. 463 of Agresti/Franklin.

(a) For a paired difference significance test, the hypotheses are H_0: $\mu_d = 0$ vs. H_0: $\mu_d \neq 0$. Press STAT, then cursor right to TESTS. Choose 2:Test. The differences are in L3, so choose Data. Supply a zero for μ_0. Specify a two-tailed alternative. With the cursor blinking on Calculate, press ENTER.

```
EDIT CALC TESTS
1:Z-Test…
2:T-Test…
3:2-SampZTest…
4:2-SampTTest…
5:1-PropZTest…
6:2-PropZTest…
7↓ZInterval…
```

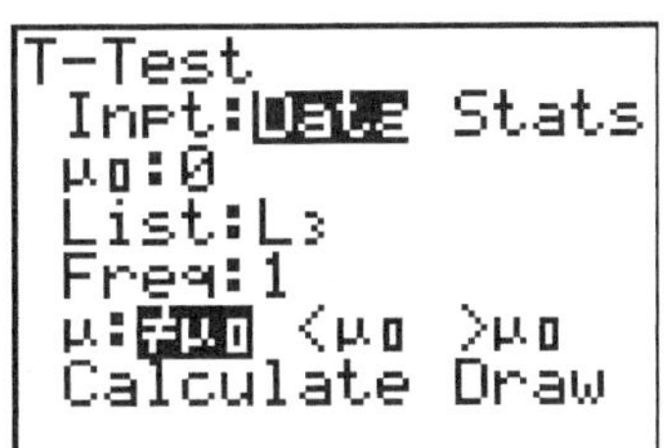

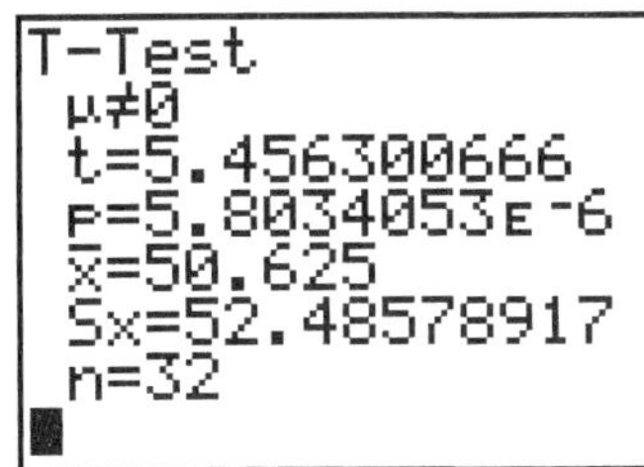

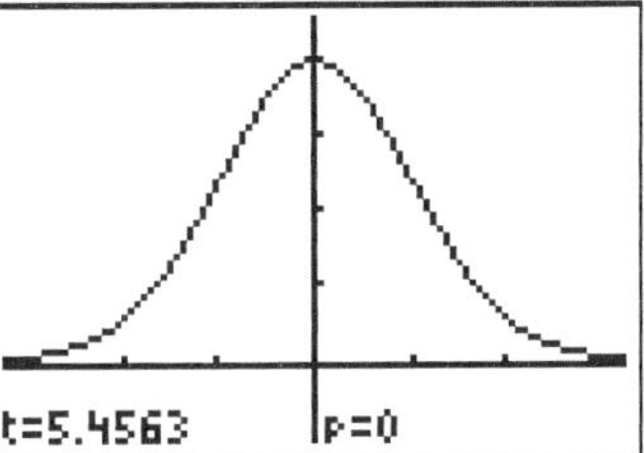

The test statistic is t = 5.46 and the p-value is .00006. We have evidence that the reaction times are different. Clearly, the reactions times are slower on average when using a cell phone.

Exercise 9.39: Does exercise help blood pressure? (pp. 467-468 of Agresti/Franklin)
Instead of calculating the before – after differences, let's make the calculator do the math. Enter the "before" blood pressure measurements into L1. Put the "after" measurements into L2. Move the cursor over so that it is on the L3 header – not the first row of L3. This way, you can define the entire list. Press 2nd 1 to specify L1, and 2nd 2 to specify L2. Press ENTER, and the three differences are calculated for you.

```
L1      L2      L3     3
150     130     ------
165     140
135     120
------  ------
L3 =L1-L2
```

```
L1      L2      L3     3
150     130     20
165     140     25
135     120     15
------  ------  ------
L3(1)=20
```

Now we can work with the differences in L3. To quickly calculate the means of the before, after, and differences as requested in part (b), we can use 1-Var Stats, or the mean function from the LIST MATH menu. Press LIST (2nd STAT), then move right to MATH. Choose 3:mean. Specify the "before" list by pressing 2nd 1 for L1, and so on. Note that the difference in means is equal to the mean of the differences.

```
NAMES OPS MATH
1:min(
2:max(
3:mean(
4:median(
5:sum(
6:prod(
7↓stdDev(
```

```
mean(L1)
             150
mean(L2)
             130
mean(L3)
              20
```

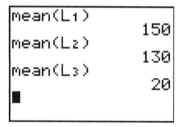

For part (c), we are asked for a confidence interval for the mean before – after difference. This is just a one-sample t-interval, so press STAT TESTS, and choose 8:TInterval.

```
EDIT CALC TESTS
2↑T-Test...
3:2-SampZTest...
4:2-SampTTest...
5:1-PropZTest...
6:2-PropZTest...
7:ZInterval...
8:TInterval...
```

Be sure to specify L3! That's where the differences are. Don't get into the habit of always assuming your data are in L1. A 95% confidence interval for the before measurements is not what we're interested in here.

The screen below shows that we have n=3 differences which have a mean of 20, and a standard deviation of 5. The interval is (7.58, 32.42). The exercise program appears to be effective in reducing blood pressure.

```
TInterval
 (7.5793,32.421)
 x̄=20
 Sx=5
 n=3
```

Exercise 9.40: Test for blood pressure (p. 468 of Agresti/Franklin)
Let's continue the previous exercise by applying a hypothesis test to the data. We could run a *one-tailed* test, since the researcher has the belief that walking will *reduce* blood pressure. Instead, let's do what researchers typically do – run a two-tailed test. If we have a significant result, we can then make note of whether the blood pressure was lowered (or raised), on average. So we shall test H_0: $\mu_d = 0$ vs. H_a: $\mu_d \neq 0$. Note that the (before – after) differences will be positive if the blood pressure went down. Follow the directions in exercise 9.39 for data entry if necessary. The differences should be in L3. From STAT TESTS, choose 2:T-Test.

```
EDIT CALC TESTS
1:Z-Test…
2:T-Test…
3:2-SampZTest…
4:2-SampTTest…
5:1-PropZTest…
6:2-PropZTest…
7↓ZInterval…
```

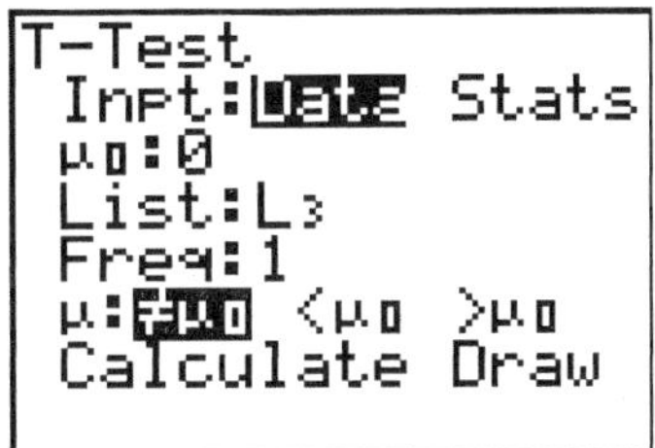

Specify "Data" and a null mean (difference) of zero. The differences are in L3, and you should leave Freq equal to 1. Indicate a two-tailed test, and press ENTER with the cursor on Calculate.

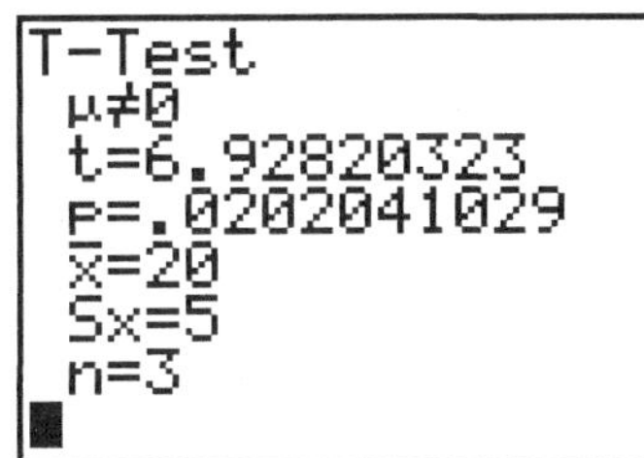

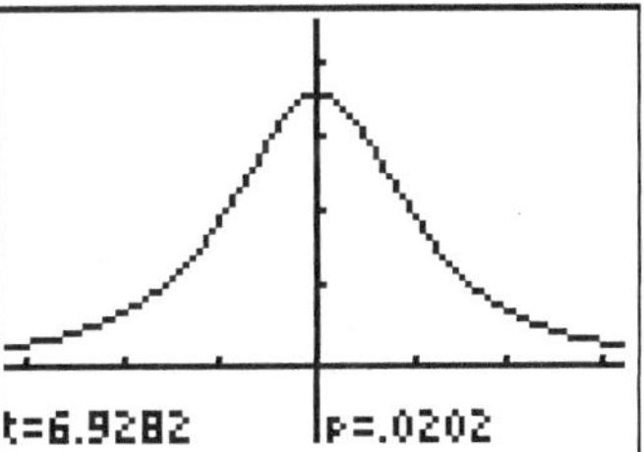

The screen above right shows a test statistic of t = 6.93 and p-value = .0202. Reject H_0 and conclude that walking was effective in reducing blood pressure, on average. The screen above right results from the "Draw" option. The test statistic is much too large to be shown on the screen.

Chapter Ten: Analyzing the Association between Categorical Variables

Chi-squared Tests of Independence: Section 10.2 of Agresti/Franklin
In this section, we will use the STAT TESTS function χ^2-Test, and the χ^2cdf command from the DISTR menu.

EXAMPLE 3: Chi-squared for happiness and family income (pp. 492-494), and EXAMPLE 4: Are happiness and income independent? (pp. 495-496 of Agresti/Franklin)
This is a continuation of a previous example. To fully understand how the chi-square test works, it is strongly recommended that you work this problem by hand first. Follow the instructions given on p. 491-492 to calculate the expected cell counts. See p. 493 for details regarding the computation of the χ^2 test statistic. Once we've done all that, we can use the χ^2cdf command to get the p-value from the χ^2 distribution. Press 2nd VARS to access the DISTR menu. Since the table has 3 rows and 3 columns, we have $(3-1)\times(3-1) = 4$ degrees of freedom (see p. 494). The p-value is the probability to the right of the test statistic, $\chi^2 = 73.4$. The χ^2cdf function works very similar to the tcdf command. Specify the lower bound (the test statistic), the upper bound (positive infinity), and the degrees of freedom. The lower and upper bounds will always follow this format for χ^2 tests. The p-value for this test is incredibly small: 4×10^{-15}. We should reject H_0 and conclude that happiness and family income are dependent.

DISTR DRAW
2↑normalcdf(
3:invNorm(
4:tpdf(
5:tcdf(
6:χ²pdf(
7:χ²cdf(
8↓Fpdf(

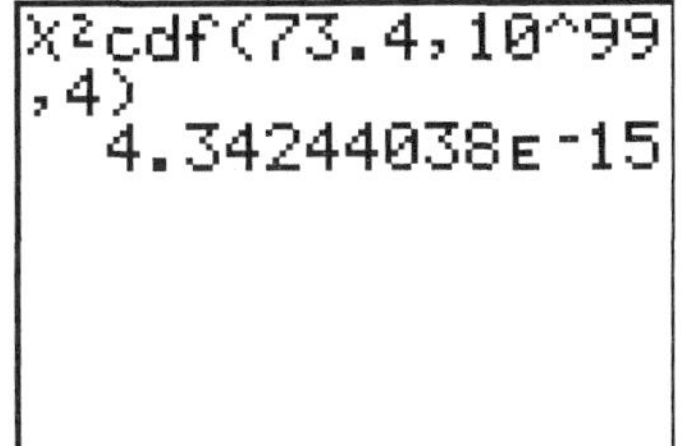

Now let's do it the "easy way". Press STAT, then move right to access the TESTS menu. Choose the χ^2-Test option.

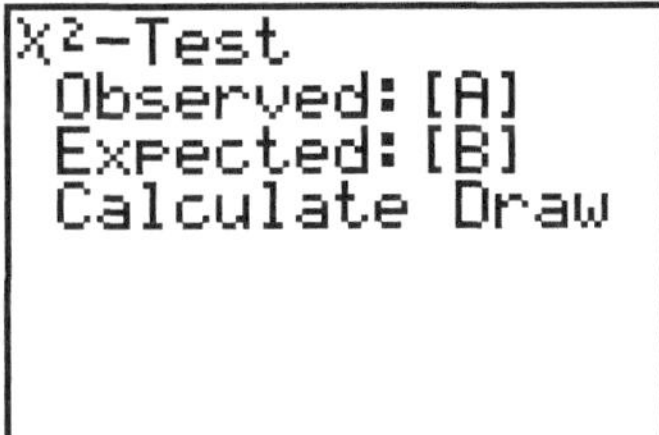

The TI-83 denotes matrices with a letter in brackets []. By default, the calculator assumes that we have entered the observed frequencies (see table 10.1 on p. 485) into matrix [A]. The calculator will compute the expected frequencies for us and place them into matrix [B], also by default. We can change these defaults to any matrices we like, [C] and [D], for example. Since we haven't entered the observed frequencies into a matrix yet, press 2nd MODE to QUIT. Press MATRX (2nd x^{-1}) to access the matrix menu. Move the cursor right to EDIT. Note that you can scroll down to choose the different matrices. With the cursor on 1: [A], press ENTER. We will place the observed frequencies from table 10.1 into matrix [A], the default. You may (or may not) have data stored in [A] already. Below right we see a 2×3 matrix from a previous problem. No matter, we will simply overwrite any pre-existing data.

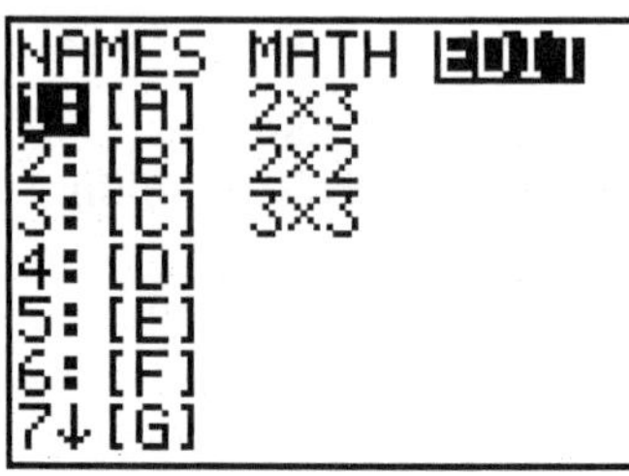

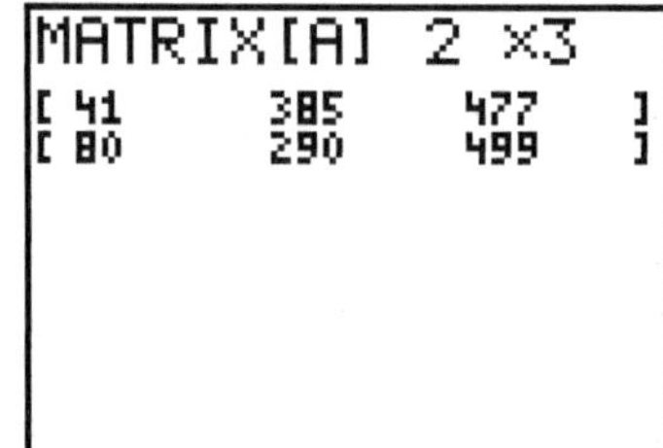

The first thing to do is to define the matrix dimension. We need to specify a 3×3 table of frequencies. Enter the observed frequencies as you see them on p. 485. Note that the row or column totals are not entered.

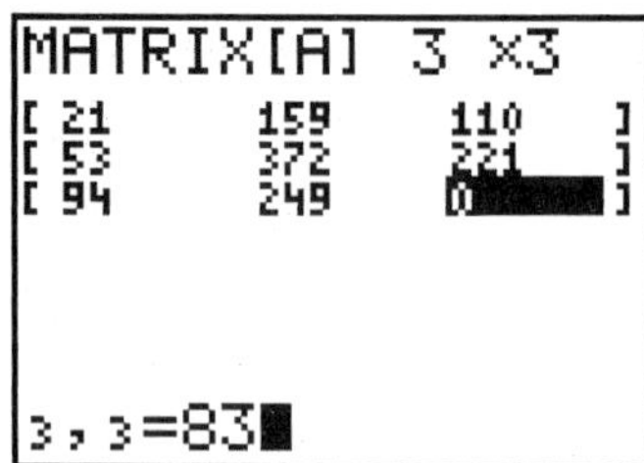

Once the observed frequencies are in matrix [A], the hard work is done. Go back to STAT TESTS and choose C: χ^2-Test.

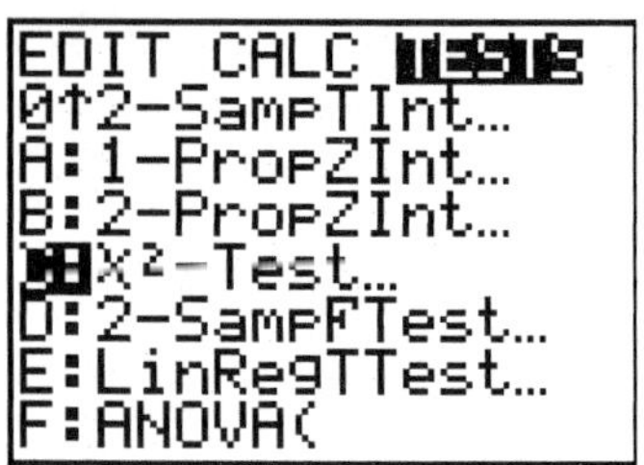

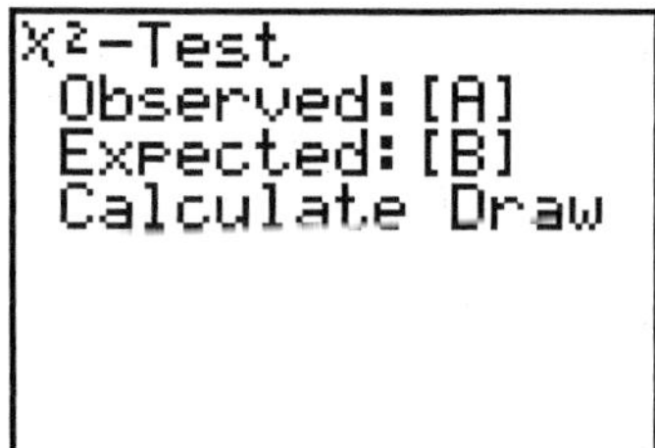

This time, we are ready to go. The expected frequencies will be computed and stored in matrix [B]. With the cursor on "Calculate", press ENTER.

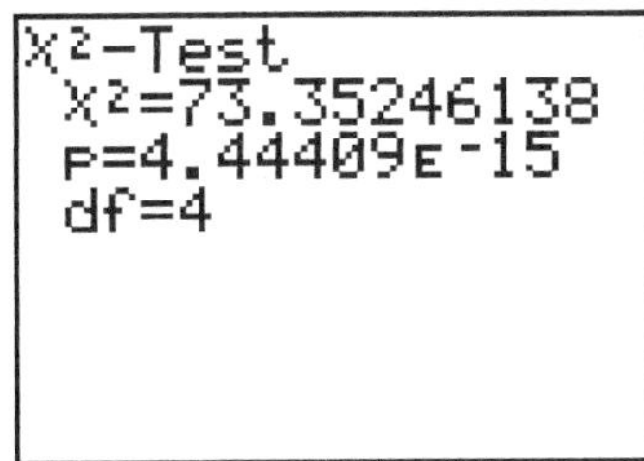

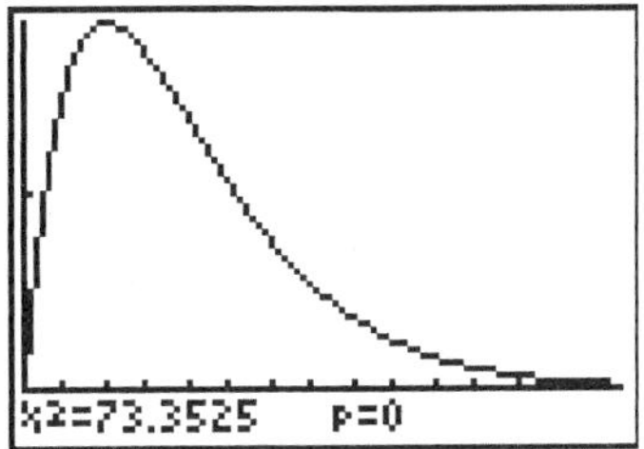

Above left, we see the test statistic and p-value we computed earlier. The image above right results from the "Draw" option. The test statistic is much too large to be shown on screen. Recall that the expected frequencies were to be stored in matrix [B]. Press MATRX (2nd x^{-1}) to access the matrix menu. Move the cursor to EDIT, then down to 2: [B]. Press ENTER. Now we are able to view the expected counts, which should match up with your hand calculations. Note that all expected cell counts are at least five, which satisfies the chi-square test assumption.

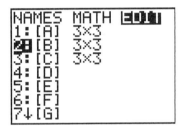

```
NAMES MATH EDIT
1:[A] 3×3
2:[B] 3×3
3:[C] 3×3
4:[D]
5:[E]
6:[F]
7↓[G]
```

```
MATRIX[B] 3 ×3
[ 35.771   166.08   88.15  ]
[ 79.683   369.96   196.36 ]
[ 52.546   243.96   129.49 ]
```

Exercise 10.9: Happiness and gender (p. 501 of Agresti/Franklin)

We initially assume that happiness and gender are independent (H_0), and will try to show that happiness depends on gender (H_a). To get the test statistic and p-value from the calculator, we can use the χ^2-Test function. First we'll need to enter the six observed frequencies into matrix [A]. Note that it is not necessary to calculate any row or column totals. Press MATRX (2nd x^{-1}) to access the matrix menu. Move the cursor to EDIT and press ENTER with the cursor on 1: [A]. No matter what data may be there, and no matter what dimension the previous matrix may been, we will overwrite it. Specify 2×3, since we have two rows and three columns.

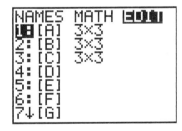

```
NAMES MATH EDIT
1:[A] 3×3
2:[B] 3×3
3:[C] 3×3
4:[D]
5:[E]
6:[F]
7↓[G]
```

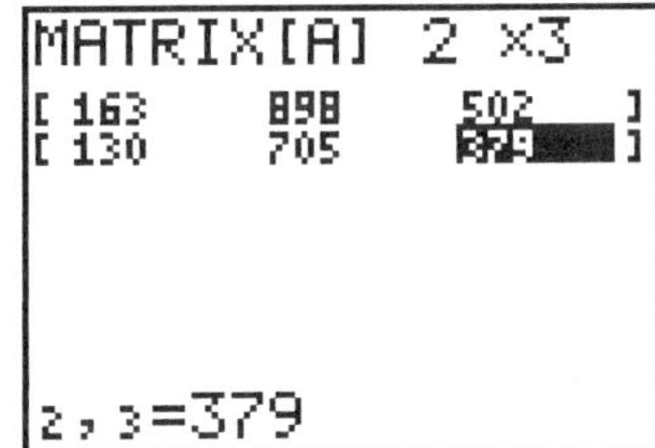

Next press STAT, move to TESTS, and choose C: χ^2-Test. Note that we could put our observed frequencies into any matrix we like, and we are likewise free to use a matrix other than [B] to store the expected counts. [A] and [B] are the default matrices used by χ^2-Test, so those will be the matrices used here.

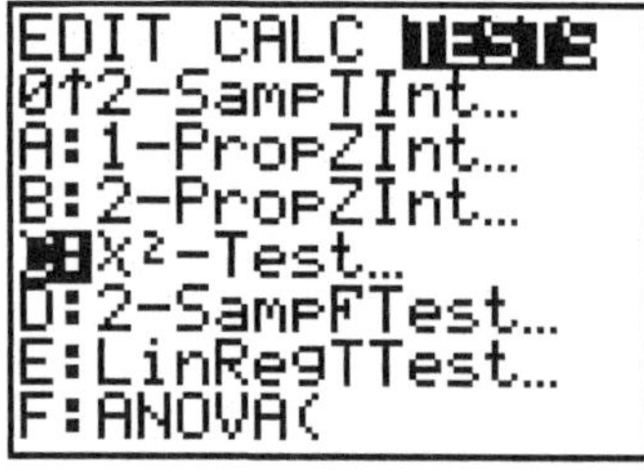

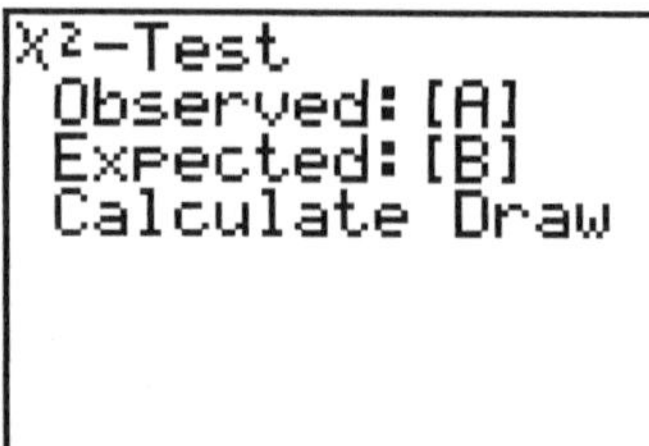

With the cursor on "Calculate", press ENTER. The screen below left is the result. The test statistic is $\chi^2 = 0.27$, and p-value = .87. Verify that df = 2 by multiplying $(r-1)\times(c-1)$ for 2 rows and 3 columns. This is a small test statistic and large p-value. Do not reject H_0 – we do not have sufficient evidence that happiness depends on gender.

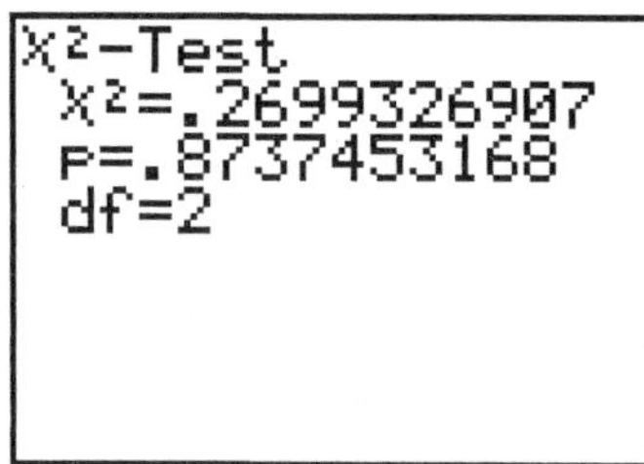

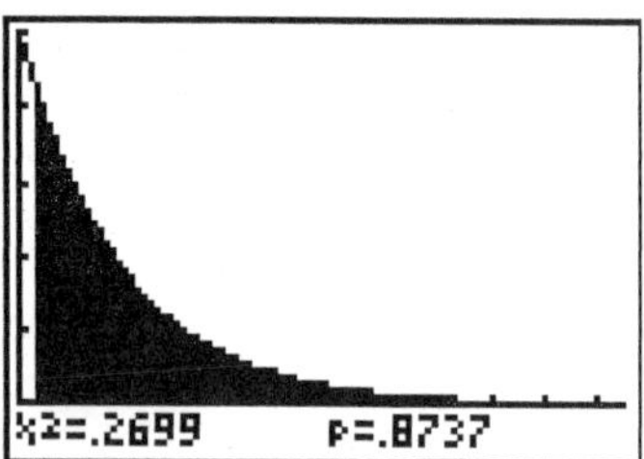

The screen above right results from the "Draw" option. Most of the curve is darkened by the p-value. Let's take a look at the expected cell counts, which have been computed and are now stored in matrix [B]. Press MATRX (2nd x^{-1}) to access the matrix menu. Move the cursor to EDIT, then down to 2: [B]. Press ENTER. All of the expected cell frequencies are at least five.

NAMES MATH EDIT
1:[A] 2×3
2:[B] 2×3
3:[C] 3×3
4:[D]
5:[E]
6:[F]
7↓[G]

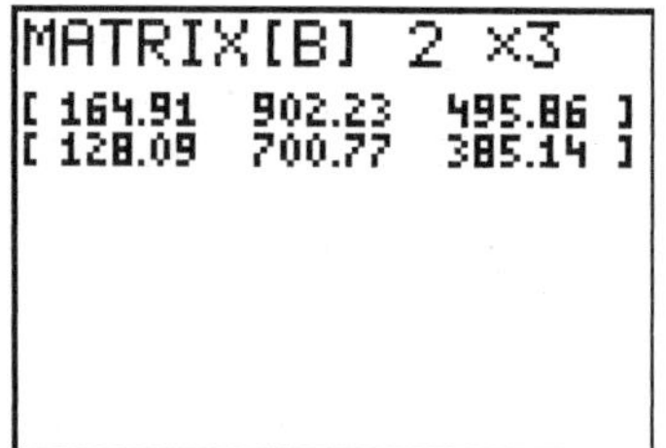

Exercise 10.10: Marital happiness and income (p.501 of Agresti/Franklin)

We initially assume that marital happiness and income are independent (H_0), and will see if the data provide evidence that marital happiness depends on income (H_a). In order to use the χ^2-Test function, we need to work with matrices. By default, the TI-83 assumes we will place the observed frequencies into matrix [A], and the expected frequencies will be calculated for us and placed in matrix [B]. Press MATRX (2nd x^{-1}) to access the matrix menu. Move the cursor to EDIT and press ENTER with the cursor on 1: [A]. Specify a 3×3 matrix and enter the observed counts as shown on p. 501.

```
NAMES MATH EDIT
1:[A] 2×3
2:[B] 2×3
3:[C] 3×3
4:[D]
5:[E]
6:[F]
7↓[G]
```

```
MATRIX[A] 3 ×3
[ 6    43    75  ]
[ 6    113   178 ]
[ 6    43    75  ]

3,3=75
```

Next press STAT, move to TESTS, and choose C: χ^2-Test.

```
EDIT CALC TESTS
0↑2-SampTInt...
A:1-PropZInt...
B:2-PropZInt...
C:χ²-Test...
D:2-SampFTest...
E:LinRegTTest...
F:ANOVA(
```

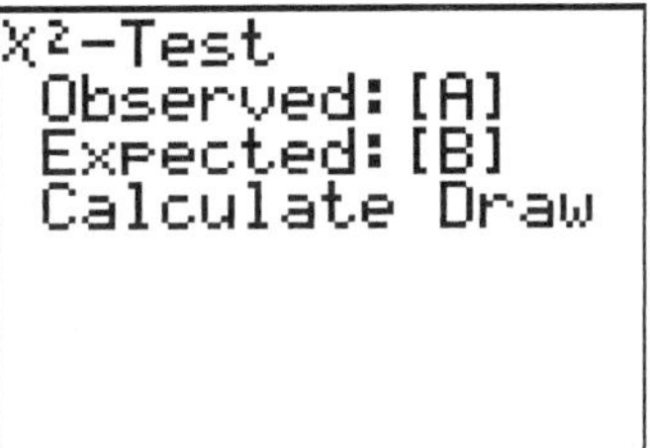

Pressing ENTER with the cursor on "Calculate" produces the screen below left. The test statistic is $\chi^2 = 3.68$, and p-value = .451. This is not significant; do not reject H_0. Marital happiness appears to be independent of income. The "Draw" option is shown below right.

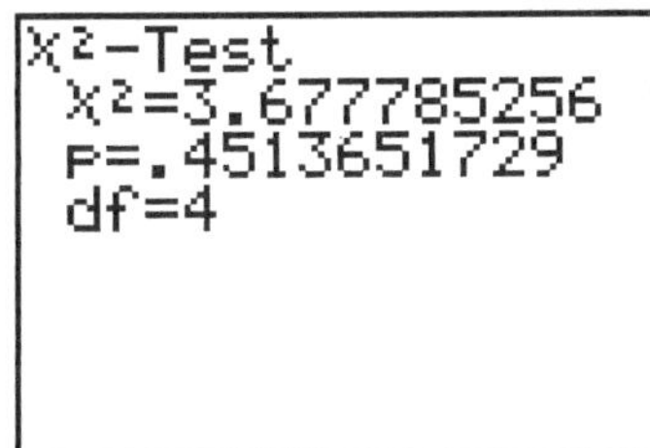

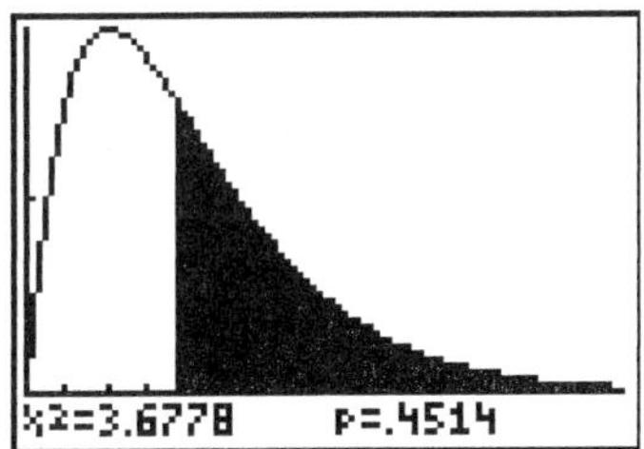

To check the assumption regarding expected cell counts, recall that the expected frequencies were to be stored in matrix [B]. Press MATRX (2nd x^{-1}) to access the matrix menu. Move the cursor to EDIT, then down to 2: [B]. Press ENTER. Note that two of the expected cell counts are less than five. This is less than 20% of the cells, so our results are probably still valid. See section 10.5 of Agresti/Franklin for a discussion of Fisher's Exact test.

```
NAMES MATH EDIT
1:[A] 3×3
2:[B] 3×3
3:[C] 3×3
4:[D]
5:[E]
6:[F]
7↓[G]
```

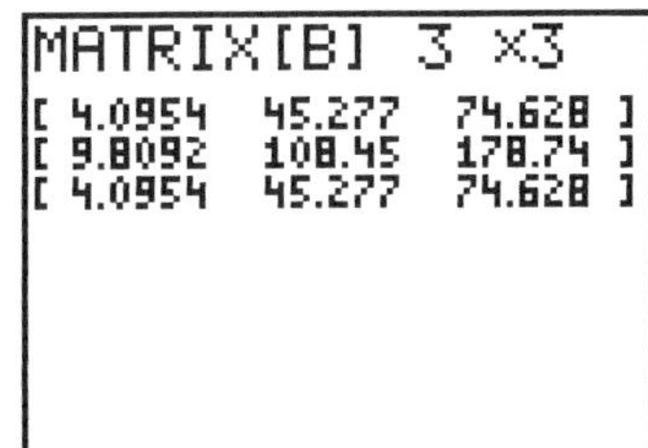

Chapter Eleven: Analyzing Association between Quantitative Variables: Regression Analysis

Recall that in chapter 3 we explored the association between pairs of quantitative variables. We created scatterplots, fit regression lines, and calculated correlations. Since then, we have learned the art of significance testing. We can apply that knowledge to regression analysis to determine if a statistically significant linear relationship exists between two quantitative variables.

Scatterplots and regression lines: Section 11.1 of Agresti/Franklin
In this section, we will revisit scatterplots and regression lines. All of the examples here use the strength study data.

EXAMPLE 1: How can you estimate a person's strength? (pp. 525-526 of Agresti/Franklin)
For this experiment, the idea was to collect data so that we could estimate the strength of female athletes. Maximum bench press was chosen for the measure of strength. We may be able to predict maximum bench press with the number of 60-pound bench presses performed. This makes maximum bench press the Y, or dependent, variable. Number of 60-pound bench presses performed will be the X, or independent, variable. We need to get these data (for 57 girls) into our calculator. If you have the CellSheet application, refer to EXAMPLE 12 in chapter 3 for downloading instructions. Otherwise, we will discuss how to download the "high school female athletes" data to your calculator from the Excel file.

Step 1. Downloading data from Excel files to your TI-83
The following instructions assume that you have a Windows PC, and a "TI Connectivity Kit", which is a cable that connects your calculator to your PC via the USB port, and the TI Connect software that accompanies the cable.

Step 2. Preparation of the Excel file
The Excel (workbook) file is called high_school_female_athletes.xls. Open this file with Excel. Save as a CSV (Comma delimited) file. Delete all columns but the two we need for this example, which are 1RM BENCH(lbs) and BRTF (60). Make note of which column is which. These instructions will assume that the two columns appear in the above order, which is Y=maximum bench press in the first column, and X=number of 60-pound bench presses in the second column. Now delete the first row of variable names. You should now have a CSV file with two columns of numeric data only. Save the file and quit Excel.

Step 3. Using the TI Connect software
Connect your calculator to your computer with the cable. Start up the TI Connect software. Activate the TI Data Editor. Pull down the File menu and choose Import (not Open). Find the file in question, possibly titled high_school_female_athletes.csv. It will appear in the TI Data Editor as a matrix. Click on the Send File icon (the last icon on the right, an arrow pointing down to a calculator). Ignore the "invalid file name" message and click OK. Verify the Device Type, and choose matrix [A] (or another matrix of your choosing) under Variable Name. Click

OK. Agree to Replace any data that may exist there. Now the high school female athletes data reside in matrix [A]. We need it in lists L1 and L2.

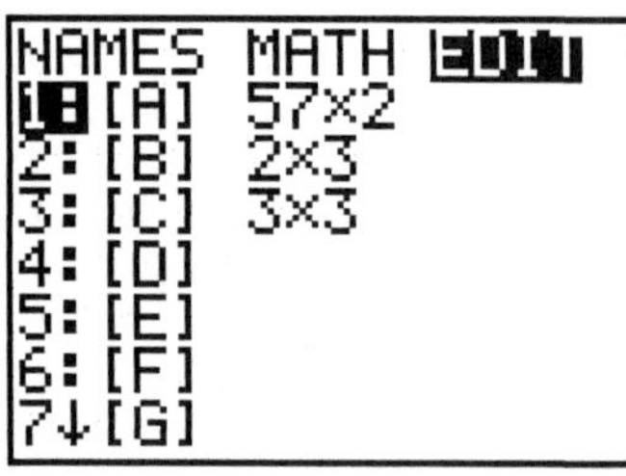

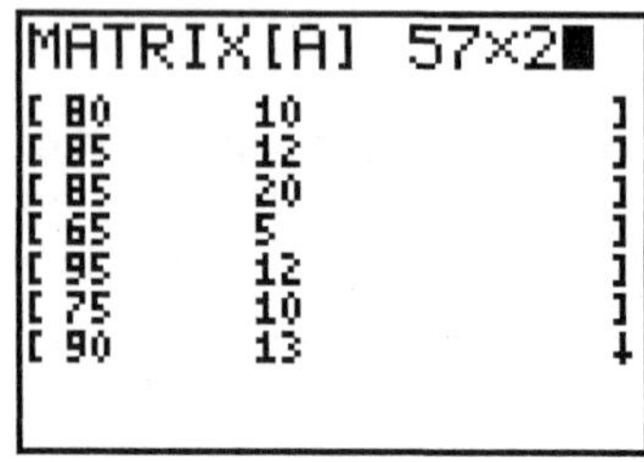

Step 4. Moving data from a matrix to lists on the TI-83
Press MATRX (2nd x^{-1}), then arrow right to MATH. Choose 8:Matr▸list.

We need to first specify the matrix [A], then the two lists, L1 and L2. From the above right screen, press MATRX (2nd x^{-1}), and with the cursor on NAMES, choose [A]. Then supply L1 and L2, separated by commas. Press ENTER.

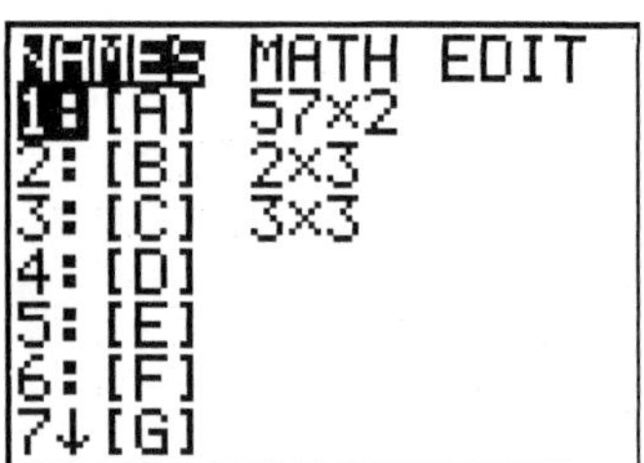

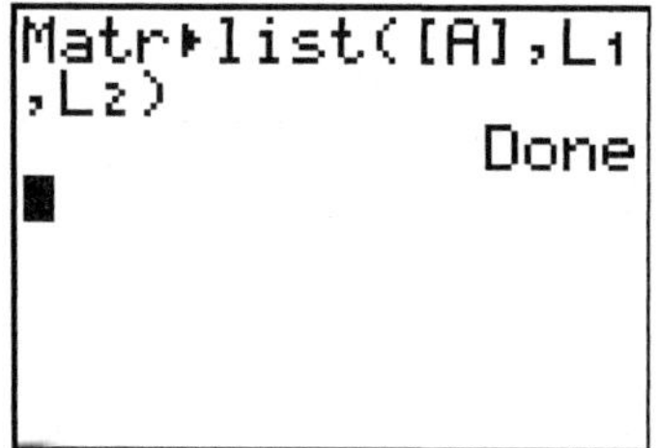

Now we can access our data with the STAT Editor.

EXAMPLE 2: What do we learn from a scatterplot in the strength study? (pp. 526-527 of Agresti/Franklin)
The strength study is introduced in EXAMPLE 1. Download the data (follow the instructions above) or type the data in to your calculator. Assuming that Y=maximum bench press is stored in L1, and X=number of 60-pound bench presses is stored in L2, let's create a scatterplot.

L1	L2	L3 1
80	10	------
85	12	
85	20	
65	5	
95	12	
75	10	
90	13	

L1(1)=80

L1	L2	L3 1
75	11	
105	30	
115	35	
80	20	
90	13	
90	15	

L1(58) =

Press STAT PLOT (2nd Y=). Make sure Plot1 is turned On, and the other two are turned Off. Choose the 1st icon, which denotes a scatterplot. Specify the X data list and the Y data list. Choose whichever of the three symbols you prefer to Mark the data points. Press ZOOM, then choose 9:ZoomStat. Remember that you can use the TRACE function to explore the graph.

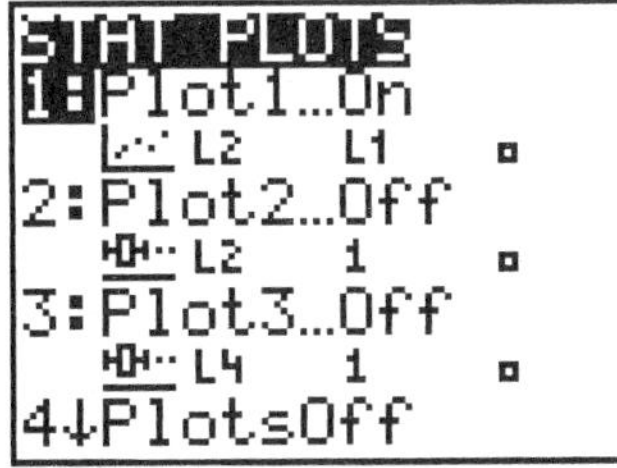

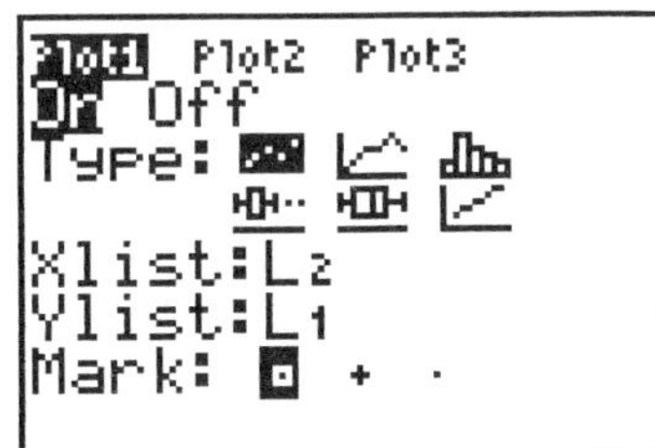

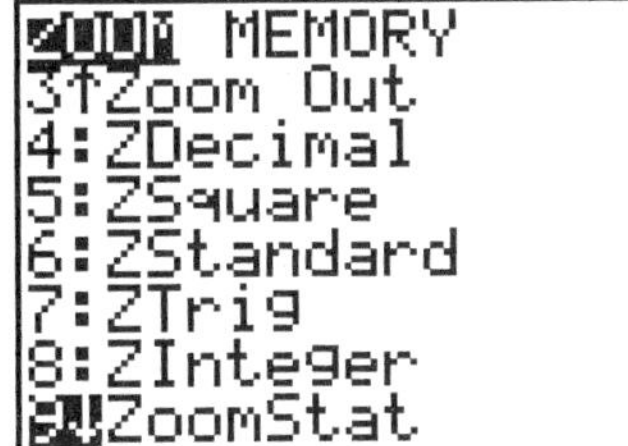

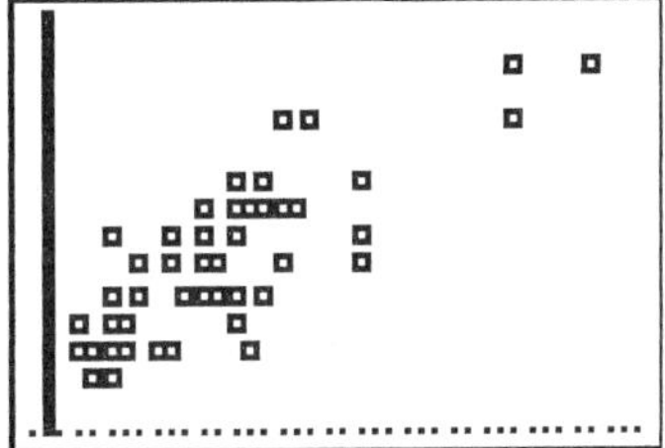

EXAMPLE 3: Which regression line predicts maximum bench press? (p. 528 of Agresti/Franklin)

To find the equation of the regression line, press STAT, then arrow right to CALC. Choose 8:LinReg(a + bx). Recall that this function matches the notation used in Agresti/Franklin much better than the 4th option in this menu. We must specify the X data list, then the Y data list. So if we have the X data stored in L2 and the Y data stored in L1 for this example, the arguments for LinReg(a+bx) should appear as you see below right.

```
EDIT CALC TESTS
2↑2-Var Stats
3:Med-Med
4:LinReg(ax+b)
5:QuadReg
6:CubicReg
7:QuartReg
8↓LinReg(a+bx)
```

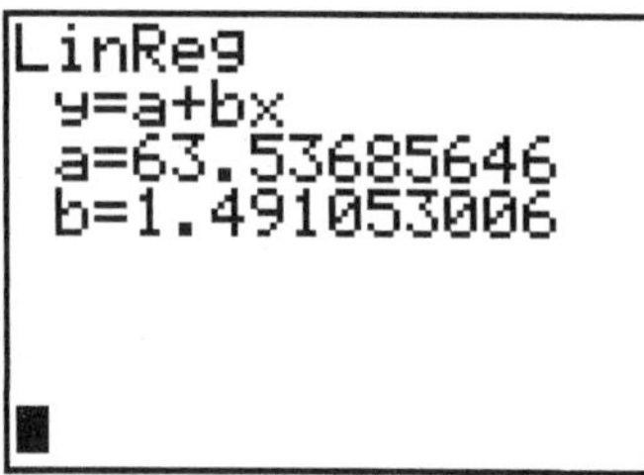

The equation for predicting Y=maximum bench press with X=number of 60-pound bench presses is $\hat{y} = 63.5 + 1.49x$. Note that you might also see r and r^2 on your screen if you have the diagnostics option turned on. See section 11.2 below.

Let's add this line to the scatterplot. From STAT CALC, choose the LinReg(a+bx) function again. This time, supply a third argument. After indicating Xlist and Ylist, give a "Y-variable" location for the TI-83 to store the equation in. The first Y-VAR is Y1, so let's use that. After the second comma (see below left), press the VARS key, then arrow right to Y-VARS. Choose 1:Function.

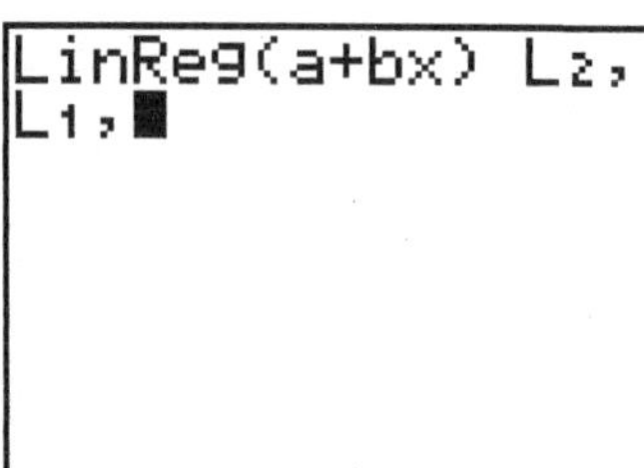

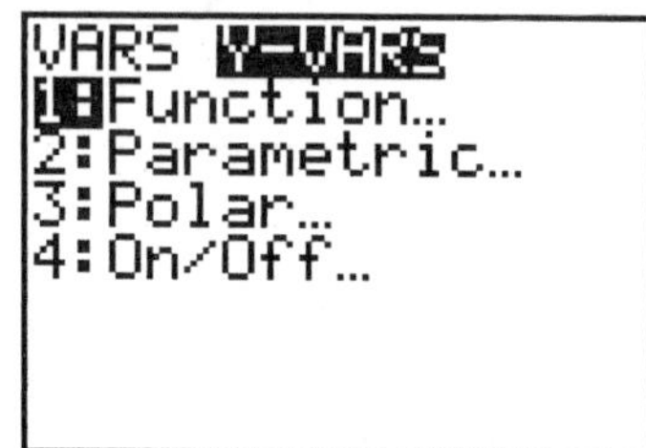

With the cursor on Y1, press ENTER. Y1 will be echoed back to the home screen.

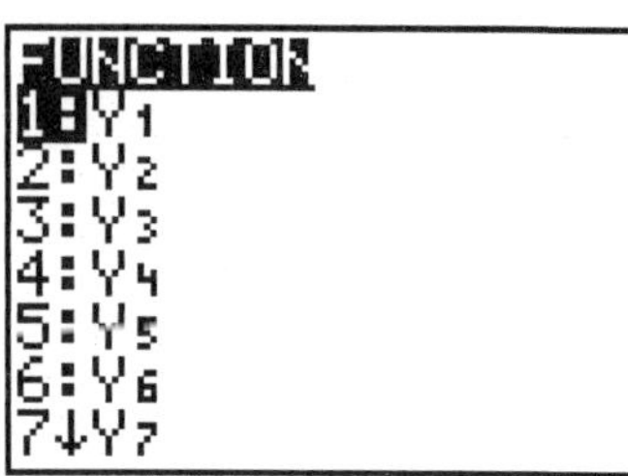

Press ENTER. This will produce the same regression results (see next page), but we also have the equation stored in Y1. Press the Y= button.

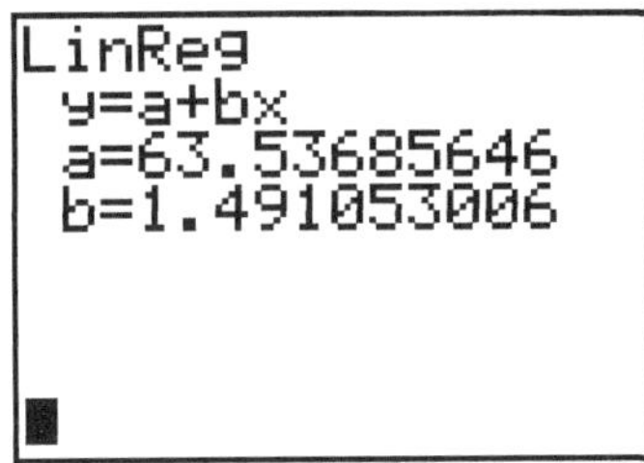

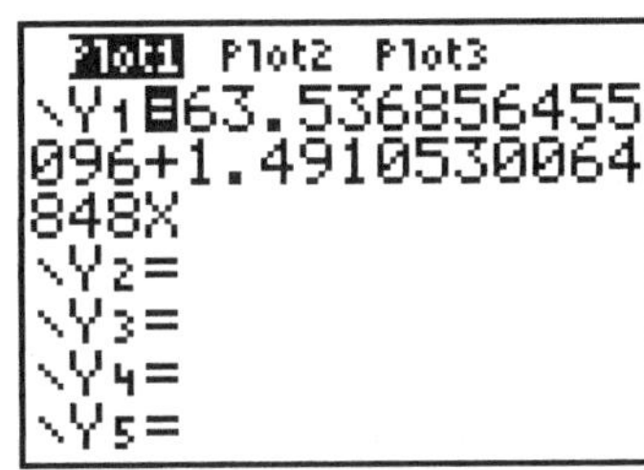

Now that the equation is stored in Y1 (by the way, we also could have typed it in there manually) it will automatically be graphed. See example 2 above for directions on setting up the scatterplot. Press GRAPH (or ZOOM/ZoomStat).

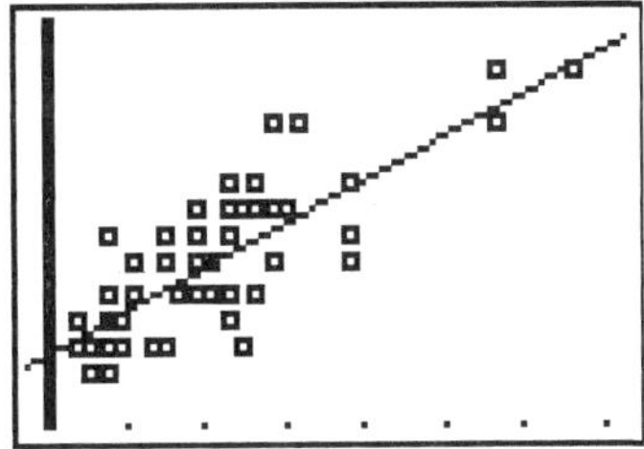

Recall from chapter 3 that the TRACE function allows us to scroll through the data points on the graph. Pressing the down arrow lets us move the cursor along the regression line.

Correlation and R^2: Section 11.2 of Agresti/Franklin

In this section, we will explore r and r^2. Again, both examples use the strength study data. Recall from chapter 3 that by default, the LinReg(a+bx) function does not produce **r**, the correlation coefficient. We must activate this option. If you need to, press CATALOG (2nd 0). The "alpha" symbol is visible, so press D (x^{-1}) to quickly find the DiagnosticOn function.

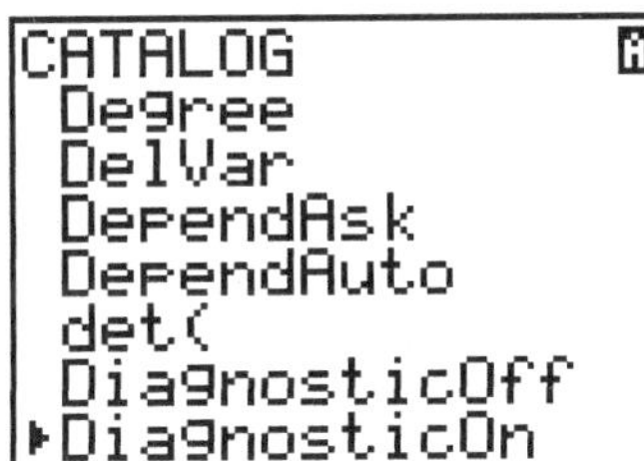

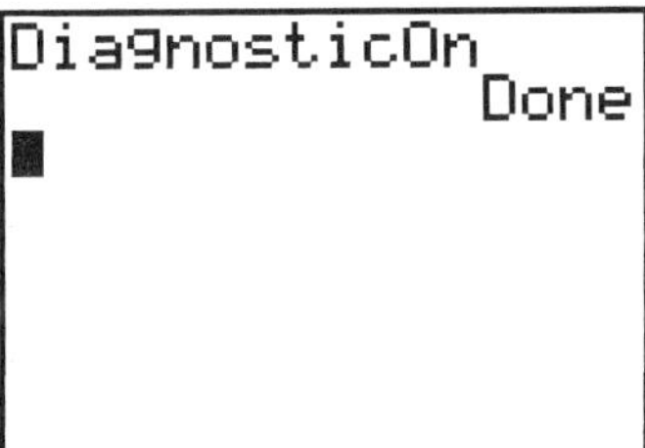

Now we will get the correlation coefficient (and R^2) as part of the LinReg(a+bx) output.

EXAMPLE 6: What's the correlation for predicting strength? (p. 537 of Agresti/Franklin)
Remember that we have the X data stored in L2 and the Y data stored in L1. Recall also that the format is LinReg(a+bx) Xlist, Ylist.

```
DiagnosticOn
            Done
LinReg(a+bx) L2,
L1
```

```
LinReg
 y=a+bx
 a=63.53685646
 b=1.491053006
 r²=.6432442672
 r=.8020251038
```

The correlation coefficient is r = .80, which indicates a strong, positive linear association.

EXAMPLE 9: What does r^2 tell us in the strength study? (p. 541 of Agresti/Franklin)
We could certainly square the above r value from example 6, or just read r^2 from the screen. Either way, we can say that 64% of the variability in Y=maximum bench press is accounted for by the variability in X=number of 60-pound bench presses.

Significance tests about the slope: Section 11.3 of Agresti/Franklin
In this section, we will make use of the LinRegTTest function, available in the STAT TESTS menu. Examples 11 and 12 both pertain to the strength study.

EXAMPLE 11: Is strength associated with 60-pound bench presses? (pp. 548-549 of Agresti/Franklin)
We initially assume that X=number of 60-pound bench presses and Y=maximum bench press are independent, and see if the data exhibit evidence of dependence. Stated in terms of the population slope, this is H_0: $\beta = 0$ vs. H_a: $\beta \neq 0$. Press STAT TESTS, then scroll *up* (it's quicker than scrolling down) to get to E:LinRegTTest. Assuming that we still have the X data stored in L2 and the Y data stored in L1, supply the Xlist and Ylist as you see below right. Freq should be one (by default), since each list entry represents one individual. Specify a two-tailed alternative. RegEQ can be left blank, or you can specify a "Y-VAR", say Y1, to store the equation in. This is similar to how we store the regression equation with the LinReg(a+bx) function. With the cursor blinking on Calculate, press ENTER.

```
EDIT CALC TESTS
0↑2-SampTInt...
A:1-PropZInt...
B:2-PropZInt...
C:χ²-Test...
D:2-SampFTest...
E:LinRegTTest...
F:ANOVA(
```

```
LinRegTTest
 Xlist:L2
 Ylist:L1
 Freq:1
 β & ρ:≠0 <0 >0
 RegEQ:Y1
 Calculate
```

```
LinRegTTest
 y=a+bx
 β≠0 and ρ≠0
 t=9.95826838
 p=6.481372E-14
 df=55
↓a=63.53685646
```

```
LinRegTTest
 y=a+bx
 β≠0 and ρ≠0
↑b=1.491053006
 s=8.003188444
 r²=.6432442672
 r=.8020251038
```

Two screens' worth of output is generated. Scroll down to see it all. The test statistic for testing zero slope (or equivalently, zero correlation) is t = 9.96. The p-value is 6×10^{-14}, which is essentially zero. Reject H_0 and conclude that a significant linear relationship exists between X and Y. Note also that df = n – 2 = 57 – 2 = 55. The intercept and slope are reported, along with r and r^2. Also included in the results is ***s***, the residual standard deviation (see p. 556 of Agresti/Franklin).

EXAMPLE 12: Estimating the slope for predicting maximum bench press (pp. 550-551 of Agresti/Franklin)

Unfortunately, the LinRegTTest function does not provide a confidence interval estimate for the slope parameter. However, the calculator is useful for finding t critical values. Recall from chapter 7 that we can use the MATH SOLVER for this. Press MATH, and choose 0:Solver. Specify 55 degrees of freedom and a tail area of .025 (for 95% confidence). With the cursor on the U= line, press SOLVE (ALPHA ENTER).

```
MATH NUM CPX PRB
4↑³√(
5:ˣ√
6:fMin(
7:fMax(
8:nDeriv(
9:fnInt(
0:Solver...
```

```
tcdf(L,U,D)-A=0
 L=-1E99
 U=-2.004044732...
 D=55
 A=.025
 bound={-1E99,1...
```

The $t_{.025}$ value is 2.00 (remember that we use the absolute value to plug into the confidence interval formula after the ± sign). Follow the steps on p. 551.

Exercise 11.35: Advertising and sales (p. 553 of Agresti/Franklin)
If you are done using the strength data (it is used again in this chapter), enter the advertising and sales data into L1 and L2, respectively. Of course, you could use L4 and L5, for example. In part (a), we are asked for the mean and standard deviation of each variable. Use the 2-Var Stats function for this. Press STAT CALC.

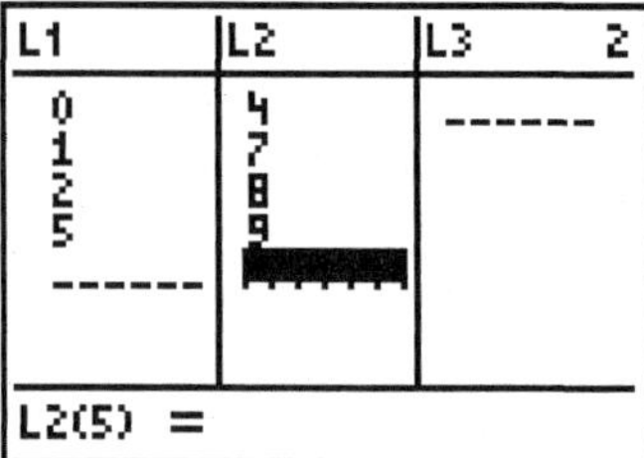

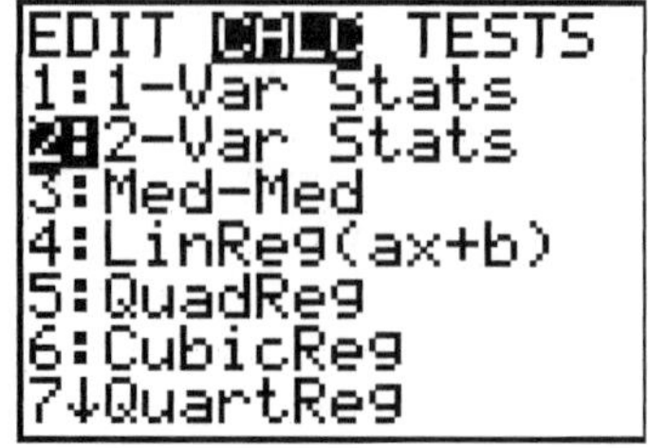

Note that X = advertising and Y = sales. We should specify the lists as you see below in order to have the variables labeled correctly in the 2-Var Stats output.

```
ClrAllLists
                Done
2-Var Stats L1,L
2
```

```
2-Var Stats
 x̄=2
 Σx=8
 Σx²=30
 Sx=2.160246899
 σx=1.870828693
↓n=4
```

The mean and standard deviation for advertising are 2 and 2.16, respectively.

```
2-Var Stats
↑ȳ=7
 Σy=28
 Σy²=210
 Sy=2.160246899
 σy=1.870828693
↓Σxy=68
```

The mean and standard deviation for advertising are 7 and 2.16, respectively. Parts (b) and (c) ask us to find the regression line, and test for a linear relationship. We can get all of the information necessary to answer both of these questions by using the LinRegTTest function. Press STAT TESTS. The X data is in L1, and the Y data is in L2.

```
EDIT CALC TESTS
0↑2-SampTInt…
A:1-PropZInt…
B:2-PropZInt…
C:χ²-Test…
D:2-SampFTest…
E:LinRegTTest…
F:ANOVA(
```

```
LinRegTTest
 Xlist:L1
 Ylist:L2
 Freq:1
 β & ρ:≠0 <0 >0
 RegEQ:Y1
 Calculate
```

```
LinRegTTest
 y=a+bx
 β≠0 and ρ≠0
 t=2.353393622
 p=.1428571427
 df=2
↓a=5.285714286
```

```
LinRegTTest
 y=a+bx
 β≠0 and ρ≠0
↑b=.8571428571
 s=1.362770288
 r²=.7346938776
 r=.8571428571
```

The regression equation is $\hat{y} = 5.29 + 0.857x$. The test statistic is t = 2.35, with p-value = .1429. We do not have sufficient evidence of a significant linear relationship. A scatterplot provides some clues regarding why not. Maybe the methods of section 11.5 would be more appropriate here.

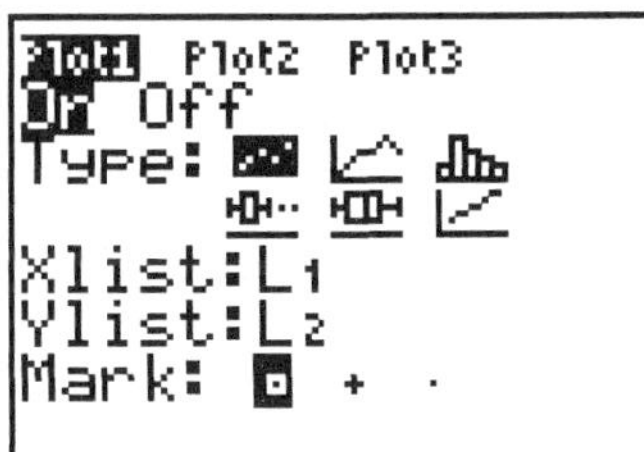

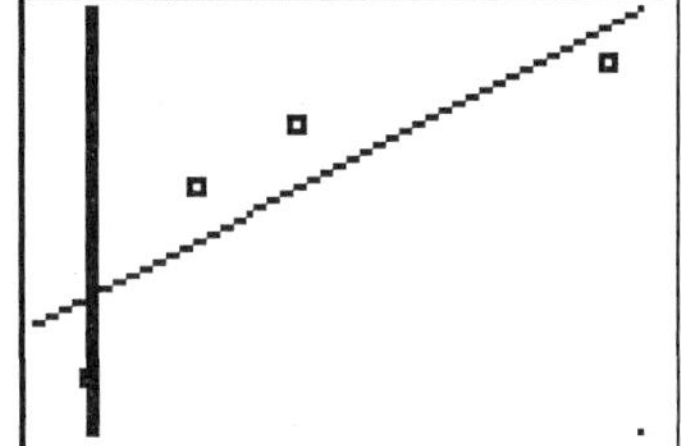

Residuals and the residual standard deviation: Section 11.4 of Agresti/Franklin

The authors explore the usefulness of standardized residuals in this section. The TI-83 does not produce standardized residuals, but it does calculate the $(y - \hat{y})$ residuals and place them in a list called RESID. Let's take a look at the residuals from the strength study. Press STAT EDIT, then 1:Edit to access the Editor.

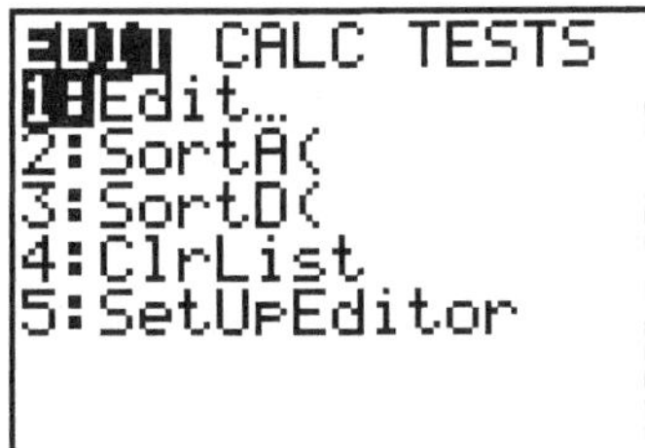

L1	L2	L3
80	10	------
85	12	
85	20	
65	5	
95	12	
75	10	
90	13	

L3 =

Maneuver the cursor so that it is on the header of L3. Press LIST (2nd STAT). Scroll up (should be quicker) and find RESID. Choose it.

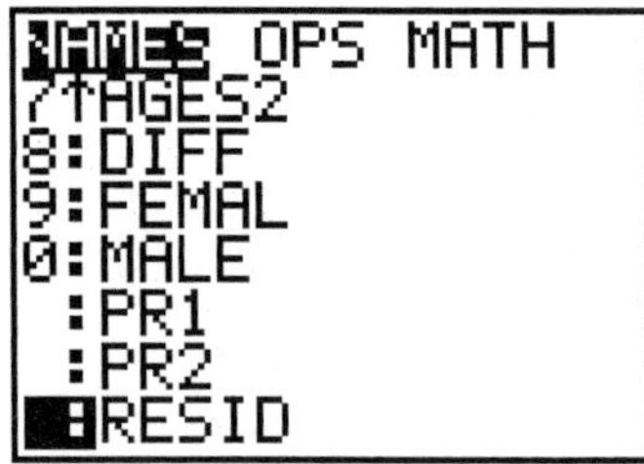

L1	L2	L3
80	10	------
85	12	
85	20	
65	5	
95	12	
75	10	
90	13	

L3 = LRESID

Press ENTER. The residuals will appear in L3.

L1	L2	L3
80	10	1.5526
85	12	3.5705
85	20	⁻8.358
65	5	⁻5.992
95	12	13.571
75	10	⁻3.447
90	13	7.0795

L3(1)=1.552613480...

The distribution of the residuals is important. See p. 556 for a brief discussion of this. Basically, we hope to see our residuals following a bell-shaped distribution with a minimum of outliers. Let's investigate this with a histogram and boxplot. Press STAT PLOT. Request a histogram of the data in L3. Press ZOOM, then choose 9:ZoomStat.

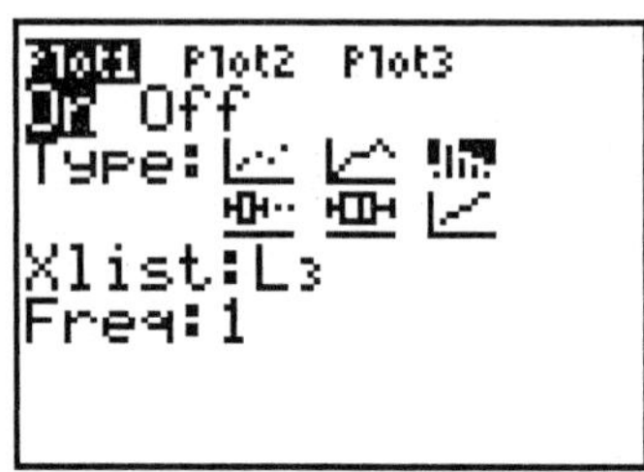

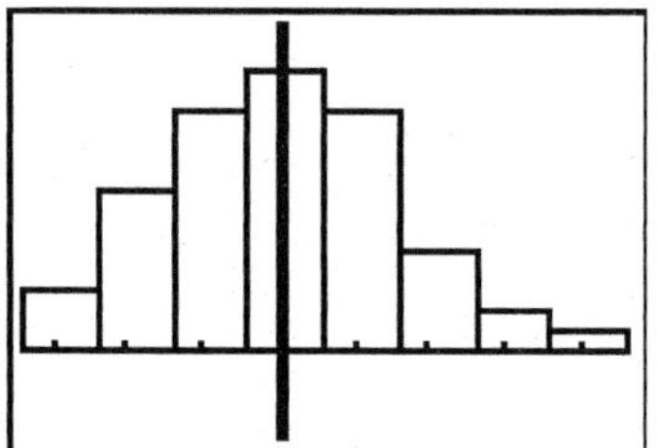

The residuals look (approximately) normally distributed. The histogram is not perfectly symmetric, but it is probably close enough to satisfy the assumptions for linear regression analysis. Note that the residuals are centered at zero, which is what we expect.

Next let's try a boxplot. Choose the icon that gives a boxplot with outliers marked with special symbols. No outliers are present.

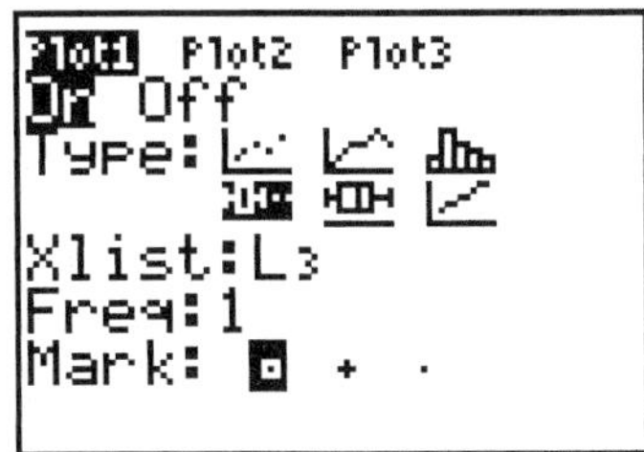

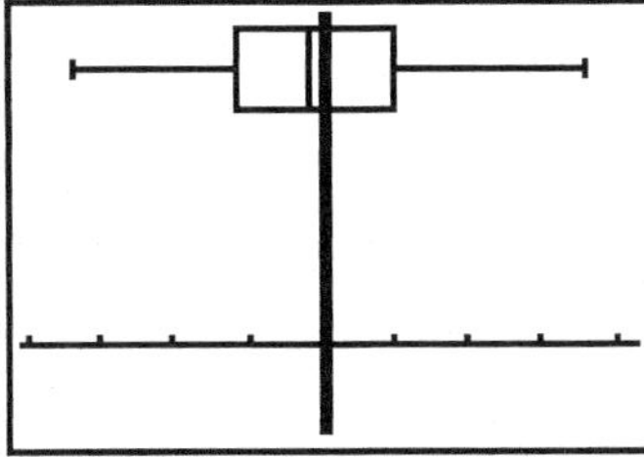

Before we leave the strength study, consider the discussion of the residual standard deviation on p. 556. The $(y - \hat{y})$ part of the formula denotes the residuals, which we have stored in L3. We can easily replicate the calculation shown at the top of p. 557. Press LIST (2nd STAT), then cursor right to MATH. Choose 5:sum. Sum up the squared residuals by squaring L3 in the sum argument. Divide by df=55 and take the square root. Recall also that s = 8.00 was given in the LinRegTTest output (see example 11).

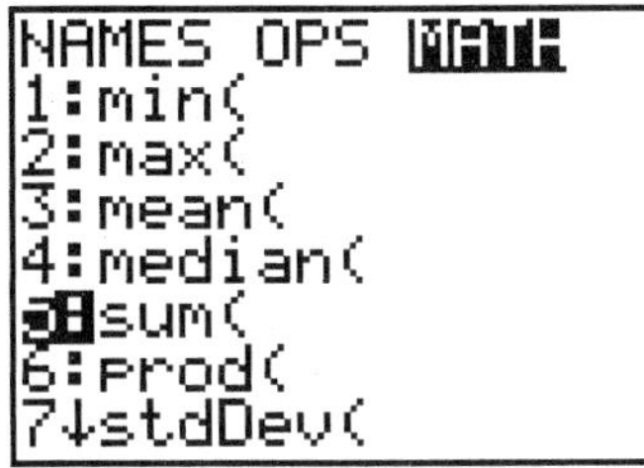

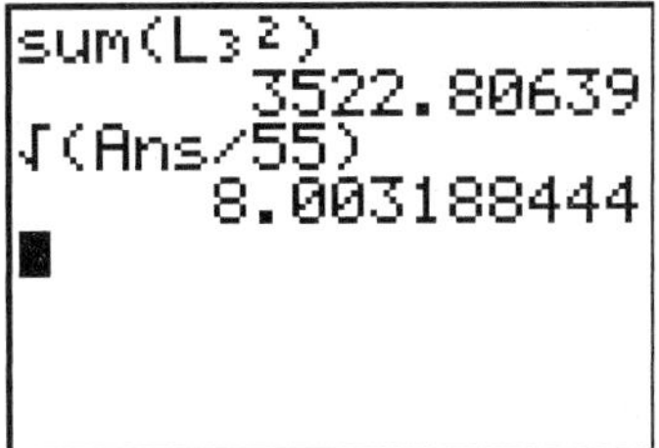

Exponential regression: Section 11.5 of Agresti/Franklin
In this section, we will explore exponential (non-linear) relationships between X and Y. We will make use of the log function.

EXAMPLE 18: Explosion in number of people using the internet (pp. 565-566 of Agresti/Franklin)
Enter the X and Y data as shown in table 11.9. Move the cursor up to the L3 header so that you can define the whole list. Press LOG, then L2 in order to place the logarithm of no. of people in L3.

L1	L2	L3 3
0	16	------
1	36	
2	76	
3	147	
4	195	
5	369	
6	513	

L3 =log(L2)

L1	L2	L3 3
0	16	1.2041
1	36	1.5563
2	76	1.8808
3	147	2.1673
4	195	2.29
5	369	2.567
6	513	2.7101

L3(1)=1.204119982...

Let's plot X vs. Y to see the exponential relationship. Press STAT PLOT. Specify as shown below left. Remember that you must clear any Y= equations before plotting.

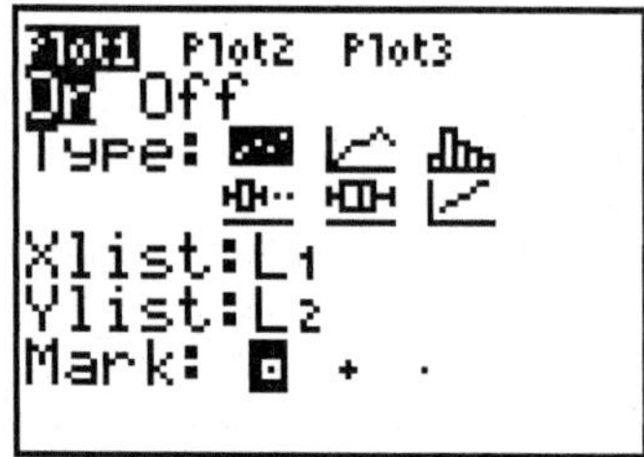

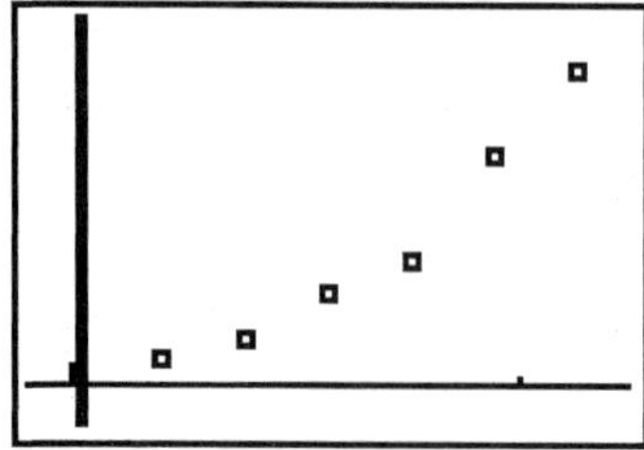

Next plot X vs. log(Y) to see if a linear relationship exists. You'll need to use ZoomStat, since the log data is on a different scale.

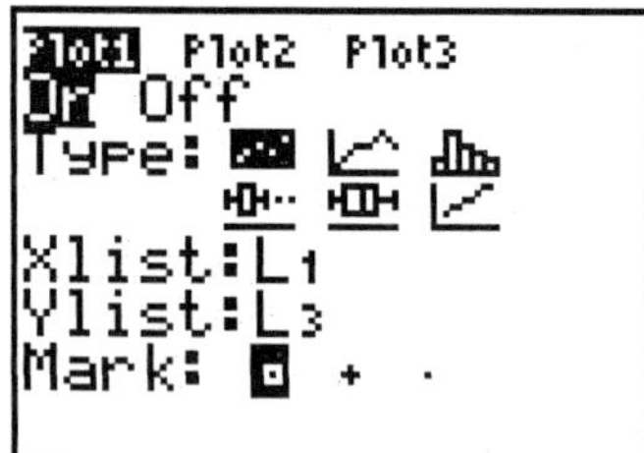

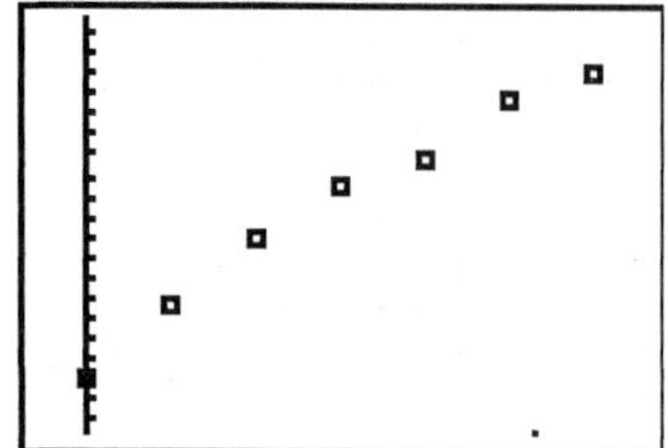

Let's find the estimates for the exponential equation given on p. 566. Press STAT CALC. Choose 0:ExpReg. The argument list is the same as LinReg(a+bx): Xlist, Ylist, RegEQ. Our X data is in L1, the Y data is in L2, and we might as well have the calculator put the equation in Y1. This last argument is optional. Recall that the "Y1" name must be pulled up by pressing VARS, Y-VARS, then 1:Function.

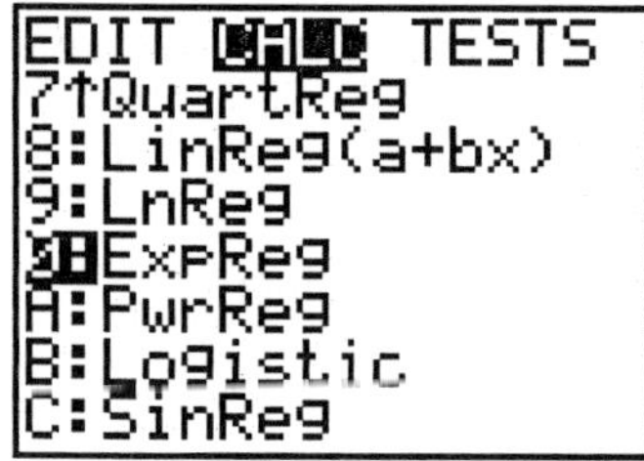

The results on the next page give the equation $\hat{y} = 20.38 \times 1.7708^x$. Go back to STAT PLOT and request a scatterplot between L1 and L2 (the original data). If you specified Y1 above, the line will be plotted as well.

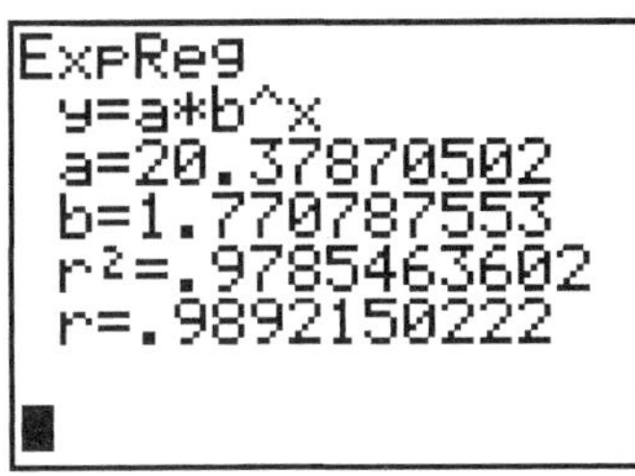

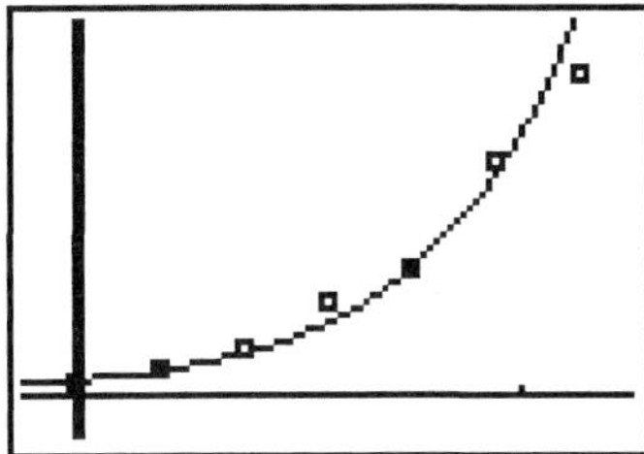

Exercise 11.55: Leaf litter decay (p. 568 of Agresti/Franklin)
Enter the X data into L1 and the Y data into L2. Plot it using STAT PLOT.

L1	L2	L3 2
7	20.4	
8	14	
9	12.3	
11	8.2	
15	3.1	
20	1.4	

L2(14) =

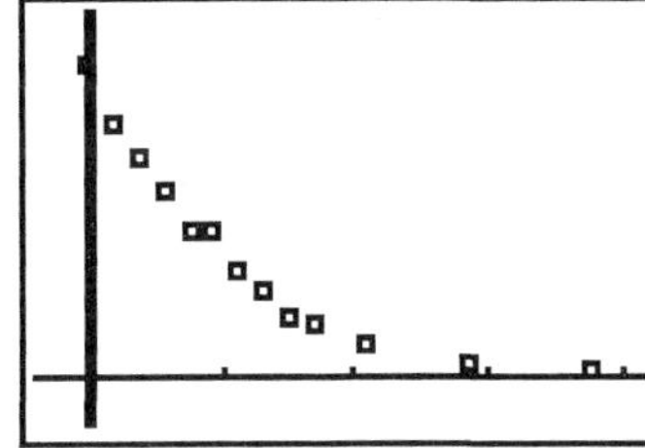

Part (b) asks us to consider a straight-line fit. Use LinReg(a+bx) from the STAT CALC menu.

EDIT CALC TESTS
2↑2-Var Stats
3:Med-Med
4:LinReg(ax+b)
5:QuadReg
6:CubicReg
7:QuartReg
8:LinReg(a+bx)

The linear coefficients as shown on p. 568 are verified. If we specify Y1 in the LinReg(a+bx) function, pressing GRAPH again will show the line, which is a poor fit.

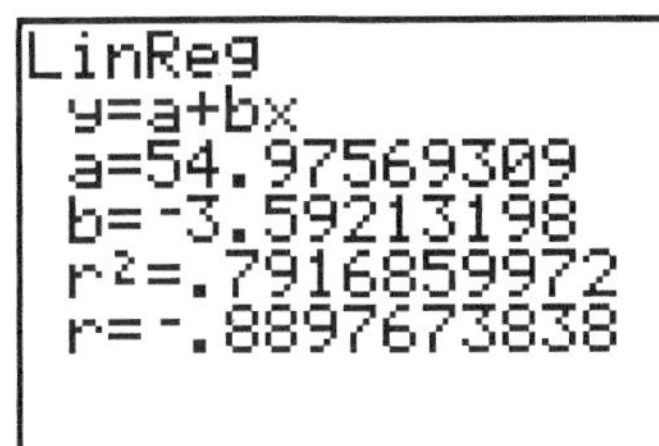

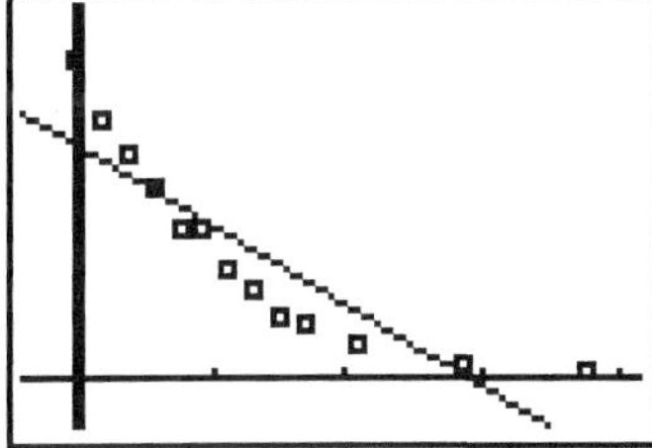

Part (b) also asks for the predicted weight for $x = 20$. Press VARS, Y-VARS, and 1:Function. Choose Y1. Supplying 20 as an argument plugs $x = 20$ into the linear equation. Since the line misses the curvature of the data, the line is below the X axis for $x = 20$, which produces a negative predicted value of weight.

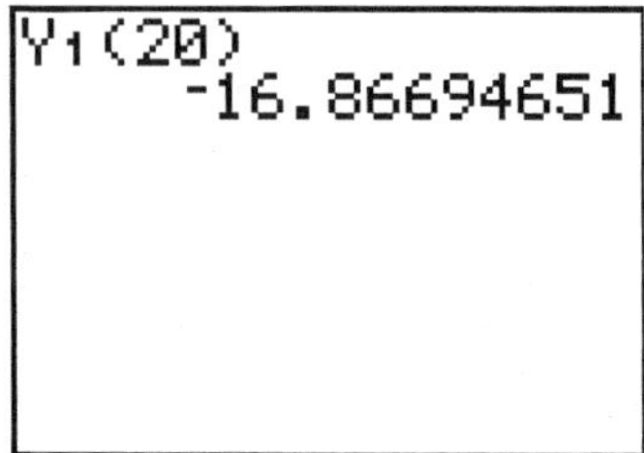

Part (c) asks for a plot of log(Y) vs. X. In STAT Editor, move the cursor to the L3 header. Press the LOG key, then supply the Y data (L2).

L1	L2	L3
0	75	------
1	60.9	
2	51.8	
3	45.2	
4	34.7	
5	34.6	
6	26.2	

L3 =log(L2)

L1	L2	L3
0	75	1.8751
1	60.9	1.7846
2	51.8	1.7143
3	45.2	1.6551
4	34.7	1.5403
5	34.6	1.5391
6	26.2	1.4183

L3(1)=1.875061263...

Now go back to STAT PLOT, and change L2 to L3. Delete the linear equation in Y1, and use ZoomStat. The plot of log(Y) vs. X does indeed look linear.

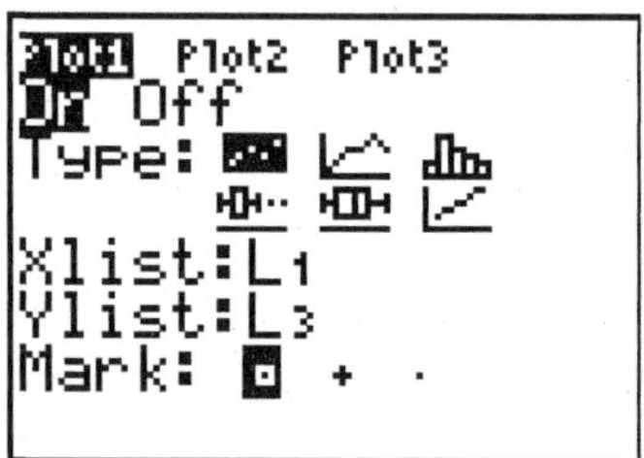

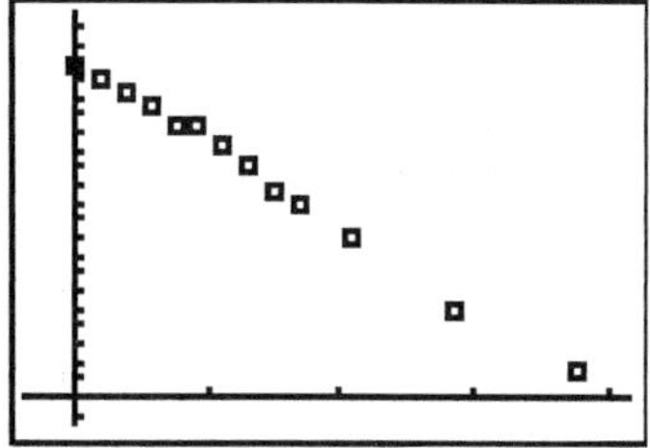

Part (d) considers fitting an exponential model to the (original) data. Press STAT CALC and choose 0:ExpReg.

Remember that the X data is in L1, and the (original) Y data is in L2. Specify Y1 to store the equation.

```
ExpReg
 y=a*b^x
 a=80.62393605
 b=.8134434131
 r²=.9935636516
 r=-.9967766308
```

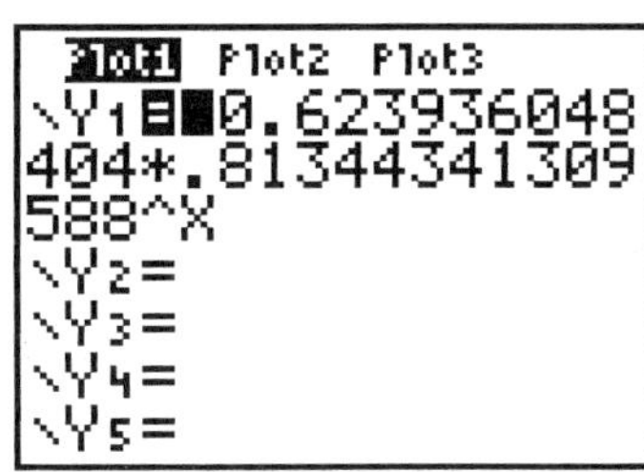

The exponential model given on p. 568 is confirmed above. Go back to STAT PLOT and change L3 back to L2. Now we can see the original data with the exponential fit.

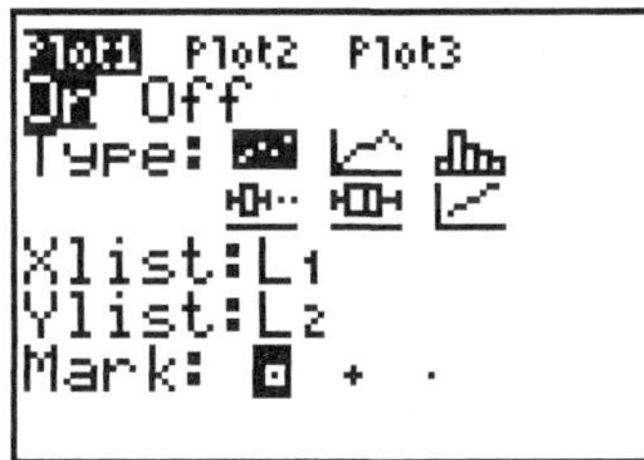

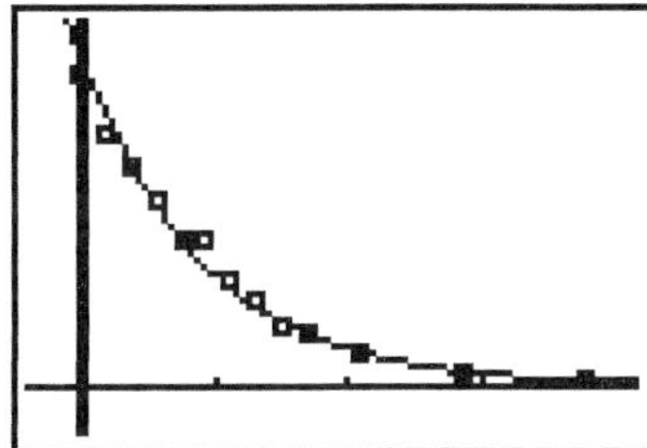

Part (d) also asks for the predicted weight for $x = 0$ and $x = 20$ using the exponential model. You could plug these values in manually, then check your work with using the Y1 equation in the calculator. Remember to press VARS, Y-VARS, 1:Function, then choose Y1. These predicted weights (80.6 kg initially and 1.30 kg after 20 weeks) make much more sense than with the linear equation.

```
Y1(0)
          80.62393605
Y1(20)
          1.297211924
```

Chapter Twelve: Multiple Regression

Unfortunately, the TI-83 does not have a pre-defined function available for regression analysis with more than one X variable. The matrix algebra capabilities of the calculator can be exploited to this end, but this method is beyond the scope of the statistics course you're taking. Programs are out there and freely available; you might try typing "multiple regression program TI-83" into an internet search engine. Users of the TI-89 that have loaded the Statistics application have multiple regression capabilities.

Chapter Fourteen: Nonparametric Statistics

Unfortunately, the same is true for chapter 14. The TI-83 does not have any pre-defined functions for nonparametric methods. Again, programs are out there and freely available; you might try typing "wilcoxon rank sum program TI-83" into an internet search engine.

Chapter Thirteen: Comparing Groups: Analysis of Variance Methods

In this chapter, we will compare the means of three or more groups. The ANOVA function from the STAT TESTS menu will be used for this. Boxplots from STAT PLOT are also useful for this method.

EXAMPLE 2: How long will you tolerate being put on hold? (pp. 627-628 of Agresti/Franklin)
We wish to compare the three recorded message groups. Enter the five holding times for the advertisement group into L1, the five holding times for the muzak group into L2, and the five holding times for the classical music group into L3. Remember that you can quickly get a "clean slate" in the STAT Editor by using the ClrAllLists command. Press MEM (2nd +), and choose 4:ClrAllLists.

L1	L2	L3 3
5	0	13
1	1	9
11	4	8
2	6	15
8	3	7
------	------	

L3(6) =

Next, let's verify the group means and standard deviations as shown in table 13.1. We can use the 1-Var Stats function from the STAT CALC menu, or the separate mean and stdDev functions from LIST MATH. Press LIST, then arrow right to MATH. Choose 3:mean, then choose 7:stdDev. Do this for all three groups.

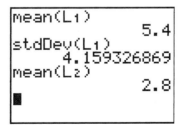

```
mean(L1)
                  5.4
stdDev(L1)
          4.159326869
mean(L2)
                  2.8
```

```
                  2.8
stdDev(L2)
       2.387467277
mean(L3)
                 10.4
stdDev(L3)
       3.435112807
```

The dotplot given on p. 628 illustrates the difference in group means, while also showing the spread of each group. We can produce a similar picture using the TI-83. Instead, we'll use boxplots to compare the groups. Press STAT PLOT (2nd Y=). Choose 5:PlotsOn. We need all three plots on for this example!

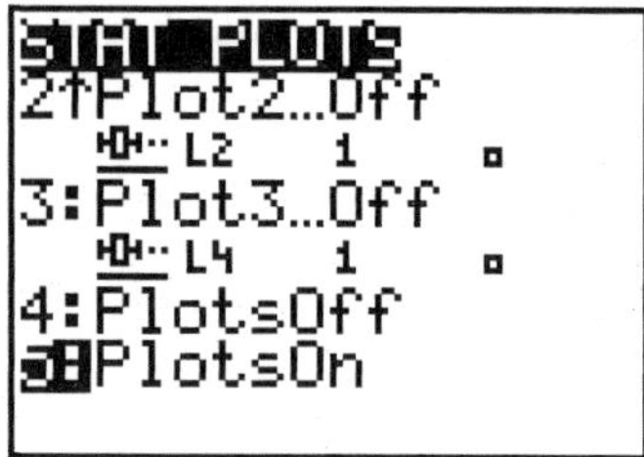

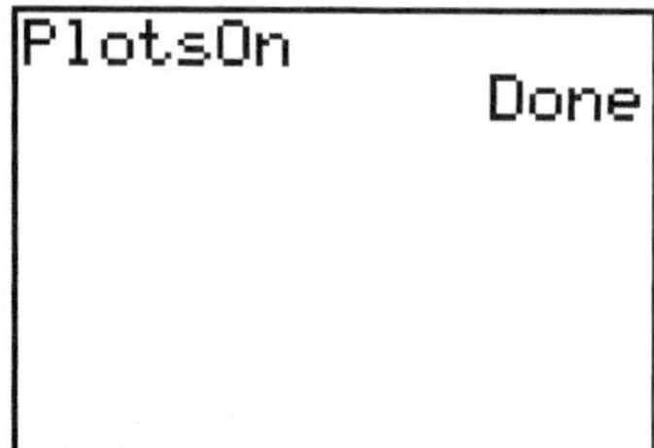

Go back to STAT PLOT. Specify a modified boxplot for L1. From the screen below right, arrow up to the top line and cursor right to Plot2. Specify a modified boxplot for L2.

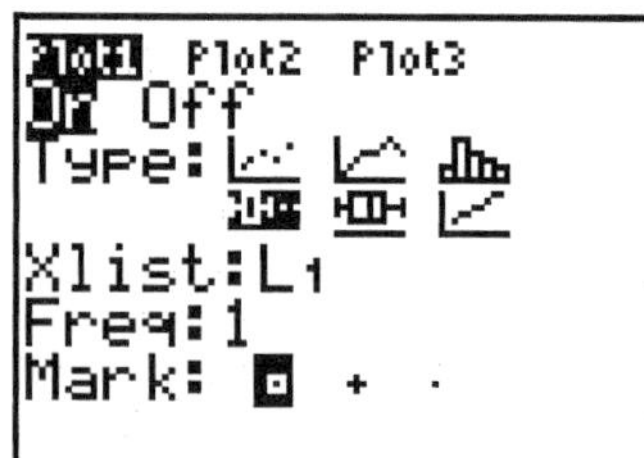

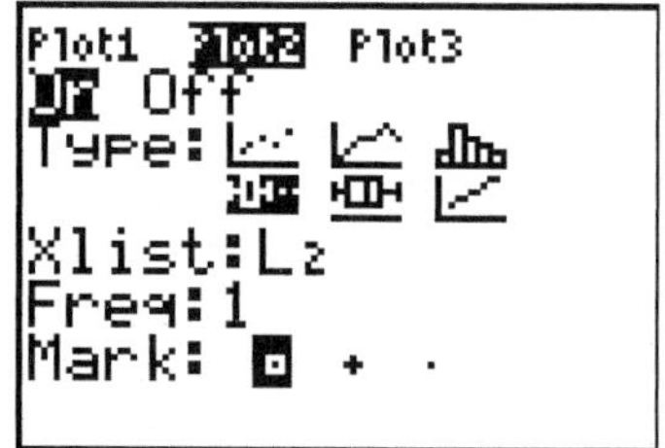

Repeat the process, specifying a modified boxplot for L3. Press ZOOM, then 9:ZoomStat. Be sure to clear out any Y= equations first.

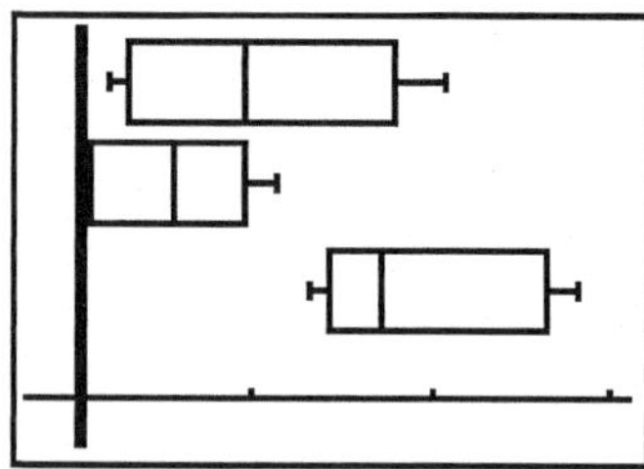

The top boxplot is L1 (advertisement), the middle boxplot is L2 (muzak), and the bottom boxplot is L3 (classical). We can see that the muzak and classical groups seem to differ, since their boxplots do not overlap. The advertisement and muzak groups overlap quite a bit – quite possibly their means do not differ significantly.

EXAMPLE 3: ANOVA for customers' holding times (p. 630 of Agresti/Franklin)
We have considered the differences and similarities among the three recorded message groups regarding average holding times graphically and with descriptive statistics (see EXAMPLE 2). Now we would like to conduct an ANOVA F-test. Initially assume that H_0: $\mu_1=\mu_2=\mu_3$. Do the data provide evidence that at least two of the means significantly differ from each other? Enter the holding times into lists L1, L2, and L3, if you haven't already. Press STAT TESTS, and scroll up once. The last option in the menu is ANOVA. Choose it. Supply the three lists as arguments.

```
EDIT CALC TESTS
0↑2-SampTInt…
A:1-PropZInt…
B:2-PropZInt…
C:χ²-Test…
D:2-SampFTest…
E:LinRegTTest…
F:ANOVA(
```

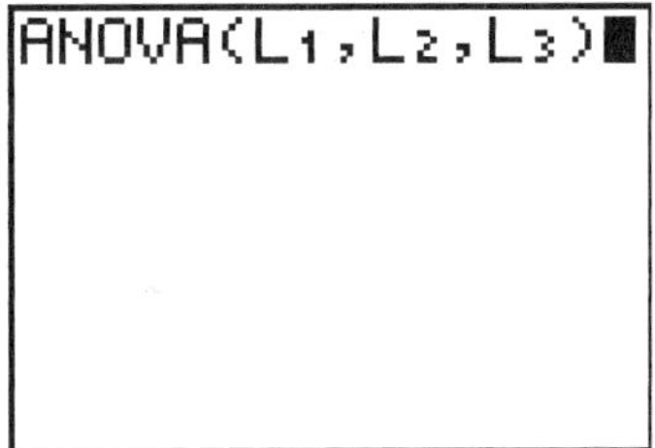

Press ENTER. Scroll down to see all of the output that is created.

```
One-way ANOVA
 F=6.431034483
 p=.0126433999
 Factor
  df=2
  SS=149.2
↓ MS=74.6
```

```
One-way ANOVA
↑ MS=74.6
 Error
  df=12
  SS=139.2
  MS=11.6
 Sxp=3.40587727
```

Note that the "Factor" df = 2. That's because we are comparing three (g = 3) groups. Factor df = g – 1 = 3 – 1 = 2. Note also that "Error" df = 12. That's because we have a total sample size of 15, and error df = N – g = 15 – 3 = 12. The test statistic is F = 6.43, which is large. The p-value is .0126, which is small. Reject H_0 and conclude that at least one difference exists among the three means. In exercise 13.14, we'll explore which of the pairs of means differ and which do not.

Following up an ANOVA F-test: Section 13.2 of Agresti/Franklin
We will consider comparing pairs of means using confidence intervals in this section. See p. 636 for the formula. Multiple comparison methods are discussed on p. 639. A Bonferroni adjustment could be made to the method shown in the next example. More likely, a data analyst would use software such as minitab to obtain Bonferroni or Tukey multiple comparison results.

Exercise 13.14: Comparing telephone holding times (p. 644 of Agresti/Franklin)
Let's take the example comparing average holding times for three recorded message groups a step further. We know (after EXAMPLE 3) that at least one difference exists among the three means, so now let's find out which groups differ, and which do not.

(a) Obtain a 95% confidence interval for the difference in mean holding times between the muzak and classical groups. Consistent with the first two examples, call muzak group 2 (data stored in L2) and classical group 3 (data stored in L3). The formula on p. 636 may be written as $\bar{y}_2 - \bar{y}_3 \pm t_{.025} s \sqrt{\frac{1}{n_2} + \frac{1}{n_3}}$. From before, we know that $\bar{y}_2 = 2.8$ and $\bar{y}_3 = 10.4$. Both sample sizes are five. The standard deviation in the formula is the square root of the mean square error from the ANOVA F-test. Refer to the last screen shot shown in EXAMPLE 3 above. The Error MS is given as 11.6, which has a square root of 3.4. This value is also given as Sxp. The t critical value has N – g = 15 – 3 = 12 df. From the t-table or MATH SOLVER, we get $t_{.025}$ = 2.18.

```
Error
 df=12
 SS=139.2
 MS=11.6
 Sxp=3.40587727
√(11.6)
       3.405877273
█
```

```
tcdf(L,U,D)-A=0
 L=-1E99
■U=-2.178812800…
 D=12
 A=.025
 bound={-1E99,1…
■left-rt=0
```

Plugging into the formula, we get $(2.8 - 10.4) \pm 2.18 \times 3.4\sqrt{\frac{1}{5} + \frac{1}{5}}$, or –7.6 ± 4.7. This gives an interval of (–12.3, –2.9) or equivalently, (2.9, 12.3). We can say with 95% confidence that those listening to classical music stayed on hold between 2.9 and 12.3 minutes longer than those subjected to muzak, on average.

(b) Look at the confidence interval formula. Since the sample sizes are all five, and we use the same t value and standard deviation regardless of which groups are being compared, the margin of error of 4.7 will not change.

(c) Since the margin of error is 4.7 for all three comparisons, this makes the next two confidence intervals easy to compute. To compare groups 1 and 2, advertising and muzak, respectively, we need $\bar{y}_1 - \bar{y}_2 \pm t_{.025} s \sqrt{\frac{1}{n_1} + \frac{1}{n_2}}$, or $(5.4 - 2.8) \pm 4.7$. 2.6 ± 4.7 gives an interval of (–2.1, 7.3). We can be 95% confident that the difference in average holding times between these groups resides in this interval, which includes zero. Therefore, we can't rule out the possibility of no difference. Recall (EXAMPLE 2) that these groups had boxplots that overlapped quite a bit.

(d) What effect would taking larger sample sizes have on the above intervals?

Performing ANOVA using regression with indicator variables: pp. 641-643 of Agresti/Franklin

Unfortunately, we cannot revisit the telephone holding times example using the regression with indicator variables method (see EXAMPLE 7). The reason is given in chapter 12 – the TI-83 does not have multiple regression capabilities.

Two-way ANOVA: Section 13.3 of Agresti/Franklin

Again, we are faced with a statistical method that the TI-83 is not equipped to perform. However, programs are available for download. The TI-89 with the Statistics application loaded has the ability to do two-way ANOVA.

EXCEL MANUAL

DIANE L. BENNER • LINDA M. MEYERS
Harrisburg Area Community College

STATISTICS

THE ART AND SCIENCE OF LEARNING FROM DATA

Agresti • Franklin

Table of Contents

INTRODUCTION TO EXCEL

INTRODUCTION: This lab session is designed to introduce you to the statistical aspects of Microsoft Excel. During this session you will learn how to enter and exit Excel, how to enter data and commands, how to print information, and how to save your work for use in subsequent sessions. As with any new skill, using this software will require practice and patience. Excel is a spreadsheet used for organizing data in columns and rows. It is an integrated part of Microsoft Office, and so data can be easily imported and exported into word processing documents, databases, graphics programs, etc. It offers a wide range of statistical functions and graphs and so is an alternative to specific statistical software.

BEGINNING AND ENDING AN EXCEL SESSION

To start Excel: Click on the Start button and choose Programs/Excel. If you have the Office shortcut bar installed, simply click on the Excel icon.

To exit Excel:

To end a Excel session and exit the program, choose **File** from the menu bar and then choose **Exit**. A dialog box will appear, asking if you want to save the changes made to this worksheet. Click **Yes** or **No**.

You can also exit Excel by clicking the X in the upper right corner of the window.

THE EXCEL WINDOWS

The Document (sheet) Window:

When you first start Excel you will be in a window titled “Microsoft Excel - Book 1” . Excel organizes itself in workbooks, each of which is made up of worksheets that are 65,536 rows by 256 columns. You can enter and edit data on several worksheets simultaneously and perform calculations based on data from multiple worksheets. When you create a chart, you can place the chart on the worksheet with its related data or on a separate chart sheet. Each of the cells within the sheet is identified by the intersection of its row and column, for example A2, or B7.

Note the three tabs at the bottom of the screen, called “sheet1”,”sheet2”, and “sheet3”. The default is a workbook with three sheets, but the number of sheets in a workbook is limited only by available memory. To add a single worksheet, click Worksheet on the Insert menu. To delete sheets from a workbook, select the sheets you want to delete. Then on the Edit menu, click Delete Sheet. To Rename a sheet, double-click the sheet tab, and type a new name over the current name.

The Application Window

When you first opened the workbook, there were three bars across the top of the screen. The top one consists of nine drop down menu items and is called the Menu Bar. It gives you access to all the Excel commands. Two other toolbars are below the Menu Bar. The first is the Standard Toolbar. This toolbar gives you easy access to the ordinary things you will be doing, such as saving, cut and paste, copying, spell check, and the **Chart Wizard**. The Chart Wizard will be discussed in the next chapter, along with the graphical display of data.

Analysis ToolPak : Microsoft Excel provides a set of data analysis tools — called the Analysis ToolPak — that you can use to save steps when you develop complex statistical or engineering analyses. You provide the data and parameters for each analysis; the tool uses the appropriate statistical or engineering macro functions and then displays the results in an output table. Some tools generate charts in addition to output tables. If the **Data Analysis** command is not on the **Tools** menu, you need to install the Analysis ToolPak. To do this, go to the **Tools** drop down menu and select **Add-ins**. When the dialog box appears, check **Analysis Toolpak** and **Analysis VBA**. Then click on **OK.**

In addition to the Data Analysis ToolPak, your textbook comes packaged with a program called PHStat2. Read the Readme file and follow the directions for loading the program on to your desktop. It contains additional macros for Excel to enable you to do advanced features.

Below the Standard Toolbar is the Formatting Toolbar. This bar gives you easy access to the tools need to format your data in a suitable way. This is where you will find bold, underlining and font choices, justifications, style, merging cells, etc.

You can choose additional toolbars by going to the **Tools** dropdown menu, selecting **Customize**, and clicking next to any toolbar you wish to display.

The Help Window in Excel

Information about Excel is stored in the program. If you forget how to use a command or need general information, you can ask Excel for help. From the

Menu Bar choose **Help .** A drop down menu will appear, giving you a choice between Microsoft Excel Help (F1 key) or the Office Assistant. You can even customize the Office assistant to a varied selection of figures. He will even travel with you from Excel into Word, or any other part of the Office package of programs.

ENTERING DATA

When a workbook is first opened, the cell A1 is outlined in black. This indicates the active cell. Move your cursor around the sheet, clicking into different cells to activate them. Note that the address changes in the box above A1. The address (row and column) of the active cell always appears here.

Let's enter data in the second column:

78 94 93 81 75 62 58 50 80 79

To do this press the down arrow key (↓) or enter key to move to the next entry position.

Let's fill the first column with the numbers 1 through 10. We can do it the same way, or we can let Excel do it for us. Enter a 1 in cell A1. Choose **Edit > Fill->Series**. In the dialog box, select **columns**, **linear**, step **1**, stop value **10**. Then click **OK.**

Column 1 should now contain the integers 1 through 10.

While you are in the sheet window, fill columns 3 and 4 with a set of ten test scores each. You should now have four columns of data.

Changing a value entered

We can edit data directly in the cell or from the formula bar at the top of the sheet. If you have not hit the Enter key yet, you can simply back space and correct your mistake. If you have entered the data, click on the cell you wish to edit to make it active. You can either retype to overwrite the data, or click into the formula bar and edit the entry.

Suppose we had inadvertently left out a value and we wish to enter it in a particular position. Place the cursor in the cell in which you wish to insert the new value. Click the Insert Cells button on the toolbar. A dialog box will appear, asking which way you wish to move the cells. A blank cell is created and the missing value can be entered. Entire rows and columns can be added the same way. You can take a short cut to this by using **Control +.**

A cell can be deleted by making the cell active, then Choose: **Edit > Delete Cells** or by using **Control -**

Copying Data

To copy the contents of one cell to another, simply activate the cell and choose **Edit > Copy** from the **Menu Bar. (Control C** will also accomplish this.) Activate the cell that you want to paste the value into and choose **Edit –>Paste** (or **Control V**) This can also be done for a range of cells. Activate the upper left cell of the range. Press shift and click the lower right corner of the range. This should highlight the entire range. You can then copy and paste as above.

Cell References:

Previously, you entered four columns of data. Click on cell B11. On the Standard Toolbar you will see a summation sign Σ. Click on it and the ten values above it will be enclosed in a box. Press enter and the sum of the ten values will be in cell B11 . Now activate cell B11, press **Control C**, highlight cells C11 and D11, and press **Control V.** This should give you the sums of columns C and D. Note what happened in the formula when you copied it. The references were changed to reflect the new column. This is called a relative reference.

If you need to preserve the value of a certain cell when copying a formula, you will have to use absolute referencing. This is accomplished by placing $ within the address. ($A$6 would keep the value in cell A6 to wherever it was copied within the worksheet.)

SAVING YOUR WORK

An Excel workbook contains all your work; the data, graphs, and all the sheets within the workbook. When you save a project, you save all of your work at once. When you open a project, you can pick up right where you left off. The contents of each sheet can be saved and printed separately from the project, in a variety of formats. You can also delete a worksheet or graph, which removes the item from the project.

RETRIEVING A FILE

You can open a wide variety of files with Excel. Choose **File Open** to select the appropriate one. There is an **Import Wizard** that will guide you through the process.

A CD ROM accompanies Agresti/Franklyn Statistics: The Art and Science of Learning from Data. This disk has data for many of the problems in the text. Follow the instructions that accompany the disk for use on your computer.
Let's use problem 1.19 to illustrate how to get the information from the CD into Excel. Under the **File** menu, click on **Open**. Use the drop down box to show the files on the CD in D drive. Open the Excel folder and click on the desired file. This should fill the worksheet with data.

PRINTING:

You have many options when it comes to printing from Excel. Go to the standard toolbar and choose the **File** drop down menu. The Set Print Area choice allows you to select the range of cells you wish to print.
The **Page SetUp** dialog box has four tabs that will help you customize your output. You can also access this dialog box through **Print Preview**. This is a good choice because it allows you to play with your selections to get the best layout for your output before you commit it to paper.

Chapter 2 Example 3 How Much Electricity Comes From Renewable Energy Sources? Creating a Pie Chart and a Bar Graph

Copy the data from the text into the worksheet.

	A	B	C
1	Source	U. S. Percentage	Canada Percentage
2	Coal	51	16
3	Hydropower	6	65
4	Natural Gas	16	1
5	Nuclear	21	16
6	Petroleum	3	1
7	Other	3	1

Choose: **Chart Wizard > Pie > Pie > Next**

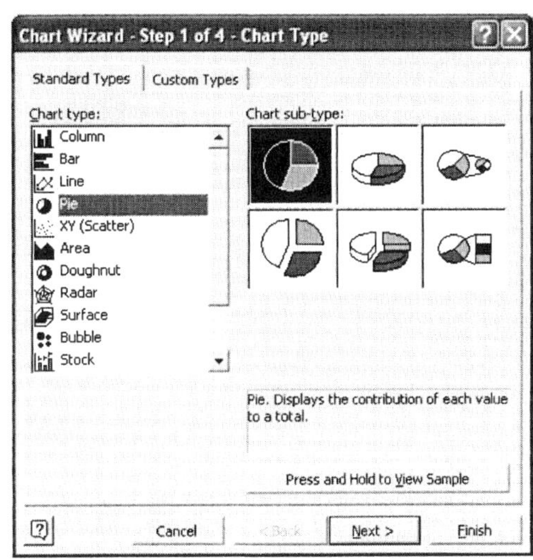

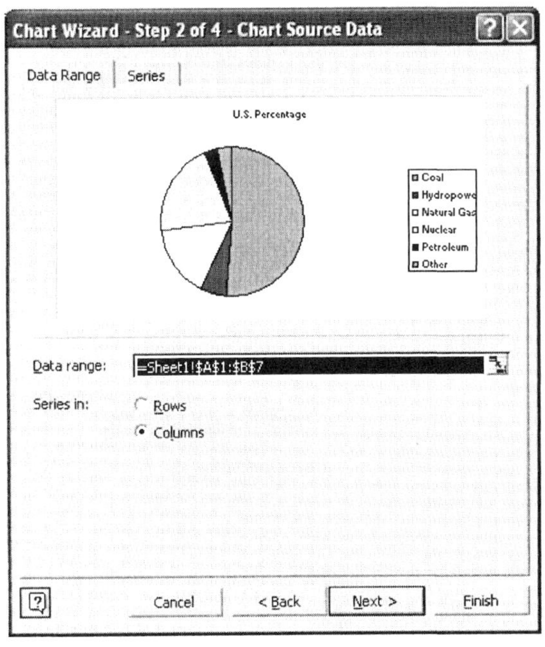

Enter: Data Range: **A1: B7 or select cells**

Select: Series in: **Columns**

Click: **Next**

Choose: **Titles** tab

Enter: Chart title: **U.S. Electricity Sources**

Choose: **Data labels** tab
Select: **Percentage > Next > Finish**

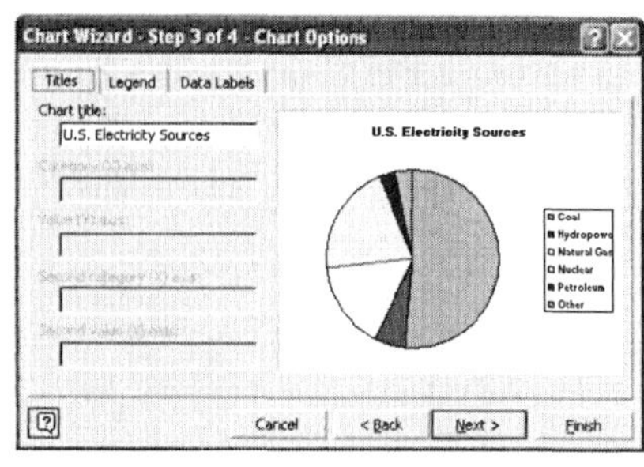

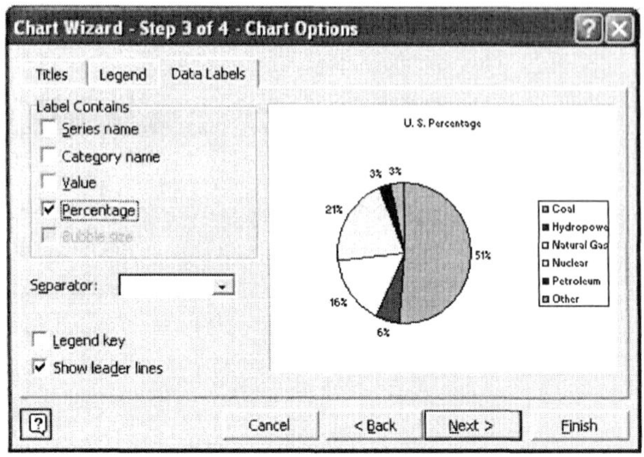

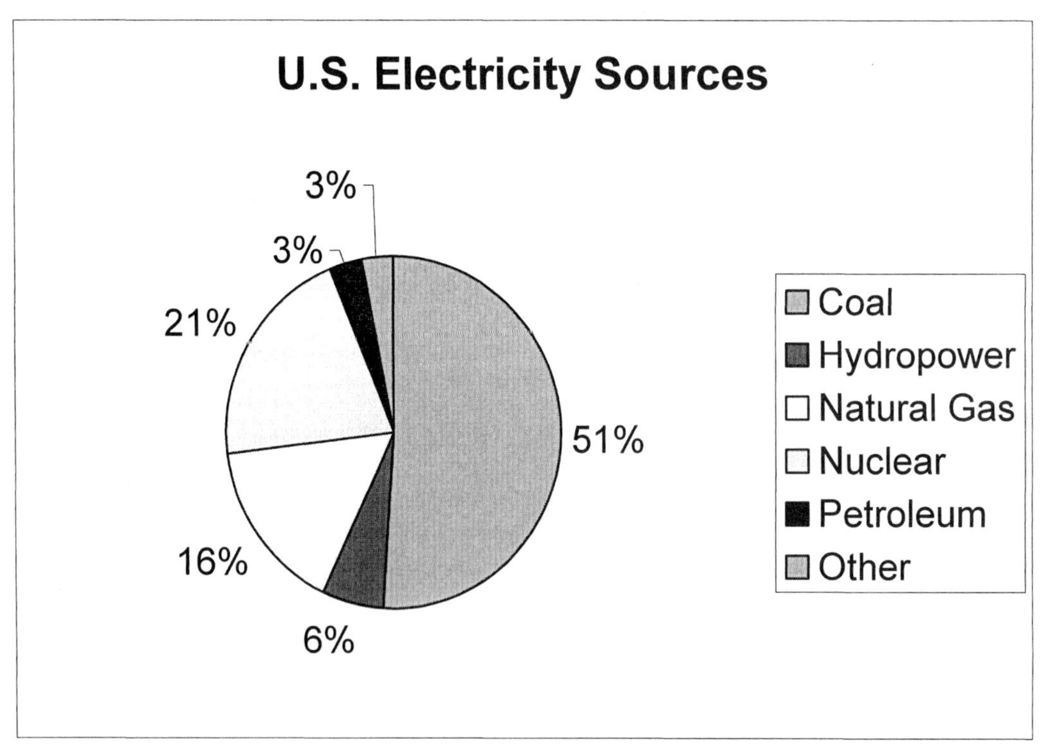

To get the bar chart:

Choose: **Chart Wizard > Column > Clustered Column > Next**

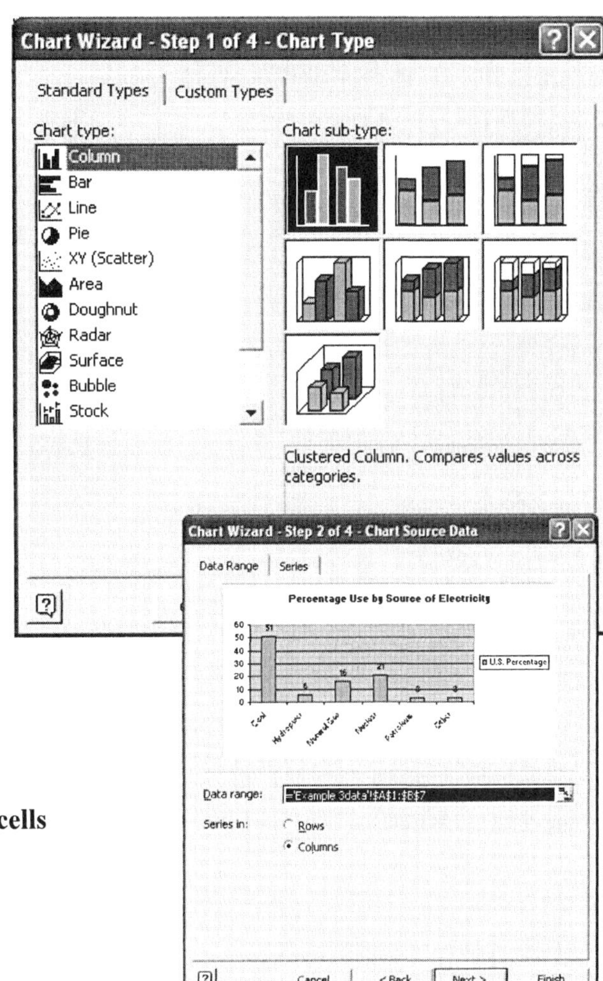

Enter: Data range: **A1:B7 or select cells**
Select: Series in: **Columns**
Click: **Next**

Choose: **Titles** tab
Enter: Chart title: **U.S. Electricity Sources**
Choose: **Legend** tab
Select: **Show legend** (to uncheck)

Click: **Next > Finish**

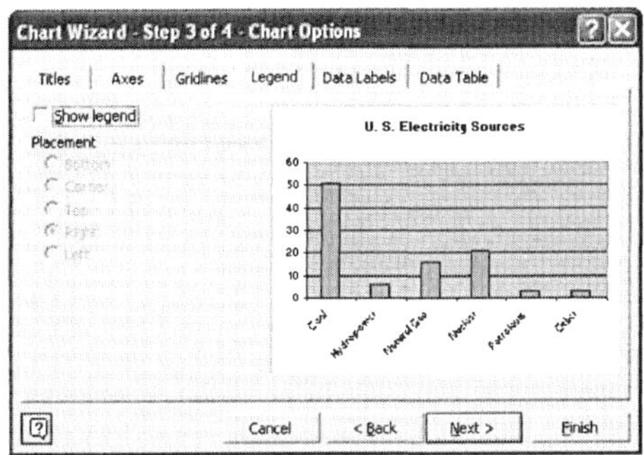

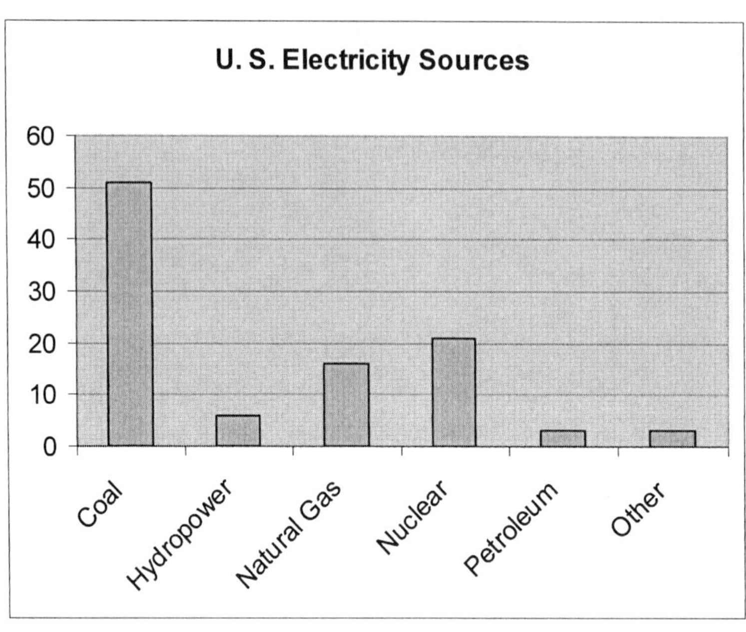

Chapter 2 Exercise 2.11 Weather Stations
Constructing a Pie Chart and Bar Graph

Use the data shown in the pie chart in the text:

	A	B
1	Region	Frequency
2	Southeast	67
3	Northeast	45
4	West	126
5	Midwest	121

Choose: **Chart Wizard > Pie > Pie > Next**
Enter: Data Range: **A1:B5 or select cells**
Check: Series in: **columns > Next**
Choose: **Titles** tab
Enter: Chart title: **Regional Distribution of Weather Stations**
> Next > Finish

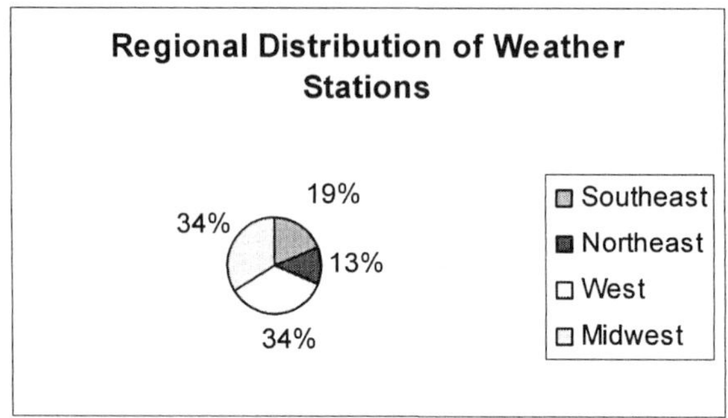

The corresponding bar chart:
Choose: **Chart Wizard > Column > Clustered Column > Next**
Enter: Data Range: **A1:B5 or select cells**
Check: Series in: **columns > Next**
Choose: **Titles**
Enter: Chart title: **Regional Distribution of Weather Stations**
> Next > Finish

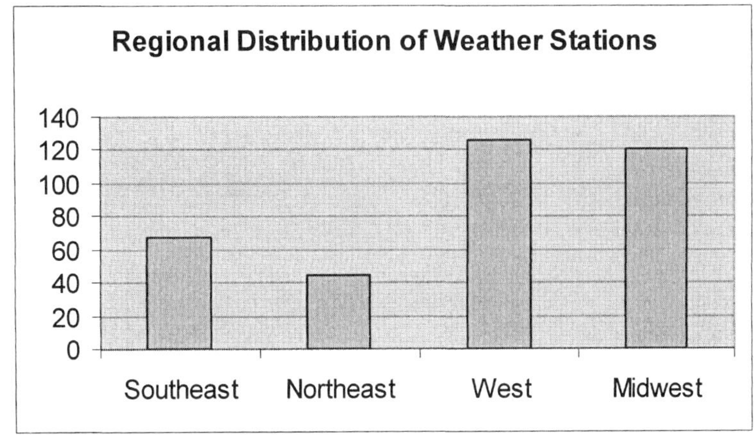

Chapter 2 Exercise 2.13 Shark Attacks Worldwide
Bar Chart and Pareto Chart

Enter the data from table 2.1 into the worksheet.

	A	B
1	Region	Frequency
2	Florida	289
3	Hawaii	44
4	California	34
5	Australia	44
6	Brazil	55
7	South Africa	64
8	Reunion Island	12
9	New Zealand	17
10	Japan	10
11	Hong Kong	6
12	Other	160

Sort the data alphabetically based on region and store the sorted columns in the original columns.
Choose: **Data > Sort >** Sort by: **Region**
Select: **Ascending >** My list has: **Header row > OK**

To create the bar chart

Choose: **Chart Wizard > Column > Clustered Column > Next**
Enter: Data Range: **A1 : B12**
Series in: **columns > Next**

Select: **Titles** tab
Enter: **appropriate title**
Select: **Legend** tab
Click: **Show legend** (to uncheck)

After reformatting the vertical axis:

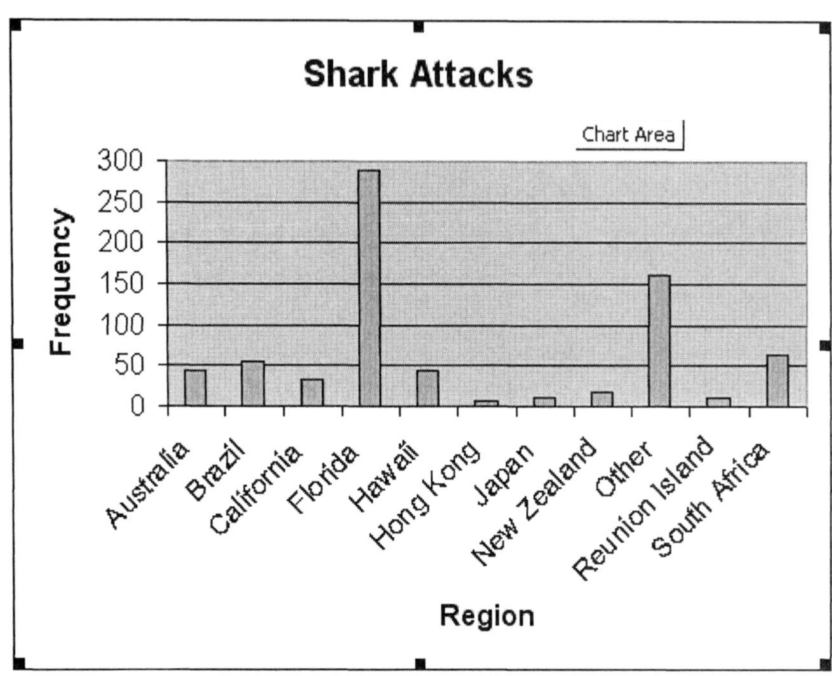

To obtain a Pareto chart for the same data, use the following menu choices

Choose: **Data > Sort >** Sort by: **Frequency**
Select: **Descending >**
Select: My list has: **Header row > OK**

Then continue with the commands necessary to create the bar graph.

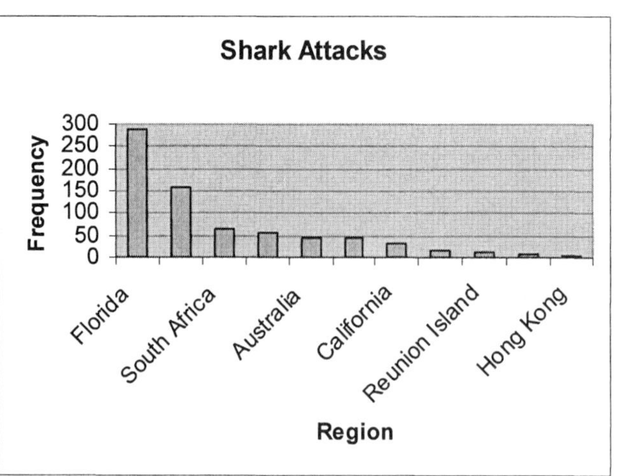

Chapter 2 Example 4 Exploring the Health Value of Cereals
Constructing a Dot Plot

Dot Plots are not available in Excel, but can be done using PHStat that comes with your text. Open the worksheet **cereal** from the Excel folder of the data disk.

Choose **PHStat > Descriptive Statistics > Dot Scale Diagram**

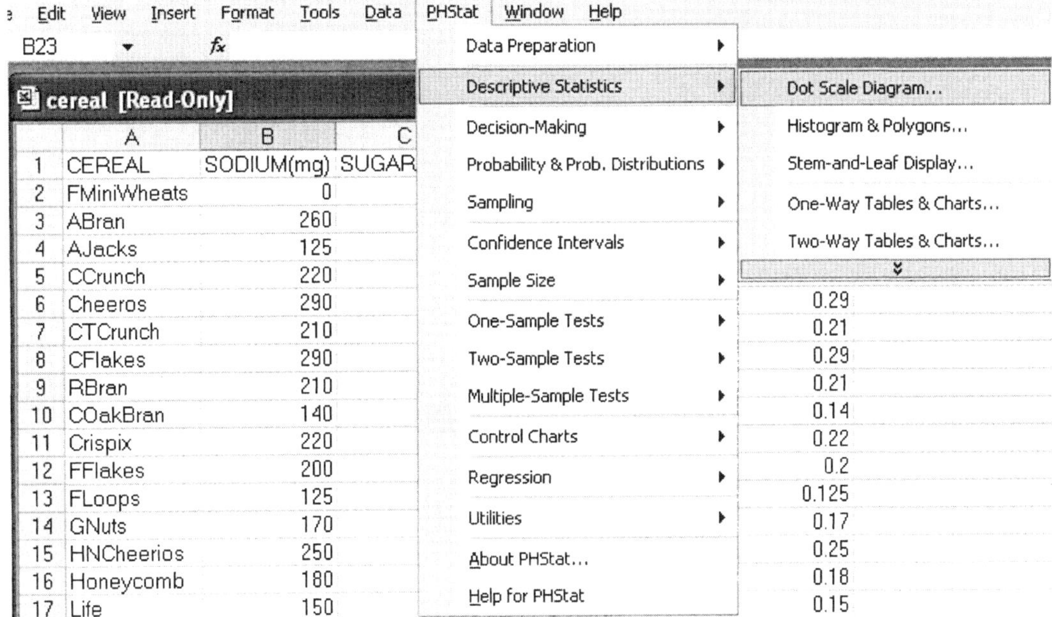

Enter the data by selecting column B.
Enter Title: **Sodium (mg)**

Click: **OK**

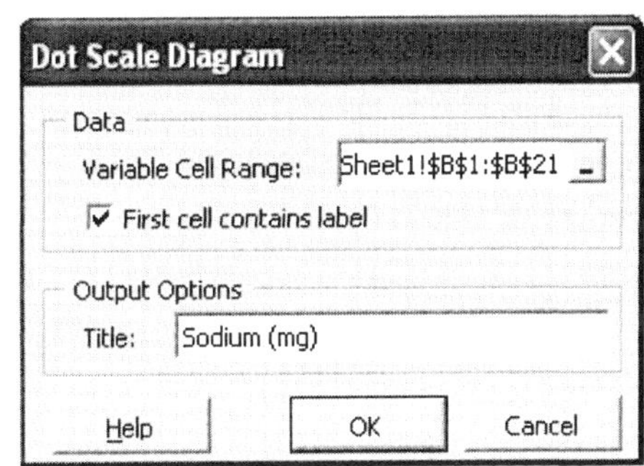

The output is shown below.

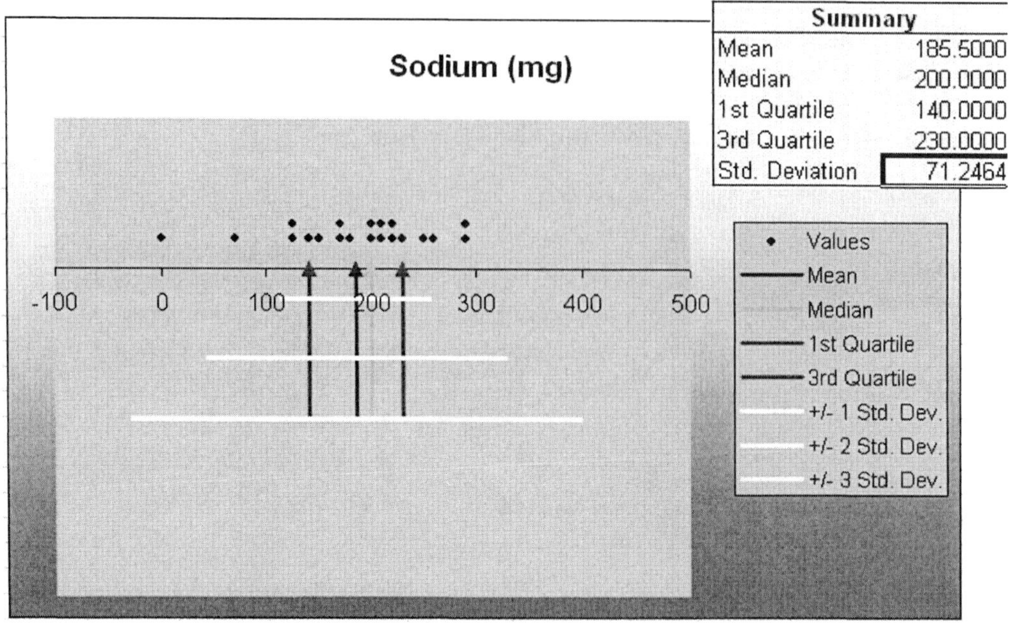

Chapter 2 Exercise 2.14 Sugar Dot Plot
Constructing a Dot Plot

Open theworksheet **cereal** from the Excel folder of the data disk.

Choose **PHStat > Descriptive Statistics > Dot Scale Diagram**

Enter the data by selecting column C.
Enter Title: **Sugar in Breakfast Cercals**
Click: **OK**

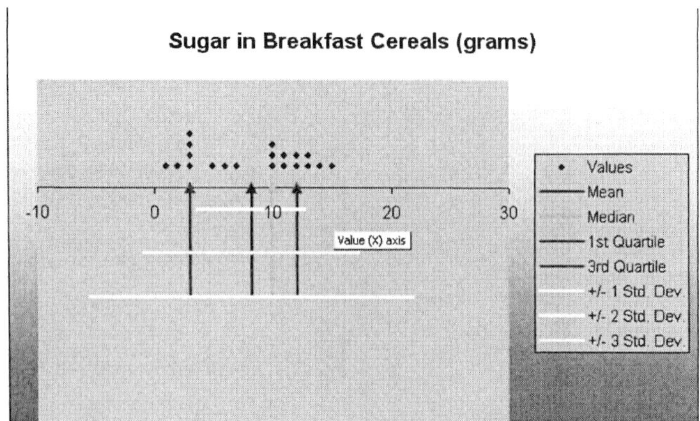

Chapter 2 Example 5 Exploring the Health Value of Cereals
Stem and Leaf Plots

Stem and Leaf Plots are not available in Excel, but can be done with PHStats.
Choose **PHStat > Descriptive Statistics > Stem-and-Leaf Display**

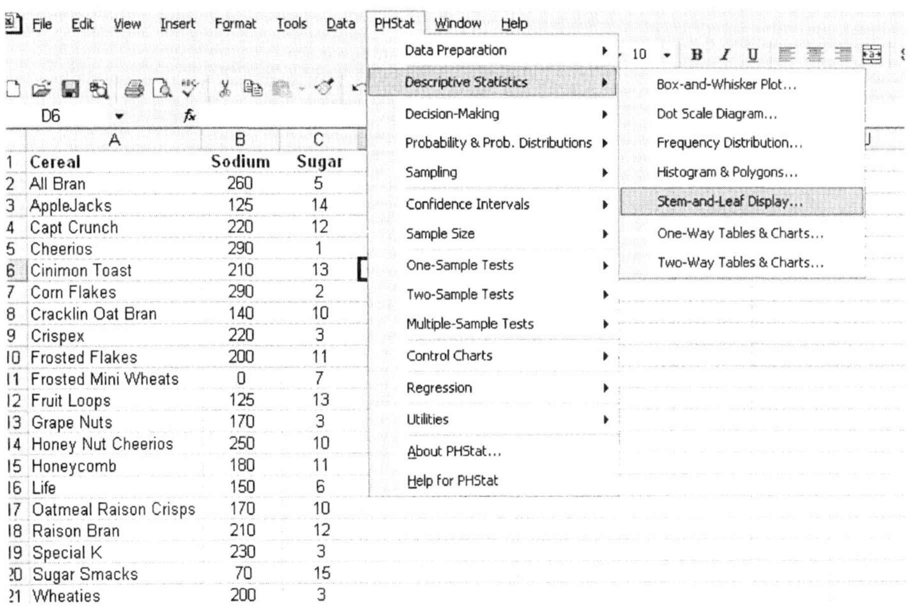

	A	B	C
1	Cereal	Sodium	Sugar
2	All Bran	260	5
3	AppleJacks	125	14
4	Capt Crunch	220	12
5	Cheerios	290	1
6	Cinimon Toast	210	13
7	Corn Flakes	290	2
8	Cracklin Oat Bran	140	10
9	Crispex	220	3
10	Frosted Flakes	200	11
11	Frosted Mini Wheats	0	7
12	Fruit Loops	125	13
13	Grape Nuts	170	3
14	Honey Nut Cheerios	250	10
15	Honeycomb	180	11
16	Life	150	6
17	Oatmeal Raison Crisps	170	10
18	Raison Bran	210	12
19	Special K	230	3
20	Sugar Smacks	70	15
21	Wheaties	200	3

Enter: Variable Cell Range: **B1-B21**
Click: **First Cell contains label**
Click: **Autocalculate stem unit**
Click: **Summary Statistics**
Click: **OK**

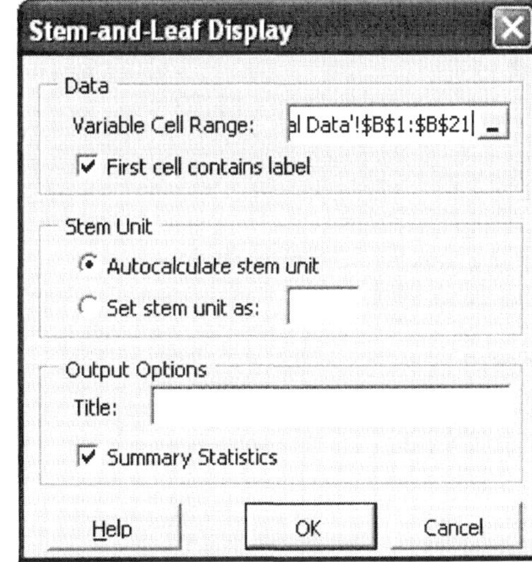

The results are shown below.

A	B	C	D	E
Stem-and-Leaf Display			Stem unit	10
			0	0
Statistics			1	
Sample Size	20		2	
Mean	185.5		3	
Median	200		4	
Std. Deviation	71.24642		5	
Minimum	0		6	
Maximum	290		7	0
			8	
			9	
			10	
			11	
			12	5 5
			13	
			14	0
			15	0
			16	
			17	0 0
			18	0
			19	
			20	0 0
			21	0 0
			22	0 0
			23	0
			24	
			25	0
			26	0
			27	
			28	
			29	0 0

Chapter 2 Exercise 2.15 eBay Prices
Stem and Leaf Plot

Choose **PHStat > Descriptive Statistics > Stem-and-Leaf Display**
Enter: **A1 : A21**
Click **Autocalculate**
Click **Summary Statistics**
OK

	A	B	C	D	E
1				eBay Prices	
2					
3				Stem unit	10
4					
5	Statistics			17	8
6	Sample Size	25		18	
7	Mean	232.16		19	9
8	Median	240		20	0
9	Std. Deviation	19.88022		21	0 0
10	Minimum	178		22	5 5 5 5 8
11	Maximum	255		23	2 5
12				24	0 0 0 5 6 6 6 9
13				25	0 0 0 5 5
14					
15					

Chapter 2 Example 7 Exploring the Health Value of Cereals
Constructing a Histogram

Open the **cereal** worksheet from the Excel folder of the data disk.

To do a histogram, you must use the Data Analysis Add-In. If Data Analysis does not show on the Tools menu:

Choose: **Tools > Add-Ins**
Select: **Analysis ToolPak**
Analysis ToolPak-VBA

To complete a Histogram, a Bin Range should be determined and entered into two columns of the worksheet. This is not required, though. If you omit the bin range, Excel creates a set of evenly distributed bins between the data's minimum and maximum values.

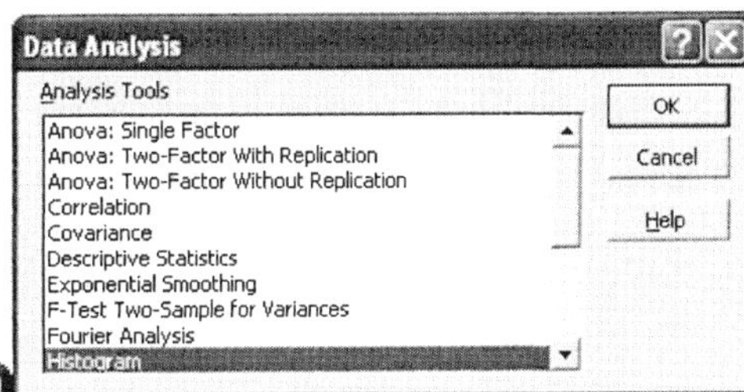

Choose: **Tools > Data Analysis > Histogram**
Click: **OK**
Enter: Input Range: **select appropriate cells**
Bin Range:
Select: **Chart Output**

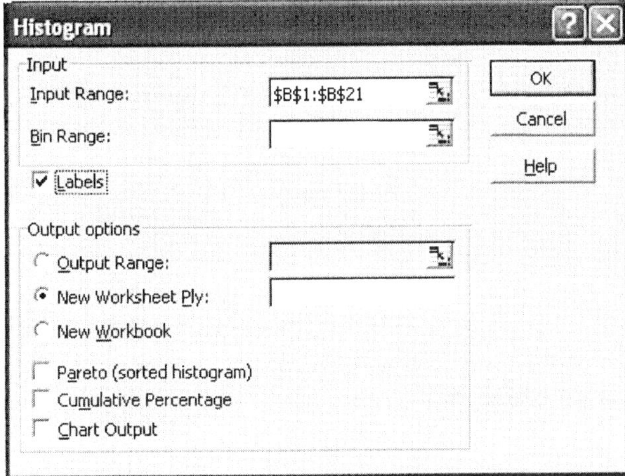

This gives you the following chart:

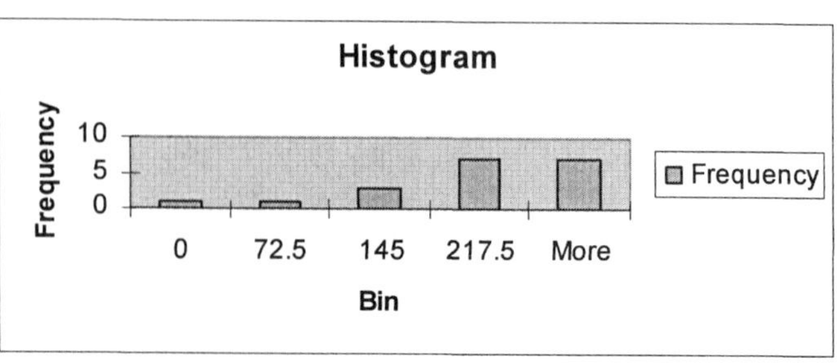

To edit the chart
Click : Anywhere within the chart on whatever it is you wish to edit

Enter an appropriate title, remove the legend, rename horizontal axis

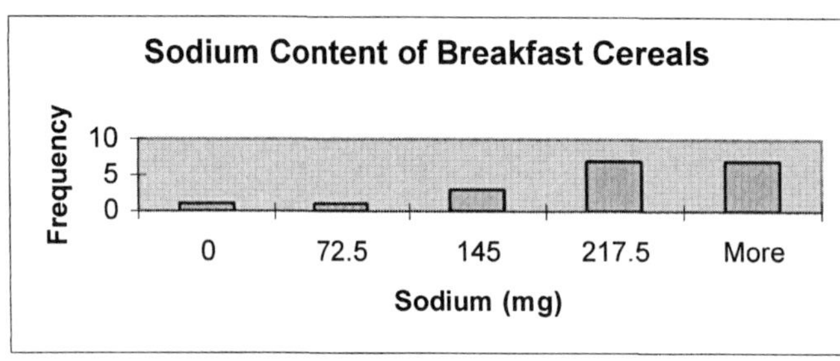

To remove gaps between bars

Right Click : **Any bar on graph**
Choose: **Format Data Series > Options**
Enter: Gap width: **0**

Click: **OK**

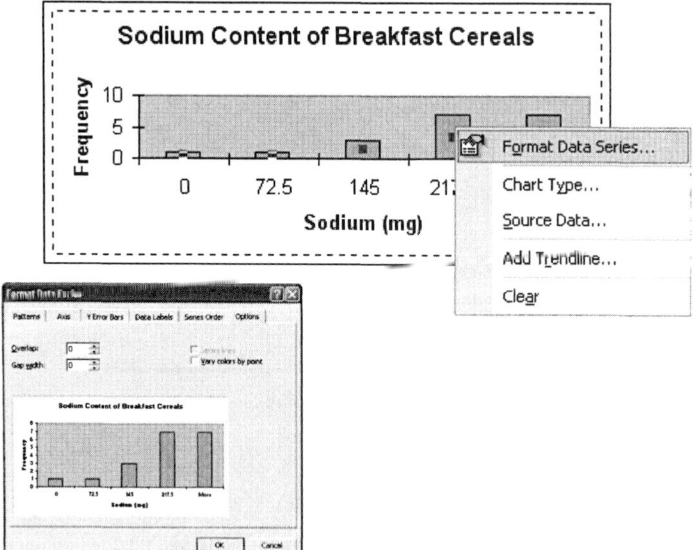

This gives you the finished chart.

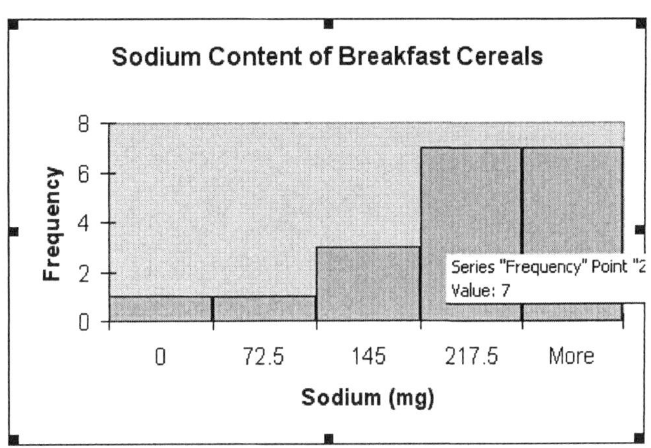

Chapter 2 Exercise 2.20 Sugar plots
Dot Plot , Stem-and-Leaf Plot and Histogram

Open the **cereal** worksheet from the Excel folder of the text data disk.

Choose **PHStat > Descriptive Statistics > Dot Scale Diagram**
Enter the data by selecting column **E**.
Enter Title: **Sugar (mg)**
OK
The chart is shown below.

Summary	
Mean	8200.0000
Median	10000.0000
1st Quartile	3000.0000
3rd Quartile	12000.0000
Std. Deviation	4560.7017

Sugar (mg)

-10000 0 10000 20000 30000

Values
Mean
Median
1st Quartile
3rd Quartile
+/- 1 Std. Dev.
+/- 2 Std. Dev.
+/- 3 Std. Dev.

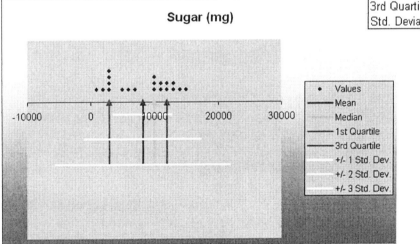

To construct the stem-and-leaf plot:
Choose **PHStat > Descriptive Statistics > Stem-and-Leaf Display**
Enter: Variable Cell Range: **E1-E21**

Click **First Cell contains label**
Click **Autocalculate**
Enter title: **Sugar (mg)**
Click **Summary Statistics**
OK

	A	B	C	D	E
1				Sugar(mg)	
2				Stem unit 1000	
3	Statistics			1	0
4	Sample Size	20		2	0
5	Mean	8200		3	0 0 0 0
6	Median	10000		4	
7	Std. Deviation	4560.702		5	0
8	Minimum	1000		6	0
9	Maximum	15000		7	0
10				8	
11				9	
12				10	0 0 0
13				11	0 0
14				12	0 0
15				13	0 0
16				14	0
17				15	0

To construct the histogram:

Choose: **Tools > Data Analysis > Histogram > OK**
Enter: Input Range: **E1 – E21**
Bin Range: **select cells - optional**
elect: **Chart Output**

Edit the chart

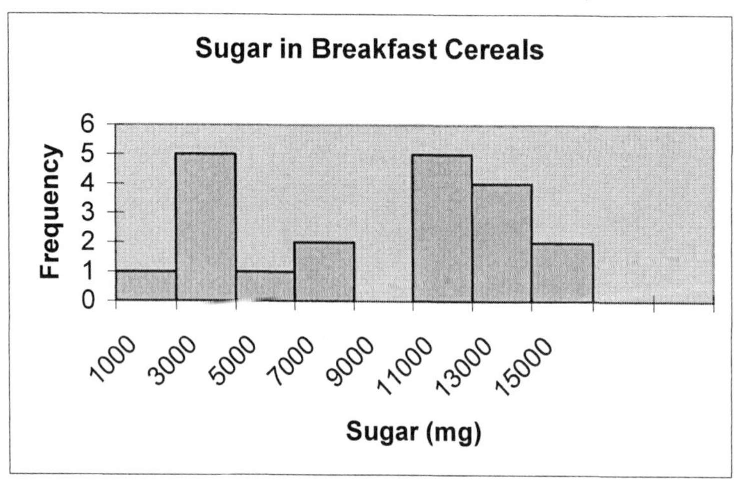

Chapter 2 Exercise 103a Temperatures in Central Park
Constructing a Histogram

Open the worksheet **central_park_yearly_temps** from the Excel folder of the data disk.

Choose: **Tools > Data Analysis > Histogram > OK**
Enter: Input Range: **(select cells)**
Bin Range: **(select cells - optional)**
Select: **Chart Output**

To edit the chart
Click: Anywhere within the chart

To remove gaps between bars
Right Click : **Any bar on graph**
Choose: **Format Data Series > Options**
Enter: Gap width: **0**
Click: **OK**

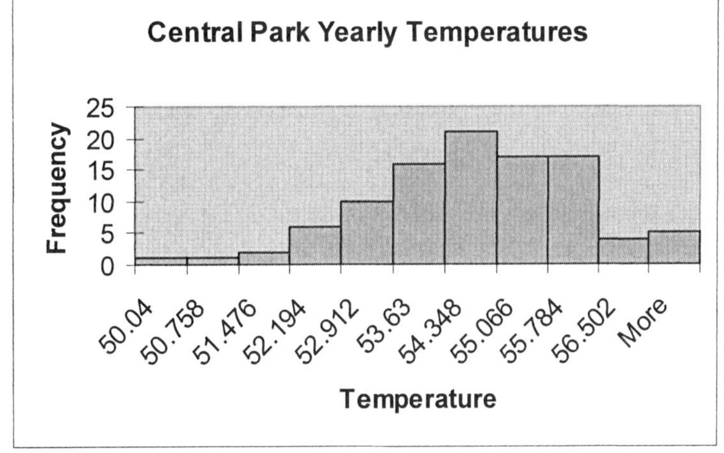

Chapter 2 Example 9 Is There a Trend Toward Warming in New York City? Constructing a Time Plot

Open the worksheet **central_park_yearly_temps** from the Excel folder of the data disk. Excel does not do Time Plots, but we can simulate one using the Chart Wizard and XY (Scatter) with data points connected by lines. Select the columns containing the data.

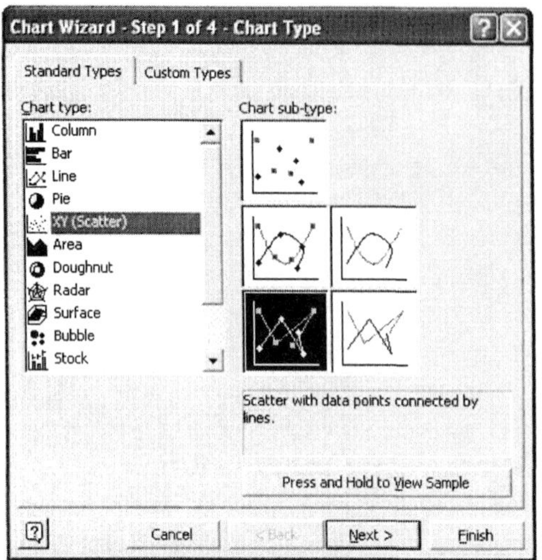

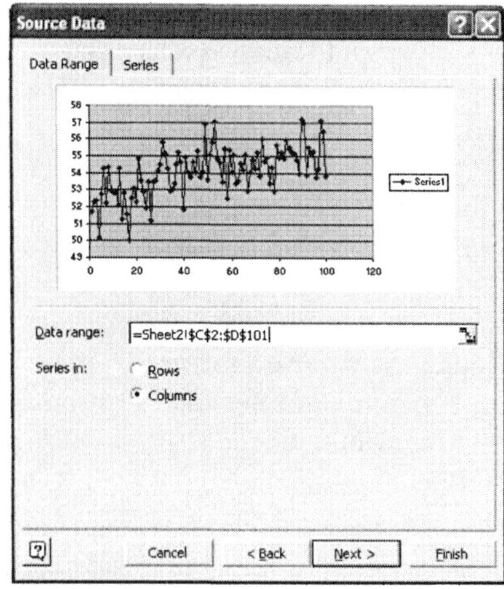

Add titles as appropriate. Place chart in new sheet.

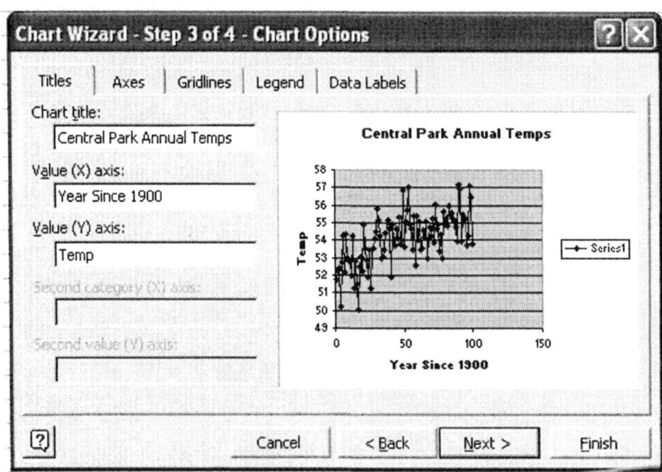

Format the graph, choosing proper axis scale and adding the trendline. You can do this by right-clicking on either axis to format axis or on a point to add the trendline.

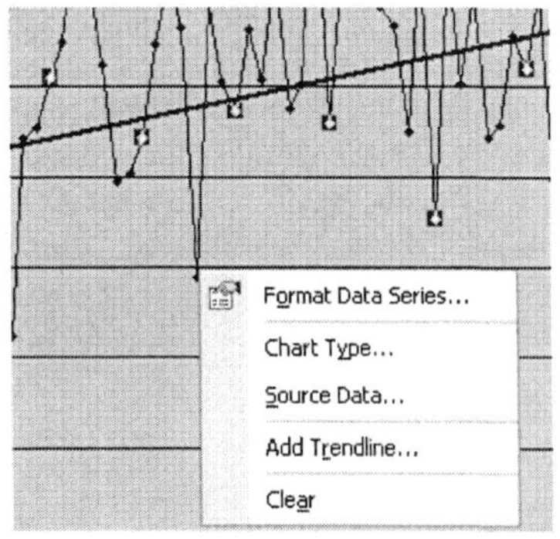

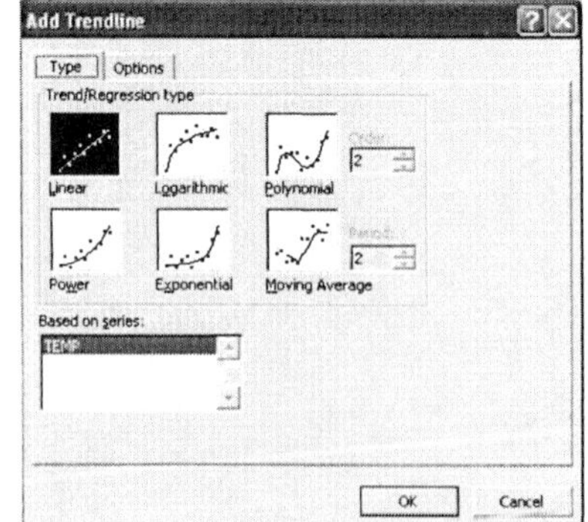

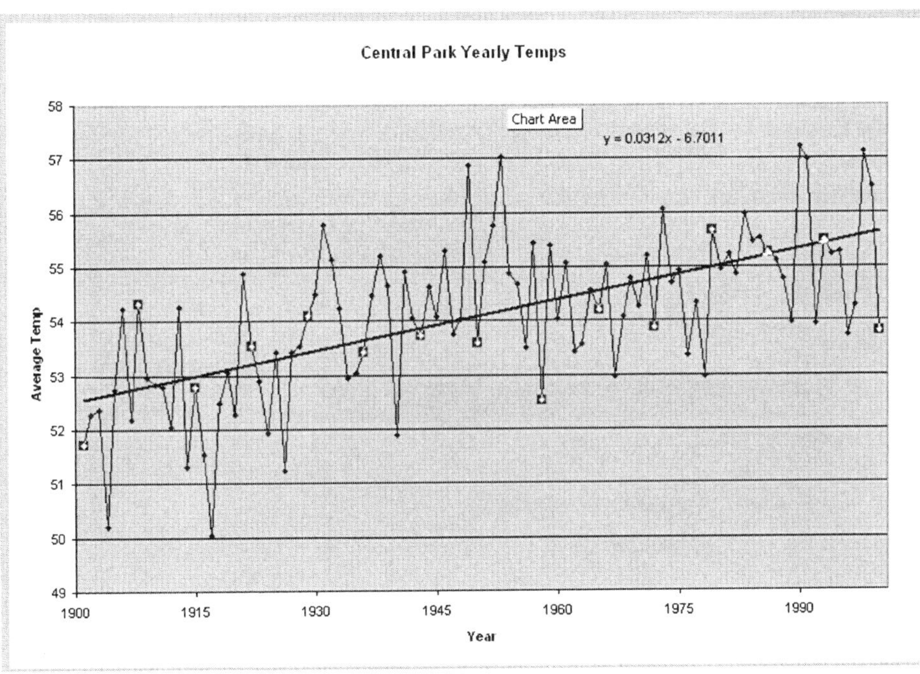

Chapter 2 Exercise 2.27 Warming in Newnan, Georgia? Constructing a Time Plot

Open the worksheet **newnan_ga_temps** from the Excel folder of the data disk. Using the Chart Wizard graph a scatterplot with connected lines. Select the columns containing the data. Add titles as appropriate. Place chart in new sheet. Format the graph, choosing proper axis scale and adding the trendline.

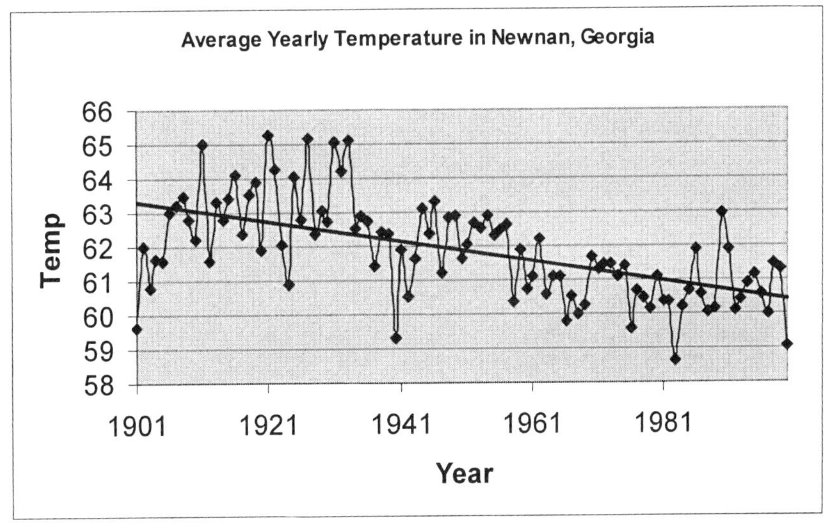

Chapter 2 Example 10 What is the Center of the Cereal Sodium Data? Determining Mean and Median

Open the Excel data file **cereal** from the Excel folder of the data disk.

Calculate the mean and median using the following commands.

Choose:
Tools > Data Analysis > Descriptive Statistics > OK

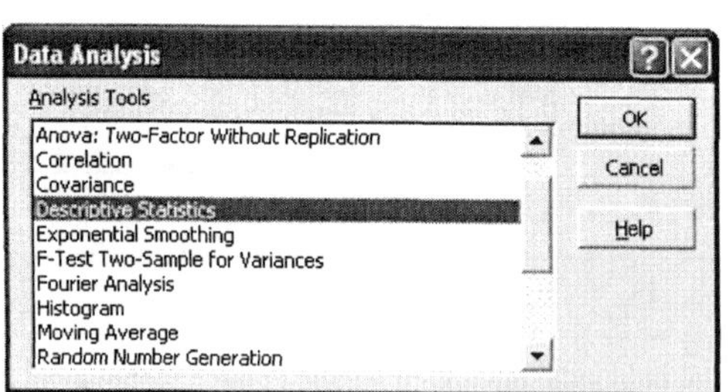

Enter: Input Range: **B1:B21**
Click: **Labels in First Row**
Click: **Summary Statistics > OK**

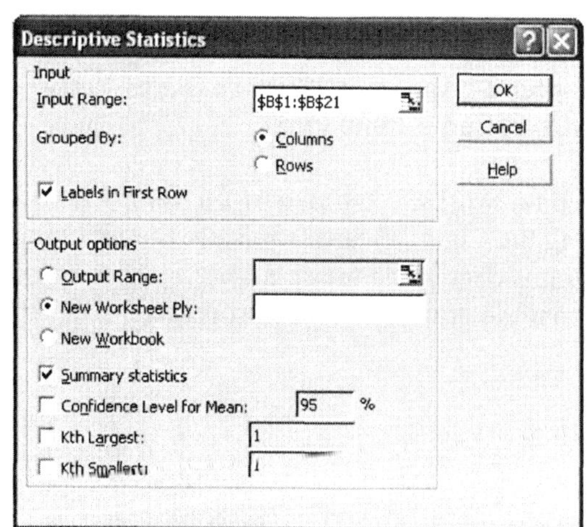

These commands generate the following output in a new worksheet ply.

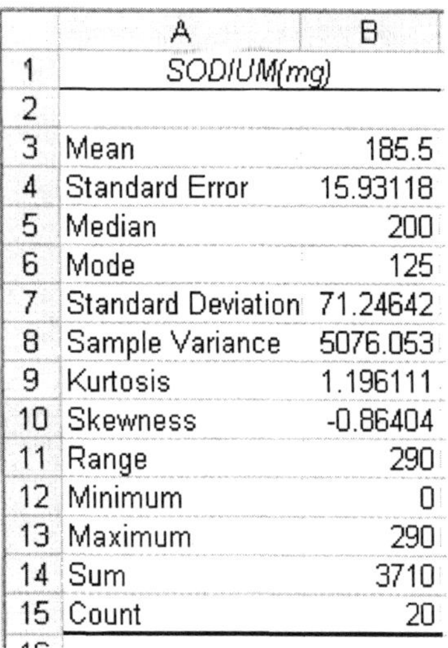

	A	B
1	SODIUM(mg)	
2		
3	Mean	185.5
4	Standard Error	15.93118
5	Median	200
6	Mode	125
7	Standard Deviation	71.24642
8	Sample Variance	5076.053
9	Kurtosis	1.196111
10	Skewness	-0.86404
11	Range	290
12	Minimum	0
13	Maximum	290
14	Sum	3710
15	Count	20

We can also find values by entering a formula and storing the result in a cell. For example, the midrange = (high + low)/2. To do this in Excel we would do the following:

Select: **a particular cell**
In the formula bar type in the expression: **(MAX(B2:B21) + MIN(B2:B21)) /2**
then press enter

Microsoft Excel - section2_3example10

File Edit View Insert Format Tools Data PHStat Window Help

A25 fx =(MAX(B2:B21)+MIN(B2:B21))/2

	A	B	C	D	E	F
1	CEREAL	SODIUM(mg)	SUGAR(g)	CODE		

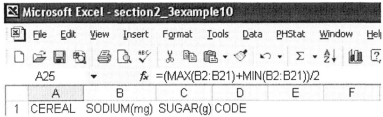

The result is placed within the cell you selected:

	A	B	C	D
1	CEREAL	SODIUM(mg)	SUGAR(g)	CODE
2	FMiniWhe:	0	7	A
3	ABran	260	5	A
4	AJacks	125	14	C
5	CCrunch	220	12	C
6	Cheeros	290	1	C
7	CTCrunch	210	13	C
8	CFlakes	290	2	A
9	RBran	210	12	A
10	COakBran	140	10	A
11	Crispix	220	3	A
12	FFlakes	200	11	C
13	FLoops	125	13	C
14	GNuts	170	3	A
15	HNCheerio	250	10	C
16	Honeycom	180	11	C
17	Life	150	6	A
18	Oatmeal R	170	10	A
19	Smacks	70	15	C
20	SpecialK	230	3	A
21	Wheaties	200	3	C
22				
23				
24	midrange			
25	145			
26				

Chapter 2 Exercise 2.29 More On CO_2 Emissions
Mean and Median

Enter the data (country and million metric tons of carbon equivalent) into the worksheet.

	A	B
1	Country	CO2 emissions
2	U.S.	1490
3	China	914
4	Russia	391
5	Japan	316
6	India	280
7	Germany	227
8	U.K.	142
9		

Calculate the mean and median using the following commands.
Choose: **Tools > Data Analysis > Descriptive Statistics > OK**
Enter: Input Range: **B1:B8**
Click: **Labels in First Row**
Click: **Summary Statistics > OK**

	A	B
1	CO2 emissions	
2		
3	Mean	537.1429
4	Standard Error	185.0784
5	Median	316
6	Mode	#N/A
7	Standard Deviation	489.6715
8	Sample Variance	239778.1
9	Kurtosis	1.78642
10	Skewness	1.58226
11	Range	1348
12	Minimum	142
13	Maximum	1490
14	Sum	3760
15	Count	7

Chapter 2 Example 15 Describing Female College Student Heights Empirical Rule

When we open the data file **heights** on the text CD, we find that the data includes male (indicated by 0) and females(indicated by 1). In this example we are only concerned with the heights of the females.

	A	B	C
1	HEIGHT		GENDER
2	70		0
3	75		0
4	67		0
5	67		1
6	73		0
7	65		0
8	73		0
9	70		0
10	65		1
11	73		0
12	71		0
13	65.5		1

To generate the histogram of female student height data only, we must specify that the data to be used should be taken from only those rows where the gender is a 1. The histogram tool under **Tools > Data Analysis > Histogram** does not provide for this. Instead, we will use the PHStat Add-in.

First, we must create the Bin Range, and bin midpoints.

D	E
Bin Cell Range	BIN midpoint
54.5	55
55.5	56
56.5	57
57.5	58
58.5	59
59.5	60
60.5	61
61.5	62
62.5	63
63.5	64
64.5	65
65.5	66
66.5	67
67.5	68
68.5	69
69.5	70
70.5	71
71.5	72
72.5	73
73.5	74
74.5	75
75.5	76
76.5	77
77.5	

To create the histogram of only the female heights:

Click: **PHStat > Descriptive Statistics > Histogram & Polygons**

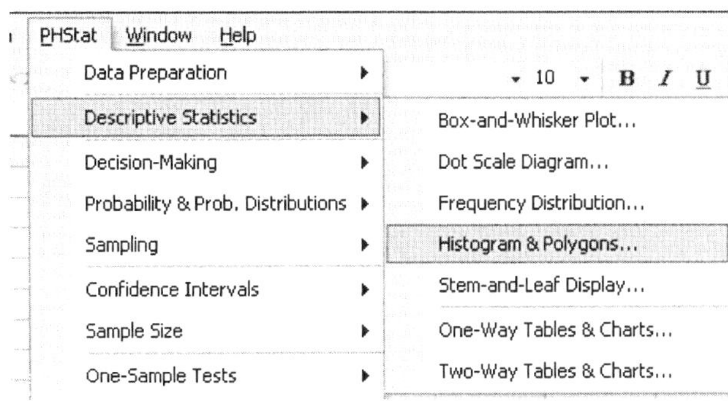

Complete the dialogue box as follows and click OK.

Histogram & Polygons
Data
Variable Cell Range: A1:A379
Bins Cell Range: Sheet1!$D$1:$D$25
Midpoints Cell Range: Sheet1!$E$1:$E$24
First cell in each range contains label
Input Options
Single Group Variable
Multiple Groups - Unstacked
Multiple Groups - Stacked
Grouping Variable Cell Range: $C$1:$C$379
Output Options
Title: Height
Histogram
Frequency Polygon
Percentage Polygon
Cumulative Percentage Polygon (Ogive)
Help OK Cancel

New worksheets are added, containing frequencies, percentages, and histograms. After some modification to the histogram, the sheet for the females appears as follows:

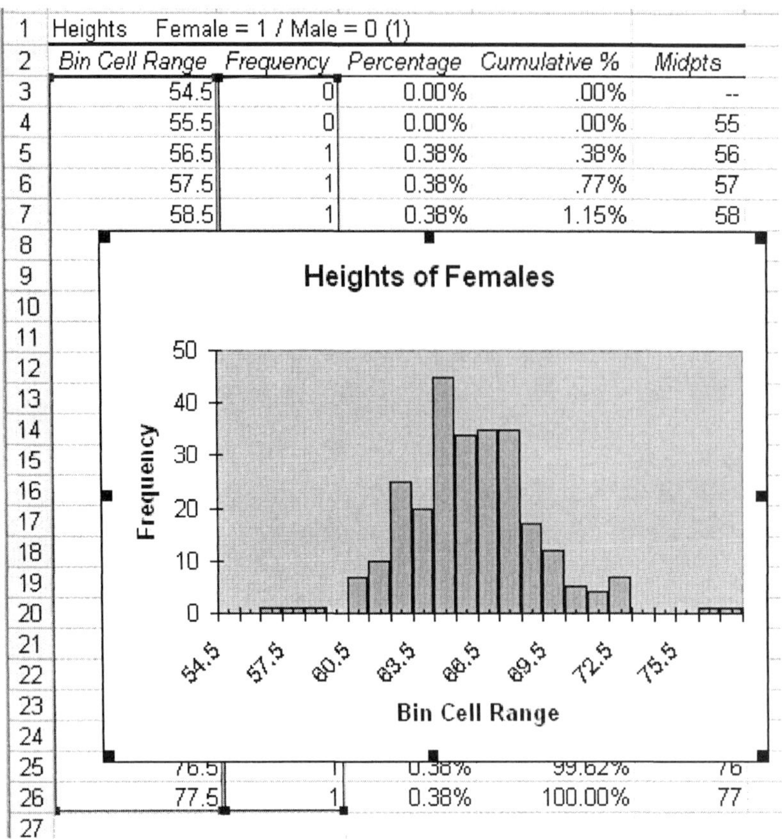

To generate the descriptive statistics for only the females, first, we will sort the data, by gender, and then by height:

Select columns A through C
Click: **Data > Sort**
Enter: Sort by **select column C** from the dropdown menu
Then by **select column A** from the dropdown menu
Click: My list has **Header row > OK**

Now, we can perform our analysis on the female heights:

Choose: **Tools > Data Analysis > Descriptive Statistics > OK**
Enter: Input Range: **A1:A262**
Click: **Labels in First Row**
Click: **Summary Statistics > OK**

	A	B
1	1	
2		
3	Mean	65.28352
4	Standard Error	0.182777
5	Median	65
6	Mode	64
7	Standard Deviation	2.952847
8	Sample Variance	8.719304
9	Kurtosis	1.330394
10	Skewness	0.40416
11	Range	21
12	Minimum	56
13	Maximum	77
14	Sum	17039
15	Count	261

	A	B	C
31	62		1
32	62		1
33	62		1
34	62		1
35	62		1
36	62		1
37	62		1
38	62		1
39	62		1
40	62		1
41	62		1
42	62		1
43	62		1
44	62		1
45	62		1
46	62.5		1
47	63		1
48	63		1
49	63		1
50	63		1
51	63		1
52	63		1
53	63		1
54	63		1
55	63		1

Now, with the lists sorted, counting the number of observations of female heights within each range is easier.

The heights between 62.3 and 68.3 lie from row 46 to row 232, which is 187 values. 72% of all female heights.

The heights between 59.3 and 71.3 lie from row 5 to row 252, which is 248 values. 95% of all female heights.

The heights between 56.3 and 74.3 lie from row 3 to row 260, which is 378 values. 99% of all female heights.

Chapter 2 Exercise 2.55 EU Data File
Empirical Rule

Open the **european_union_unemployment** data file on the text CD.

Generating the histogram, we see that the distribution is skewed, and not bell shaped.

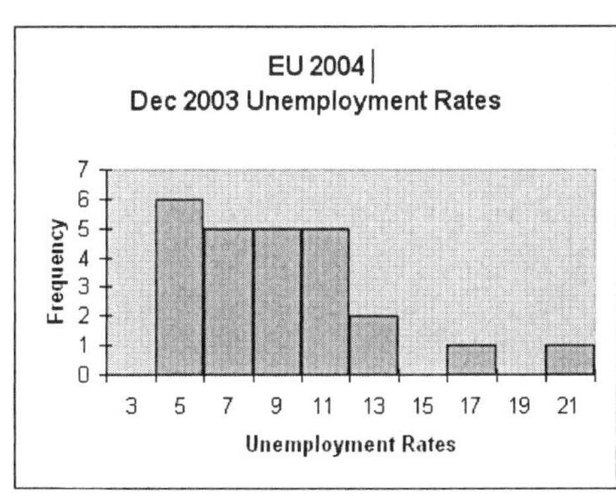

Generating basic statistics

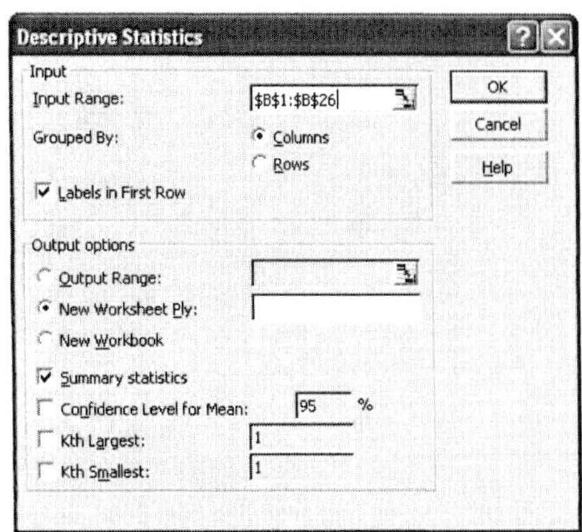

	A	B
1	*unemployment*	
2		
3	Mean	8.288
4	Standard Error	0.738948
5	Median	8.3
6	Mode	6
7	Standard Deviation	3.69474
8	Sample Variance	13.6511
9	Kurtosis	2.364485
10	Skewness	1.38482
11	Range	15.2
12	Minimum	3.9
13	Maximum	19.1
14	Sum	207.2
15	Count	25
16		

To determine the number of data values that fall in the ranges
$\bar{x} \pm 1 \cdot s = (4.593, 11.983), \bar{x} \pm 2 \cdot s = (.898, 15.678), \bar{x} \pm 3 \cdot s = (-2.797, 19.373)$
sort the unemployment rates

1	country	unemployment
2	Luxembourg	3.9
3	Netherlands	4.4
4	Austria	4.5
5	Ireland	4.6
6	Cyprus	4.7
7	United Kingdom	4.8
8	Hungary	5.9
9	Denmark	6
10	Sweden	6
11	Slovenia	6.4
12	Portugal	6.7
13	Czech Republic	8.1
14	Belgium	8.3
15	Italy	8.5
16	Malta	8.6
17	Finland	8.9
18	Germany	9.2
19	Greece	9.3
20	France	9.5
21	Estonia	9.6
22	Latvia	10.5
23	Spain	11.2
24	Lithuania	11.9
25	Slovakia	16.6
26	Poland	19.1

We determine that there are 20, 23, and 25 data values within the three ranges, respectively, which is 80%, 92 %, and 100%. Notice that theses percentages don't even begin to match the 68%, 95% and 99.7% of the Empirical Rule. This provides further evidence that the distribution of unemployment rates is not bell shaped.

Chapter 2 Example 16 What Are the Quartiles for the Cereal Sodium Data? Quartiles

Open the **cereal** data file on the text CD.

In this example, we wish to determine the quartiles for the sodium values.
Choose: **PHStats > Descriptive Statistics > Box-and-Whisker Plot**

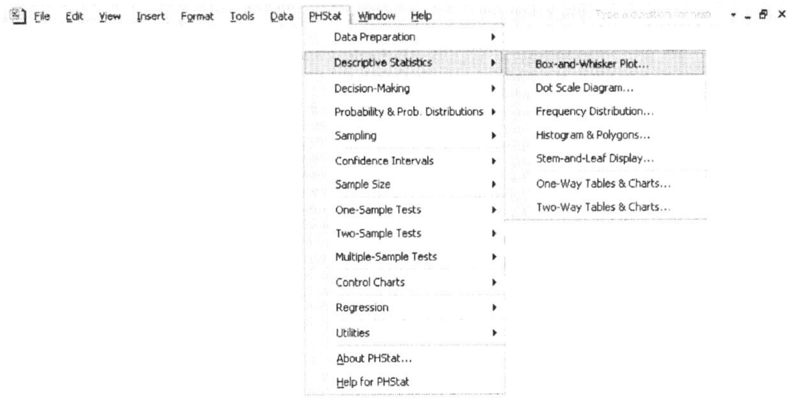

Click: **Five-Number Summary**
Make selections as indicated and click **OK**

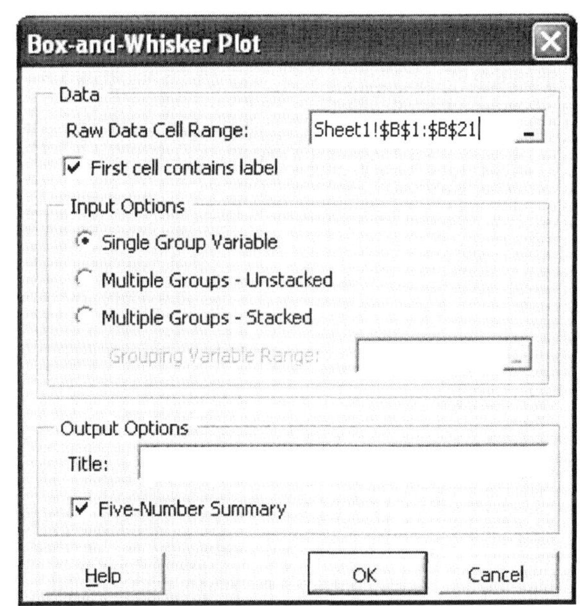

A new worksheet ply named FiveNumbers is created, containing the needed information.

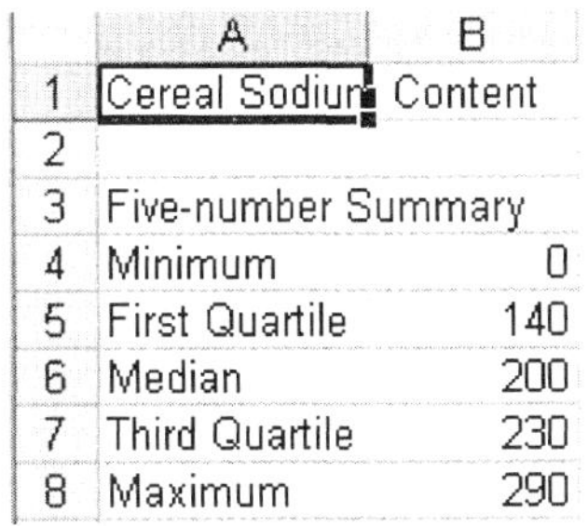

	A	B
1	Cereal Sodium Content	
2		
3	Five-number Summary	
4	Minimum	0
5	First Quartile	140
6	Median	200
7	Third Quartile	230
8	Maximum	290

Note that the results differ from those in the text. Not all statistical programs use the same method for determining the quartiles. You should explore how and why PHStat generates Q1=140 and Q3 = 230.

More general percentiles are very easy to generate with Excel. Let's say we're interested in the 30th, 65th and 98th percentiles.

Select a cell in the worksheet to store the response. When that cell is activated, you will type in the formula bar **PERCENTILE(array,k)** where array is the array or range of data that defines relative standing and k is the percentile value in the range 0..1, inclusive.

=PERCENTILE(B2:B21,0.3)

B	C	D
SODIUM(n	SUGAR(g)	CODE
0	7	A
260	5	A
125	14	C
220	12	C
290	1	C
210	13	C
290	2	A
210	12	A
140	10	A
220	3	A
200	11	C
125	13	C
170	3	A
250	10	C
180	11	C
150	6	A
170	10	A
70	15	C
230	3	A
200	3	C

30th percentile
164

65th percentile
213.5

98th percentile
290

Chapter 2 Exercise 2.58 European Unemployment Quartiles

Enter the data into columns A and B of the worksheet. Be sure to label the columns.

	A	B
1	Country	Unemployment rate
2	Belgium	8.3
3	Denmark	6
4	Germany	9.2
5	Greece	9.3
6	Spain	11.2
7	France	8.5
8	Portugal	6.7
9	Netherlands	4.4
10	Luxembourg	3.9
11	Ireland	4.6
12	Italy	8.5
13	Finland	8.9
14	Austria	4.5
15	Sweden	6
16	U.K.	4.8

Click **PHStats > Descriptive Statistics>Box-and-Whisker Plot**
Enter: Raw Data Cell Range: **select appropriate cells**
Enter: Title: **an appropriate title**
Click: **Five-Number Summary > OK**

	A	B
1	EU Unemployment Rates	
2		
3	Five-number Summary	
4	Minimum	3.9
5	First Quartile	4.6
6	Median	6.7
7	Third Quartile	9.2
8	Maximum	11.2

Chapter 2 Example 17 Box Plot for the Breakfast Cereal Sodium Data
Box Plots

Open worksheet **cereal** from the Excel folder of the data disk. To construct a box plot (Box-and-Whisker Plot) for the breakfast cereal sodium data, use the following commands:

Click: **PHStats > Descriptive Statistics > Box-and-Whisker Plot**

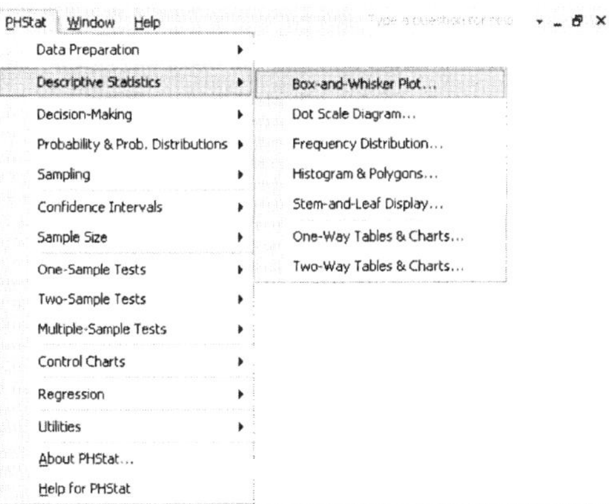

In the Box-and-Whisker Plot dialogue box, enter the **Raw Data Cell Range**, the **Title** and click **OK**.

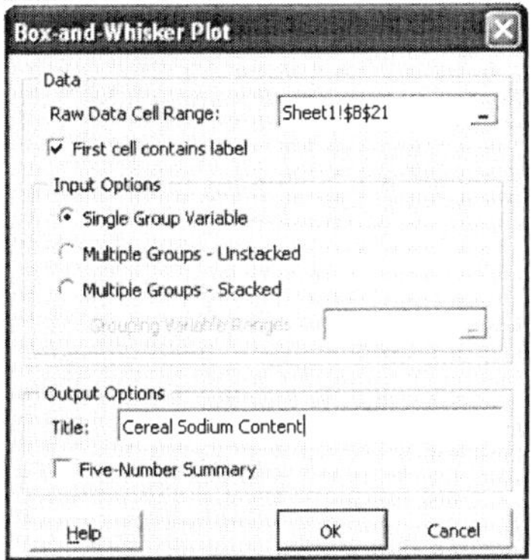

A new worksheet named BoxWhiskerPlot is created.

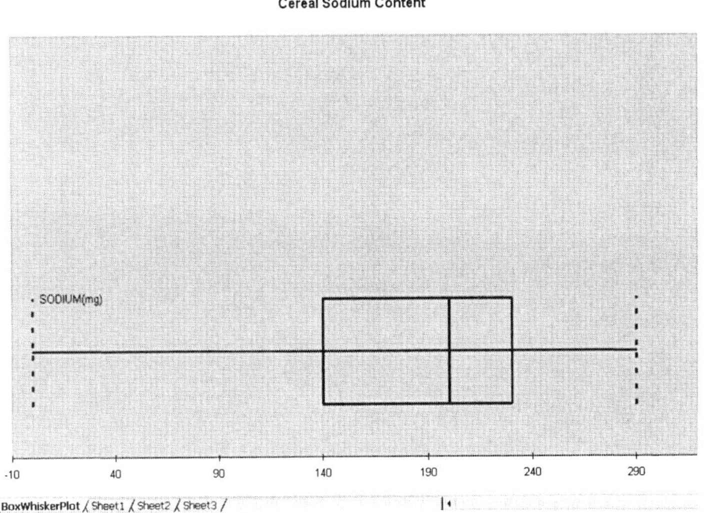

This macro does not highlight potential outliers with an asterisk, as MINITAB and the TI83/84 will do. Examination of whether values should be considered outliers would have to be carried out by you, using the 1.5 X IQR criterion.

Chapter 2 Exercise 2.69 European Union Unemployment Rates
Box Plots

Enter the data from exercise 2.58 into columns A and B.
Choose: **PHStats > Descriptive Statistics > Box-and-Whisker Plot**

In the Box-and-Whisker Plot dialogue box
Enter: Raw Data Cell Range: **appropriate cells**
Title: **an appropriate title**
Click: **OK**.

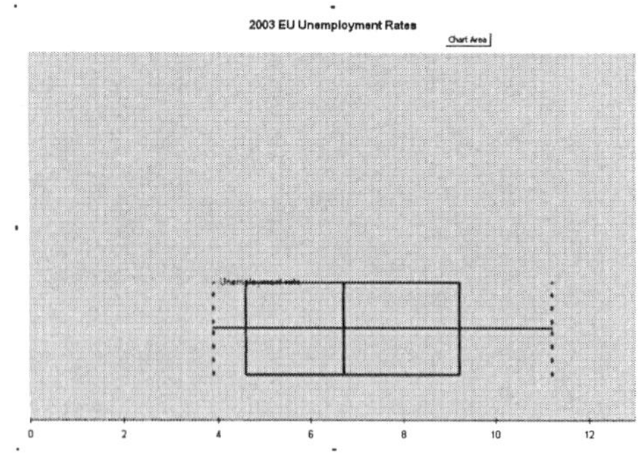

Chapter 2 Exercise 2.73 Florida Students Again
Box Plots

Open worksheet **fla_student_survey** from the Excel folder of the data disk.

Choose: **PHStats > Descriptive Statistics > Box-and-Whisker Plot**

In the Box-and-Whisker Plot dialogue box
Enter: Raw Data Cell Range: **appropriate cells**
Title: **an appropriate title**
Click: **OK**

TV Viewing (hours per week)

TV

-10 0 10 20 30 40

BoxWhiskerPlot / Sheet1 / Sheet2 / Sheet3 /

Chapter 2 Exercise 2.103b Temperatures in Central Park
Box Plots

Open worksheet **central_park_yearly_temps** from the Excel folder of the data disk.

Choose: **PHStats > Descriptive Statistics > Box-and-Whisker Plot**

In the Box-and-Whisker Plot dialogue box
Enter: Raw Data Cell Range: **appropriate cells**
Title: **an appropriate title**
Click: **OK**

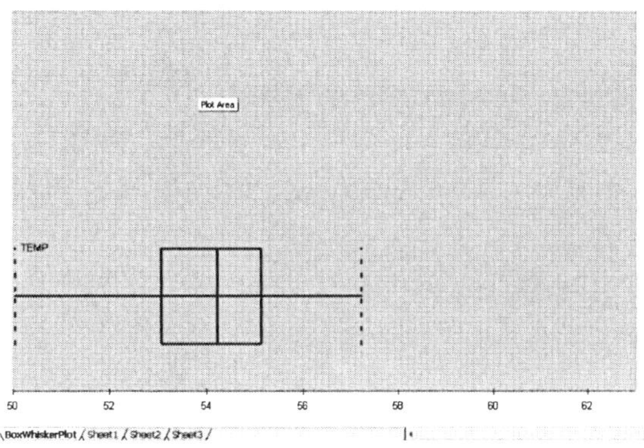

We can compare this to the histogram produced for part (a) of this exercise.

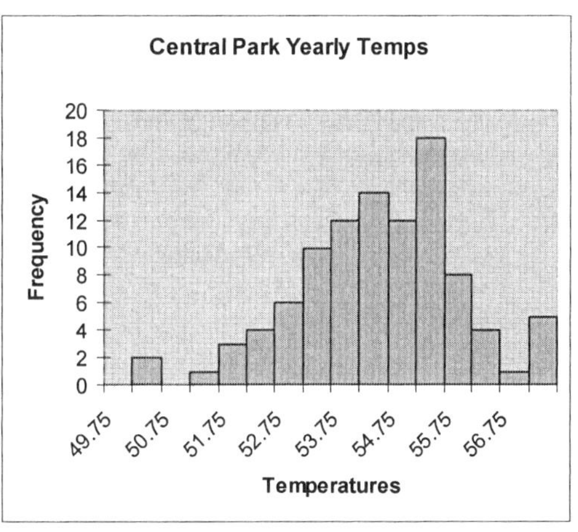

Chapter 2 Exercise 2.113 How Much is Spent on Haircuts?
Side-by-Side Plots

Open worksheet **georgia_student_survey** from the Excel folder of the data disk. For this exercise we wish to compare how much males and females spend on a haircut. First note that gender is indicated in column B as 0 for male and 1 for female. Many of the graphs illustrated in this chapter can be done as side-by-side plots, graphing with a separate plot for the males and the females, using the same scale, so that comparison of the distributions can be made

	A	B	C	D	E
1	Height	Gender	Haircut	Job	Studytime
2	65	1	25	1	7
3	71	0	12	0	2
4	68	1	4	0	4
5	64	1	0	1	3.5
6	64	1	50	0	4.5
7	66	0	10	1	3

Generating histograms with the same scale (by using the same bin ranges)

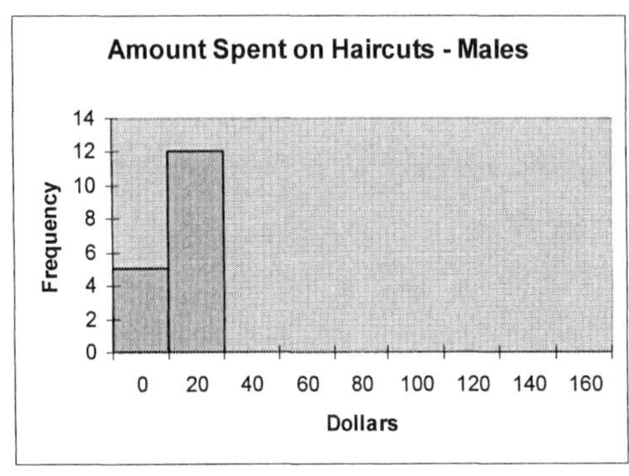

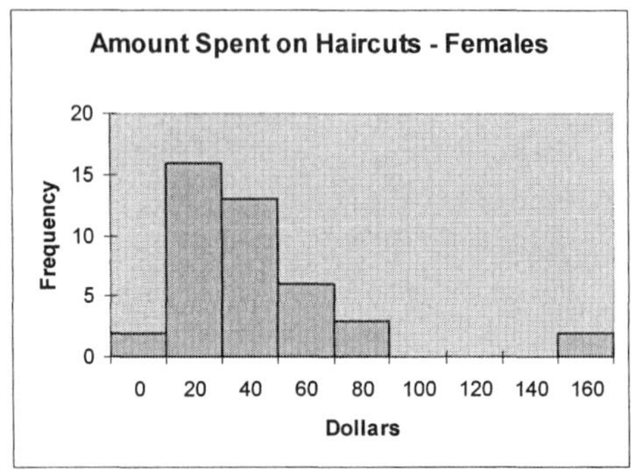

Side-by-side box plots:

Select: **PHStat > Descriptive Statistics > Box-and-Whisker Plot**

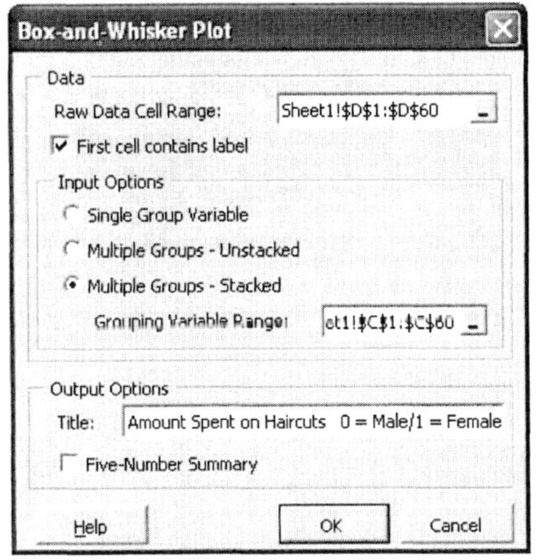

Complete dialogue box as indicated.

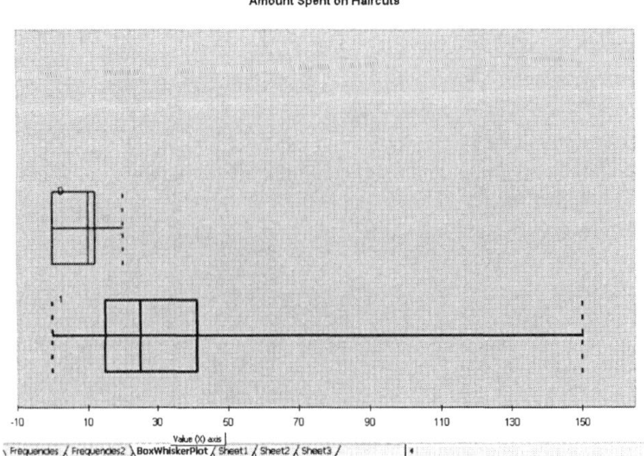

Chapter 3 Example 3 How Can We Compare Pesticide Residues For the Food Types Graphically?
Side-by-Side Bar Chart

Enter the conditional proportions (Table 3.2) into the worksheet.

	A	B	C
1	Food Type	Pesticide Present	Pesticide Not Present
2	Organic	0.23	0.77
3	Conventional	0.73	0.27
4			

To generate a single bar graph that shows side-by-side bars to compare the conditional proportion of pesticide residues in conventionally grown and organic foods:

Choose: **Insert > Chart**
Select: Chart Type: **Column**
Chart sub-type: **Clustered Column**
Click: **Next**
Select: Data range: **Click and drag to select appropriate cells**

Click: **Next**

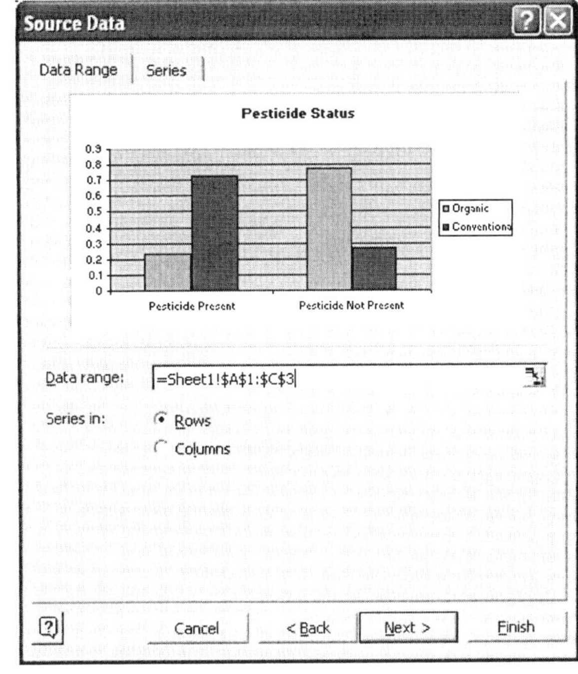

On the Titles tab
Enter: Chart title: **Enter an appropriate title**

Click: **Next**
Click: **Finish**

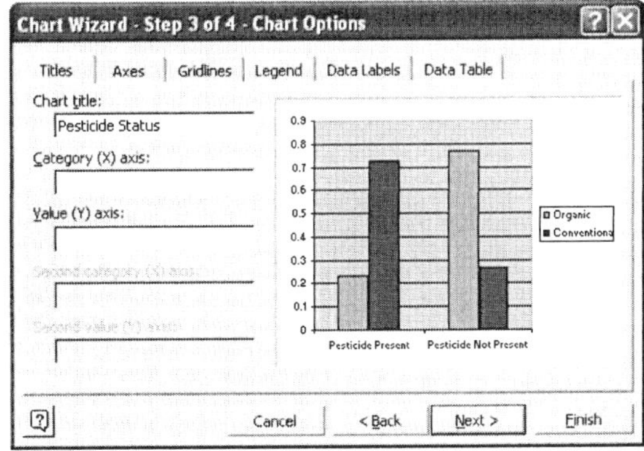

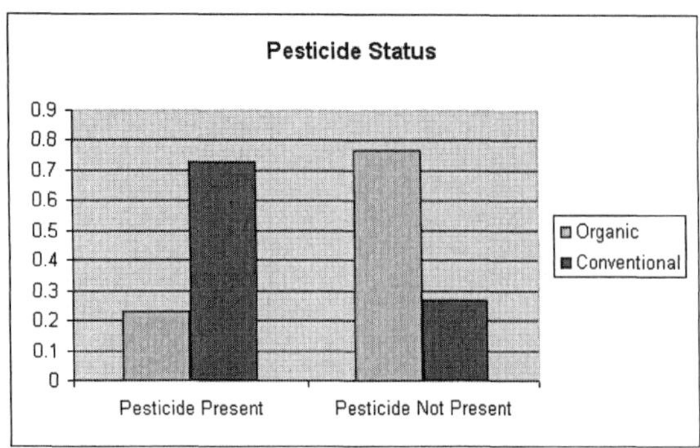

Chapter 3 Exercise 3.4 Religious Activities
Side-by-Side Bar Chart

Enter the data into the worksheet. Since we are comparing unequal numbers of males and females, we need to enter the data as proportions on hours of home religious activity within each category of gender.

	A	B	C	D	E	F
1	Gender	0	1 to 9	10 to 19	20 to 39	40 or more
2	Female	0.29896	0.38773	0.11488	0.13446	0.06397
3	Male	0.43533	0.38328	0.09306	0.06309	0.02524

Choose: **Insert > Chart**
Select: Chart Type: **Column**
Chart sub-type: **Clustered Column**
Click: **Next**
Select: Data range: **Click and drag to select appropriate cells**

Click: **Next**
On the titles tab
Enter: Chart title: **Enter an appropriate title**
Click: **Next**
Click: **Finish**

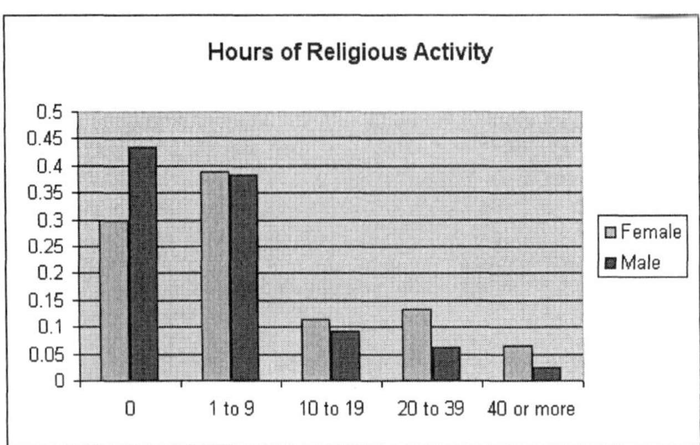

Chapter 3 Example 5 Constructing a Scatterplot for Internet Use and GDP
Constructing a Scatterplot

Open the worksheet **human_development**, which is found in the Excel folder of the data disk.

	A	B	C	D	E	F	G
1	C1-T	INTERNET	GDP	CO2	CELLULAR	FERTILITY	LITERACY
2	Algeria	0.65	6.09	3	0.3	2.8	58.3
3	Argentina	10.08	11.32	3.8	19.3	2.4	96.9
4	Australia	37.14	25.37	18.2	57.4	1.7	100
5	Austria	38.7	26.73	7.6	81.7	1.3	100
6	Belgium	31.04	25.52	10.2	74.7	1.7	100
7	Brazil	4.66	7.36	1.8	16.7	2.2	87.2
8	Canada	46.66	27.13	14.4	36.2	1.5	100

Since we are interested in how Internet use depends on GDP, column C GDP is the x-variable, and column B INTERNET is the y-variable.

To create a scatterplot of the data
Choose: **Insert > Chart**
In the Chart Type dialog box
Select: **XY (Scatter)**
Select: **Scatter**
Click: **Next**

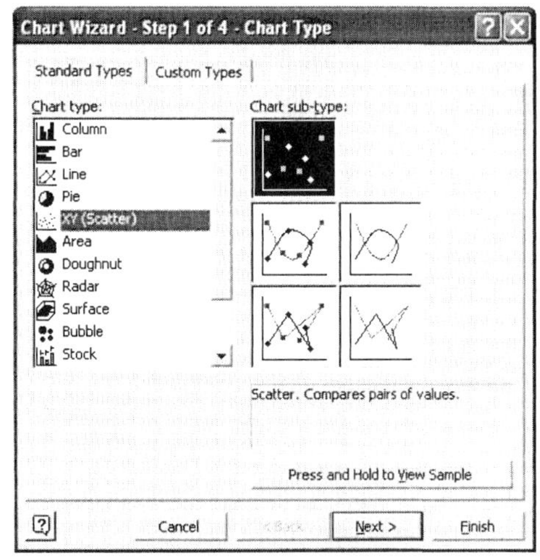

In the Chart Source Data dialog box
Enter: Data Range: **select the appropriate cells**

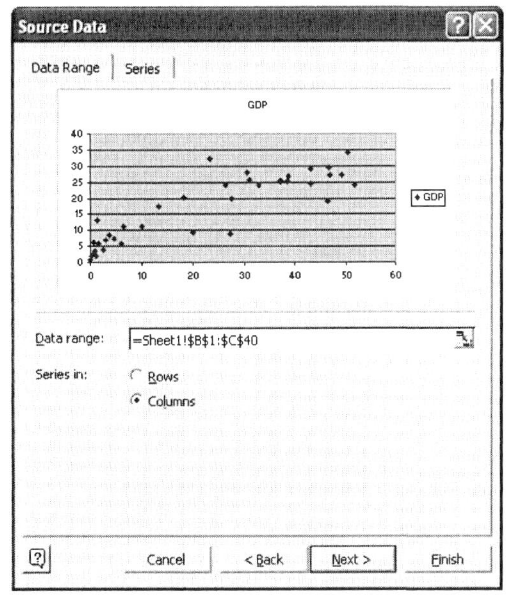

Click: the **Series** tab

Note: by default, Column B Internet is selected as the x and column C GDP is selected as the y

We need to change this to
X Values: **Sheet1!$C$2:$C$40**
Y Values: **Sheet1!$B$2:$B$40**

Click: **Next**

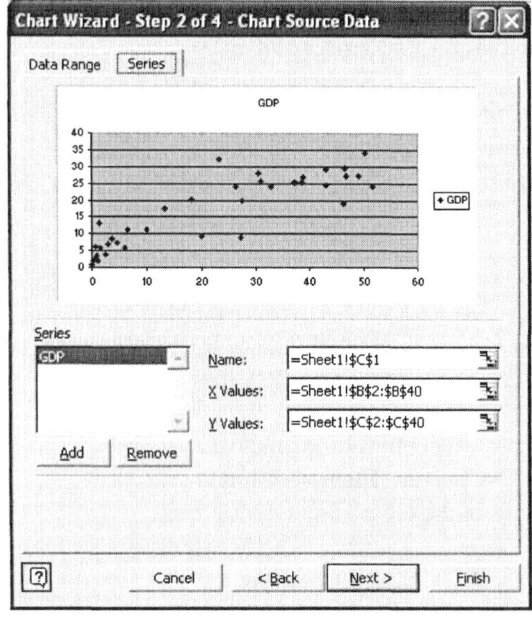

In the Chart Options dialog box
on the Titles tab
Enter: Chart title: **GDP vs Internet Usage**
Value (X) axis: **Gross Domestic Product (thousands of $)**
Value (Y) axis: **% of adults who are internet subscribers**

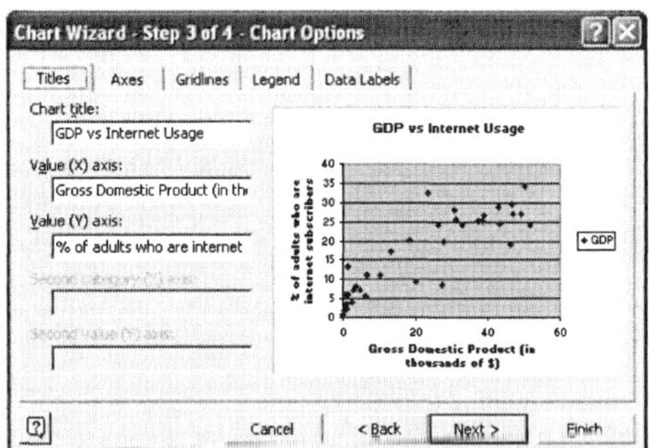

on the Legend tab
Click: **Show Legend**
(so that a checkmark does not appear there)

Click: **Next**

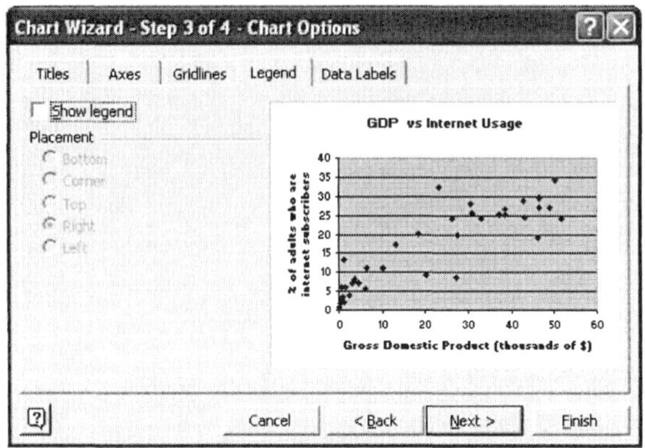

in the Chart Location dialog box
Select: **either of the two options**
Click: **Finish**

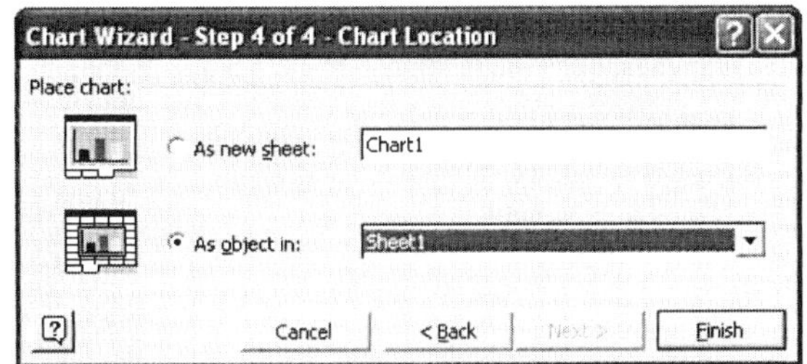

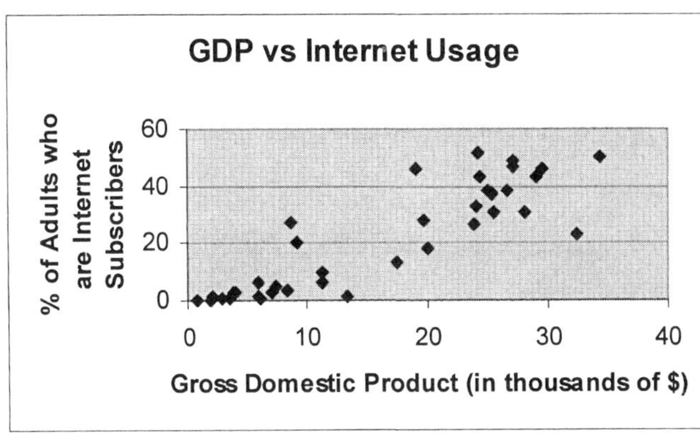

Chapter 3 Example 7 What's the Correlation Between Internet Use and GDP?
Computing Correlation

Open the worksheet **human_development**, which is found in the Excel folder of the data disk.

Since we are interested in how Internet use depends on GDP, column C, GDP, is the x-variable, and Column B, Internet, is the y-variable.

To find the correlation :
Click: **within an empty cell**

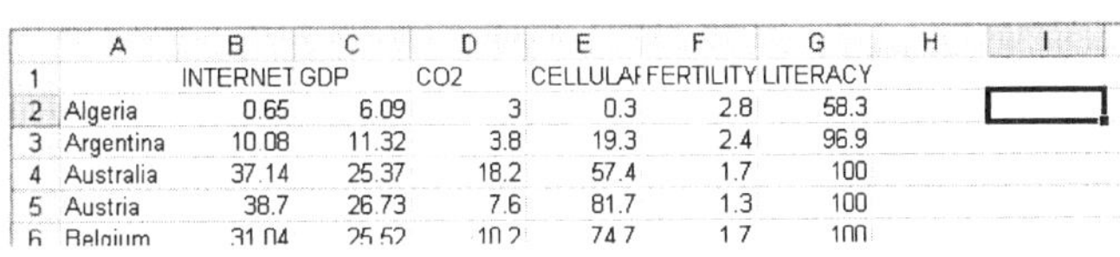

	A	B	C	D	E	F	G	H	I
1		INTERNET	GDP	CO2	CELLULAF	FERTILITY	LITERACY		
2	Algeria	0.65	6.09	3	0.3	2.8	58.3		
3	Argentina	10.08	11.32	3.8	19.3	2.4	96.9		
4	Australia	37.14	25.37	18.2	57.4	1.7	100		
5	Austria	38.7	26.73	7.6	81.7	1.3	100		
6	Belgium	31.04	25.52	10.2	74.7	1.7	100		

Choose: **Insert > Function**

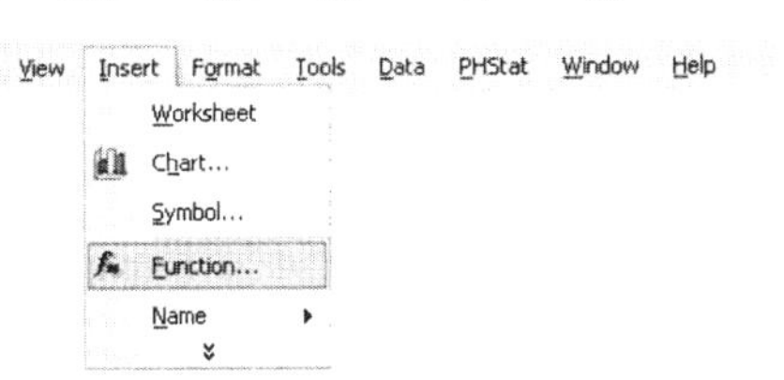

Select: Or select a category: **Statistical**
Select: **CORREL**
Click: **OK**

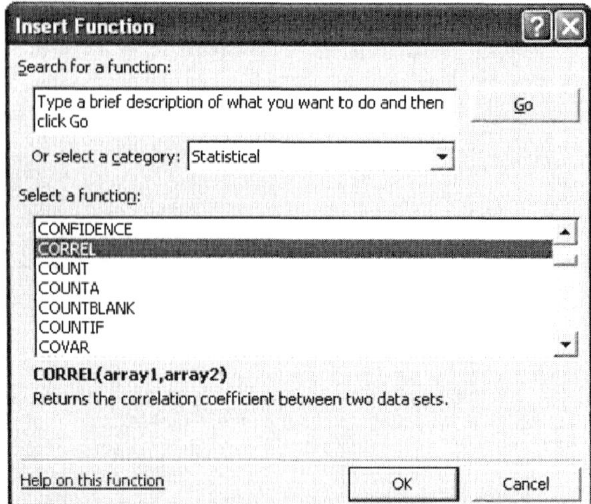

Complete the Function Arguments dialog box as shown. Click **OK**.

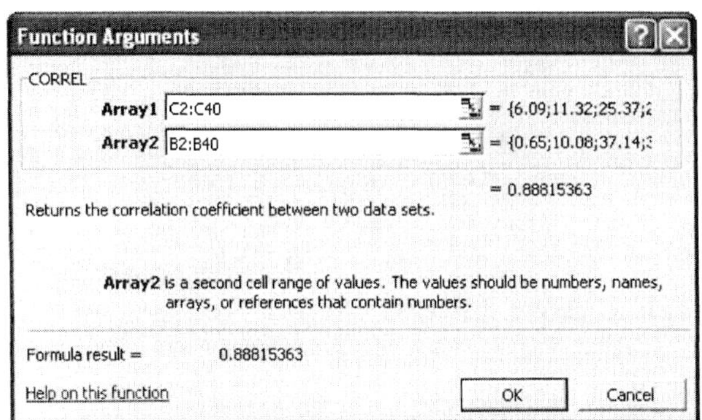

Your output should appear as shown.

	A	B	C	D	E	F	G	H	I
2	Algeria	0.65	6.09	3	0.3	2.8	58.3		0.888154
3	Argentina	10.08	11.32	3.8	19.3	2.4	96.9		
4	Australia	37.14	25.37	18.2	57.4	1.7	100		
5	Austria	38.7	26.73	7.6	81.7	1.3	100		
6	Belgium	31.04	25.52	10.2	74.7	1.7	100		

Chapter 3 Exercise 3.19 Which Mountain Bike to Buy?
Computing Correlation

Open the worksheet **mountain_bike**, which is found in the Excel folder of the data disk.

Since we are interested in whether and how weight affects the price, column B, weight, is the x-variable, and column A, price, is the y-variable.

To create a scatterplot of the data

Choose: **Insert > Chart**
In the Chart Type dialog box
Select: **XY (Scatter)**

Select: **Scatter**
Click: **Next**

In the Chart Source Data dialog box
Enter: Data Range: **select the appropriate cells**
Click: **Next**

Notice, by default, that the values in column A (price) are being treated as the x-variable. This needs to be corrected.

Select: the **Series** tab
Enter: X values: **B2:B13**
Y values: **A2:A13**

Click: **Next**

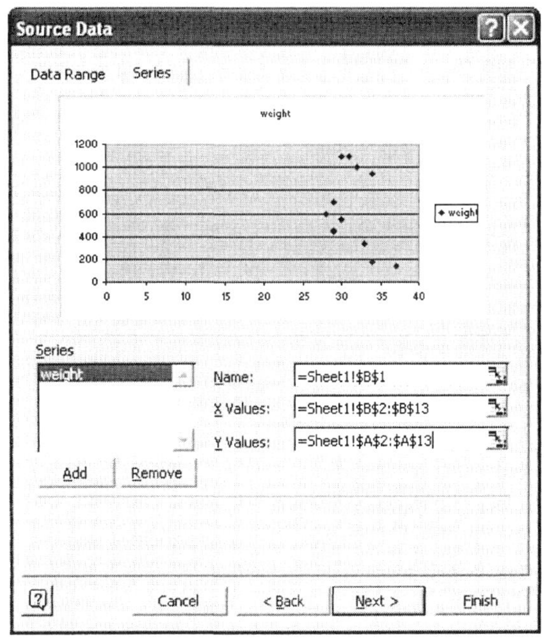

In the Chart Options dialog box
on the Title tab
Enter: Chart title: **Mountain Bikes / Weight vs Price**
Value (X) axis: **Weight (lb)**
Value (Y) axis: **Price ($)**

on the Legend tab
Click: **Show Legend** (so that a checkmark does not appear there)
Click: **Next**

Notice that now we also need to adjust the scale on the x-axis. Place your cursor within the chart on the x-axis, so that the mouseover reads "Value (X) axis", and double-click.

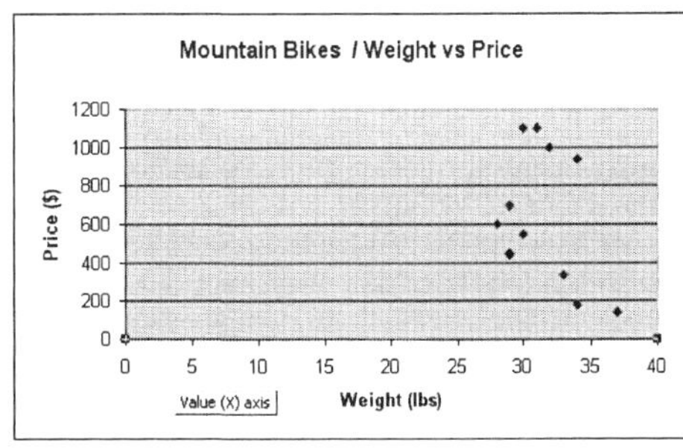

in the Format Axis window, on the Scale tab

Enter: Minimum: **27**

Click: **OK**

in the Chart Location dialog box

Select: **either of the two options**

Click: **Finish**

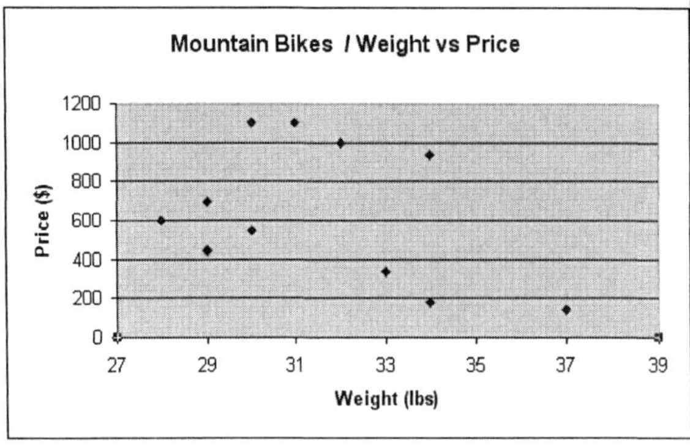

To find the correlation :

Click: **within an empty cell**

Choose: **Insert > Function**

Select: Or select a category: **Statistical**

Select: **CORREL**

Click: **OK**

Complete the Function Arguments dialog box as shown.

Click **OK**.

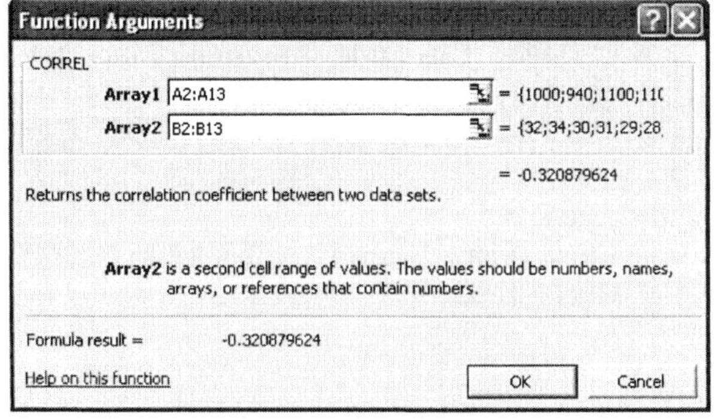

Your output should appear as shown.

	A	B	C	D
1	price	weight		
2	1000	32		-0.32088
3	940	34		
4	1100	30		
5	1100	31		

Chapter 3 Exercise 3.21 Buchanan Vote
Computing Correlation

Open the worksheet **Buchanan_and_the_butterfly_ballot** which is in the Excel folder of the data disk.

Creating a box plot of Gore and Buchanan:

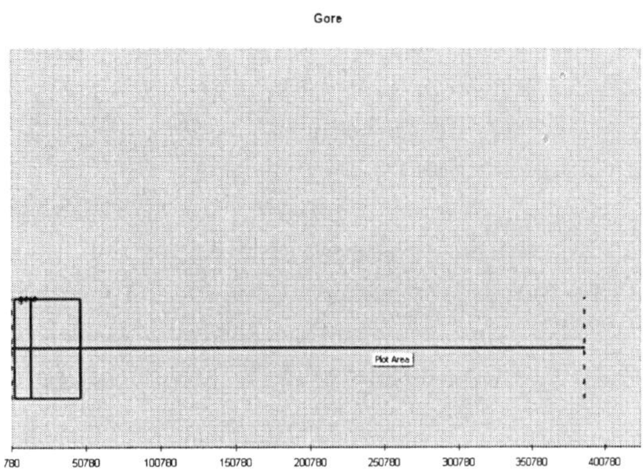

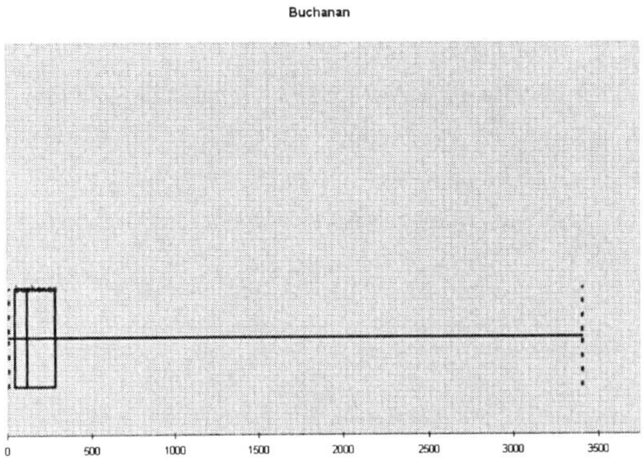

Buchanan

0 500 1000 1500 2000 2500 3000 3500

Note the different scales when comparing the two distributions

To create a scatterplot of the data

Choose: **Insert > Chart**

In the Chart Type dialog box

Select: **XY (Scatter)**

Select: **Scatter**

Click: **Next**

In the Chart Source Data dialog box

Enter: **select the appropriate cells**

(the cells we wish to use are not directly next to each other, so select column C, type a comma (,) and select column E**)**

Click: **Next**

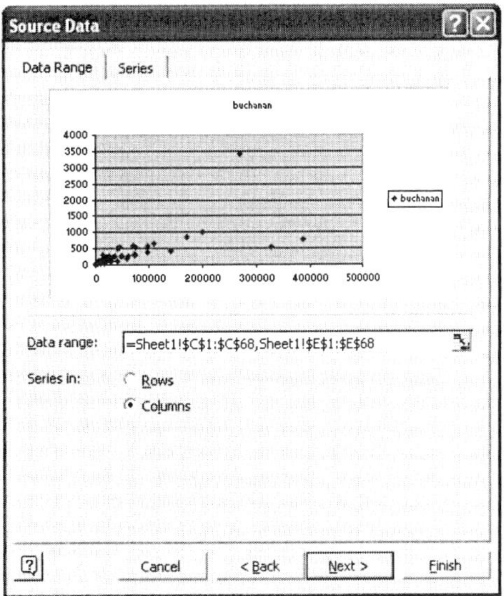

In the Chart Options dialog box
on the Title tab
Enter: Chart title: **The Gore Vote vs The Buchanan Vote**
Value (X) axis: **Gore**
Value (Y) axis: **Buchanan**

on the Legend tab
Click: **Show Legend** (so that a checkmark does not appear there)
Click: **Next**

We will now adjust the scale on each axis. Place your cursor within the chart on the x-axis, so that the mouseover reads "Value (X) axis", and double-click.

in the Format Axis window, on the Scale tab
Enter: Maximum: **400000**
Click: **OK**

Place your cursor within the chart on the y-axis, so that the mouseover reads "Value (Y) axis", and double-click.

in the Format Axis window, on the Scale tab
Enter: Maximum: **3500**
Click: **OK**

in the Chart Location dialog box
Select: **either of the two options**
Click: **Finish**

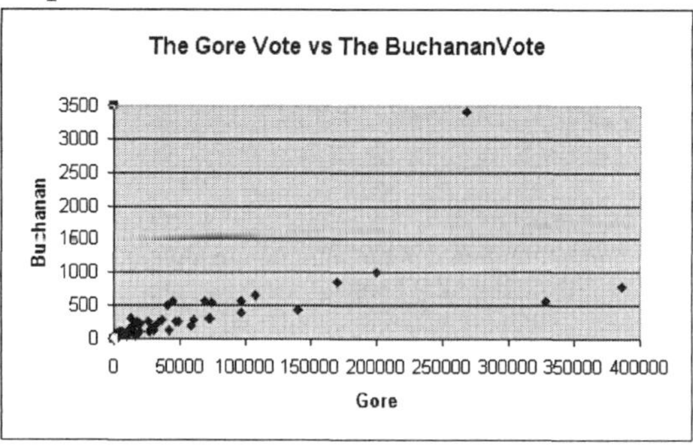

To find the correlation :
Click: **within an empty cell**

Choose: **Insert > Function**
Select: Or select a category: **Statistical**
Select: **CORREL**
Click: **OK**

Complete the Function Arguments dialog box as shown. Click **OK**.

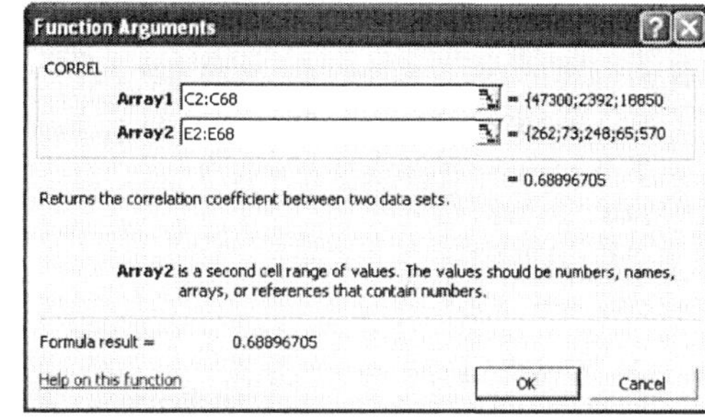

Your output should appear as shown.

	A	B	C	D	E	F	G
1	county	perot	gore	bush	buchanan		
2	Alachua	8072	47300	34062	262		0.688967
3	Baker	667	2392	5610	73		
4	Bay	5922	18850	38637	248		
5	Bradford	819	3075	5414	65		

Chapter 3 Example 9 How Can We Predict Baseball Scoring Using Batting Average? Generating the Regression Equation

Open the worksheet **al_team_statistics** which is in the Excel folder of the data disk.

Choose:	**Chart Wizard > XY(Scatter) > Scatter > Next**
Enter:	Data Range: **Select cells > Next**
Choose:	Titles
Enter:	Chart title: **Batting Average vs Team Scoring**
	Value (x) axis: **Batting Average**
	Value (y) axis: **Scoring Average > Finish**

Adjust the scale on each axis.

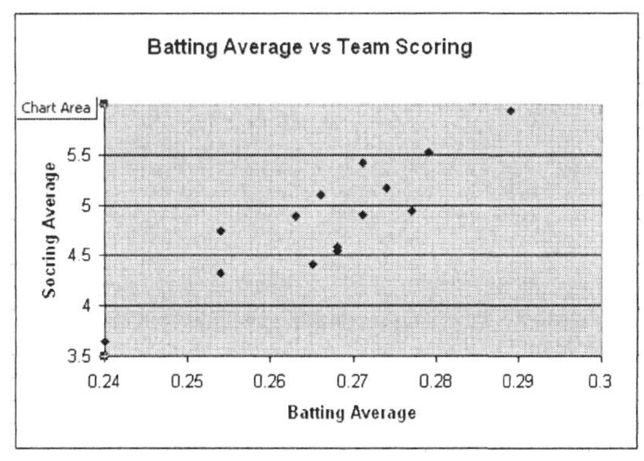

Now we will calculate the regression equation:

Choose: **Tools>Data Analysis**

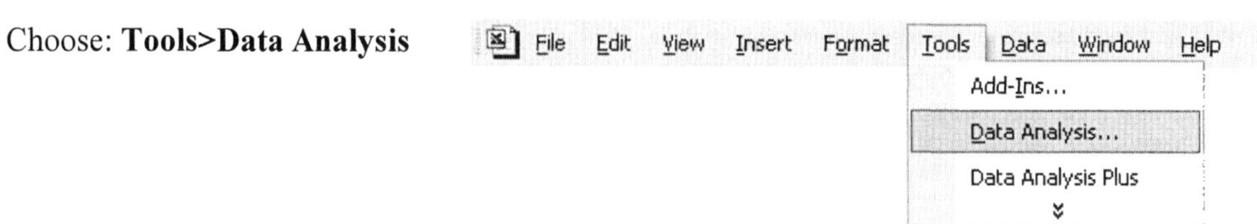

Select: **Regression**
Click: **OK**

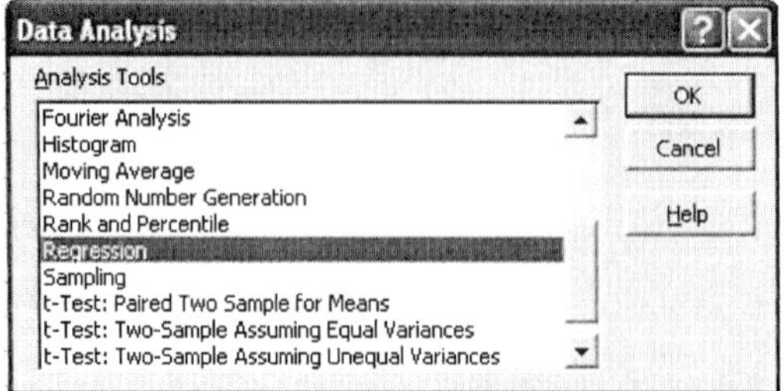

Indicate the location of the data, as appropriate:

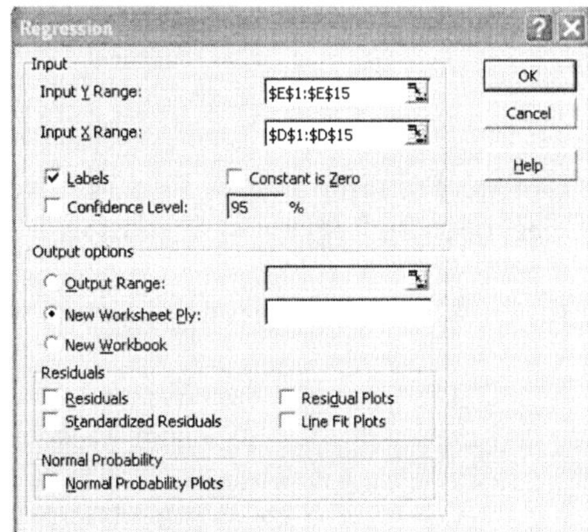

Here is the default output generated by the **Regression** command for Example 9.
Notice that a great deal of information is generated, but at this point we would need only the coefficients circled
.

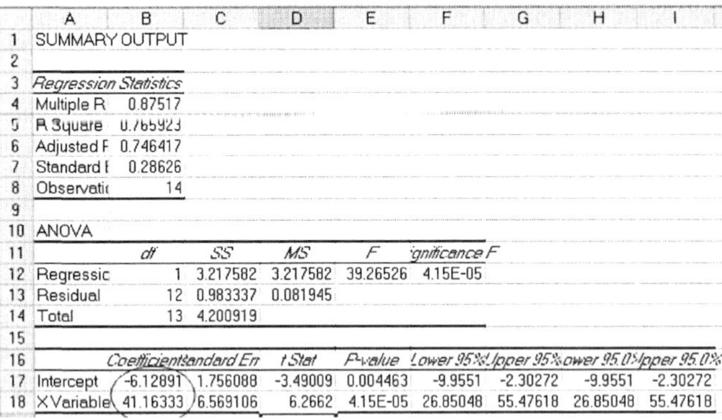

	A	B	C	D	E	F	G	H	I
1	SUMMARY OUTPUT								
2									
3	*Regression Statistics*								
4	Multiple R	0.87517							
5	R Square	0.765923							
6	Adjusted F	0.746417							
7	Standard I	0.28626							
8	Observati	14							
9									
10	ANOVA								
11		*df*	*SS*	*MS*	*F*	*gnificance F*			
12	Regressic	1	3.217582	3.217582	39.26526	4.15E-05			
13	Residual	12	0.983337	0.081945					
14	Total	13	4.200919						
15									
16		*Coefficients*	*tandard Err*	*t Stat*	*P-value*	*Lower 95%*	*Upper 95%*	*ower 95.0%*	*pper 95.0%*
17	Intercept	-6.12891	1.756088	-3.49009	0.004463	-9.9551	-2.30272	-9.9551	-2.30272
18	X Variable	41.16333	6.569106	6.2662	4.15E-05	26.85048	55.47618	26.85048	55.47618

The least squares line can be added to the plot, along with its equation and the value of r^2. Right click on one of the data points shown in the scatter plot. A drop-down menu will appear.

Select: **Add Trendline**
Select: Type : **Linear**
Select: Options,
then check **Display equation on chart and Display R-squared on chart > OK**

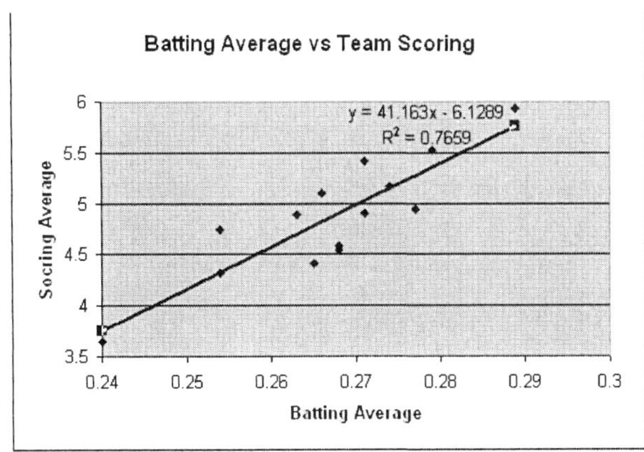

Use the regression equation to do the prediction.

Chapter 3 Exercise 3.35 Mountain Bikes Revisited
Generating a Regression Equation

Construct the scatterplot and include the regression line

Choose:**Chart Wizard > XY(Scatter) > Scatter > Next**
Enter: Data Range: **Select cells > Next**
Choose: Titles
Enter: Chart title: **Weight vs Price of Mountain Bikes**
Value (x) axis: **Weight**
Value (y) axis: **Price > Finish**

After formatting the chart and adding the trendline we get:

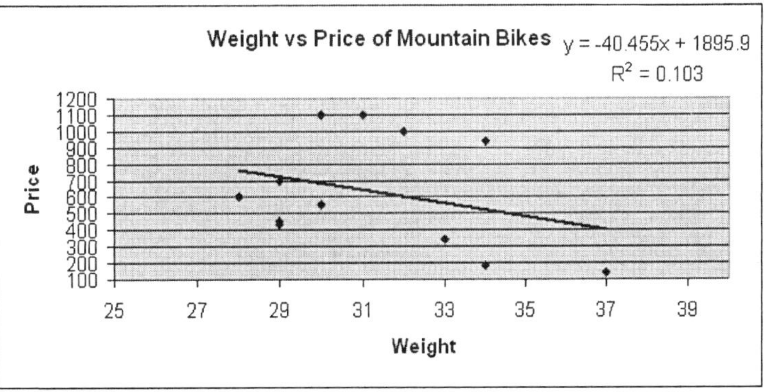

Chapter 3 Exercise 3.36 Mountain Bike and Suspension Type
Generating Regression Equations

Do a scatterplot designating front end and full suspensions as different series. Include both regression equations on the chart.

Choose: **Chart Wizard > XY(Scatter) > Scatter > Next**
Enter: Data Range: **Select appropriate cells for Series 1 and Series 2 and name each**
Click: **Next**

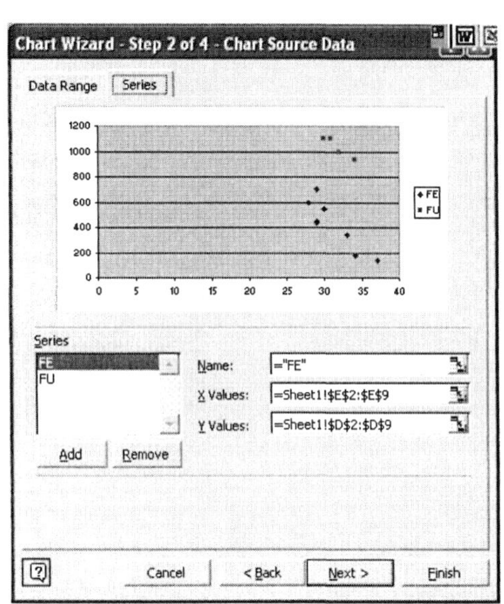

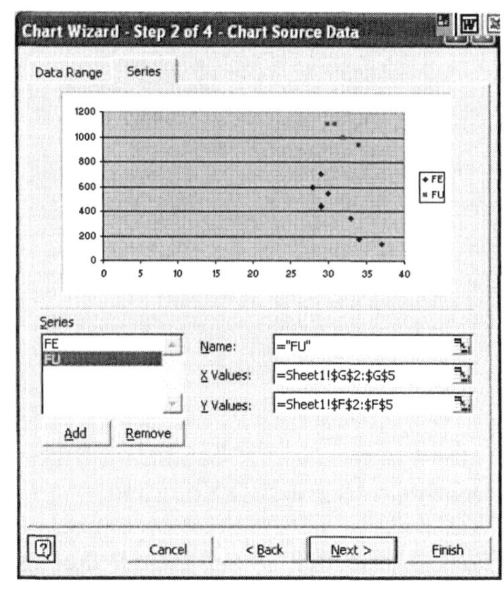

Choose: Titles:
Enter: Chart title: **Full Suspension & Front-end Suspension**
Value (x) axis: **Weight**
Value (y) axis: **Price**
Click: **Finish**

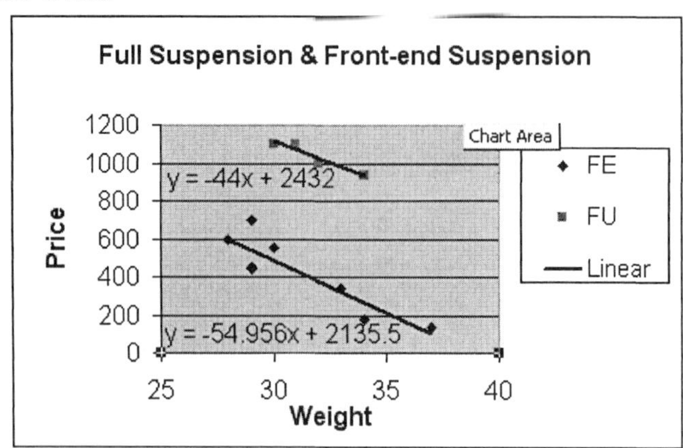

Find the correlations for the two types separately:

	A	B	C	D
1		weight_FU	price_FU	
2	weight_FU	1		
3	price_FU	-0.95178	1	
4				
5				

	A	B	C	D
1		weight_FE	price_FE	
2	weight_FE	1		
3	price_FE	-0.88842	1	
4				
5				

What conclusions can you draw?

Chapter 3 Exercise 3.79 High School Graduation Rates and Health Insurance Scatterplot, Correlation, and Regression

Open the worksheet **hs_graduation_rates** from the Excel folder of the data disk.

Choose:**Chart Wizard > XY(Scatter) > Scatter > Next**
Enter: Data Range: S**elect cells > Next**
Choose: Titles
Enter: Chart title: **HS Grad rates vs Health Insurance**
Value (x) axis: **Grad rates**
Value (y) axis: **Health Ins > Finish**

After formatting the chart and adding the trendline we get:

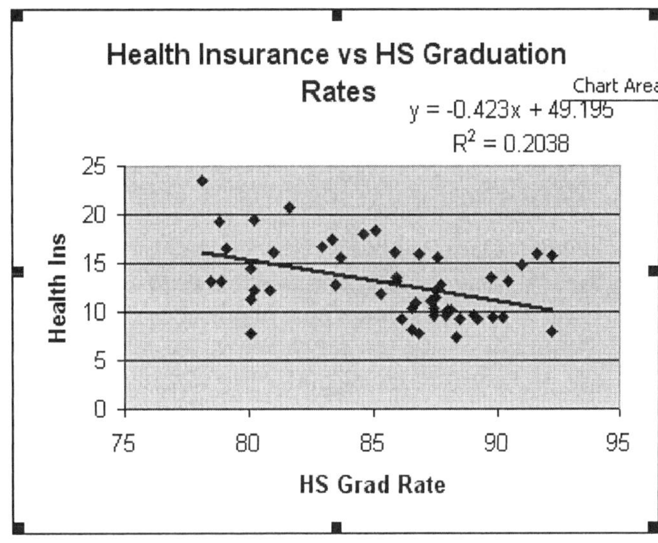

Computing correlation: Click: **within an empty cell**
Choose: **Insert > Function**
Select: Or select a category: **Statistical**
Select: **CORREL**
Click: **OK**

Complete the Function Arguments dialog box as appropriate. Click **OK**.

	A	B	C	D
1		HS Grad Rate	Health Ins	
2	HS Grad Rate	1		
3	Health Ins	-0.451437877	1	
4				

To do the regression: Choose: **Tools>Data Analysis >Regression >OK**
Enter: Input Y Range: **D2:D52**
Input X Range: **C2:C52**
Click: **OK**

SUMMARY OUTPUT

Regression Statistics	
Multiple R	0.451438
R Square	0.203796
Adjusted F	0.187547
Standard E	3.358214
Observatio	51

ANOVA

	df	*SS*	*MS*	*F*	*gnificance F*
Regressio	1	141.444	141.444	12.54203	0.000884
Residual	49	552.6023	11.2776		
Total	50	694.0463			

	Coefficients	*tandard Err*	*t Stat*	*P-value*	*Lower 95%*	*Upper 95%*	*ower 95.0%*	*Upper 95.0%*
Intercept	49.19483	10.225	4.811228	1.47E-05	28.64692	69.74273	28.64692	69.74273
X Variable	-0.42299	0.11944	-3.54147	0.000884	-0.66301	-0.18297	-0.66301	-0.18297

Chapter 3 Example 10 How Can We Detect an Unusual Vote Total? Looking at Residuals

Open the worksheet **Buchanan_and_the_butterfly_ballot** from the Excel folder of the data disk.

Select: **Tools > Data Analysis > Regression**
Enter: Input Y Range: **select appropriate cells**
Input X Range: **select appropriate cells**
Select: **Residuals**

Click: **OK**

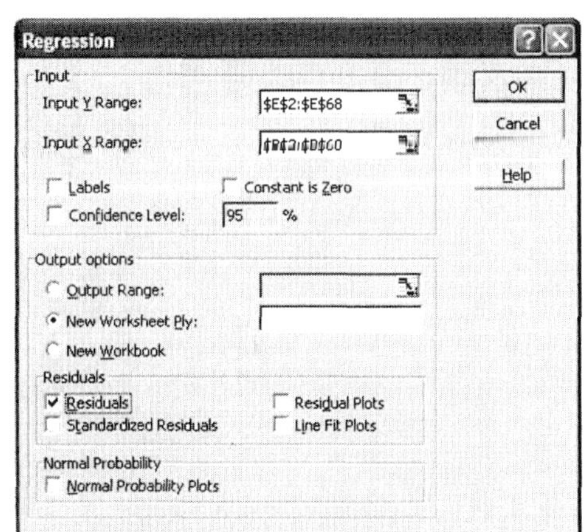

Next, you will copy the residuals to the worksheet containing the data so that you can use them in a histogram of the residuals. Click and drag over the residuals, so that the range is highlighted. Right click in the highlighted section and select copy. Paste into data worksheet.

22	RESIDUAL OUTPUT		
23			
24	Observation	Predicted Y	Residuals
25	1	289.5312	-27.5312
26	2	24.91641	48.08359
27	3	212.7018	35.29824
28	4	30.34807	34.65193
29	5	903.3445	-333.344
30	6	1393.445	-604.445
31	7	23.59423	66.40577
32	8	279.2039	-97.2039
33	9	259.9429	10.05709
34	10	118.3267	67.67333
35	11	226.9241	-104.924
36	12	71.4786	17.5214

	A	B	C	D	E	F
1	county	perot	gore	bush	buchanan	residuals
2	Alachua	8072	47300	34062	262	-27.5312
3	Baker	667	2392	5610	73	48.08359
4	Bay	5922	18850	38637	248	35.29824
5	Bradford	819	3075	5414	65	34.65193
6	Brevard	25249	97318	115185	570	-333.344

Generating a histogram of the residuals:

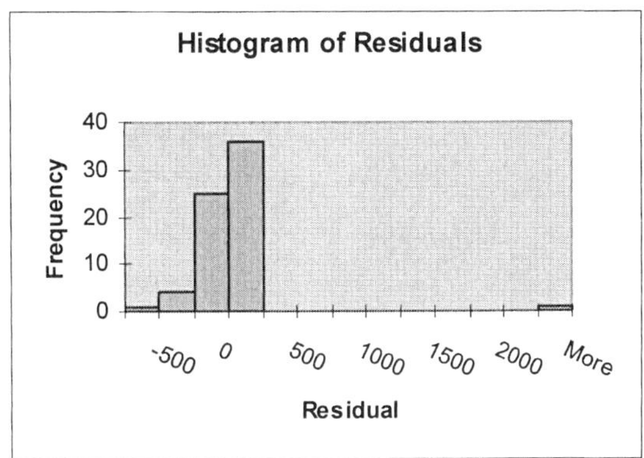

Chapter 3 Exercise 3.29 Regression Between Cereal Sodium and Sugar Residuals

Open the worksheet **CEREAL** from the Excel folder of the data disk. Perform the regression, checking Residuals in the dialog box.

Select: **Tools > Data Analysis > Regression**
Enter: Input Y Range: **select appropriate cells**
Input X Range: **select appropriate cells**
Select: **Residuals**
Click: **OK**
Copy the residuals to the worksheet containing the data, and construct a histogram of the residuals.

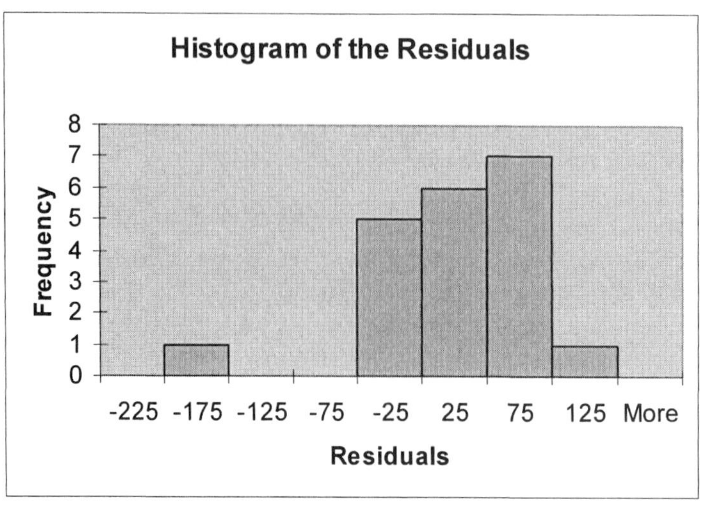

Chapter 3 Example 12 How Can We Forecast Future Global Warming? Regression on a Time Series Plot

Open the worksheet **central_park_yearly_temps** from the Excel folder of the data disk.

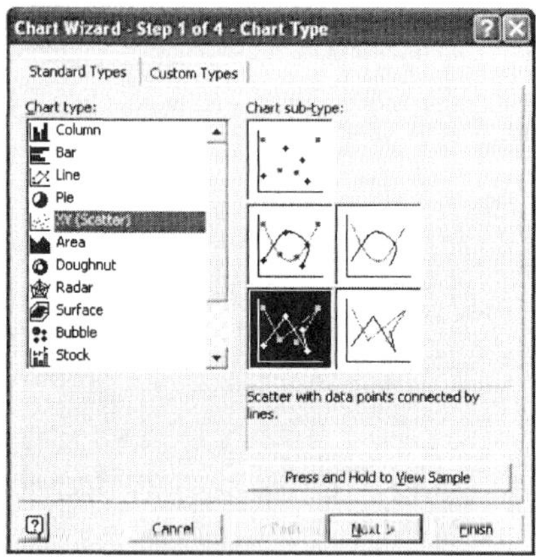

Because our x-variable is years, the scatterplot can be interpreted as a time plot.
Using the Chart Wizard select: **XY(Scatter) > Scatter with data points connected by lines.**

Select the columns containing the data.
Add titles as appropriate.

Format the graph, choosing proper axis scale and adding the trendline.

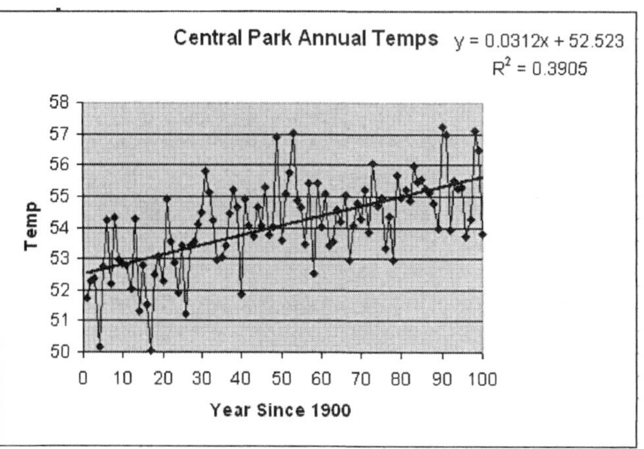

Chapter 3 Example 13 Is Higher Education Associated with Higher Murder Rates? Influential Outliers

Open the worksheet **us_statewide_crime** from the Excel folder of the data disk.

Choose:**Chart Wizard > XY(Scatter) > Scatter > Next**
Enter: Data Range: **Select cells > Next**
Choose: Titles
Enter: Chart title:
Value (x) axis: **% with College Education**
Value (y) axis: **Murder rate**

Click: **Finish**

After formatting the chart and adding the trendline we get:

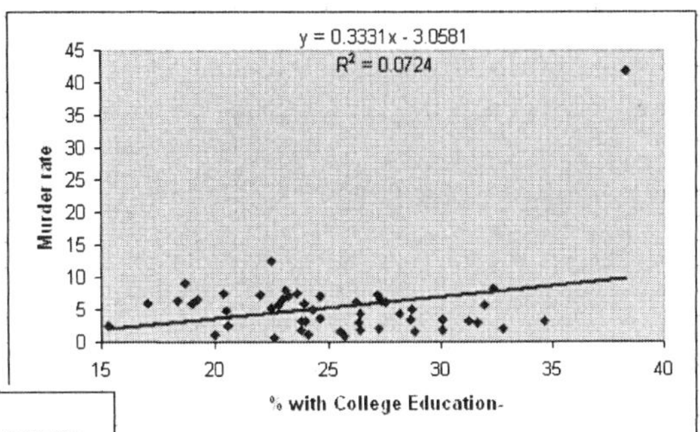

Redoing the entire process, excluding the data for D.C. we get:

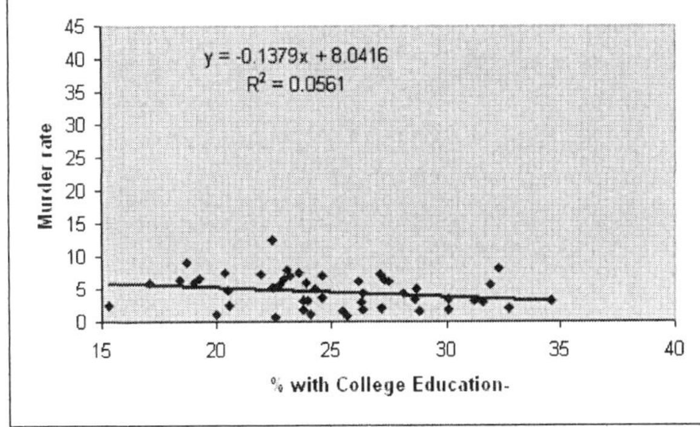

Chapter 3 Exercise 3.41 Murder and Education Application of the Regression Model

Using the Excel results of the previous problem

a) In the equation $y = 0.3331\ x - 3.0581$ we substitute 15 and 40 for x, we obtain 1.89 and 10.26 for y, respectively.

b) In the equation $y = -.1379\ x + 8.0416$ we substitute 15 and 40 for x, we obtain 5.9 and 2.4 for y, respectively.

Exercise 3.45 Regression Between Sodium and Sugar Influential Outliers

Open the worksheet **CEREAL** from the Excel folder of the data disk. Let x = sodium (mg) and y = sugar (g). Using the chart wizard construct a scatterplot and include the regression line.

Choose:**Chart Wizard > XY(Scatter) > Scatter > Next**
Enter: Data Range: **Select appropriate cells > Next**
Choose: Titles
  Enter: Chart title: **Sodium vs Sugar**
    Value (x) axis: **Sodium (mg)**
    Value (y) axis: **Sugar (mg)**
  Click: **Finish**

After formatting the chart and adding the trendline we get:

Note the outlier at (7000, 0)

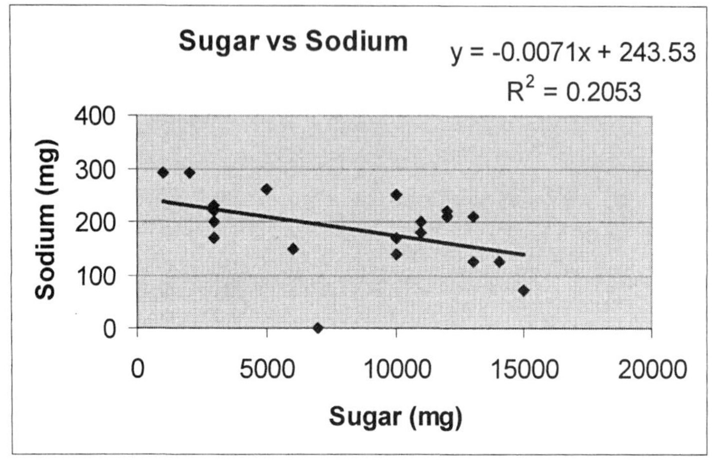

Computing correlation:
Click: **within an empty cell**
Choose: **Insert > Function**
  Select: Or select a category: **Statistical**
  Select: **CORREL**
  Click: **OK**
Complete the Function Arguments dialog box as appropriate. Click **OK**

	A	B	C	D
1		*SODIUM(m*	*SUGAR(g)*	
2	SODIUM(n	1		
3	SUGAR(g)	-0.45305	1	
4				

Then without the outlier:

	A	B	C	D
1		*SODIUM(m*	*SUGAR(g)*	
2	SODIUM(n	1		
3	SUGAR(g)	-0.62255	1	
4				

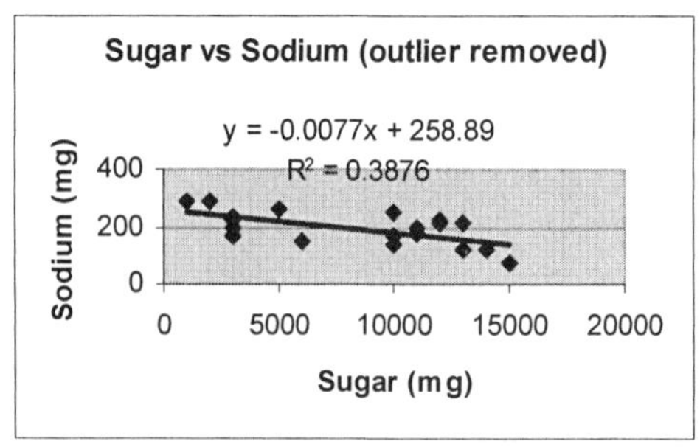

What conclusions can you draw?

Chapter 4 Example 5 Auditing the Accounts of a School District
Random Selection

To use random numbers within Excel to select 10 accounts to audit in a school district that has 60 accounts we will use the PHStat add-in.

Within a blank Excel worksheet,
Click: **PHStat**
Select: **Sampling >**
Random Sample Generation

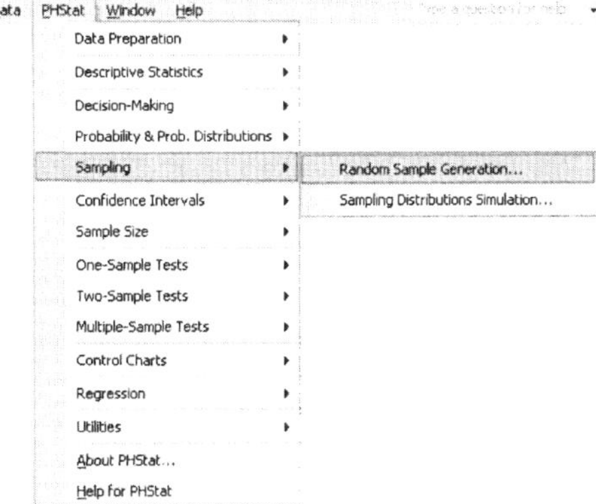

Complete the dialog box as shown:

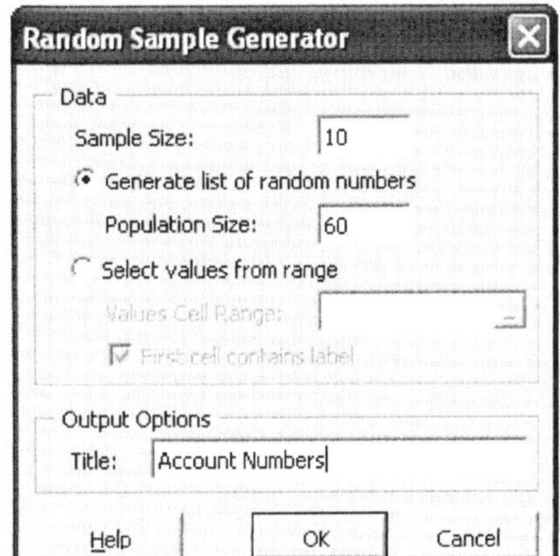

The output is displayed in a new worksheet.

that

	A
1	Account Numbers
2	51
3	50
4	30
5	54
6	52
7	7
8	18
9	43
10	42
11	26

Because the numbers were randomly generated, it is not likely your output will be exactly the same.

Chapter 4 Exercise 4.18 Auditing Accounts
Random Selection

To use random numbers within Excel to select 10 accounts to audit in a school district that has 60 accounts we will use the PHStat add-in.

Within a blank Excel worksheet,
Click: **PHStat**
Select: **Sampling >**
Random Sample Generation

Complete the dialog box as shown:

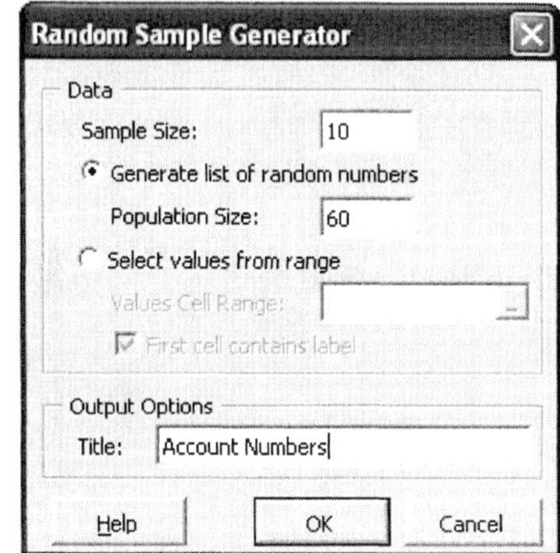

The output is displayed in a new worksheet.

Because we are using a random number generator, each time this command is issued, different accounts will be selected.

	A
1	Account #s
2	35
3	12
4	41
5	60
6	18
7	43
8	47
9	57
10	46
11	16

Chapter 6 Example 7 What IQ Do You Need to Get Into MENSA?
Determining Normal Probabilities

Stanford-Binet IQ scores are approximately normally distributed with $\mu = 100$ and $\sigma = 16$. To be eligible for MENSA, you must rank at the 98th percentile (i.e. you must score in the top 2%). Find the lowest IQ score that still qualifies for Mensa membership.

Choose: **Insert > Function**

File Edit View Insert Format Tools Data PHStat Window
- Worksheet
- Chart...
- Symbol...
- Function...
- Name

Select: Or select a category: **Statistical**
Select: Select a function: **NORMINV**

Click: **OK**

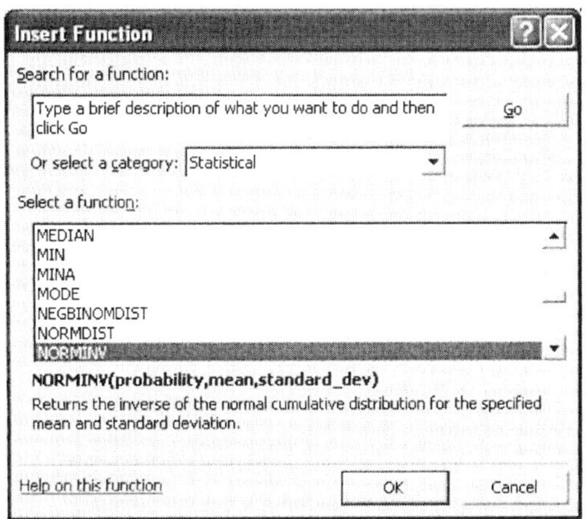

Function Arguments
Enter: Probability: **.98**
(area to the left of the value we wish to find)
Mean : **100**
Standard deviation: **16**

Click: **OK**

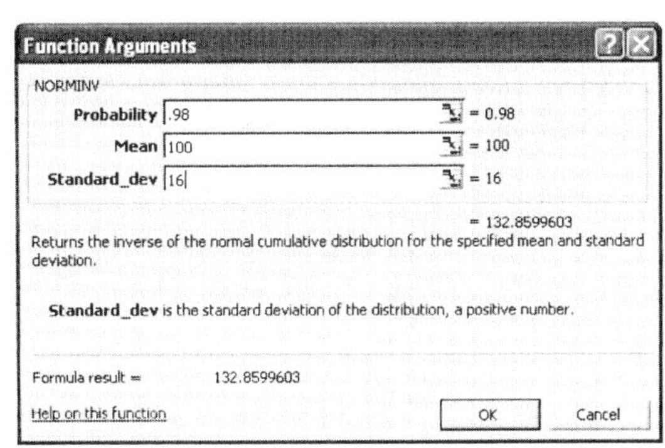

The X-value should appear in the worksheet.

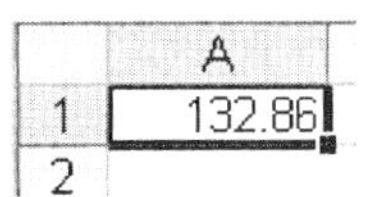

Notice that the test score needed to join Mensa is 133.

Chapter 6 Example 8 Finding Your Relative Standing on the SAT
Determining Normal Probabilities

SAT scores are approximately normally distributed with $\mu = 500$ and $\sigma = 100$. If one of your SAT scores was 650, what percentage of SAT scores were higher than yours?

Choose: **Insert > Function**
Select: Or select a category: **Statistical**
Select: Select a function: **NORMDIST**

Click: **OK**

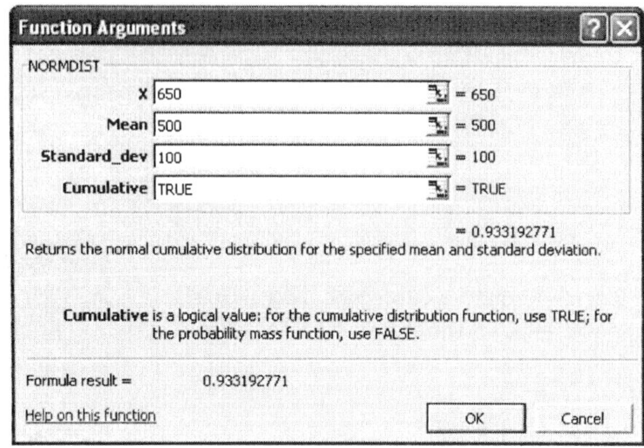

Enter: X **650**
Mean **500**
Standard_dev **100**
Click: **OK**

0.933193

Notice that the result displayed in the worksheet is the probability that X is less than 650.

Since we were asked to determine the percentage of SAT scores higher than 650 we must calculate 1 - 0.9332 = 0.0668. Only about 7% of SAT scores fall above 650.

Chapter 6 Example 9 What Proportion of Students Get a Grade of B?
Determining Normal Probabilities

On the midterm exam, an instructor always gives a grade of B to students who score between 80 and 90. One year, the scores on the exam have an approximately normal distribution with $\mu = 83$ and $\sigma = 5$. About what proportion of students get a B?

We will have to compute two cumulative probabilities, one for X = 80 and one for X = 90, and then subtract the two probabilities.

Choose: **Insert > Function > NORMDIST**
Enter: X **80** then repeat using **90**
Mean **83**
Standard_dev **5**
Click: **OK**

	A
1	0.274253
2	0.919243

It follows that about 0.9192 – 0.2743 = 0.6449, or about 64%, of the exam scores were in the B range.

Chapter 6 Exercise 6.21 Blood pressure
Determining Normal Probabilities

In Canada, systolic blood pressure readings has are normally distributed with $\mu = 121$ and $\sigma = 16$. A reading above 140 is considered to be high blood pressure. What proportion of Canadians suffers from high blood pressure?

Choose: **Insert > Function > NORMDIST**
Enter: X **140**
Mean **121**
Standard_dev **16**
Click: **OK**

0.882485

It follows that about 1 – 0.8825 = 0.1175, or about 12%, of Canadians suffer from high blood pressure.

We are also asked to determine the proportion of Canadians having systolic blood pressure in the range from 100 to 140.

We have already computed the $P(X < 140) = 0.8825$, so we just need to determine $P(X < 100)$ and subtract the two results.

0.094676

Approximately 0.8825 – 0.0947 = 0.7878, or 78% of Canadians have systolic blood pressures in the range from 100 to 140.

Chapter 6 Exercise 6.23 Mental Development Index
Determining Normal Probabilities

The Mental Development Index of the Bayley Scales of Infant Development is a standardized measure used in observing infants over time. MDI is approximately normally distributed with $\mu = 100$ and $\sigma = 16$.

What proportion of children has MDI of at least 120?

Choose: **Insert > Function > NORMDIST**
Enter: X **120**
Mean **100**
Standard_dev **16**
Click: **OK**

0.89435

It follows that about 1 – 0.8944 = 0.1056, or about 11%, of children have an MDI of at least 120.

We are also asked to determine the proportion of children having an MDI of at least 80.

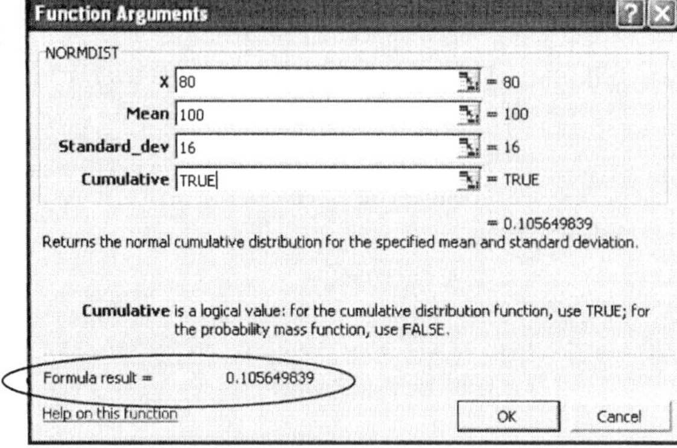

Note the Formula Result = 0.105649639

So, approximately 1.0 – 0.1057 = 0.8943, or 89% of children have an MDI of at least 80.

To determine the MDI score that is the 99th percentile:

Choose: **Insert >Function > NORMINV**
Enter: Probability **.99**
Mean **100**
Standard_dev **16**
Click: **OK**

Function Arguments
NORMINV
Probability .99 = 0.99
Mean 100 = 100
Standard_dev 16 = 16
= 137.221552
Returns the inverse of the normal cumulative distribution for the specified mean and standard deviation.
Standard_dev is the standard deviation of the distribution, a positive number.
Formula result = 137.221552
Help on this function
OK
Cancel

An MDI of 138 is the 99th percentile.

To determine the MDI such that only 1% of the population has an MDI below it, we will use the probability **.01**

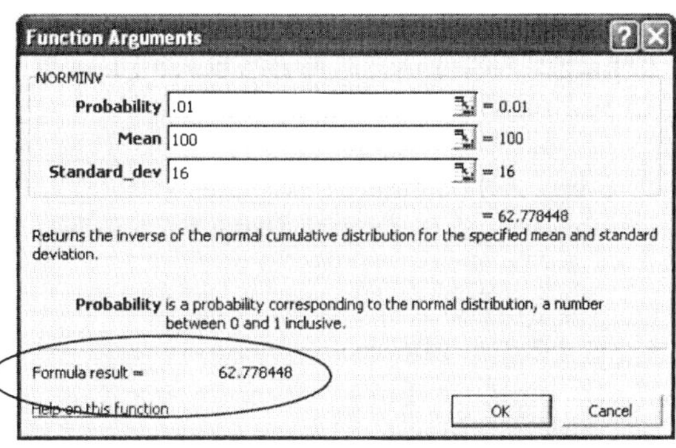

Approximately 1% of the population has an MDI less than 63.

Chapter 6 Exercise 6.98 Psychomotor Development Index
Determining Normal Probabilities

Choose: **Insert > Function > NORMDIST**
Enter: X **97** then again with **103**
Mean **100**
Standard_dev **15**
Click: **OK**

This gives the following results:

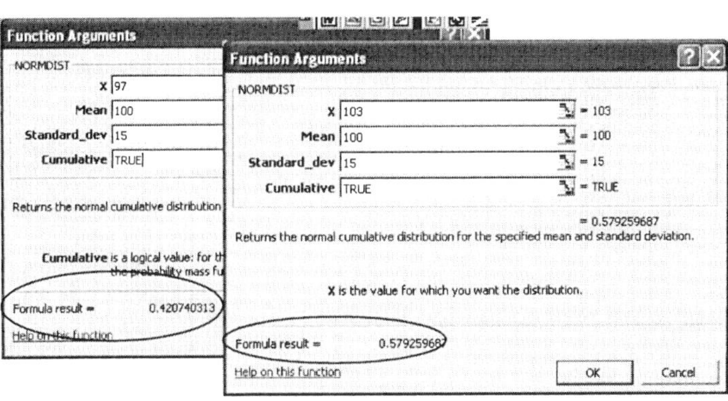

a) The probability P(PDI > 103) = 1 - 0.57962 = .42038
b) The probability P(97 ≤ PDI ≤ 103) = .57926 - .420740 = .15852
c) To calculate the z score for X = 90, using Excel
Select: **Insert > Function > STANDARDIZE**
Click: **OK**

Enter: X **90**
Mean **100**

Standard_dev **15**

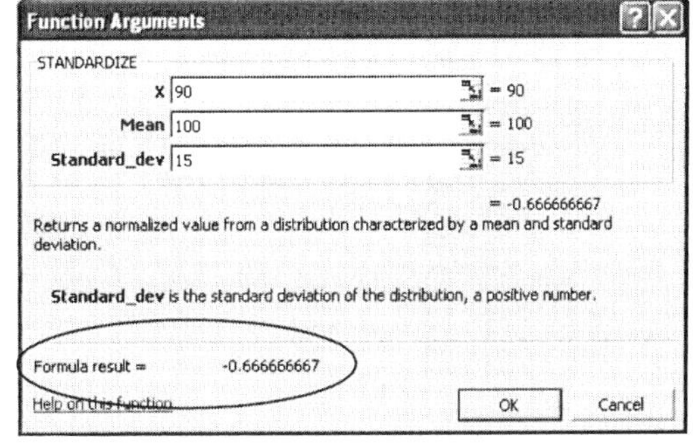

The z-score is -0.6667

Chapter 6 Example 12 Are Women Passed Over for Managerial Training? Determining Binomial Probabilities

Let X denote the number of females selected in a random sample of ten employees (X can have any value 0, 1, 2, 3, …, 10). Enter column labels "x" and "P(X=x)" into cells A1 and B1

Enter: **0** (in cell A2)
Choose: **Edit > Fill > Series**
Select: **columns** **linear** step **1** stop value **10**
Click: **OK**

Select cell B2.
Select: **Insert > Function > BINOMDIST**

Click: **OK**

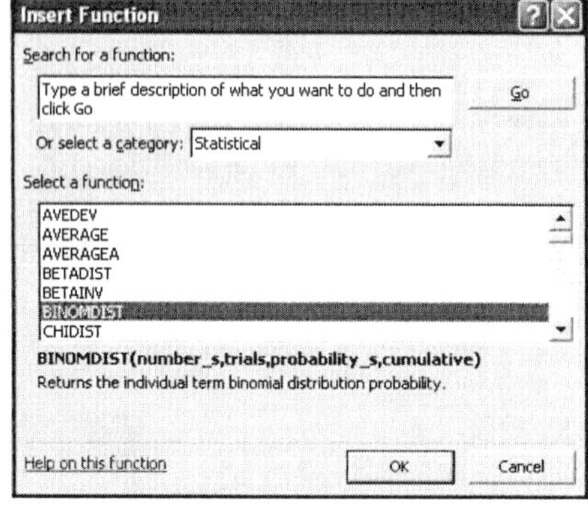

Enter: Number_s **A2**
Trials **10**
Probability_s **.5**
Cumulative: **FALSE**
Click: **OK**

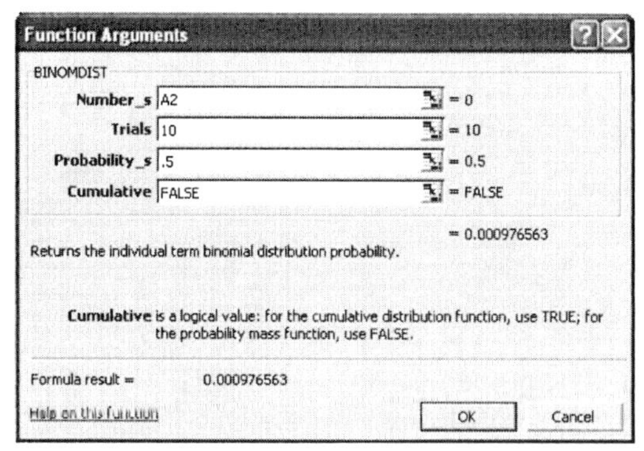

Drag to fill the remainder of B3 to B12

	A	B
1	x	P(X=x)
2	0	0.000977
3	1	0.009766
4	2	0.043945
5	3	0.117188
6	4	0.205078
7	5	0.246094
8	6	0.205078
9	7	0.117188
10	8	0.043945
11	9	0.009766
12	10	0.000977

This will generate the distribution in the column B of the worksheet

Chapter 6 Example 14 How Can We Check for Racial Profiling?
Determining Binomial Probabilities

Let X denote the number of police car stops where the driver is African-American (X can have any value 0, 1, 2, 3, …, 262). Enter column labels "x" and "P(X=x)" into cells A1 and B1

Enter: **0** (in cell A2)
Choose: **Edit > Fill > Series**
Select: **columns** **linear** step **1** stop value **262**
Click: **OK**

Select cell B2.
Select: **Insert > Function > BINOMDIST**

Click: **OK**

Enter: Number_s **A2**
Trials **262**
Probability_s **.422**
Cumulative: **FALSE**
Click: **OK**

Drag to fill the remainder of B3 to B264

Note the probabilities for X >=207

x	P(X=x)
207	3.94491E-34
208	7.61589E-35
209	1.43666E-35
210	2.64724E-36
211	4.76321E-37
212	8.36601E-38
213	1.43382E-38
214	2.39696E-39
215	3.90704E-40
216	6.20693E-41
217	9.60637E-42
218	1.44777E-42
219	2.1237E-43
220	3.03056E-44
221	4.20499E-45
222	5.66997E-46
223	7.42541E-47
224	9.4389E-48

Chapter 6 Exercise 6.36 Exit Poll
Mean and Standard Deviation of the Binomial Random Variable X

The data is binary (vote for or against), the voters were randomly selected, and each voter is separate and independent from another voter

N = 3160 p = .5 1- p = .5

We could generate the probabilities and use
(recall that the mean and standard deviation for a general probability distribution is)

$\mu = \sum xP(x)$ and $\sigma = \sqrt{\sum (x-u)^2 P(x)}$

But, for the binomial distribution we can use the following expressions

$\mu = n * p$ and $\sigma = \sqrt{n \cdot p \cdot (1-p)}$

Select an empty cell, and in the Function box, type in the required calculation.

fx =3160 * 0.5

	C	D
		mean
0		1580

fx =SQRT(3160*0.5*(1-0.5))

	C	D	E
		mean	std dev
0		1580	28.10694

Chapter 6 Exercise 6.37 Jury Duty
Determining Binomial Probabilities

a) Check the three criteria for being binomial
b) Let n = 12, p = .4 . Enter 0 through 12 into A2:A14, then
   Select cell **B2**
   Choose: **Insert > Function > BINOMDIST**

   Enter: Number_s **A2**
   Trials: **12**
   Probability_s: **.4**

   Cumulative: **FALSE**
   Click: **OK**

Click and drag to fill B3:B14

	A	B
1	x	P(X=x)
2	0	0.002177
3	1	0.017414
4	2	0.063852
5	3	0.141894
6	4	0.212841
7	5	0.22703
8	6	0.176579
9	7	0.100902
10	8	0.042043
11	9	0.012457
12	10	0.002491
13	11	0.000302
14	12	1.68E-05

Chapter 6 Example 16 Exit Poll of California Voters Revisited
The Mean and Standard Deviation of the Sampling Distribution of a Proportion

Neither Excel not PhStat can handle the calculations in determining the binomial probabilities for our sample size of 3160. See the Minitab section of this manual.

Chapter 6 Exercise 6.47 Other Scenario for Exit Poll
The Sampling Distribution of Sample Proportion

This problem repeats Example 16, but uses p = .55 instead of .5 . See the Minitab section of this manual.

Chapter 6 Class Exploration 6.124
Simulating a Sampling Distribution for a Sample Mean

Open the worksheet **heads_of_households** from the Excel folder of the data disk. Construct a histogram.

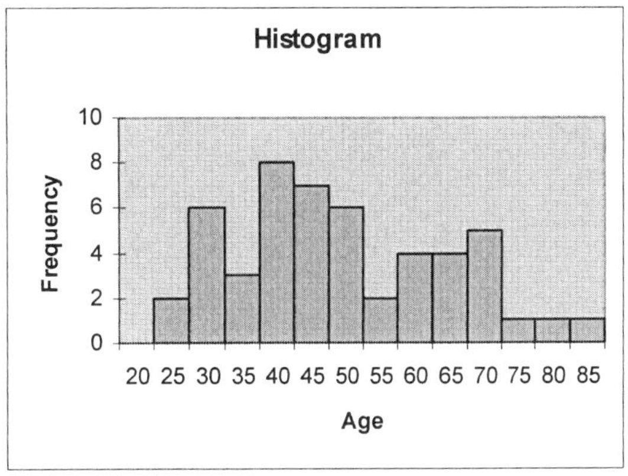

b) Lets assume a class size of 30 students. Each student can collect random samples of size 9. We will store the samples in nine columns of 30 cells each.

Tools > Data Analysis > Sampling > OK

Enter: Input Range: **select appropriate cells**
Select: **Random**
Enter: Number of Samples: **30**
Column **C3**
Enter: Output range**: enter a column**
Click: **OK**

Note you will have to do this nine times, naming a new column each time.

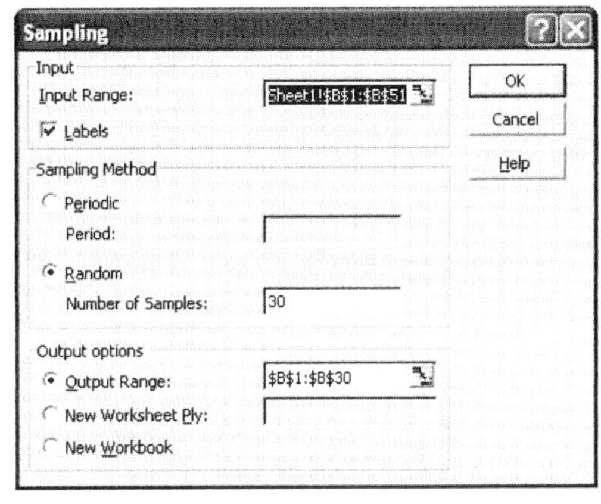

To find the mean of each sample :

Select cell J1. Use formula bar to specify the mean of A1 through I1.

fx =AVERAGE(A1:I1)

C	D	E	F	G	H	I	J
76	67	30	39	66	29	63	50.55556

Click and drag from J1 down to J30.

Now, find the mean and standard deviation of column J.

Tools > Data Analysis > Descriptive Statistics > OK

Complete the dialogue box as indicated, specifying the column containing the sample means, and where you want the summary statistics placed.

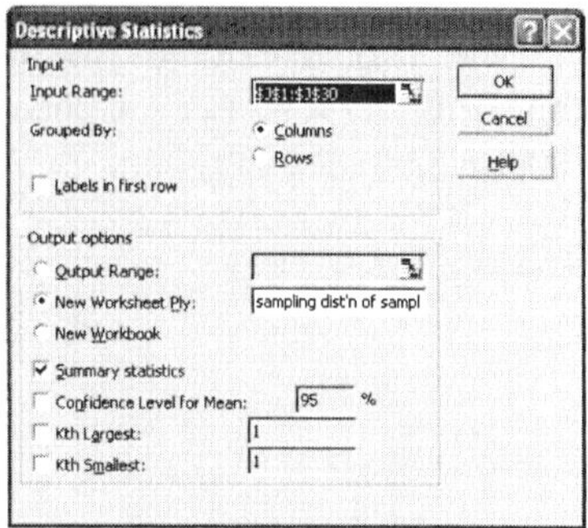

Note the mean of all thirty sample means is 48.5, and the standard deviation of all thirty sample means in 4.8547

	A	B
1	*Column1*	
2		
3	Mean	48.5
4	Standard Error	0.886343
5	Median	49.22222
6	Mode	47.22222
7	Standard Deviation	4.8547
8	Sample Variance	23.56811
9	Kurtosis	-0.6634
10	Skewness	-0.10158
11	Range	19.66667
12	Minimum	39.33333
13	Maximum	59
14	Sum	1455
		30

What conclusions can you draw?

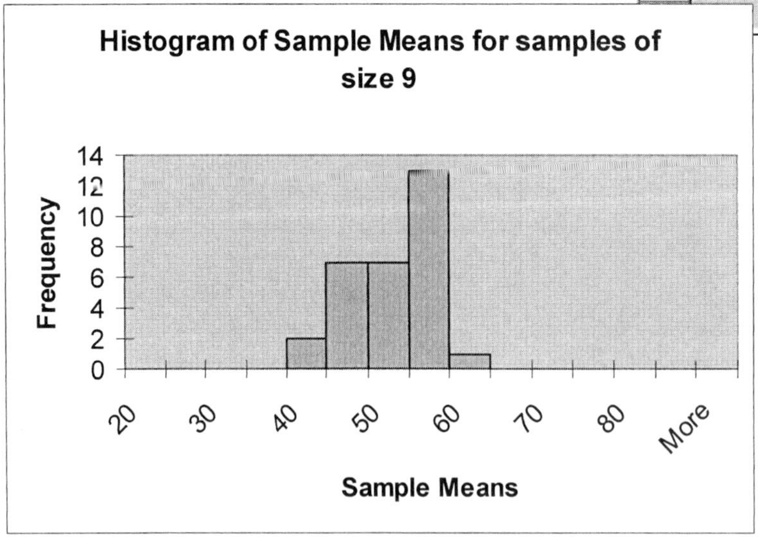

Histogram of Sample Means for samples of size 9

Frequency

Sample Means

Compare this histogram to the histogram of the population that we did at the start of the exercise.

Chapter 7 Example 2 Should a Wife Sacrifice Her Career For Her Husband's? Constructing the Confidence Interval Estimate for a Population Proportion

From Chapter 6 we recall that 95% of a normal distribution falls within two standard deviations of the mean. So if we look at the mean $\pm$ 1.96 standard deviations, we would find 95% of a normal distribution. So in this problem, if we look at 1.96(.01) we get 0.02. we can then construct the interval as the sample proportion $\pm$ 0.02, or .19 $\pm$ 0.02 which gives us (0.17, 0.21).

To do this in Excel:
Choose **PHStat > Confidence Interval > Estimate for a Population Proportion**

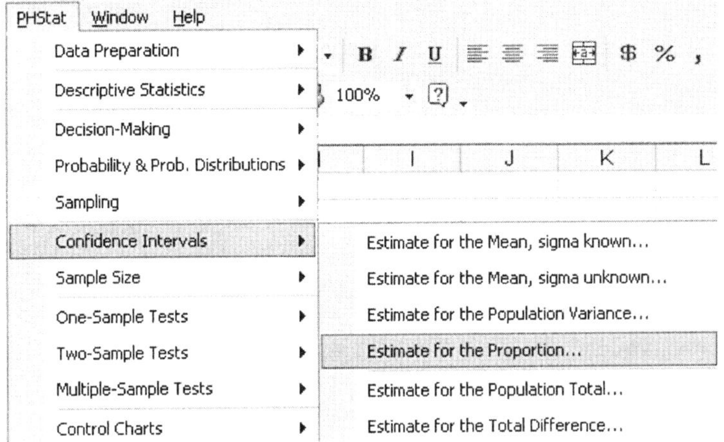

Enter Sample size **1823**
Number of Successes **346**
Confidence Level: **95**
Output Options:
Title: **C.I. for Proportion**
OK

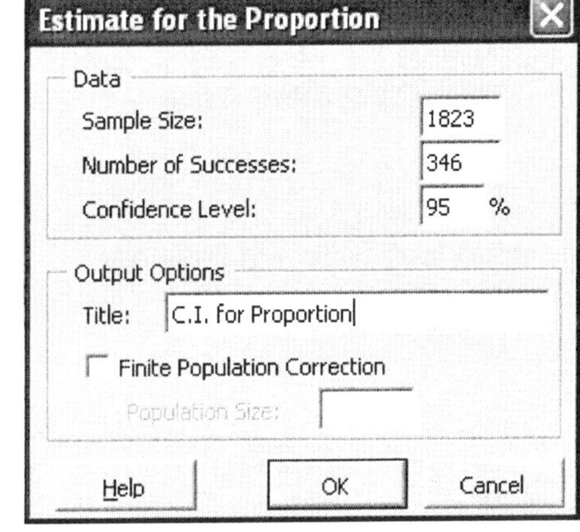

The output follows in a separate sheet:

	A	B
1	**C.I. for Proportion**	
2		
3	**Data**	
4	**Sample Size**	**1823**
5	**Number of Successes**	**346**
6	**Confidence Level**	**95%**
7		
8	Intermediate Calculations	
9	Sample Proportion	0.189797038
10	Z Value	-1.95996279
11	Standard Error of the Proportion	0.009184347
12	Interval Half Width	0.018000979
13		
14	**Confidence Interval**	
15	**Interval Lower Limit**	**0.171796059**
16	**Interval Upper Limit**	**0.207798017**

Note the point estimate (sample proportion) is .19 and the interval is (.17, .21).
The margin of error is .02. Excel calls this the Interval Half Width.

Chapter 7 Exercise 7.7 Believe in Heaven?
Constructing the confidence interval estimate for a proportion

Choose **PHStat > Confidence Interval > Estimate for the Proportion**
Enter Sample size **1158**
Number of Successes **996**
Confidence Level: **95** **Note:** This is 86% of the sample
Output Options:
Title: **C.I. for Proportion**
OK

The output appears in a separate sheet.

	A	B
1	**CI for Proportion**	
2		
3	**Data**	
4	**Sample Size**	**1158**
5	**Number of Successes**	**996**
6	**Confidence Level**	**95%**
7		
8	Intermediate Calculations	
9	Sample Proportion	0.860103627
10	Z Value	-1.95996279
11	Standard Error of the Proportion	0.010193524
12	Interval Half Width	0.019978927
13		
14	**Confidence Interval**	
15	**Interval Lower Limit**	**0.8401247**
16	**Interval Upper Limit**	**0.880082554**

Note the sample proportion, or point estimate = .86
The interval is (.84, .88)
And the error is .02

Chapter 7 Exercise 7.21 Exit Poll Predictions
Constructing the Confidence Interval Estimate for a Population Proportion

Choose **PHStat > Confidence Interval > Estimate for the Proportion**
Enter: Sample size **1400**
Number of Successes **660**
Confidence Level**: 95**
Output Options:
Title: **95% C.I. for Proportions**
OK

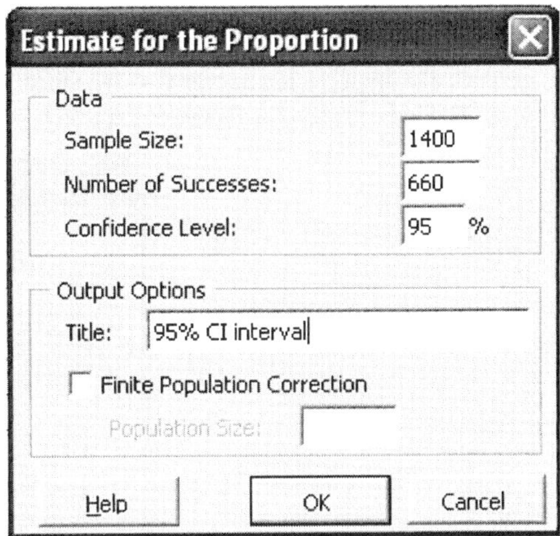

The output is in a separate sheet:

	A	B
1	**95% CI interval**	
2		
3	**Data**	
4	**Sample Size**	**1400**
5	**Number of Successes**	**660**
6	**Confidence Level**	**95%**
7		
8	Intermediate Calculations	
9	Sample Proportion	0.471428571
10	Z Value	-1.95996279
11	Standard Error of the Proportion	0.013341227
12	Interval Half Width	0.026148308
13		
14	**Confidence Interval**	
15	**Interval Lower Limit**	**0.445280263**
16	**Interval Upper Limit**	**0.49757688**

The interval is (.45, .50) with a point estimate of .47.

Repeat the procedure for the 99% Confidence Interval to get a (.44, .51) Interval with a point estimate of .47

What conclusions can you draw?

	A	B
1	**99% CI interval**	
2	**Data**	
3	**Sample Size**	**1400**
4	**Number of Successes**	**660**
5	**Confidence Level**	**99%**
6	Intermediate Calculations	
7	Sample Proportion	0.471428571
8	Z Value	-2.57583134
9	Standard Error of the Proportion	0.013341227
10	Interval Half Width	0.034364751
11	**Confidence Interval**	
12	**Interval Lower Limit**	**0.437063821**
13	**Interval Upper Limit**	**0.505793322**

Chapter 7 Example 7: eBay Auctions of Palm Handheld Computers
Constructing a Confidence Interval Estimate for a Population Mean?

Open the data file **ebay_ auctions** from the Excel folder of the data disk.
To get the descriptive statistics for the set of data
Choose **Tools > Data Analysis > Descriptive Statistics > OK**

Enter Input Range **A1 – B16**
Click Grouped By **Columns**
Click **Summary Statistics**
OK

The output will appear in cell A22 or you may choose a new worksheet.

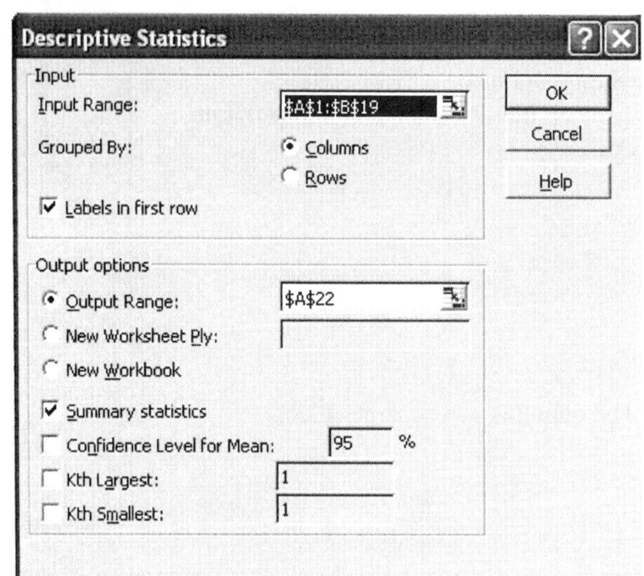

Note the mean and standard deviations.

22	*Buy-It-Now*		*Bidding*	
23				
24	Mean	233.5714286	Mean	231.6111111
25	Standard Error	5.532833352	Standard Error	5.170390112
26	Median	235	Median	240
27	Mode	225	Mode	246
28	Standard Deviation	14.63850109	Standard Deviation	21.93610746
29	Sample Variance	214.2857143	Sample Variance	481.1928105
30	Kurtosis	-0.65488889	Kurtosis	0.582289202
31	Skewness	-0.38824558	Skewness	-1.13783237
32	Range	40	Range	77
33	Minimum	210	Minimum	178
34	Maximum	250	Maximum	255
35	Sum	1635	Sum	4169
36	Count	7	Count	18

You can also compare the selling prices by doing a dotplot of the data values. Choose **PHStat > Descriptive Statistics > Dot Scale Diagram** for each column of data. The output will also give you some descriptive statistics. Only the Buy It Now is shown here.

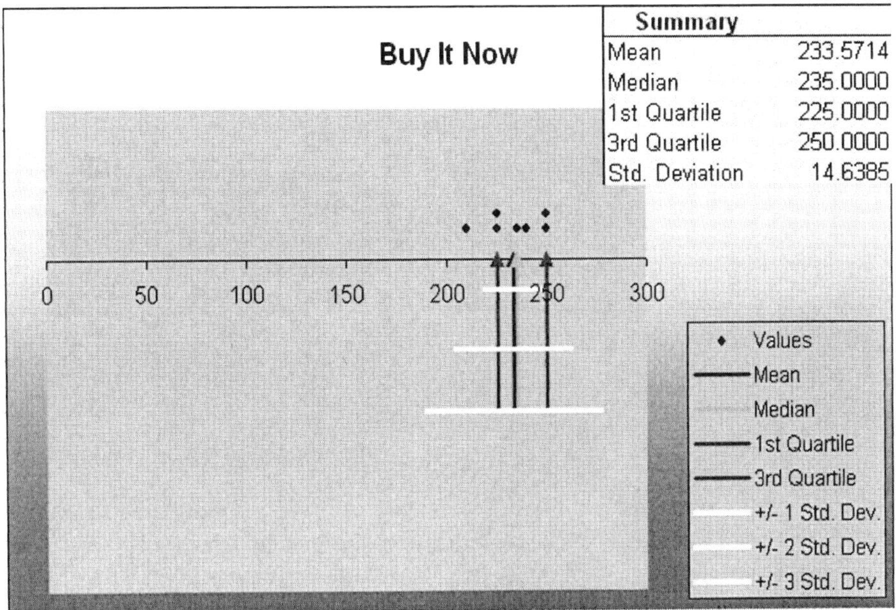

To get the Confidence interval around the mean
Choose > **PHStat > Confidence intervals > Estimate for the Mean Sigma Unknown**
Click **Sample Statistics Unknown**
Sample Cell Range: **A1-A8**
Click: **First cell contains label**
Enter Title: **Buy It Now**
OK

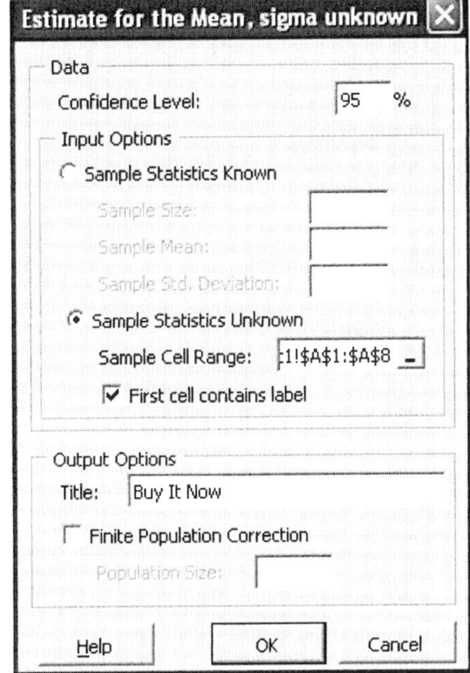

The output is placed in a separate sheet.

The interval is (220.03, 247.11)

	A	B
1	**Buy It Now**	
2		
3	**Data**	
4	**Sample Standard Deviation**	**14.63850109**
5	**Sample Mean**	**233.5714286**
6	**Sample Size**	**7**
7	**Confidence Level**	**95%**
8		
9	Intermediate Calculations	
10	Standard Error of the Mean	5.532833352
11	Degrees of Freedom	6
12	*t* Value	2.446913641
13	Interval Half Width	13.5383654
14		
15	**Confidence Interval**	
16	**Interval Lower Limit**	**220.03**
17	**Interval Upper Limit**	**247.11**

Using the same data, and now constructing the confidence interval for the Bidding column we get the following output. (This is exercise 7.29). Note that $178 seems to be an outlier. Repeat the procedure after deleting $178. Compare the two outputs.

	A	B
1	**Bidding**	
2		
3	**Data**	
4	**Sample Standard Deviation**	**21.93610746**
5	**Sample Mean**	**231.6111111**
6	**Sample Size**	**18**
7	**Confidence Level**	**95%**
8		
9	Intermediate Calculations	
10	Standard Error of the Mean	5.170390112
11	Degrees of Freedom	17
12	*t* Value	2.109818524
13	Interval Half Width	10.90858484
14		
15	**Confidence Interval**	
16	**Interval Lower Limit**	**220.70**
17	**Interval Upper Limit**	**242.52**

	A	B
1	**Bidding**	
2		
3	**Data**	
4	**Sample Standard Deviation**	**17.91831958**
5	**Sample Mean**	**234.7647059**
6	**Sample Size**	**17**
7	**Confidence Level**	**95%**
8		
9	Intermediate Calculations	
10	Standard Error of the Mean	4.345830838
11	Degrees of Freedom	16
12	*t* Value	2.119904821
13	Interval Half Width	9.212747743
14		
15	**Confidence Interval**	
16	**Interval Lower Limit**	**225.55**
17	**Interval Upper Limit**	**243.98**

Chapter 7 Exercise 7.51 Do You Like Tofu?
Confidence Interval Estimate of a Population Proportion with a Small Sample Size

In this problem, we have a very small sample of five students. All say they like tofu. We will use the plus four method in this situation, giving us seven successes out of nine trials. Using the formulas in the text:

We have N = 9, $\hat{p} = 7/9 = 0.7777$ and
$$se = \sqrt{\hat{p}(1-\hat{p})/n} = \sqrt{.7777(1-.7777)/9} = .1386.$$

The resulting interval = $\hat{p} \pm 1.96(se) = 0.7777 \pm (1.96)(0.1386)$.

This gives us the interval (0 .506, 1.05). Note that the right end of the interval is 1.05. We cannot have a value greater than 1.0 for a proportion, so we would report the interval as (0.506, 1.00).

We can do this using Excel.
Choose **PHStat > Confidence Intervals > Estimate for the Proportion** and enter the fields as shown.

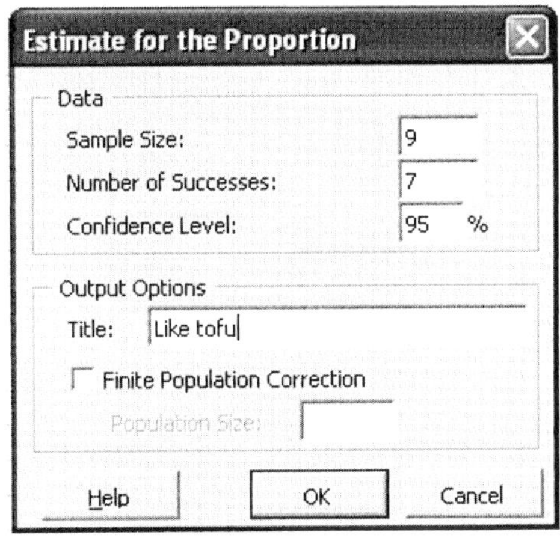

We are using the plus 4 method, adding 4 to the sample size, 2 of which are considered successes.

The results are shown in a separate sheet.

Note that the upper limit is greater than 1, so we state the interval as (.51 , 1.0).

	A	B
1	**Like tofu**	
2		
3	**Data**	
4	**Sample Size**	9
5	**Number of Successes**	7
6	**Confidence Level**	95%
7		
8	Intermediate Calculations	
9	Sample Proportion	0.777777778
10	Z Value	-1.95996279
11	Standard Error of the Proportion	0.138579903
12	Interval Half Width	0.271611453
13		
14	**Confidence Interval**	
15	**Interval Lower Limit**	**0.506166324**
16	**Interval Upper Limit**	**1.049389231**

Chapter 8 Example 4 Dr. Dog: Can Dogs Detect Cancer By Smell?
Hypothesis Testing For a Proportion

We are asked to test H_0: p = 1/7 vs H_1: p > 1/7. The sample evidence presented is that in the total of 54 trials, the dogs made the correct decision 22 times.

Choose **PHStat > One Sample Tests > z Test for the Proportion**

Enter Null Hypothesis: **.143**
Level of Significance: **.05**
Number of Successes: **22**
Sample Size: **54**
Select: **Upper Tail Test**
Enter Title: **Dr Dog Test for the Proportion**
OK

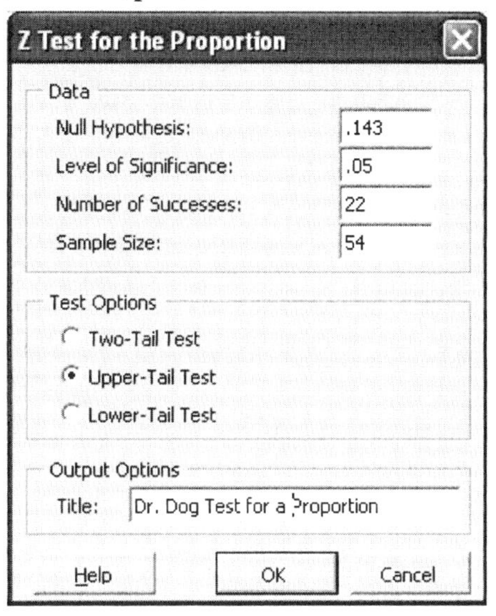

	A	B
1	**Dr. Dog Test for a Proportion**	
2		
3	**Data**	
4	**Null Hypothesis** *p*=	**0.143**
5	**Level of Significance**	**0.05**
6	**Number of Successes**	**22**
7	**Sample Size**	**54**
8		
9	Intermediate Calculations	
10	Sample Proportion	0.407407407
11	Standard Error	0.047638881
12	**Z Test Statistic**	**5.550243898**
13		
14	**Upper-Tail Test**	
15	**Upper Critical Value**	**1.644853476**
16	***p*-Value**	**1.43002E-08**
17	**Reject the null hypothesis**	

Note the test statistic (z= 5.5) and the p-value (0.000) With such a small p-value we will reject the null hypothesis.

Chapter 8 Example 6: Can TT Practitioners Detect a Human Energy Field?
Hypothesis Testing For A Proportion

We are asked to test H_0: p = .5 vs H_1: p > .5 . The sample evidence presented is that in the total of 150 trials, the TT practitioners were correct with 70 of their predictions.

Choose **PHStat > One Sample Tests > z Test for the Proportion**

Enter Null Hypothesis: **.5**
Level of Significance: **.05**
Number of Successes: **70**
Sample Size: **150**
Select: **Upper Tail Test**
Enter Title: **TT Practitioners Test for the Proportion**
OK

The results appear in a separate worksheet.

With such a large p-value (p-value = 0.793) the decision is Fail To Reject.

	A	B
1	TT Practitionors Test for the Proportio	
2		
3	Data	
4	Null Hypothesis p=	0.5
5	Level of Significance	0.05
6	Number of Successes	70
7	Sample Size	150
8		
9	Intermediate Calculations	
10	Sample Proportion	0.466666667
11	Standard Error	0.040824829
12	Z Test Statistic	-0.816496581
13		
14	Upper-Tail Test	
15	Upper Critical Value	1.644853476
16	p-Value	0.792891972
17	Do not reject the null hypothesis	

Chapter 8 Exercise 8.15 Another Test of Therapeutic Touch
Hypothesis Testing For A Proportion

We are asked to test H_0: p = .5 vs H_1: p > .5 . The sample evidence presented is that in the total of 130 trials, the TT practitioners were correct with 53 of their predictions.

Choose **PHStat > One Sample Tests > z Test for the Proportion**

Enter Null Hypothesis: **.5**
Level of Significance: **.05**
Number of Successes: **53**
Sample Size: **130**
Select: **Upper Tail Test**
Enter Title: **TT Practitioners Test for the Proportion**
OK

The results appear in a separate worksheet.

	A	B
1	TT Practitionors Test for the Proportic	
2		
3	Data	
4	Null Hypothesis p=	0.5
5	Level of Significance	0.05
6	Number of Successes	53
7	Sample Size	130
8		
9	Intermediate Calculations	
10	Sample Proportion	0.407692308
11	Standard Error	0.043852901
12	Z Test Statistic	-2.104939246
13		
14	Upper-Tail Test	
15	Upper Critical Value	1.644853476
16	p-Value	0.982351764
17	Do not reject the null hypothesis	

Chapter 8 Exercise 8.17 Gender Bias In Selecting Managers
Hypothesis Testing For A Proportion

We are asked to test H_0: p = .6 vs H_1: p $\neq$ .6 . The sample evidence presented is that in the total of 40 employees chosen for management training, 28 were male.

Choose **PHStat > One Sample Tests > z Test for the Proportion**

Enter Null Hypothesis: **.6**
Level of Significance: **.05**
Number of Successes: **28**
Sample Size: **40**
Select: **Two Tail Test**
Enter Title: **Gender Bias Test for the Proportion**
OK

The results appear in a separate worksheet.

	A	B
1	**Gender Bias Test for the Proportion**	
2		
3	**Data**	
4	**Null Hypothesis** *p*=	**0.6**
5	**Level of Significance**	**0.05**
6	**Number of Successes**	**28**
7	**Sample Size**	**40**
8		
9	Intermediate Calculations	
10	Sample Proportion	0.7
11	Standard Error	0.077459667
12	Z Test Statistic	**1.290994449**
13		
14	**Two-Tail Test**	
15	**Lower Critical Value**	**-1.959962787**
16	**Upper Critical value**	**1.959962787**
17	***p*-Value**	**0.196705733**
18	**Do not reject the null hypothesis**	

Chapter 8 Example 7 Mean Weight Change in Anorexic Girls
Significance Test About a Mean

Open the worksheet **anorexia,** which is found in the Excel folder of the data disk. We wish to perform a two-tailed test on $H_0 : \mu = 0$ versus Ha: $\mu \neq 0$ using the weight gains found in Column G cogchange.

Choose **PHStat >One Sample Tests > t-test for the Mean, sigma unknown**

Enter:
Null Hypothesis: **0**
Level of Significance: **.05**
Choose **Sample Statistics Unknown**
Enter Sample Range: **G1 – G30**
Check: **First cell contains label**
Enter Title: **Mean Wt Change**
OK

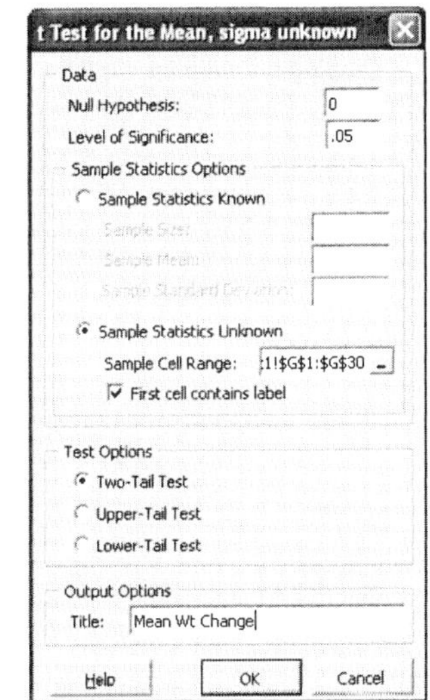

The output appears in a separate worksheet.

	A	B
1	**Mean Wt Change**	
2		
3	**Data**	
4	**Null Hypothesis** μ=	0
5	**Level of Significance**	0.05
6	**Sample Size**	29
7	**Sample Mean**	3.006896552
8	**Sample Standard Deviation**	7.308504392
9		
10	Intermediate Calculations	
11	Standard Error of the Mean	1.357155195
12	Degrees of Freedom	28
13	***t* Test Statistic**	2.215587844
14		
15	**Two-Tail Test**	
16	**Lower Critical Value**	-2.048409442
17	**Upper Critical Value**	2.048409442
18	***p*-Value**	0.035022597
19	**Reject the null hypothesis**	

Chapter 8 Exercise 8.35 Anorexia in Teenage girls
Significance Test About a Mean

We wish to perform a two-tailed test on $H_0 : \mu = 0$ versus Ha: $\mu \neq 0$ using the weight gains given. First we will plot the data with a box plot. There are no outliers indicated, and the data distribution is not highly skewed.

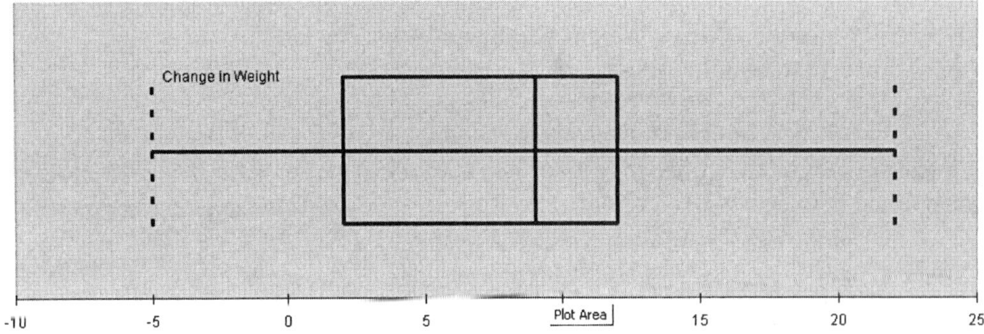

Performing the significance test about $H_0 : \mu = 0$ versus Ha: $\mu \neq 0$
Choose **PHStat >One Sample Tests > t-test for the Mean, sigma unknown**

Enter:

Null Hypothesis: **0**

Level of Significance: **.05**

Choose **Sample Statistics Unknown**
Enter Sample Range: **A1-A18**
Check: **First cell contains label**
Enter Title: **Change in Weight**
OK

The output appears in a separate worksheet.

	A	B
1	**Change in Weight**	
2		
3	**Data**	
4	**Null Hypothesis** μ=	**0**
5	**Level of Significance**	**0.05**
6	**Sample Size**	**17**
7	**Sample Mean**	**7.294117647**
8	**Sample Standard Deviation**	**7.183006908**
9		
10	Intermediate Calculations	
11	Standard Error of the Mean	1.74213507
12	Degrees of Freedom	16
13	***t* Test Statistic**	**4.186884113**
14		
15	**Two-Tail Test**	
16	**Lower Critical Value**	**-2.119904821**
17	**Upper Critical Value**	**2.119904821**
18	***p*-Value**	**0.000697365**
19	**Reject the null hypothesis**	

Chapter 9 Example 4 Confidence Interval Comparing Heart Attack Rates for Aspirin and Placebo – Confidence Interval for Difference Between Two Proportions

To construct a confidence interval $p_1 - p_2$, note that for the placebo group X = 189 and n = 11034, and for the aspirin group X = 104 and n = 11037.

Select **PHStat > Two Sample Tests > z test for Differences in Two Proportions.**

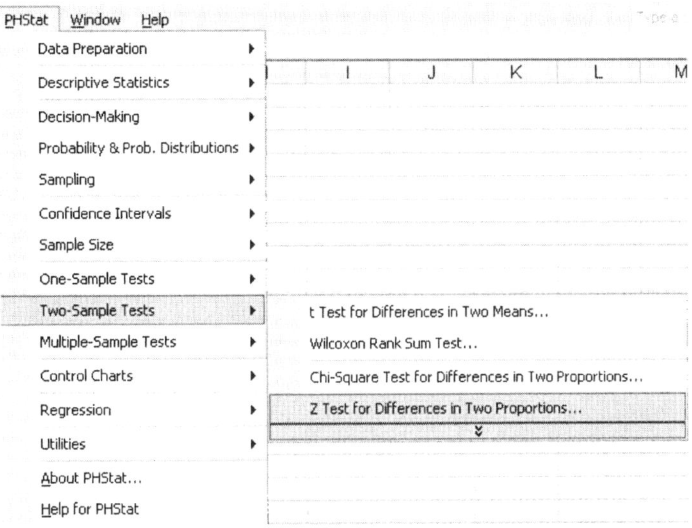

Enter Hypothesized Difference: **0**
Level of Significance: **.05**
Population 1 Sample
Number of Successes: **189**
Sample Size: **11034**
Population 2 Sample
Number of Successes: **104**
Sample Size: **11037**
Test Options
Click **Two Tail Test**
Output Options:
Title: **Placebo vs Aspirin**

OK

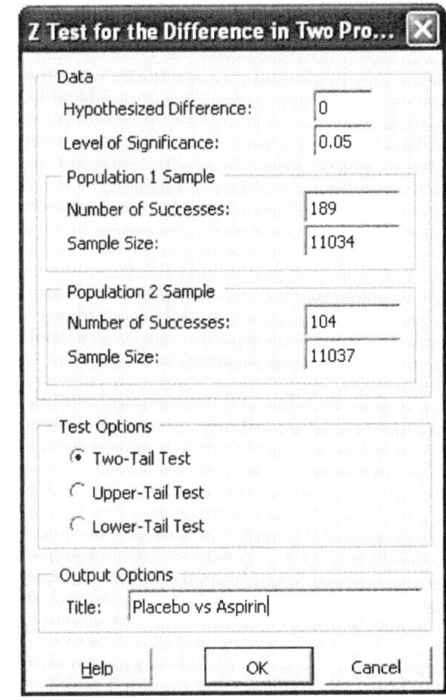

The results appear in a new worksheet.

This is not a confidence interval for a difference in two proportions. Excel and PHStat cannot produce a confidence interval. However by using this two-sided test with a level of significance = .05, we can find the z statistic and the point estimate. We would still need to calculate the margin of error to determine the interval.

	A	B
1	Placebo vs Aspirin	
2		
3	Data	
4	Hypothesized Difference	0
5	Level of Significance	0.05
6	Group 1	
7	Number of Successes	189
8	Sample Size	11034
9	Group 2	
10	Number of Successes	104
11	Sample Size	11037
12		
13	Intermediate Calculations	
14	Group 1 Proportion	0.017128874
15	Group 2 Proportion	0.00942285
16	Difference in Two Proportions	0.007706024
17	Average Proportion	0.013275339
18	Z Test Statistic	5.001388204
19		
20	Two-Tail Test	
21	Lower Critical Value	-1.959962787
22	Upper Critical Value	1.959962787
23	*p*-Value	5.70091E-07
24	Reject the null hypothesis	

Chapter 9 Example 5 Is TV Watching Associated With Aggressive Behavior? Significance Test for Difference Between Two Proportions

To perform a significance test for H_0: p1 – p2 = 0 versus H_a: p1 – p2 ≠ 0 note that for the "less than 1 hour of TV per day" group X = 5 and n = 88 and for the " at least 1 hour of TV per day group X = 154 and n = 619.

Select **PHStat > Two Sample Tests > z test for Differences in Two Proportions.**

Enter Hypothesized Difference: **0**
Level of Significance: **.05**
Population 1 Sample
Number of Successes: **5**
Sample Size: **88**
Population 2 Sample
Number of Successes: **154**
Sample Size: **619**
Test Options
Click **Two Tail Test**
Output Options:
Title: **TV Watching with Aggressive Behavior**
OK

The results appear in a new worksheet.

	A	B
1	TV Watching with Agressive Behavior	
2		
3	Data	
4	Hypothesized Difference	0
5	Level of Significance	0.05
6	Group 1	
7	Number of Successes	5
8	Sample Size	88
9	Group 2	
10	Number of Successes	154
11	Sample Size	619
12		
13	Intermediate Calculations	
14	Group 1 Proportion	0.056818182
15	Group 2 Proportion	0.248788368
16	Difference in Two Proportions	-0.191970187
17	Average Proportion	0.224893918
18	Z Test Statistic	-4.035908513
19		
20	Two-Tail Test	
21	Lower Critical Value	-1.959962787
22	Upper Critical Value	1.959962787
23	*p*-Value	5.44182E-05
24	Reject the null hypothesis	

Chapter 9 Exercise 9.9 Drinking and Unplanned Sex
Significance Test for the Difference Between Two Populations

To perform a significance test for H_0: p1 – p2 = 0 versus H_a: p1 – p2 ≠ 0 note that for the 2001 group X = 1871 (.213*8783) and n = 8783, and for the 1993 group X = 2440 (.192*12708) and n = 12708

Select **PHStat > Two Sample Tests > z test for Differences in Two Proportions.**

Enter Hypothesized Difference: **0**
Level of Significance: **.05**
Population 1 Sample
Number of Successes: **1871**
Sample Size: **8783**
Population 2 Sample
Number of Successes: **2440**
Sample Size: **12708**
Test Options
Click **Two Tail Test**
Output Options:
Title: **Drinking and Unplanned Sex**

OK

	A	B
1	**Drinking and Unplanned Sex**	
2		
3	**Data**	
4	**Hypothesized Difference**	**0**
5	**Level of Significance**	**0.05**
6	**Group 1**	
7	**Number of Successes**	**1871**
8	**Sample Size**	**8783**
9	**Group 2**	
10	**Number of Successes**	**2440**
11	**Sample Size**	**12708**
12		
13	Intermediate Calculations	
14	Group 1 Proportion	0.213025162
15	Group 2 Proportion	0.192005036
16	Difference in Two Proportions	0.021020126
17	Average Proportion	0.200595598
18	**Z Test Statistic**	**3.782884498**
19		
20	**Two-Tail Test**	
21	**Lower Critical Value**	**-1.959962787**
22	**Upper Critical Value**	**1.959962787**
23	***p*-Value**	**0.000155072**
24	**Reject the null hypothesis**	

Chapter 9 Example 8 Nicotine – How Much More Addicted are Smokers than Ex-Smokers? -- Confidence Interval for Difference Between Population Means

To construct a confidence interval for $\mu_1 - \mu_2$, note that for the smoker group, $\bar{X} = 5.9$ and s = 3.3 for n = 75 and for the ex-smoker group $\bar{X} = 1.0$ and s = 2.3 for n = 257.

Choose **PHStat> Two Sample Tests > t Test for Differences in Two Means**

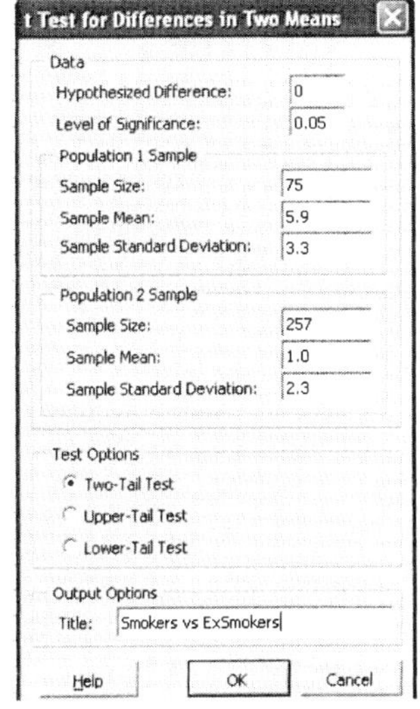

Enter
Hypothesized Difference **0**
Level of Significance **.05**
Population 1
Sample Size **75**
Sample Mean **5.9**
Sample Standard Deviation **3.3**
Population 2

Sample Size **257**
Sample Mean **1.0**
Sample Standard Deviation **2.3**

Choose **Two Tail Test**

OK

The results appear in a new worksheet. As noted before, Excel and PHStat cannot produce a confidence interval. However by using this two-sided test with a level of significance = .05, we can find the z statistic and the point estimate. We would still need to calculate the margin of error to determine the interval.

	A	B
1	**Smokers vs ExSmokers**	
2		
3	**Data**	
4	**Hypothesized Difference**	**0**
5	**Level of Significance**	**0.05**
6	**Population 1 Sample**	
7	**Sample Size**	**75**
8	**Sample Mean**	**5.9**
9	**Sample Standard Deviation**	**3.3**
10	**Population 2 Sample**	
11	**Sample Size**	**257**
12	**Sample Mean**	**1**
13	**Sample Standard Deviation**	**2.3**
14		
15	Intermediate Calculations	
16	Population 1 Sample Degrees of Freedom	74
17	Population 2 Sample Degrees of Freedom	256
18	Total Degrees of Freedom	330
19	Pooled Variance	6.545758
20	Difference in Sample Means	4.9
21	***t* Test Statistic**	**14.59299**
22		
23	**Two-Tail Test**	
24	**Lower Critical Value**	**-1.96718**
25	**Upper Critical Value**	**1.967178**
26	***p*-Value**	**1.45E-37**
27	**Reject the null hypothesis**	

Chapter 9 Example 9 Does Cell Phone Use While Driving Impair Reaction Time? Significance Test for Comparing Two Population Means

To perform a significance test for H_0: $\mu_1 - \mu_2 = 0$ versus : Ha: $\mu_1 - \mu_2 \neq 0$, note that for the "cell phone" group , $\bar{X} = 585.2$ and s = 89.6 for n = 32 and for the "control" group $\bar{X} = 533.7$ and s = 65.3 for n = 32.

Choose **PHStat> Two Sample Tests > t Test for Differences in Two Means**

Enter

Hypothesized Difference **0**
Level of Significance **.05**

Population 1

Sample Size **32**
Sample Mean **585.2**
Sample Standard Deviation **89.6**

Population 2

Sample Size **32**
Sample Mean **533.7**
Sample Standard Deviation **65.3**

Choose **Two Tail Test**

OK

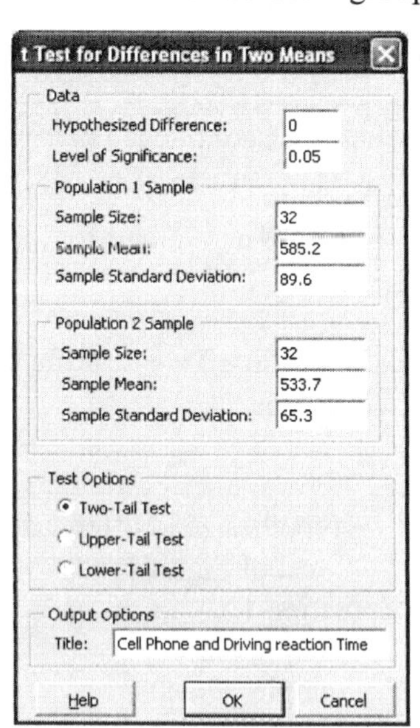
t Test for Differences in Two Means
Data
Hypothesized Difference: 0
Level of Significance: 0.05
Population 1 Sample
Sample Size: 32
Sample Mean: 585.2
Sample Standard Deviation: 89.6
Population 2 Sample
Sample Size: 32
Sample Mean: 533.7
Sample Standard Deviation: 65.3
Test Options
Two-Tail Test
Upper-Tail Test
Lower-Tail Test
Output Options
Title: Cell Phone and Driving reaction Time
Help OK Cancel

The results appear in a new worksheet.

	A	B
1	Cell Phone and Driving reaction Time	
2		
3	Data	
4	Hypothesized Difference	0
5	Level of Significance	0.05
6	Population 1 Sample	
7	Sample Size	32
8	Sample Mean	585.2
9	Sample Standard Deviation	89.6
10	Population 2 Sample	
11	Sample Size	32
12	Sample Mean	533.7
13	Sample Standard Deviation	65.3
14		
15	Intermediate Calculations	
16	Population 1 Sample Degrees of Freedom	31
17	Population 2 Sample Degrees of Freedom	31
18	Total Degrees of Freedom	62
19	Pooled Variance	6146.125
20	Difference in Sample Means	51.5
21	*t* Test Statistic	2.627644
22		
23	Two-Tail Test	
24	Lower Critical Value	-1.99897
25	Upper Critical Value	1.998969
26	*p*-Value	0.010822
27	Reject the null hypothesis	

Chapter 9 Exercise 9.21 Females or Males More Nicotine Dependant Significance Test for Comparing Two Population Means

To perform a significance test for H$_0$: $\mu_1 - \mu_2 = 0$ versus : Ha: $\mu_1 - \mu_2 \neq 0$, note that for the females , $\bar{X} = 2.8$ and s = 3.6 for n = 150 and for the males $\bar{X} = 1.6$ and s = 2.9 for n = 182.

Choose **PHStat> Two Sample Tests > t Test for Differences in Two Means**
Enter

Hypothesized Difference **0**
Level of Significance **.05**
Population 1
Sample Size **150**
Sample Mean **2.8**
Sample Standard Deviation **3.6**
Population 2
Sample Size **182**
Sample Mean **1.6**
Sample Standard Deviation **2.9**

Choose **Two Tail Test**
OK

	A	B
1	Nicotine dependance	
2		
3	Data	
4	Hypothesized Difference	0
5	Level of Significance	0.05
6	Population 1 Sample	
7	Sample Size	150
8	Sample Mean	2.8
9	Sample Standard Deviation	3.6
10	Population 2 Sample	
11	Sample Size	182
12	Sample Mean	1.6
13	Sample Standard Deviation	2.9
14		
15	Intermediate Calculations	
16	Population 1 Sample Degrees of Freedom	149
17	Population 2 Sample Degrees of Freedom	181
18	Total Degrees of Freedom	330
19	Pooled Variance	10.46439
20	Difference in Sample Means	1.2
21	*t* Test Statistic	3.363849
22		
23	Two-Tail Test	
24	Lower Critical Value	-1.96718
25	Upper Critical Value	1.967178
26	*p*-Value	0.000859
27	Reject the null hypothesis	

Chapter 9 Exercise 9.23 TV Watching and Gender
Significance Test for Comparing Two Population Means

To perform a significance test for H_0: $\mu_1 - \mu_2 = 0$ versus : Ha: $\mu_1 - \mu_2 \neq 0$, note that for the females , $\overline{X} = 3.06$ and s = 2.12 for n = 506 and for the males $\overline{X} = 2.88$ and s = 2.63 for n = 399.

Choose **PHStat> Two Sample Tests > t Test for Differences in Two Means**
Enter

Hypothesized Difference	**0**
Level of Significance	**.05**
Population 1	
Sample Size	**506**
Sample Mean	**3.06**
Sample Standard Deviation	**2.12**
Population 2	
Sample Size	**399**
Sample Mean	**2.88**
Sample Standard Deviation	**2.63**

Choose **Two Tail Test**
OK

Note that we are using PHStat for this test. Data Analysis in Excel has a test for comparing two means, but not for summarized data.

	A	B
1	TV Watching and Gender	
2		
3	Data	
4	Hypothesized Difference	0
5	Level of Significance	0.05
6	Population 1 Sample	
7	Sample Size	506
8	Sample Mean	3.06
9	Sample Standard Deviation	2.12
10	Population 2 Sample	
11	Sample Size	399
12	Sample Mean	2.88
13	Sample Standard Deviation	2.63
14		
15	Intermediate Calculations	
16	Population 1 Sample Degrees of Freedom	505
17	Population 2 Sample Degrees of Freedom	398
18	Total Degrees of Freedom	903
19	Pooled Variance	5.562124
20	Difference in Sample Means	0.18
21	*t* Test Statistic	1.13996
22		
23	Two-Tail Test	
24	Lower Critical Value	-1.96259
25	Upper Critical Value	1.962594
26	*p*-Value	0.254605
27	Do not reject the null hypothesis	

Chapter 9 Example 10 Is Arthroscopic Surgery Better Than Placebo
Significance Test for Comparing Two Population Means Assuming Equal Population Standard Deviations

To perform a significance test for H_0: $\mu_1 - \mu_2 = 0$ versus : Ha: $\mu_1 - \mu_2 \neq 0$, note that for the "placebo" group , $\overline{X} = 51.6$ and s = 23.7 for n = 60 and for the "lavage arthroscopic" group $\overline{X} = 53.7$ and s = 23.7 for n = 61.

Choose **PHStat> Two Sample Tests > t Test for Differences in Two Means**

Enter
- Hypothesized Difference **0**
- Level of Significance **.05**

Population 1
- Sample Size **60**
- Sample Mean **51.6**
- Sample Standard Deviation **23.7**

Population 2
- Sample Size **61**
- Sample Mean **53.7**
- Sample Standard Deviation **23.7**

Choose **Two Tail Test**
OK

Note that we are using PHStat for this test. Data Analysis in Excel has a test for comparing two means, but not for summarized data.

	A	B
1	Arthroscopic Surgery	
2		
3	**Data**	
4	**Hypothesized Difference**	0
5	**Level of Significance**	0.05
6	**Population 1 Sample**	
7	**Sample Size**	60
8	**Sample Mean**	51.6
9	**Sample Standard Deviation**	23.7
10	**Population 2 Sample**	
11	**Sample Size**	61
12	**Sample Mean**	53.7
13	**Sample Standard Deviation**	23.7
14		
15	Intermediate Calculations	
16	Population 1 Sample Degrees of Freedom	59
17	Population 2 Sample Degrees of Freedom	60
18	Total Degrees of Freedom	119
19	Pooled Variance	561.69
20	Difference in Sample Means	-2.1
21	*t* **Test Statistic**	-0.48733
22		
23	**Two-Tail Test**	
24	**Lower Critical Value**	-1.9801
25	**Upper Critical Value**	1.980097
26	*p*-**Value**	0.626924
27	**Do not reject the null hypothesis**	

Chapter 9 Example 13 Matched Pairs Analysis of Cell Phone Impact on Driver Reaction Time
Comparing Means With Matched Pairs

Enter the data from Table 9.9. Since neither Excel nor PHStat will do a matched pairs test, we will perform a t test on the difference column and get the same results.

Choose **PHStat> One Sample Tests > t Test for the Mean Sigma Unknown**
Enter
- Null Hypothesis: **0**
- Level of Significance: **.05**
- Select **Sample Statistics Unknown**
- Sample Cell Range: **D1-D33**

Check **First cell contains label**
Check **Two Tail Test**
Enter Title: **Cell Phone Impact on Driver Reaction Time**
OK

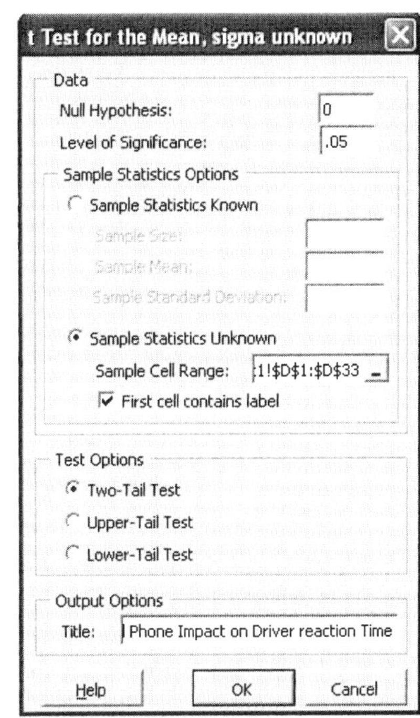

The results appear in a separate worksheet.

	A	B
1	**Cell Phone Impact on Driver reaction Time**	
2		
3	**Data**	
4	**Null Hypothesis** μ=	**0**
5	**Level of Significance**	**0.05**
6	**Sample Size**	**32**
7	**Sample Mean**	**50.625**
8	**Sample Standard Deviation**	**52.48578917**
9		
10	Intermediate Calculations	
11	Standard Error of the Mean	9.278264359
12	Degrees of Freedom	31
13	***t* Test Statistic**	**5.456300666**
14		
15	**Two-Tail Test**	
16	**Lower Critical Value**	**-2.039514584**
17	**Upper Critical Value**	**2.039514584**
18	***p*-Value**	**5.80341E-06**
19	**Reject the null hypothesis**	

Chapter 9 Exercise 9.41 Matched Pairs Analysis of Movies and Sports
Comparing Means With Matched Pairs

Enter the data from table in the problem. Form the difference and place it in the next column. Since neither Excel nor PHStat will do a matched pairs test, we will perform a t test on the difference column and get the same results.

	A	B	C	D
1	Student	Movies	Sports	Difference
2	1	10	5	5
3	2	4	0	4
4	3	12	20	-8
5	4	2	6	-4
6	5	12	2	10
7	6	7	8	-1
8	7	45	12	33
9	8	1	25	-24
10	9	25	0	25
11	10	12	12	0

Choose **PHStat> One Sample Tests > t Test for the Mean Sigma Unknown**
Enter
- Null Hypothesis: **0**
- Level of Significance: **.05**
- Select **Sample Statistics Unknown**
- Sample Cell Range: **D1-D33**

Check **First cell contains label**
Check **Two Tail Test**
Enter Title: **Movies VS Sports**
OK

	A	B
1	**Movies vs Sports**	
2		
3	**Data**	
4	**Null Hypothesis** μ=	**0**
5	**Level of Significance**	**0.05**
6	**Sample Size**	**10**
7	**Sample Mean**	**4**
8	**Sample Standard Deviation**	**16.16580754**
9		
10	Intermediate Calculations	
11	Standard Error of the Mean	5.112077203
12	Degrees of Freedom	9
13	***t* Test Statistic**	**0.782460796**
14		
15	**Two-Tail Test**	
16	**Lower Critical Value**	**-2.262158887**
17	**Upper Critical Value**	**2.262158887**
18	***p*-Value**	**0.454037837**
19	**Do not reject the null hypothesis**	

We can also get a confidence interval for the mean difference.

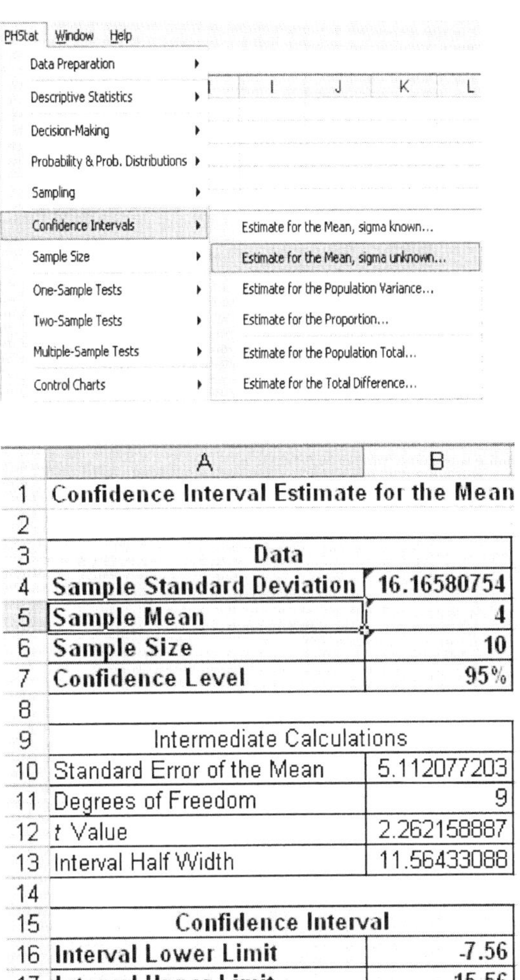

	A	B
1	**Confidence Interval Estimate for the Mean**	
2		
3	**Data**	
4	**Sample Standard Deviation**	16.16580754
5	**Sample Mean**	4
6	**Sample Size**	10
7	**Confidence Level**	95%
8		
9	Intermediate Calculations	
10	Standard Error of the Mean	5.112077203
11	Degrees of Freedom	9
12	*t* Value	2.262158887
13	Interval Half Width	11.56433088
14		
15	**Confidence Interval**	
16	**Interval Lower Limit**	-7.56
17	**Interval Upper Limit**	15.56

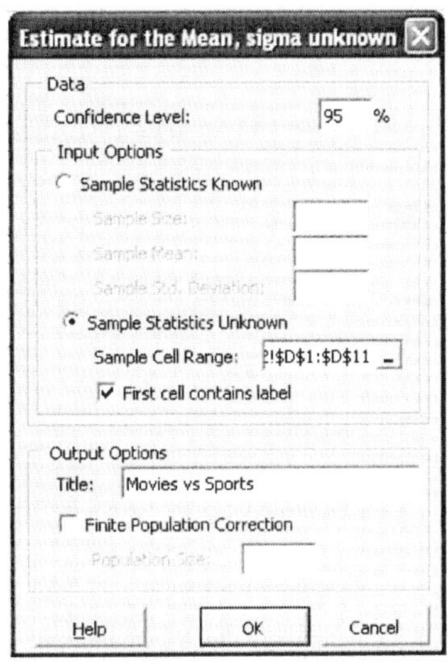

Chapter 9 Example 15 Inferences Comparing Beliefs in Heaven and Hell Comparing Proportions with Dependent Samples

We will create the 95% confidence interval for population mean difference $p_1 - p_2$ using summary statistics.

First note that the sample mean of the 1120 difference scores equals

$[(0)(833) + (1)(125) + (-1)(2) + (0)(160)]/1120 = 123/1120 = 0.109821$

and the standard deviation of the 1120 difference scores equals

$$\sqrt{\{[(0-.109821)^2*(833)+(1-.109821)^2*(125)+(-1-.109821)^2*(2)+(0-.109821)^2*(160)]/(1120-1)\}}$$

$= 0.318469$

Choose **PHStat > Confidence Intervals > Estimate for the Mean, Sigma Unknown**

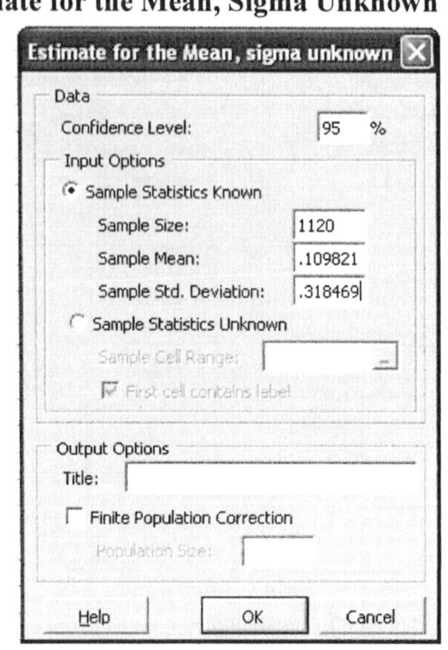

Enter: Confidence level **95**
Choose Sample Statistics Known

Enter: Sample Size **1120**
Sample Mean: **0.19821**
Sample Std. Deviation: **.318469**

OK

The results appear in a new worksheet.

	A	B
1	**Confidence Interval Estimate for the Mean**	
2		
3	**Data**	
4	**Sample Standard Deviation**	**0.318469**
5	**Sample Mean**	**0.109821**
6	**Sample Size**	**1120**
7	**Confidence Level**	**95%**
8		
9	Intermediate Calculations	
10	Standard Error of the Mean	0.009516081
11	Degrees of Freedom	1119
12	*t* Value	1.962084752
13	Interval Half Width	0.018671358
14		
15	**Confidence Interval**	
16	**Interval Lower Limit**	**0.09**
17	**Interval Upper Limit**	**0.13**

Chapter 9 Exercise 9.49 Heaven and Hell
Comparing Proportions for Dependent Samples

A point estimate for the difference between the population proportions believing in heaven and believing in hell is $\hat{p}_1 - \hat{p}_2 = .631 - .496 = .135$
Because we do not know the sample size used, but can assume dependent sampling, we'll use McNemar's test. Since McNemar's test is not included in Excel Data Analysis nor in PHStat, we will just do the computation in the worksheet.

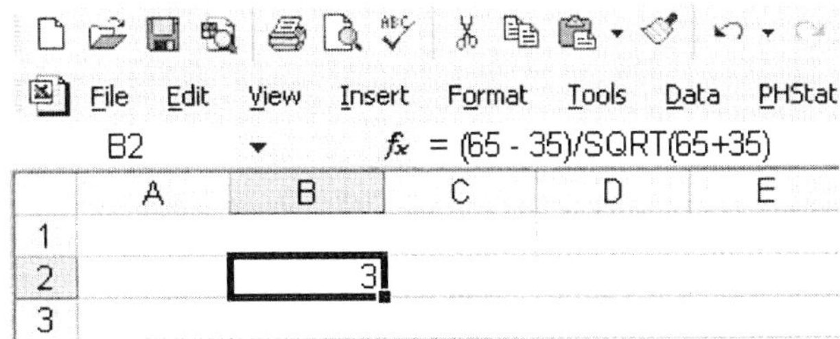

Chapter 10 Exercise 10.1 Gender Gap in Politics?
Contingency Tables

Excel does not normally do contingency tables, so all the work must be done manually. First enter the data in the spreadsheet . Using the Σ tool enter the row and column totals.

	A	B	C	D	E
1	**Gender**	**Democrat**	**Independent**	**Republican**	**All**
2	**Female**	567	534	395	1496
3	**Male**	356	460	369	1185
4	**All**	923	994	764	2681
5					

Now we will add the row and column percentages.

6	Row percents		Column Percents		
7	**Gender**	**Democrat**	**Independent**	**Republican**	**All**
8	**Female**	567	534	395	1496
9		0.38	0.36	0.26	1.00
10		0.61	0.54	0.52	
11	**Male**	356	460	369	1185
12		0.30	0.39	0.31	1.00
13		0.39	0.46	0.48	
14	**All**	923	994	764	2682
15		1.00	1.00	1.00	

Note the row percents are in rows 9 and 12 and the column percents are in rows 10 and 13.

To show bar graphs of this data use the Chart Wizard as described in Chapter 2.

Choose **Chart Wizard,> Column Chart** and follow the prompts

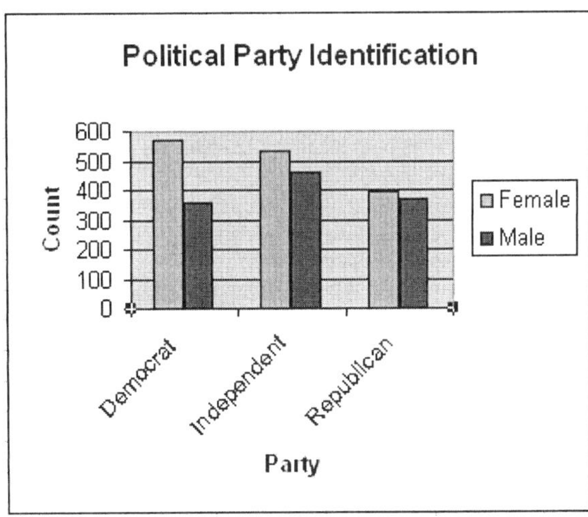

Chapter 10 Exercise 10.5 Marital Happiness and Income Contingency Table

Following the directions in Activity1 at the end of section 10.2, you can access the data in the GSS data base for this problem. Enter the fields as shown below .

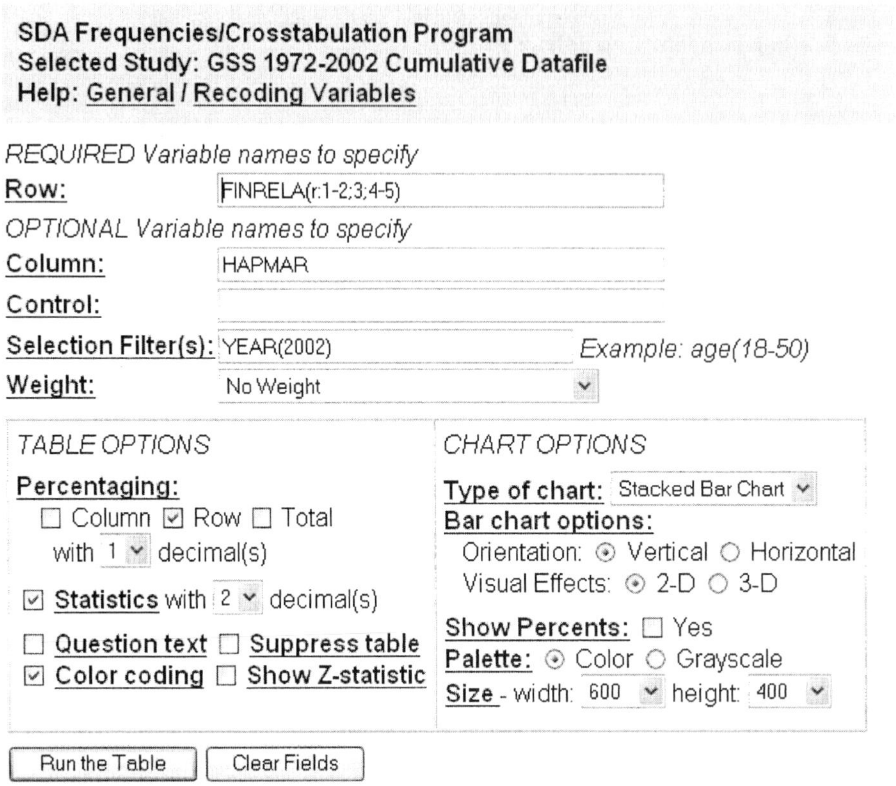

SDA Frequencies/Crosstabulation Program
Selected Study: GSS 1972-2002 Cumulative Datafile
Help: General / Recoding Variables

REQUIRED Variable names to specify

Row: FINRELA(r:1-2;3;4-5)

OPTIONAL Variable names to specify

Column: HAPMAR

Control:

Selection Filter(s): YEAR(2002) *Example: age(18-50)*

Weight: No Weight

TABLE OPTIONS

Percentaging:
☐ Column ☑ Row ☐ Total
with 1 decimal(s)

☑ Statistics with 2 decimal(s)

☐ Question text ☐ Suppress table
☑ Color coding ☐ Show Z-statistic

CHART OPTIONS

Type of chart: Stacked Bar Chart

Bar chart options:
Orientation: ⦿ Vertical ○ Horizontal
Visual Effects: ⦿ 2-D ○ 3-D

Show Percents: ☐ Yes
Palette: ⦿ Color ○ Grayscale
Size - width: 600 height: 400

Run the Table | Clear Fields

After you have accessed the data, enter it into an Excel worksheet as follows:

	A	B	C	D
1		Happily married		
2	Income	very	pretty	not
3	below	75	43	6
4	average	178	113	6
5	above	117	57	6

Using the Σ tool enter the row and column totals and add the row and column percentages.

	A	B	C	D	E
1			Happily married		
2	Income	very	pretty	not	Totals
3	below	75	43	6	124
4		0.60	0.35	0.05	1.00
5		0.20	0.20	0.33	
6	average	178	113	6	297
7		0.60	0.38	0.02	1.00
8		0.48	0.53	0.33	
9	above	117	57	6	180
10		0.65	0.32	0.03	1.00
11		0.32	0.27	0.33	
12	Totals	370	213	18	601
13		1.00	1.00	1.00	

Chapter 10 Example 3 Chi-Squared for Happiness and Family Income

The table in the margin gives the data for this problem. To test the null hypothesis that Happiness and Family Income are independent, we will perform a Chi Square Test.

Choose **PHStat > Multiple Sample tests> Chi-Square Test > OK**

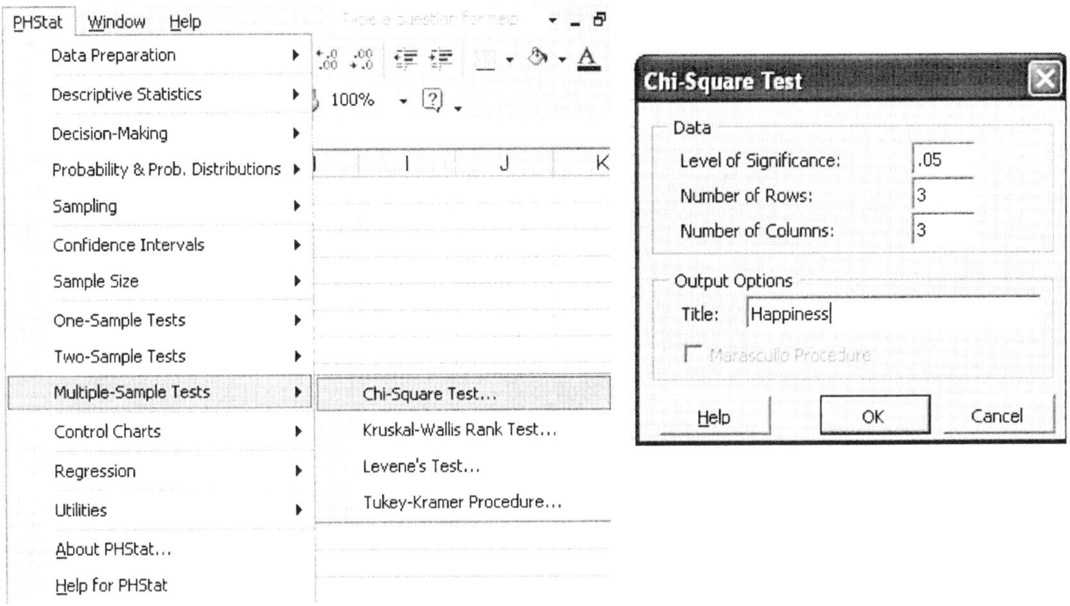

A chart template will pop up. Fill in the labels and counts in the appropriate columns. As you do, the rest of the chart will fill in according to pre-entered formulas.

Note the Chi-Square statistic and p-value in rows 27 and 28.

	A	B	C	D	E	F	G	H	I
1	**Happiness**								
2									
3		**Observed Frequencies**							
4			**Happiness**				Calculations		
5	**income**	**not**	**pretty**	**very**	**Total**			fo-fe	
6	**above**	**21**	**159**	**110**	**290**		-14.7709	-7.0793	21.85022
7	**average**	**53**	**372**	**221**	**646**		-26.6828	2.044053	24.63877
8	**below**	**94**	**249**	**83**	**426**		41.45374	5.035242	-46.489
9	**Total**	**168**	**780**	**414**	**1362**				
10									
11		Expected Frequencies							
12			Happimess						
13	income	not	pretty	very	Total			(fo-fe)^2/fe	
14	above	35.77093	166.0793	88.14978	290		6.099373	0.301762	5.416147
15	average	79.68282	369.9559	196.3612	646		8.935086	0.011294	3.091592
16	below	52.54626	243.9648	129.489	426		32.70286	0.103923	16.69042
17	Total	168	780	414	1362				
18									
19	**Data**								
20	**Level of Significance**	**0.05**							
21	Number of Rows	3							
22	Number of Columns	3							
23	Degrees of Freedom	4							
24									
25	**Results**								
26	**Critical Value**	**9.487728**							
27	**Chi-Square Test Statistic**	**73.35246**							
28	**p-Value**	**4.44E-15**							
29	**Reject the null hypothesis**								
30									
31	***Expected frequency assumption***								
32	***is met.***								

Chapter 10 Example 4 Are Happiness and Income Independent?
Chi Squared Test of Independence

In the chart above we noted that the $X^2 = 73.4$ and the degrees of freedom = 4. The p-value reported in the table is 0 to very many decimal places. We therefore can reject H_0 and conclude that there is an association between happiness and family income.

Exercise 10.11 Life After Death and Gender
Contingency Tables

The problem gives data that 425 of 563 males and 550 of 648 females believe in life after death.

Choose **PHStat > Multiple Sample tests > Chi-Square Test > OK**

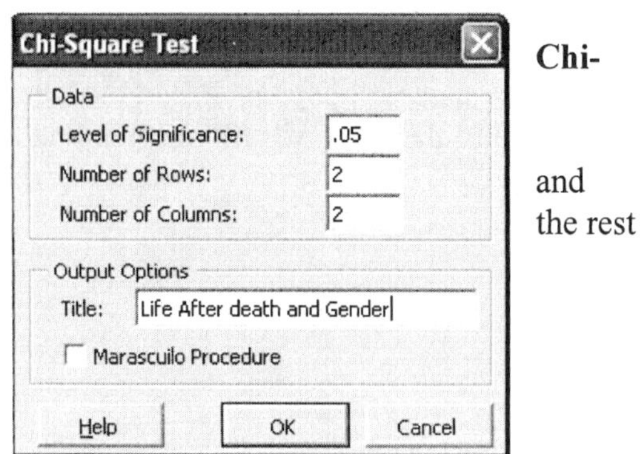

A chart template will pop up. Fill in the labels and counts in the appropriate columns. As you do, the rest of the chart will fill in according to pre-entered formulas.

Note the Chi-Square statistic and p-value.

	A	B	C	D	E	F	G
1	**Life After death and Gender**						
2							
3	**Observed Frequencies**						
4		**Life After Death**				Calculations	
5	**Gender**	**Yes**	**No**	**Total**		fo-fe	
6	**Male**	425	138	563		-28.2824	28.28241
7	**Female**	550	98	648		28.28241	-28.2824
8	**Total**	975	236	1211			
9							
10	Expected Frequencies						
11		Life After Death					
12	Gender	Yes	No	Total		(fo-fe)^2/fe	
13	Male	453.2824	109.7176	563		1.764672	7.290488
14	Female	521.7176	126.2824	648		1.533195	6.334174
15	Total	975	236	1211			
16							
17	**Data**						
18	**Level of Significance**	0.05					
19	Number of Rows	2					
20	Number of Columns	2					
21	Degrees of Freedom	1					
22							
23	**Results**						
24	**Critical Value**	3.841455					
25	**Chi-Square Test Statistic**	16.92253					
26	***p*-Value**	3.89E-05					
27	**Reject the null hypothesis**						
28							
29	*Expected frequency assumption*						
30	*is met.*						

Exercise 10.15 Help The Environment
Chi Squared Test for Different Levels

The data for this problem is given in the margin of the text. It is a 2 row by 5 column chart. Choose **PHStats > Multiple Sample tests> Chi-Square Test > OK**
Enter the size of the chart and the title and the chart will pop up. Fill in the chart. for a significance level of .05 Note the calculations on the right and the expected cell counts in rows 12 and 13.

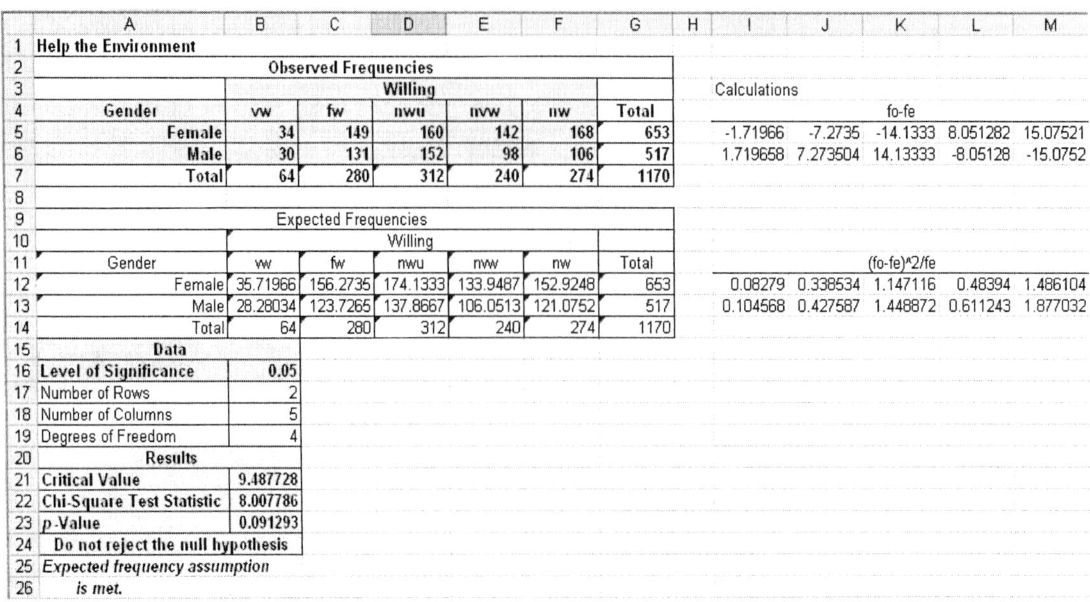

Help the Environment

Observed Frequencies						
	Willing					
Gender	vw	fw	nwu	nvw	nw	Total
Female	34	149	160	142	168	653
Male	30	131	152	98	106	517
Total	64	280	312	240	274	1170

Expected Frequencies						
	Willing					
Gender	vw	fw	nwu	nvw	nw	Total
Female	35.71966	156.2735	174.1333	133.9487	152.9248	653
Male	28.28034	123.7265	137.8667	106.0513	121.0752	517
Total	64	280	312	240	274	1170

Data	
Level of Significance	0.05
Number of Rows	2
Number of Columns	5
Degrees of Freedom	4
Results	
Critical Value	9.487728
Chi-Square Test Statistic	8.007786
p-Value	0.091293
Do not reject the null hypothesis	

Expected frequency assumption is met.

Calculations

fo-fe				
-1.71966	-7.2735	-14.1333	8.051282	15.07521
1.719658	7.273504	14.13333	-8.05128	-15.0752

(fo-fe)^2/fe				
0.08279	0.338534	1.147116	0.48394	1.486104
0.104568	0.427587	1.448872	0.611243	1.877032

Repeat the chart. For a significance level of .1 . Note the differences

Help the Environment

Observed Frequencies						
	Willing					
Gender	vw	fw	nwu	nvw	nw	Total
Female	34	149	160	142	168	653
Male	30	131	152	98	106	517
Total	64	280	312	240	274	1170

Expected Frequencies						
	Willing					
Gender	vw	fw	nwu	nvw	nw	Total
Female	35.71966	156.2735	174.1333	133.9487	152.9248	653
Male	28.28034	123.7265	137.8667	106.0513	121.0752	517
Total	64	280	312	240	274	1170

Data	
Level of Significance	0.1
Number of Rows	2
Number of Columns	5
Degrees of Freedom	4
Results	
Critical Value	7.779434
Chi-Square Test Statistic	8.007786
p-Value	0.091293
Reject the null hypothesis	

Expected frequency assumption is met.

Calculations

fo-fe				
-1.71966	-7.2735	-14.1333	8.051282	15.07521
1.719658	7.273504	14.13333	-8.05128	-15.0752

(fo-fe)^2/fe				
0.08279	0.338534	1.147116	0.48394	1.486104
0.104568	0.427587	1.448872	0.611243	1.877032

Chapter 11 Example 2 What Do We Learn From a Scatterplot in The Strength Study? Review of Producing a Scatterplot

Open the data file **high_school_female_athletes** from the Excel folder of the data disk.

To create a scatterplot of the data with column O BP(60) as the x-variable, and column L BP as the y-variable

Click: **Chart Wizard > Scatter > Scatter > Next**
On the Series Tab, complete the dialogue box as illustrated
Click: **Next**

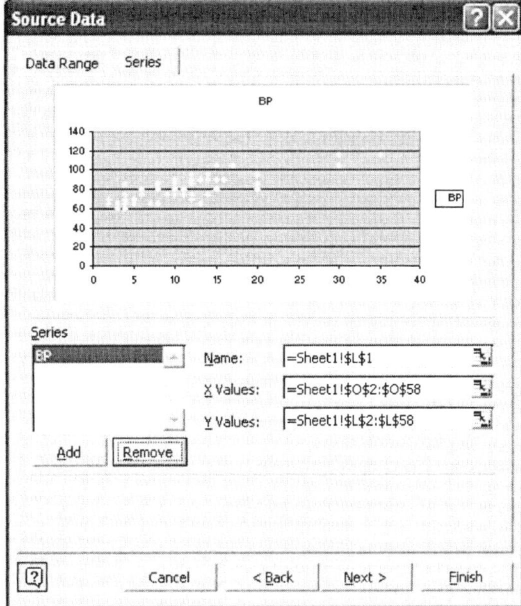

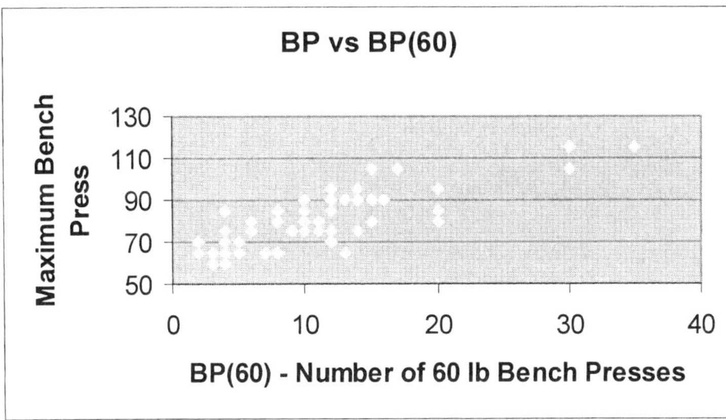

The scatterplot shows that female athletes with higher numbers of 60-pound bench presses also tended to have higher vales for the maximum bench press.

Chapter 11 Example 3 Which Regression Line Predicts Maximum Bench Press? Review of Generating a Regression Line

Open the data file **high_school_female_athletes** from the Excel folder of the text CD.

To generate the least squares regression line

Click: **Tools > Data Analysis > Regression**

Enter: Input Y Range: **L1:L58**
Input S Range: **O1:O58**
Click: **OK**

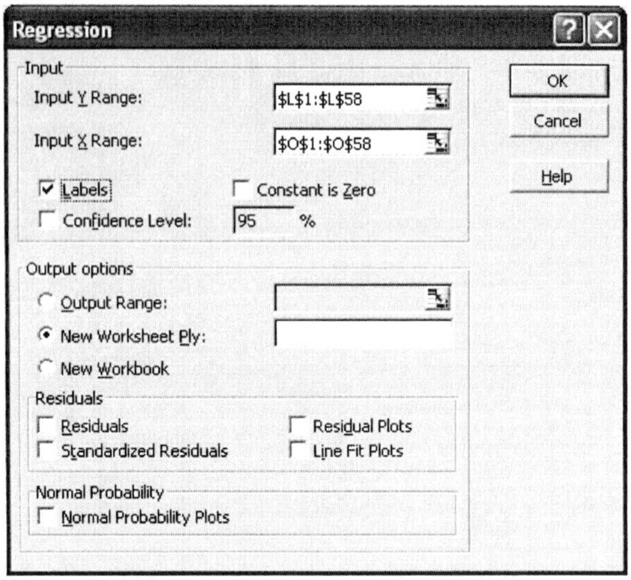

Regression
Input
Input Y Range: $L$1:$L$58
Input X Range: $O$1:$O$58
Labels
Constant is Zero
Confidence Level: 95 %
OK
Cancel
Help
Output options
Output Range:
New Worksheet Ply:
New Workbook
Residuals
Residuals
Standardized Residuals
Residual Plots
Line Fit Plots
Normal Probability
Normal Probability Plots

The results are displayed in a new worksheet.

SUMMARY OUTPUT

Regression Statistics	
Multiple R	0.802025
R Square	0.643244
Adjusted F	0.636758
Standard E	8.003188
Observatic	57

ANOVA

	df	*SS*	*MS*	*F*	*gnificance F*
Regressic	1	6351.755	6351.755	99.16711	6.48E-14
Residual	55	3522.806	64.05103		
Total	56	9874.561			

	Coefficient	*tandard Err*	*t Stat*	*P-value*	*Lower 95%*	*Upper 95%*	*ower 95.0%*	*pper 95.0%*
Intercept	63.53686	1.956469	32.47528	1.44E-37	59.61601	67.45771	59.61601	67.45771
BP (60)	1.491053	0.14973	9.958268	6.48E-14	1.190987	1.791119	1.190987	1.791119

To add the regression line to the scatterplot produced for chapter 11 example 2, right-click on one of the data points then select **Add Trendline, Linear, Display equation on chart , and Display R-squared on chart, OK.**

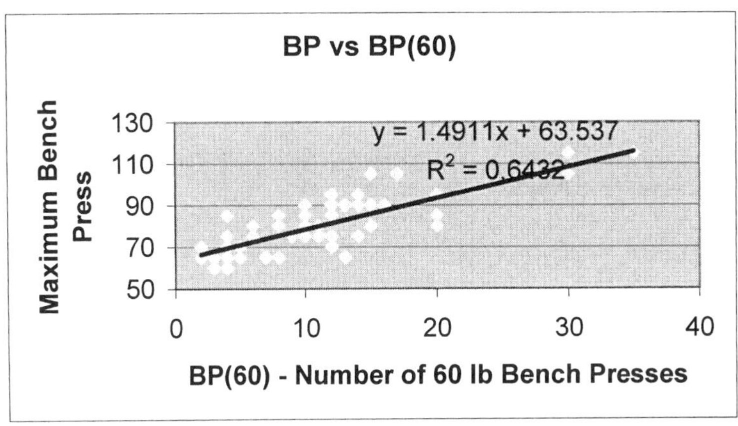

Chapter Exercise 11.1 Car Mileage and Weight
Generating a Regression Line

Open the data file **car_weight_and_mileage** from the Excel folder of the text CD.

To generate the least squares regression line

Click: **Tools > Data Analysis > Regression**

Enter: Input Y Range: **C1:C26**
Input S Range: **B1:B26**

Click: **OK**

SUMMARY OUTPUT

Regression Statistics	
Multiple R	0.866579
R Square	0.750959
Adjusted F	0.740131
Standard E	3.016149
Observatio	25

ANOVA

	df	*SS*	*MS*	*F*	*gnificance F*
Regressic	1	630.9254	630.9254	69.35415	2.14E-08
Residual	23	209.2346	9.097155		
Total	24	840.16			

	Coefficient	*Standard Err*	*t Stat*	*P-value*	*Lower 95%*	*Upper 95%*	*ower 95.0%*	*pper 95.0%*
Intercept	45.64536	2.602758	17.5373	8.3E-15	40.26115	51.02957	40.26115	51.02957
weight	-0.00522	0.000627	-8.32791	2.14E-08	-0.00652	-0.00392	-0.00652	-0.00392

Chapter Exercise 11.7 Predicting College GPA
Generating a Regression Line

Open the data file **georgia_student_survey** from the Excel folder of the text CD. We are interested in predicting college GPA based on high school GPA.

explanatory: high school GPA (column H)
response: college GPA (column I)

To generate the scatterplot:
Click: **Chart Wizard > Scatter > Scatter > Next**

On the Series Tab, complete the dialogue box as illustrated

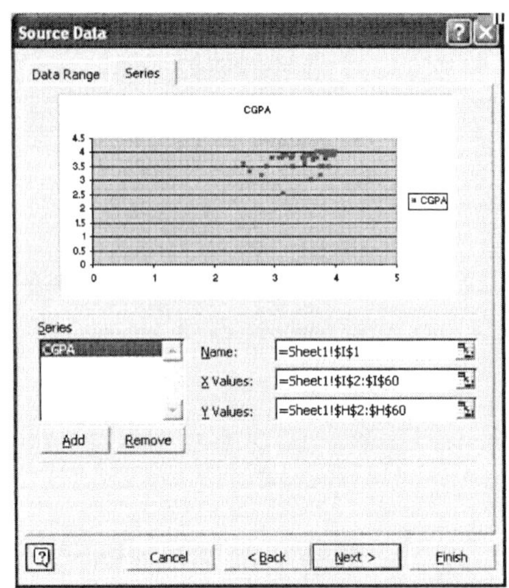

Click: **Next**

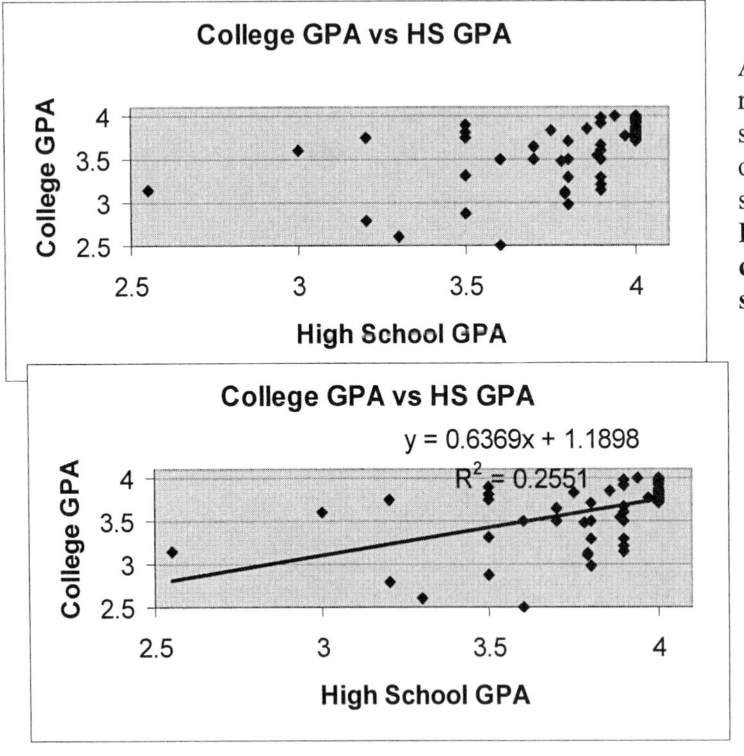

Add the least squares regression line to the scatterplot. Right-click on one of the data points then select **Add Trendline, Linear, Display equation on chart , and Display R-squared on chart, OK.**

Chapter 11 Example 6 What's the Correlation for Predicting Strength?
Review of Determining Correlation

Open the data file **high_school_female_athletes** from the Excel folder of the text CD. We have already identified column L BP as the response variable and column O BP(60) as the explanatory variable.

To determine the correlation using Excel:
Click: **within an empty cell**
Choose: **Insert > Function**
Select: Or select a category: **Statistical**
Select: **CORREL**
Click: **OK**

Complete the Functions Arguments dialog box as shown. Click **OK.**

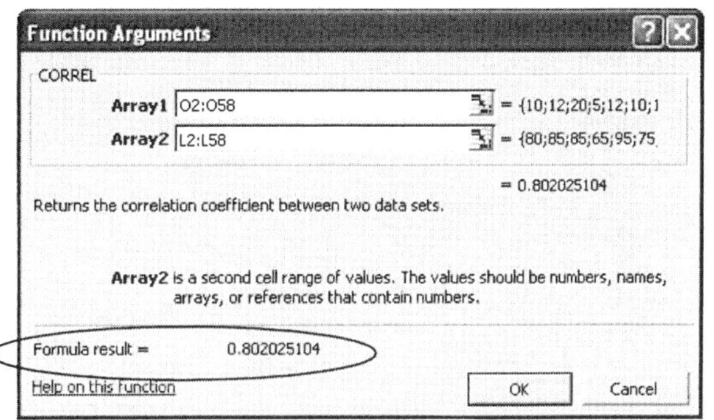

The results are displayed within the dialog box. r = 0.802

Chapter 11 Example 9 What Does r^2 Tell Us In The Strength Study?
Determining r^2

Open the data file **high_school_female_athletes** from the Excel folder of the text CD. We have already identified column L BP as the response variable and column O BP(60) as the explanatory variable.

To determine the value of r^2 using Excel:
Select: **PHStat > Regression > Simple Linear Regression**
Enter: Y Variable Cell Range: **O1:O58**
X Variable Cell Range: **L1:L58**
Select: **Regression Statistics Table**
Click: **OK**

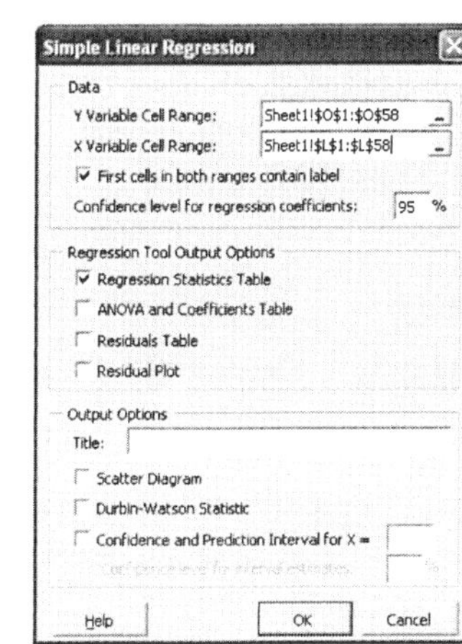

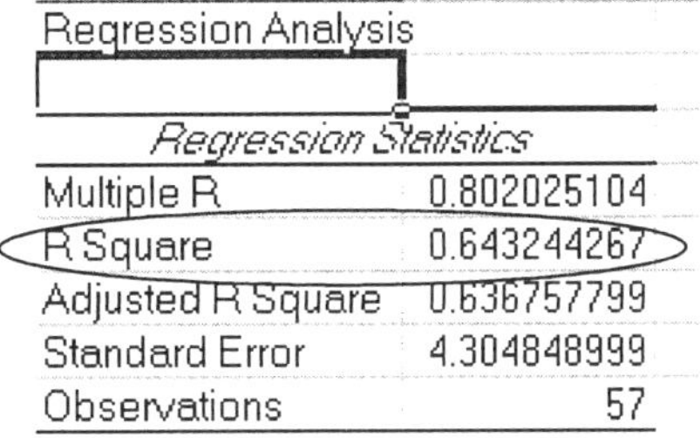

Regression Analysis	
Regression Statistics	
Multiple R	0.802025104
R Square	0.643244267
Adjusted R Square	0.636757799
Standard Error	4.304848999
Observations	57

Chapter 11 Exercise 11.13 Sit-ups and the 40-yard dash
Determining the Value of r^2

Open the data file **high_school_female_athletes** from the Excel folder of the text CD. For this exercise we have column H SIT-UP as the explanatory variable and column I 40-YD (sec) as the response variable.

To determine the value of r^2 using Excel:
Select: **PHStat > Regression > Simple Linear Regression**
Enter: Y Variable Cell Range: **O1:O58**
X Variable Cell Range: **L1:L58**
Select: **Regression Statistics Table**
Select: **ANOVA and Coefficients Table**
Click: **OK**

Regression Analysis

Regression Statistics	
Multiple R	0.459309491
R Square	0.210965208
Adjusted R Square	0.196619121
Standard Error	0.327208191
Observations	57

Note the value of $R^2 = 21.1\%$

ANOVA

	df	*SS*	*MS*	*F*	*Significance F*
Regression	1	1.574438532	1.574438532	14.70541804	0.00032572
Residual	55	5.88858603	0.107065201		
Total	56	7.463024561			

	Coefficients	*Standard Error*	*t Stat*	*P-value*	*Lower 95%*	*Upper 95%*
Intercept	6.706527218	0.177891008	37.70020353	5.5637E-41	6.350025759	7.063028678
SIT-UP	-0.024346063	0.006348777	-3.834764405	0.00032572	-0.037069293	-0.011622833

Note the regression equation is $\hat{y} = 6.71 - 0.0243x$. Using this equation we can predict time in the 40-yard dash for any subject who can do a given number of sit-ups.

Chapter 11 Exercise 11.15 Student ideology
Determining the Value of r^2

Open the data file **fla_student_survey** from the Excel folder of the text CD. For this exercise we have column J newspapers as the explanatory variable and column N political_ideology as the response variable.

Doing a complete analysis of the association of these two variables, we start with the scatterplot:
Select: **Chart Wizard > Scatter > Scatter > Next**

Complete the dialog box as indicated.

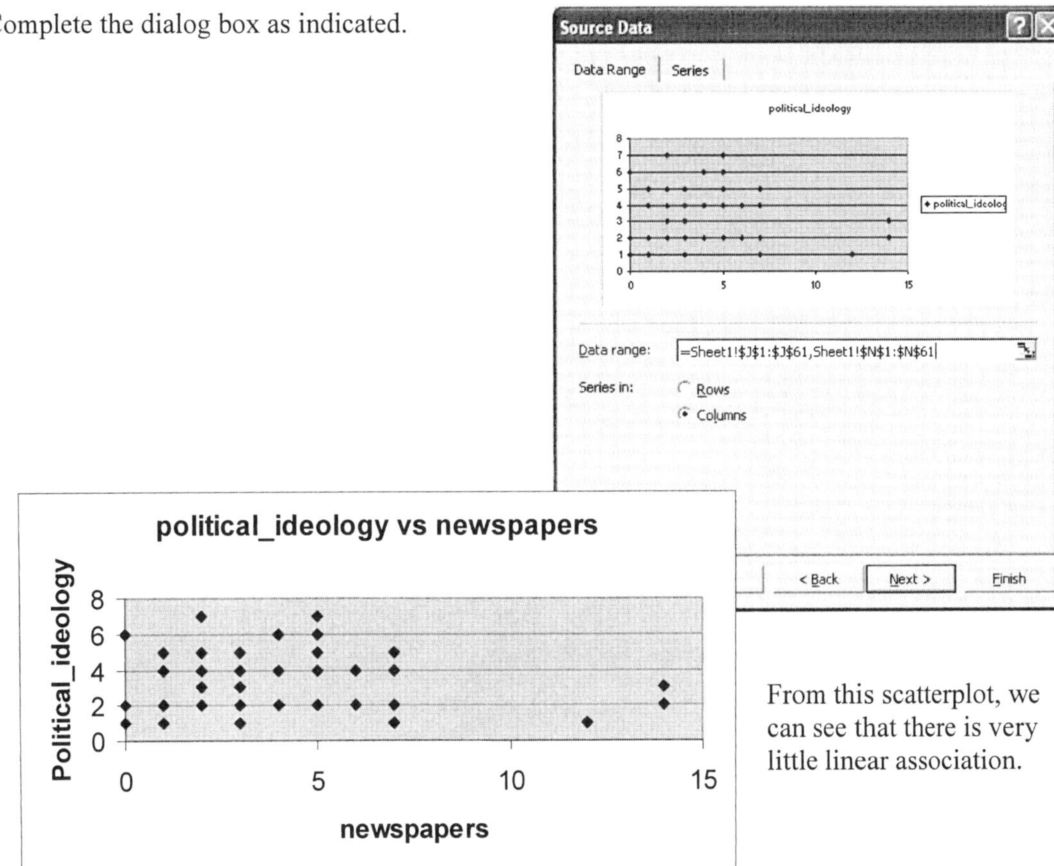

From this scatterplot, we can see that there is very little linear association.

Determining the correlation:

Select: **Insert > Function > CORR**

Complete the dialog box as indicated

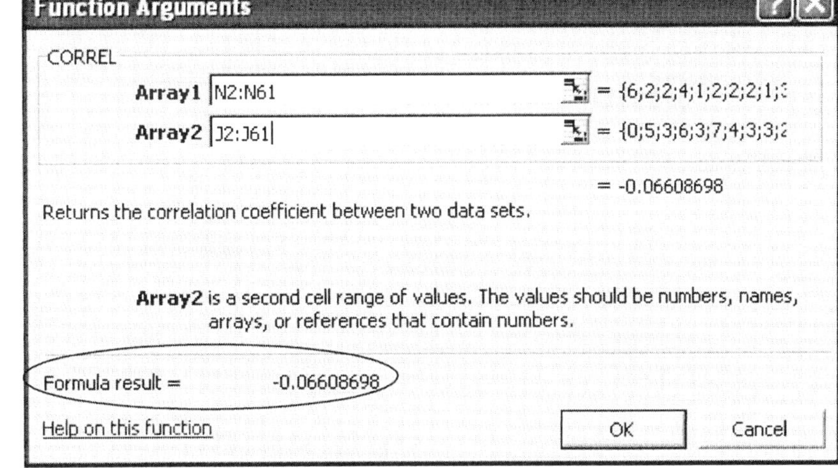

Again, based on the value of r = -0.066, we see very little evidence of a linear association between these two variables.

Determining the value of r^2:
Select: **PHStat > Regression > Simple Linear Regression**
Enter: Y Variable Cell Range: **N1:N61**
X Variable Cell Range: **J1:J61**
Select: **Regression Statistics Table**
Select: **ANOVA and Coefficients Table**

A	B	C	D	E	F	G
Regression Analysis						
Regression Statistics						
Multiple R	0.06608698					
R Square	0.004367489					
Adjusted R Square	-0.012798589					
Standard Error	1.646540581					
Observations	60					
ANOVA						
	df	*SS*	*MS*	*F*	*Significance F*	
Regression	1	0.689772075	0.689772075	0.254425555	0.615886951	
Residual	58	157.2435613	2.711095884			
Total	59	157.9333333				
	Coefficients	*Standard Error*	*t Stat*	*P-value*	*Lower 95%*	*Upper 95%*
Intercept	3.180284775	0.360640175	8.818442854	2.64656E-12	2.458385572	3.902183979
newspapers	-0.035988108	0.071347482	-0.504406141	0.615886951	-0.178805504	0.106829288

Note the displayed value of R-Sq = .4%

Chapter 11 Example 11 Is Strength Associated with 60-pound Bench Presses? A Significance Test of Independence

Open the worksheet **high_school_female_student_athletes,** which is found in the Excel folder. The number of bench presses before fatigue, column O BP(60), is the x-variable, and maximum bench press, column L BP, is the y-variable. The regression output includes the t test statistic and the p-value.

Regression Analysis						
Regression Statistics						
Multiple R	0.802025104					
R Square	0.643244267					
Adjusted R Square	0.636757799					
Standard Error	8.003188444					
Observations	57					
ANOVA						
	df	*SS*	*MS*	*F*	*Significance F*	
Regression	1	6351.755014	6351.755014	99.16710914	6.48137E-14	
Residual	55	3522.80639	64.05102527			
Total	56	9874.561404				
	Coefficients	*Standard Error*	*t Stat*	*P-value*	*Lower 95%*	*Upper 95%*
Intercept	63.53685646	1.956468584	32.4752756	1.43832E-37	59.61600676	67.45770615
BP (60)	1.491053006	0.149730149	9.95826838	6.48137E-14	1.190987157	1.791118856

Notice the t test statistic ($t = 9.96$) appears under the column heading "tStat" in the row for the predictor BP (60), and the p-value (p = 0.000) for the two-sided alternative H_a: $\beta \neq 0$ appears under the heading "P-value" in the row for the predictor BP (60).

Chapter 11 Exercise 11.29 Predicting House Prices
Significance Test of Independence and a Confidence Interval for Slope

Open the worksheet **house_selling_prices**, which is found in the Excel folder. The size of the house is in column H size, and selling price is in column G price.

Select: **PHStat > Regression > Simple Linear Regression**
Enter: Y Variable cell range: **G1:G101**
X Variable cell range: **H1:H101**
Select: **Regression Statistics Table**
Select: **ANOVA and Coefficients Table**
Click: **OK**

Regression Analysis

Regression Statistics	
Multiple R	0.761262112
R Square	0.579520003
Adjusted R Square	0.57522939
Standard Error	36730.2057
Observations	100

ANOVA

	df	*SS*	*MS*	*F*	*Significance F*
Regression	1	1.8222E+11	1.8222E+11	135.0669725	3.83804E-20
Residual	98	1.32213E+11	1349108011		
Total	99	3.14433E+11			

	Coefficients	*Standard Error*	*t Stat*	*P-value*	*Lower 95%*	*Upper 95%*
Intercept	9161.158864	10759.78616	0.851425737	0.396608629	-12191.28619	30513.60392
size	77.00769255	6.626123525	11.62183172	3.83804E-20	63.85836631	90.15701878

Notice the t test statistic ($t = 11.62$), and the p-value ($p = 0.000$) for the two-sided alternative H_a: $\beta \neq 0$. Since p-value = 0.000 is less than a 0.05 significance level, there is evidence that these two variables are not independent, and that the sample association between these two variables is not just random variation.

The 95% confidence interval for the population slope is (63.859, 90.157)

This interval does not support the builder's claim that selling price increases $100, on the average, for every extra square foot. (The interval does not contain 100.)

Chapter 11 Exercise 11.35 Advertising and Sales
A Significance Test of Independence

Enter the data into the Excel worksheet. Enter the x's into column A and name the column Advertising. Enter the y's into column B and name the column Sales.

Select: **PHStat > Regression > Simple Linear Regression**
Enter: Y Variable cell range: **B1:B5**
X Variable cell range: **A1:A5**

Regression Analysis

Regression Statistics	
Multiple R	0.857142857
R Square	0.734693878
Adjusted R Square	0.602040816
Standard Error	1.362770288
Observations	4

ANOVA

	df	*SS*	*MS*	*F*	*Significance F*
Regression	1	10.28571429	10.28571429	5.538461538	0.142857143
Residual	2	3.714285714	1.857142857		
Total	3	14			

	Coefficients	*Standard Error*	*t Stat*	*P-value*	*Lower 95%*	*Upper 95%*
Intercept	5.285714286	0.997445717	5.299250068	0.033814111	0.994048759	9.577379812
Advertising	0.857142857	0.36421568	2.353393622	0.142857143	-0.709951822	2.424237536

Notice the t test statistic ($t = 2.35$), and the p-value (p = 0.143) for the two-sided alternative H_a: $\beta \neq 0$. Since p-value = 0.143 is not less than significance level 0.05, there is not evidence that these two variables are independent. A word of caution though, were the assumptions validated? This is a very small sample, so results are suspect.

Chapter 11 Example 13 Detecting an Underachieving College Student
How Data Vary Around the Regression Line

Open the worksheet **georgia_student_survey**, found in the Excel folder. High school GPA is found in column H and college GPA is in column I.

As part of the regression analysis, PHStat can be instructed to include a table of the residuals. However, PHStat does not highlight any "unusual observations."

Select: **PHStat > Regression > Simple Linear Regression**
Enter: Y Variable cell range: **I1:I60**
X Variable cell range: **H1:H60**
Select: **Regression Statistics Table**
Select: **ANOVA and Coefficients Table**
Select: **Residuals Table**
Click: **OK**

Here is a portion of the residual table produced.

Observation	*Predicted CGPA*	*Residuals*
1	3.673924508	-0.373924508
2	3.603860484	-0.473860484
3	3.100673406	0.499326594
4	3.673924508	-0.173924508
5	3.482840807	0.017159193
6	3.22806254	0.52193746
7	3.597491028	-0.127491028

Chapter 11 Example 16 Predicting Maximum Bench Press and Estimating Its Mean Confidence Interval for Population Mean of y and Prediction Interval for a Single y-value

Open the worksheet **high_school_female_athletes**, found in the Excel folder.

Proceed with the regular regression command:
Select: **PHStat > Regression > Simple Linear Regression**

Enter: Y Variable Cell Range: **O1:O58**
X Variable Cell Range: **L1:L58**
Select: **Regression Statistics Table**
Enter: Output Options: Confidence and Prediction Interval for X = **11**
Enter: Confidence level for interval estimates: **95** %
Click: **OK**

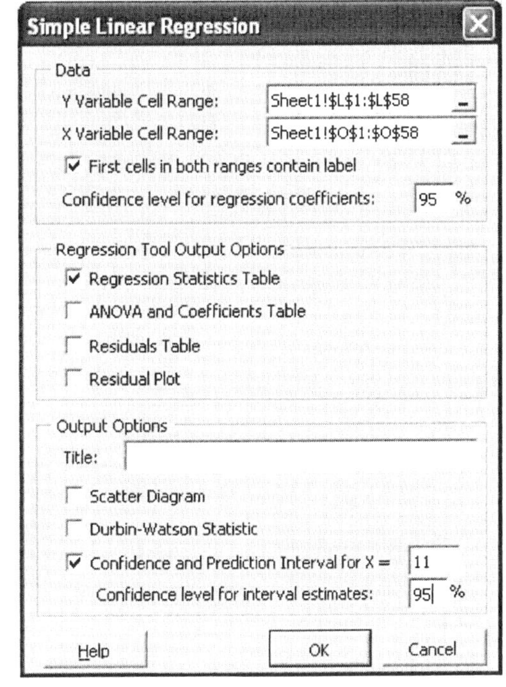

Data	
X Value	**11**
Confidence Level	**95%**
Intermediate Calculations	
Sample Size	57
Degrees of Freedom	55
t Value	2.004044
Sample Mean	10.98246
Sum of Squared Difference	2856.982
Standard Error of the Estimate	8.003188
h Statistic	0.017544
Predicted Y (YHat)	79.93844
For Average Y	
Interval Half Width	2.12439
Confidence Interval Lower Limi	**77.814**
Confidence Interval Upper Limi	**82.0628**
For Individual Response Y	
Interval Half Width	16.17882
Prediction Interval Lower Limit	**63.7596**
Prediction Interval Upper Limit	**96.1173**

Chapter 11 Exercise 11.37 Poor predicted strengths
How Data Vary Around the Regression Line

Open the worksheet **high_school_female_athletes**, found in the MINITAB folder.

Select: **PHStat > Regression > Simple Linear Regression**
Enter: Y Variable Cell Range: **O1:O58**
X Variable Cell Range: **L1:L58**
Select: **Regression Statistics Table**
Select: **ANOVA and Coefficients Table**
Select: **Residuals Table**
Click: **OK**

Regression Analysis

Regression Statistics	
Multiple R	0.802025104
R Square	0.643244267
Adjusted R Square	0.636757799
Standard Error	8.003188444
Observations	57

ANOVA

	df	*SS*	*MS*	*F*	*Significance F*
Regression	1	6351.755014	6351.755014	99.16710914	6.48137E-14
Residual	55	3522.80639	64.05102527		
Total	56	9874.561404			

	Coefficients	*Standard Error*	*t Stat*	*P-value*	*Lower 95%*	*Upper 95%*
Intercept	63.53685646	1.956468584	32.4752756	1.43832E-37	59.61600676	67.45770615
BP (60)	1.491053006	0.149730149	9.95826838	6.48137E-14	1.190987157	1.791118856

A portion of the residuals table:

RESIDUAL OUTPUT

Observation	*Predicted BP*	*Residuals*
1	78.44738652	1.55261348
2	81.42949253	3.570507467
3	93.35791658	-8.357916585
4	70.99212149	-5.992121488

Chapter 11 Exercise 11.43 ANOVA Table for Leg Press

Open the worksheet **high_school_female_student_athletes,** which is found in the Excel folder. The number of leg presses before fatigue, column M LP (200), is the x-variable, and maximum leg press, column N LP , is the y-variable

Proceed with the regular regression command:
Select: **PHStat > Regression > Simple Linear Regression**
Enter: Y Variable Cell Range: **M1:M58**
X Variable Cell Range: **N1:N58**
Select: **Regression Statistics Table**
Select: **ANOVA and Coefficients Table**

Select: **Residuals Table**
Click: **OK**

A	B	C	D	E	F	G
Regression Analysis						
Regression Statistics						
Multiple R	0.792504645					
R Square	0.628063612					
Adjusted R Square	0.621301132					
Standard Error	36.10702319					
Observations	57					
ANOVA						
	df	*SS*	*MS*	*F*	*Significance F*	
Regression	1	121082.4003	121082.4003	92.87474879	2.0619E-13	
Residual	55	71704.44179	1303.717123			
Total	56	192786.8421				
	Coefficients	*Standard Error*	*t Stat*	*P-value*	*Lower 95%*	*Upper 95%*
Intercept	233.8878981	13.06444129	17.90263303	1.32112E-24	207.7061792	260.0696171
LP (200)	5.271025893	0.546948359	9.6371546	2.0619E-13	4.174917157	6.367134629

From the ANOVA table, we can use SS Residual Error = 71704 to determine the residual standard deviation of y-values. Note that $df = 57 - 2 = 55$.

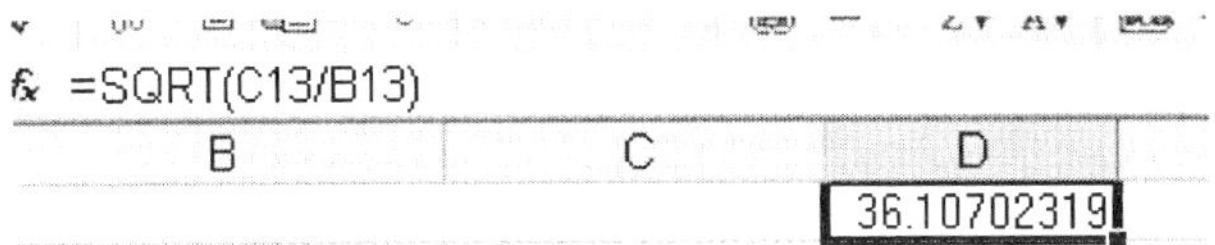

At any fixed value x of number of 200-pound leg presses, we estimate that the maximum leg press values have a standard deviation of 36.1 pounds.

For female athletes with $x = 22$, we would estimate the mean maximum leg press to be `LP = 234 + 5.27 (22)` = 349.94 pounds and the variability of their maximum leg press values to be 36.1 pounds.

If y-values are approximately normal, then 95% of the y-values would fall in the interval approximately $\hat{y} \pm 2s = 349.94 \pm 2(36.1) = (277.74, 422.14)$
Using Excel to generate the prediction interval, notice we get similar results.

For Individual Response Y	
Interval Half Width	72.99248
Prediction Interval Lower Limit	**276.858**
Prediction Interval Upper Limit	**422.843**

Chapter 11 Example 18 Explosion in Number of People Using the Internet
Exponential Regression

Chapter 11 Exercise 11.55 Leaf Litter Decay
Exponential Regression

Excel and PHStat do not include an exponential regression tool. See the MINITAB section.

Chapter 12 Example 2 Predicting Selling Price Using House and Lot Sizes Multiple Regression and Plotting the Relationships

Open the worksheet **house_selling_prices** , which is found in the Excel folder. The price is found in G, house size is found in H, and lot size is found in I
Select:
PHStat > Regression > Multiple Regression

Enter:
Y Variable Cell Range: **select appropriate cells**
X Variables Cell Range: **select appropriate cells**
Select: **Regression Statistics Table**
Select: **ANOVA and Coefficients Table**
Click: **OK**

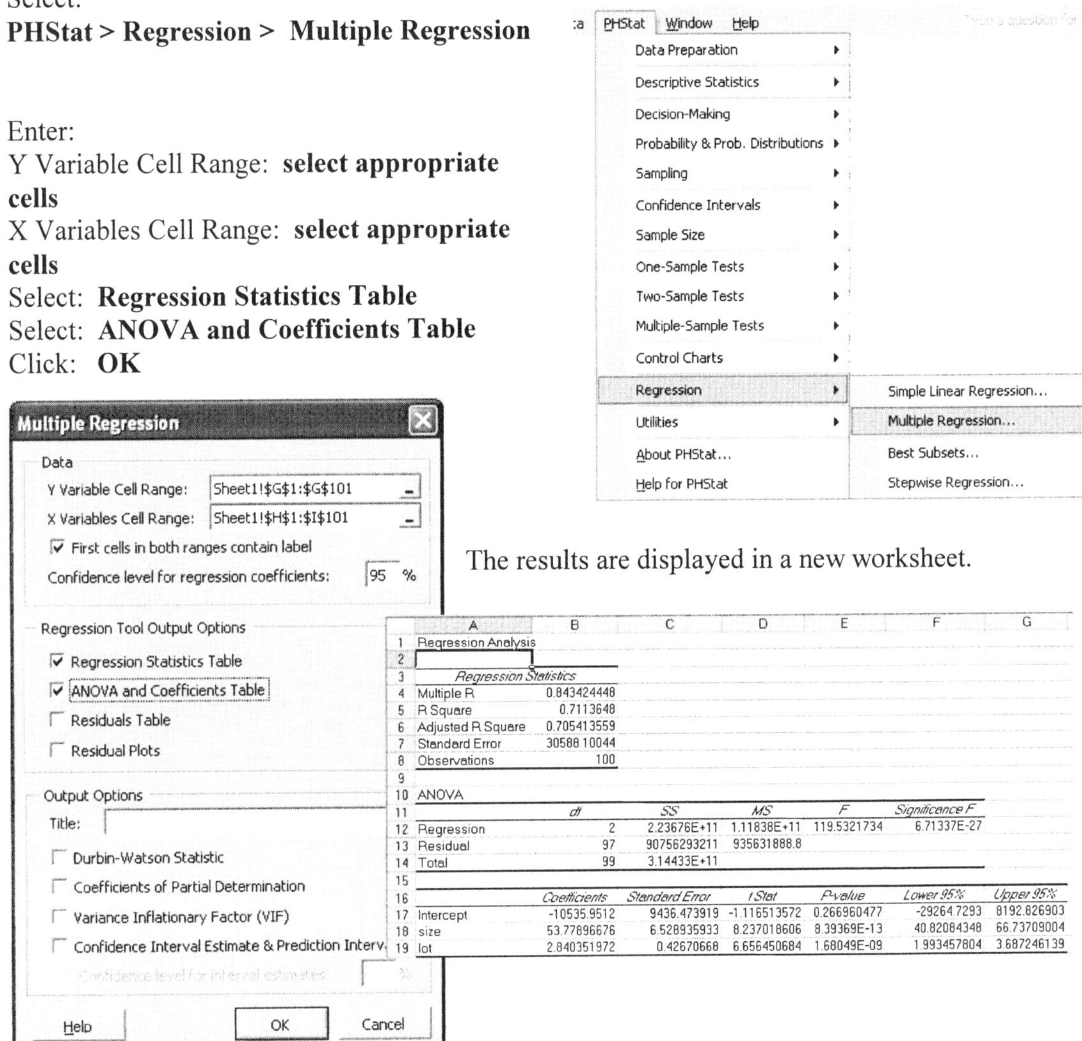

The results are displayed in a new worksheet.

Regression Analysis

Regression Statistics	
Multiple R	0.843424448
R Square	0.7113648
Adjusted R Square	0.705413559
Standard Error	30588.10044
Observations	100

ANOVA

	df	SS	MS	F	Significance F
Regression	2	2.23676E+11	1.11838E+11	119.5321734	6.71337E-27
Residual	97	90756293211	935631888.8		
Total	99	3.14433E+11			

	Coefficients	Standard Error	t Stat	P-value	Lower 95%	Upper 95%
Intercept	-10535.9512	9436.473919	-1.116513572	0.266960477	-29264.7293	8192.826903
size	53.77896676	6.528935933	8.237018606	8.39369E-13	40.82084348	66.73709004
lot	2.840351972	0.42670668	6.656450684	1.68049E-09	1.993457804	3.687246139

Excel and PHState cannot be used to generate a scatterplot matrix. See MINITAB section.

Chapter 12 Exercise 12.5 Does more education cause more crime? Multiple Regression

Open the worksheet **fla_crime**, which is found in the Excel folder. The crime rate is found in B, education is found in C, and urbanization is found in D.

Select: **PHStat > Regression > Multiple Regression**
Enter: Response: **B1:B68**
Predictors: **C1:C68, D1:D68** Select: **Results**
Select: **Regression Statistics Table**
Select: **ANOVA and Coefficients Table**
Click: **OK**

The results are displayed in the new worksheet.

	A	B	C	D	E	F	G
1	Regression Analysis						
2							
3	*Regression Statistics*						
4	Multiple R	0.686599834					
5	R Square	0.471419331					
6	Adjusted R Square	0.454901186					
7	Standard Error	20.81558241					
8	Observations	67					
9							
10	ANOVA						
11		*df*	*SS*	*MS*	*F*	*Significance F*	
12	Regression	2	24731.65726	12365.82863	28.53948225	1.37899E-09	
13	Residual	64	27730.46215	433.2884711			
14	Total	66	52462.1194				
15							
16		*Coefficients*	*Standard Error*	*t Stat*	*P-value*	*Lower 95%*	*Upper 95%*
17	Intercept	59.11806772	28.36531056	2.084167828	0.04114165	2.451896833	115.7842306
18	education	-0.583377348	0.472459127	-1.234767868	0.22143116	-1.527222104	0.360467407
19	urbanization	0.682501422	0.12321259	5.539218224	6.1108E-07	0.4363562	0.928646644

Using the regression equation generated by Excel,

```
crime = 59.1 - 0.583 education + 0.683 urbanization
```

we can predict crime rates for a county that has 0% in an urban environment by substituting 0 for "urbanization" and using the resulting equation crime = 59.1 – 0.583 education . With education set equal to 70% we determine the crime rate to be 59.1 – 0.583(70) = 18.29. For an 80% high school graduation rate, the crime rate is predicted to be 59.1 – 0.583(80) = 12.46.

Chapter 12 Example 3 How Well Can We Predict House Selling Prices? ANOVA and R

Open the worksheet **house_selling_prices** , which is found in the Excel folder. The price is found in G, house size is found in H, and lot size is found in I. The regression command in PHStat includes output of the value of R-squared. We have done this numerous times.

Select: **Stat > Regression > Multiple Regression**
Enter: Y Variable Cell Range: **G1:G101**
X Variables Cell Range: **H1:H101, I1:I101**
Select: **Regression Statistics Table**
Select: **ANOVA and Coefficients Table**
Click: **OK**

	A	B
1	Regression Analysis	
2		
3	*Regression Statistics*	
4	Multiple R	0.843424448
5	R Square	0.7113648
6	Adjusted R Square	0.705413559
7	Standard Error	30588.10044
8	Observations	100

Note the value of R-squared is 71.1%. The multiple correlation between selling price and the two explanatory variables is $R = \sqrt{R^2} = \sqrt{.711} = 0.84$.

Chapter 12 Example 4 What Helps Predict a Female Athlete's Weight? Significance Test and Confidence Interval about a Multiple Regression Parameter β

Open the worksheet **college_athletes**, which is found in the Excel folder. The total body weight (TBW) is found in A, height (HGT) in inches is found in B, the percent of body fat (%BF) is found in C, and age (AGE) is found in K. PHStat will only allow selection of "X Variable Cell Ranges" that are contiguous. Copy columns HGT and %BF to columns L and M.

Select: **Stat > Regression > Multiple Regression**
Enter: Y Variable Cell Range: **A1:A65**
X Variable Cell Ranges: **K1:M65**
Select: **Regression Statistics Table**
Select: **ANOVA and Coefficients Table**
Click: **OK**

The results are displayed in the new worksheet.

Note the value of R-squared is 66.9%. The predictive power is good.

	A	B
1	Regression Analysis	
2		
3	*Regression Statistics*	
4	Multiple R	0.818101194
5	R Square	0.669289564
6	Adjusted R Square	0.652754042
7	Standard Error	10.10860657
8	Observations	64

ANOVA

	df	SS	MS	F	Significance F
Regression	3	12407.94877	4135.982922	40.47586592	1.9772E-14
Residual	60	6131.03561	102.1839268		
Total	63	18538.98438			

	Coefficients	Standard Error	t Stat	P-value	Lower 95%	Upper 95%
Intercept	-97.69378154	28.78521964	-3.393886959	0.001226151	-155.272775	-40.11478809
AGE	-0.960088375	0.648277817	-1.480982922	0.143843722	-2.256836658	0.336659908
HGT	3.428473996	0.367899254	9.319056661	2.87856E-13	2.692566158	4.164381834
%BF	1.364265481	0.312552565	4.364915328	5.09641E-05	0.739067469	1.989463493

To test whether age helps us to predict weight, if we already know height and percent body fat, we perform a significance test on $H_0 : \beta_3 = 0$ versus $H_a: \beta_3 \neq 0$. The *t* test statistic is reported in the table as -1.48 and the p-value is .144. This p-value does not give much evidence against the null hypothesis. Age does not significantly predict weight, if we already know height and percentage of body fat.

Chapter 12 Example 5 What's Plausible for the Effect of Age on Weight? Confidence Interval about a Multiple Regression Parameter β

Open the worksheet **college_athletes**, which is found in the Excel folder. The total body weight (TBW) is found in A, height (HGT) in inches is found in B, the percent of body fat (%BF) is found in C, and age (AGE) is found in K.

Using the results from the previous example:

ANOVA

	df	SS	MS	F	Significance F
Regression	3	12407.94877	4135.982922	40.47586592	1.9772E-14
Residual	60	6131.03561	102.1839268		
Total	63	18538.98438			

	Coefficients	Standard Error	t Stat	P-value	Lower 95%	Upper 95%
Intercept	-97.69378154	28.78521964	-3.393886959	0.001226151	-155.272775	-40.11478809
AGE	-0.960088375	0.648277817	-1.480982922	0.143843722	-2.256836658	0.336659908
HGT	3.428473996	0.367899254	9.319056661	2.87856E-13	2.692566158	4.164381834
%BF	1.364265481	0.312552565	4.364915328	5.09641E-05	0.739067469	1.989463493

The 95% confidence interval for the population slope β_3 is (-2.26, 0.34)
At fixed values of height and percent body fat, we infer that the population mean weight changes very little (and may not change at all, since this interval includes 0), making these results consistent with what we found in example 4.

Chapter 12 Example 7 The F Test For Predictors of Athletes' Weight
The F Test that All Multiple Regression Parameters β = 0

Open the worksheet **college_athletes**, which is found in the Excel folder. The total body weight (TBW) is found in A, height (HGT) in inches is found in B, the percent of body fat (%BF) is found in C, and age (AGE) is found in K.

Once again using the results from example 4:

Regression Analysis

Regression Statistics	
Multiple R	0.818101194
R Square	0.669289564
Adjusted R Square	0.652754042
Standard Error	10.10860657
Observations	64

ANOVA

	df	*SS*	*MS*	*F*	*Significance F*
Regression	3	12407.94877	4135.982922	40.47586592	1.9772E-14
Residual	60	6131.03561	102.1839268		
Total	63	18538.98438			

	Coefficients	*Standard Error*	*t Stat*	*P-value*	*Lower 95%*	*Upper 95%*
Intercept	-97.69378154	28.78521964	-3.393886959	0.001226151	-155.272775	-40.11478809
AGE	-0.960088375	0.648277817	-1.480982922	0.143843722	-2.256836658	0.336659908
HGT	3.428473996	0.367899254	9.319056661	2.87856E-13	2.692566158	4.164381834
%BF	1.364265481	0.312552565	4.364915328	5.09641E-05	0.739067469	1.989463493

The ANOVA table for the multiple regression output includes the value of the F test statistic (F = 40.48) and the p-value (p = 0.000) for testing H_0: $\beta_1 = \beta_2 = \beta_3 = 0$. We can reject H_0 and conclude that at least one predictor has an effect on weight.

Chapter 12 Exercise 12.20 Study time help GPA?
Hypothesis Test and Confidence Interval About β

Open the worksheet **georgia student survey**, which is found in the Excel folder.
Performing a multiple regression for college GPA based on high school GPA and study time:

Select: **Stat > Regression > Multiple Regression**
Enter: Y Variable Cell Range: **I1:I60**
X Variables Cell Range:Predictors: **H1:H60, E1:E60**
Select: **Regression Statistics Table**
Select: **ANOVA and Coefficients Table**
Click: **OK**

Regression Analysis

Regression Statistics	
Multiple R	0.508122099
R Square	0.258188067
Adjusted R Square	0.231694784
Standard Error	0.318824302
Observations	59

ANOVA

	df	*SS*	*MS*	*F*	*Significance F*
Regression	2	1.981222325	0.990611162	9.745416	0.000233468
Residual	56	5.692340387	0.101648935		
Total	58	7.673562712			

	Coefficients	*Standard Error*	*t Stat*	*P-value*	*Lower 95%*	*Upper 95%*
Intercept	1.126227293	0.568970902	1.979411054	0.05269364	-0.013557629	2.266012214
HSGPA	0.643428775	0.14576196	4.41424342	4.67363E-05	0.351432675	0.935424875
Studytime	0.007757836	0.016136202	0.480772131	0.632551508	-0.02456684	0.040082513

To test whether study time helps us to college GPA, if we already know high school GPA, we perform a significance test on $H_0 : \beta_2 = 0$ versus H_a: $\beta_2 \neq 0$. The *t* test statistic is reported in the table as 0.48 and the p-value is 0.633. This p-value does not give evidence against the null hypothesis. Study time does not significantly predict college GPA if we already know high school GPA.

The 95% confidence interval for the population slope β_2 : (-0.0247, 0.0399)
At a fixed value of high school GPA, we infer that the population mean college GPA changes very little (and may not change at all, since this interval includes 0), making these results consistent with the significance test we have already completed.

Chapter 12 Exercise 12.29 More predictors for house price
The F Test that All Multiple Regression Parameters β = 0

Open the worksheet **house_selling_prices**, which is found in the Excel folder. Performing multiple regression for predicting house selling price using size of home, lot size, and real estate tax (copy column B to column J):

Select: **Stat > Regression > Multiple Regression**
Enter: Y Variable Cell Range: **G1:G101**
X Variables Cell Range: **H1:J101**
Select: **Regression Statistics Table**
Select: **ANOVA and Coefficients Table**
Click: **OK**

Regression Analysis

Regression Statistics	
Multiple R	0.870016326
R Square	0.756928408
Adjusted R Square	0.749332421
Standard Error	28215.98487
Observations	100

ANOVA

	df	*SS*	*MS*	*F*	*Significance F*
Regression	3	2.38003E+11	79334302193	99.64845705	2.25232E-29
Residual	96	76429613021	796141802.3		
Total	99	3.14433E+11			

	Coefficients	*Standard Error*	*t Stat*	*P-value*	*Lower 95%*	*Upper 95%*
Intercept	6305.319269	9567.27338	0.65905081	0.511440892	-12685.58281	25296.22135
size	34.51179153	7.543276837	4.575172338	1.42238E-05	19.5384939	49.48508915
lot	1.594361818	0.491127242	3.246331463	0.001610374	0.619481202	2.569242433
Taxes	22.03549329	5.194517429	4.242067447	5.10402E-05	11.72444981	32.34653678

The ANOVA table for the multiple regression output includes the value of the F test statistic (F = 99.65) and the p-value (p = 0.000) for testing $H_0\colon\ \beta_1 = \beta_2 = \beta_3 = 0$. We can reject H_0 and conclude that at least one predictor has an effect on price.

The p-values of 0.000, 0.002, 0.000 for *t*-tests for each of the explanatory variables, respectively, indicate that each predictor (house size, lot size, and taxes) does significantly predict price, if we already know the values of the other two variables.

Chapter 12 Example 9 Another Residual Plot for House Selling Prices
Plots of Residuals Against Explanatory Variables

Open the worksheet **house_selling_prices** , which is found in the Excel folder. The price is found in G, house size is found in H, and lot size is found in I .

To produce a histogram of standardized residuals for the multiple regression model predicting selling price by the house size and the lot size:

Within the **PHStat > Regression > Multiple Regression** command, select **Residuals Table.** Although PHStat does not allow production of a histogram of the residuals, we can use the Histogram command from **Tools > Data Analysis > Histogram** to create a histogram of the residuals included in the Multiple Regression output, a portion of which is shown here.

RESIDUAL OUTPUT

Observation	*Predicted price*	*Residuals*
1	107766.7244	37233.27564
2	82070.99553	-14070.99553
3	107418.5374	7581.462637
4	90352.13611	-21352.13611
5	135017.8588	27982.1412
6	81020.78	-11120.78
7	63684.51979	-13684.51979
8	116402.7511	20597.24891
9	116410.7462	4889.25382
10	66145.24984	3854.750164

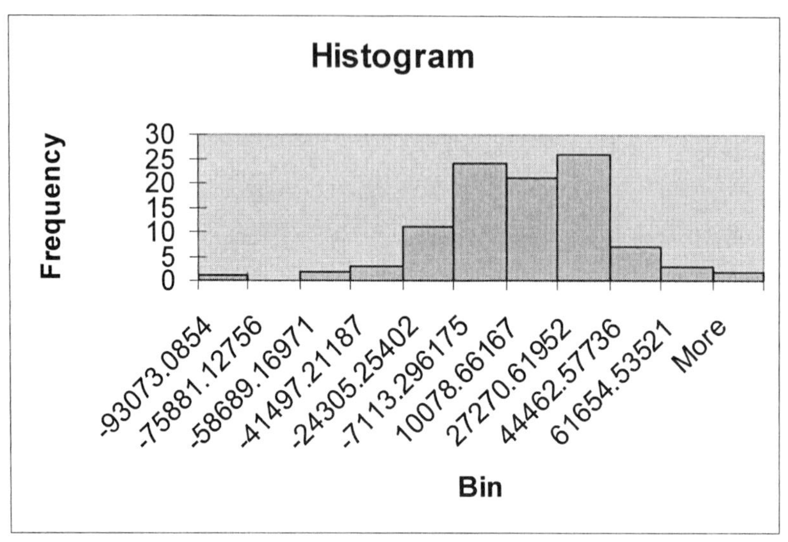

To plot the residuals against the explanatory variable house size:
Select: **Residual Plots** from within the PHStat Multiple Regression dialogue box.

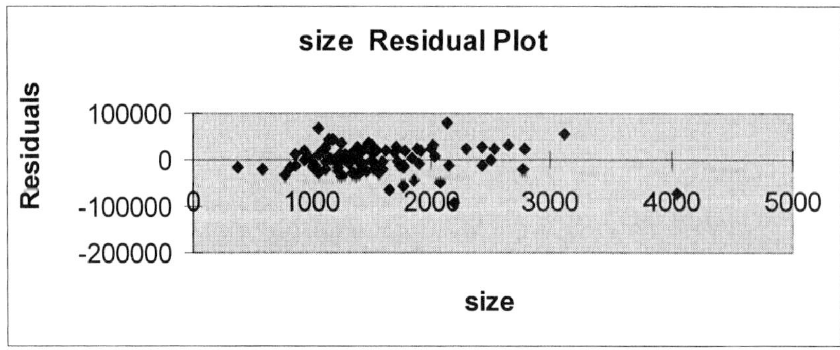

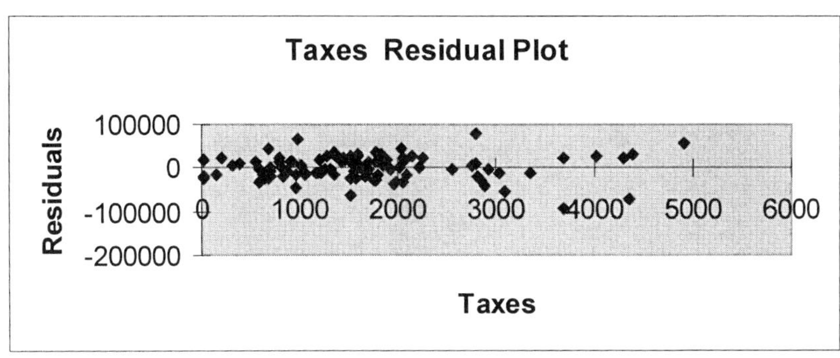

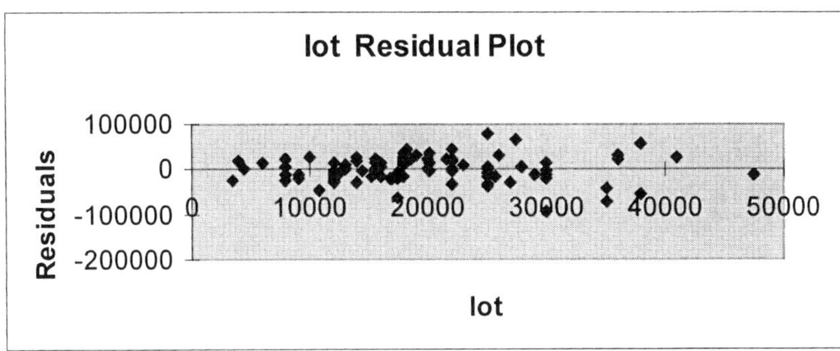

Chapter 12 Example 11 Comparing Winning High Jumps For Men and Women Including Categorical Predictors in Regression

Open the worksheet **high_jump** , which is found in the Excel folder. The following screen shot indicates the organization of the data file.

	A	B	C	D
1	Men_Meters	Year_Men	Year_Women	Women_Meters
2	1.81	1896		
3	1.9	1900		
4	1.8	1904		
5	1.905	1908		
6	1.93	1912		
7	1.935	1920		
8	1.98	1924		
9	1.94	1928	1928	1.59
10	1.97	1932	1932	1.657
11	2.03	1936	1936	1.6
12	1.98	1948	1948	1.68
13	2.04	1952	1952	1.67

Note that 1928 is the first year that women participated in the high jump.

We need to modify the data file so that each winning high jump (whether for male or female) is listed in one column and add columns to indicate male/female (1 = male, 0 = female) and number of years since 1928. Since we're only interested in years starting at 1928 we will not use the data for the men for prior years.

E	F	G
Meters	Male/Female	Years Since
1.94	1	0
1.97	1	4
2.03	1	8
1.98	1	20
2.04	1	24
2.12	1	28
2.16	1	32
2.16	1	36
2.24	1	40
2.23	1	44
2.25	1	48
2.36	1	52
2.35	1	56
2.36	1	60
2.34	1	64
2.39	1	68
2.36	1	72
1.59	0	0
1.657	0	4
1.6	0	8

Performing multiple regression for predicting winning height (in meters) as a function of number of years since 1928 and gender (1 = male, 0 = female)

Select: **PHStat > Regression > Multiple Regression**
Enter: Y Variable Cell Range:**E1:E35**
X Variables Cell Range: **F1: G35**
Click: **OK**

Regression Analysis

Regression Statistics	
Multiple R	0.983971376
R Square	0.968199669
Adjusted R Square	0.966148035
Standard Error	0.04302062
Observations	34

ANOVA

	df	*SS*	*MS*	*F*	*Significance F*
Regression	2	1.746820632	0.873410316	471.9163174	6.13325E-24
Residual	31	0.057373985	0.001850774		
Total	33	1.804194618			

	Coefficients	*Standard Error*	*t Stat*	*P-value*	*Lower 95%*	*Upper 95%*
Intercept	1.586768443	0.016747139	94.74862872	9.3886E-40	1.552612409	1.620924477
Male/Female	0.341352941	0.014755951	23.1332393	4.07317E-21	0.311257964	0.371447918
Years Since	0.006862708	0.000339469	20.21598066	2.06556E-19	0.006170355	0.007555061

To produce a scatterplot that includes the categorical distinction:
Select: **Chart Wizard > Scatterplot > Scatterplot > OK**

On the Series tab, create a series for the men, selecting appropriate cells, and create a series for the women.

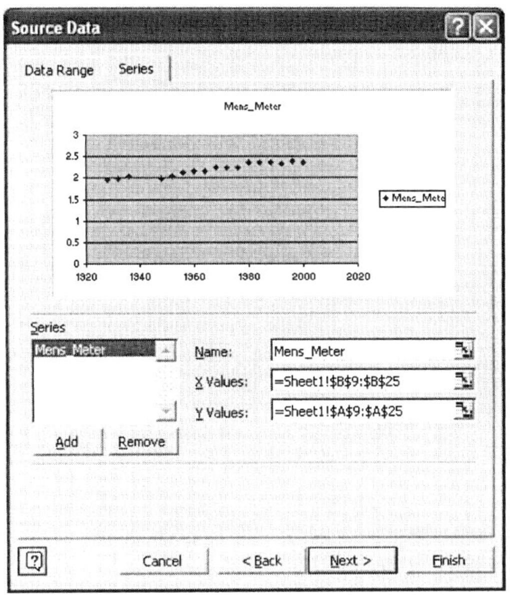

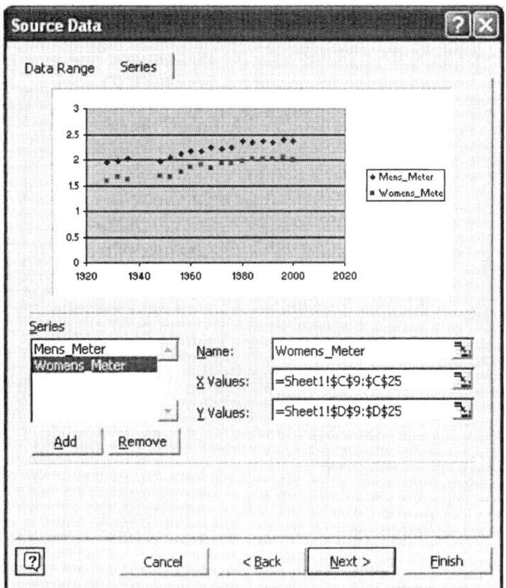

After modifying the axes:

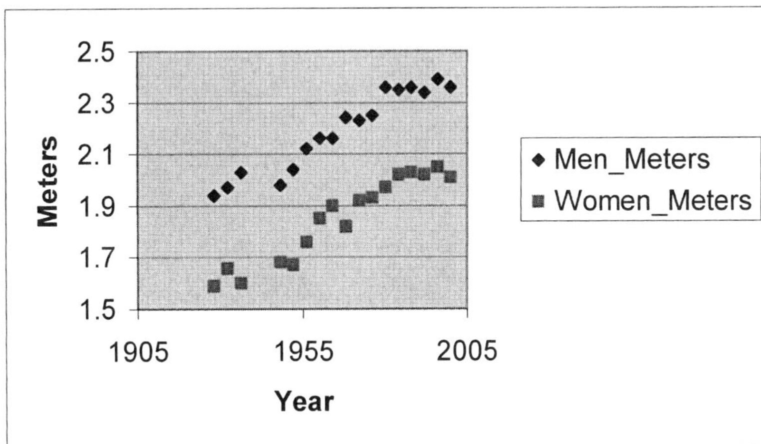

Chapter 12 Exercise 12.45 Houses, tax, and NW
Including Categorical Predictors in Regression

Open the worksheet **house_selling_prices** , which is found in the Excel folder. The following screen shot indicates the organization of the data file.

	A	B	C	D	E	F	G	H	I
1	House	Taxes	Bedrooms	Baths	Quadrant	NW	price	size	lot
2	1	1360	3	2	NW	1	145000	1240	18000
3	2	1050	1	1	NW	1	68000	370	25000
4	3	1010	3	1.5	NW	1	115000	1130	25000
5	4	830	3	2	SW	0	69000	1120	17000
6	5	2150	3	2	NW	1	163000	1710	14000
7	6	1230	3	2	NW	1	69900	1010	8000
8	7	150	2	2	NW	1	50000	860	15300

The data file includes column F NW where 1 = NW and 0 = other.

Performing multiple regression for predicting price as a function of real estate tax and whether the home is in the NW region:

Select: **PHStat > Regression > Multiple Regression**
Enter: Y Variable Cell Range: **select appropriate cells**
X Variables Cell Range: **select appropriate cells**
Click: **OK**

Regression Analysis

Regression Statistics	
Multiple R	0.827887418
R Square	0.685397577
Adjusted R Square	0.678910929
Standard Error	31934.40967
Observations	100

ANOVA

	df	*SS*	*MS*	*F*	*Significance F*
Regression	2	2.15511E+11	1.07756E+11	105.6628305	4.37986E-25
Residual	97	98921232511	1019806521		
Total	99	3.14433E+11			

	Coefficients	*Standard Error*	*t Stat*	*P-value*	*Lower 95%*	*Upper 95%*
Intercept	43014.39208	7830.66244	5.493071934	3.17562E-07	27472.70347	58556.08069
Taxes	45.3021067	3.215981488	14.08655705	3.41226E-25	38.91927725	51.68493616
NW	10814.17803	7458.343929	1.449943598	0.150299978	-3988.561816	25616.91787

From the coefficient for NW we see that the selling price increases by $10,814 just because the house is located in the NW region.

Chapter 12 Example 12 Annual Income and Having a Travel Credit Card
Logistic Regression

Chapter 12 Example 14 Estimating Proportion of Students Who Have Used Marijuana
Logistic Regression

Logistic Regression is not available in Excel or PHStat. See the MINITAB section.

Chapter 13 Example 3 Customers' Telephone Holding Times ANOVA

Enter the data into three separate columns as found in Example 2

	A	B	C
1	Advertisement	Muzak	Classical
2	5	0	13
3	1	1	9
4	11	4	8
5	2	6	15
6	8	3	7

Choose **Tools> Data Analysis > ANOVA – Single Factor**
Enter Input Range: **A1 – C6**
Click Grouped by : **Columns**
Click **Labels in first row**
Enter Alpha: **.05**
Output range: **A10 – E15 or new worksheet**
OK

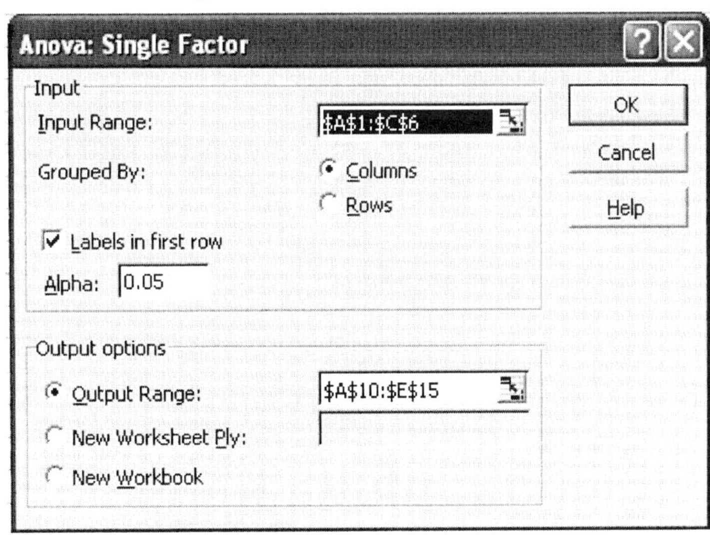

Note the p-value and the F statistic. The critical value is also given as 3.88.

10	Anova: Single Factor						
11							
12	SUMMARY						
13	*Groups*	*Count*	*Sum*	*Average*	*Variance*		
14	Advertisement	5	27	5.4	17.3		
15	Muzak	5	14	2.8	5.7		
16	Classical	5	52	10.4	11.8		
17							
18							
19	ANOVA						
20	*ource of Variatic*	*SS*	*df*	*MS*	*F*	*P-value*	*F crit*
21	Between Group	149.2	2	74.6	6.431034	0.012643	3.88529
22	Within Groups	139.2	12	11.6			
23							
24	Total	288.4	14				

Chapter 13 Exercise 13.3 What's The Best Way to Learn French? ANOVA

Enter the data given into three separate columns in a new worksheet.

	A	B	C
1	Group I	Group II	Group III
2	4	1	9
3	6	5	10
4	8		5

Choose **Tools> Data Analysis > ANOVA – Single Factor**
Enter Input Range: **A1 – C4**
Click Grouped by**: Columns**
Click **Labels in first row**
Enter Alpha: **.05**
Output range: **A10 or you may choose new worksheet**
OK

The output is shown below.

10	Anova: Single Factor						
11	SUMMARY						
12	*Groups*	*Count*	*Sum*	*Average*	*Variance*		
13	Group1	3	18	6	4		
14	Group 2	2	6	3	8		
15	Group 3	3	24	8	7		
16							
17	ANOVA						
18	*Source of Variation*	*SS*	*df*	*MS*	*F*	*P-value*	*F crit*
19	Between Groups	30	2	15	2.5	0.1767767	5.786148449
20	Within Groups	30	5	6			
21	Total	60	7				

Chapter 13 Example 7: Regression Analysis of Telephone Holding Times

We will be using the data from Examples 1 – 4, but to do the regression analysis we will be setting it up a little differently. We will use two independent variables, x_1 and x_2. Let $x_1 = 1$ for advertisement and 0 for Muzak and classical. Let $x_2 = 1$ for Muzak and 0 for advertisement and classical. The dependant variable is the holding times. Observe the way the data is laid out.

	A	B	C	D	E	F	G	H
1	Advertisement	Muzak	Classical	Holding Time		x1	x2	
2	5	0	13	5		1	0	
3	1	1	9	1		1	0	
4	11	4	8	11		1	0	
5	2	6	15	2		1	0	
6	8	3	7	8		1	0	
7				0		0	1	
8				1		0	1	
9				4		0	1	
10				6		0	1	
11				3		0	1	
12				13		0	0	
13				9		0	0	
14				8		0	0	
15				15		0	0	
16				7		0	0	

To perform the regression analysis choose **Tools > Data Analysis > Regression > OK**

Enter Input Y range: **D1 – D16**
Input X range: **F1 – G16**
Check Labels
Confidence Intervals **95%**
Choose Output Range: **A20**
OK

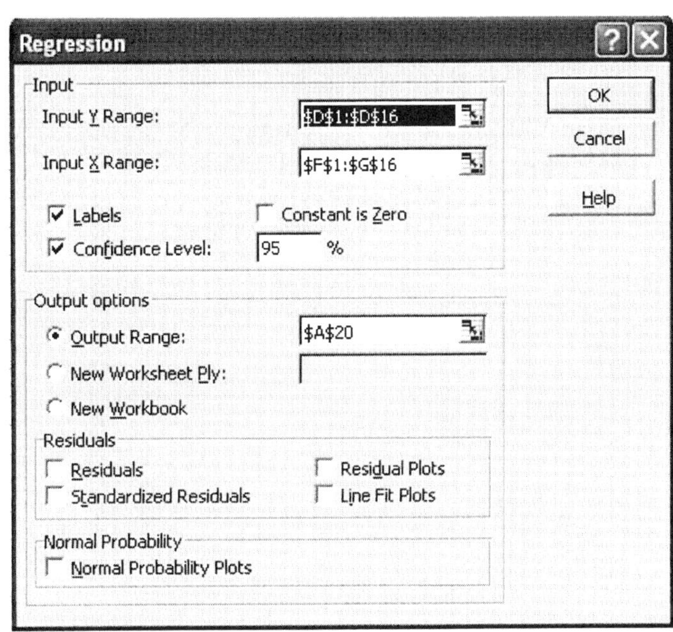

The output will be placed starting in cell A20.

	A	B	C	D	E	F	G	H	I
20	SUMMARY OUTPUT								
21									
22	*Regression Statistics*								
23	Multiple R	0.71926145							
24	R Square	0.51733703							
25	Adjusted R Sq	0.4368932							
26	Standard Error	3.40587727							
27	Observations	15							
28									
29	ANOVA								
30		*df*	*SS*	*MS*	*F*	*Significance F*			
31	Regression	2	149.2	74.6	6.4310345	0.0126434			
32	Residual	12	139.2	11.6					
33	Total	14	288.4						
34									
35		*Coefficients*	*Standard Error*	*t Stat*	*P-value*	*Lower 95%*	*Upper 95%*	*Lower 95.0%*	*Upper 95.0%*
36	Intercept	10.4	1.523154621	6.827934509	1.829E-05	7.081331227	13.71866877	7.081331227	13.718669
37	x1	-5	2.154065923	-2.321191727	0.0386835	-9.693306388	-0.30669361	-9.693306388	-0.3066936
38	x2	-7.6	2.154065923	-3.528211425	0.0041601	-12.29330639	-2.90669361	-12.29330639	-2.9066936

Note the coefficients, F statistic and P-value shown in the above chart.

These form the regression equation

$$\hat{y} = 10.4 - 5.1x_1 - 7.6x_2$$

Chapter 13 Example 11 Testing the Main Effects for Corn Yield
Two Way ANOVA allowing for interaction

Enter the data in the worksheet as shown.

	A	B	C
1	Fertilizer		Manure
2		High	Low
3	High	13.7	16.4
4		15.8	12.5
5		13.9	14.1
6		16.6	14.4
7		15.5	12.2
8	Low	15	12.4
9		15.1	10.6
10		12	13.7
11		15.7	8.7
12		12.2	10.9

To perform a Two-Way ANOVA:
Choose **Tools > Data Analysis > ANOVA: Two Factor with Replication**

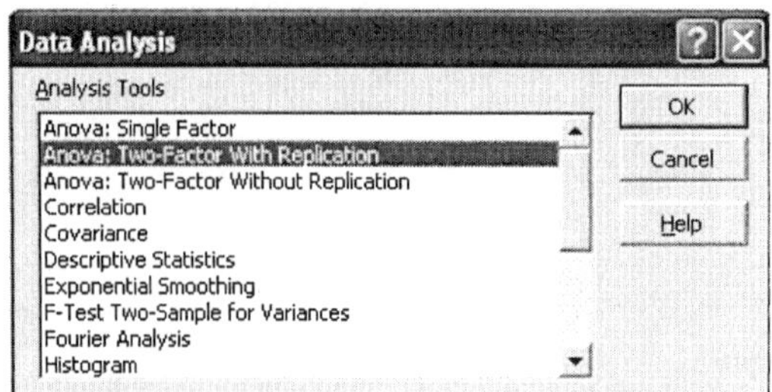

Enter Input Range: **A2-C12**
Rows per Sample: **5**
Alpha: **.05**
Output Range: **A17**
OK

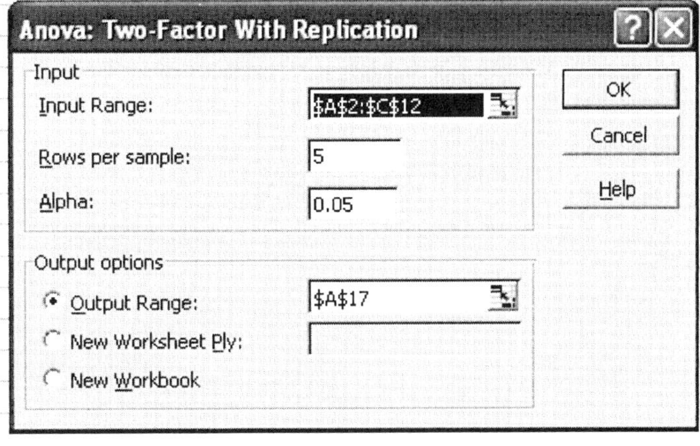

The output is extensive. Note the highlighted areas.

	A	B	C	D	E	F	G
17	Anova: Two-Factor With Replication						
18							
19	SUMMARY	High	Low	Total			
20	*High*						
21	Count	5	5	10			
22	Sum	75.5	69.6	145.1			
23	Average	15.1	13.92	14.51			
24	Variance	1.575	2.847	2.352111			
25							
26	*Low*						
27	Count	5	5	10			
28	Sum	70	56.3	126.3			
29	Average	14	11.26	12.63			
30	Variance	3.085	3.593	5.053444			
31							
32	*Total*						
33	Count	10	10				
34	Sum	145.5	125.9				
35	Average	14.55	12.59				
36	Variance	2.407222	4.827667				
37							
38							
39	ANOVA						
40	*Source of Variation*	*SS*	*df*	*MS*	*F*	*P-value*	*F crit*
41	Sample	17.672	1	17.672	6.368288	0.022578	4.493998
42	Columns	19.208	1	19.208	6.921802	0.018162	4.493998
43	Interaction	3.042	1	3.042	1.096216	0.310658	4.493998
44	Within	44.4	16	2.775			
45							
46	Total	84.322	19				

Chapter 13 Example 10: Regression Modeling to Estimate and Compare Mean Corn Yields

Enter the data in three columns, using 1 to represent high, and 0 to represent low for the levels of manure and fertilizer. The corn yield values will be the dependent variable and the manure and fertilizer levels will be the independent variables.

H	I	J
f	**m**	**yield**
1	1	13.7
1	1	15.8
1	1	13.9
1	1	16.6

To do the multiple regression we use PHStat.

Choose **PHStat > Regression > Multiple regression**
Enter Y Variable Cell Range: **J1 – J21**
X Variable Cell Range **H1 – I21**
Click **First cells contain labels**
Click **Regression Statistics Table**
Click **ANOVA and Coefficients Table**
OK

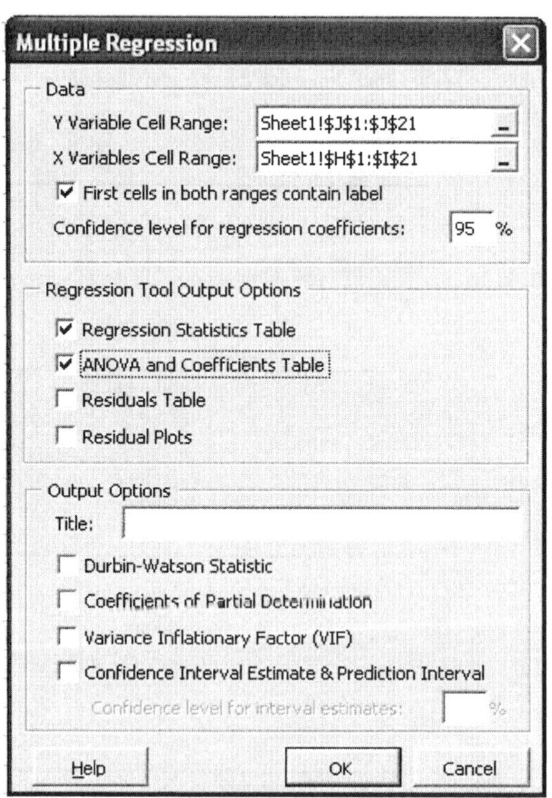

The output is extensive. Note the coefficients for the regression equation.

	A	B	C	D	E	F	G
1	Corn Yield						
2							
3	*Regression Statistics*						
4	Multiple R	0.661340328					
5	R Square	0.43737103					
6	Adjusted R Square	0.371179387					
7	Standard Error	1.670540596					
8	Observations	20					
9							
10	ANOVA						
11		*df*	*SS*	*MS*	*F*	*Significance F*	
12	Regression	2	36.88	18.44	6.607647232	0.007531609	
13	Residual	17	47.442	2.790705882			
14	Total	19	84.322				
15							
16		*Coefficients*	*Standard Error*	*t Stat*	*P-value*	*Lower 95%*	*Upper 95%*
17	Intercept	11.65	0.646997591	18.00624943	1.65275E-12	10.2849525	13.0150475
18	f	1.88	0.747088466	2.516435583	0.022188041	0.303778914	3.456221086
19	m	1.96	0.747088466	2.623517948	0.017792196	0.383778914	3.536221086
20							

Chapter 13 Exercise 13.33 Diet and Weight Gain
Two Way ANOVA, with and without interaction

Enter the data as presented in the problem. A small sample is shown to the right.

	A	B	C
1	Source	Level	
2		High	Low
3	Beef	73	90
4		102	76
5		118	90
6		104	64
7		81	86
8		107	51
9		100	72
10		87	90
11		117	95
12		111	78
13	Cereal	98	107
14		74	95
15		56	97
16		111	80

Choose **Tools > Data Analysis > ANOVA: Two Factor without Replication**

Enter:
Input Range: **A2-C32**
Check Box for Labels
Output range: **M2**
OK

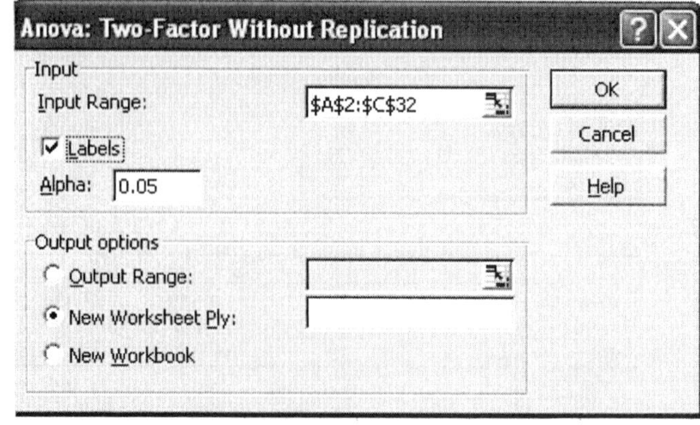

The output is extensive. Only part is shown here.

39	ANOVA						
40	*Source of Variation*	*SS*	*df*	*MS*	*F*	*P-value*	*F crit*
41	Rows	5597.933	29	193.0322	0.753146	0.775036	1.860812
42	Columns	3168.267	1	3168.267	12.3615	0.001462	4.182965
43	Error	7432.733	29	256.3011			
44							
45	Total	16198.93	59				

To repeat this and allow for interaction
Choose **Tools > Data Analysis > ANOVA: Two Factor with Replication**

Enter
Input range**: A2 – C32**
Rows per sample **10**
Alpha: **.05**
Output range: **F2**
OK

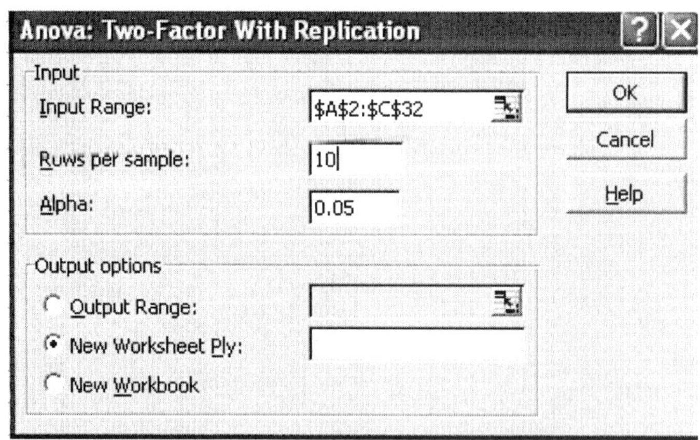

Again the output is extensive and only part is shown here.

29	ANOVA						
30	*Source of Variation*	*SS*	*df*	*MS*	*F*	*P-value*	*F crit*
31	Sample	266.5333	2	133.2667	0.621129	0.541132	3.168246
32	Columns	3168.267	1	3168.267	14.76665	0.000322	4.01954
33	Interaction	1178.133	2	589.0667	2.74552	0.073188	3.168246
34	Within	11586	54	214.5556			
35							
36	Total	16198.93	59				

Note the interaction line in this output that does not appear in the one above. Take note of the other differences and similarities.

Chapter 14 Example 4 Estimating the Difference Between Median Reaction Times Wilcoxon Rank Sum Test

Excel does not provide Nonparametric tests, but we can perform some of them by using the add-in program PHStat. Open the worksheet **cell_phones** from the Excel folder of the data disk. A portion is shown below.

Microsoft Excel - Book1

File Edit View Insert Format Tools Data PHStat

E11 fx

	A	B	C	D	E
1	Cell phone	Control	rankcellph	rankcontrol	
2	636	557	27	20	
3	623	572	25	22	
4	615	457	24	5	
5	672	489	29	12	
6	601	532	22	17	

Choose **PHStat > Two Sample Test > Wilcoxon Rank Sum Test.**

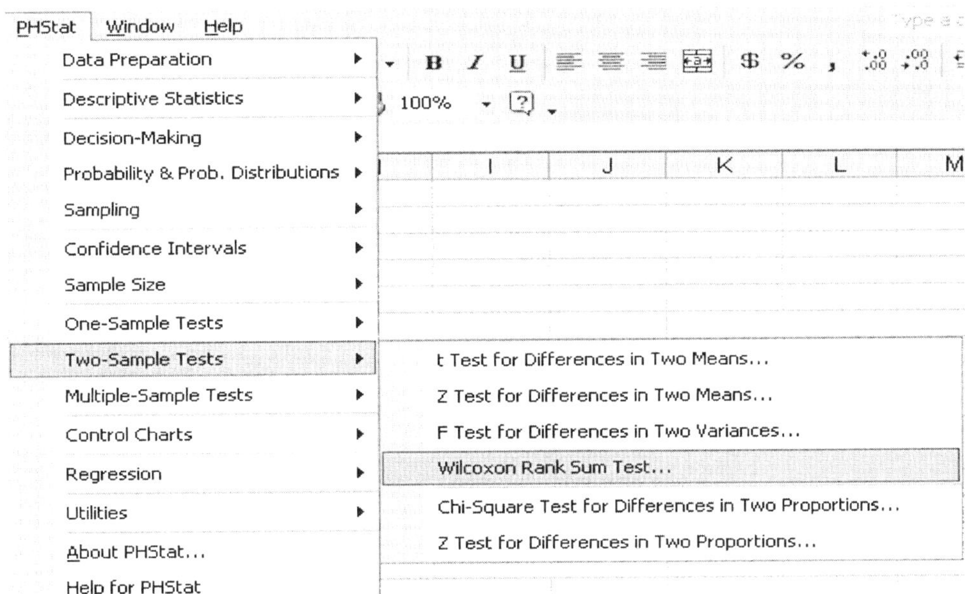

Enter: Level of Significance **.05**
Population 1 Sample Cell Range: **B1 to B33**
Population 2 Sample Cell Range: **A1 – A33**
Check box for First Cell contains label
Check **Two Tail Test**
Enter Title: **Cell Phone Study of Reaction Time**

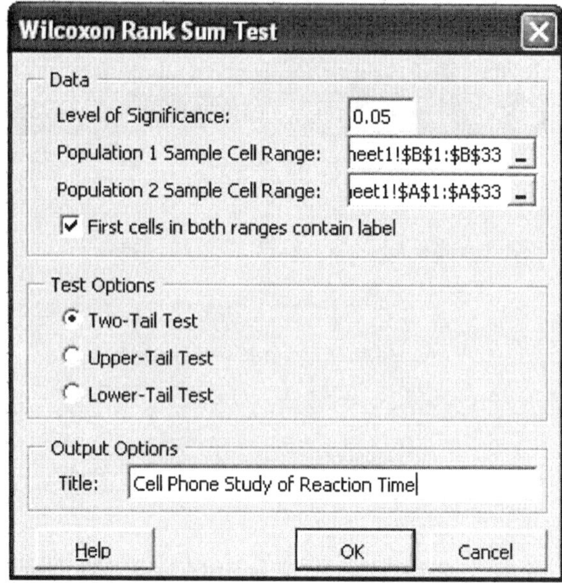

Notes:
PHStat performs the Wilcoxon rank sum test for differences between two medians in a new worksheet. This procedure also generates a second worksheet that contains the ranks of the sample data from the two populations.

	A	B	C
1	**Cell Phone Study of Reaction Time**		
2			
3	**Data**		
4	**Level of Significance**	**0.05**	
5			
6	Population 1 Sample		
7	Sample Size	32	
8	Sum of Ranks	864	
9	Population 2 Sample		
10	Sample Size	32	
11	Sum of Ranks	1216	
12			
13	Intermediate Calculations		
14	Total Sample Size n	64	
15	*T*1 Test Statistic	864	
16	*T*1 Mean	1040	
17	Standard Error of *T*1	74.4759469	
18	**Z Test Statistic**	**-2.363179084**	
19			
20	**Two-Tail Test**		
21	**Lower Critical Value**	**-1.959962787**	
22	**Upper Critical Value**	**1.959962787**	
23	***p*-Value**	**0.018118881**	
24	**Reject the null hypothesis**		

Chapter 14 Exercise 14.4 Anticipation of Hypnosis
One Sided Wilcoxon Rank Sum Test

Enter the data from the text into Columns A and B.
Choose **PHStat > Two Sample Test > Wilcoxon Rank Sum Test**.
Enter: Level of Significance **.05**
Population 1 Sample Cell Range**: A1 –A 9**
Population 2 Sample Cell Range: **B1- B9**
Check box for First Cell contains label
Check **Lower Tail Test**
Enter Title: **Anticipation of Hypnosis**
OK

	A	B	C
1	**Anticipation of Hypnosis**		
2			
3	**Data**		
4	**Level of Significance**	**0.05**	
5			
6	Population 1 Sample		
7	Sample Size	8	
8	Sum of Ranks	49	
9	Population 2 Sample		
10	Sample Size	8	
11	Sum of Ranks	87	
12			
13	Intermediate Calculations		
14	Total Sample Size n	16	
15	*T*1 Test Statistic	49	
16	*T*1 Mean	68	
17	Standard Error of *T*1	9.521905	
18	**Z Test Statistic**	**-1.9954**	
19			
20	**Lower-Tail Test**		
21	**Lower Critical Value**	**-1.64485**	
22	***p*-Value**	**0.023**	
23	**Reject the null hypothesis**		

Chapter 14 Exercise 14.7 Teenage Anorexia
Wilcoxon Rank Sum Test

Open the file **anorexia** from the Excel folder of the data disk.

Choose **PHStat > Two Sample Test > Wilcoxon Rank Sum Test**.
Enter: Level of Significance **.05**
Population 1 Sample Cell Range **G1 –G30**
Population 2 Sample Cell Range: **H1- H27**
Check box for First Cell contains label
Check **Two Tail Test**
Enter Title: **Anorexia**
OK

Level of Significance	**0.05**
Population 1 Sample	
Sample Size	29
Sum of Ranks	907
Population 2 Sample	
Sample Size	26
Sum of Ranks	633
Intermediate Calculations	
Total Sample Size n	55
*T*1 Test Statistic	633
*T*1 Mean	728
Standard Error of *T*1	59.31835
Z Test Statistic	**-1.60153**
Two-Tail Test	
Lower Critical Value	**-1.95996**
Upper Critical Value	**1.959963**
***p*-Value**	**0.10926**
Do not reject the null hypothesis	

Chapter 14 Example 5 Does Heavy Dating Affect College GPA?
Kruskal-Wallis Test

PHStat performs the Kruskal-Wallis rank sum test of hypothesis for differences between medians from multiple independent sample groups and places the analysis in a new worksheet. It also generates a second worksheet that contains the ranks of the sample data from the multiple independent samples. Any non-numeric values in the sample data cell range (excluding the first cells of columns if First cells contain label is selected) are treated as zero values for the purposes of ranking.

Enter the data in three separate columns

	A	B	C	D
1	Rare	Occasional	Regular	
2	1.75	2	2.4	
3	3.15	3.2	2.95	
4	3.5	3.44	3.4	
5	3.68	3.5	3.67	
6		3.6	3.7	
7		3.71	4	
8		3.8		

Choose PHStat> Multiple Sample Tests > Kruskal Wallis Rank Test

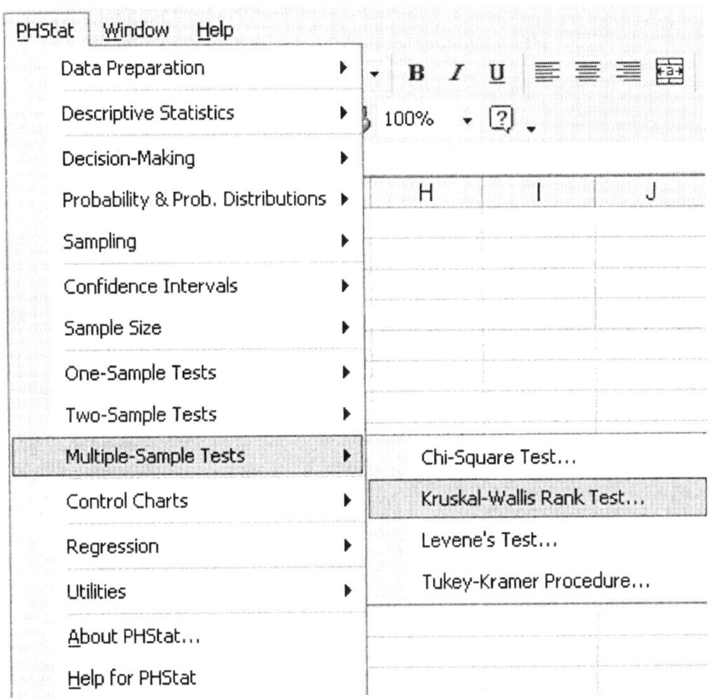

Enter Level of Significance: **.05**
Sample Data Cell Range: **A1 – C8**
Check box for First Cells contain labels
Title : **GPA vs Dating**
OK

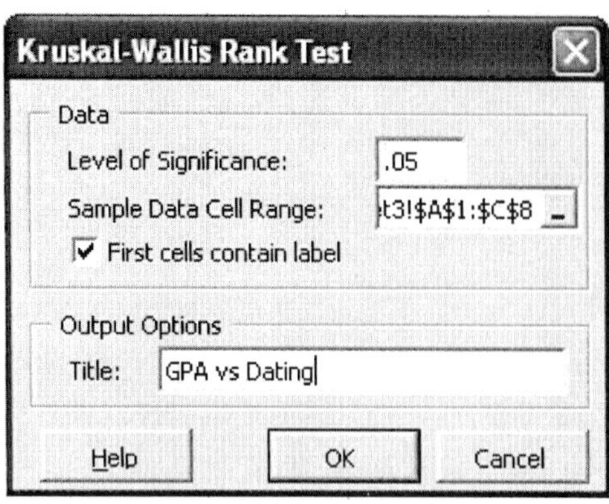

The results appear in a separate worksheet.

	A	B	C	D	E	F	G	H
1	GPA vs Dating							
2								
3	**Data**							
4	**Level of Significance**	**0.05**		Group	Sample Size	Sum of Ranks	Mean Ranks	
5				1	4	28.5	7.125	
6	Intermediate Calculations			2	7	67.5	9.64285714	
7	Sum of Squared Ranks/Sample Size	1395.455		3	6	57	9.5	
8	Sum of Sample Sizes	17						
9	Number of Groups	3						
10								
11	**Test Result**							
12	**H Test Statistic**	**0.723739**						
13	**Critical Value**	**5.991476**						
14	**p-Value**	**0.696373**						
15	**Do not reject the null hypothesis**							

Chapter 14 Example 6 Spend More Time Browsing the Internet or Watching TV? Sign Test for Matched Pairs

Excel does not have a nonparametric sign test, but we can do the calculations manually. Retreive the data for the **georgia_student_survey** from the data disk. We are only interested in the columns for Internet and TV watching, so you can hide all the others. Create a column named DIFF for the differences between the two columns.
Use the COUNTIF function to count the number of entries greater than zero, which would be considered your successes.

Choose **Insert > Function > Statistical > COUNTIF**

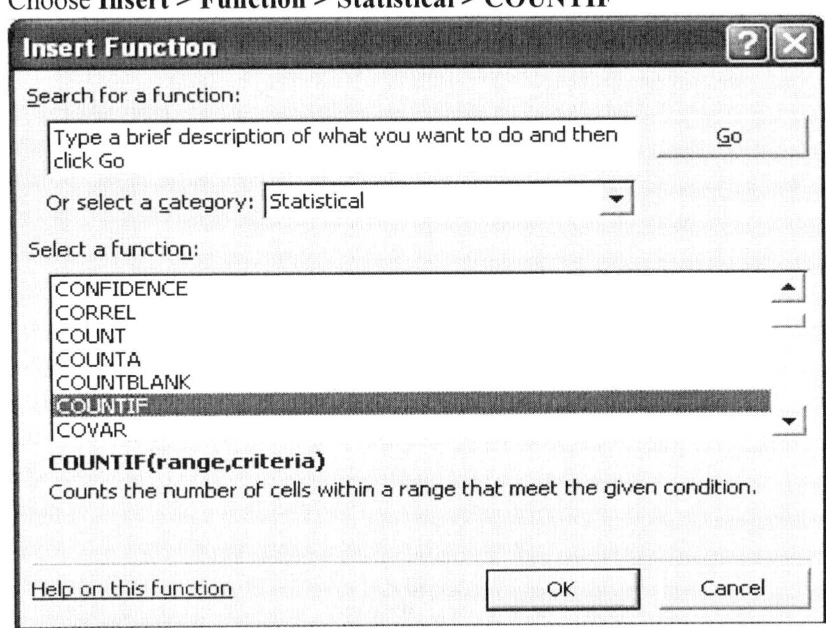

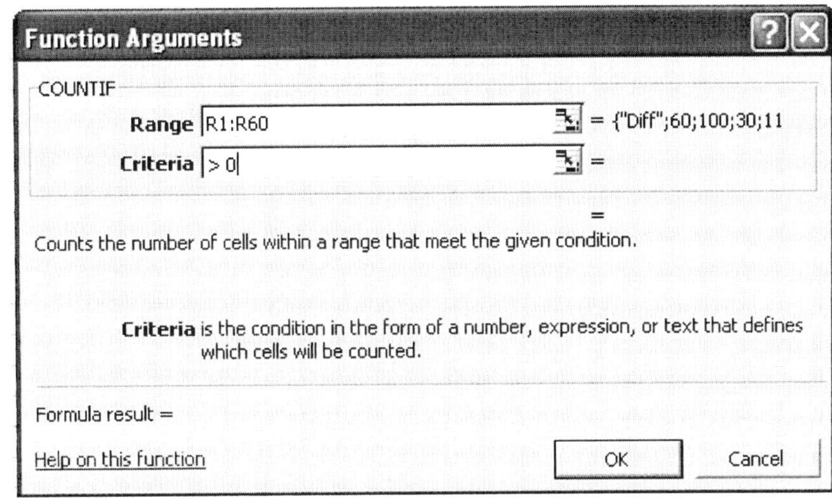

This will enter the number 35 into your worksheet. Then perform a z-test for the proportion by choosing **PHStat > One-Sample test > Z test for the proportion**

Enter Null Hypothesis : **.5**
Level of Significance **.05**
Number of Successes**: 35**
Sample Size**: 54**
Test Option: **Two Tail Test**
Title: **Internet vs TV**

OK

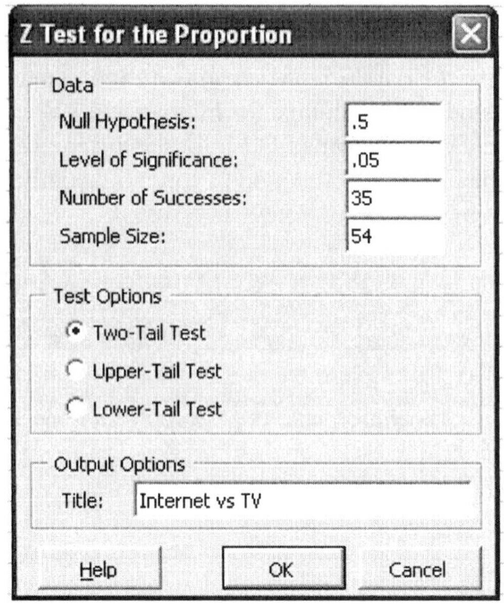

The result will appear in a new worksheet.

	A	B
1	**Internet vs TV**	
2		
3	**Data**	
4	**Null Hypothesis** $p=$	**0.5**
5	**Level of Significance**	**0.05**
6	**Number of Successes**	**35**
7	**Sample Size**	**54**
8		
9	Intermediate Calculations	
10	Sample Proportion	0.648148148
11	Standard Error	0.068041382
12	**Z Test Statistic**	**2.177324216**
13		
14	**Two-Tail Test**	
15	**Lower Critical Value**	**-1.959962787**
16	**Upper Critical value**	**1.959962787**
17	***p*-Value**	**0.029456281**
18	**Reject the null hypothesis**	

Chapter 14 Exercise 14.8 How Long Do You Tolerate Being On Hold
Kruskal-Wallis Test

Enter the data in three separate columns.

	A	B	C
1	Muzak	Advertisement	Classical
2	0	5	13
3	1	1	9
4	4	11	8
5	6	2	15
6	3	8	7

Choose PHStat> Multiple Sample Tests > Kruskal Wallis Rank Test
Enter Level of Significance: **.05**
Sample Data Cell Range: **A1 – C6**
Check box for First Cells contain labels
Title : **Holding Time vs Group**
OK

The results appear in a separate worksheet.

	A	B	C	D	E	F	G
1	Holding Time vs Group						
2							
3	Data						
4	Level of Significance	0.05		Group	Sample Size	Sum of Ranks	Mean Ranks
5				1	5	22.5	4.5
6	Intermediate Calculations			2	5	37	7.4
7	Sum of Squared Ranks/Sample Size	1107.1		3	5	60.5	12.1
8	Sum of Sample Sizes	15					
9	Number of Groups	3					
10							
11	Test Result						
12	H Test Statistic	7.355					
13	Critical Value	5.991476					
14	p-Value	0.025286					
15	Reject the null hypothesis						

Chapter 14 Exercise 14.10 Sports vs TV
Sign Test

Open the **fla_student_survey** file. We are only interested in the Sports and TV columns, so hide the others. Create a Difference column. A small sample of the file is shown.

fla_student_survey [Read-Only]

	H	I	U
1	TV	sports	diff
2	3	5	-2
3	15	7	8
4	0		-4

Use the COUNTIF function to count the successes (ie > 0)
Choose **Insert > Function > Statistical > COUNTIF**
Enter: Range: **T1-T61**
Criteria: **>0**
OK
The result should be 30. This represents the number of successes.

Then perform a z-test for the proportion:
Choose **PHStat > One-Sample test > Z test for the proportion**
Enter Null Hypothesis : **.5**
Level of Significance **.05**
Number of Successes: **30**
Sample Size:
Test Option: **Two Tail Test**
Title: **Sport vs TV**
OK

The result will appear in a new worksheet

	A	B
1	**Sports vs TV**	
2		
3	**Data**	
4	**Null Hypothesis** *p*=	**0.5**
5	**Level of Significance**	**0.05**
6	**Number of Successes**	**30**
7	**Sample Size**	**54**
8		
9	Intermediate Calculations	
10	Sample Proportion	0.555555556
11	Standard Error	0.068041382
12	**Z Test Statistic**	**0.816496581**
13		
14	**Two-Tail Test**	
15	**Lower Critical Value**	**-1.959962787**
16	**Upper Critical value**	**1.959962787**
17	***p*-Value**	**0.414216057**
18	**Do not reject the null hypothesis**	

SPSS MANUAL

DEBRA HYDORN
University of Mary Washington

STATISTICS

THE ART AND SCIENCE OF LEARNING FROM DATA

Agresti • Franklin

Acknowledgments

I would like to thank my colleague, Wyatt Mangum, for working through each example in his "spare" time. He found several cases where I had omitted steps and also made suggestions for maintaining consistency throughout the manual.

I would also like to thank my husband Michael for carefully reviewing most of the manual. His suggestions about how differences in operating systems might affect the output and SPSS windows were very helpful. His questions about why I discussed some options and not others helped me to anticipate questions students might have while using the manual.

Finally, I would like to thank the individuals at Prentice Hall who gave me the opportunity to write this manual. It gave me the chance to "play" with SPSS and in the process I "discovered" some new things about SPSS!

Debra L. Hydorn

Table of Contents

Chapter 1 Getting Started with SPSS

SPSS is a Windows-based program with drop-down menus for you to access commands for manipulating and analyzing your data. In SPSS, you select the appropriate command from one of the menus and typically fill in the details about the analysis you want in a dialog box. Most of the commands you will be using have a dialog box. If you have used drop-down menus and dialog boxes before then you will find SPSS very easy to use, and, if you haven't used them before, don't worry – this manual will provide step-by-step instructions! The purpose of this manual is to get you started with using SPSS for your data analyses. SPSS includes commands for conducting many more advanced methods than will be covered here. And, for most of the commands demonstrated in this manual you will find some options that are not discussed as they are beyond the scope of an introductory statistics course.

The examples for this manual were produced using version 13.0 of SPSS; if you are using an earlier version you might notice some differences in the screens or output that you encounter. For commands where version 13.0 is drastically different, examples using version 11.0 will also be provided.

Depending on the computer you are using, you may access SPSS through "All Programs" found on that computer's **Start** button, or by double clicking on an SPSS icon on the desk top. If you access it through the Start button, make sure to select SPSS for Windows (and not an SPSS Production Facility, for example). When you first start SPSS you will see the following dialog box that appears for you to indicate what you want to do:

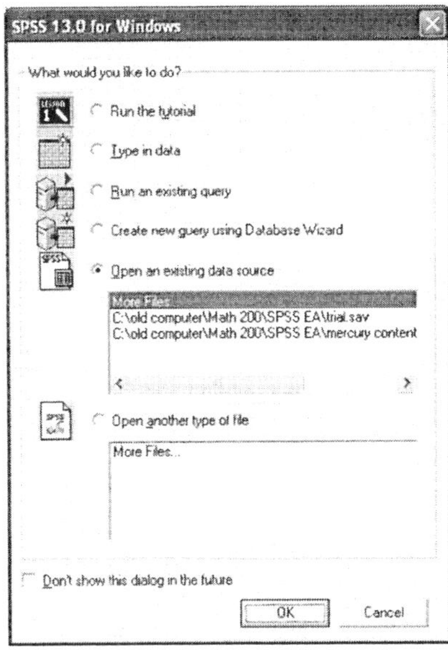

The default setting in this dialog box is to "Open an existing data source." The CD that comes with your textbook includes a number of SPSS data files that you will access later on using this option. This dialog box also has an option for accessing output files that have been saved. For a simple introduction to SPSS, though, click on "Type in data" and then on the **OK** button. You should now see the following display:

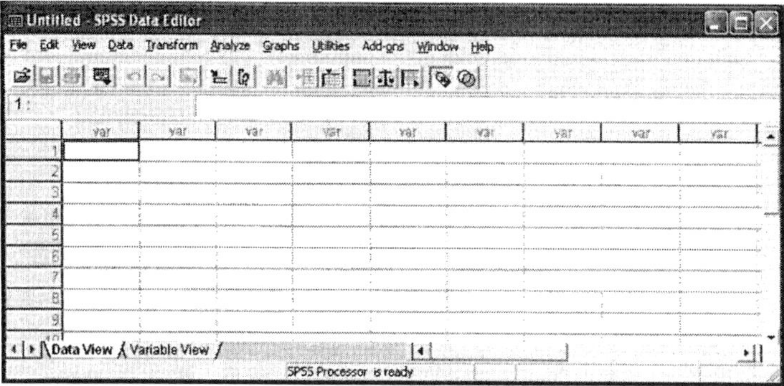

This is the first of the two windows that you will be using in SPSS. This is the **SPSS Data Editor** window, and it is here that you will enter new data to analyze. It resembles spreadsheets you have probably seen before and it has some of the same properties. Unlike some other spreadsheets, however, the Data Editor in SPSS is for data only and all output you produce will appear in the **Output Viewer** window, which is the second SPSS window you will be using. To see the Output Viewer you will have to first enter some data, but let's explore the Data Editor a little more first.

The menu bar is immediately below the title bar. Once you enter some data and save it you will see the name on that bar change. The menus you will be using most are:

File – commands for opening and saving data files and for exiting SPSS
Edit – commands for modifying SPSS data files or output (e.g., copy and paste)
Data – commands for adding or deleting variables and for manipulating the data set (e.g., select cases and split file)
Transform – commands for creating or re-coding new variables using variables that are already in the data set
Analyze – commands for producing univariate and bivariate descriptive statistics as well as for conducting inferential statistics
Graphs – commands for producing various graphical displays

Below the menu bar is a series of icons representing a few common commands, for those who prefer that approach. The directions in this manual use the drop-down menus only. Notice the two tabs at the bottom of the window, **Data View** and **Variable View**. The Data View is for data entry and has columns for the variables and rows for the observations. The columns are currently each labeled "var" but these labels will change when you enter data. The Variable View is for defining and labeling your variables. You can enter the data in the Data View and then move to the Variable View to name the variables, or you can define your variables first in the Variable View and then enter the data in the Data View. If you click on the Variable View tab you will see that the labels on the columns of the spreadsheet change, as shown below.

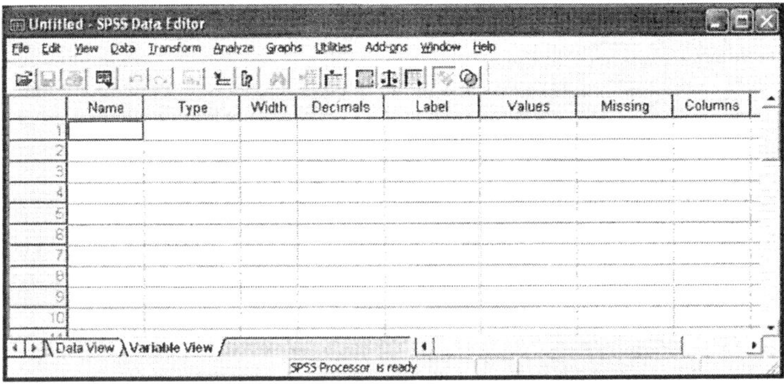

Information about the variables in your data set is entered using the columns in this new spreadsheet, some of which have obvious functions. In the "Name" column you can enter a one word name for your variable; SPSS does not allow the use of spaces or some special characters in this column. For more descriptive names you can use the underscore _ symbol or enter longer descriptions in the "Label" column. Try to avoid using variable names such as var1 or X as these names do not provide any useful information about the data. If you click on the far right side of the cells under some of the columns a dialog box will appear with options for your variables. For example, if you click on the right side of the first cell under "Type" the following dialog box will appear.

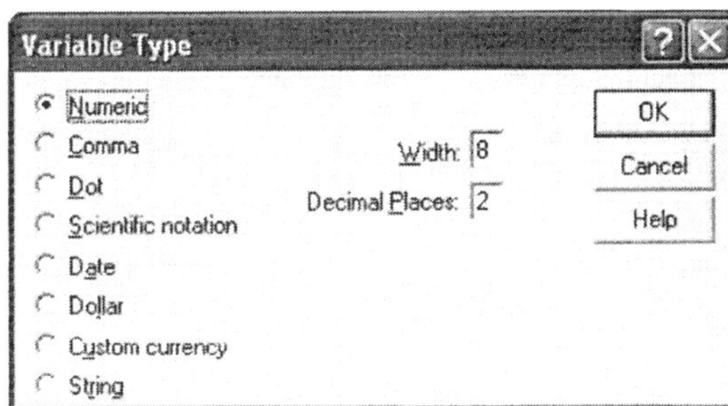

Notice that "Numeric" is the default variable type; if you want to enter character data you will need to change the type to "String." You can increase or decrease the total number of digits or characters by changing the entry under "Width," while the number of decimals to the right of the decimal point can be altered under "Decimals." For both of these variable characteristics up and down arrows will appear when you click on the right side of a cell in those columns. If you have categorical data that you wish to code numerically, this can be done using the "Values" column. This option will be demonstrated in the example below.

Entering Data

To introduce you to data entry in SPSS, let's start with the University of Florida Student Survey example in the first chapter of your text. The complete data set is available on the text CD but you will enter only a portion of it here. In addition to a student identification number, the variables you will enter are:

Gender (f=female, m=male)
Racial-ethnic group (b=black, h=Hispanic, w=white)
Age (in years)
College GPA (scale 0 to 4)
Average number of hours per week watching TV
Whether a vegetarian (yes, no)
Political party (dem=Democrat, rep=Republican, ind=Independent)
Marital status (1=married, 0=unmarried)

Some of these variables are categorical and some are numeric. For entering categorical variables you have two choices: you can enter it as a string variable or you can enter it as a numeric variable with codes to identify the categories. Of the categorical variables, all but "Marital status" will be entered as string variables. Marital status is entered using a numerical coding scheme (but the variable is still categorical). How you enter categorical data will determine which analyses will be available in SPSS for you to use. You will learn more about this later on.

Let's start by defining the variables in the Variable View and then entering the data in the Data View. The following display shows the variable names entered and which variables are numeric and which are string. It also shows the number of decimals for the numerical variables decreased to match the data to be entered. Notice the up and down arrows on the right side of the cell in the "Decimals" column of the "married" variable. These were used to decrease the number of decimals from the default setting of 2 to 0. To match the variable names as they are given in the text, variable labels have been included.

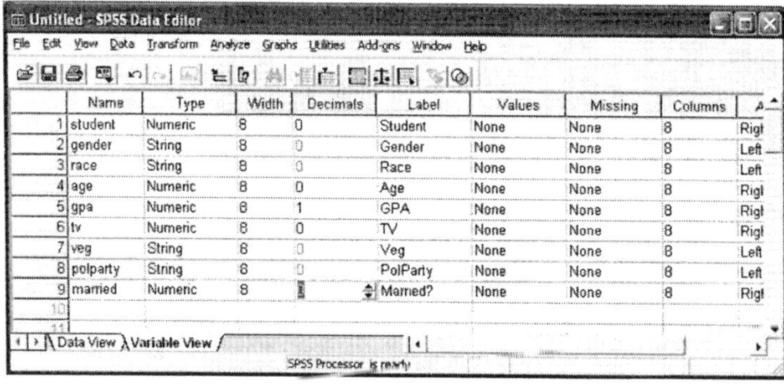

The last thing you need to do is enter the coding for the "married" variable. Click on the right hand side of the cell under the "Values" column in the "married" variable row. The following dialog box will appear:

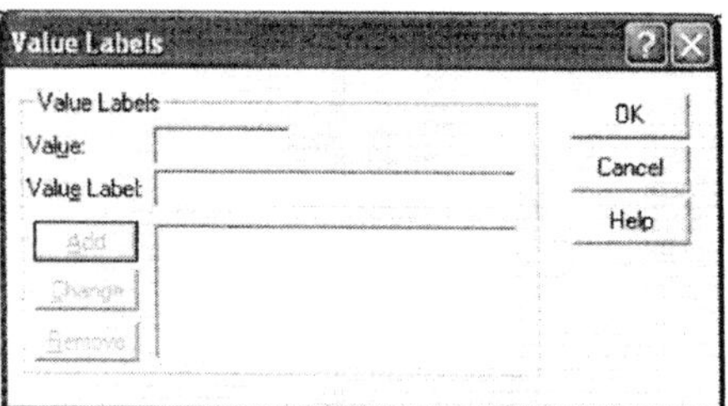

Enter the value "1" in the "Value" box and then click on the "Value Label" box and enter "Married" (but without the quotes). Notice that the "Add" button becomes highlighted as soon as you start entering a label. Click on **Add** so that the dialog box looks like:

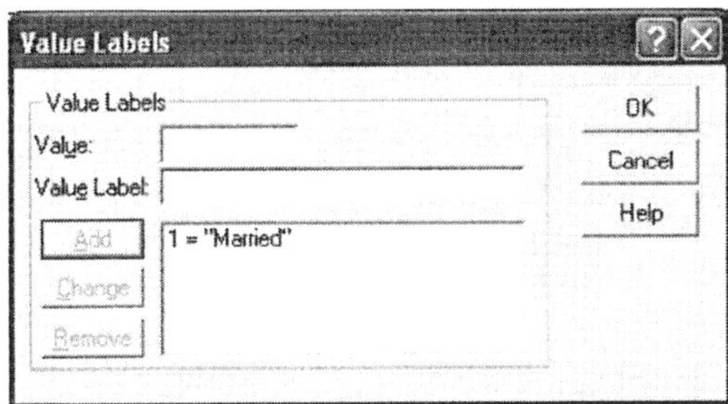

Now enter "0" in the "Value" box and "Unmarried" in the "Value Label" box and click on **Add** again. Click on **OK** to complete the process and close this dialog box. The Variable View should now look like:

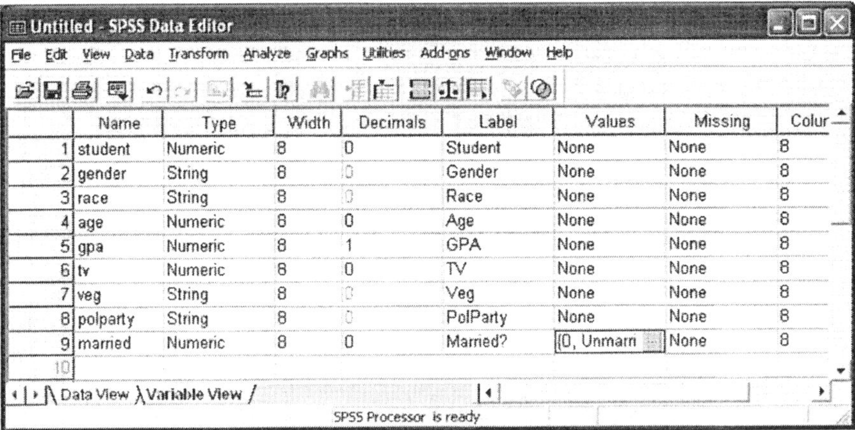

Now click on the Data View tab. Notice that the variable names now replace "var" at the top of each column. The data to be entered is shown in the following table and represents just a portion of the data set available on your text CD in the "fla_student_survey" data file.

Student	Gender	Race	Age	GPA	TV	Veg	PolParty	Married?
1	m	w	32	3.5	3	no	rep	1
2	f	w	23	3.5	15	yes	dem	0
3	f	w	27	3.0	0	yes	dem	0
4	f	h	35	3.2	5	no	ind	1
5	m	w	23	3.5	6	no	ind	0
6	m	w	39	3.5	4	yes	dem	1
7	m	b	24	3.7	4	no	ind	0
8	f	h	31	3.0	4	no	ind	1

Enter the data so that your Data Editor appears as the one shown below.

Untitled - SPSS Data Editor

File Edit View Data Transform Analyze Graphs Utilities Add-ons Window Help

9 : married

	student	gender	race	age	gpa	tv	veg	polparty	married
1	1	m	w	32	3.5	3	no	rep	1
2	2	f	w	23	3.5	15	yes	dem	0
3	3	f	w	27	3.0	0	yes	dem	0
4	4	f	h	35	3.2	5	no	ind	1
5	5	m	w	23	3.5	6	no	ind	0
6	6	m	w	39	3.5	4	yes	dem	1
7	7	m	b	24	3.7	5	no	ind	0
8	8	f	h	31	3.0	5	no	ind	1

Data View / Variable View

SPSS Processor is ready

To view the value labels for the "married" variable, instead of the values 0 and 1, click on the "View" menu and select "Value labels." A check mark will appear next to "Value labels" and the labels you entered for each value should now appear in the Data Editor in place of the values 0 and 1. Once you have the data entered it is important that you now check for any data entry errors. Possible errors to look for are misplaced decimals or omitted cases. You can easily correct any mistakes by using the arrow keys on your keyboard to move to the cell you need to correct or by clicking on that cell with the mouse. Make a correction and press **Enter**. Note: If you first enter data in the Data View before defining your variables in the Variable View, the variable names at the top of each column you use will automatically change to "var00001" and "var00002," for example, which you can then change to appropriate names in the Variable View.

Note: If a data set has "missing" values, these observations are entered in SPSS as a "." (without the quotes). Many of the SPSS commands you will be using include a table indicating if there were any missing values omitted from an analysis.

Saving Data

Now that you have entered this data set in SPSS (and it is free from errors), let's see how to save it. Select the "Save as…" command under the "File" menu. A dialog box will appear in which you can enter where you want the data file to be saved and what name you want it saved as. In the display below the data file is being saved into a folder called "Intro Statistics" with the name "fla student survey" (again, without the quotes). Note that it is OK to have spaces in the names of SPSS data files.

Click on **Save** to complete the process. The data file will be saved with the ".sav" file-type extension. Notice that once the data has been saved with a name, the title in the bar at the top of the Data Editor now shows that data file name.

Producing Output

To demonstrate how SPSS works, let's print a table of this data and then find some descriptive statistics and produce a graph for the "age" variable. The command to print a table of your data is the "Case Summaries" option under the "Reports" command from the Analyze menu. Click on the "Analyze" menu and then scroll down to "Reports." Then click on "Case Summaries" from the list of options that appears. Let's abbreviate this action with select **Analyze → Reports → Case Summaries...** (menu name → command name → subcommand or option). A triangle to the right of a command name indicates that there are several options from which you can choose. The ellipsis (...) following a subcommand indicates that a dialog box will appear for you to enter information about the analysis you want SPSS to do. The following dialog box will appear. All of the variables in the data set are listed in the box on the left.

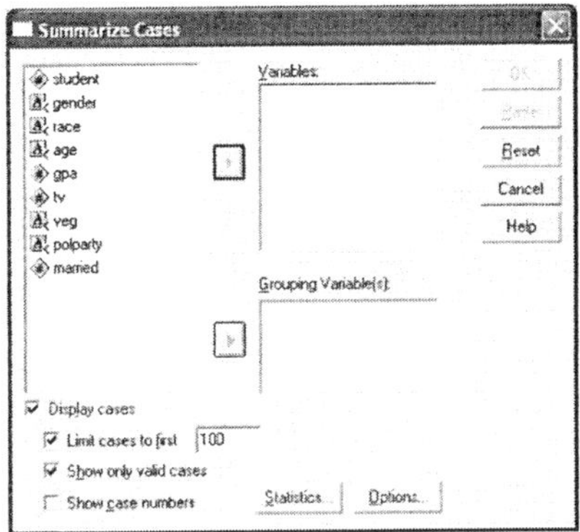

To select a variable to be included in the output from this command, click once on the variable name and then once on the triangle button between the two boxes. Or, just double click on the variable name. This double-clicking option doesn't work all of the time, but only when there is only one place to enter a variable name. The dialog box below shows the variables selected.

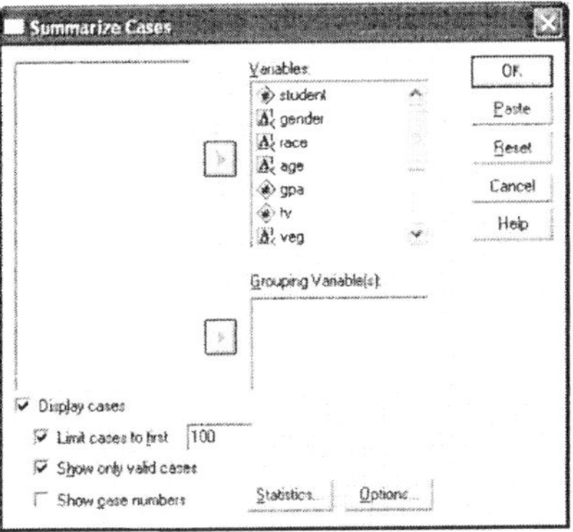

To un-select a variable, click on it once in the "Variables" box and then on the triangle button. In the dialog box below, the variable gender has been selected. Notice that the triangle button is pointing the other way.

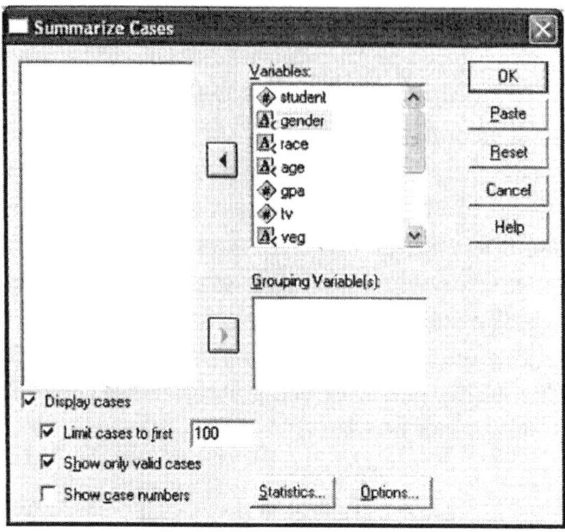

Notice also that the **OK** button has become highlighted. This is how you will know when you have entered enough information in a dialog box for SPSS to be able to execute the command. Click on **OK** to produce the output. Now you will see the SPSS Output Viewer. Notice that the Output Viewer also has a list of menus at the top and is divided into two parts. On the right you will see the actual output produced, in this case, a table showing the data that has been entered. A flow chart, or **Output Navigator**, showing the commands you have used and the parts of the output produced, is shown on the left. The list of menus in the menu bar of the Output Viewer is slightly different from those available in the Data Editor window, but the Analyze and Graphs menus are in both menu bars so you can produce additional output while either the Data Editor or the Output Viewer is the active window. To move back and forth between the Data Editor and the Output Viewer you can use the "Window" menu or the window name tags that are at the bottom of your computer screen.

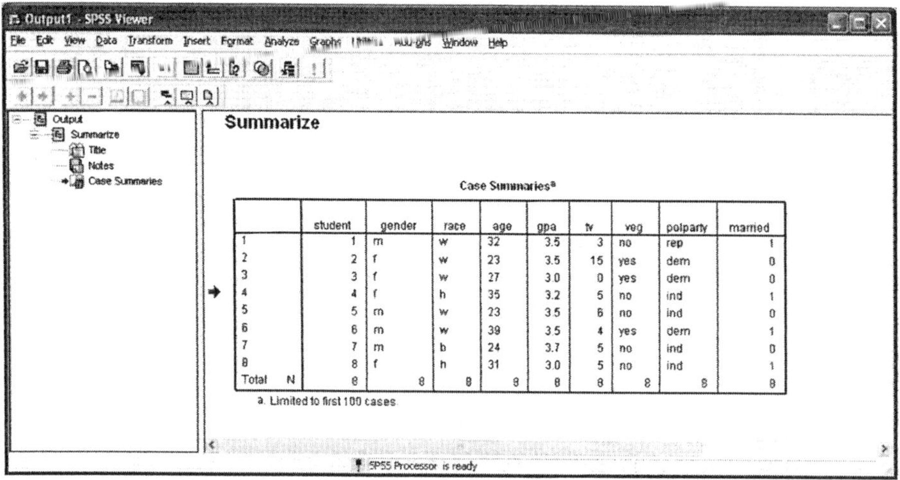

Case Summaries[a]

	student	gender	race	age	gpa	tv	veg	polparty	married
1	1	m	w	32	3.5	3	no	rep	1
2	2	f	w	23	3.5	15	yes	dem	0
3	3	f	w	27	3.0	0	yes	dem	0
4	4	f	h	35	3.2	5	no	ind	1
5	5	m	w	23	3.5	6	no	ind	0
6	6	m	w	39	3.5	4	yes	dem	1
7	7	m	b	24	3.7	5	no	ind	0
8	8	f	h	31	3.0	5	no	ind	1
Total N	8	8	8	8	8	8	8	8	8

a. Limited to first 100 cases.

This output is typical of what SPSS produces. It includes a title, some notes that are currently "hidden," a "Case Processing Summary," and the actual output. The "Case Processing Summary" was omitted in the output above. To delete a piece of output just click on it once and press the "Delete" key on your keyboard. You can also "hide" a piece of output by double clicking on its icon in the Output Navigator. To "un-hide" it just double click on the icon again. Notice that the title of the output is the name as the command you used. You will see how to modify the title so that it is more descriptive after making a histogram for "age." Within the output, SPSS uses the variable names and labels that you entered in the Variable View of the Data Editor. If you forget to name a variable, you will see "var00001" or something similar. You will need to go into the Variable View to name the variable and then redo the output. Unlike some other data analysis programs, any changes you make to the data after producing output are not incorporated into output that has already been produced. This is also true for correcting any data entry errors – SPSS does not automatically update output.

One SPSS command that finds a simple collection of descriptive statistics is the "Descriptive Statistics" command under the "Analyze" menu. Select **Analyze → Descriptive Statistics → Descriptives...** and the following dialog box will appear.

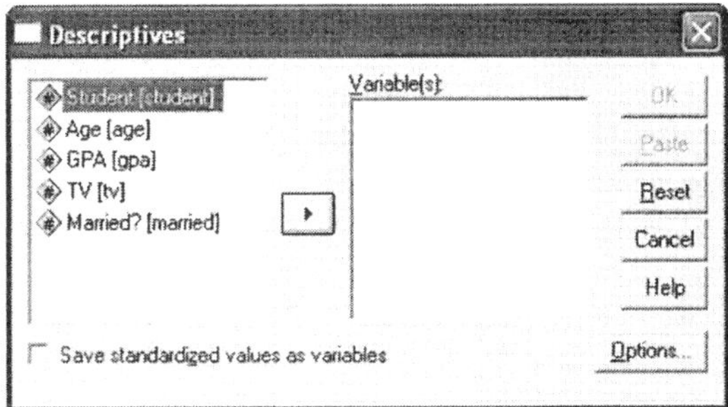

Notice that only the numeric variables are listed on the left. Also notice that each variable is listed with the variable label first followed by the variable name in parentheses. Note that some SPSS dialog boxes, like this one, allow you to select more than one variable to analyze at the same time. Others only allow you to select and analyze one variable at a time. Select the "age" variable so that it appears in the "Variable(s)" list as shown below.

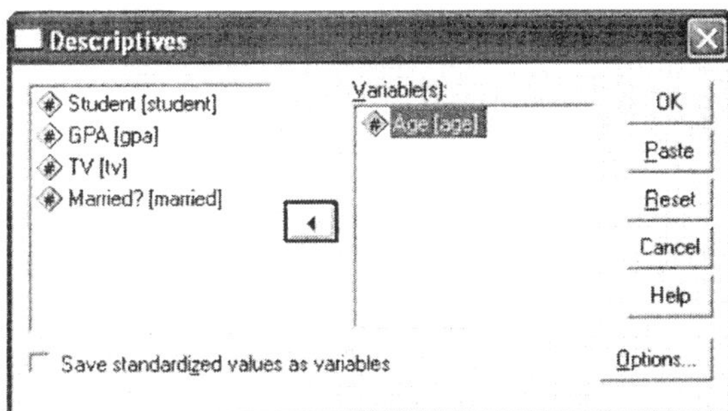

Click on **OK** to produce the output

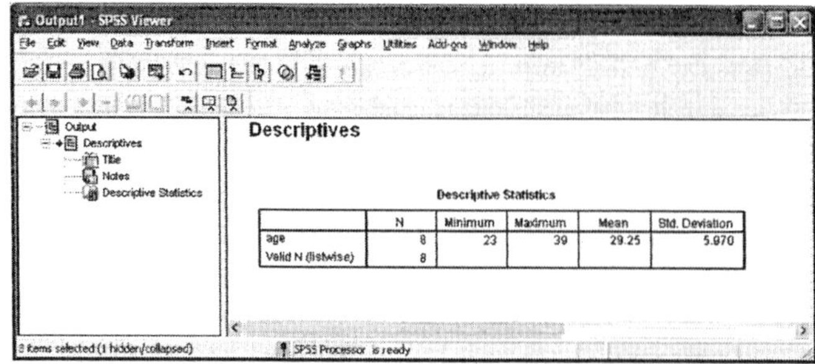

To make a histogram, select **Graphs → Histogram...** The following dialog box will appear:

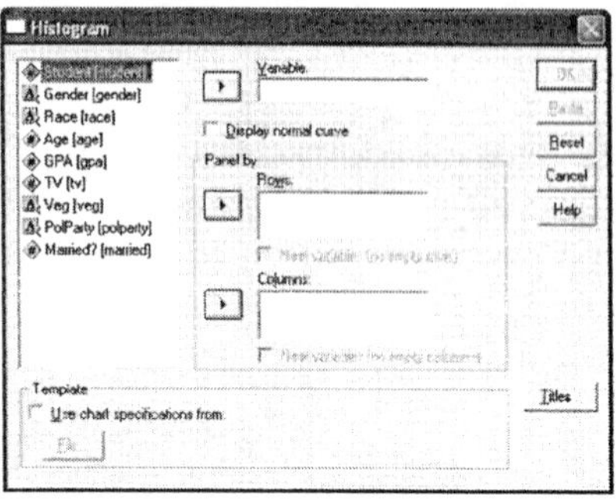

The "Histogram" command is an example of a command that allows you to analyze only one variable at a time, unlike the "Descriptives" command that allows you to analyze more than one variable at a time. Select the variable "age" so that it appears in the Variable box, and click on **OK**. Depending on the size of your computer screen, your Output Viewer might now look like the following display.

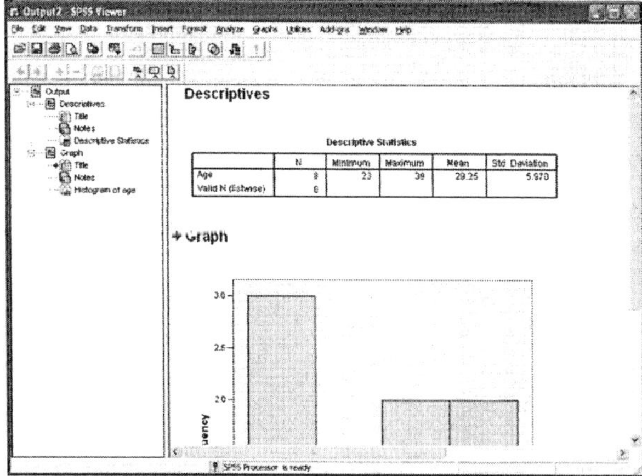

To view the entire histogram you can use the scroll bar on the right. Notice that the Output Navigator shows the additional output and its parts, and that the title for the histogram is just "Graph," which like the "Descriptives" title is not very informative. But, it is very easy to modify SPSS output so that it is more useful!

Modifying output

To modify any part of any output produced by SPSS all you have to do is double click on it to open an Editor. For example, double click on the "Descriptives" title and you will notice that a "broken-edged" box appears around the title and a curser is within that box. Make the changes you want, such as changing the title to "Descriptive Statistics for Age." For longer titles, press **Enter** when you want to start a new line. SPSS should automatically increase the size of the title box. If it doesn't, follow the process described below for changing the size of output. Click anywhere outside the box to close the Editor. Using the same process, change the title of the histogram to something like "Histogram for Age."

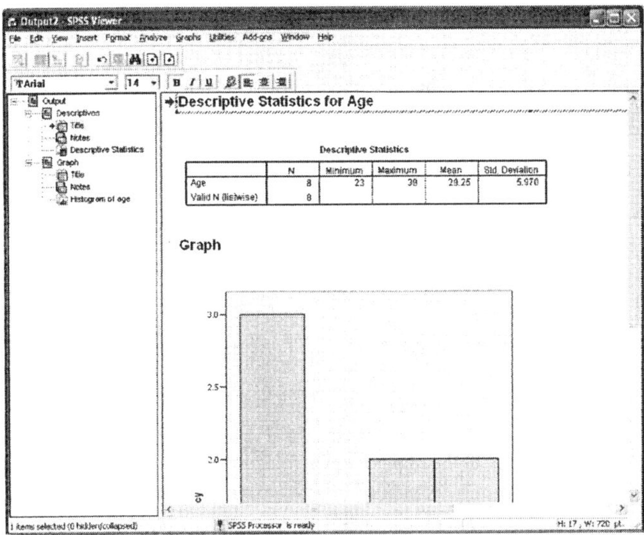

The "Descriptives" output is actually a text table, but some of the horizontal and vertical bars are hidden. You can change the height and width of the cells in the table so that the table takes up less space. Just double click on the output to open a text Editor. Two boxes appear, one called "Pivoting Trays" and the other "Formatting Toolbar," but you don't need them. Move the cursor around within the table until it changes to a double arrow. Then, drag the column or row to the desired width or height. Double click anywhere outside that output to close the Editor.

You will see how to modify a Histogram and other graphs using the Chart Editor in the next chapter, but one thing you might want to do is resize it so that the entire graph appears in the Output Viewer window. To do this, click once on the histogram so that a solid-line box appears around the graph. The solid-line box will have smaller boxes on the corners and edges that you can "grab" using the mouse and drag to resize the graph. This may be hard to do if the lower edge is right at the bottom of the window. The following display shows the modifications described above. Click anywhere in the right side of the window but off of any output to remove the sold-line box.

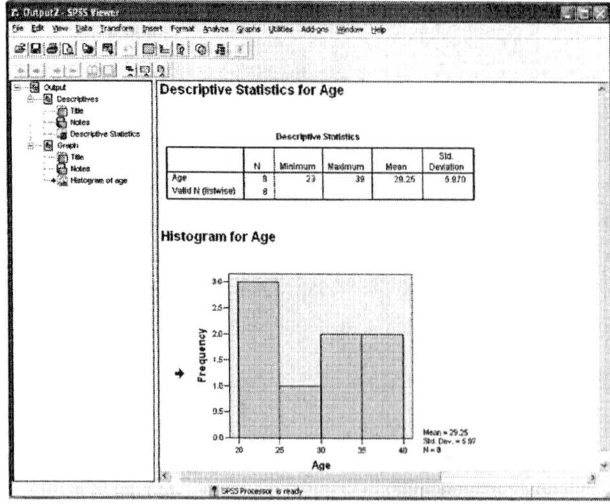

Printing Output

Before selecting the "Print" command from the "File" menu make sure that all of the output you want to print has been selected. For example, after you have changed a title or another piece of output, only that piece of output is selected unless you click once somewhere else in the output (but not on another piece of output). To determine which pieces of output, if any, are currently selected look at the Output Navigator on the left of the Output Viewer. In the display below, notice that only the Histogram is selected and only it would be printed if the print command was executed.

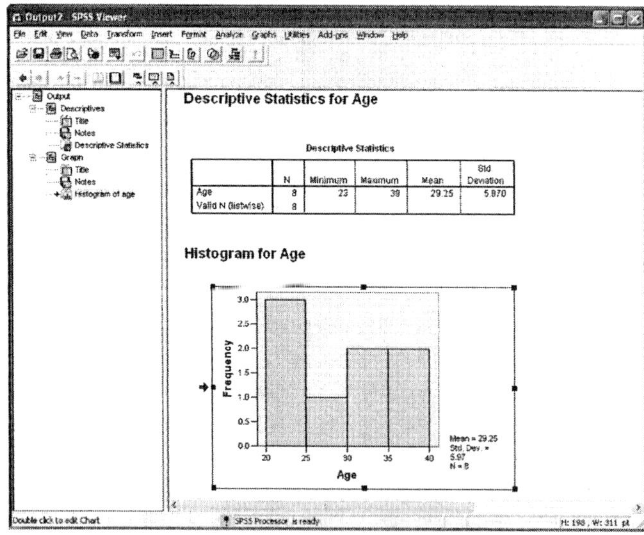

To print all of the output you can either have nothing selected or everything selected. If nothing is selected, SPSS will automatically print all of your output. To select everything click on the word "Output" at the top of the Output Navigator on the left. All of the pieces of output in the Output Viewer will then be selected. To select nothing click anywhere in the output window off of the actual output. If you only want to print a portion of your output, just select those items in the Output Navigator. If the pieces of output you want to print are not next to each other you can hide the output in between them, or press the "Ctrl" button on your keyboard while selecting output you want to print. To print your output, select **File → Print**. Another option for printing output is described below under Saving Output.

Saving Output

You have two options for saving SPSS output. The first is to select **File → Save As...** which will save your output as an SPSS output file with the ".spo" file-type extension. If you resize graphs and insert page breaks (available under the "Insert" menu) so that titles are not separated from their output, this approach works well for producing output that might appear as an appendix to a report, for example. Another option is to save the output into a word processing document. This is a good option for when you want to include output as part of your report, rather than in an appendix, and is also the second option for printing output. Open a word processing document along with your SPSS output, then click on a piece of output to copy. In SPSS Version 13.0 the "Copy" command under the "Edit" menu works for all output but if you are using an earlier version of SPSS you might have to use the "Copy Objects" command instead for some text output. After selecting "Copy" move to the word processing document and click where you want the output to appear and select "Paste" from the "Edit" menu. For example, the output from the "Descriptives" command completed above has been copied to this document below using this process. With this option you can save your SPSS output and print it later, even on a computer that doesn't have SPSS installed on it. Output saved in a .spo file can only be printed from a computer that has SPSS installed on it.

Descriptive Statistics for Age

Descriptive Statistics

	N	Minimum	Maximum	Mean	Std. Deviation
Age	8	23	39	29.25	5.970
Valid N (listwise)	8				

Exiting SPSS

Before you leave SPSS make sure that you have saved any data or output that you might want to have access to later on. This is especially important for larger data sets! To quit this or any SPSS session select **File → Exit**. If you haven't saved your data or output at this point a dialog box will appear asking if you want to save it now.

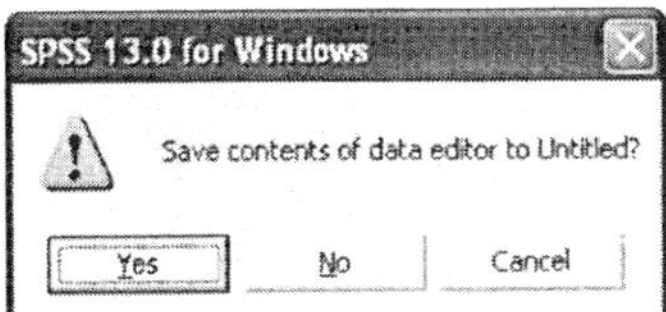

Additional Comments

In the example provided above you saw how typical SPSS commands operate and the output those commands produce. Every command you might need for an introductory statistics course is found in one of the drop-down menus and most commands have dialog boxes similar to those shown above. The rest of this manual will continue to use the convention used above for identifying the menu, command and option, that is, select **menu → command → subcommand**.

Because there are different options for entering data, some SPSS commands include options for producing the same or similar output. For example, the "married" variable in the above example could have been entered as a string variable as were the other categorical variables in that data set. As another example, if we wanted to compare the ages of the male and female students in a class, we could enter the data as two separate variables, one for the ages of the male students and the other for the ages of the female students. Or, we could enter the data as in the example data set, using one variable for the ages and another for the students' gender. Where appropriate, the different options for data entry will be described in this manual and the corresponding command options will be demonstrated.

Hints for Effective Data Analysis

The quality of the output you produce will go a long way toward helping you to communicate the information available in your data. Producing descriptive statistics and graphs is only half of the process involved in data analysis. The second half involves presenting that output to others and describing what it shows. Any output you produce should be created by keeping in mind that others will need to be able to examine it and understand what is being displayed. Here are some hints as you begin learning how to analyze data using SPSS:

1. Use descriptive variable names and variable labels, where appropriate.
2. Use value labels for numerically coded categories.
3. Include descriptive titles on all output.
4. Provide meaningful labels on all axes of graphs and include units and a starting point.
5. Provide a legend if needed.
6. Avoid any features that distract from the data, such as unusual colors or shading patterns in graphs.
7. Use graphs that are appropriate for the type of data and how it was collected.
8. Use the same scales on graphs that will be compared.
9. Use proportions instead of frequencies for comparing groups that are of different sizes.
10. Use multiple statistics and graphs to showcase your data – each statistic and graph provides different information about the data.

Chapter 2 Exploring Data with Graphs and Numerical Summaries

Working with Categorical Data

Recall from the previous chapter that you have two options for entering categorical data. You can enter the data as a "String" variable, or you can enter it as a coded "Numeric" variable. In this section you will explore both options.

Pie Charts for Tabulated Data

When data has already been tabulated, as in Exercises 2.11 and 2.13, it is very easy to produce a pie or bar chart using SPSS. The key is to use a "weight" variable to let SPSS know how many observations are in each category. The data for Example 2.11 is in the form of a pie chart that shows the regional distribution of weather stations and can be summarized in the following table:

Region	Count	Percent
Southeast	67	18.7
Northeast	45	12.5
West	126	35.1
Midwest	121	33.7

We will enter the data in SPSS in two columns, one for the region and the other for the count in each region.

Data Entry Option 1: Category as a string variable

In a new SPSS Data Editor create two variables, "region" as a string variable and "count" as a numeric variable. You will have to increase the Width for "region" from 8 to 9 (under the Variable View) if you want to type "Southeast" and "Northeast" entirely. Enter the data from the table above so that your Data Editor appears as the one in the following display.

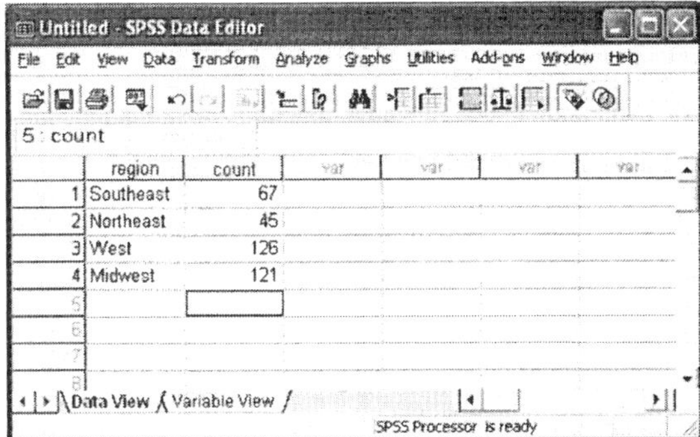

To let SPSS know that the variable "count" represents the number of observations in each category defined by the variable "region," select **Data → Weight Cases....** In the dialog box that appears, click on "Weight cases by" and enter the "count" variable in the "Frequency Variable" box.

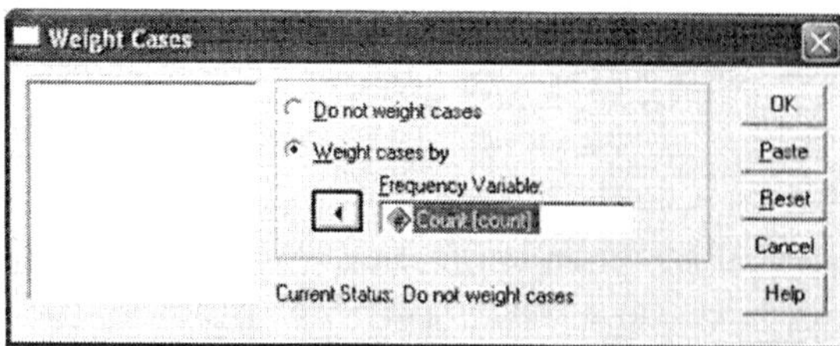

Click on **OK** to complete the process. To produce a pie chart for this data, select **Graphs → Pie...** and in the dialog box that appears select the "Summaries for groups of cases" option and click on **Define**. Note: The "Values of individual cases" option will be discussed in the next section.

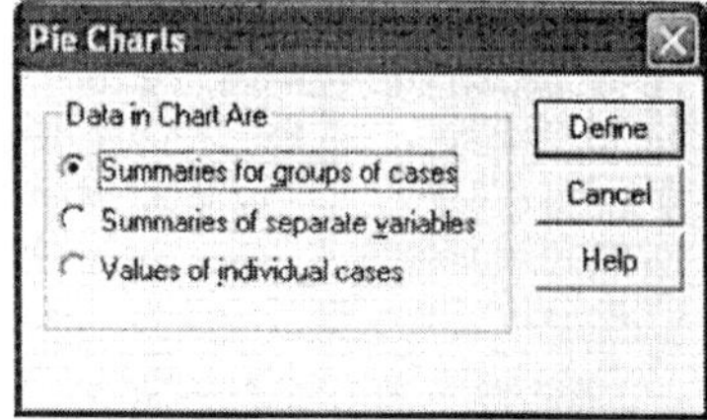

Enter "region" in the "Define Slices by" box and click on **OK**.

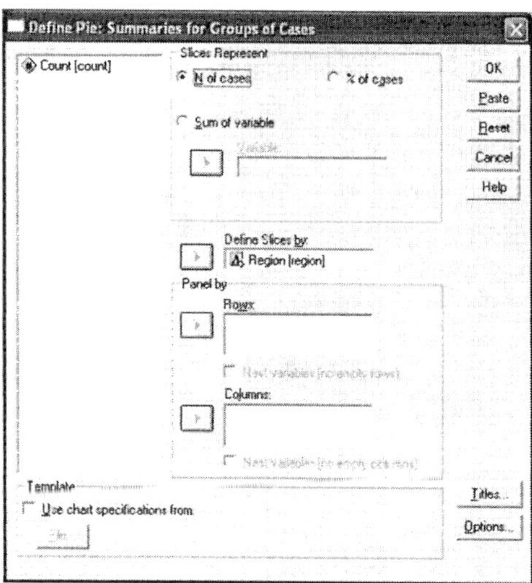

The Output View will show the following pie chart. The size of this chart was decreased in the display below so that it would fit in the window and the title has also been changed.

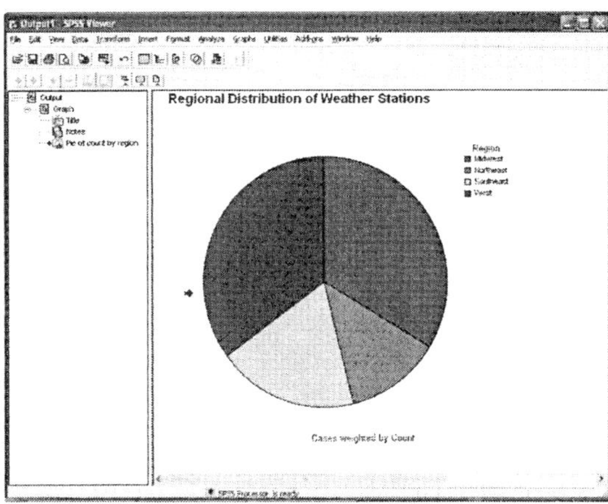

If you don't like the color scheme selected by SPSS it is very easy to change. You can also add text boxes that indicate the count and percentage of each category. Double click anywhere on the pie chart to open the **Chart Editor**. A smaller version of the pie chart will appear in a window with new menu options.

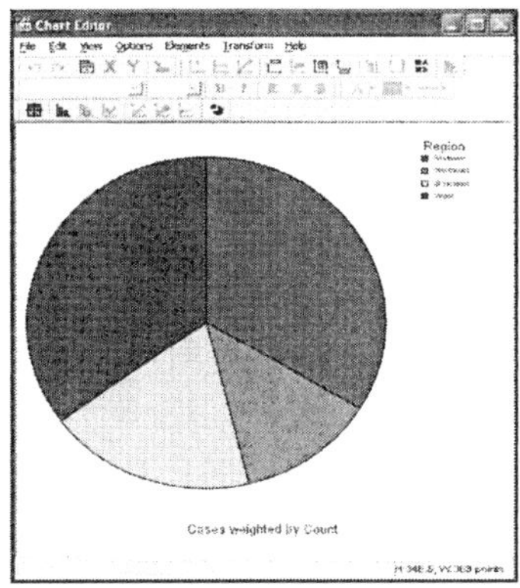

Select **Elements → Show Data Labels** and the following dialog box will appear.

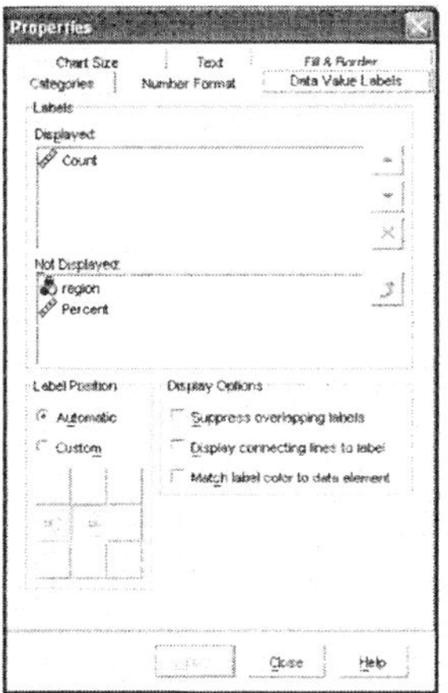

The current option is to display the "Count" in each category ("Count" is the only item listed in the "Displayed" box). If you click on "Percent" in the "Not Displayed" box and then on the curved arrow to the right of that box, then both "Count" and "Percent" will be displayed. If you select "region" to also be displayed then you won't need a legend included with the graph. Click on **Apply** and then **Close** when you have made your selections. You will see the items you selected now displayed in the Chart Editor version of the pie chart. The changes you make to a graph in the Chart Editor are not applied to the actual graph until you close the Chart Editor. To close the Chart Editor, select **File → Close** (or use the close box in the upper right corner).

To change the color of one or more of the "slices" of a pie chart, click once anywhere on the pie chart in the Chart Editor to select it (you will see a circle outline appear around the edges of the pie chart) and then click once on the slice you want to change. Just the slice you want to change should now be outlined.

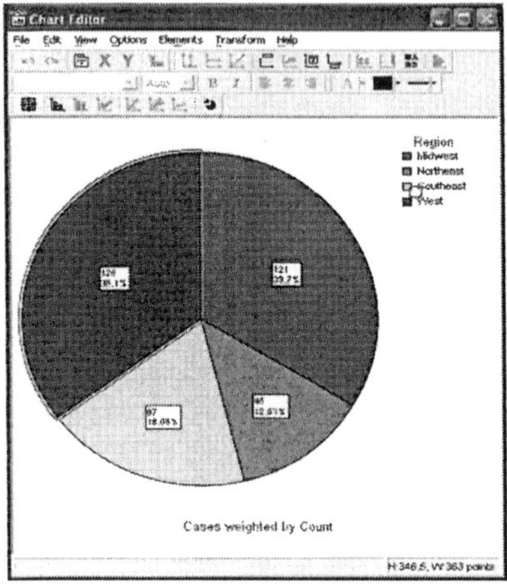

To change the color of the slice you selected, select **Edit → Properties**. Click on the "Fill and "Border" tab.

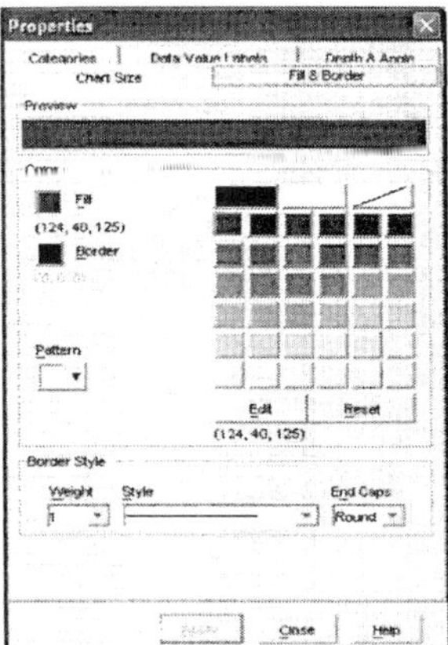

The current color of the slice appears in the "Preview" box. Select a new color from the options that appear then click on **Apply** and **Close**. The following display shows the same pie chart with each of the slices changed to a new color. If you don't have a color printer you might want to change each of the slices to white or a light shade of gray.

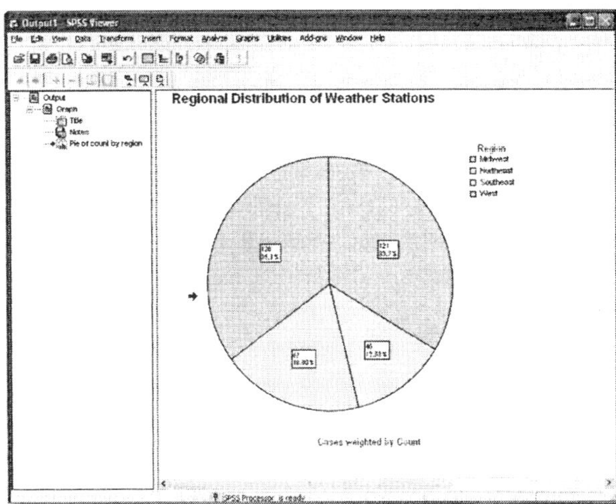

Data Entry Option 2: Category as a coded numeric variable

To code "region" numerically, use the "Numeric" type in the Variable View and enter the following value labels following the process described in the first chapter of this manual.

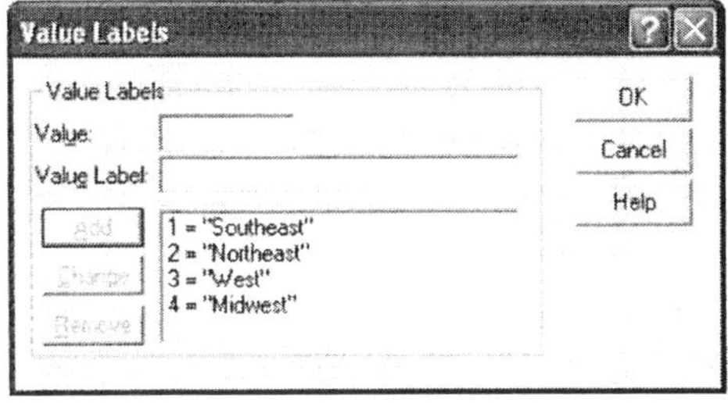

Click on the Data View tab and enter the data the same as above, except enter the values 1, 2, 3 and 4 for each region. Select **View → Value Labels** so that the Data View appears as follows:

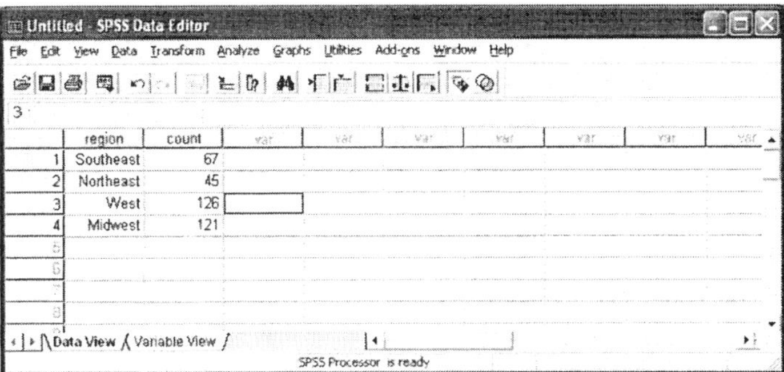

	region	count
1	Southeast	67
2	Northeast	45
3	West	126
4	Midwest	121

The data set appears exactly the same as it did for Option 1 but the process for producing a pie chart is slightly different. Select **Graph → Pie...** but this time select the "Values of individual cases" option.

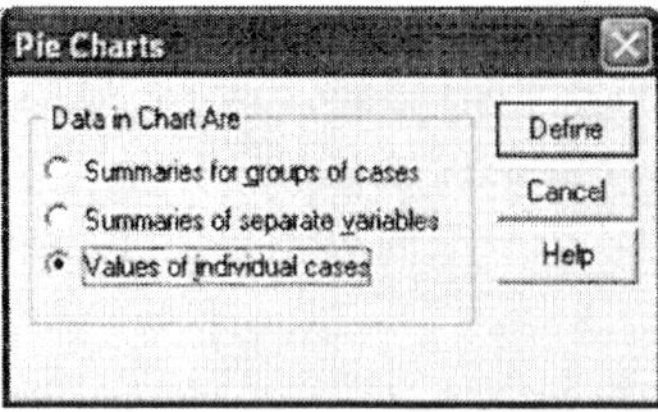

Click on **Define**. Enter "count" in the "Slices Represent:" box and then click on "Variable" under "Slice Labels." Enter "region" in the box below "Variable" and click on **OK**.

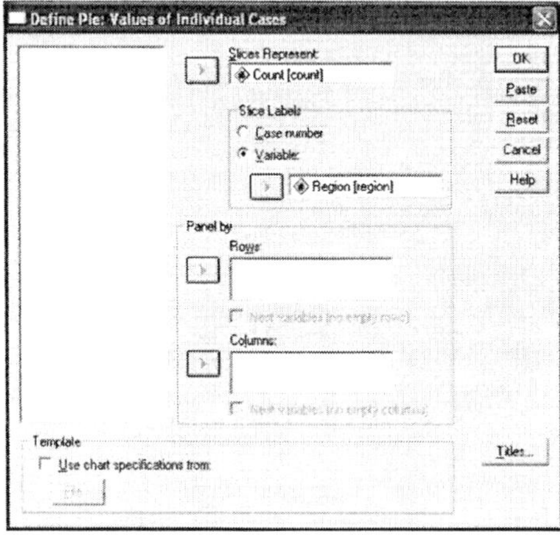

The resulting pie chart differs from the first pie chart in the order that the slices are presented. When "region" was a string variable, SPSS ordered the slices alphabetically, but when it is coded numerically the slice coded 1 (Northeast) appears first. The following display shows the pie chart for this data set with each of the slices colored white. The legend was removed by selecting **Options → Hide Legend** from the Chart Editor.

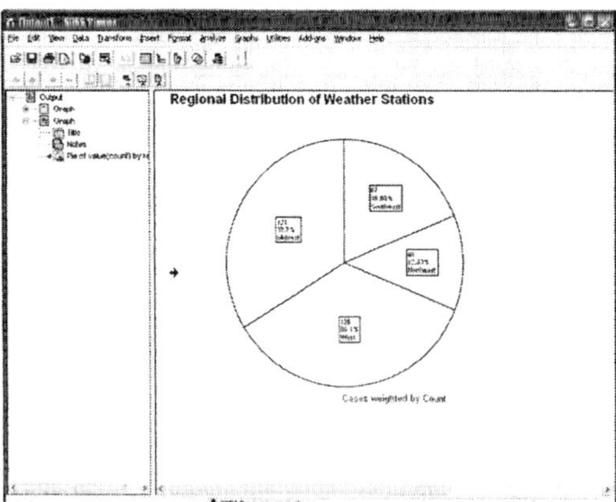

Which option to use?

Both options are easy to use. Indicating that the slices represent the "count" in each category when you code the data numerically serves the same purpose as indicating that "count" is a weighting variable when a variable is entered as a string variable. If the category names are long, data entry might be easier using numerical coding. The graphs are nearly identical so the appearance of the output isn't really an issue. If you want the slices to appear in a non-alphabetical order that is very easy to do by using an appropriate coding scheme. For example, if you want the slices to appear largest to smallest clockwise around the circle, then you would code Midwest = 1, West = 2, Southeast = 3, and West = 4. Both data entry options include the same options for modifying the color, size and labeling of the resulting graph. However, for some analyses, for example, for comparing two groups, some SPSS commands will only work if the variable used to define the groups is coded numerically. You will learn more about this later.

Pie Charts in SPSS Version 11.0

If you are using an earlier version of SPSS than 13.0, most of the directions given above will be the same. The display below shows a pie chart made for the same data set as the example above.

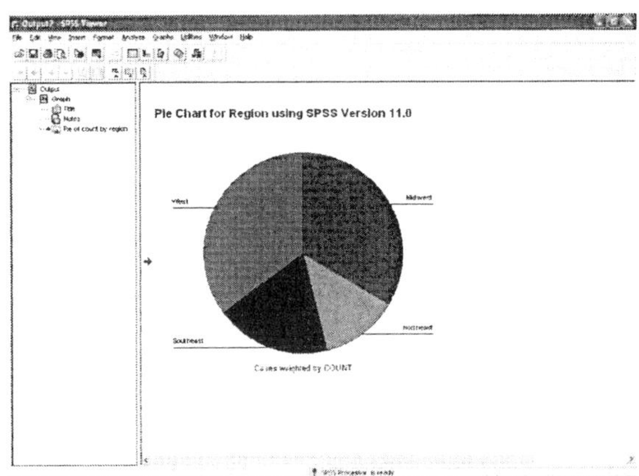

The Chart Editor is a little different in Version 11.0.

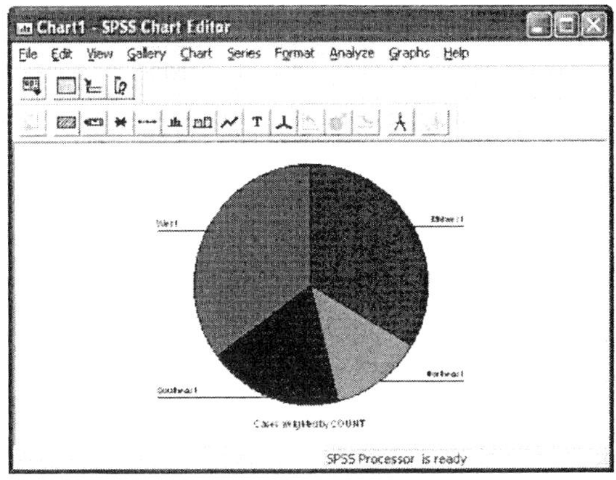

To change the color of one of the slices, click on that slice and then select **Format → Color...** Another dialog box will appear.

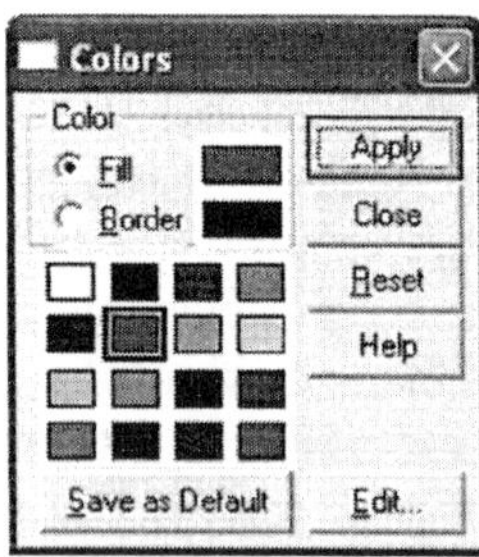

Choose the color you want, then click on **Apply** and then **Close**. To add the counts or percents associated with each slice, double click on one of the section labels. In the dialog box that appears, click on "Values" and/or "Percents" under "Labels" and then on **OK**.

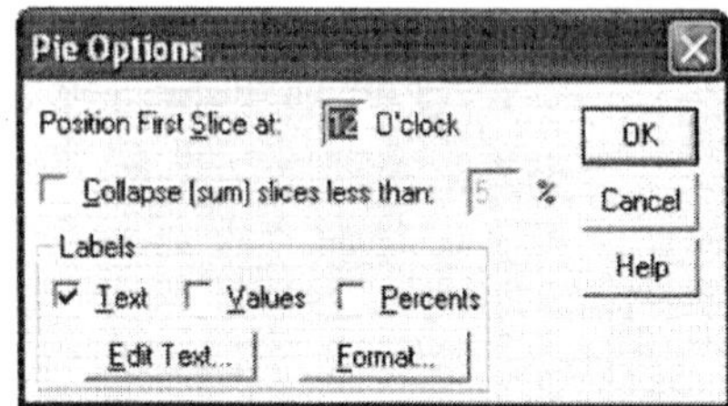

Bar Charts and Pareto Charts for Tabulated Data

The commands for making a bar chart or Pareto chart operate exactly the same as the command for making a pie chart. If you enter a categorical variable using the string option choose the "Summaries of groups of cases" option and use the "Weight Cases" command to indicate which variable provides the recorded counts. If you code it numerically choose the "Values of individual cases" option. Let's make a bar chart for the data from Exercise 2.13 on the frequency of shark attacks worldwide. The data for this exercise is given in the table below:

Region	Frequency
Florida	289
Hawaii	44
California	34
Australia	44
Brazil	55
South Africa	64
Reunion Island	12
New Zealand	17
Japan	10
Hong Kong	6
Other	160

Enter the data set in SPSS using either option described above. To clear the Data Editor, select **File → New → Data.** A dialog box will ask you if you want to save the data from the previous example if you have not already saved it. After you have entered the new data set, select **Graph → Bar...** and then select the appropriate option for how you entered the data. For "Region" as a string variable, SPSS will produce the following output, with the categories listed alphabetically:

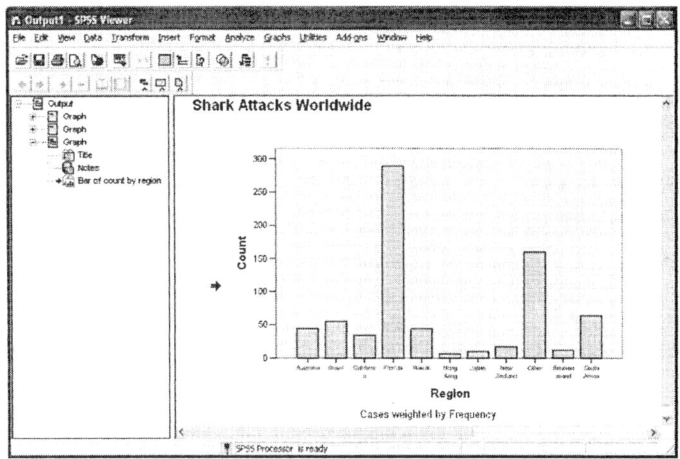

The size of the bar chart above was decreased to make the display fit the size of the window. Note that the vertical axis is labeled "Count;" if you coded the data numerically, the vertical axis will be labeled "Value Frequency" and the order of the bars will depend on the coding scheme you used. You can change the label on any axis by opening the Chart Editor and clicking on the label you want to change. Here is the same bar chart output with "Frequency" as the label for the vertical axis and the color of the bars changed to white. You can also change each bar to a different color just by clicking on each one individually in the Chart Editor and selecting a new color.

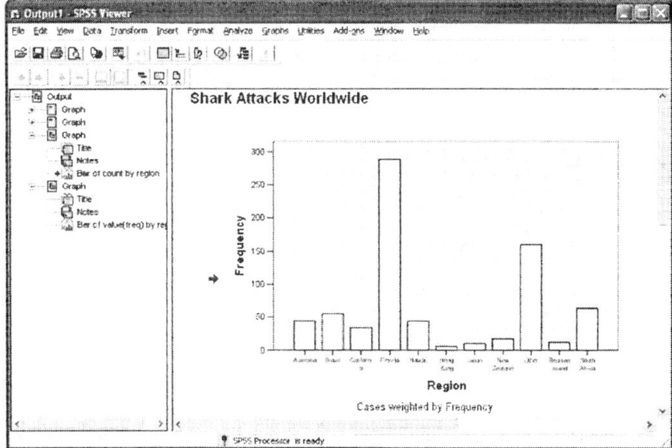

To make a Pareto chart, select **Graphs → Pareto...** Click on **Define** and enter "Region" in the "Category Axis" box. Click on **OK**. The default graph is shown below.

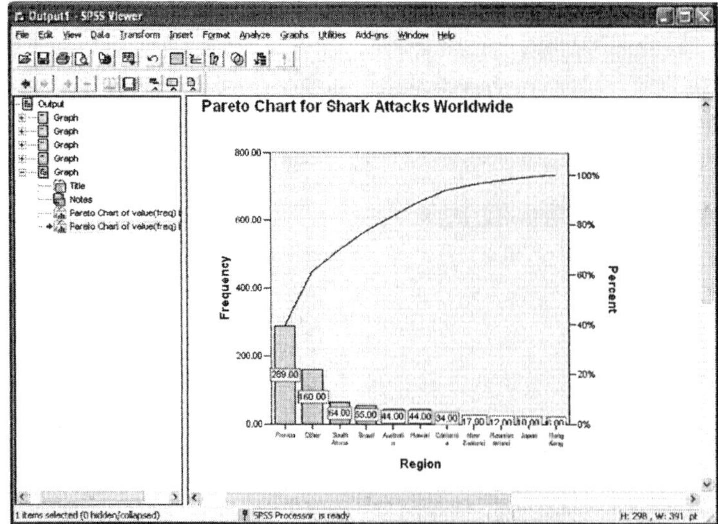

SPSS includes the count in each category on top of each bar and a cumulative percent line above the bars. The counts can be removed in the Chart Editor. Click once on one of the count labels in the graph to highlight the counts and then select **Elements → Hide Data Labels**. The cumulative percent line can also be removed by clicking on it once to highlight it and pressing the "Delete" button on your keyboard.

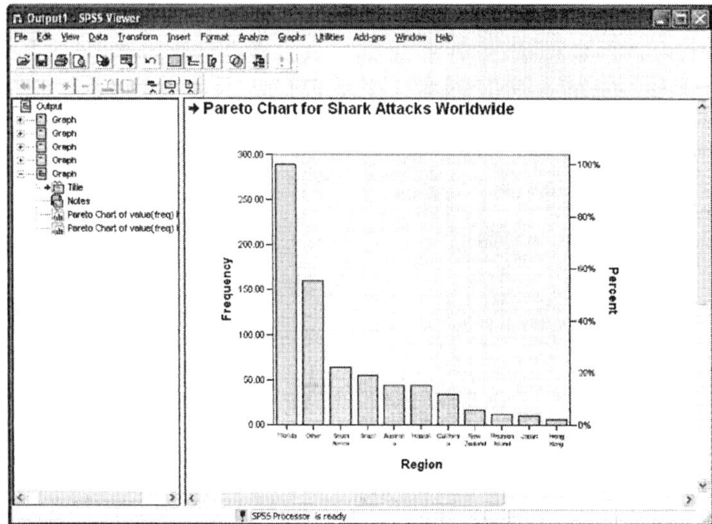

Frequency Tables and Graphs for Non-Tabulated Categorical Data

SPSS can also organize and display categorical data that hasn't already been tabulated. To demonstrate this, let's use the "gender" variable from the "fla_student_survey" data set from your text CD. Access this data file in SPSS by selecting **File → Open → Data** and locating your text CD in the "Look in" box. Find the file among those listed and double click on it. A portion of this data set is displayed below.

	subject	gender	age	high_sch_GPA	college_GPA	distance_home	distance_residence	TV	sports	news
1	1.0	m	32.0	2.20	3.50	.00	5.00	3.00	5.0	
2	2.0	f	23.0	2.10	3.50	1200.00	.30	15.00	7.0	
3	3.0	f	27.0	3.30	3.00	1300.00	1.50	.00	4.0	
4	4.0	f	35.0	3.50	3.20	1500.00	8.00	5.00	5.0	
5	5.0	m	23.0	3.10	3.50	1600.00	10.00	6.00	6.0	
6	6.0	m	39.0	3.50	3.50	350.00	3.00	4.00	5.0	
7	7.0	m	24.0	3.60	3.70	.00	.20	5.00	12.0	
8	8.0	f	31.0	3.00	3.00	5000.00	1.50	5.00	3.0	
9	9.0	m	34.0	3.00	3.00	5000.00	2.00	7.00	5.0	
10	10.0	m	28.0	4.00	3.10	900.00	2.00	1.00	1.0	
11	11.0	m	23.0	2.30	2.60	253.00	1.50	10.00	15.0	

To find the frequency table for this variable, select **Analyze → Descriptive Statistics → Frequencies...**, then select "gender" and click on **OK**. The following output will appear:

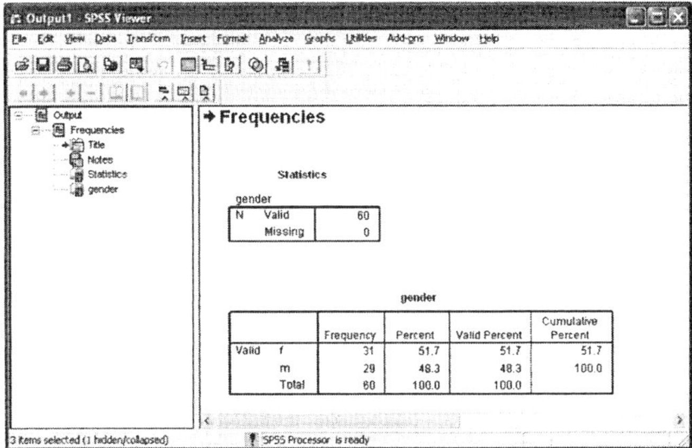

Frequencies

Statistics

gender

N	Valid	60
	Missing	0

gender

		Frequency	Percent	Valid Percent	Cumulative Percent
Valid	f	31	51.7	51.7	51.7
	m	29	48.3	48.3	100.0
	Total	60	100.0	100.0	

This output has three parts that are currently showing. The first is the title "Frequencies," which you should change to something like "Frequency Table for Gender." The second is a "Statistics" portion where SPSS reports how many "valid" or non-missing observations were analyzed. This part of the output can be omitted if there aren't any missing values in the data set. Or, you can temporarily "hide" it by double clicking on its icon in the Output Navigator. Here we see that 51.7% of the students surveyed were female. Since "gender" was entered as a String variable, the categories are listed alphabetically in this output.

You can make a pie or bar chart using the "Summaries for groups of cases" option of these graph commands, or you can use an option available in the "Frequencies" command. You might have noticed the "Charts" button in the Frequencies dialog box, shown below.

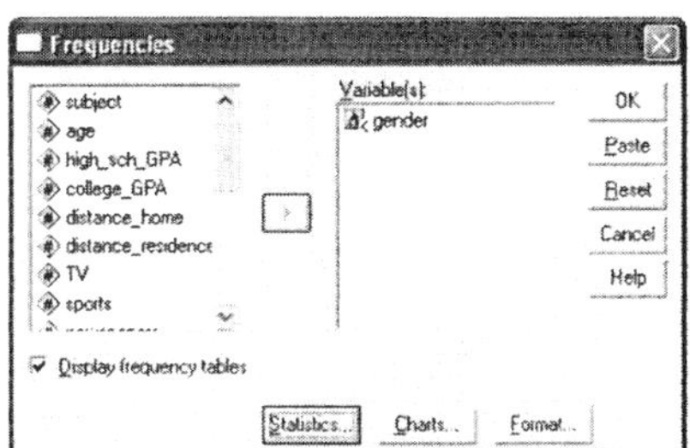

If you click on **Charts** a dialog box will appear that allows you to produce a bar chart or a pie chart along with the frequency table. Some of the commands under the "Analyze" menu also include options for producing graphs. The "Statistics" button in the "Frequencies" dialog box allows you to produce some statistics along with a frequency distribution of a data set.

Note that each of these commands will also work if the categorical variable is entered as a coded numeric variable.

Working with Quantitative Data

For quantitative data, you will now see how to make a dot plot, a stem-and-leaf plot, and histograms. If the data was collected over time you can also make a time plot.

Dot Plot

SPSS Version 13.0 includes a command for producing dot plots. Earlier versions of SPSS may not have this option. For example, Version 11.0 does not include this command. To demonstrate how to make a dot plot in SPSS we will use the "SODIUMmg" variable of the "cereal" data set from your text CD. Access this data and select **Graphs → Scatter/Dot...** In the dialog box that appears click on **Simple Dot** and then on **Define**.

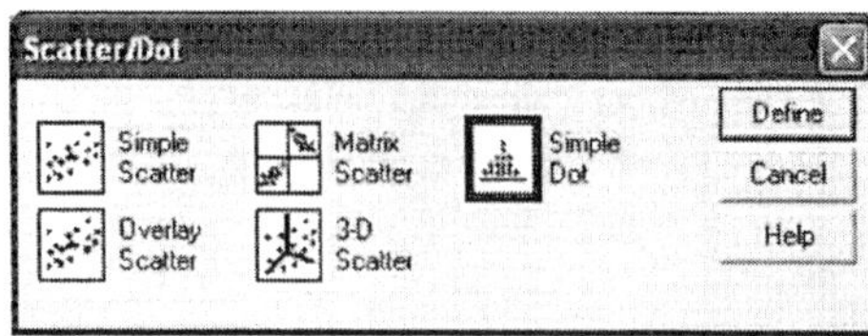

Enter "SODIUMmg" in the "X-Axis Variable" box.

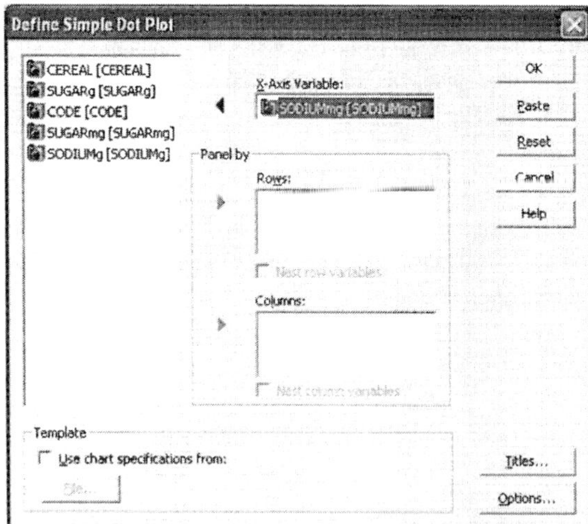

Click on **OK** to produce the output.

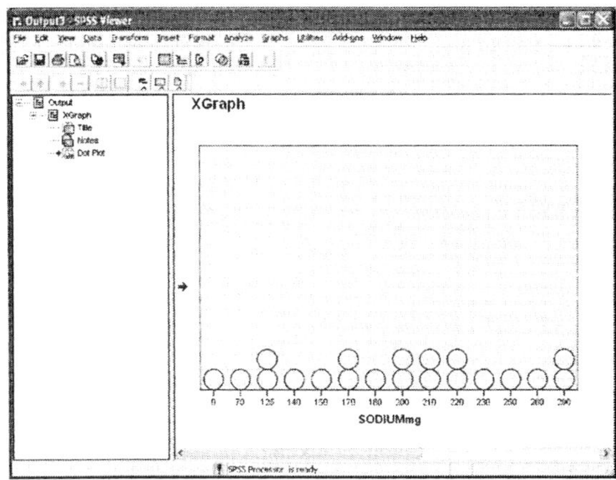

The size and fill of the circles can be modified using the Chart Editor. With the Chart Editor open, click once on one of the circles in the graph, then select **Edit → Properties**... Click on the "Marker" tab and enter a new size in the "Size" box. Click on the box next to "Fill" under "Color" and then on the black box to fill in the circles.

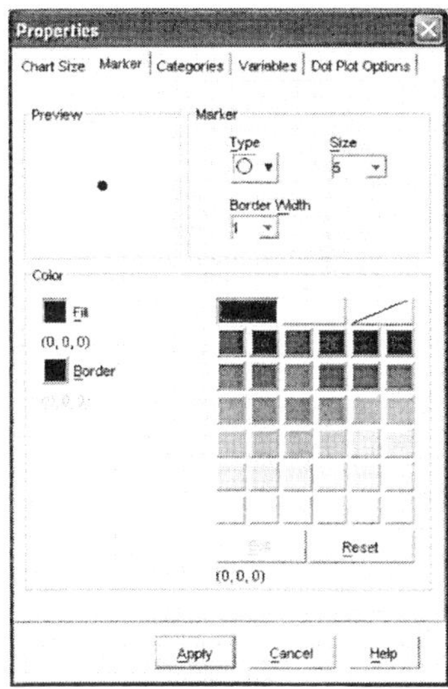

Click on **Apply** and then on **Close** to complete the change, then close the Chart Editor.

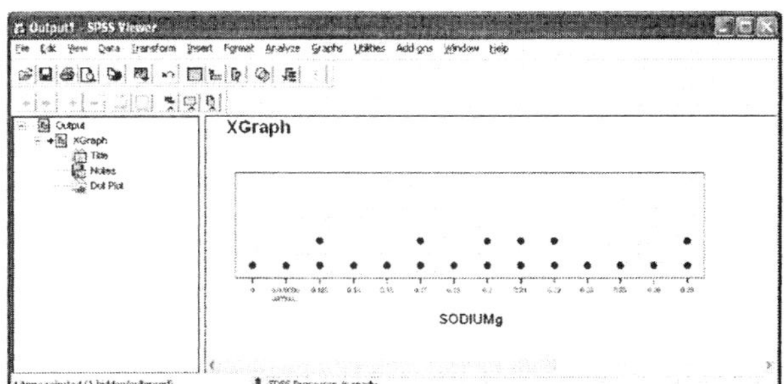

Stem-and-Leaf Plot

For this example we will use the data for Exercise 2.15 of final eBay prices (in dollars). The data is shown below and a portion of it is shown in the SPSS Data Editor below that.

235, 225, 225, 240, 250, 250, 210, 250, 249, 255, 200, 199,
240, 228, 255, 232, 246, 210, 178, 246, 240, 245, 225, 246, 225

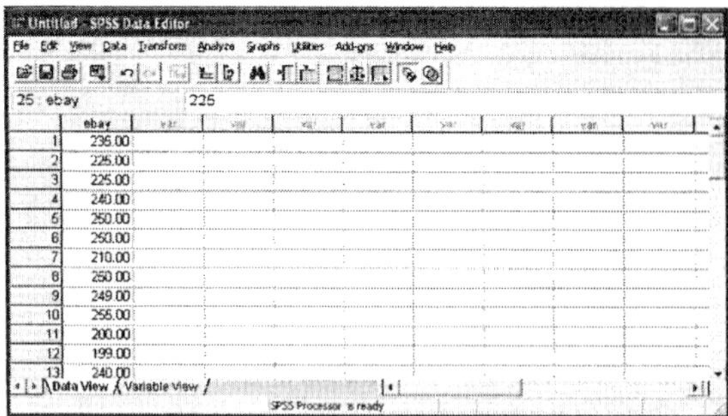

SPSS does not include stem-and-leaf plots with the other graphical methods under the "Graphs" menu. Instead it is an option under the "Explore" command under the "Analyze" menu. You will learn more about this very useful command later on.

To make a stem-and-leaf plot select **Analyze → Descriptive Statistics → Explore...** In the dialog box that appears, enter "eBay" in the "Dependent" list, and then click on **Plots** underneath "Display" in the lower left corner.

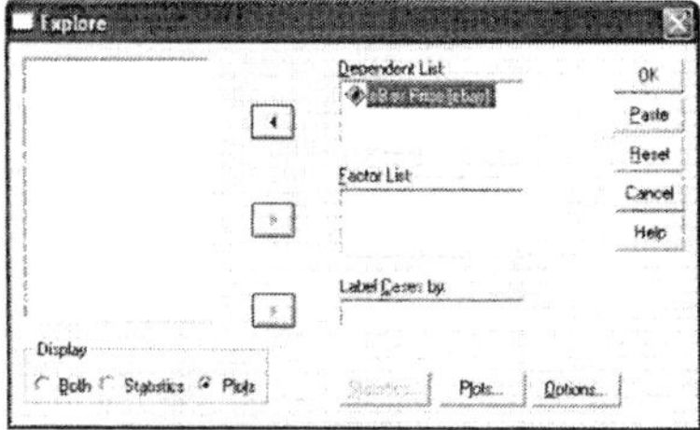

One of the default graphs produced by this command is a stem-and-leaf plot. The other default graph is a box plot. To produce only the stem-and-leaf plot, click on the "Plots" button on the lower right side of the dialog box and select "None" underneath "Boxplots."

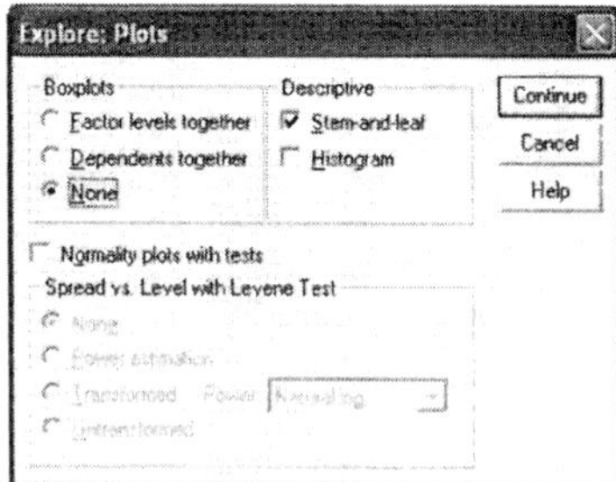

Click on **Continue** and then on **OK** to produce the stem-and-leaf plot. The following output will appear.

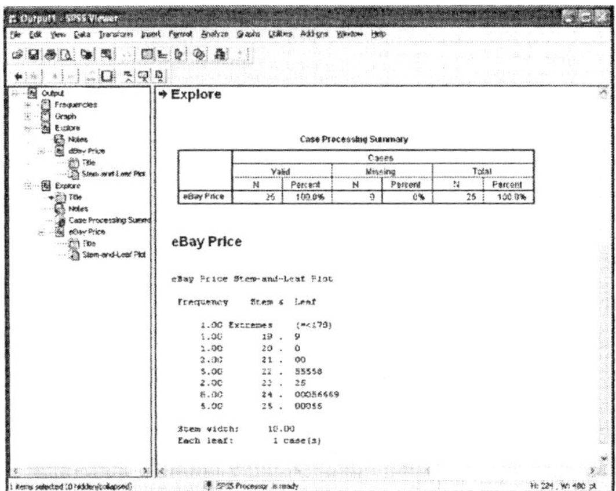

Explore

Case Processing Summary

	Cases					
	Valid		Missing		Total	
	N	Percent	N	Percent	N	Percent
eBay Price	25	100.0%	0	0%	25	100.0%

eBay Price

```
eBay Price Stem-and-Leaf Plot

 Frequency    Stem &  Leaf

     1.00 Extremes    (=<178)
     1.00       19 .  9
     1.00       20 .  0
     2.00       21 .  00
     5.00       22 .  55558
     2.00       23 .  25
     8.00       24 .  00056669
     5.00       25 .  00055

 Stem width:      10.00
 Each leaf:        1 case(s)
```

The "Explore" command includes a "Case Processing Summary" that reports the number of valid observations included in the analysis. Like the "Statistics" portion of the "Frequencies" command output, the "Case Processing Summary" may be omitted if there aren't any missing values in the data set. The stem-and-leaf plot is the last piece of output produced. The plot is actually text output that can be modified by double clicking on it to open a text editor. The first column on the left indicates how many "leaves" are associated with each "stem." Notice that the first stem is listed as "Extremes" with an observation "=< 178" and that the stem 18 is omitted since there aren't any observations between 180 and 189 in value.

Histograms

The command for making a histogram is one of the options available under the "Graphs" menu. You might have also noticed that it is an option under the "Frequencies" and "Explore" commands. Let's use the "Histogram" command under the "Graphs" menu to make a histogram of the sugar content of cereals for Exercise 2.20. The data for this exercise is the "cereal" data set on the text CD.

cereal.sav - SPSS Data Editor

1 : CEREAL FMiniWheats

	CEREAL	SODIUMmg	SUGARg	CODE	SUGARmg	SODIUMg
1	FMiniWheats	.00	7.0	A	7000.00	.00
2	ABran	260.00	5.0	A	5000.00	.26
3	AJacks	125.00	14.0	C	14000.00	.13
4	CCrunch	220.00	12.0	C	12000.00	.22
5	Cheeros	290.00	1.0	C	1000.00	.29
6	CTCrunch	210.00	13.0	C	13000.00	.21
7	CFlakes	290.00	2.0	A	2000.00	.29
8	RBran	210.00	12.0	A	12000.00	.21
9	COakBran	140.00	10.0	A	10000.00	.14
10	Crispix	220.00	3.0	A	3000.00	.22
11	FFlakes	200.00	11.0	C	11000.00	.20
12	FLoops	125.00	13.0	C	13000.00	.13
13	GNuts	170.00	3.0	A	3000.00	.17

Select **Graphs → Histogram…** and enter the "SUGARg" variable in the "Variable" box.

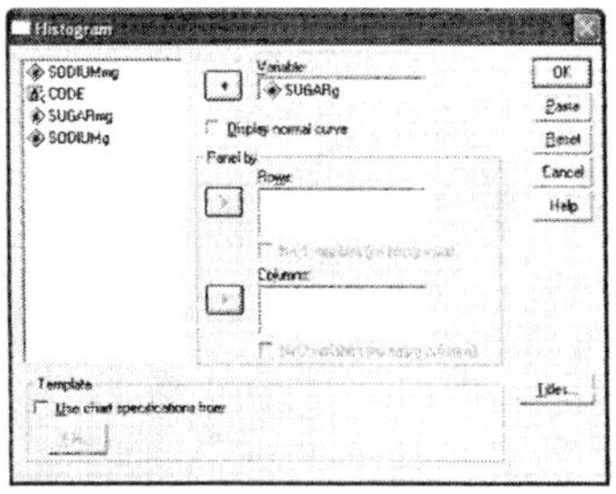

Click on **OK** to produce the graph.

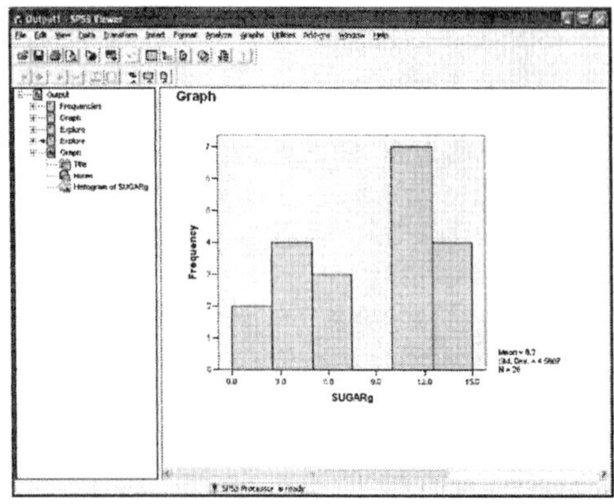

Changing the Number of Intervals and the Scale of a Histogram

The histogram above is an example of a "default" histogram made by SPSS. The "Histogram" command has a "rule" created by the SPSS programmers for determining how many intervals to produce, how wide they are, and where to start the first interval. But, this may not be the histogram you want or the "best" histogram for displaying your data. It is usually a good idea to make more than one histogram of the same data set to explore how changing the number of intervals and how wide they are affects the apparent shape of the distribution. The easiest way to make additional histograms is to make a copy of the default histogram. You can then modifying the copy using the Chart Editor. To make a copy of any piece of output click once on it and then select **Edit → Copy** and then **Edit → Paste After**.

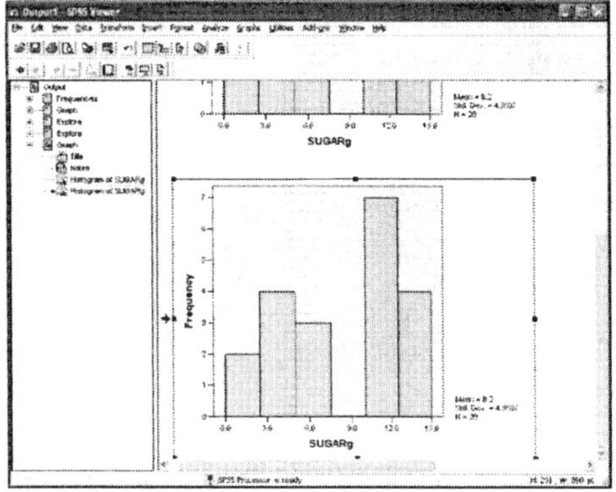

You now have two copies of this histogram, the original one to leave as is and a second one to modify. Double click on the second graph to open the Chart Editor. To change the number of bars double click anywhere on the scale of the horizontal axis. In the dialog box that appears, click on the "Histogram Options" tab at the top.

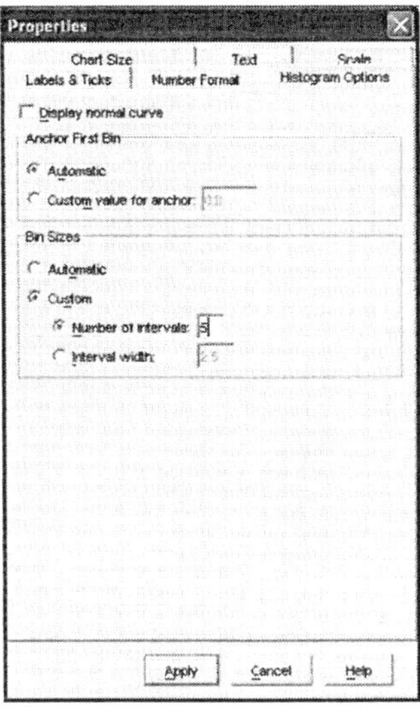

Click on "Custom" underneath "Bin Sizes," and then on the "Number of Intervals" box. Enter "5," click on **Apply** and then on **Close**. The histogram now appears as the one shown below.

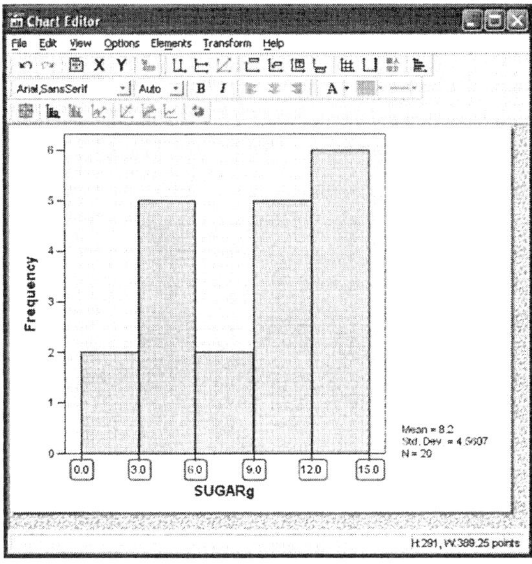

To change the scale on the horizontal axis, click on the "Scale" tab in the same dialog box as the "Histogram Options" tab you used above. Under "Range" click on the items you want to change and enter new values. For example, you could make the scale start at a "Minimum" of 0 and extend to a "Maximum" of 16, with tic marks ("Major Increment") 2 units apart. These choices are shown in the following dialog box.

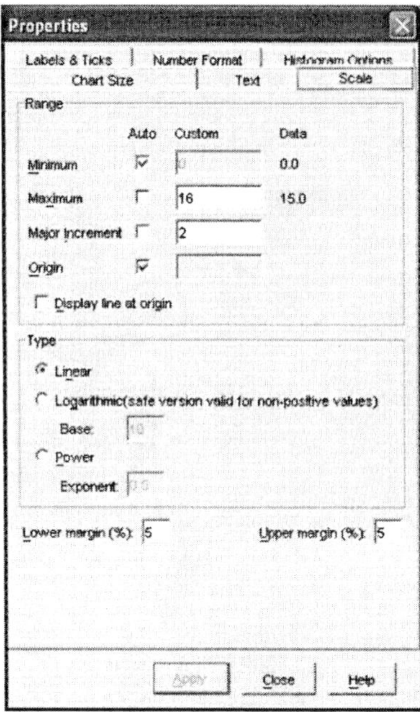

Click on **Apply** and then on **Close** to make the change.

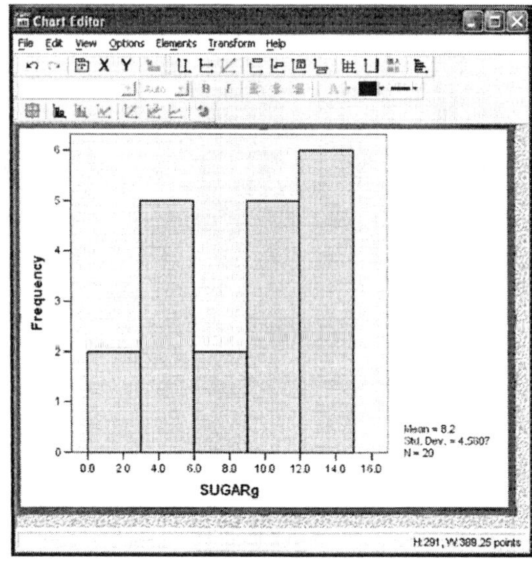

You might now change the interval width under the "Histogram Options" tab to also be 2 so that the tic marks and intervals line up. This change is shown in the following dialog box.

Click on **Apply** and then on **Close** to make the change. The process for changing the scale on the vertical axis is similar to the process described above. Just double click on the vertical axis and select the "Scale" tab in the dialog box that appears.

As another example, let's make some histograms for the data associated with Exercise 2.103a, the "central_park_yearly_temps" data set on your text CD. Access this data set and select **Graphs → Histogram**. Enter the variable "TEMP" in the "Variable" box and click on **OK**.

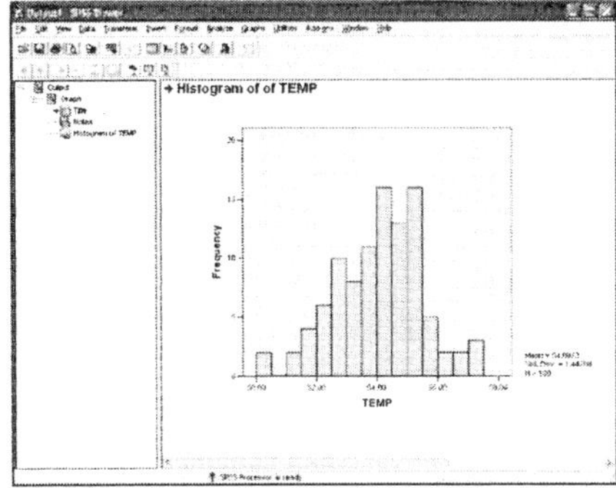

The default histogram for this data set has 15 intervals. Let's make one with only 10. As in the example above, double click on the histogram to open the Chart Editor, then double click on the scale for the horizontal axis. Click on the "Histogram Options" tab, then on "Custom" under "Bin Sizes" and enter 10 in the corresponding box. Click on **Apply** and then on **Close**.

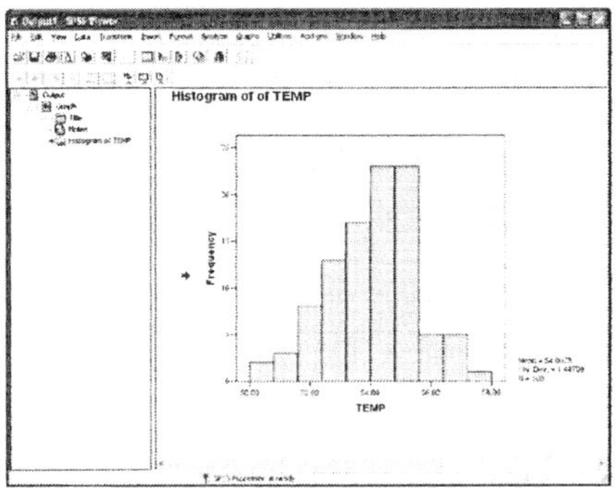

This histogram has a smoother shape with no empty intervals. Now decrease the number of intervals to 5 and compare the results to the other two histograms.

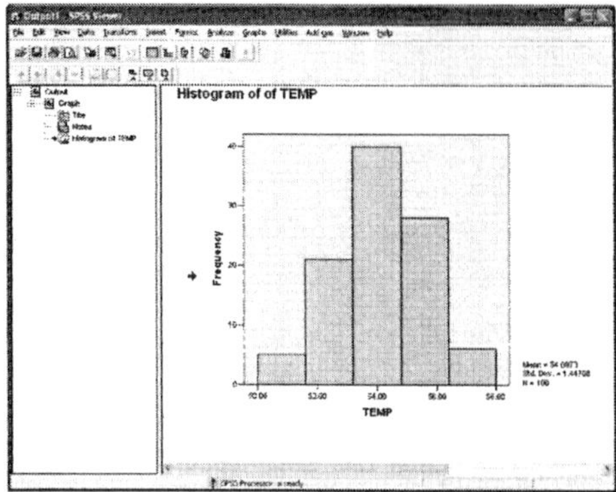

By decreasing the number of intervals we are producing a histogram with a smoother shape, but be careful not to "over smooth" by using too few intervals!

Histograms using SPSS Version 11.0

As with the pie and bar charts, the Chart Editor for modifying a histogram is a little different in earlier versions of SPSS. For the Central Park temperatures data set SPSS Version 11.0 would produce the following default histogram.

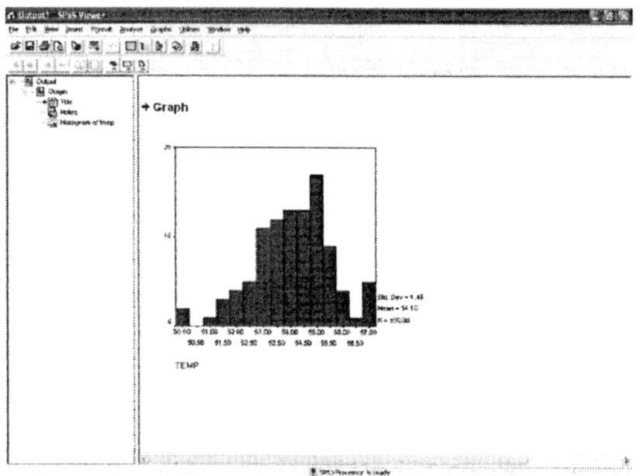

To change the color of the bars, use the same directions given above for pie charts in SPSS Version 11.0. To change the number of intervals, double click on the scale below the horizontal axis within the Chart Editor. The following dialog box will appear.

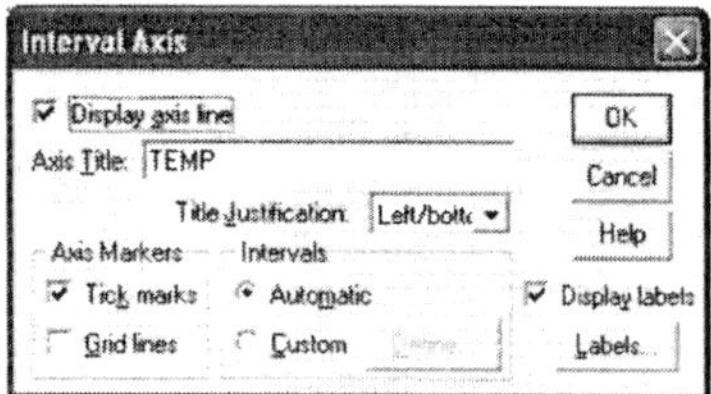

Click on "Custom" underneath "Intervals" and then on **Define**.

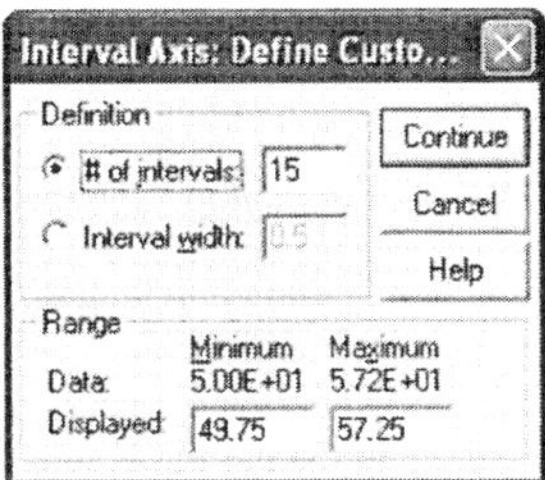

Enter a new value for the number of intervals or click on "Interval width" and enter a new value there. Then click on **Continue** and then on **OK**.

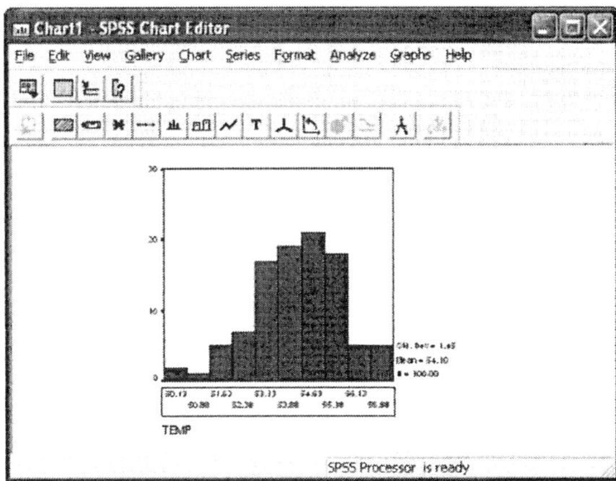

Notice that the scale automatically adjusts to the new number of bars. The values on the horizontal axis correspond to the midpoints of each interval. If you want to have the ranges of each interval displayed instead, just click on "Labels" in the lower right corner of the "Interval Axis" dialog box to make the change. Notice that the vertical axis doesn't have a label on it. To add one, double click on that axis and enter a title in the "Axis Title" box of the dialog box that appears.

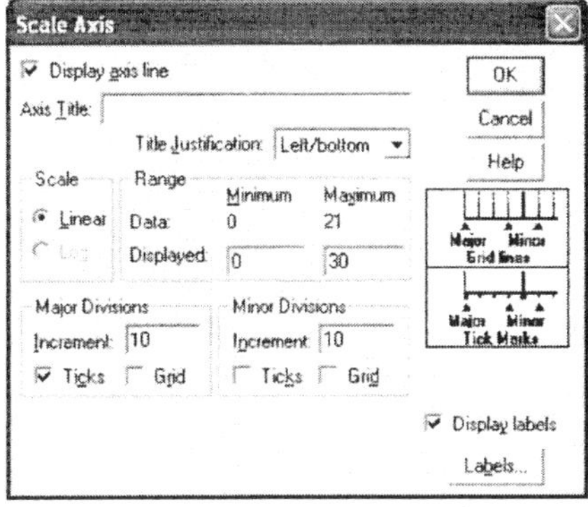

Click on **OK** to make the change. A revised histogram is shown below.

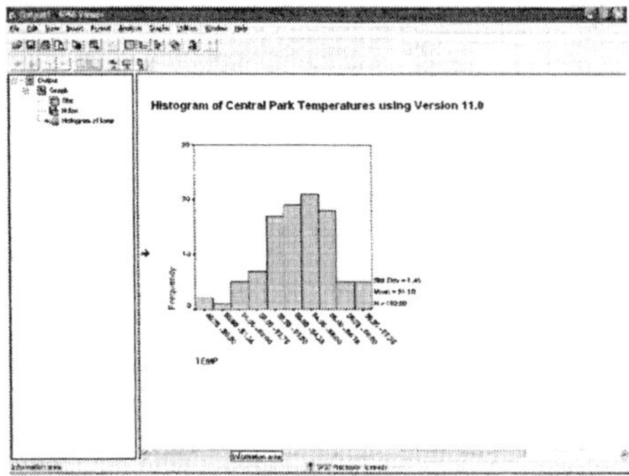

Time Plots

When data is collected over time (or some other dimension such as distance) a time plot can be useful for showing trends or other changes in the "process" that is producing the data. The "Graphs" menu includes two commands for producing a time plot, the "Line" command and the "Sequence" command. We will use the "newnan_temps" data file for Exercise 2.27 from your text CD for this demonstration. Access this data file and select **Graphs → Line**... The following dialog box will appear.

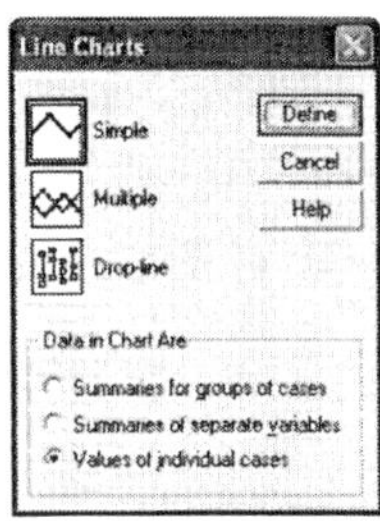

You can make a time plot for this data set using either the "Summaries for groups of cases" option or the "Values of individual cases" option. The "Summaries for groups of cases" option graphs the average of observations collected at the same time while the "Values of individual cases" option assumes the data consists of one observation at each time period.

For the "Summaries for groups of cases" option, enter "Year" in the "Category Axis" box. Then click on "Other Statistic" under "Line Represents" and enter "TEMP" in the "Variable" box. Click on **OK** to produce the graph.

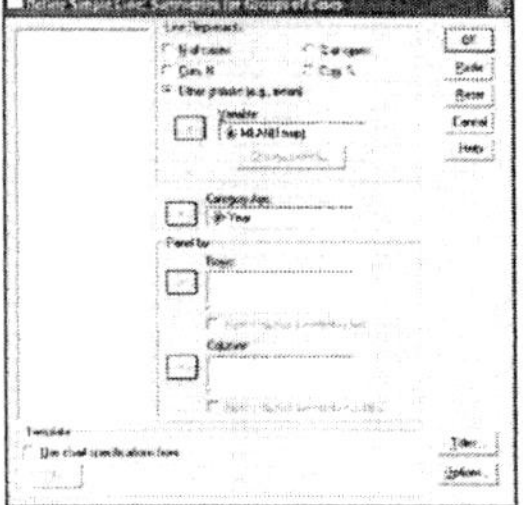

For the “Values of individual cases” option, enter “TEMP” under “Line Represents” then click on “Variable” under “Category Labels” and enter “Year” in the box beneath “Variable.” This option is different from the “Summaries for groups of cases” option in that it doesn’t require a variable for the horizontal axis (“Category Labels”). This is a nice option for graphing data that is in an ordered sequence but no time (or order) variable is included in the data set. The horizontal axis will just be the observation number. Click on **OK** to produce the plot.

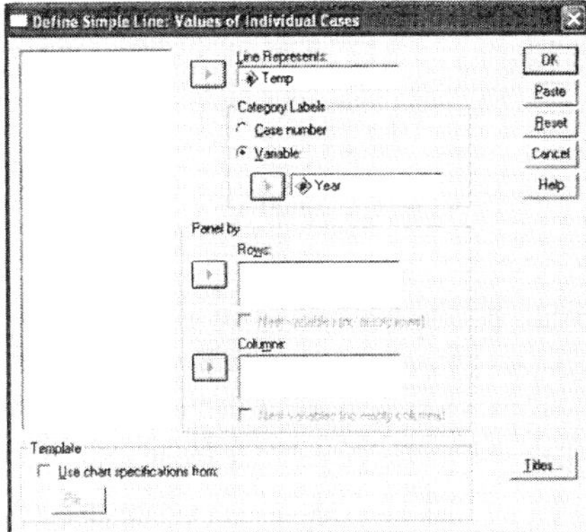

The graph will appear the same using either option, except that the vertical axis is labeled “Mean Temp” for the first option and “Value Temp” for the second.

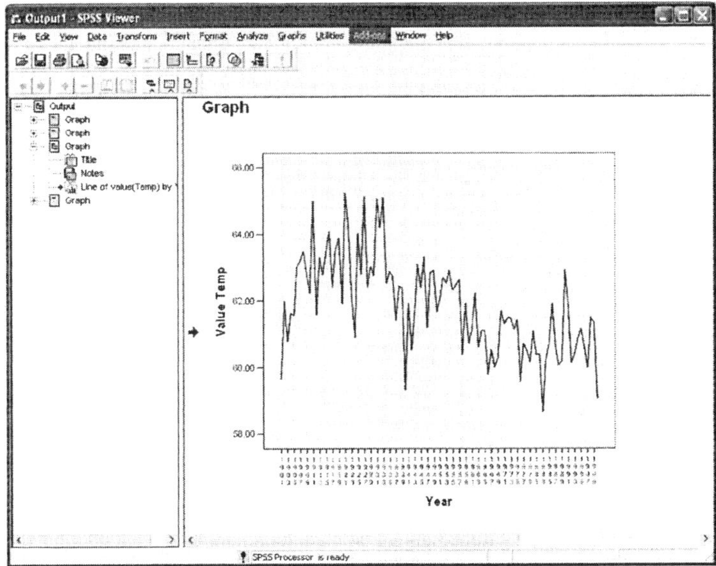

To change the scale on the horizontal axis so that every five years is listed, rather than each year, double click on the graph to open the Chart Editor, double click on the horizontal axis and then click on the “Labels & Ticks” tab. Click on “Custom” underneath “Category Label Placement” and enter “5” in the corresponding box.

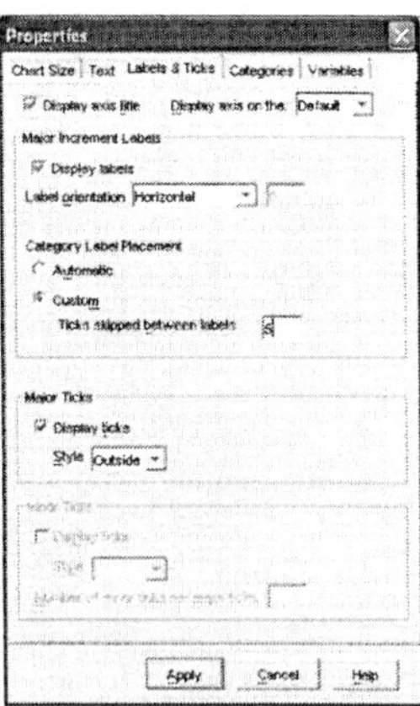

Click on **Apply** and then on **Continue** to make the change.

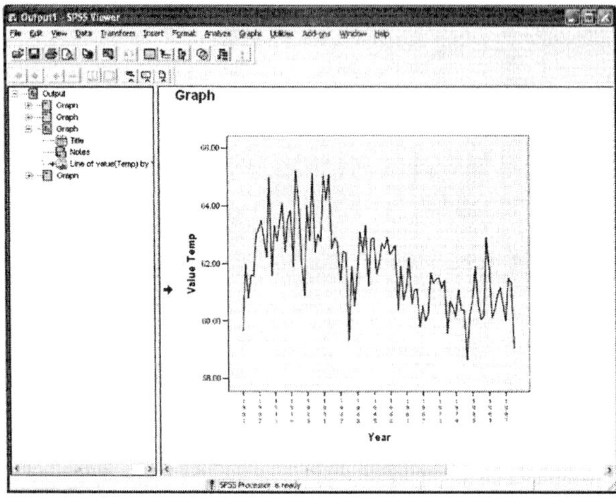

To produce a time plot using the "Sequence" command, select **Graphs → Sequence...**, then enter "TEMP" in the "Variables" box and "YEAR" in the "Time Axis Labels" box. This command is like the "Values of individual cases" option for the "Line" command in that it doesn't require a variable for the horizontal axis ("Time Axis Labels").

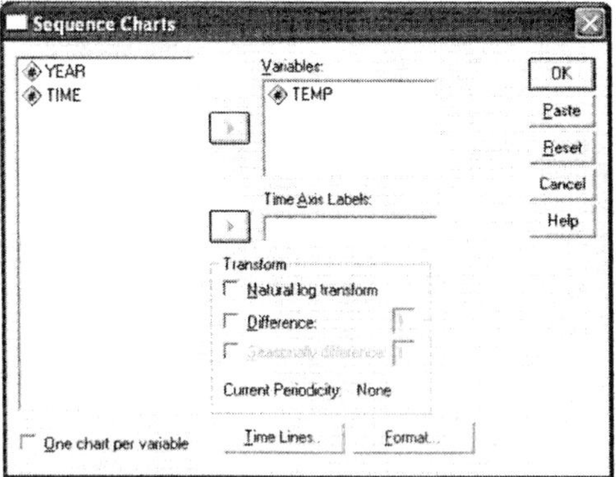

You can also make another version of a time plot using the "Scatterplot" command under the "Graphs" menu. However, this method will only work if you have a time variable in your data set for the horizontal axis. Select **Graphs → Scatterplot...**, then select the "Simple" option and click on **Define**.

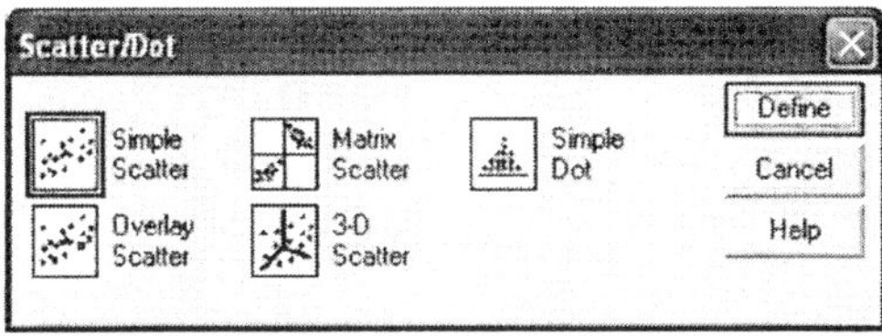

Enter "Temp" as the Y-variable and "Year" as the X-variable and click on **OK** to produce the plot.

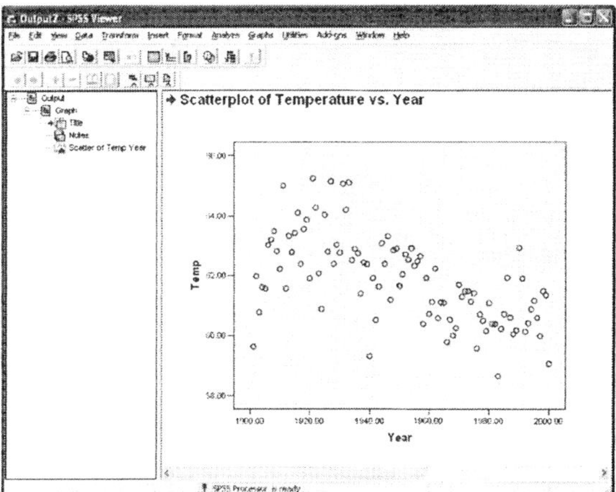

Double click on the graph to open the Chart Editor, then select **Element → Interpolation Line**. Click on **Close** to see that the points are now connected as in the graph produced by the Line graph command. This option is useful if you want to add a regression line to the graph. You will see how to do this in the next chapter of this manual.

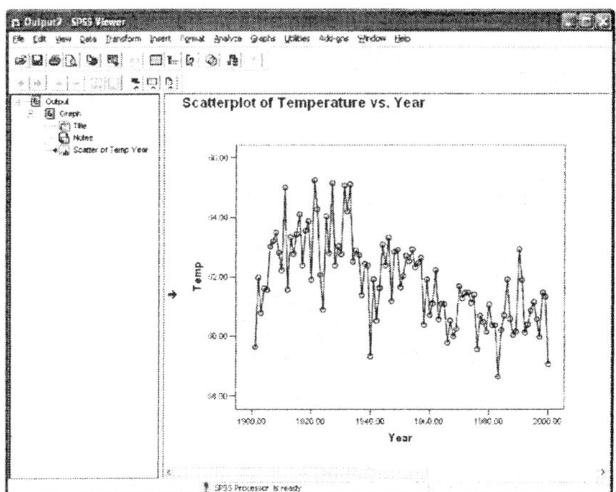

Time Plots in SPSS Version 11.0

The procedure for producing a time plot is exactly the same in earlier versions of SPSS, but the appearance of the graph is slightly different.

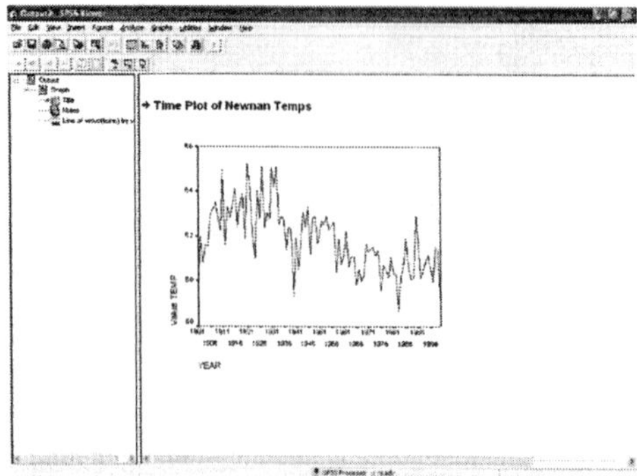

Descriptive Statistics

You might have noticed that the output from the "Histogram" command includes both the mean and the standard deviation of the variable being graphed. To find these and other descriptive statistics, you also have three options under the "Descriptive Statistics" command from the "Analyze" menu: Frequencies, Descriptives, and Explore. "Descriptives" is the most basic command and was demonstrated in the first chapter; the default output for this command includes only the minimum, maximum, mean, and standard deviation. Under the "Options" button in the dialog box for this command you can produce some additional statistics.

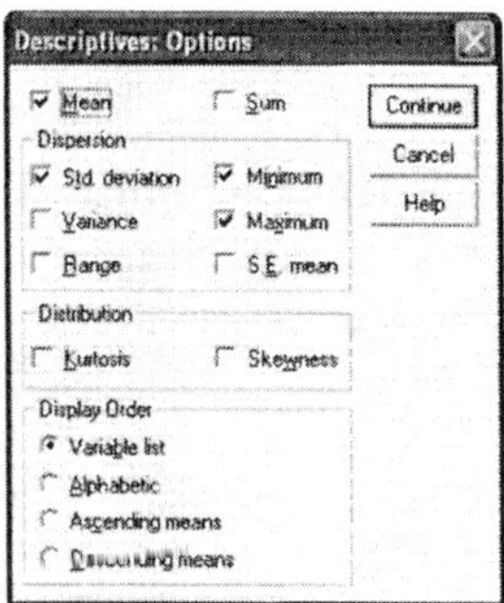

Click on any of the default statistics (the ones that are already checked) that you don't want to find to un-select them. The "Frequencies" and "Explore" commands offer more choices than "Descriptives" does.

Mean and Median

Let's demonstrate the "Frequencies" command using the data from Exercise 2.29. The data for this exercise consists of the 1997 carbon dioxide emissions (in million metric tons of carbon equivalent) from fossil fuel combustion for seven countries with the highest emissions and is given in the following table:

Country	U.S.	China	Russia	Japan	India	Germany	U.K.
CO2	1490	914	391	316	280	227	142

Enter this data in SPSS as shown below.

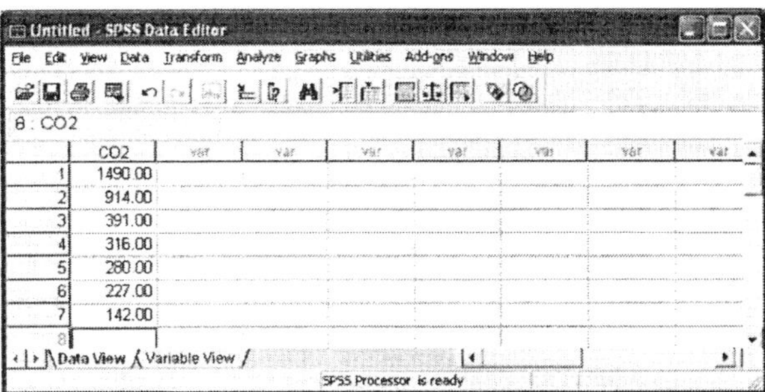

Select **Analyze → Descriptive Statistics → Frequencies…** and enter CO2 in the "Variable(s)" box. Click on the "Statistics" button and select the statistics you want to produce. For this example, we will find just the mean and median for CO2.

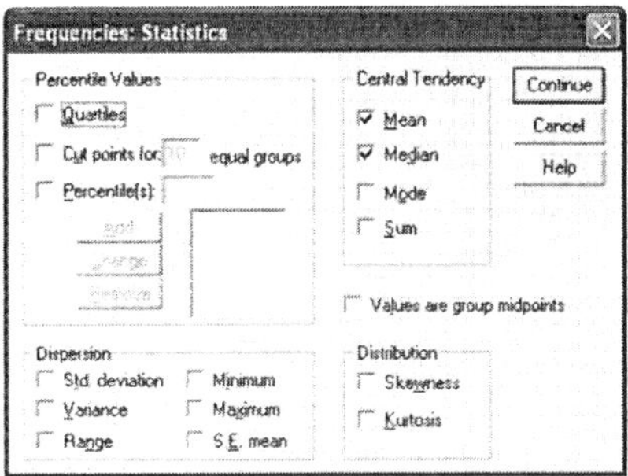

Click on **Continue** and then click on the "Display frequency tables" box so that the frequency distribution won't be produced. With only 7 observations, the frequency distribution will not be useful here.

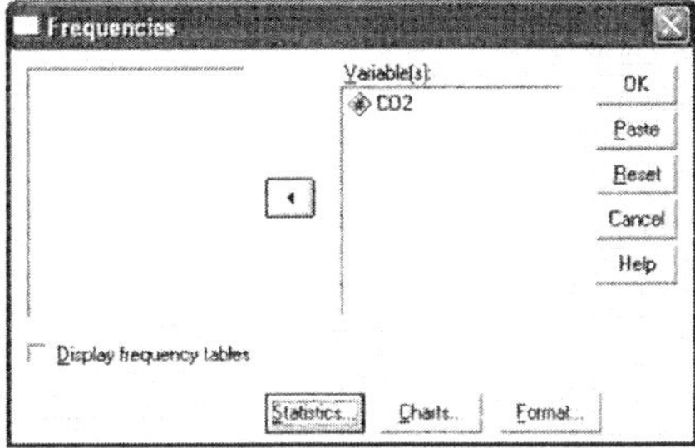

Note that you could click on the "Charts" button to also produce a histogram for this data set. Click on **OK** to produce the output. The following display shows the default output from the "Descriptives" command along with the mean and median output from the "Frequencies" command for comparison.

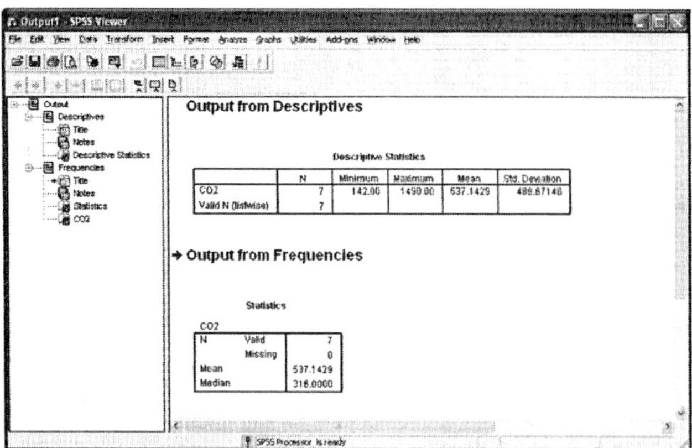

Standard Deviation and Range

The standard deviation and range can be found using any one of “Frequencies,” “Descriptives,” and “Explore.” Let’s find the standard deviation for the “European_union_unemployment” data of Exercise 2.55 using the “Descriptives” command. Access this data file from your text CD.

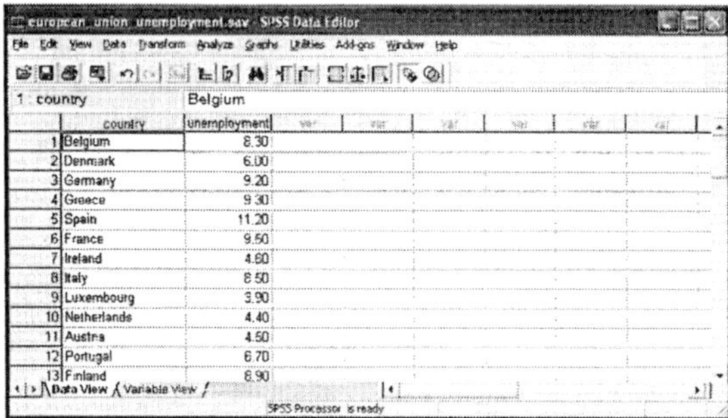

Select **Analyze → Descriptive Statistics → Descriptives…** and enter “unemployment” in the “Variable(s)” list. Click on the “Options” button and select “Range” from the list of possible statistics to produce. The standard deviation is one of the default statistics produced by this command.

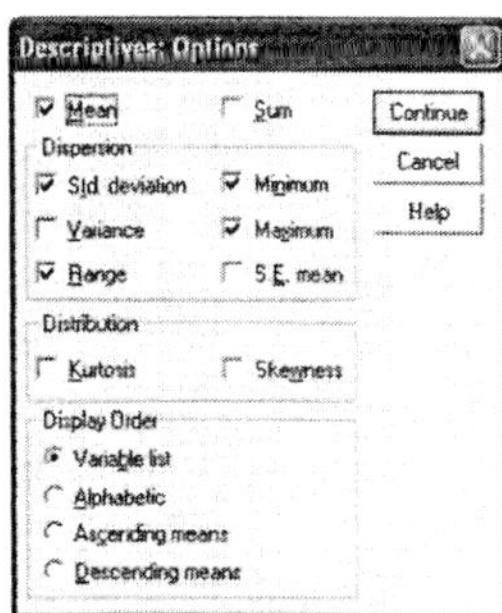

Click on **OK** to produce the default output.

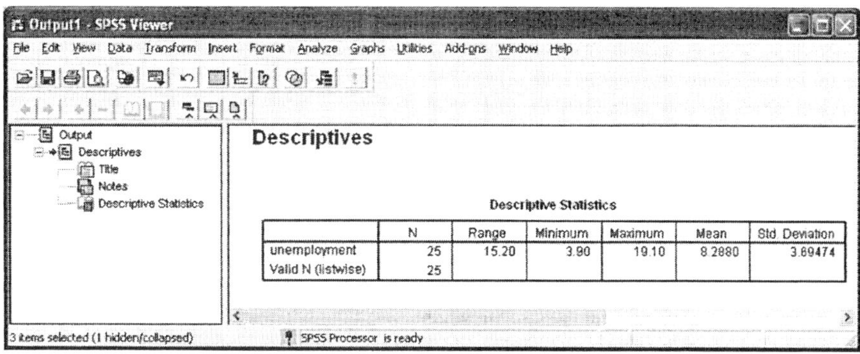

Quartiles and Other Percentiles

Both the "Frequencies" and "Explore" commands will produce quartiles and some percentiles. You actually have more options with the "Frequencies" command; the "Explore" command will only produce the 5th, 10th, 25th, 50th, 75th, 90th and 95th percentiles. Let's find the quartiles for the data from Exercise 2.58 using "Frequencies." The data is the unemployment rate for the 15 nations of the European Union in 2003 and is shown in the following table:

Belgium	8.3	France	9.5	Italy	8.5
Denmark	6.9	Portugal	6.7	Finland	8.9
Germany	9.2	Netherlands	4.4	Austria	4.5
Greece	9.3	Luxembourg	3.9	Sweden	6.0
Spain	11.2	Ireland	4.6	U.K.	4.8

Enter the data in SPSS as shown below.

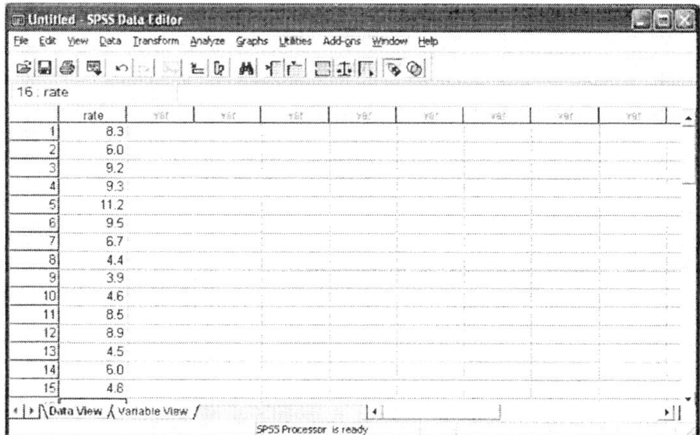

Select **Analyze → Descriptive Statistics → Frequencies...** and enter the unemployment rate variable in the "Variables" box, then click on the "Statistics" button. Click on "Quartiles" under "Percentile Values" and then on **Continue**.

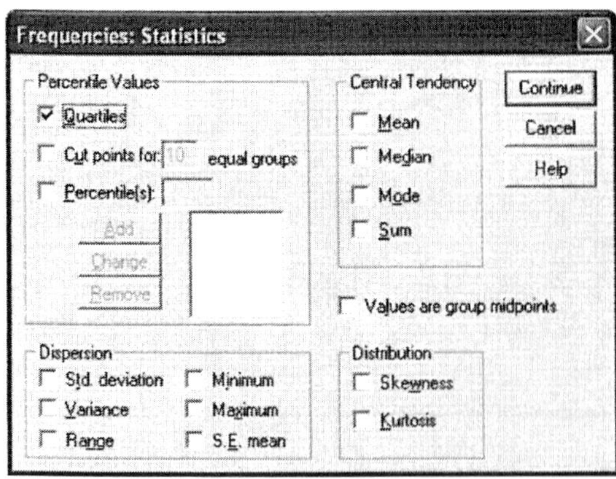

Click on "Display frequency tables" so that the frequency distribution will not be produced.

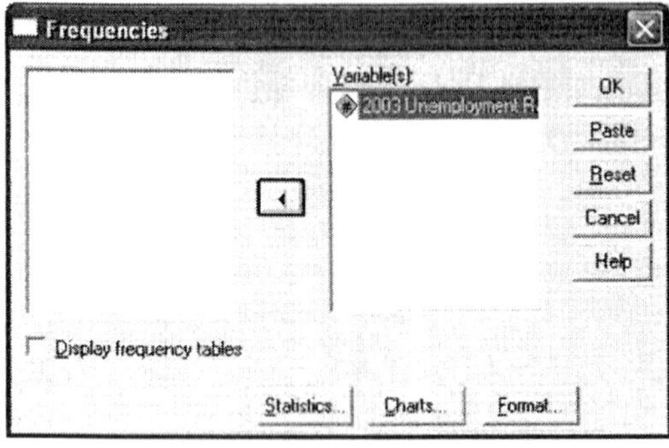

Click on **OK** to produce the output.

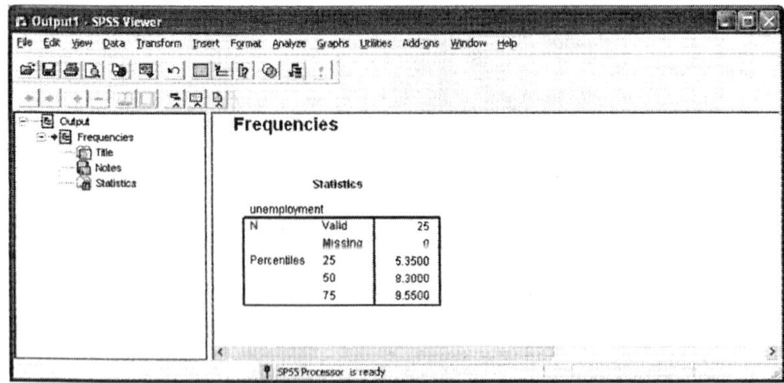

Frequencies

Statistics

unemployment

N	Valid	25
	Missing	0
Percentiles	25	5.3500
	50	8.3000
	75	9.5500

To produce other percentiles, you would click on the "Percentile(s)" button in the "Frequencies: Statistics" dialog box and then enter the value for the percentile you want in the box that opens up. Click on the "Add" button and repeat to request additional percentiles. Let's demonstrate this process with the "cereal" data set on your text CD. Access this data set, then select **Analyze → Descriptive Statistics → Frequencies...** and enter "SUGARmg" in the "Variables" box. Click on the "Statistics" button and then on "Percentiles." The following example shows a request for the 30th and 70th percentiles.

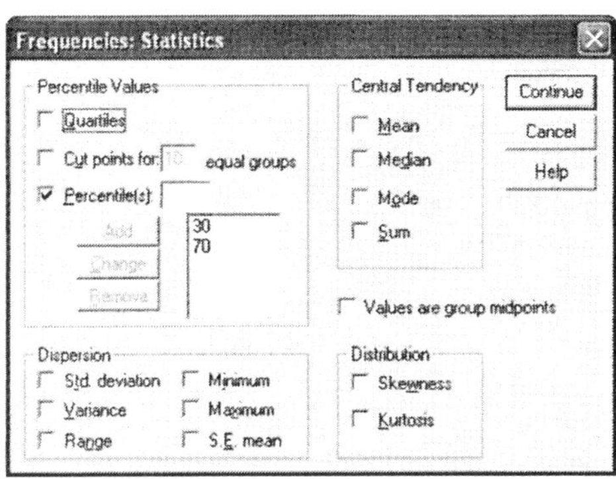

Click on **Continue** and then on **OK** to produce the output.

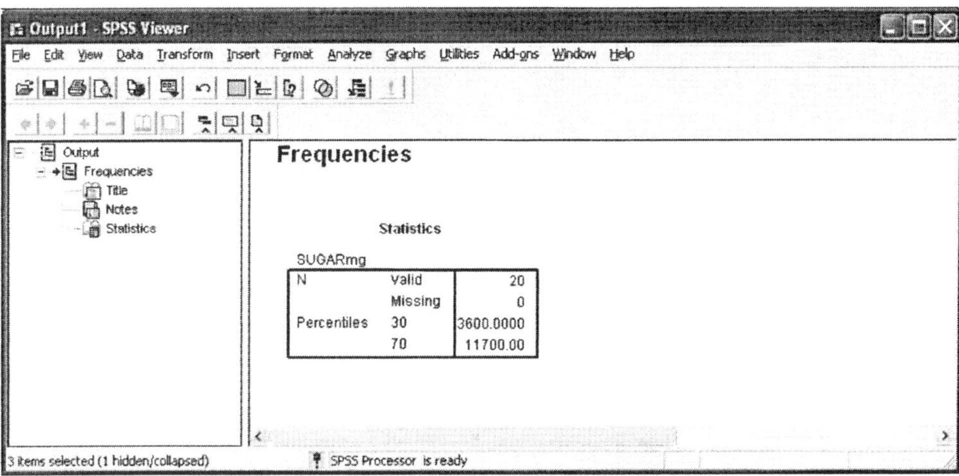

N	Valid	20
	Missing	0
Percentiles	30	3600.0000
	70	11700.00

To produce percentiles using the "Explore" command, select **Analyze → Descriptive Statistics → Explore…**, click on the "Statistics" button and then on "Percentiles." If you click on "Descriptives" to remove the check mark next to it, then SPSS will only produce the percentiles output.

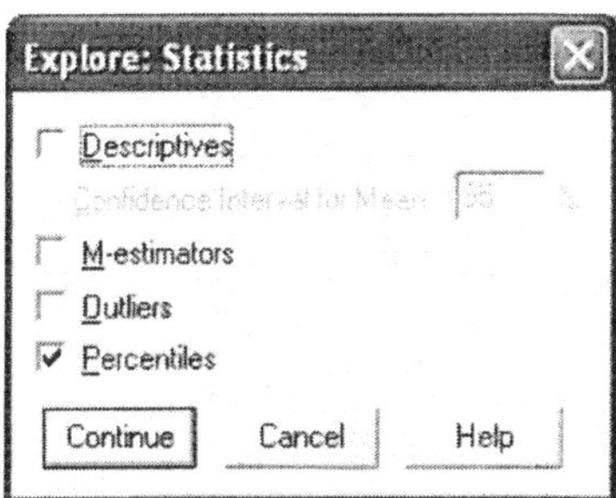

In the output below, the column widths of the Percentiles table were decreased so that all of the output would appear in the output window, and so it would all print on one page.

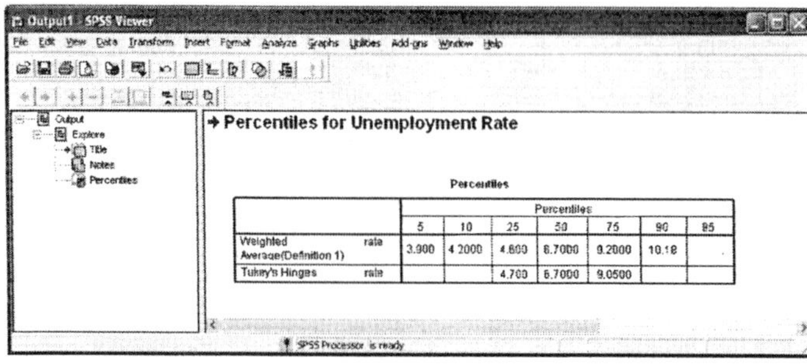

Percentiles

		Percentiles						
		5	10	25	50	75	90	95
Weighted Average(Definition 1)	rate	3.900	4.2000	4.800	6.7000	9.2000	10.18	.
Tukey's Hinges	rate			4.700	6.7000	9.0500		

Box Plots

The "Graphs" menu includes a command for producing a box plot, but it is also an option under the "Explore" command. Let's produce a box plot for the unemployment rate data from the previous example. Select **Graphs → Boxplot...** The following dialog box will appear:

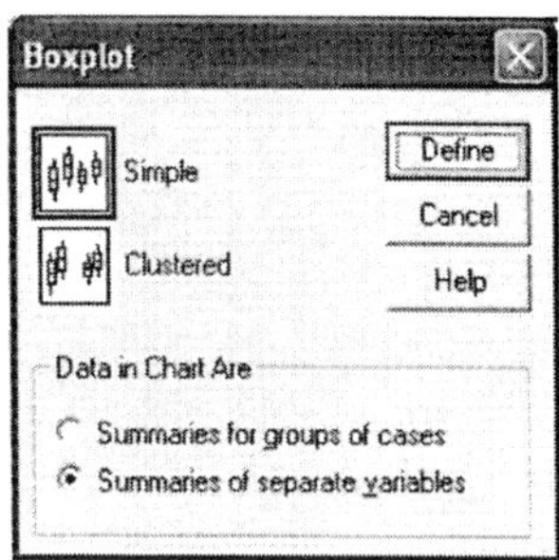

Click on "Simple" and "Summaries of separate variables" and then on **Define**. The "Summaries for groups of cases" option is for creating side-by-side box plots for a variable observed on two or more groups. You will learn about this option in another example. Enter the unemployment rate variable in the "Boxes represent" box and click on **OK**. The following output will be produced. Note that the Case Processing Summary part of this output has already been omitted.

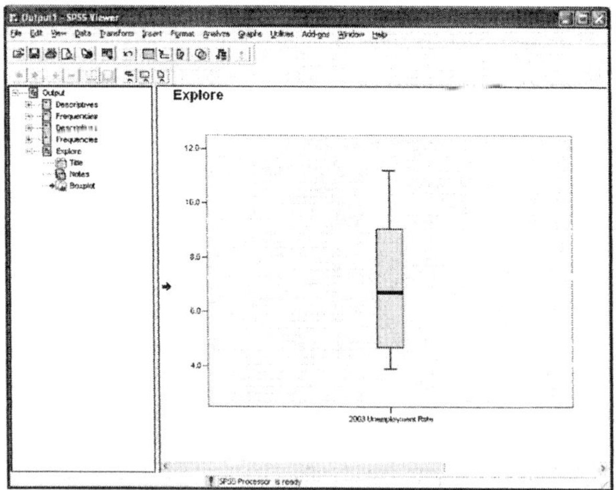

If you prefer to view box plots horizontally, double click on the graph to open the "Chart Editor" and then select **Options → Transpose Chart**. The box plot will now be displayed horizontally instead of vertically. Close the "Chart Editor" to complete the change.

Let's explore box plots with some more examples. Open the "fla_student_survey" data set from your text CD and make a box plot of the number of hours of TV watching.

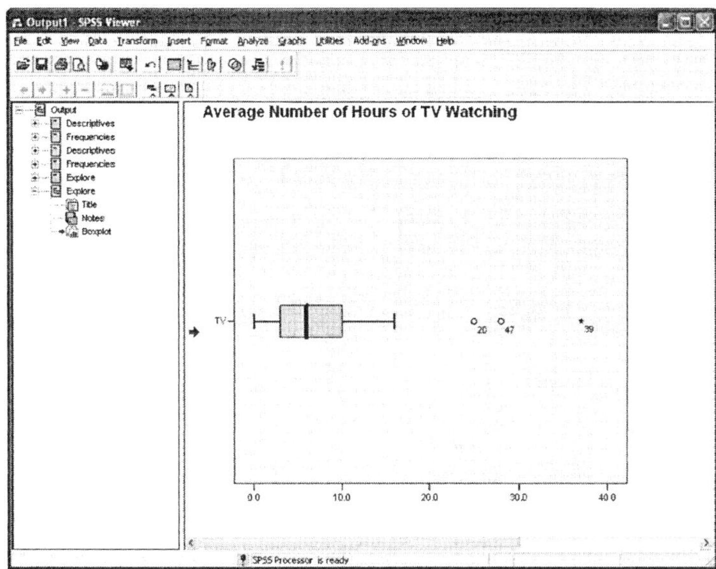

Notice that the right-hand "whisker" does not extend all the way out to the maximum value but is stopped at a "fence;" any observations beyond the fence are considered to be outliers. Notice also that there are three outliers in this graph. Different programs have different rules for locating upper and lower fences and for identifying outliers. Two possible rules are to extend the upper whisker out to a distance of 1.5 × IQR from the third quartile, or to the next larger data value below this distance. Any points beyond the fence are displayed as outliers. SPSS appears to be using the later rule. Similar rules are applied for the left-hand whisker if there are outliers on that side of the distribution. One of the outliers in this box plot is designated as an "extreme" outlier (more than 3 × IQR from the third quartile) by the symbol used to identify it (*).

The values below each outlier are the observation numbers of each of these points within the data set, not the actual data value. Knowing the observation number makes it easier to locate those values in case they are the result of a data entry error. To eliminate the observation numbers from the graph, click on an outlier in the "Chart Editor" and select **Elements → Hide Data Labels.**

Side-By-Side Box Plots

To see how to make side-by-side box plots of a quantitative variable for groups defined by a categorical variable, open the "georgia_student_survey" data set from your text CD. Select

Graphs → Boxplot... and then select the "Summaries for groups of cases" option. Click on **Define** and enter "Haircut" in the "Variable" box and "Gender" in the "Category Axis" box.

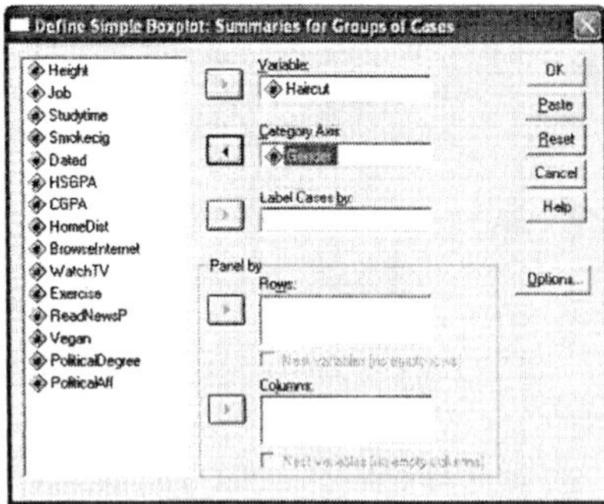

Click on **OK** to produce the output. In the following display some of the output has been omitted and a legend was added to the title to identify the groups, since value labels were not assigned to the "Gender" variable within the data set.

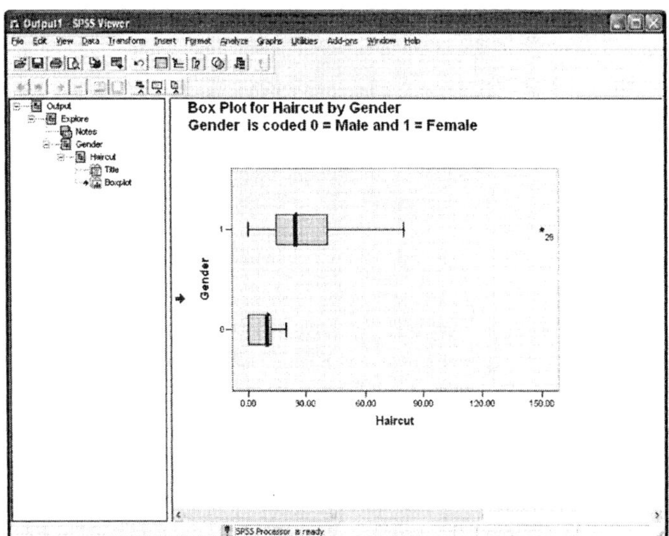

Side-by-side box plots are very useful for comparing the distributions of two or more groups. As we can see in the above output, the price spent on hair cuts by the male students typically varies less than the price spent by the female students, and the median price spent by the female students appears to be larger than the maximum spent by any of the male students.

Box Plots in SPSS Version 11.0

As with the other graphs we've looked at, modifying box plots in earlier versions of SPSS is slightly different, although the process for producing a box plot is the same. To display a box plot horizontally, select **Format → Swap Axes** in the "Chart Editor**.**" The color of the box part of the plot can be changed by clicking once inside the box (but not on the median line) and then selecting **Format → Color.** Select the color you want, then click on **Apply** and then on **Close**. Close the "Chart Editor" to complete the changes.

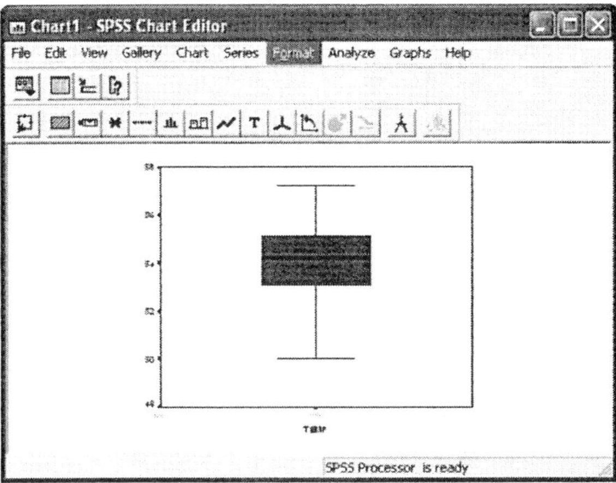

The Explore Command

We used the "Explore" command above to produce a stem-and-leaf plot and percentiles. Each time we used that command, however, we only produced a small portion of what this command can do all at once. With this one command you can not only find descriptive statistics and percentiles, but also produce a histogram, stem-and-leaf plot and a box plot! To produce this output make, sure that "Both" is selected under "Display" and that both "Descriptives" and "Percentiles" are selected under "Statistics." Click on the "Plots" button and select both "Histogram" and "Stem-and-leaf" under "Descriptive," along with "Factor levels together" under "Boxplots." Click on **Continue** and then on **OK** to see all the output this one command will produce!

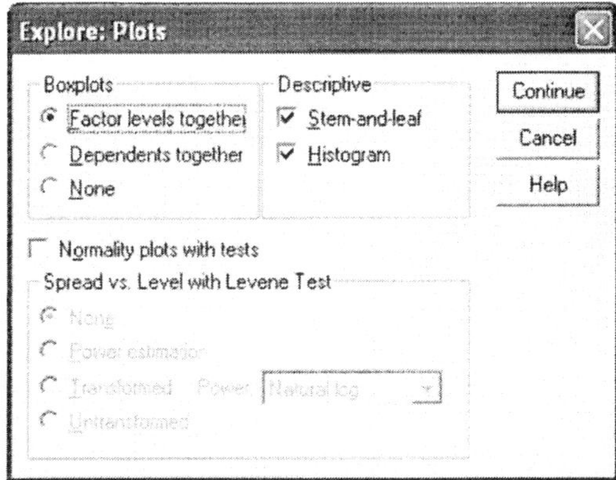

Comparison of Frequencies, Descriptives, and Explore

The following table will help you determine which of these three commands under "Descriptive Statistics" is the most useful for the analyses you want to do. Notice that "Frequencies" can be used on both categorical and quantitative data, while "Descriptives" and "Explore" only operate on quantitative data. Also, "Descriptives" is the only one of these commands that does not include the option to produce at least one type of graph.

Command	**Variable Type**	**Statistics**	**Graphs**
Frequencies	Categorical	Frequency Table	Bar and pie charts
	Quantitative	Mean, Median, Mode Standard Deviation, Variance, Range Minimum, Maximum, S.E. Mean Quartiles, Percentiles	Histogram
Descriptives	Quantitative	Mean Standard Deviation, Variance, Range Minimum, Maximum, S.E. Mean	None
Explore	Quantitative	Mean, Median Confidence Interval for the Mean Standard Deviation, Variance, Range, IQR Minimum, Maximum, S.E. Mean Some Percentiles	Stem-and-leaf Histogram Box plot

Chapter 3 Association: Contingency, Correlation and Regression

In the previous chapter you saw how to produce descriptive statistics and graphs for one categorical or quantitative variable. This chapter will introduce you to statistical methods for analyzing two variables at the same time.

Two Categorical Variables

When working with two categorical variables, you will want to display the joint distribution in a contingency table. You might also want to produce some conditional proportions or maybe a side-by-side bar chart to make comparisons.

Contingency Tables

To produce a contingency table from data that has already been tabulated, the process is similar to how we dealt with tabulated data for a single variable in the previous chapter. In that case, we entered two variables: one for the categories and one for count or frequency of observations in each category. For a contingency table, we will need an additional column, one for the categories of the second categorical variable.

As an example, let's look at the data for Exercise 3.4. This data set has two categorical variables, "Gender" and "Number of Hours of Home Religious Activity," and is shown in the table below.

Religious Activity by Gender

	Number of Hours of Home Religious Activity				
Gender	0	1-9	10-19	20-39	40 or more
Female	229	297	88	103	49
Male	276	243	59	40	16

Open a new SPSS Data Editor and enter the data in three columns, as shown below. Notice that there are 10 "cases" in this table, one for each combination of the two categories for "Gender" and the five categories for "hours." You can enter "Gender" and "hours" as string-type variables or as numeric variables with value labels. The "hours" variable has the label "Number of Hours of Home Religious Activity" added to it in the Variable View of the Data Editor.

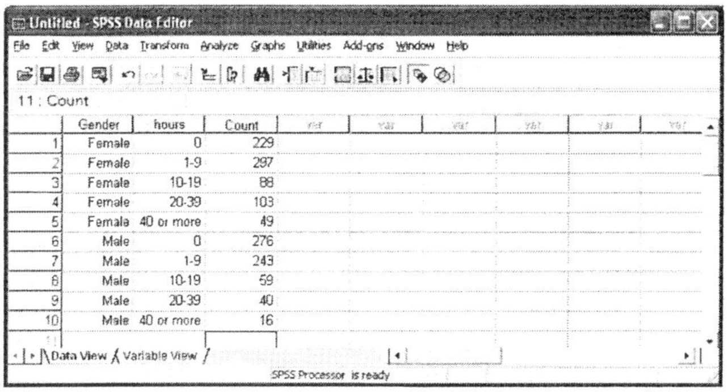

Untitled - SPSS Data Editor

11 : Count

	Gender	hours	Count
1	Female	0	229
2	Female	1-9	297
3	Female	10-19	88
4	Female	20-39	103
5	Female	40 or more	49
6	Male	0	276
7	Male	1-9	243
8	Male	10-19	59
9	Male	20-39	40
10	Male	40 or more	16

As with the examples from the previous chapter with one categorical variable, you will have to indicate to SPSS that the data in the variable "Count" represents the number of observations in each of the 10 different combinations of the categories of the "Gender" and "hours" variables. Select **Data → Weight Cases...** and enter "Count" as the "Frequency Variable" to "Weight cases by."

To produce a contingency table for this data using SPSS, select **Analyze → Descriptive Statistics → Crosstabs...** and enter "Gender" as the "Row Variable" and "hours" as the "Column Variable" and click on **OK**.

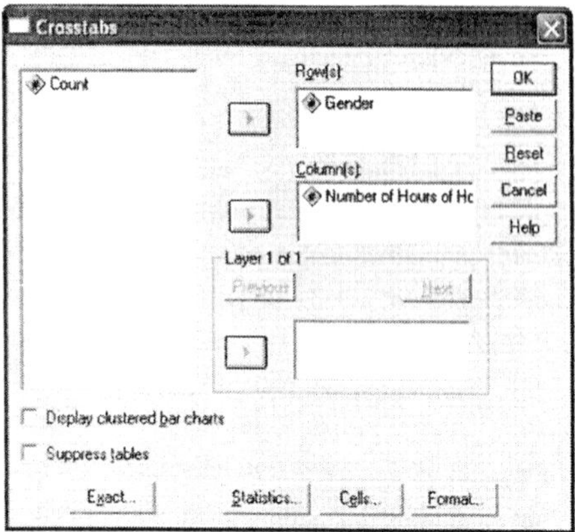

The output is shown below, without the "Case Processing Summary."

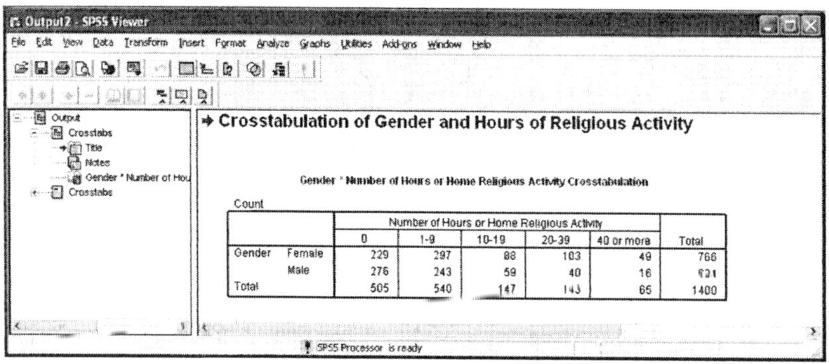

Crosstabulation of Gender and Hours of Religious Activity

Gender * Number of Hours or Home Religious Activity Crosstabulation

Count

		Number of Hours or Home Religious Activity					Total
		0	1-9	10-19	20-39	40 or more	
Gender	Female	229	297	88	103	49	766
	Male	276	243	59	40	16	631
Total		505	540	147	143	65	1400

Both of the categorical variables were entered as Numeric variables in this example. If you entered the "hours" variable as a string variable, you will find that SPSS may reorder the categories in the output produced. For this example the "1-9" and "10-19" columns would be switched in order if the "hours" variable was entered as a String variable.

Conditional Proportions

If all you needed was the contingency table for this example, then there wouldn't be any need to have entered it in SPSS. But, with the data entered this way you can now find conditional proportions and produce a side-by-side bar chart allowing you to compare the "hours" variable by "Gender." To find the conditional proportions for "hours" select **Analyze → Descriptive Statistics → Crosstabs...** again, but this time click on the "Cells" button. In the dialog box, click on "Row" underneath Percentages.

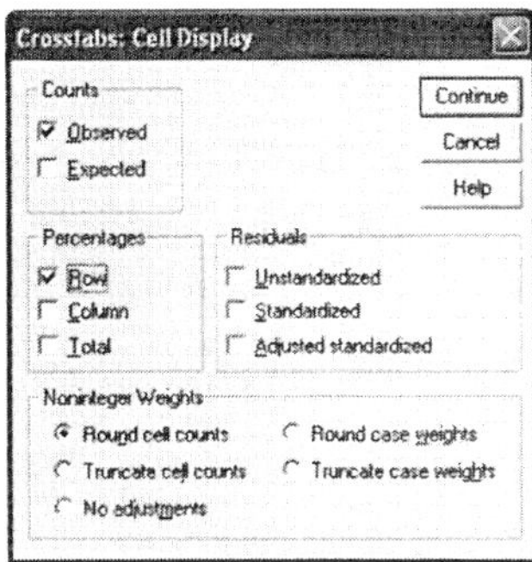

Click on **Continue** and then click on **OK** to produce the following output. If you also select "Column" and "Total" under "Percentages" then these results would be added to the output.

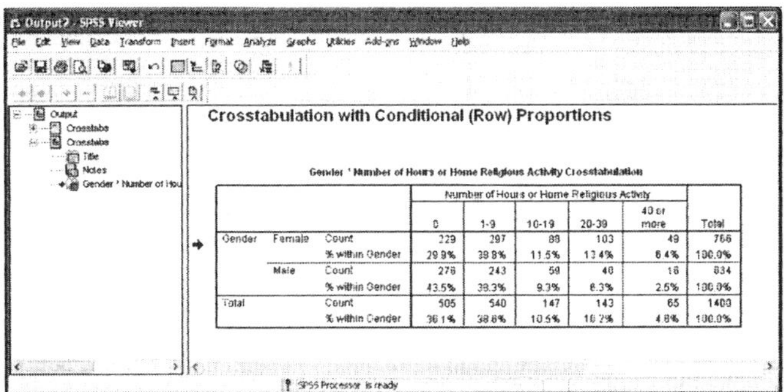

Crosstabulation with Conditional (Row) Proportions

Gender * Number of Hours or Home Religious Activity Crosstabulation

			Number of Hours or Home Religious Activity					
			0	1-9	10-19	20-39	40 or more	Total
Gender	Female	Count	229	297	88	103	49	766
		% within Gender	29.9%	38.8%	11.5%	13.4%	6.4%	100.0%
	Male	Count	276	243	59	40	16	634
		% within Gender	43.5%	38.3%	9.3%	6.3%	2.5%	100.0%
Total		Count	505	540	147	143	65	1400
		% within Gender	36.1%	38.6%	10.5%	10.2%	4.6%	100.0%

Side-By-Side Bar Charts

To produce side-by-side bar charts for this data, select **Graphs → Bar...** and click on the "Summaries for groups of cases" option. Click on "Define," then enter "Gender" in the "Category Axis" and "hours" under "Columns." Click on "Other Statistics" under "Bars Represent" and enter "Count." Click on **OK**.

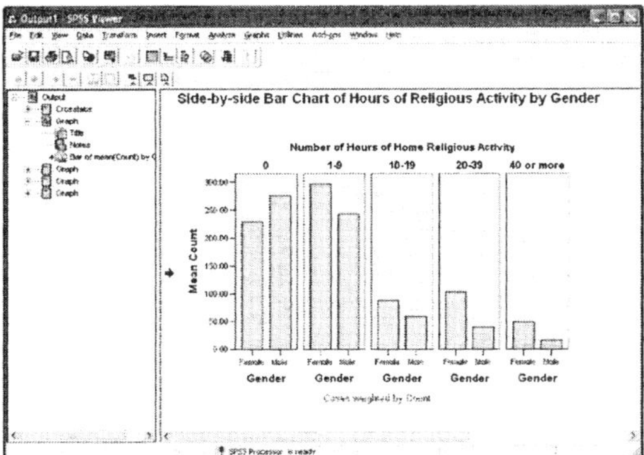

You can put the "hours" variable on the horizontal axis instead of "Gender" by switching their roles in the dialog box. If you put "Gender" under "Columns" and "hours" under "Category Axis" then the following graph will be produced instead.

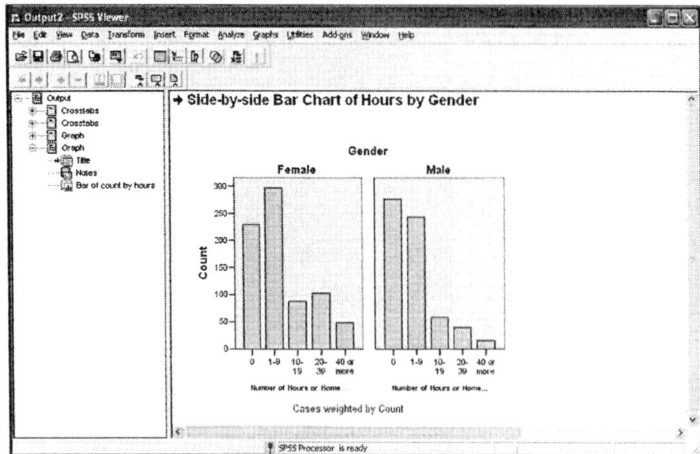

If you place a variable in the "Rows" box instead of the "Columns" box then you will get the same graphs but with the graphs stacked one on top of another, rather than side-by-side.

Just as in the one-variable case, there is another option for producing this graphical output. For this option, do not weight the cases by the variable "Count." (If the cases are weighted, then select **Data → Weight Cases...** and click on "Do not weight cases.") Select **Graphs → Bar...** and click on the "Values of individual cases" option. Click on **Define** and enter "Count" under "Bars Represent." Click on "Variable" under "Category Labels" and enter "Gender." Enter "hours" under "Columns" and click on **OK**. The following graph will be produced.

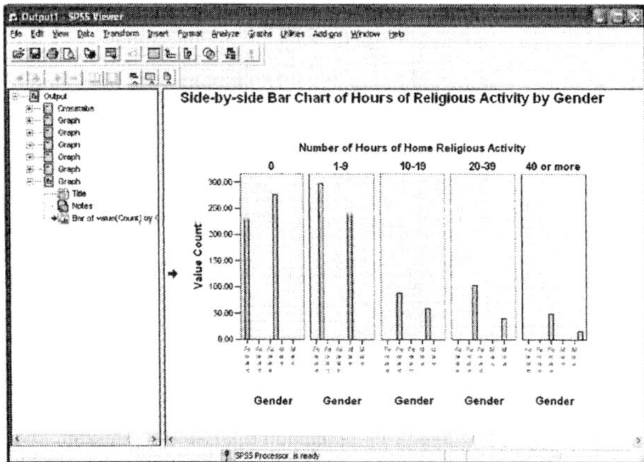

The graph is similar to the first side-by-side bar chart produced above except the bars are narrower and the axes are labeled differently. You can also switch the roles of "Gender" and "hours" as above to produce a graph similar to the second side-by-side bar chart produced above.

Contingency Tables for Non-Tabulated Data

What if the data hasn't been tabulated yet? This is actually an easier situation to deal with since you won't have to bother with a weighting variable. Let's demonstrate this using the variables "gender" and "vegetarian" from the "fla_student_survey" data set from your text CD. Once you have accessed this data file, select **Analyze → Descriptive Statistics → Crosstabs...** and enter "gender" as the row variable and "vegetarian" as the column variable.

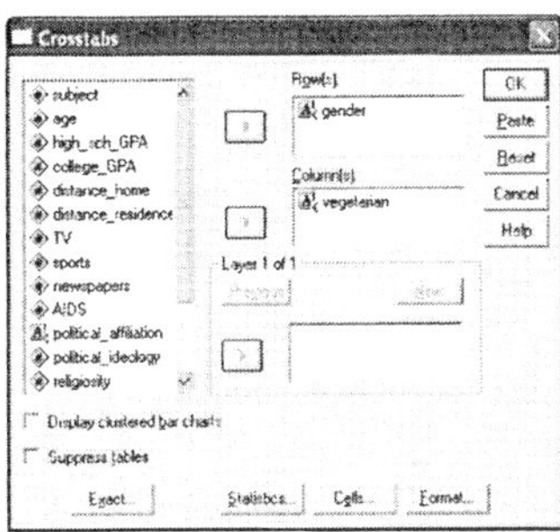

Click on **OK** to produce the contingency table.

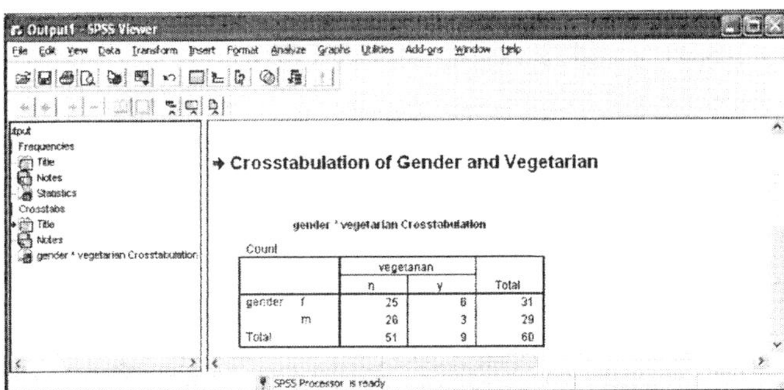

gender * vegetarian Crosstabulation

Count

		vegetarian		Total
		n	y	
gender	f	25	6	31
	m	26	3	29
Total		51	9	60

Let's produce side-by-side bar charts using the same process as described above. For example, if you select **Graphs → Bar...** and enter "gender" under "Category Axis" and "vegetarian" under "Columns represent" the following graph will result.

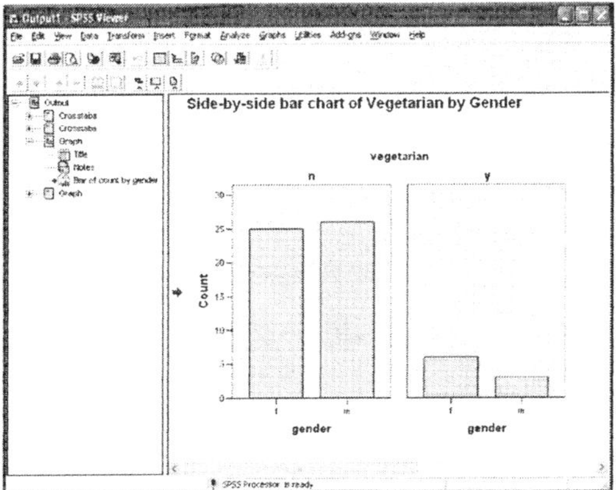

One Categorical Variable and One Quantitative Variable

You actually already dealt with this case in the previous chapter. For this situation you would want to produce descriptive statistics and graphs of the quantitative variable for each category of the categorical variable. The easiest way to do this is to use "Explore" command. To demonstrate this, let's use the "fla_student_survey" data set.

Select **Analyze → Descriptive Statistics → Explore...** and enter "TV" in the "Dependent List" and "vegetarian" in the "Factor List." Click on **OK**. Among the output produced is a table with various descriptive statistics for average TV watching for the vegetarian and non-vegetarian students and side-by-side box plots, both shown below.

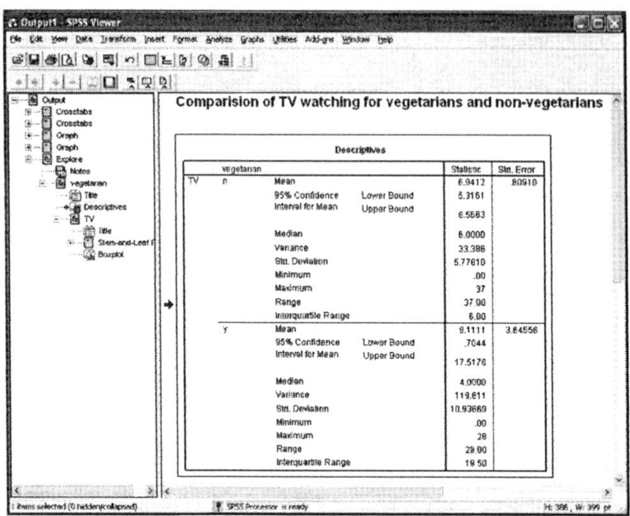

Comparision of TV watching for vegetarians and non-vegetarians

Descriptives

	vegetarian			Statistic	Std. Error
TV	n	Mean		6.9412	.80910
		95% Confidence Interval for Mean	Lower Bound	5.3161	
			Upper Bound	8.5563	
		Median		6.0000	
		Variance		33.363	
		Std. Deviation		5.77610	
		Minimum		.00	
		Maximum		37	
		Range		37.00	
		Interquartile Range		6.00	
	y	Mean		9.1111	3.64556
		95% Confidence Interval for Mean	Lower Bound	.7044	
			Upper Bound	17.5178	
		Median		4.0000	
		Variance		119.611	
		Std. Deviation		10.93669	
		Minimum		.00	
		Maximum		28	
		Range		28.00	
		Interquartile Range		19.50	

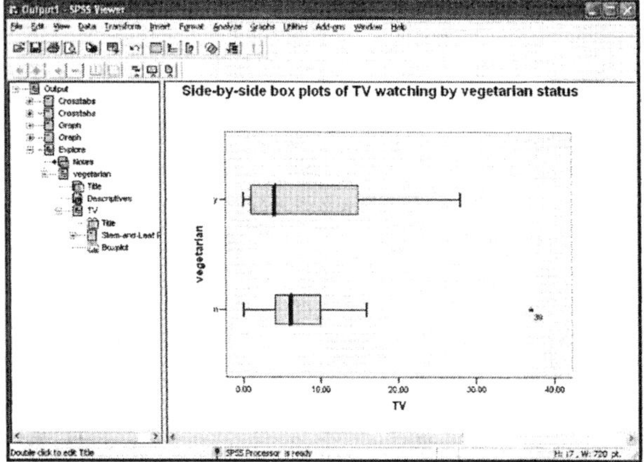

Another option for comparing groups according to a quantitative variable is the "Compare Means" command. Let's demonstrate this command with the same data set as above. Select **Analyze → Compare Means → Means...** and enter the "TV" variable in the "Dependent List" and the "vegetarian" variable in the "Independent list."

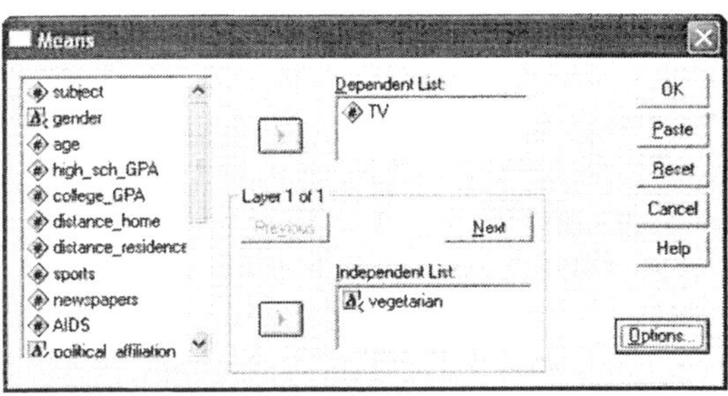

Click on **OK** to produce the output.

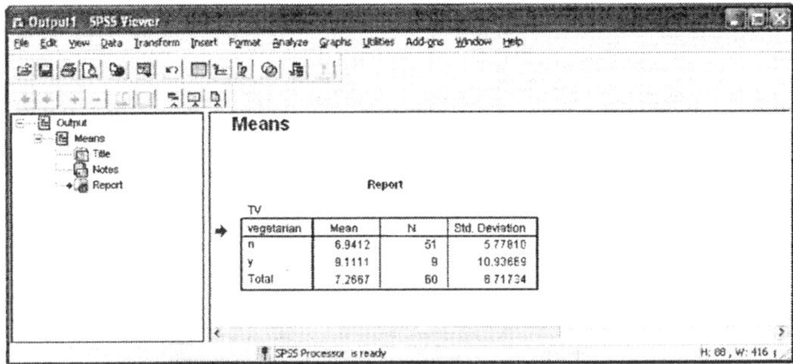

vegetarian	Mean	N	Std. Deviation
n	6.9412	51	5.77810
y	9.1111	9	10.93689
Total	7.2667	60	6.71734

The "Options" button in the "Means" dialog box has options for producing some additional statistics for each group.

A third alternative for producing statistics and graphs of a quantitative variable for groups defined by a categorical variable is to use the "split file" command from the "Data" menu. Continuing with the same data set, let's compare the TV watching habits for the groups defined by political affiliation. Select **Data → Split File...** and click on "Organize output by groups." Click on the "political affiliation" variable and then on the triangle button to enter that variable in the "Groups Base on:" box. Click on **OK**.

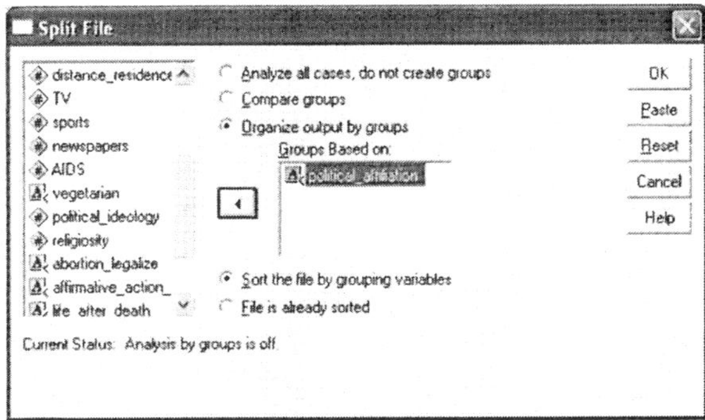

All of the commands you do after this will be done separately for each group defined by the grouping variable. The "Compare groups" option will do the same thing but it will organize the output to make comparisons easier. The output from the "Descriptives..." command for the TV watching variable produced after the "Split File..." command was done is shown below.

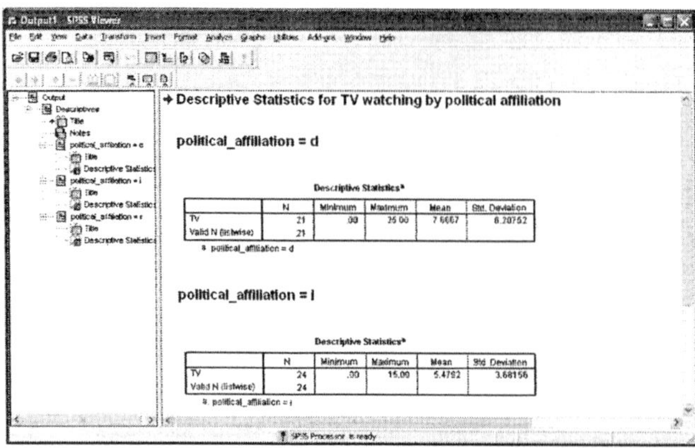

When you are done using the split file, select **Data → Split File…** again and select "Analyze all cases, do not create groups." Click on **OK** to complete the command.

Two Quantitative Variables

For two quantitative variables you will want to display the association in a scatterplot. If the association appears linear, the correlation coefficient and regression line will also be useful.

Scatterplots

To produce a scatterplot to investigate the association between two quantitative variables, use the "Scatter/Dot" command under the "Graphs" menu. Let's demonstrate this using the "GDP" and "Internet Use" variables from the "human_development" data set on your text CD. Access this data file, then select **Graphs → Scatter/Dot…** and click on "Simple Scatter" and then on **Define**.

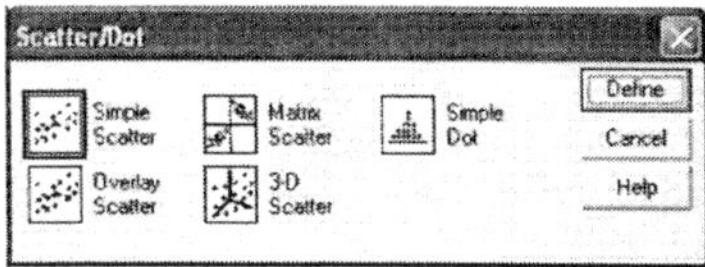

Enter "INTERNET" as the "Y Axis:" variable and "GDP" as the "X axis:" variable as shown below.

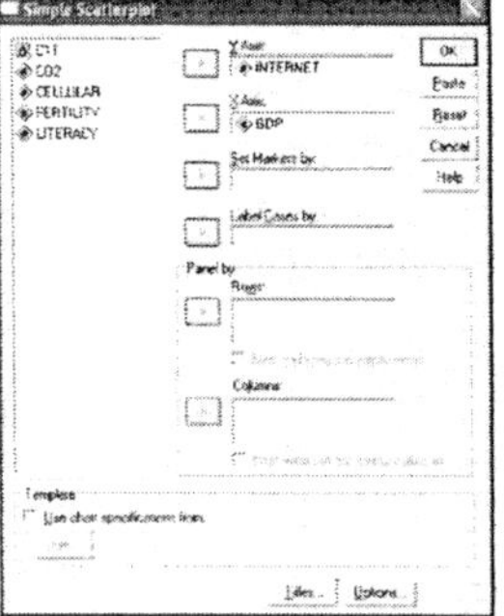

Click on **OK** to produce the output. As demonstrated in Chapter 2 for the dot plot, you can change the size, shape and color of the plotting symbol. This process will also be described again in the section below on how to add the regression line to a scatter plot.

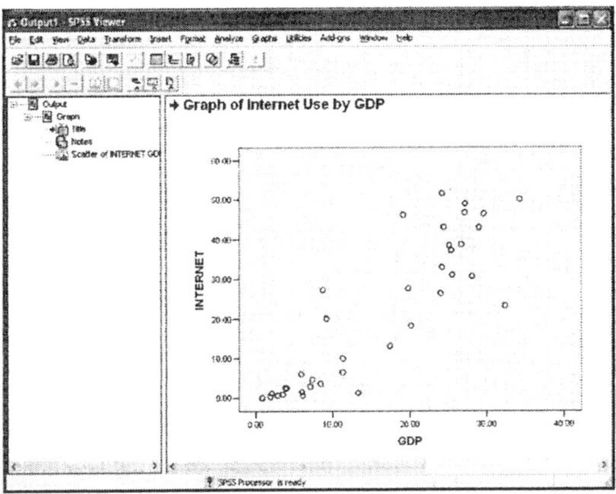

The Correlation Coefficient

To find the correlation coefficient for the association between these two variables select **Analyze → Correlate → Bivariate…** and enter both variables in the "Variables" box.

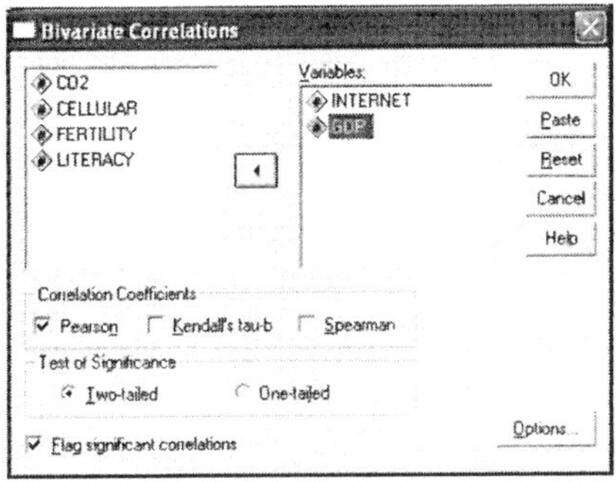

Click on **OK** to produce the output.

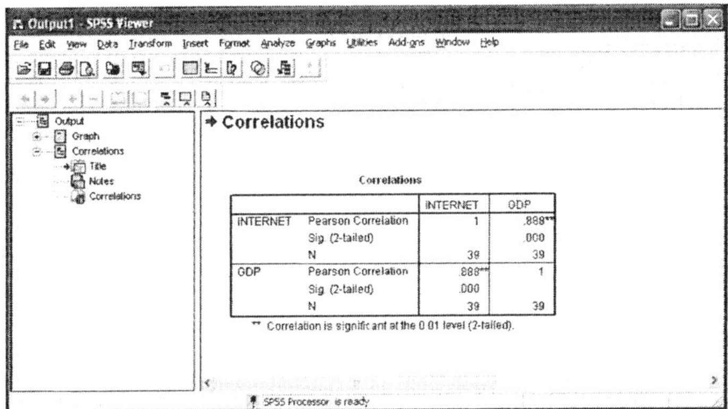

Correlations

		INTERNET	GDP
INTERNET	Pearson Correlation	1	.888**
	Sig. (2-tailed)		.000
	N	39	39
GDP	Pearson Correlation	.888**	1
	Sig. (2-tailed)	.000	
	N	39	39

** Correlation is significant at the 0.01 level (2-tailed).

SPSS produces a "correlation matrix" that shows the correlation of each variable with itself (1) along with the correlation between "GDP" and "INTERNET" and the correlation between "INTERNET" and "GDP" (.888). SPSS also includes the results ("Sig") of a hypothesis test to determine whether or not this correlation is "significantly different" from zero.

The Regression Line

The command for finding the regression equation also finds the correlation coefficient as well. To demonstrate this command, let's use the "Batting Average" and "Team Scoring" variables in the "al_team_statistics" data set from you text CD. Access this data file and select **Analyze → Regression → Linear…**

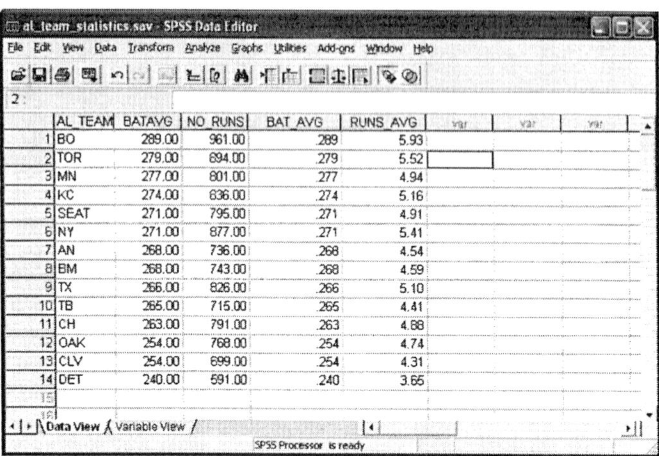

	AL_TEAM	BATAVG	NO_RUNS	BAT_AVG	RUNS_AVG
1	BO	289.00	961.00	.289	5.93
2	TOR	279.00	894.00	.279	5.52
3	MN	277.00	801.00	.277	4.94
4	KC	274.00	836.00	.274	5.16
5	SEAT	271.00	795.00	.271	4.91
6	NY	271.00	877.00	.271	5.41
7	AN	268.00	736.00	.268	4.54
8	BM	268.00	743.00	.268	4.59
9	TX	266.00	826.00	.266	5.10
10	TB	265.00	715.00	.265	4.41
11	CH	263.00	791.00	.263	4.88
12	OAK	254.00	768.00	.254	4.74
13	CLV	254.00	699.00	.254	4.31
14	DET	240.00	591.00	.240	3.65

Enter "BAT_AVG" as the "Independent Variable" and "RUNS_AVG" as the "Dependent Variable."

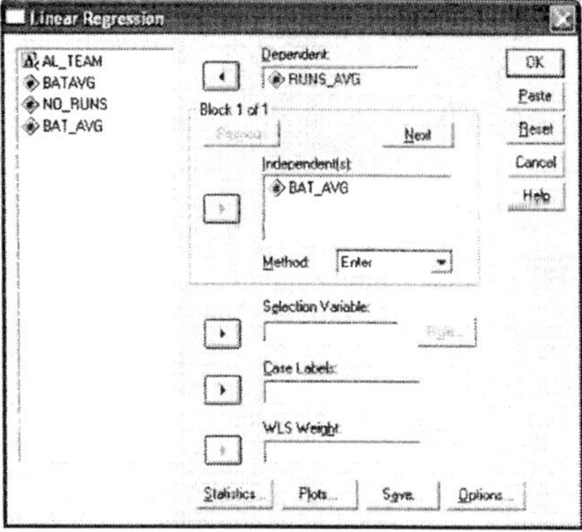

Click on **OK** to produce the output. The resulting output has five components, some of which you will learn about later on. The parts that you will need right now are the title, the Model Summary, and the Coefficients. The other parts have been omitted from the output shown below.

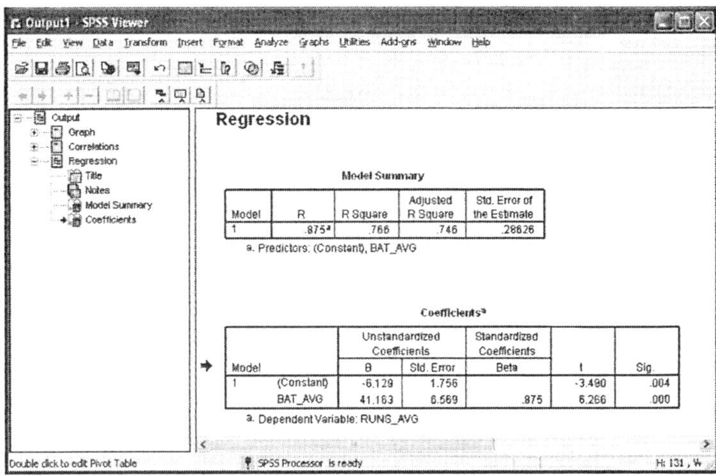

Model Summary

Model	R	R Square	Adjusted R Square	Std. Error of the Estimate
1	.875[a]	.766	.746	.28626

a. Predictors: (Constant), BAT_AVG

Coefficients[a]

Model		Unstandardized Coefficients		Standardized Coefficients		
		B	Std. Error	Beta	t	Sig.
1	(Constant)	-6.129	1.756		-3.490	.004
	BAT_AVG	41.163	6.569	.875	6.266	.000

a. Dependent Variable: RUNS_AVG

The "Model Summary" provides the absolute value of the correlation coefficient under "R." Since the association is positive we know that r is +.875. The intercept and slope can be found in the "B" column under "Unstandardized Coefficients" in the "Coefficients" part of the output. The intercept can be found next to "(Constant)" (value = -6.129) and the slope is next to the independent variable, "BAT_AVG" in this case (value = 41.163). So the regression line is $\hat{Y} = -.6129 + 41.163X$. Note: Most of the rest of the statistics included in the regression output are covered in a later chapter in your text and in this manual.

If you are interested in finding some descriptive statistics for the regression variables just click on the "Statistics" button before you click on **OK**. Click on "Descriptives," then on **Continue**, and finally on **OK**.

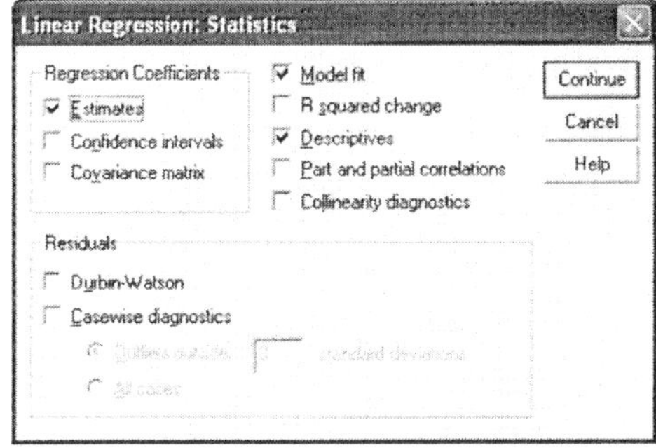

Adding the Regression Line to a Scatterplot

The "Scatter/Dot" command has an option for you to add the regression line to a scatterplot. Make a scatterplot of "RUNS_AVG" vs. "BAT_AVG" as shown below.

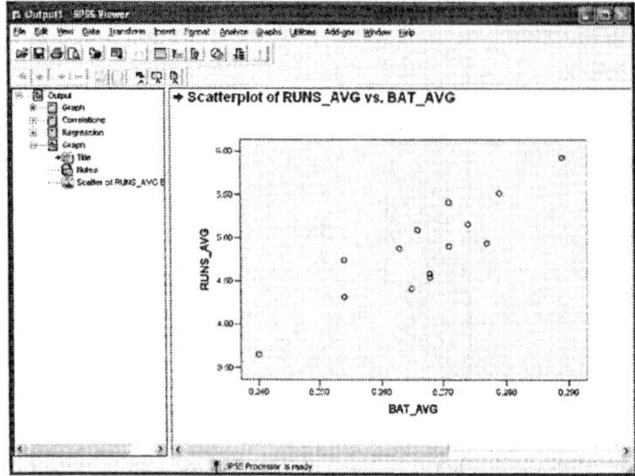

To add the regression line to the plot, double click on the graph to open the Chart Editor, then select **Elements → Fit Line at Total**. The line will be added to the graph, along with a statistic called "R-Squared." With the Chart Editor open you can also change the size and symbol used for the data points. To do this, click once on one of the points so that a circle appears around each point.

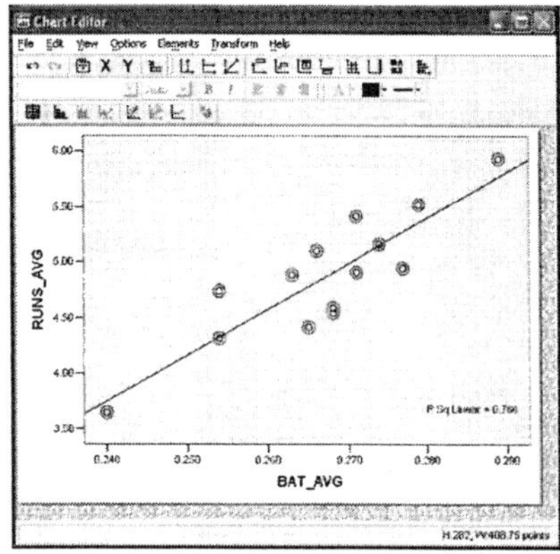

Select **Edit → Properties...** and click on the "Marker" tab. Select a new symbol by clicking on the down arrow next to the box under "Type." To produce a closed symbol rather than an open symbol, click on the "Fill" box under "Color."

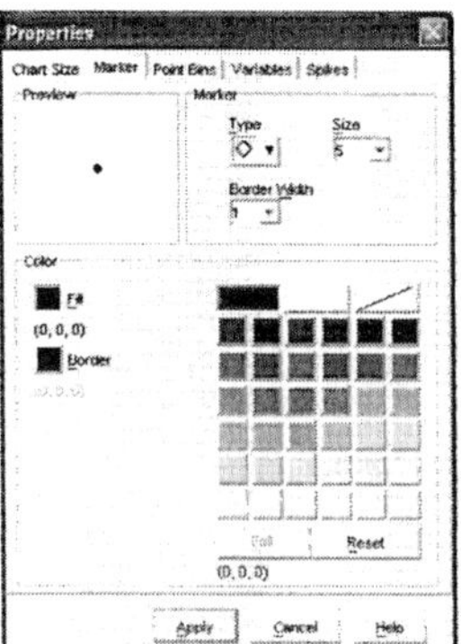

Click on **Apply** and then on **Close** to complete the changes.

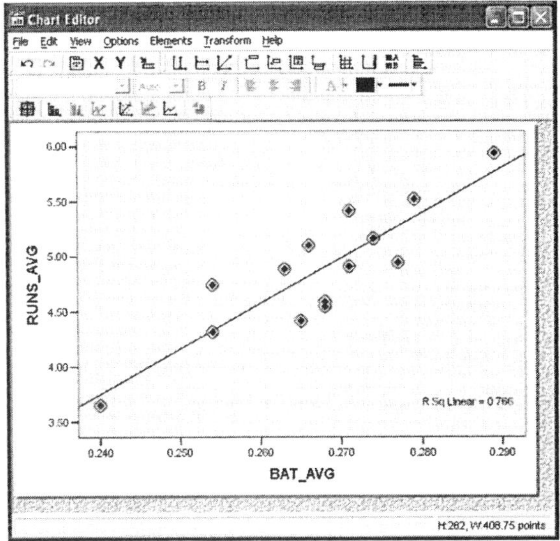

Scatterplots, Correlation and Regression in SPSS Version 11.0

Each of these commands works the same in Version 11.0, as they do in Version 13.0. But, the process for adding the regression line to a scatterplot and changing the plotting symbol are a little different in earlier versions of SPSS. For example, in Version 11.0, the Chart Editor appears as below.

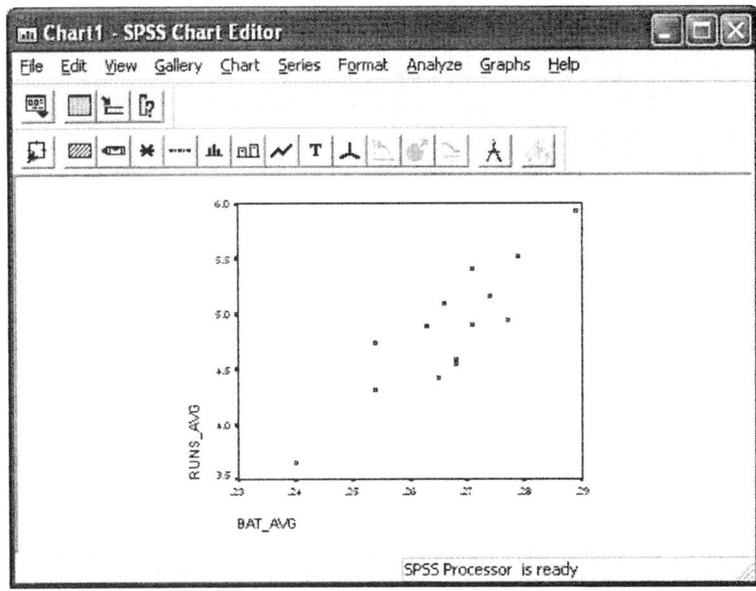

To add the regression line select **Chart → Options...** and click on "Total" under "Fit Line." Click **OK** to add the line to the graph.

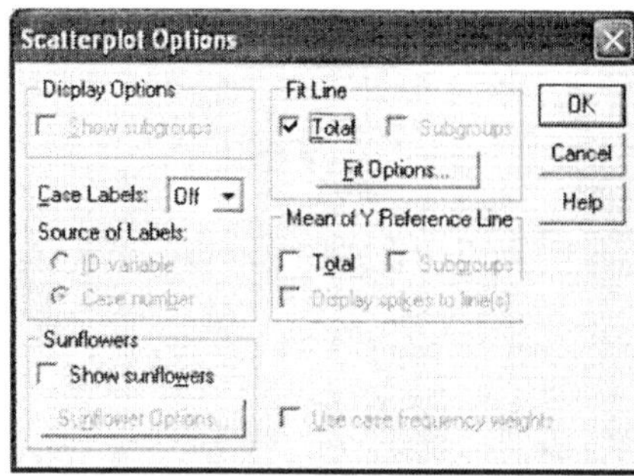

To change the plot symbol click once on one of the data points so that a collection of the points is highlighted. Then select **Format → Marker...** and select a new symbol and size from those listed and click on **Apply** and then on **Close**. Graphs made with tiny, open circles can be hard to read, so it is a good idea to increase the size of the circles and fill them in.

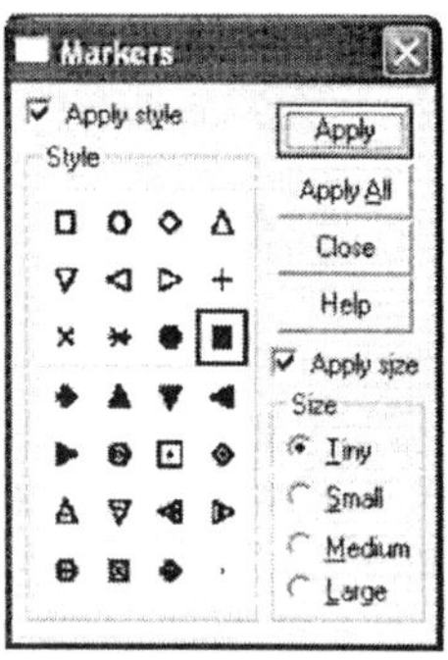

Saving and Graphing Regression Residuals

The "Regression" command in SPSS has an option to save the regression residuals to the Data Editor. The residuals can then be graphed to look for outliers, for example. To demonstrate this, let's use the "buchanan_and_the_butterfly_ballot" data set from Example 6 in Chapter 3, which can be found on the text CD. Access this data set and perform the regression of "Buchanan" (Y-variable) on "Perot" (X-variable), but click on the "Save" button and select "Unstandardized" under "Residuals."

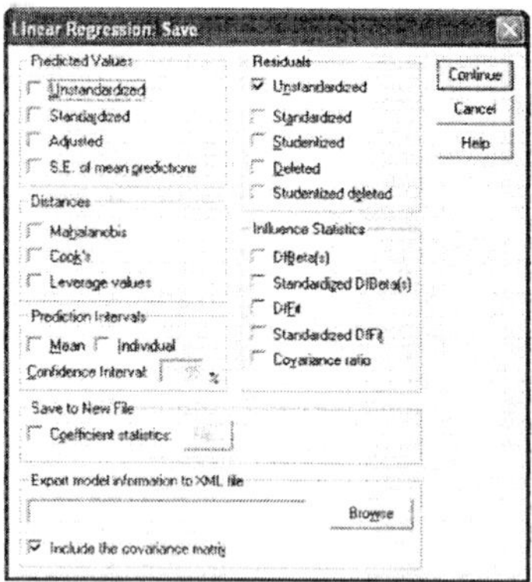

Click on **Continue** and then on **OK**. If you look at the Data Editor now you will see that a new variable has been added to the data set, "RES_1," that contains the residuals. Residuals from subsequent regression analyses would be saved as "RES_2," "RES_3," etc.

buchanan_and_the_butterfly_ballot.sav - SPSS Data Editor

1 : RES_1 -27.5311599239134

	county	perot	gore	bush	buchanan	RES_1
1	Alachua	8072.00	47300.00	34062.00	262.00	-27.53116
2	Baker	667.00	2392.00	5610.00	73.00	48.08359
3	Bay	5922.00	18850.00	38637.00	248.00	35.29824
4	Bradford	819.00	3075.00	5414.00	65.00	34.65193
5	Brevard	25249.00	97318.00	115185.0	570.00	-333.34446
6	Broward	38964.00	386518.0	177279.0	789.00	-604.44456
7	Calhoun	630.00	2155.00	2873.00	90.00	66.40577
8	Charlotte	7783.00	29645.00	35426.00	182.00	-97.20386
9	Citrus	7244.00	25525.00	29766.00	270.00	10.05709
10	Clay	3281.00	14632.00	41736.00	186.00	67.67333
11	Collier	6320.00	29918.00	60433.00	122.00	-104.92413
12	Columbia	1970.00	7047.00	10964.00	89.00	17.52140
13	Desoto	965.00	3322.00	4256.00	36.00	.43467
14	Dixie	652.00	1826.00	2697.00	29.00	4.61961
15	Duval	13844.00	107864.0	152098.0	652.00	156.20870
16	Escambia	8587.00	40958.00	73029.00	504.00	196.06552

Data View / Variable View

SPSS Processor is ready

One graph that is useful for analyzing residuals is the histogram. Select **Graphs →** **Histogram…** and enter "RES_1" in the "Variable" box. Click on **OK** to make the graph.

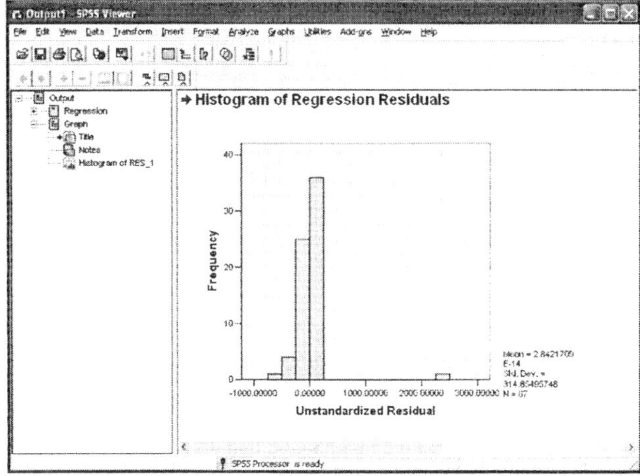

This default histogram produced by SPSS is a little different from the one in your text but the outlier shows up just the same.

Adding the Regression Line to a Time Plot

If you produce a time plot using the "Scatter/Dot" command instead of the "Line" command, then you can also insert the regression line on this graph. Let's demonstrate this using the "central_park_yearly_temps" data file from Example 12 in Chapter 3, which is on your text CD. Access this data file then select **Graphs → Scatter/Dot…** and select the "Simple Scatter" option. Click on **Define** and then enter "TEMP" as the Y-variable and "YEAR" as the X-variable. Click on **OK** to produce the plot. Double click to open the Chart Editor and select

Elements → Interpolation Line to connect the points. Click on **Close**. Next, select **Elements → Fit Line at Total** to add the regression line. Click on **Close** to complete the change.

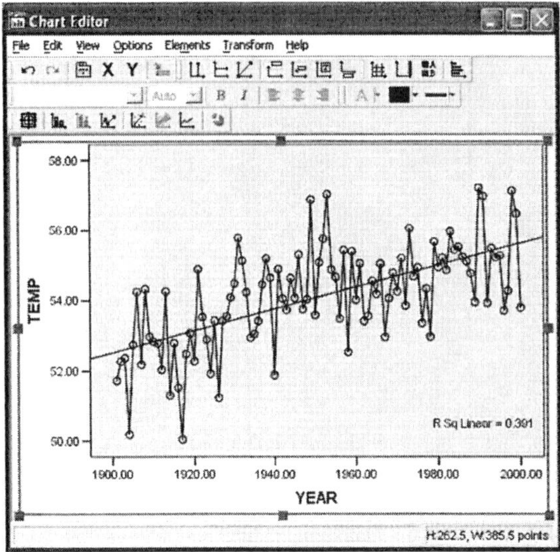

Influential Points

If you discover a potential influential point, it is very easy to temporarily eliminate this observation and repeat the analysis, to assess the impact of that one observation. To demonstrate how to do this, let's take a look at the "us_statewide_crime" data from Example 13 in Chapter 3, which is on your text CD. Make a scatterplot of Y = murderrate vs. X = college.

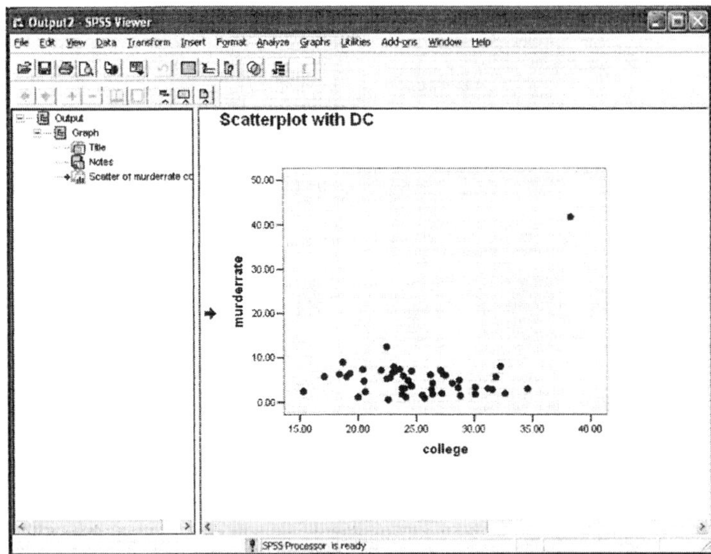

Notice the outlier in the upper right corner that is also a potential influential point. This is the observation for Washington DC. Insert the regression line in the plot.

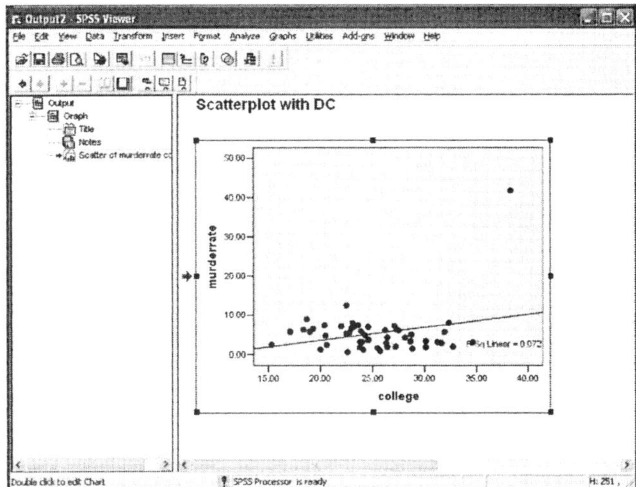

The line does appear to be "pulled up" toward the Washington DC data point. To produce the same graph without Washington DC, select **Data → Select Cases...** and click on "If condition is satisfied."

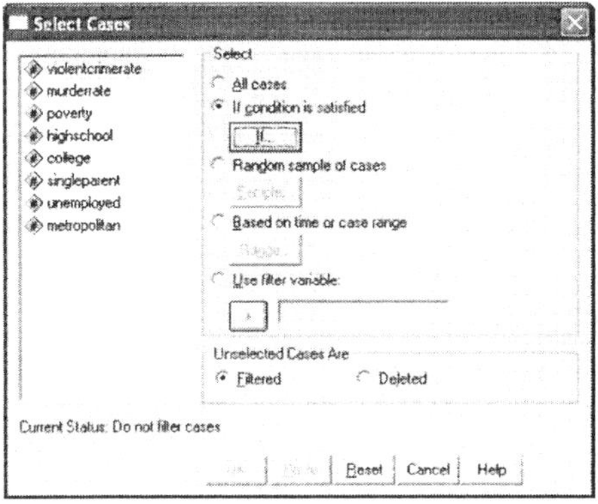

Click on the "If..." button and enter the expression "murderrate < 40" in the box at the top. Since the murder rate for DC is so much larger than it is for the other states, this is an easy way to eliminate the DC observation from the analysis without actually deleting the observation from the data set.

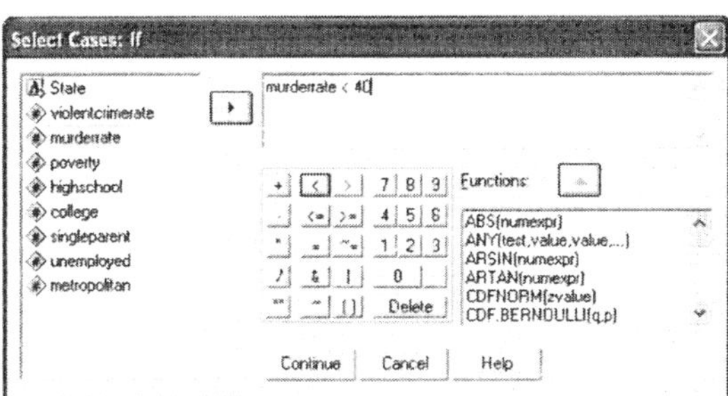

Click on **Continue** and then on **OK**. If you look at the Data Editor now you will see that the number to the left of "District of Columbia" is crossed out. SPSS also adds a "filter" variable to the data set with a value of 1 for cases to include in subsequent analyses and 0 for those to exclude. All of the cases except DC will have a value of 1 for the "filter," while DC has the value 0.

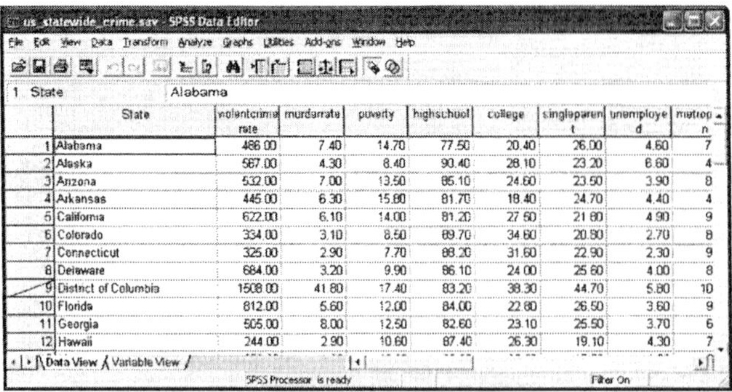

us_statewide_crime.sav - SPSS Data Editor

	State	violentcrimerate	murderrate	poverty	highschool	college	singleparent	unemployed	metropn
1	Alabama	486.00	7.40	14.70	77.50	20.40	26.00	4.60	7
2	Alaska	567.00	4.30	8.40	90.40	28.10	23.20	6.60	4
3	Arizona	532.00	7.00	13.50	85.10	24.60	23.50	3.90	8
4	Arkansas	445.00	6.30	15.80	81.70	18.40	24.70	4.40	4
5	California	622.00	6.10	14.00	81.20	27.50	21.80	4.90	9
6	Colorado	334.00	3.10	8.50	89.70	34.60	20.80	2.70	8
7	Connecticut	325.00	2.90	7.70	88.20	31.60	22.90	2.30	9
8	Delaware	684.00	3.20	9.90	86.10	24.00	25.60	4.00	8
9	District of Columbia	1508.00	41.80	17.40	83.20	38.30	44.70	5.80	10
10	Florida	812.00	5.60	12.00	84.00	22.80	26.50	3.60	9
11	Georgia	505.00	8.00	12.50	82.60	23.10	25.50	3.70	6
12	Hawaii	244.00	2.90	10.60	87.40	26.30	19.10	4.30	7

Make the scatter plot of murder rate vs. college again and add the regression line.

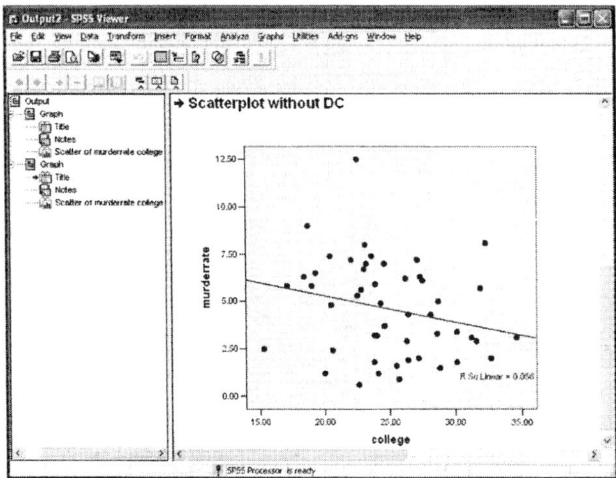

Instead of a positive slope, as in the plot with DC, this graph shows a regression line with a negative slope. The value of R-Square is also changed.

Chapter 4 Sampling

Most of the time that you will be using SPSS you will probably be doing data analyses similar to the examples given in the previous chapters of this manual. Occasionally, though, you might need to generate a random sample from a data set or from a list of numbers that represent the individuals in a population. These tasks are easy to do without SPSS, of course, but if you don't have access to a random number table, SPSS has a command for selecting samples that will do the job for you!

Random Samples

If you want to select a random sample from within a data set, you can do this very easily using SPSS. You can also use this feature to generate a sequence of random numbers. First we will demonstrate how to select a random sample from an existing data set. Access the "fla_student_survey" data set from your text CD. Suppose we want to select a random sample of 5 students from the 60 students in this data set. Select **Data → Select Cases...** and click on "Random sample of cases."

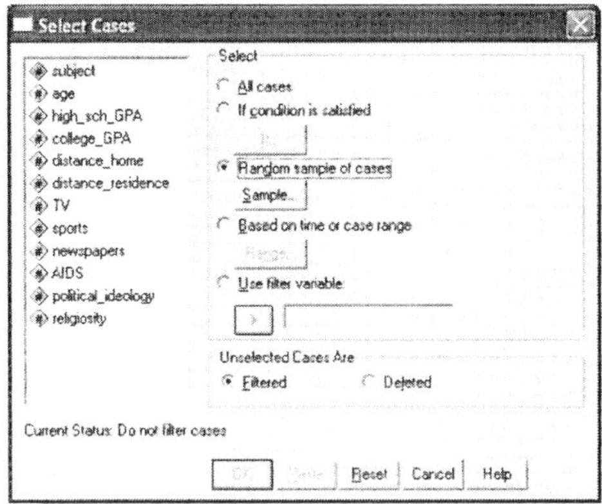

Click on the "Sample" button and in the dialog box that appears click on "Exactly" and enter "5" and "60" as in the example dialog box below (so that it reads "Exactly 5 cases from the first 60 cases").

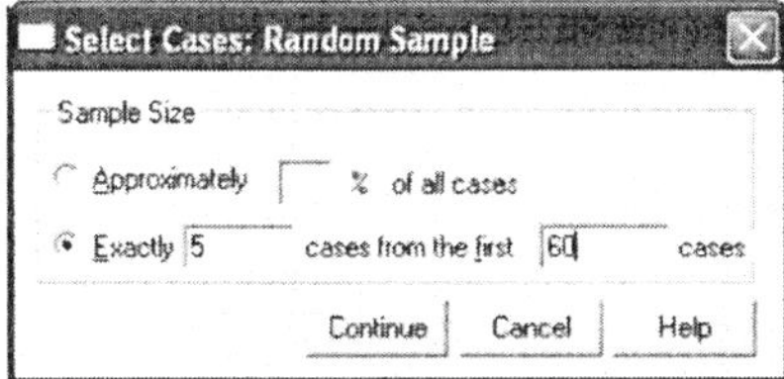

Click on **Continue** and then on **OK**. You will see that most of the case numbers now have a diagonal slash indicating that they are not included in the random sample.

fla_student_survey.sav - SPSS Data Editor

	subject	gender	age	high_sch_GPA	college_GPA	distance_home	distance_residence	TV	sports	newspapers
1	1.0	m	32.0	2.20	3.50	.00	5.00	3.00	5.0	.00
2	2.0	f	23.0	2.10	3.50	1200.00	.30	15.00	7.0	6.00
3	3.0	f	27.0	3.30	3.00	1300.00	1.50	.00	4.0	3.00
4	4.0	f	35.0	3.50	3.20	1500.00	8.00	5.00	5.0	6.00
5	5.0	m	23.0	3.10	3.50	1600.00	10.00	6.00	6.0	3.00
6	6.0	m	39.0	3.50	3.50	350.00	3.00	4.00	5.0	7.00
7	7.0	m	24.0	3.60	3.70	.00	.20	5.00	12.0	4.00
8	8.0	f	31.0	3.00	3.00	5000.00	1.50	5.00	3.0	3.00
9	9.0	m	34.0	3.00	3.00	5000.00	2.00	7.00	5.0	3.00
10	10.0	m	28.0	4.00	3.10	900.00	2.00	1.00	1.0	2.00
11	11.0	m	23.0	2.30	2.60	253.00	1.50	10.00	15.0	1.00

Data View / Variable View — SPSS Processor is ready — Filter On

In the example above we can see that subject #9 was selected for the sample. To see which cases were selected you can scroll through the data set or print a list using the "Case Summaries" command. Select **Analyze → Reports → Case Summaries...** and enter the variables you want to review for your sample. In the example below, the variables "subject," "age," and "TV" have been selected.

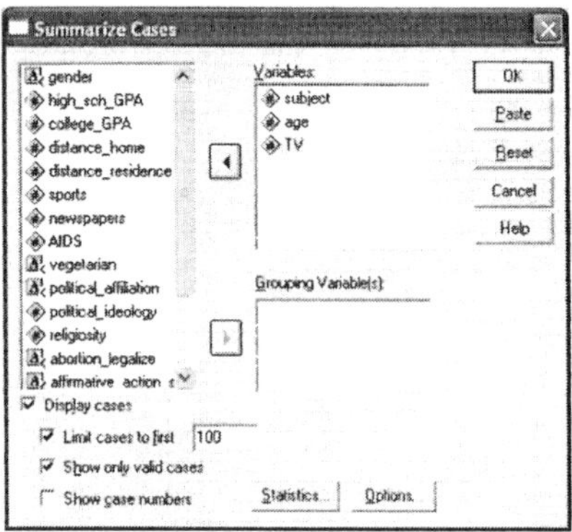

Click on **OK** to produce the output. In the output that is produced, we can see that observations 9, 12, 20, 29 and 59 were selected. If you use the "Select Cases" command again, you will get a different random sample each time.

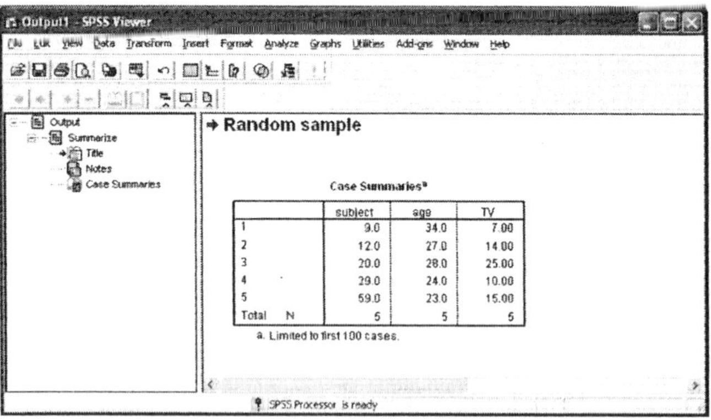

Random sample

Case Summaries[a]

		subject	age	TV
1		9.0	34.0	7.00
2		12.0	27.0	14.00
3		20.0	28.0	25.00
4		29.0	24.0	10.00
5		59.0	23.0	15.00
Total	N	5	5	5

a. Limited to first 100 cases.

Generating a Sequence of Random Numbers

If you are not working with an existing SPSS data file and you would like to generate a sequence of random numbers to identify a random sample, as in Example 5 of Chapter 4 of your text, SPSS can generate the numbers for you. First, you will have to enter the numbers corresponding to the items you will be sampling from. In Example 5, the items are 60 account numbers. Enter the numbers 1 through 60 in the first column of the Data Editor. You can name the variable "account" if you want to, but it is not necessary.

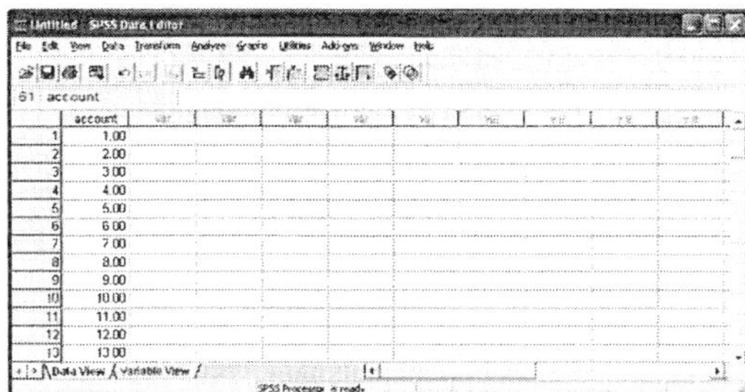

Select a random sample as above. Select **Data → Select Cases...** then click on "Random sample of cases." Click on "Sample" and enter 10 for the number of cases to select and 60 for the total number of cases. Click on "Continue" and then on "OK.'

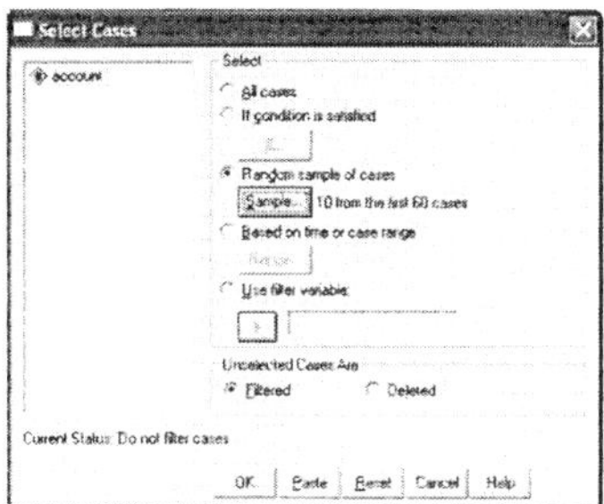

Scroll through the list to see which numbers were selected or use the "Case Summaries" command as above.

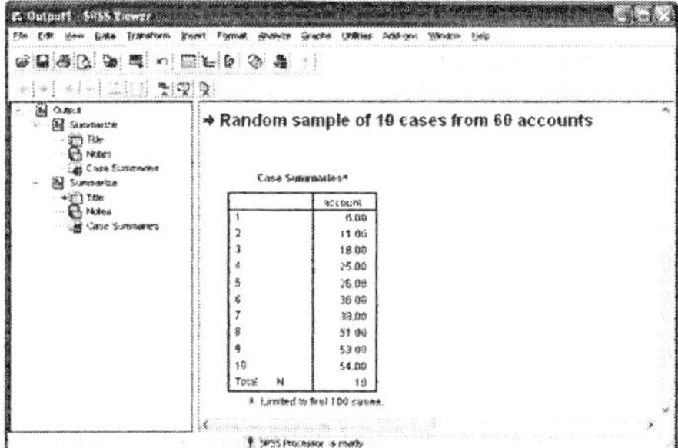

For this example, accounts 6, 11, 18, 25, 26, 36, 39, 51, 53 and 54 were selected for the random sample.

Chapter 5 Probability Calculations

Your main goal for learning to use SPSS is to learn which commands produce the output you need, but you can also use SPSS to do probability calculations. It is fairly easy to find probabilities using tables, but it is nice to know that you can use SPSS to do this when you don't have access to the right table!

The binomial and normal distributions are among the discrete and continuous probability distributions available in SPSS. To find a probability, you will need a Data Editor with at least one variable and at least one observation already in it – the commands for calculating probabilities (or for producing any output) won't work in an empty Data Editor. Open a new Data Editor in SPSS and enter 0 (or any other value) in the first row of the first column, just so there is at least one observation entered. SPSS will name the variable "VAR00001" but there is no need to rename this variable because you are not going to use it for anything.

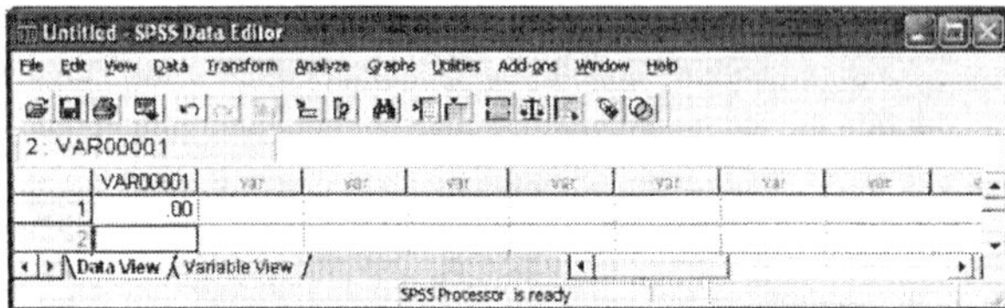

Binomial Probability Calculations

To find the binomial probability of no women in 10 employees chosen for managerial training, as in Example 12 of Chapter 6 in your text, select **Transform → Compute...** and enter "prob1" (or some other variable name) in the "Target Variable" box. Click on the box under "Numeric Expression" and scroll through the list under "Function Group" to find "PDF & Noncentral PDF." (PDF stands for "Probability Density Function.") Click on this to reveal the list of available probability distributions under "Functions and Special Variables. Scroll through the list under "Functions and Special Variables" to find "PDF.Binom" and double click on it.

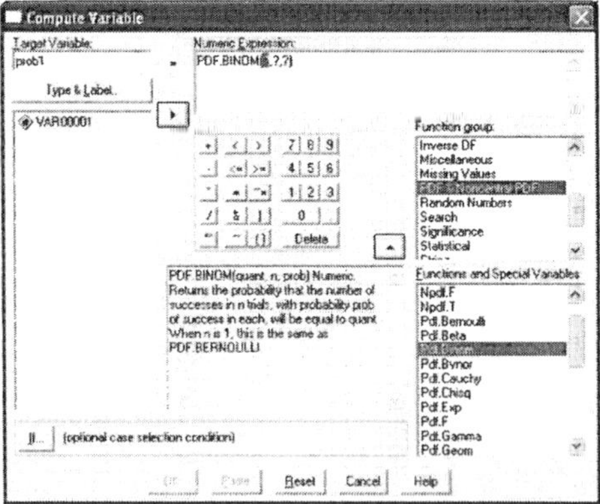

The expression "PDF. Binom(?,?,?) will appear in the "Numeric Expression" box. The first question mark is for the value that you want to calculate the probability for, in this case 0. The other two question marks are for the sample size (n) and the probability of a success (p), respectively. We want the probability of no (0) successes in a sample of n=10 when the probability of a success is p=0.50. This value of p represents the case when there are an equal proportion of men and women from the population from which the sample is taken. Edit the three question marks by clicking on the expression in the "Numeric Expression" box, then erasing the three questions marks and typing in "0,10,.5" between the parentheses.

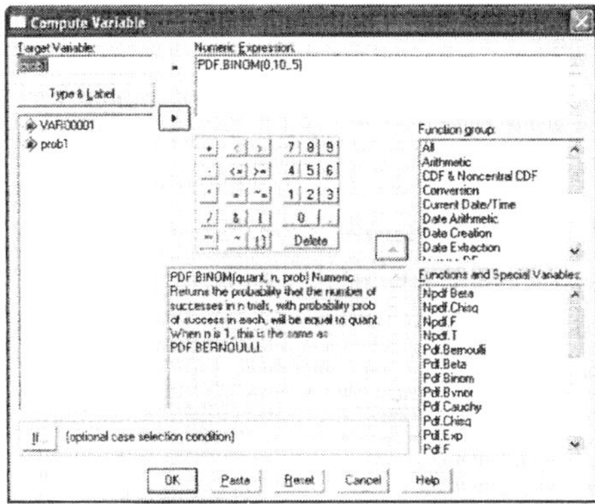

Click on **OK** to complete the calculation. Notice that no output is produced from this command. Instead, the probability will appear in the Data Editor in a new column with variable name "prob1." You will have to increase the number of decimal places in the Variable View to see the probability value.

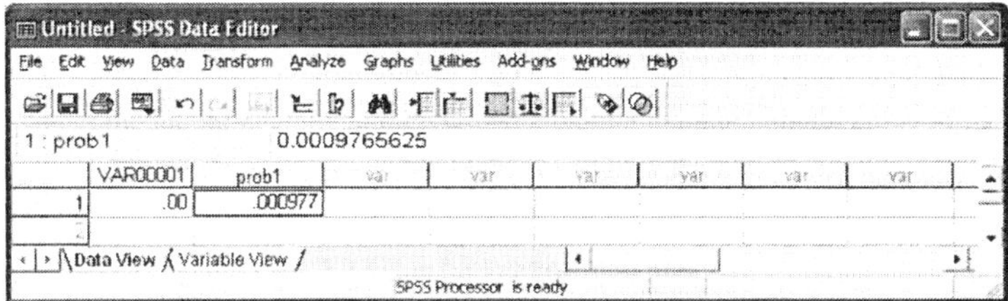

So, the probability of no women selected in a sample of 10 when the proportion of women is .5 is .000977, or approximately .001. To find the entire probability distribution for n=10 and p=0.50, just repeat this process and change the first value entered to the desired value. For example, to find the probability of selecting 3 women in 10 trials when the proportion of women is .50 the "Numeric Expression" would be "PDF.Binom(3,10,.5)." SPSS will replace the first probability calculation with this new result unless you enter a new variable name, such as prob2 in the "Target Variable" box.

Normal Probability Calculations

The process for calculating a probability for a normal distribution is similar to that for finding a binomial probability except you select "CDF & Noncentral CDF" under "Function group" instead of the PDF option. (CDF represents "Cumulative Distribution Function.") Select **Transform → Compute...** and scroll through the list of available distributions under "Functions and Special Variables to find "Cdf.Normal" and double click on it.

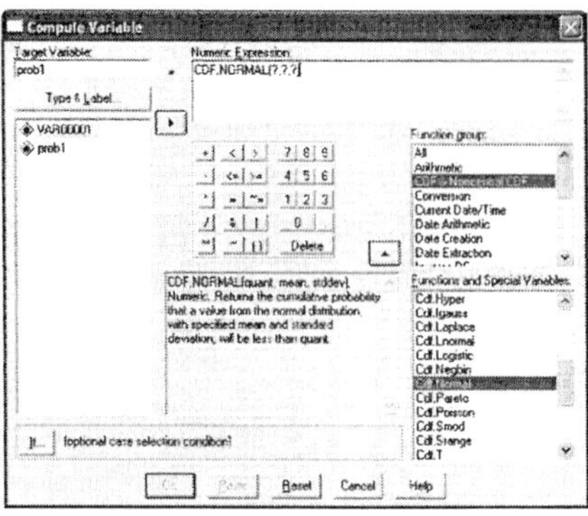

The expression "CDF.Normal(?,?,?)" will appear in the "Numeric Expression" box. Here the three question marks represent a value from a normal distribution that you want to find the cumulative probability for, and the mean and standard distribution of that distribution, respectively. SPSS will report the probability of observing a normal random variable at or below the value you enter. Let's use Example 8 of Chapter 6 in your text to demonstrate how to find normal probabilities. In this example, you are asked to find the probability of an SAT score higher than 650 when the mean SAT score is 500 and the standard deviation is 100. Replace the three question marks, as above, with "650,500,100" and click on **OK**. If you are using the same Data Editor and same variable name that you used to calculate the binomial probability above, then SPSS will ask you if you want to change the existing variable.

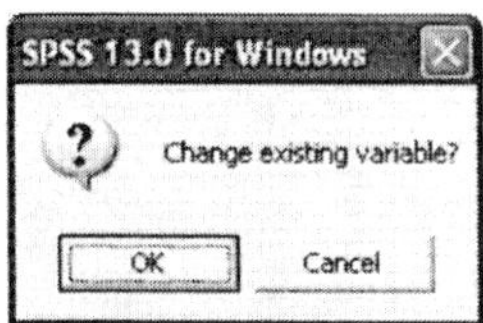

Click on **OK**. The probability shown in the Data Editor is .933193, but this is the probability of an SAT score *below* 650, not above. You can find the probability of an SAT score larger than 650 by subtracting .933193 from 1, of course, or you could enter the expression "1 – CDF.Normal(650,500,100)" in the "Numeric Expression" box and let SPSS do the calculation for you.

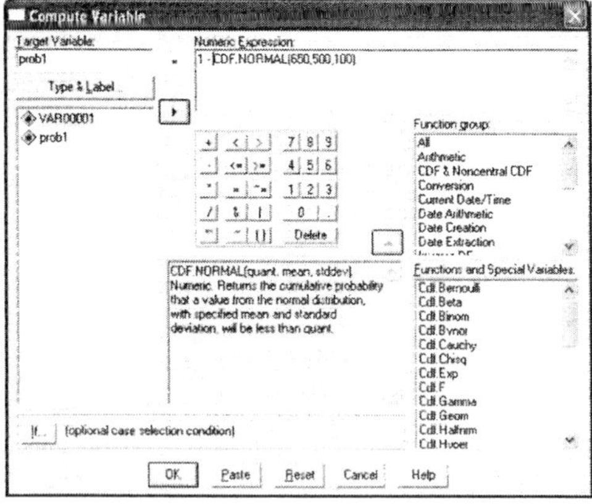

Finding a Standard Normal P-value

The function for finding normal probabilities may be useful for you later on when you need to find something called a "p-value" for a standard normal Z-score called a "test statistic." A p-value is a probability associated with an observed Z-score. For these calculations, the mean and standard deviation are 0 and 1, respectively, because the corresponding distribution is standard normal. For example, suppose we have an observed Z-score of 1.97 and the p-value we need corresponds to the probability of getting a value larger than 1.97. (Don't worry – the reason for needing this probability will be explained in your statistics course. Here you will just see how to use SPSS to find the probability.) Select **Transform → Compute...**, enter "pvalue" as the "Target Variable" and enter the expression "1-CDF.Normal(1.97,0,1)" in the "Numeric Expression" box.

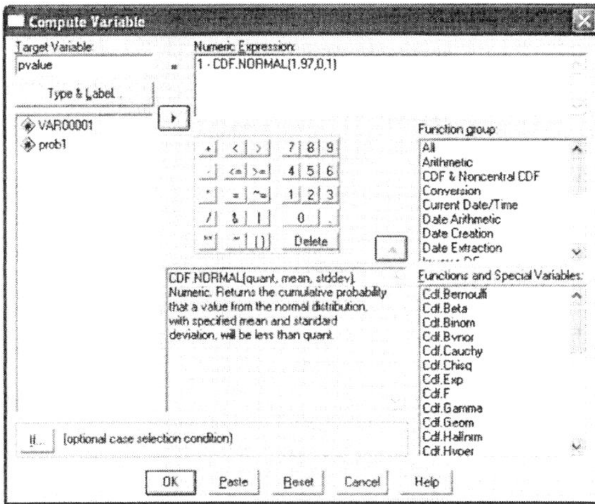

Click on **OK** to find the p-value.

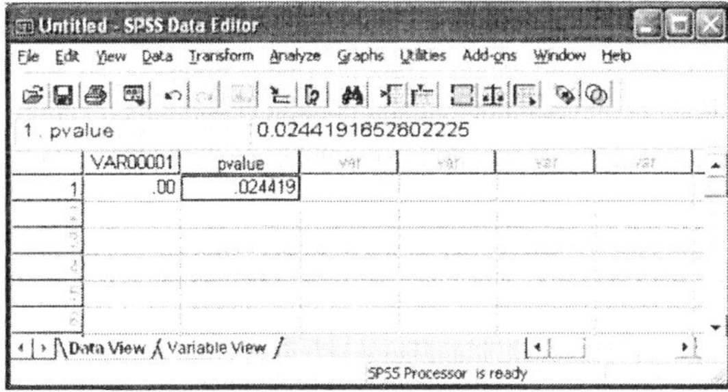

The p-value is .024419.

If you scroll through the list of Cdf functions available you will see some additional distribution functions that you may need to use to find p-values later on in your statistics course. These include the chi-square distribution (Cdf.chisq), the F distribution (Cdf.F), and the T distribution (Cdf.T).

Inverse Distribution Functions

In some situations you may need to find the z-score value for the standard normal distribution that has a specified proportion of the distribution below it in value. For example, to construct a 95% confidence interval for a proportion, as in Example 2 in Chapter 7, you need the z-score that has 97.5% of the distribution below it in value. The easiest way to find such a z-score is to use a standard normal table. But, in a situation where you don't have access to a table you can use SPSS to find the z-score you need. Select **Transform → Compute...** and enter "zvalue" (or some other variable name) in the "Target Variable" box. Scroll through the list under "Function group" to find "Inverse DF" and click on it. Scroll through the list under "Functions and Special Variables" to find "Idf.Normal" and double click on it. The expression "IDF.NORMAL(?,?,?)" will appear in the "Numeric Expression" box. This command works similarly to the normal distribution Cdf function above in that the last two question marks are for the mean and standard deviation of a normal distribution, respectively. The first question mark is for the appropriate probability value. If you want to find the zscore for a 95% confidence interval, replace the three question marks with ".975,0,1". You have to enter .975 instead of .95 because the "Inverse DF" is the inverse of the cumulative distribution function. So, to find the middle 95% you need the value that has 97.5% below it.

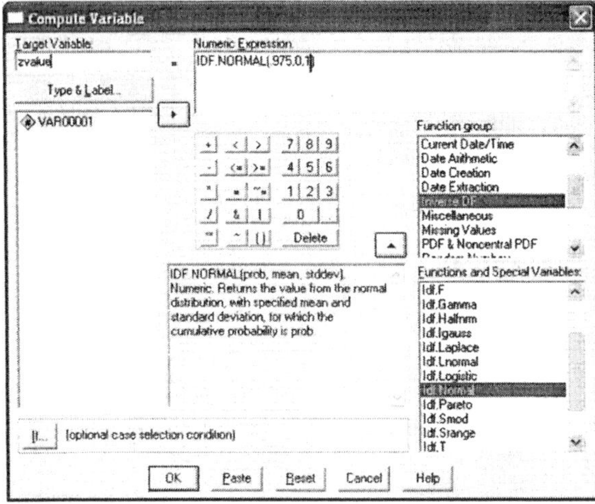

Click on **OK** to complete the calculation. Just like the Pdf and Cdf functions, no output will be produced. Instead the zscore value will appear as the value of a new variable created by the "Transform" command.

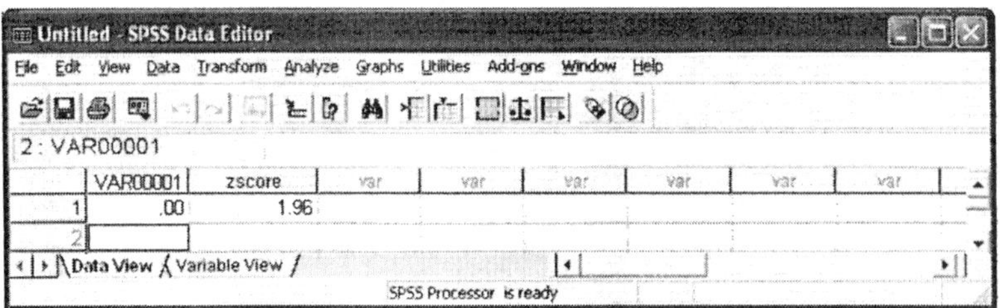

Simulating a Sampling Distribution

SPSS does not have a built-in command for simulating a sampling distribution but the process for doing this is fairly easy and takes only a few steps. Let's demonstrate this using the data for Exercise 6.124, the ages of all 50 heads of households in a small fishing village in Nova Scotia. This data is provided below.

50	45	23	28	67	62	41	68	37	60	41	70	47	66	51	57	40	36	38	81	27	37	56	71	39
46	49	30	28	31	45	43	43	54	62	67	48	32	42	33	36	25	29	57	39	50	64	76	63	29

Enter the ages in SPSS, then select **Data → Select Cases...** and click on "Random sample of Cases." Click on the "Sample" button and fill in the dialog box so that SPSS will select 9 of the 50 cases in the data set. Click on **Continue** and then on **OK**.

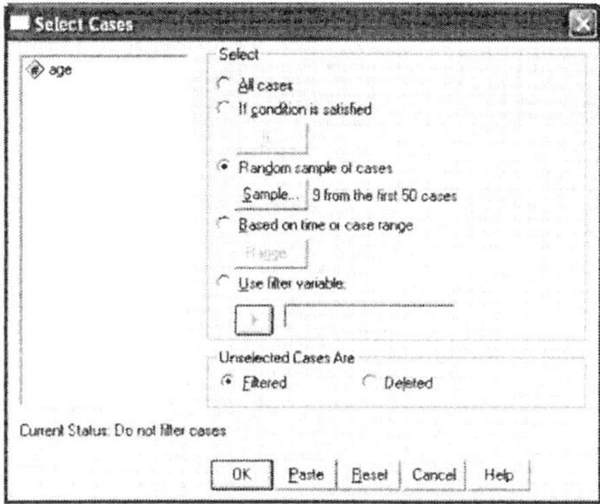

The next step is to create a new variable that has only the values of "age" for the 9 selected households for the first sample. To do this, select **Transform → Compute...** and enter "sample1" in the "Target Variable" box. Enter the expression "age" in the "Numeric Expression" box, then click on the "If..." button in the lower left corner of the dialog box. Click on "Include if case satisfies condition" and enter "filter_$>0" in the box.

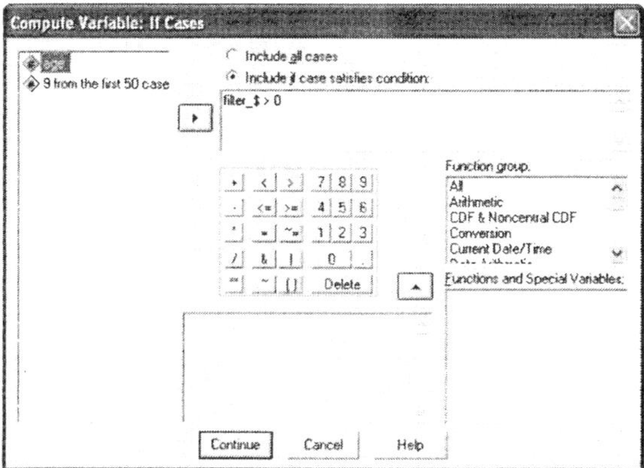

Click on **Continue**. Your dialog box should be filled in as the one below.

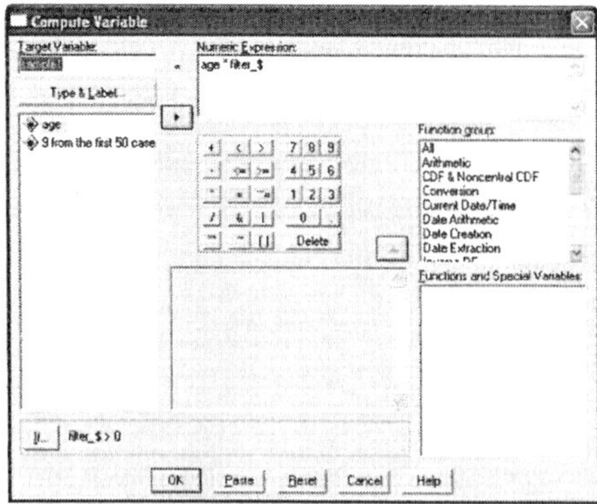

Click on **OK** to create the sample.

nova scotia ages.sav - SPSS Data Editor

1 : filter_$ 0

	age	filter_$	sample1
1	50	0	.
2	45	0	.
3	23	1	23.00
4	28	0	.
5	67	0	.
6	62	0	.
7	41	0	.
8	68	0	.
9	37	0	.
10	60	0	.
11	41	0	.

For this sample, the ages 23, 40, 36, 46, 45, 32, 39, 50, and 29 were selected. To create a second random sample, select **Data → Select Cases...** again and just click on **OK** since the dialog box is already set for selecting a new sample of 9 cases. Now select **Transform → Compute...** again and change the "Target Variable" to "sample2" and click on **OK**.

nova scotia ages.sav - SPSS Data Editor

1 : filter_$ 0

	age	filter_$	sample1	sample2
1	50	0	.	.
2	45	0	.	.
3	23	0	23.00	.
4	28	0	.	.
5	67	0	.	.
6	62	1	.	62.00
7	41	0	.	.
8	68	0	.	.
9	37	1	.	37.00
10	60	0	.	.
11	41	1	.	41.00

For this second sample, ages 62, 37, 41, 27, 56, 71, 39, 30, and 31 were selected. Repeat this process, changing the target variable name each time, for as many samples as you wish to create. Once you have the number of samples you need, you can find the mean of each sample by using one of the options under the "Descriptive Statistics" command. But, first you will need to "turn off" the "Select Cases" option under the Data menu. Select **Data → Select Cases...**, click on "All cases" under "Select," and then on **OK**. The following display shows the means for twenty samples of size nine chosen from this distribution.

Descriptive Statistics for 20 random samples of size 9

Descriptive Statistics

	N	Minimum	Maximum	Mean	Std. Deviation
sample1	9	23.00	50.00	37.7778	9.71461
sample2	9	27.00	71.00	43.7778	15.54953
sample3	9	28.00	50.00	38.8889	6.43126
sample4	9	23.00	68.00	43.6667	14.69843
sample5	9	23.00	66.00	47.5556	14.54799
sample6	9	31.00	81.00	48.8889	15.43506
sample7	9	33.00	68.00	52.5556	13.27806
sample8	9	23.00	81.00	51.1111	17.86388
sample9	9	25.00	68.00	45.6667	14.83240
sample10	9	25.00	85.00	45.0000	18.00000
sample11	9	37.00	81.00	59.4444	15.45744
sample12	9	27.00	81.00	55.7778	18.13671
sample13	9	28.00	76.00	48.8889	16.67566
sample14	9	29.00	66.00	48.3333	13.55544
sample15	9	29.00	67.00	50.7778	13.54417
sample16	9	30.00	70.00	40.7778	11.78624
sample17	9	28.00	66.00	44.6667	14.33527
sample18	9	25.00	86.00	49.5556	14.85613
sample19	9	29.00	67.00	44.4444	12.90456
sample20	9	27.00	64.00	45.5556	13.38143
Valid N (listwise)	0				

You can now enter these means in the same data file under a new variable named "mean."

	sample15	sample16	sample17	sample18	sample19	sample20	mean
1	50.00						37.78
2					45.00		43.78
3						23.00	38.89
4				28.00			43.67
5	67.00			67.00			47.56
6				62.00			48.89
7		41.00		41.00	41.00		52.56
8				68.00			51.11
9		37.00				37.00	45.67
10							45.00
11		41.00	41.00		41.00	41.00	59.44
12		70.00					55.78
13				47.00			48.89
14	66.00		66.00				48.33
15							50.78
16	57.00		57.00	57.00			40.78
17			40.00			40.00	44.67
18		36.00					49.56
19	38.00				38.00		44.44
20							45.56

To investigate the sampling distribution, find the mean and standard deviation of the variable "mean" and produce a histogram, as shown below.

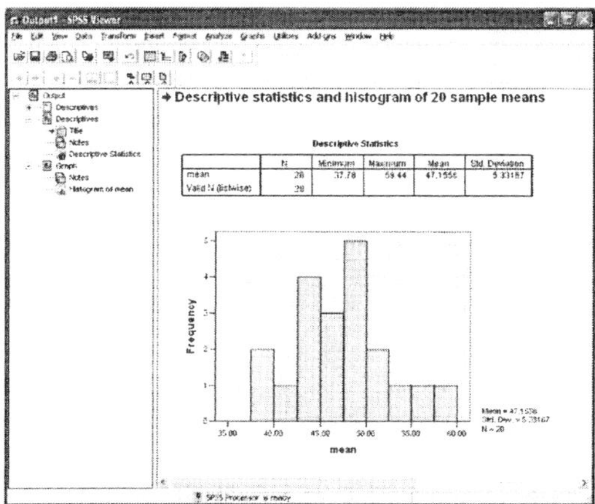

	N	Minimum	Maximum	Mean	Std. Deviation
mean	20	37.78	59.44	47.1556	5.33167
Valid N (listwise)	20				

For these twenty samples, the mean of the sample means is 47.1556, which is very close to the population mean of 47.18. The standard deviation of the sample means is 5.33167, which is a little more than the expected value of 14.74/3, or 4.9133. The histogram also shows a relatively mounded and symmetric shape, as expected.

Chapter 6 Confidence Intervals

In this chapter we will demonstrate SPSS commands for producing a confidence interval for a mean. SPSS does not have a command for producing a confidence interval for a proportion. You can use SPSS to find the appropriate critical z-score for the level of confidence you are using if you don't have access to a normal distribution table. This process is described in the previous chapter in the section on Inverse Distribution Functions.

Confidence Interval for a Mean

Two of the commands under the "Analyze" menu will provide you with the confidence interval for a mean, "Explore" under the "Descriptive Statistics" command and "One-sample T Test" under the "Compare Means" command. We will demonstrate the "Explore" command in this section using the eBay auction data from Example 7 in Chapter 7, and then see how the "One-Sample T Test" command operates in the next chapter. The data is shown below and is the final prices (in dollars) of an item sold using two methods, "buy-it-now" and "bidding only."

Buy-it-now	235, 225, 225, 240, 250, 250, 210
Bidding only	250, 249, 255, 200, 199, 240, 228, 255, 232, 246, 210, 178, 246, 240, 245, 225, 246, 225

Data Entry Options

You have two options for entering the data in SPSS. One way would be to enter the data in two separate columns, one for each method used to sell the item. The data would appear as below.

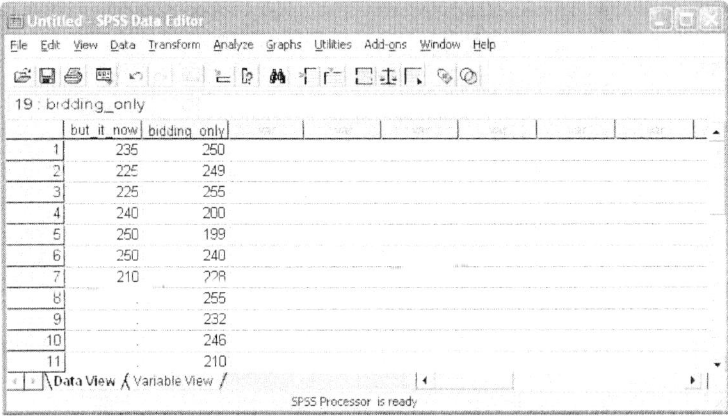

The other option would be to enter all of the selling prices in one variable and then to use a second variable to identify which selling method was used. The first variable would be a numeric variable, named something like "final_price," and the second one could be either numeric or string and named "selling_method," for example. Entered using this second method the data would appear as below.

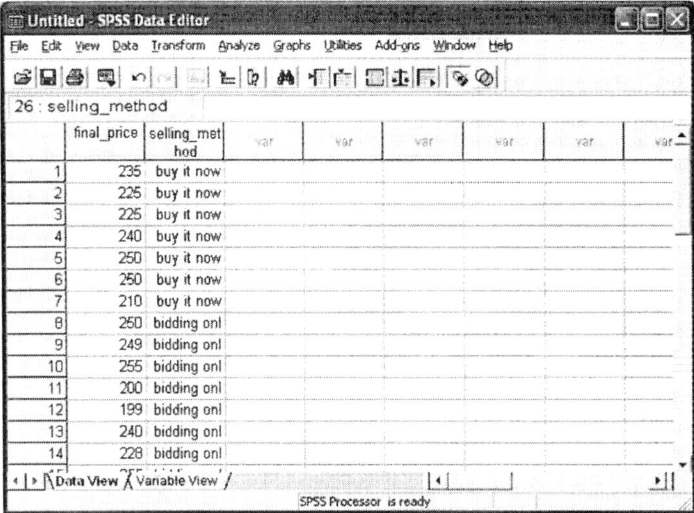

For finding the confidence interval for the mean of each sample you can use either data entry method. But, if you want to use a method for comparing the means of two or more independent samples, you will need to use the second data entry method. And, if you want to use a method for comparing the means of two or more dependent samples, you will need to use the first data entry method. Note that the data for this example is available on your text CD entered using the first data entry method in the file "ebay_auctions."

To find the confidence interval for a mean using the first data entry method, select **Analyze → Descriptive Statistics → Explore...** and enter both variable names in the box under "Dependent List." If you do not want any graphs, click on "Statistics" under "Display."

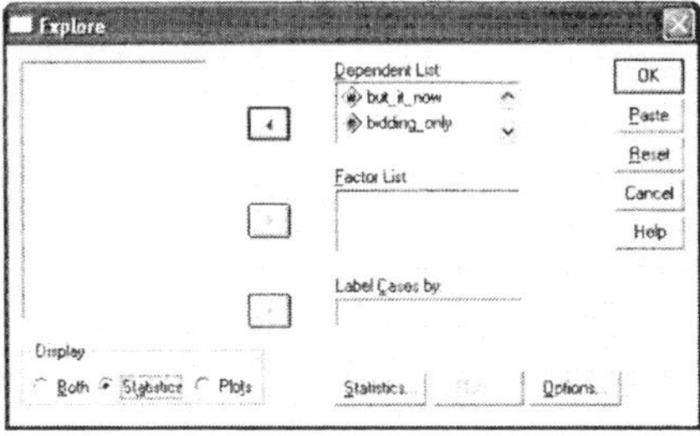

Click on **OK** to produce the output. Some of the statistics produced by SPSS have been deleted from the output display below.

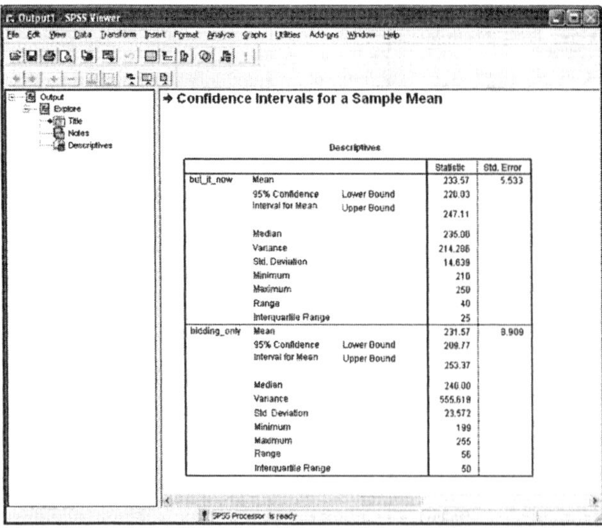

The confidence interval for the “buy it now” method is (220.03, 247.11) and for the “bidding only” method it is (209.77, 253.37).

To find confidence intervals for a mean using the second data entry method select **Analyze → Descriptive Statistics → Explore…** and enter the variable with the prices in the “Dependent List” box and the variable with the selling method in the “Factor List” box. Click on **OK**. Except for some labels, the output will appear exactly the same as the output produced above.

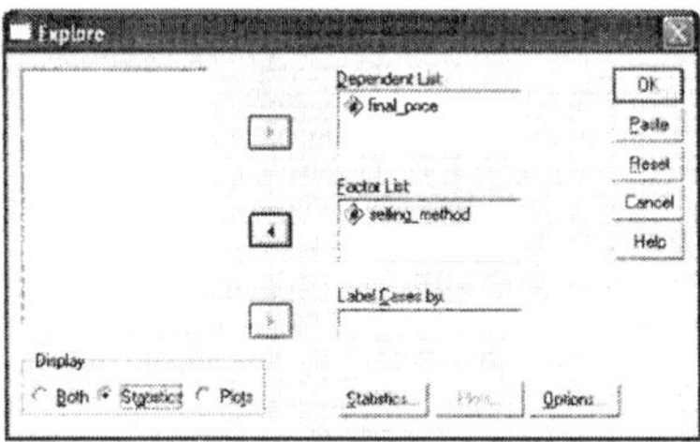

Chapter 7 Significance Tests about Hypotheses

In this section you will be introduced to a number of SPSS commands that you can use to conduct a hypothesis test.

Hypothesis Test for a Proportion

SPSS has two commands that you can use to do a test on the proportion of "successes" in a population. The first test is called the "Binomial Test" and it will give you approximately the same results as the Z-test in your text. The second test is a Chi-square "Goodness-of-fit" test; although the results for this test are equivalent to the Z-test for large samples, the process for doing this test is a little more complicated than it is for the Binomial Test. Both of these tests are available under the "Nonparametric Tests" command under the "Analyze" menu and both tests will be demonstrated below. You will learn about nonparametric tests in the last chapter in your text and in this manual. Remember, though, that the Z-test for a proportion is easy to calculate by hand and you can use the method described in Chapter 5 of this manual to find the corresponding standard normal p-value.

The Binomial Test

Let's demonstrate this test using the data from Exercise 4 in Chapter 8. In this example, dogs were trained to identify patients with bladder cancer, so *p* is the proportion of dogs that make the correct selection on a given trial. In this study, the dogs made the correct selection 22 times in 54 trials. Enter this data in SPSS using two variables, one to indicate the result of each trial ("correct" = 1 or "incorrect" = 0) and one to indicate the count for each result. The "Result" variable must be entered as a coded numeric variable and you must code the response that corresponds to the proportion of interest with the value 1. Don't forget to select **Data → Weight Cases...** and enter "count" as the weighting variable.

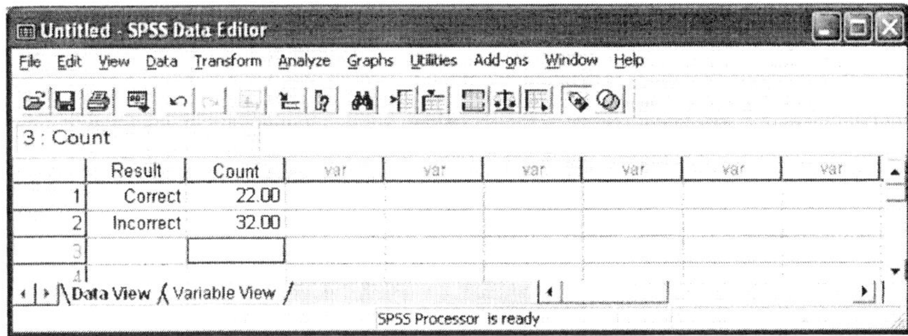

The null hypothesis is that the proportion of correct results is 1/7 and the alternate hypothesis is that it is different from 1/7. To conduct the test, select **Analyze → Nonparametric Tests → Binomial...** Enter "Result" in the "Test Variable List" and .143 (=1/7) in the "Test Proportion" box.

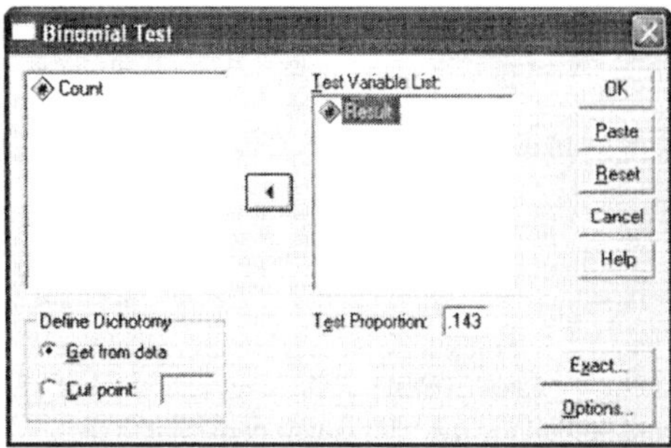

Click on **OK** to produce the output.

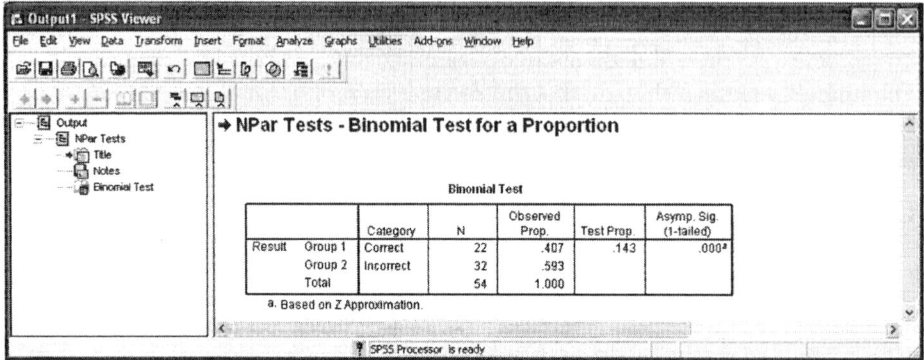

Binomial Test

		Category	N	Observed Prop.	Test Prop.	Asymp. Sig. (1-tailed)
Result	Group 1	Correct	22	.407	.143	.000[a]
	Group 2	Incorrect	32	.593		
	Total		54	1.000		

a. Based on Z Approximation.

SPSS reports the observed proportion of each result (correct and incorrect) as well as the test (null) proportion for the correct response. SPSS also reports the p-value for the two-sided alternative hypothesis, but not the actual value of the test statistic. The p-value isn't .000, however. SPSS only reports p-values to three decimal places, so we can tell that the p-value is less than .0005. With such a small p-value (less than .0005) you can't tell that the results for this Binomial Test differ slightly from the results for the Z-test. You will be able to see this, though, with the next example.

For Example 6 in Chapter 8, p is the proportion of correct predictions made by therapeutic touch practitioners. In this example the alternative hypothesis is that p is more than $p_0 = .50$. The sample results were that 70 out of 150 therapeutic touch practitioners correctly identified which of their hands was closer to the hand of a researcher. Enter this data in SPSS using the same format as above. (If you have just done the previous example, all you will have to do is replace the counts with the appropriate counts for this example.)

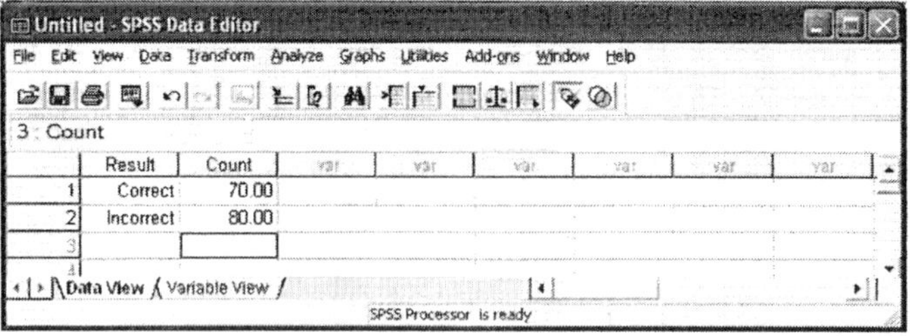

	Result	Count
1	Correct	70.00
2	Incorrect	80.00

To conduct the test, select **Analyze → Nonparametric Tests → Binomial...** Enter "Result" in the "Test Variable List" and 0.50 in the "Test Proportion" box.

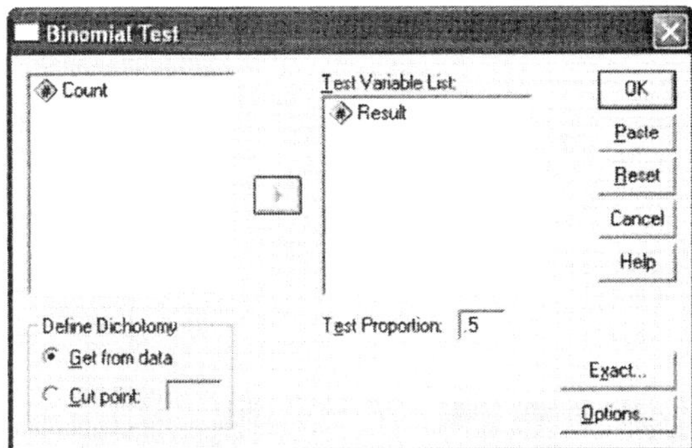

Click on **OK** to complete the test.

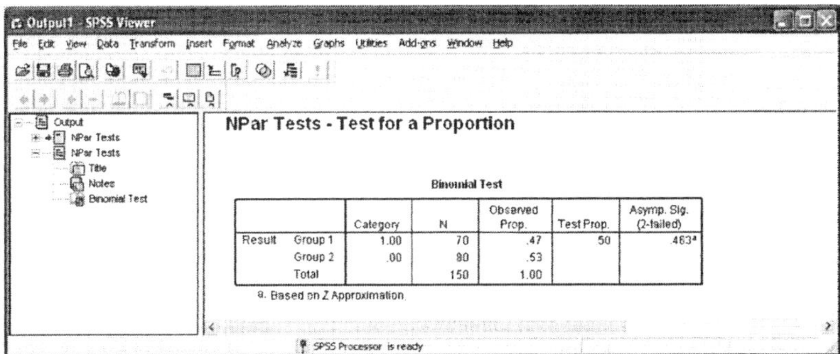

		Category	N	Observed Prop.	Test Prop.	Asymp. Sig. (2-tailed)
Result	Group 1	1.00	70	.47	.50	.463[a]
	Group 2	.00	80	.53		
	Total		150	1.00		

a. Based on Z Approximation

Since we are using the "greater than" alternate hypothesis and the observed proportion is less than the hypothesize proportion, the p-value for the Binomial Test is 1-0.463/2 = 0.7685, which is a little less than the p-value of 0.79 reported in your text. For most situations, the results for the Binomial Test and the Z-test will agree; that is, if you reject the null hypothesis using one test then you will also reject it using the other test.

Chi-Square Goodness-of-Fit Test

This test will compare the observed frequencies of observations in two or more categories to the expected frequencies assuming a specified distribution. When there are just two categories, the chi-square test statistic is equal to the square of the Z-statistic for this situation. Although calculated using a chi-square distribution instead of the standard normal distribution, the p-value reported by SPSS for this test turns out to be the same as the p-value for the Z-statistic for the two-sided alternative hypothesis. Your instructor may not want you to use this chi-square test, so it is best that you check first before learning this alternate method.

Let's try this approach using both of the examples we used above for the Binomial Test. Recall that for Example 4 there were 22 correct responses and 32 incorrect responses.

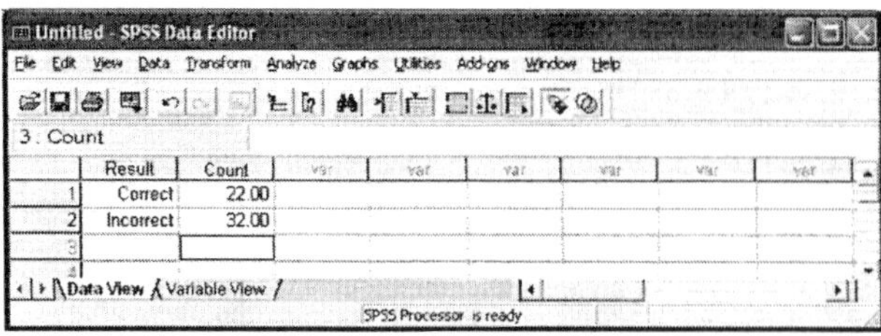

To conduct the test, select **Analyze → Nonparametric Tests → Chi-square...** Enter "Result" in the "Test Variable List" and then click on "Values" below "Expected Values." This test is based on comparing the actual counts to the expected counts, instead of comparing the sample proportion to the hypothesized proportion. To do this test, you have to tell SPSS what the expected frequencies would be for the two groups (correct and incorrect) under the null hypothesis. The hypothesized value for p is 1/7 and there were 54 trials, so we would expect 54/7 or 7.71 correct trials and 54 - 7.71 = 46.29 incorrect trials. Click on the box next to "Values" and enter "7.71" and click on "Add." Then click on that box again, enter "46.29" and click on "Add" again. We had to round these values because SPSS only allows a limited number of decimal places in the "Expected Values" box. At this point, the "OK" button should be highlighted.

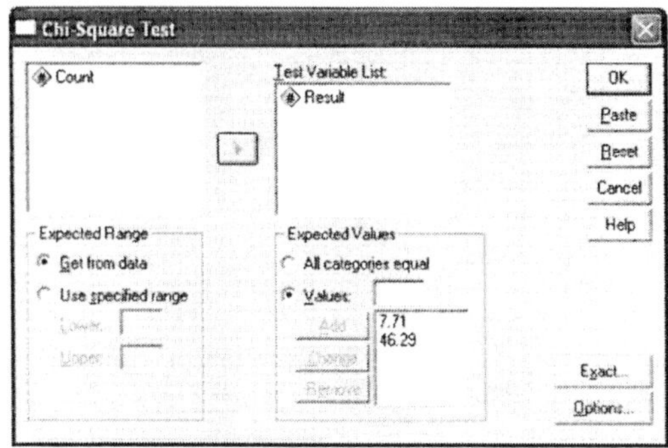

Click on **OK** to complete the command.

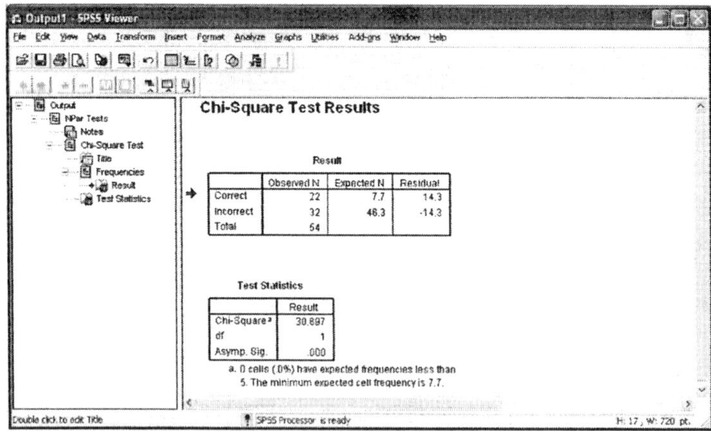

The value of the test statistic is 30.897. The Z test statistic in your text is calculated as $Z = \dfrac{.407 - .143}{\sqrt{\dfrac{.143(1-.143)}{54}}} = 5.555$, where $\hat{p} = \dfrac{22}{54} = .407$ and $p_0 = \dfrac{1}{7} = .143$. You can check that the square of 5.555 is roughly 30.897, the value of the Chi-Square test statistic reported by SPSS. The p-value for the test is the value in the output next to "Asymp. Sig." and is less than .0005.

Now let's try this with Example 6 where there were 70 correct responses and 80 incorrect, and the alternative hypothesis is one-sided. Enter the data in SPSS as shown below.

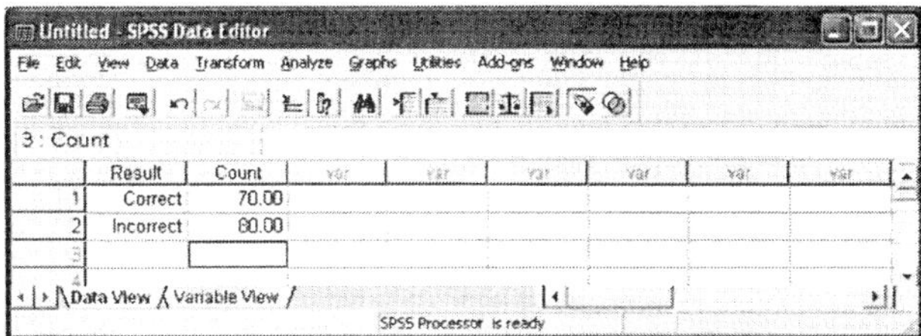

Select **Analyze → Nonparametric Tests → Chi-square...** Enter "Result" in the "Test Variable List" and then click on "All categories equal."

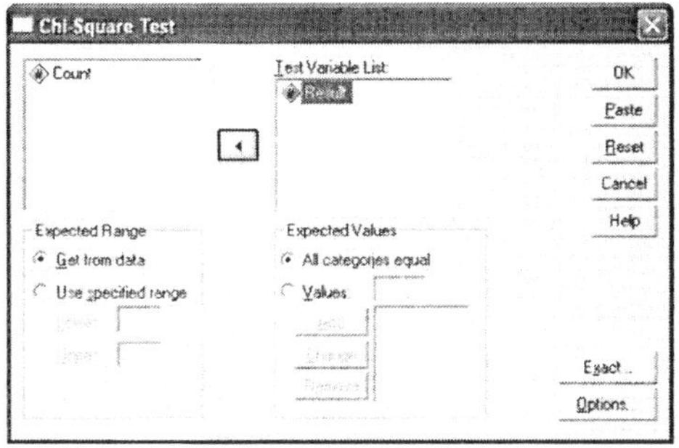

Click on **OK** to complete the command.

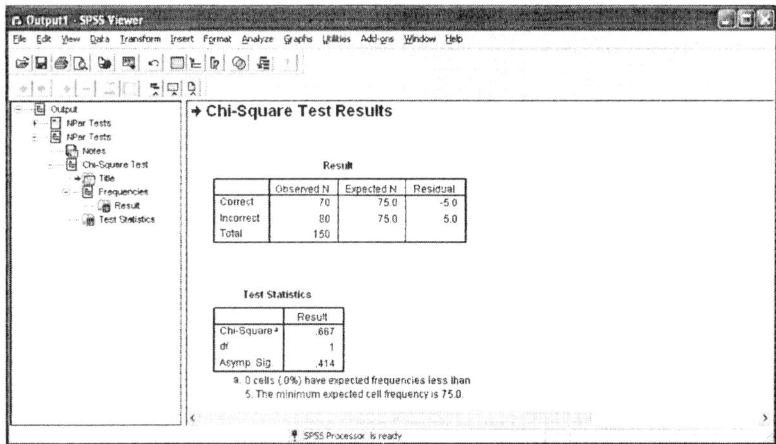

Result

	Observed N	Expected N	Residual
Correct	70	75.0	-5.0
Incorrect	80	75.0	5.0
Total	150		

Test Statistics

	Result
Chi-Square[a]	.667
df	1
Asymp. Sig.	.414

a. 0 cells (.0%) have expected frequencies less than 5. The minimum expected cell frequency is 75.0.

Since we know that Z = -0.82, we expect that the chi-square statistic will be $(-0.82)^2 = 0.67$. The chi-square statistic reported by SPSS is .667 and the p-value is .414. This p-value represents two times the "tail area" above 0.817. Since the sample proportion is less than the expected proportion we know that the test statistic is -0.817 and the p-value should be 1 - .414/2, or .793. These results agree more closely with the results of the Z-test; however, you have to convert the sample and null proportions to sample and null (expected) counts and, although the p-values are the same, the value of the chi-square test statistic is the square of the Z-statistic, which might be a little confusing for the beginning statistician!

One-sample Test for a Mean

To demonstrate how to conduct a test for a mean, we will use the anorexia data set in Example 7 of Chapter 8. This data set consists of the weights of 29 anorexia patients both before and after a treatment program, along with their change in weights. This data is shown in the table below.

	Weight				Weight				Weight		
Girl	Before	After	Change	Girl	Before	After	Change	Girl	Before	After	Change
1	80.5	82.2	1.7	11	85.0	96.7	11.7	21	83.0	81.6	-1.4
2	84.9	85.6	0.7	12	89.2	95.3	6.1	22	76.5	75.7	-0.8
3	81.5	81.4	-0.1	13	81.3	82.4	1.1	23	80.2	82.6	2.4
4	82.6	81.9	-0.7	14	76.5	72.5	-4.0	24	87.8	100.4	12.6
5	79.9	76.4	-3.5	15	70.0	90.9	20.9	25	83.3	85.2	1.9
6	88.7	103.6	14.9	16	80.6	71.3	-9.3	26	79.7	83.6	3.9
7	94.9	98.4	3.5	17	83.3	85.4	2.1	27	84.5	84.6	0.1
8	76.3	93.4	17.1	18	87.7	89.1	1.4	28	80.8	96.2	15.4
9	81.0	73.4	-7.6	19	84.2	83.9	-0.3	29	87.4	86.7	-0.7
10	80.5	82.1	1.6	20	86.4	82.7	-3.7				

We would like to determine if the after weights are "significantly" different from the before weights, on average. Access the data set on your text CD, but notice that the data set also includes the same variables for a second group of anorexia patients. We want only the first 29 patients (those that were given the "cognitive" treatment"). To eliminate the other patients select **Data → Select Cases...** and select only those subjects in this group. Alternatively, you could just enter the data in the table above, using three columns, as in the display below. Note that for this demonstration we don't need the before and after weights, so you could just enter the data for the change variable.

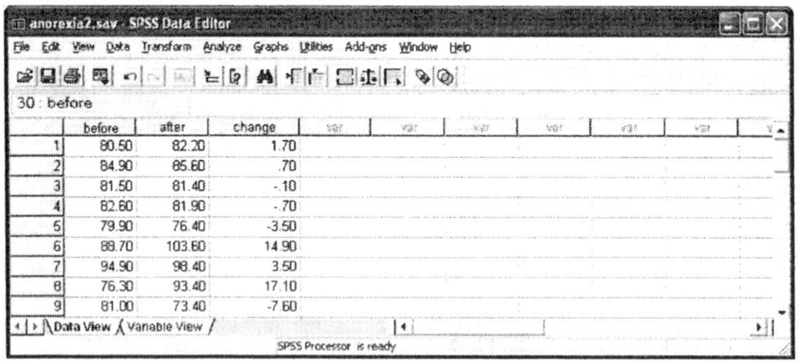

anorexia2.sav - SPSS Data Editor

File Edit View Data Transform Analyze Graphs Utilities Add-ons Window Help

30 : before

	before	after	change
1	80.50	82.20	1.70
2	84.90	85.60	.70
3	81.50	81.40	-.10
4	82.60	81.90	-.70
5	79.90	76.40	-3.50
6	88.70	103.60	14.90
7	94.90	98.40	3.50
8	76.30	93.40	17.10
9	81.00	73.40	-7.60

Data View / Variable View

SPSS Processor is ready

To do the test, select **Analyze → Compare Means → One-Sample T Test...** and enter the "change" variable in the "Test Variable(s)" box.

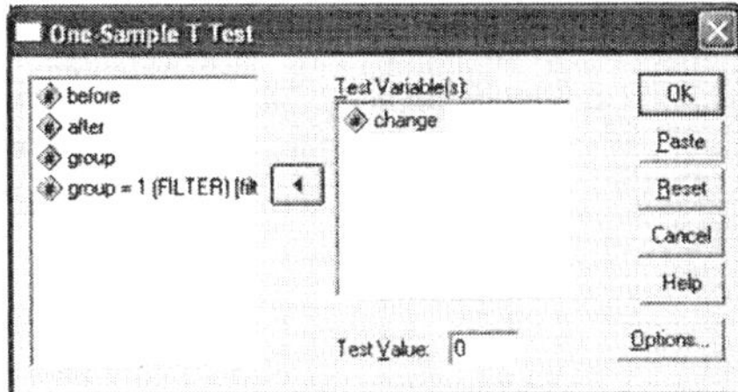

The dialog box includes an option for you to change the "null" or test value. For this example, we want to test if the change is different from zero, which is the default value. Click on **OK** to produce the output.

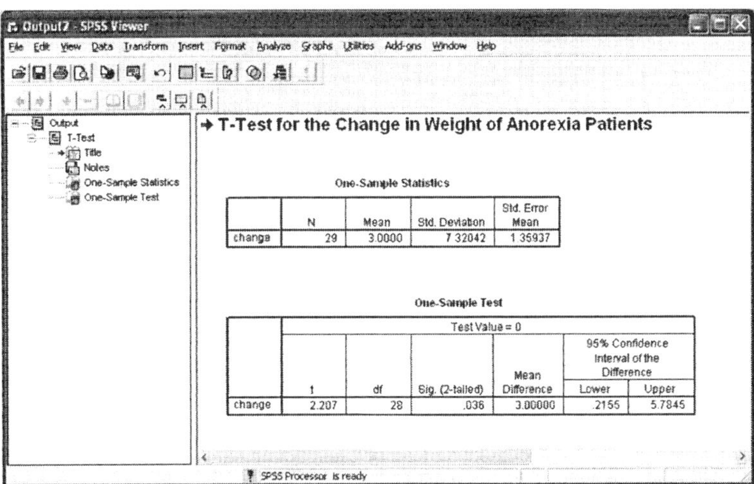

One-Sample Statistics

	N	Mean	Std. Deviation	Std. Error Mean
change	29	3.0000	7.32042	1.35937

One-Sample Test

	Test Value = 0					
					95% Confidence Interval of the Difference	
	t	df	Sig. (2-tailed)	Mean Difference	Lower	Upper
change	2.207	28	.036	3.00000	.2155	5.7845

The output consists of two parts. The first part provides some descriptive statistics for the "change" variable, including the mean and standard deviation. The second part provides the value of the test statistic, the degrees of freedom for the associated T-distribution, and the p-value for the two-sided (or two-tailed) alternate hypothesis. The p-values is listed as "Sig. (2-tailed)" in the output. SPSS does not include an option for finding one-sided p-values. If you are doing a one-sided test you can find the p-value as one-half of Sig or 1 minus one-half of Sig, depending on the direction of your alternate hypothesis and the value (positive or negative) of the T-statistic. The output also includes a confidence interval for the mean.

Chapter 8 Comparing Two Groups

In this chapter we will look at methods for comparing the means or proportions for two samples. While there are no commands for producing just confidence intervals for these situations, most of these commands either include a confidence interval as part of the default output or they include an option for producing a confidence interval.

Comparing the Proportions of Two Independent Samples

SPSS includes an option under the "Nonparametric Tests" command for doing a test to compare the proportions of "successes" from two independent samples. Unlike the Binomial Test, the results of this test will agree with the test shown in your test. SPSS also includes a command for conducting a chi-square "Contingency Table" test that is equivalent to the Z-test that will be discussed in the next chapter.

Let's demonstrate this test using the data from Example 5 in Chapter 9. In this example, 5 of 88 teenagers who watched TV for less than 1 hour per day reported an aggressive act while 154 of 465 teenagers who watched TV at least 1 hour per day reported an aggressive act. Enter this data in SPSS using three variables, one for the response variable (aggressive act), one for the group (TV watching habits) and one for the counts in each group with each response. As we saw in the previous chapter, both the response variable and the group variable must be entered as coded numeric variables. Make sure to weight the cases using the "count" variable.

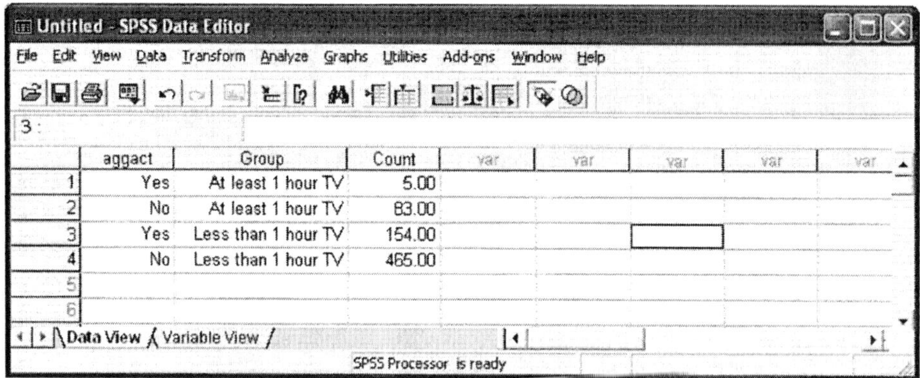

To do the test, select **Analyze → Nonparametric Tests → 2 Independent Samples...** and enter the "aggressive act" variable in the "Test Variable List" box and the "group" variable in the "Grouping Variable:" box. Click on "Define Groups" and enter the two numeric values you used to define the groups. The data for this example was entered using "1" for the "less than 1 hour of TV" group and "2" for the "at least 1 hour of TV" group.

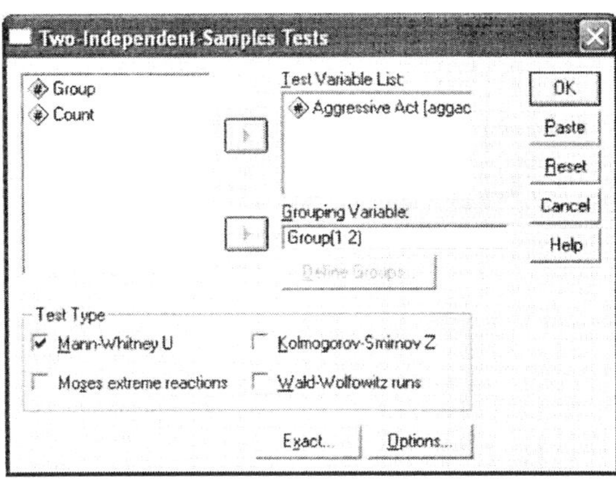

Click on **OK** to produce the output.

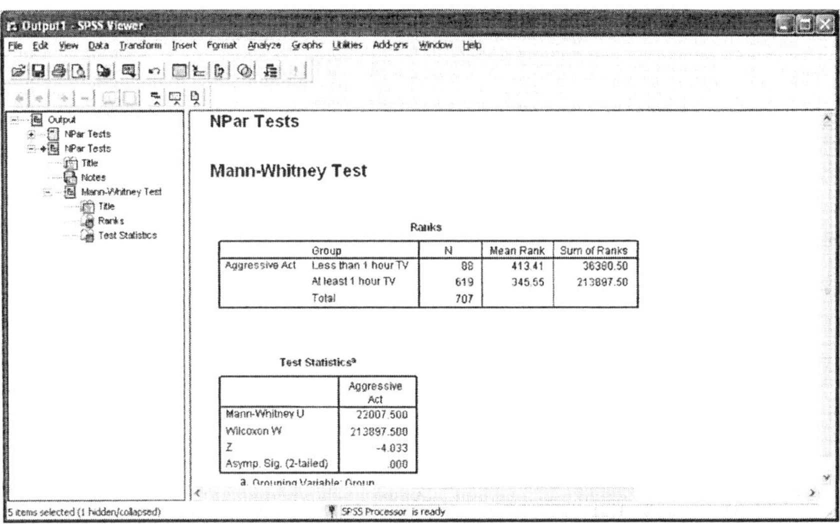

Ranks

	Group	N	Mean Rank	Sum of Ranks
Aggressive Act	Less than 1 hour TV	88	413.41	36380.50
	At least 1 hour TV	619	345.55	213897.50
	Total	707		

Test Statistics[a]

	Aggressive Act
Mann-Whitney U	22007.500
Wilcoxon W	213897.500
Z	-4.033
Asymp. Sig. (2-tailed)	.000

a. Grouping Variable: Group

The value listed next to "Z" in the output is the value of the two-sample Z-statistic shown in your text, and the value next to "Asymp. Sig" is the two-sided p-value for this test.

Comparing the Means of Two Independent Samples

The "Analyze" menu in SPSS includes both parametric and nonparametric methods for comparing means. When the samples to be compared are independent, the parametric methods are produced by "Independent-Samples T Test" under the "Compare Means" command. As mentioned in Chapter 6, the data must be entered using the second data entry method; that is, you must have one variable for the variable of interest (the dependent variable) and a second variable to identify the sample or group (the independent variable). The "Independent-Samples T Test" command will work if the variable identifying the groups is a string variable or a numerically coded variable. The nonparametric commands will be discussed in the last chapter of this manual.

Let's demonstrate these commands using the "newspaper" variable from the "fla_student_survey" data set on your text CD. Access this data file, then select **Analyze → Compare Means → Independent-Samples T Test...** Enter "newspaper" in the "Test Variable(s)" box and "gender" in the "Grouping Variable" box. Click on "Define Groups" and enter "m" in the "Group 1" box and "f" in the "Group 2" box. If you are working with a data set in which the variable that defines the groups is coded numerically, you would enter the appropriate numeric values instead of "m" and "f."

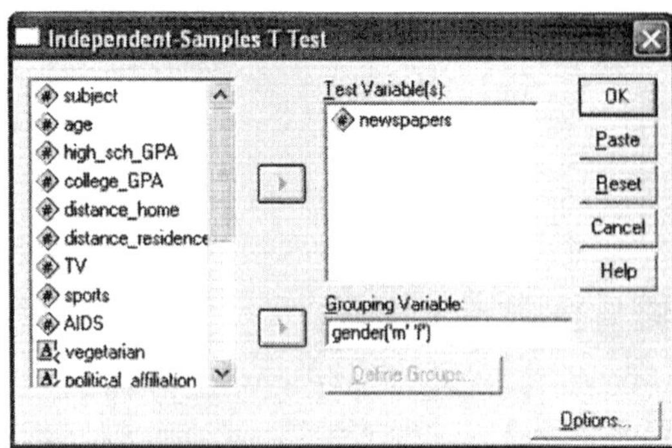

Click on **Continue** and then on **OK**.

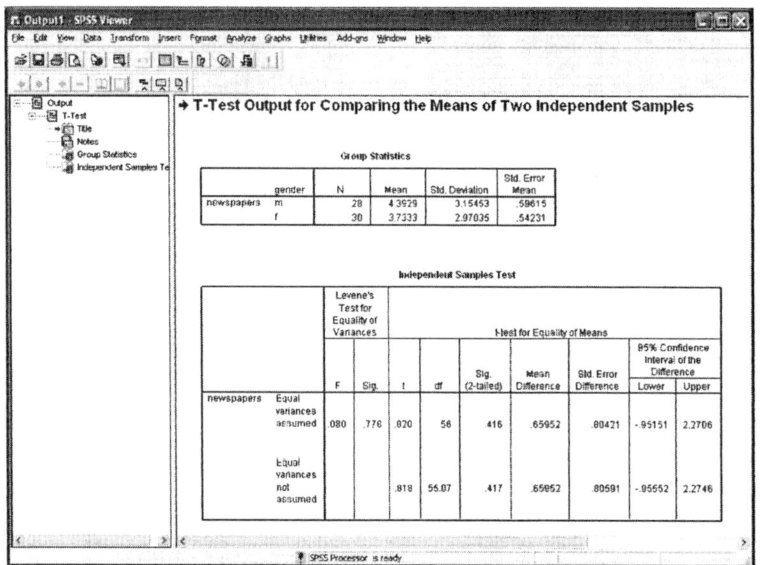

T-Test Output for Comparing the Means of Two Independent Samples

Group Statistics

	gender	N	Mean	Std. Deviation	Std. Error Mean
newspapers	m	28	4.3929	3.15453	.59615
	f	30	3.7333	2.97035	.54231

Independent Samples Test

		Levene's Test for Equality of Variances		t-test for Equality of Means						
									95% Confidence Interval of the Difference	
		F	Sig.	t	df	Sig. (2-tailed)	Mean Difference	Std. Error Difference	Lower	Upper
newspapers	Equal variances assumed	.080	.778	.820	56	.416	.65952	.80421	-.95151	2.2706
	Equal variances not assumed			.818	55.07	.417	.65952	.80591	-.95552	2.2746

The first part of the output provides some descriptive statistics for the variable "newspaper" for each gender. The second part includes the results of a test to compare the variances of "newspaper" of the two genders (Levene's Test for Equality of Variances) and the results of tests to compare the means of "newspaper" of the two genders (t-test for Equality of Means). We are interested in the results of the test to compare two means. SPSS reports the results for both the equal variances and unequal variances versions of the test statistic for comparing means. Your text book presents the equal variances test statistic as an alternative that can be used when it is reasonable to assume, under the null hypothesis, that both the means and the variances are the same. For this data set, the t-statistic assuming equal variances is .82, with a two-sided p-value of .416, while the t-statistic assuming unequal variances is .818, with a two-sided p-value of .417. We would make the same conclusion regardless of which test statistic we used for this data set. The evidence suggests no difference in the average newspaper reading habits between this sample of male and female students.

Your text mentions an "F" test for comparing the variances of two groups that is not "robust" to the assumption of normal data. Levene's test, however, is a robust test and can be used to decide whether it is better to use the equal variances t-statistic or the unequal variances t-statistic. The hypothesis of equal variances would be rejected if the p-value is small. For this data set, the p-value ("Sig.") for Levene's test is .778, indicating that there is no evidence to support the conclusion that the variances are different. In this case, the equal variances test statistic could be used instead of the unequal variances test statistic. Since the unequal variances test statistic is always valid, many text book authors present only this statistic.

Comparing the Means of Two Dependent Samples

As indicated in Chapter 6, to compare the means of two dependent samples, the data must be entered with each of the paired variables in a separate column. We will demonstrate the methods for comparing means of dependent samples using the cell phone data from Example 12 in Chapter 9. The reaction times are shown in the table below.

	Using cell phone?			Using cell phone?	
Student	No	Yes	Student	No	Yes
1	604	636	17	525	626
2	556	623	18	508	501
3	540	615	19	529	574
4	522	672	20	470	468
5	459	601	21	512	578
6	544	600	22	487	560
7	513	542	23	515	525
8	470	554	24	499	647
9	556	543	25	448	456
10	531	520	26	558	688
11	599	609	27	589	679
12	537	559	28	814	960
13	619	595	29	519	558
14	536	565	30	462	482
15	554	573	31	521	527
16	467	554	32	543	536

Enter the data in SPSS in two columns labeled "no" and "yes." We will use SPSS to calculate the differences for us. Select **Transform → Compute...** and enter "difference" in the "Target Variable" box. Then, enter the expression "yes – no" in the "Numeric Expression" box.

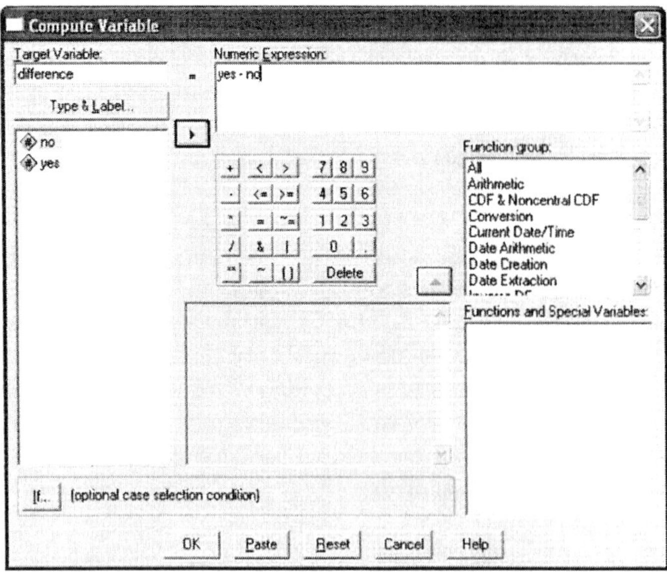

Click on **OK** and the new variable "difference" will be added to the SPSS Data Editor.

Paired-Samples T-Test

To compare the means of these two samples, select **Analyze → Compare Means → Paired-Samples T Test …** and click once on "no" and once on "yes" in the variable list on the left.

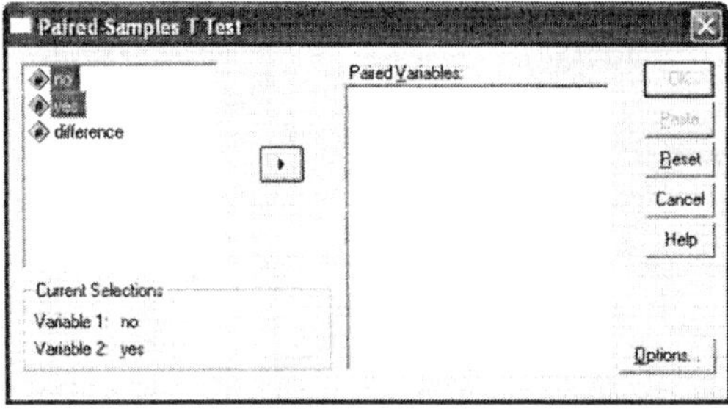

Click on the triangle button to enter the pair "no-yes" in the "Paired Variables" box.

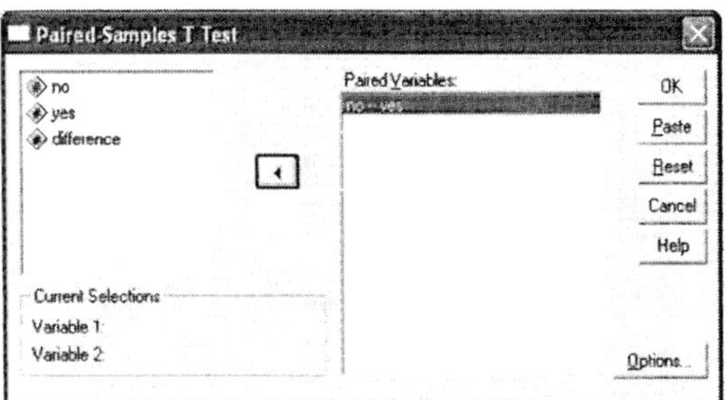

Click on **OK** to produce the output. SPSS will calculate the differences as "no – yes" instead of "yes – no." If you want the difference to be "yes – no" you will have to enter the "Yes" column to the left of the "No" column in the Data Editor. It doesn't make any difference, though, for the analysis because the results will just be the negative of the result if the differences were calculated as "yes – no."

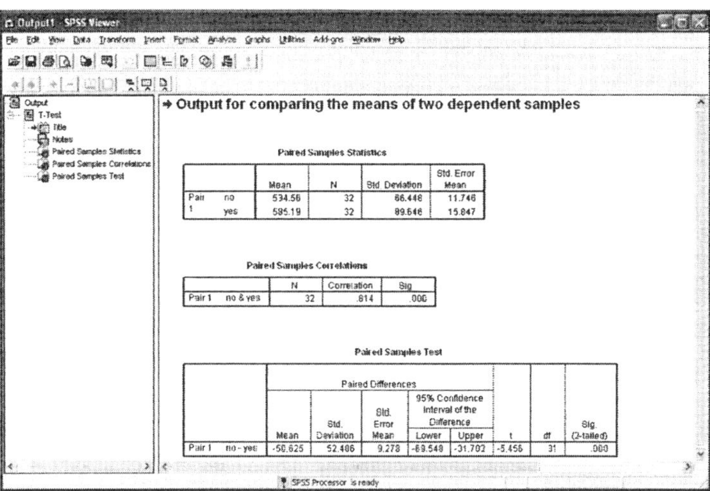

Output for comparing the means of two dependent samples

Paired Samples Statistics

		Mean	N	Std. Deviation	Std. Error Mean
Pair 1	no	534.56	32	66.448	11.746
	yes	585.19	32	89.646	15.847

Paired Samples Correlations

		N	Correlation	Sig.
Pair 1	no & yes	32	.814	.000

Paired Samples Test

		Paired Differences					t	df	Sig. (2-tailed)
		Mean	Std. Deviation	Std. Error Mean	95% Confidence Interval of the Difference Lower	Upper			
Pair 1	no - yes	-50.625	52.406	9.278	-69.548	-31.702	-5.456	31	.000

Similar to the output produced for the independent samples methods, SPSS produces some descriptive statistics of each sample and then the results for the paired t-test on the mean difference between the paired observations, including a confidence interval for the mean difference. SPSS also produces the correlation coefficient for the paired variables. Note that since we had SPSS calculate the differences you could have done a one-sample T-test on the differences instead of doing the paired T-test command. If you use a test value of zero for the one-sample test, the results will be the same as for the paired t-test.

To look at the distribution of the differences, you could make a box plot. Select **Graphs → Boxplot...** and select the "Summaries of separate variables" option.

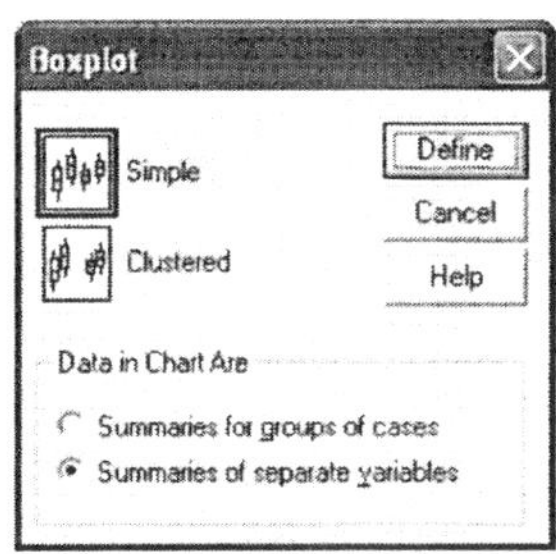

Click on **Define** and enter the difference variable in the "Boxes represent" box. Click on **OK** to produce the graph.

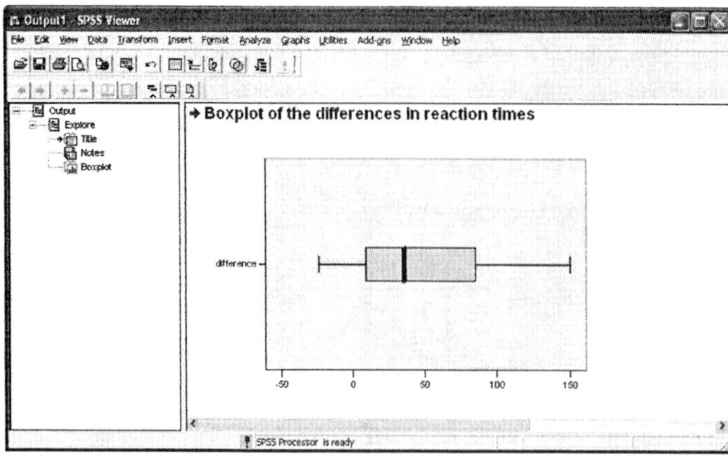

McNemar's Test

Although SPSS does not have a command for comparing proportions between two independent samples is does have a command for conducting McNemar's test on dependent samples when the data is categorical. To demonstrate this command, we will use the data from Example 15 in Chapter 9. This data is from the General Social Survey and is shown in the following table.

	Belief in Hell	
Belief in Heaven	Yes	No
Yes	833	125
No	2	160

Enter the data in SPSS using three variables, "heaven," "hell," and "count" and weight the cases by "count." Both "heaven" and "hell" must be entered as coded numeric variables in order to do this test.

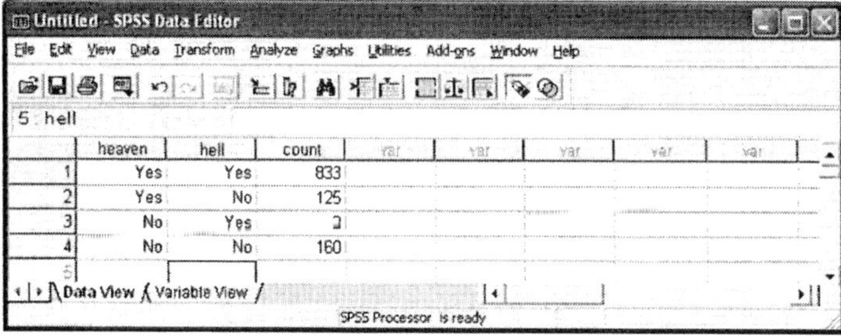

To complete the test, select **Analyze → Nonparametric Tests → 2 Related Samples…** and click on the "Statistics" button. Click on the paired variables as you did for the paired t-test above and then click on the arrow button. Click on "McNemar" and then on **Continue**.

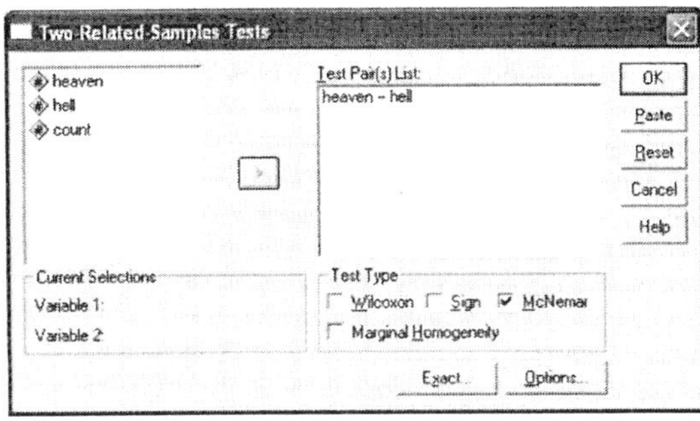

Click on **OK** to produce the output.

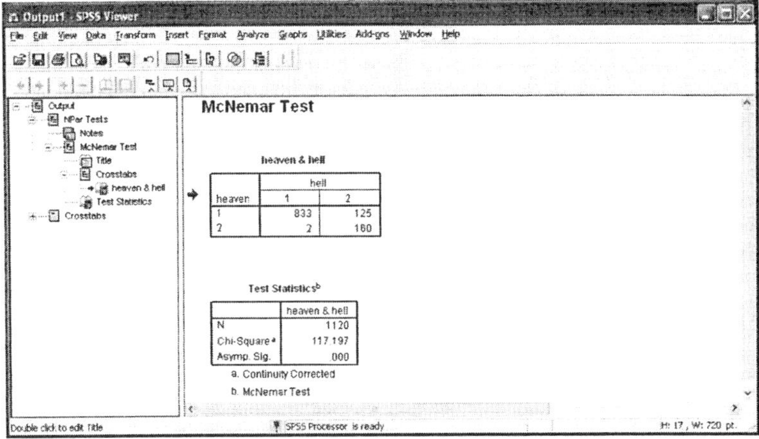

heaven & hell

heaven	hell 1	hell 2
1	833	125
2	2	160

Test Statistics[b]

	heaven & hell
N	1120
Chi-Square[a]	117.197
Asymp. Sig.	.000

a. Continuity Corrected
b. McNemar Test

There is also an option for doing McNemar's test under the "Crosstabs" command. Select **Analyze → Descriptive Statistics → Crosstabs...** and click on the "Statistics" button. Click on "McNemar" and then on **Continue**.

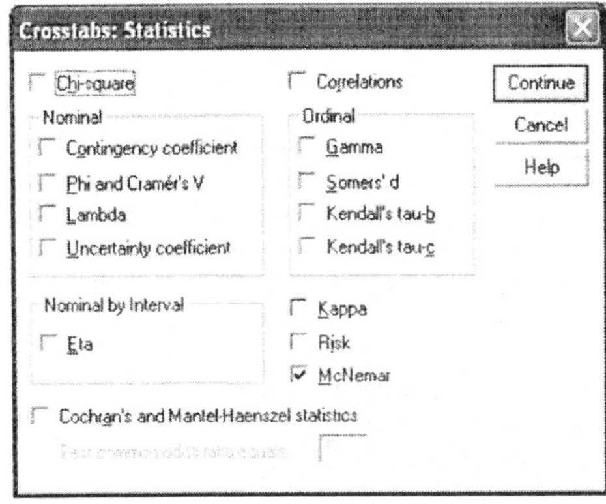

Enter "hell" as the "Column" variable and "heaven" as the "Row" variable. Click on **OK** to produce the output.

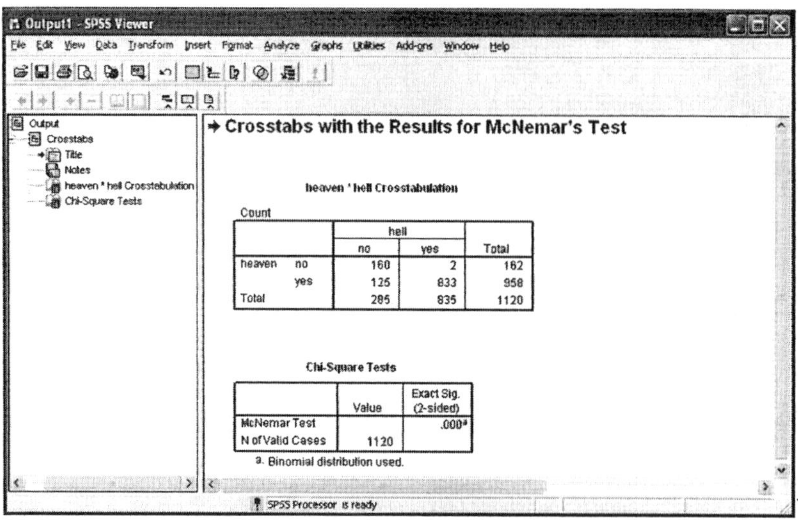

heaven * hell Crosstabulation

Count

		hell		
		no	yes	Total
heaven	no	160	2	162
	yes	125	833	958
Total		285	835	1120

Chi-Square Tests

	Value	Exact Sig. (2-sided)
McNemar Test		.000[a]
N of Valid Cases	1120	

a. Binomial distribution used.

SPSS reports only the p-value for this test.

Chapter 9 Analyzing the Association between Categorical Variables

In chapter 7 of this manual you saw how to use a chi-square test that is equivalent to a one-sample Z-test for a proportion. There is also an equivalent chi-square test for comparing a proportion between two independent samples when the populations have just two groups. Depending on the situation, this test is sometimes called a test for independence (or association) or a test for homogeneity.

Contingency Table Tests

To demonstrate the chi-square test we will use the data provided in Example 3 in Chapter 10. This data is from the General Social Survey and shows the joint distribution of the variables "Happiness" and "Income." The data is shown in the following table.

		HAPPINESS	
INCOME	Not too Happy	Pretty Happy	Very Happy
Above average	21	159	110
Average	53	372	221
Below average	94	249	83

Enter the data in SPSS using three columns, one for each of the response variables, HAPPINESS and INCOME, and one for the count in each cell of the table. Remember that HAPPINESS and INCOME can be entered as string or numerically coded variables. Don't forget to select **Data → Weight Cases...** and enter "count" as a weighting variable.

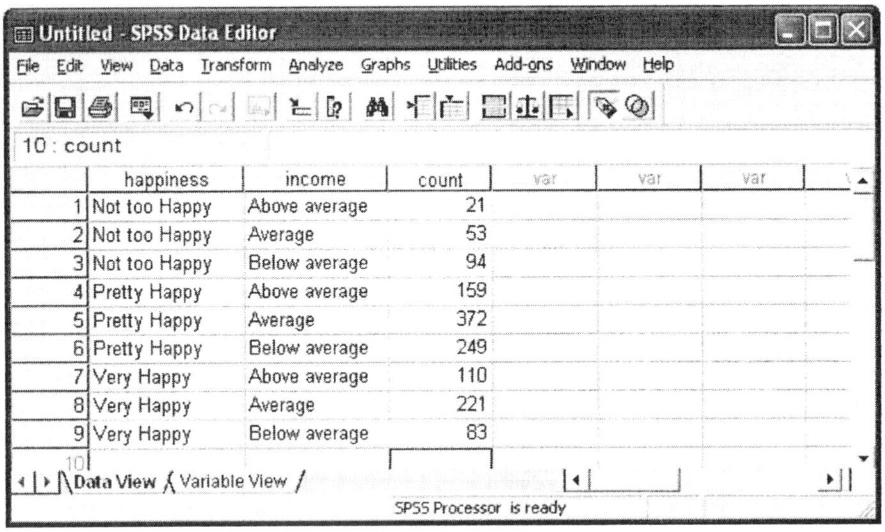

Untitled - SPSS Data Editor

File Edit View Data Transform Analyze Graphs Utilities Add-ons Window Help

10 : count

	happiness	income	count	var	var	var
1	Not too Happy	Above average	21			
2	Not too Happy	Average	53			
3	Not too Happy	Below average	94			
4	Pretty Happy	Above average	159			
5	Pretty Happy	Average	372			
6	Pretty Happy	Below average	249			
7	Very Happy	Above average	110			
8	Very Happy	Average	221			
9	Very Happy	Below average	83			
10						

Data View / Variable View

SPSS Processor is ready

To conduct the test select **Analyze → Descriptive Statistics → Crosstabs**... and enter "income" as the "Row" variable and "happiness" as the "Column" variable.

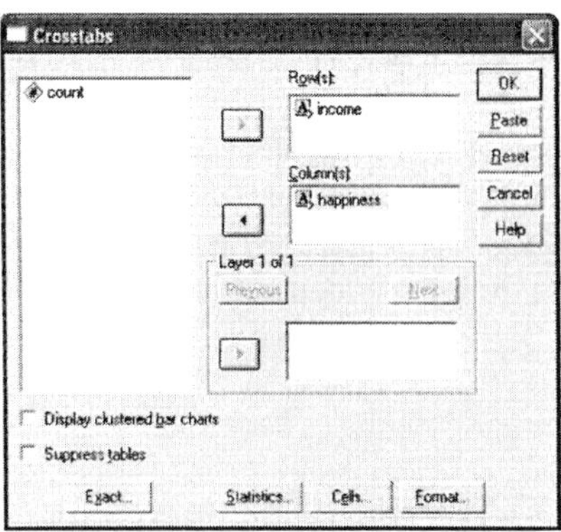

Click on the "Statistics" button and then on "Chi-square."

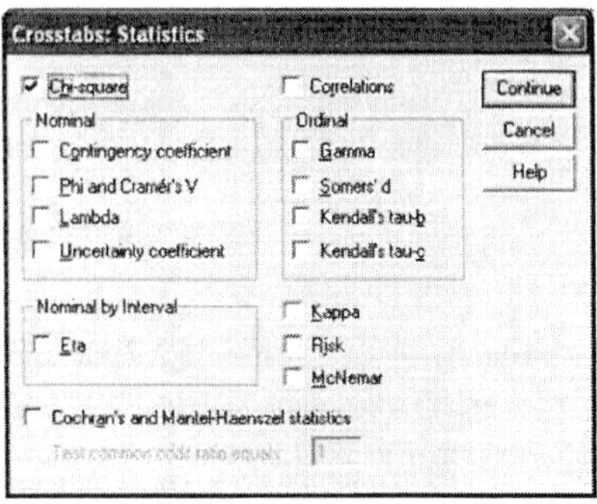

Click on **Continue** and then on **OK**.

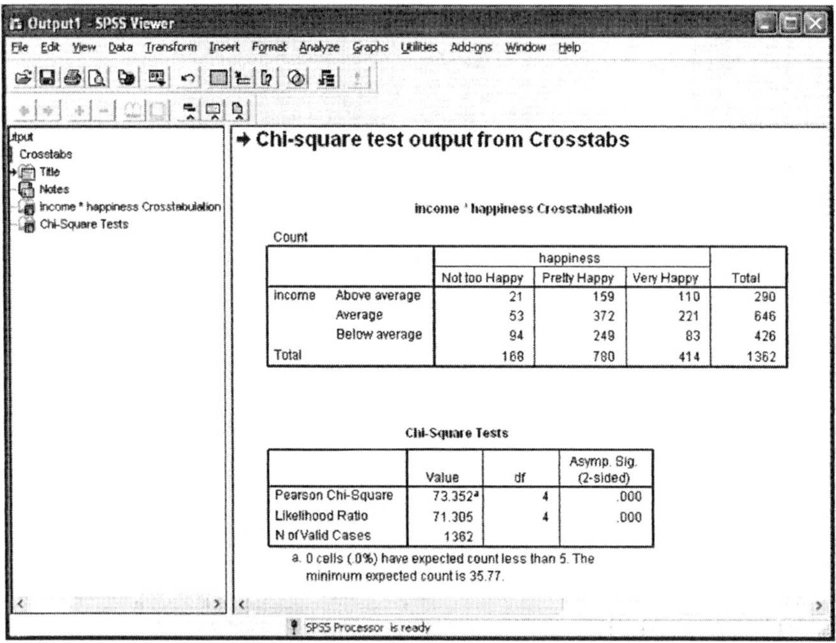

Chi-square test output from Crosstabs

income * happiness Crosstabulation

Count

		happiness			Total
		Not too Happy	Pretty Happy	Very Happy	
income	Above average	21	159	110	290
	Average	53	372	221	646
	Below average	94	249	83	426
Total		168	780	414	1362

Chi-Square Tests

	Value	df	Asymp. Sig. (2-sided)
Pearson Chi-Square	73.352[a]	4	.000
Likelihood Ratio	71.305	4	.000
N of Valid Cases	1362		

a. 0 cells (.0%) have expected count less than 5. The minimum expected count is 35.77.

In addition to the value of the chi-square statistic and its associated p-value, under "Asymp. Sig.", SPSS also reports the number of cells with expected cell count less than 5. Note that the results of the test would be the same if "income" was entered as the "Column" variable and "happiness" was the "Row" variable.

You may recall from chapter 2 of this manual that "Crosstabs" includes options for calculating the conditional (row or column) proportions for a contingency table. It also has an option for calculating the standardized residual for each cell in the table. Select **Analyze → Descriptive Statistics → Crosstabs**… again and click on the "Cells…" button. Click on "Expected" under "Counts" ("Observed" should already be selected) and also on "Adjusted standardized" under "Residuals." These are the residuals that your text refers to as "standardized."

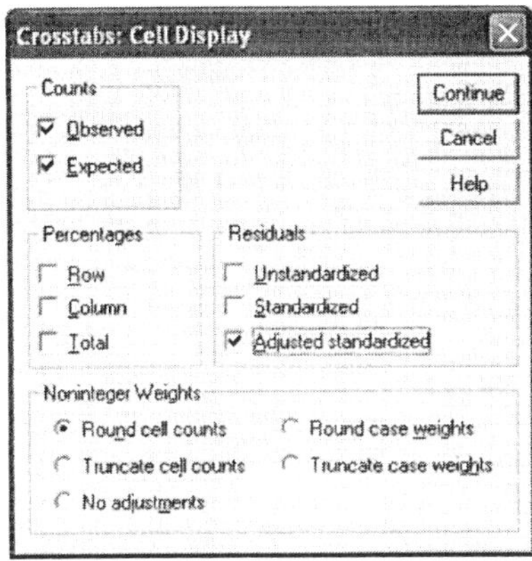

Click on **Continue** and then on **OK**.

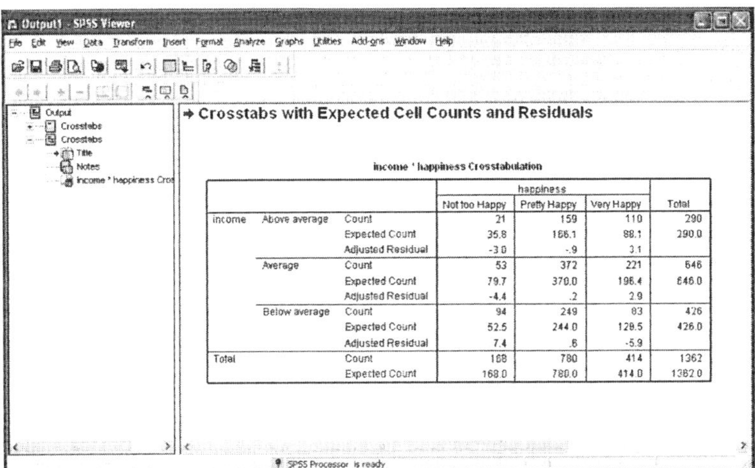

Crosstabs with Expected Cell Counts and Residuals

income * happiness Crosstabulation

			happiness			Total
			Not too Happy	Pretty Happy	Very Happy	
income	Above average	Count	21	159	110	290
		Expected Count	35.8	166.1	88.1	290.0
		Adjusted Residual	-3.0	-.9	3.1	
	Average	Count	53	372	221	646
		Expected Count	79.7	370.0	196.4	646.0
		Adjusted Residual	-4.4	.2	2.9	
	Below average	Count	94	249	83	426
		Expected Count	52.5	244.0	129.5	426.0
		Adjusted Residual	7.4	.6	-5.9	
Total		Count	168	780	414	1362
		Expected Count	168.0	780.0	414.0	1362.0

Comparing Two Proportions

As with the chi-square goodness-of-fit test, you should check with your instructor before learning how to use this alternate method for comparing two proportions.

Let's demonstrate this test using Example 5 from Chapter 9. In this example, p is the proportion that showed aggressive behavior and the two groups are determined by their amount of TV watching. The results are shown in the following table.

	Aggressive Act	
TV Watching	Yes	No
Less than 1 hour per day	5	83
At least 1 hour per day	154	465

Enter the data in SPSS using three variables, one for each of the categorical variables and one for the count in each cell of the table. Make sure to weight the cases by the "count" variable.

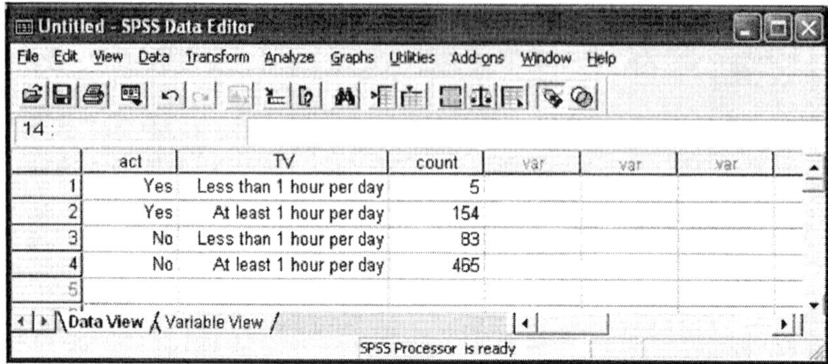

To do the test select **Analyze → Descriptive Statistics → Crosstabs**... and enter the "hours" variable in the "Row(s)" box and the "TV" variable in the "Column(s)" box. Click on the "Statistics" button and select the "Chi-Square" statistic.

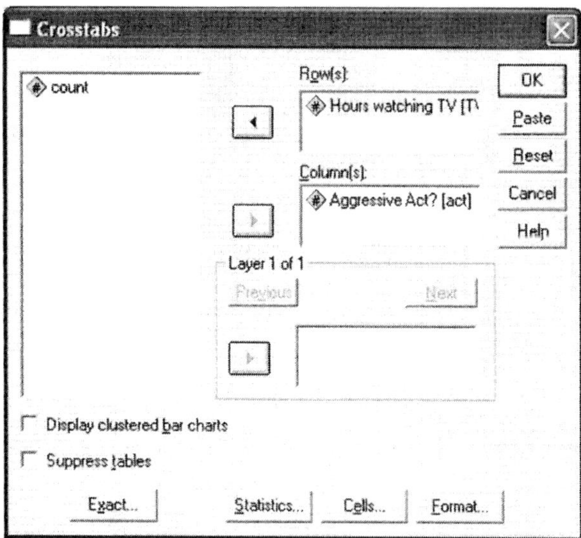

Click on **Continue** and then on **OK**.

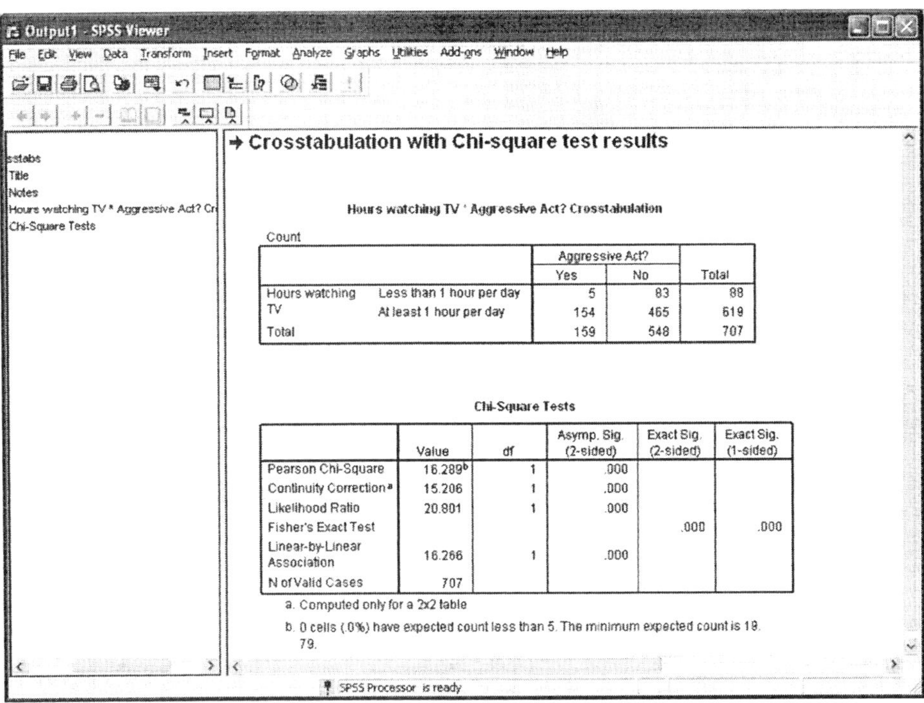

→ Crosstabulation with Chi-square test results

Hours watching TV * Aggressive Act? Crosstabulation

Count

		Aggressive Act? Yes	No	Total
Hours watching TV	Less than 1 hour per day	5	83	88
	At least 1 hour per day	154	465	619
Total		159	548	707

Chi-Square Tests

	Value	df	Asymp. Sig. (2-sided)	Exact Sig. (2-sided)	Exact Sig. (1-sided)
Pearson Chi-Square	16.289[b]	1	.000		
Continuity Correction[a]	15.206	1	.000		
Likelihood Ratio	20.801	1	.000		
Fisher's Exact Test				.000	.000
Linear-by-Linear Association	16.266	1	.000		
N of Valid Cases	707				

a. Computed only for a 2x2 table

b. 0 cells (.0%) have expected count less than 5. The minimum expected count is 19.79.

The value of the chi-square test statistic is 16.289, which is the same as the square of the value of the Z-statistic given in your text. (If you don't round your calculations at intermediate steps when calculating the Z-statistic, you will get a value of -4.036 and the square of -4.036 is 16.289.) The p-value for this test (and for the corresponding Z-statistic) is less than .0005.

Fisher's Exact Test

SPSS also includes a command to conduct Fisher's Exact Test, which is useful when dealing with a small sample size. To demonstrate this test we will use the data from Example 10 in Chapter 10. This data is shown in the following table.

	PREDICTION	
ACTUAL	Milk	Tea
Milk	3	1
Tea	1	3

This data set is from an experiment where a taster made a prediction of whether tea or milk was poured in a cup first. The experiment consisted of 4 trials where milk was poured first and 4 where tea was poured first.

Enter the data in SPSS using three variables, "actual," "prediction" and "count," then select **Data → Weight Cases...** and enter "count" as the weighting variable. Select **Analyze → Descriptive Statistics → Crosstabs...**, then click on the "Exact" button and then on "Exact."

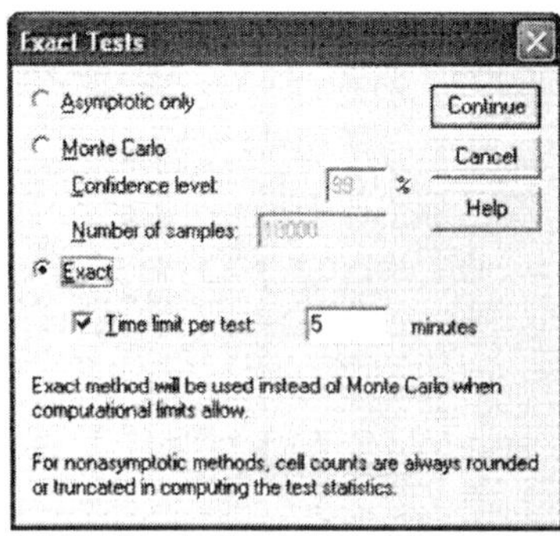

Click on **Continue** and enter "actual" as the "Row" variable and "predicted" as the "Column" variable.

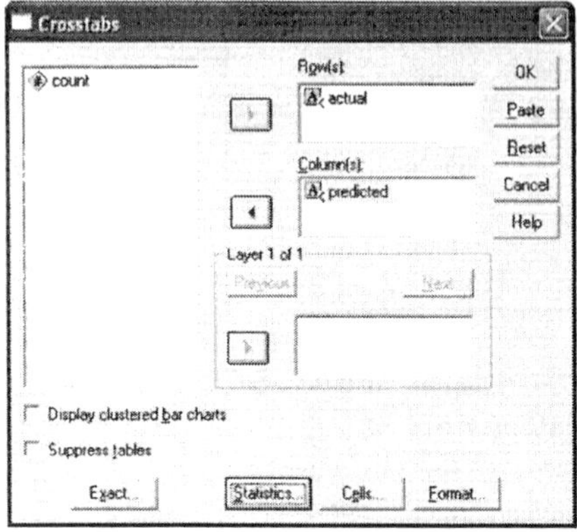

Click on the "Statistics" button and select "Chi square." Click on Continue and then on **OK** to produce the output.

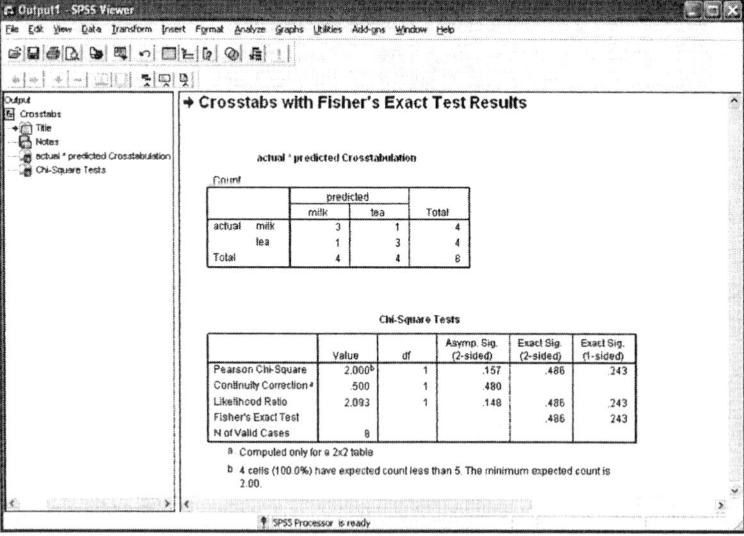

Crosstabs with Fisher's Exact Test Results

actual * predicted Crosstabulation

Count

		predicted		Total
		milk	tea	
actual	milk	3	1	4
	tea	1	3	4
Total		4	4	8

Chi-Square Tests

	Value	df	Asymp. Sig. (2-sided)	Exact Sig. (2-sided)	Exact Sig. (1-sided)
Pearson Chi-Square	2.000[b]	1	.157	.486	.243
Continuity Correction[a]	.500	1	.480		
Likelihood Ratio	2.093	1	.148	.486	.243
Fisher's Exact Test				.486	.243
N of Valid Cases	8				

a. Computed only for a 2x2 table

b. 4 cells (100.0%) have expected count less than 5. The minimum expected count is 2.00.

Chapter 10 Regression Analysis

You probably noticed in the regression examples in Chapter 3 of this manual that the regression command in SPSS does a lot more than just find the slope and intercept of the best fit line. Some of the output that SPSS produces for this command was omitted back in Chapter 3. We will explore that output more in this chapter

Testing for the significance of the slope

In Chapter 3 we used SPSS to find the correlation coefficient between two variables and to find the best fit line, but we didn't address whether or not the association was "significant" or "meaningful." In the regression output SPSS provides the results for a test to determine if the slope of the best fit line is significantly different from zero. We will explore this output using the "high school female athletes" data set from your text CD. Access this data set and select **Analyze → Regression → Linear...** Enter Maximum Bench Press (@1RMBENCHlbs) as the "Dependent" variable and Number of 60-Pound Bench Presses (BRTF_60) as the "Independent" variable.

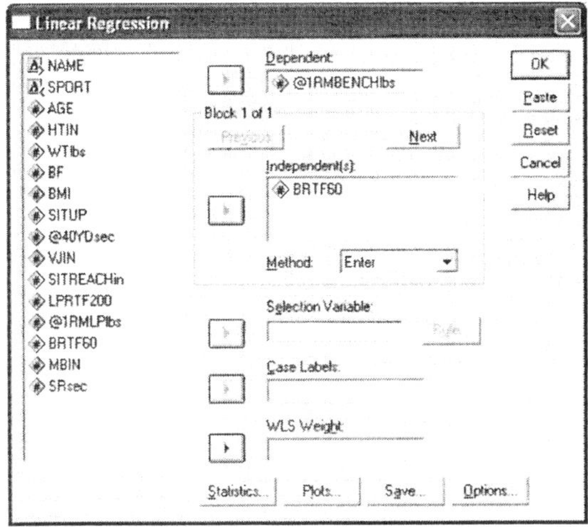

Click on **OK** to produce the output. The Model Summary and Coefficients portions of the regression output are shown below. The best fit line is $\hat{y} = 63.5 + 1.49x$, as shown in your text and verified by the "B" values under Unstandardized Coefficients in the output. Next to the estimates of the intercept and slope you will see their corresponding standard errors. On the far right of the "Coefficients" output you will see columns titled "t" and "Sig." These columns are for conducting two separate hypothesis tests, one to determine if the y-intercept is significantly different from zero and the other to determine if the slope is significantly different from zero. For this example we're only interested in the test concerning the slope.

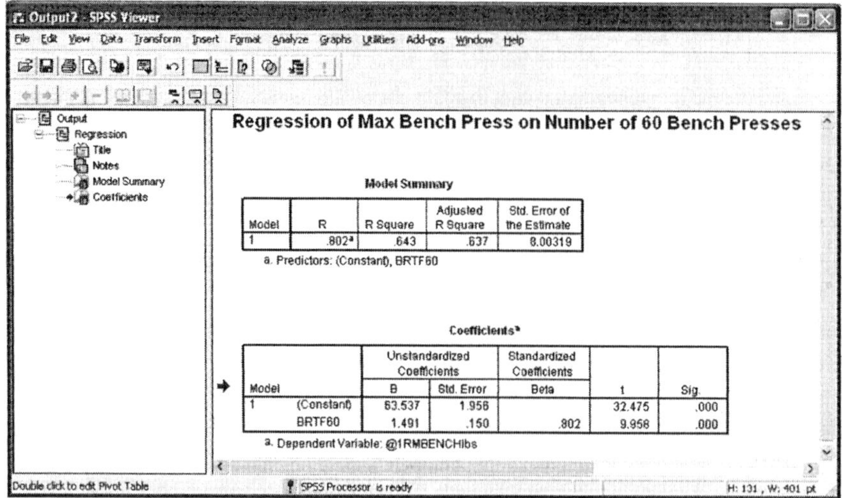

Regression of Max Bench Press on Number of 60 Bench Presses

Model Summary

Model	R	R Square	Adjusted R Square	Std. Error of the Estimate
1	.802[a]	.643	.637	8.00319

a. Predictors: (Constant), BRTF60

Coefficients[a]

Model		Unstandardized Coefficients		Standardized Coefficients	t	Sig.
		B	Std. Error	Beta		
1	(Constant)	63.537	1.956		32.475	.000
	BRTF60	1.491	.150	.802	9.958	.000

a. Dependent Variable: @1RMBENCHlbs

In your text book the test statistic for testing the significance of the slope is given as $t = \frac{b-0}{se}$, where b is the estimate of the slope and se is the standard error of b. For this example, $t = \frac{1.491-0}{.15} = 9.96$, which agrees with the t-statistic provided by SPSS. The p-value for the two-sided test is given by "Sig," which is listed as .0000. This result just means that the p-value is less than .00005.

The Squared Correlation and Residual Standard Deviation

Included in the Model Summary portion of the regression output is a value listed as "R-Square." This value tells us the proportion reduction in error in using $\hat{y}$ to predict y instead of $\bar{y}$. For a regression analysis with one Y-variable and one X-variable, the proportional reduction in error can easily be found by squaring the value of r. We will explore the actual formula for doing this calculation in the next chapter, but for now it is easy to see for the example above that $(.802)^2 =$ .643, so the error in predicting the maximum number of bench presses is 64.3% smaller using the regression equation than the error using the average number of bench presses.

The residual standard deviation is given in the Model Summary under "Std. Error of the Estimate." For this example, the residual standard deviation is 8.003. This value can be used to create a prediction interval for y as described in your text.

Confidence Interval for the Slope

The Regression command in SPSS includes an option for also calculating a confidence interval for the slope. Select **Analyze → Regression → Linear...** and click on the "Statistics" button. Click on "Confidence Intervals" under "Regression Coefficients" and then on **Continue**.

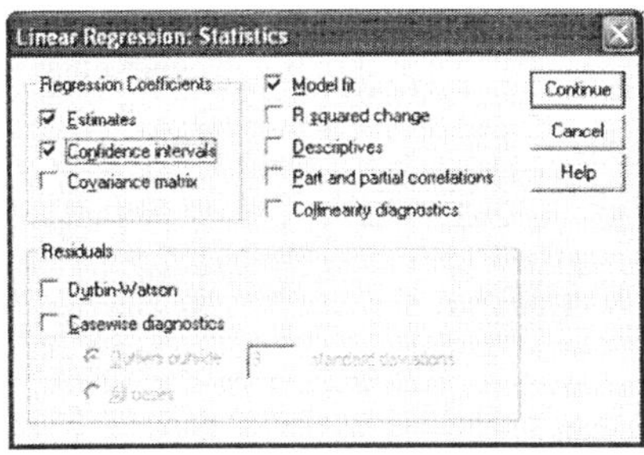

Click on **OK** to produce the output. Notice the additional columns for the lower and upper bounds in the "Coefficients" portion of the output.

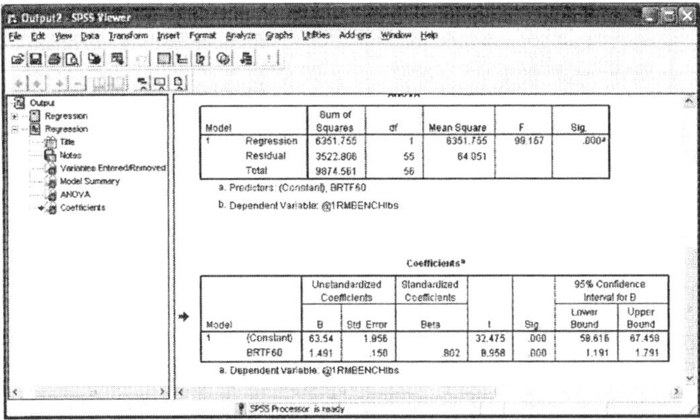

Model		Sum of Squares	df	Mean Square	F	Sig.
1	Regression	6351.755	1	6351.755	99.167	.000a
	Residual	3522.806	55	64.051		
	Total	9874.561	56			

a. Predictors: (Constant), BRTF60

b. Dependent Variable: @1RMBENCHlbs

Coefficients[a]

Model		Unstandardized Coefficients		Standardized Coefficients	t	Sig.	95% Confidence Interval for B	
		B	Std. Error	Beta			Lower Bound	Upper Bound
1	(Constant)	63.54	1.956		32.475	.000	59.616	67.458
	BRTF60	1.491	.150	.802	9.958	.000	1.191	1.791

a. Dependent Variable: @1RMBENCHlbs

Detecting Unusual Observations

The Regression command in SPSS includes an option for printing a table of observations with large residuals. To print this table, select **Analyze → Regression → Linear...** and click on the "Statistics" button. Click on "Casewise Diagnostics" under "Residuals" and enter the value "2" in the "standard deviations" box. Click on **Continue** and then on **OK**.

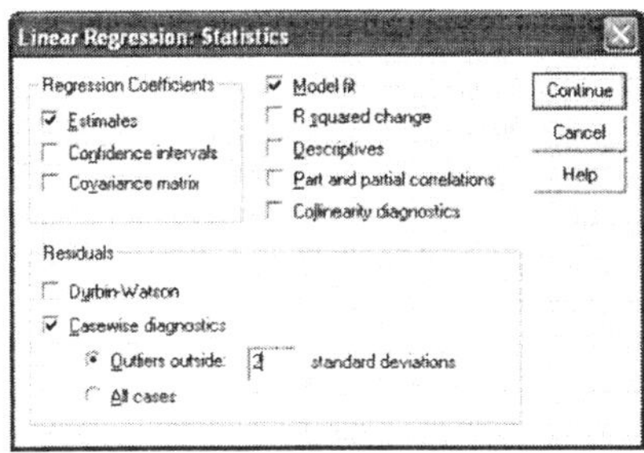

The output is shown below. This data set has two observations with standardized residuals larger than 2. The output in your text lists three observations instead of two, because different statistical programs use different "rules" to identify unusual cases. Case #59, for example, represents a student whose residual is 3.104 standard errors below zero, indicating that this student's actual college GPA is far below the value predicted by the best-fit regression line.

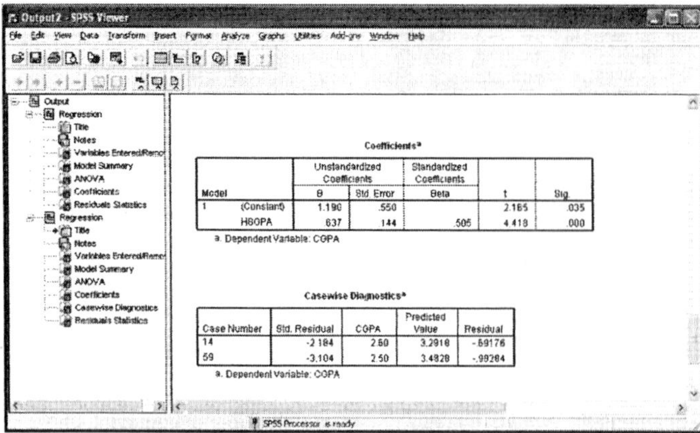

Exponential Regression

SPSS has an option for fitting an exponential model instead of a linear model. To demonstrate this we will use the data for Example 18 in Chapter 11 in your text. This data is shown in the table below:

Year	Years since 1995	Number of People
1995	0	16
1996	1	36
1997	2	76
1998	3	147
1999	4	195
2000	5	369
2001	6	513

The data has been entered into the SPSS Data Editor shown below.

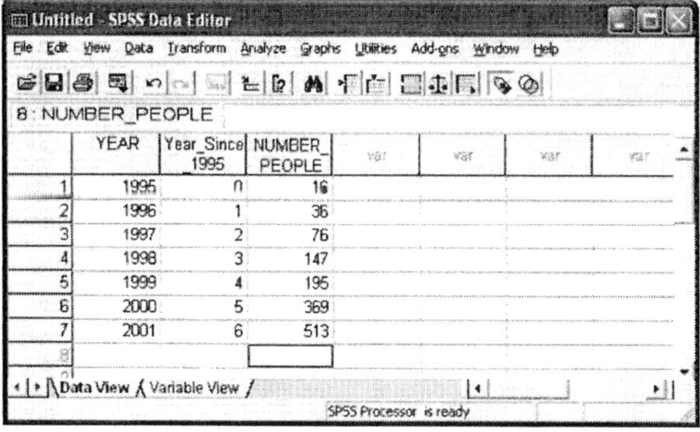

Select **Analyze → Regression → Curve Estimation…** and enter "Number_People' as the dependent variable and "Year since 1995" as the independent variable. Click on "Exponential" under "Models." You will notice that "Linear" has already been selected. SPSS will produce both the linear and exponential fits to this data set. If you don't want to see the linear results, click on "Linear" to un-select it.

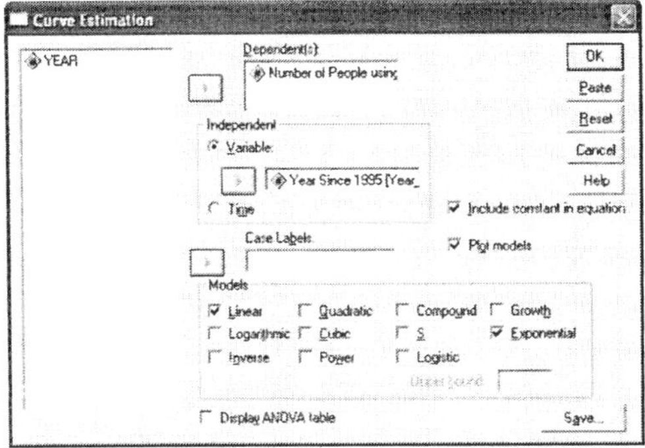

Click on **OK** to produce the output.

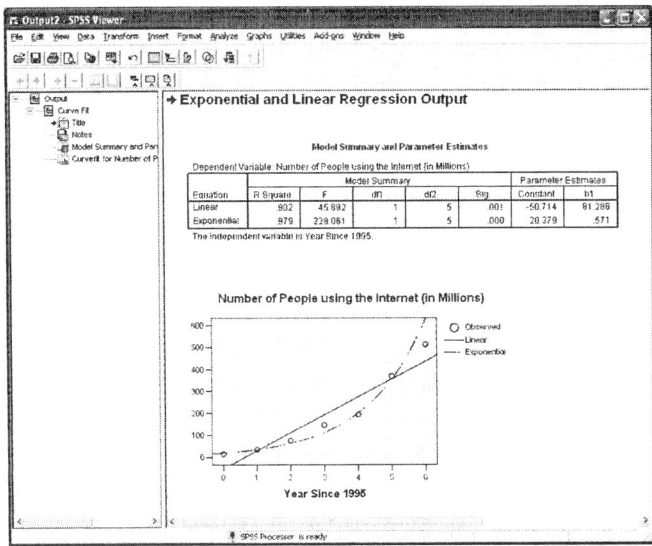

Model Summary and Parameter Estimates

Dependent Variable: Number of People using the Internet (in Millions)

Equation	Model Summary					Parameter Estimates	
	R Square	F	df1	df2	Sig.	Constant	b1
Linear	.902	45.892	1	5	.001	-50.714	81.286
Exponential	.979	228.061	1	5	.000	20.379	.571

The independent variable is Year Since 1995.

Some of the output produced by SPSS has been omitted in the display above. Unlike the example in your text, SPSS does the exponential regression using the logarithm with base *e* (instead of base 10), so the base for the exponent *x* in the exponential model appears different than it does in your text. To convert the coefficient b1 to log base 10, raise *e* to the power given by the value of b1 from the output. That is, $e^{.571} = 1.77$. So the exponential regression model is $\hat{y} = 20.38 \times 1.77^x$.

Another way to find the exponential regression model is to use SPSS to calculate the logarithm of the number of people using the internet, as shown in your text. Select **Transform → Compute...** and enter the variable name "log_people" in the Target Variable box. Then select "Arithmetic" under "Function Group" and scroll through the "Functions and Special Variables" list to find "Lg10." Double click on this and then replace the question mark in the expression "LG10(?)" with the variable "Number_People."

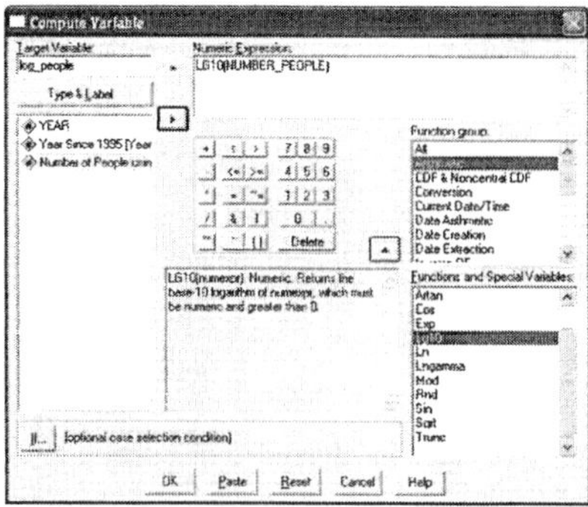

Click on **OK**. The data editor will now include the base 10 logarithm of the number of people using the Internet as a new variable.

	YEAR	Year_Since_1995	NUMBER_PEOPLE	log_people
1	1995	0	16	1.20
2	1996	1	36	1.56
3	1997	2	76	1.88
4	1998	3	147	2.17
5	1999	4	195	2.29
6	2000	5	369	2.57
7	2001	6	513	2.71

To do the exponential regression using this approach, select **Analyze → Regression → Linear…** and enter "log_people" as the "Dependent" variable and "Year_Since_1995" as the "Independent" variable. Click on **OK**.

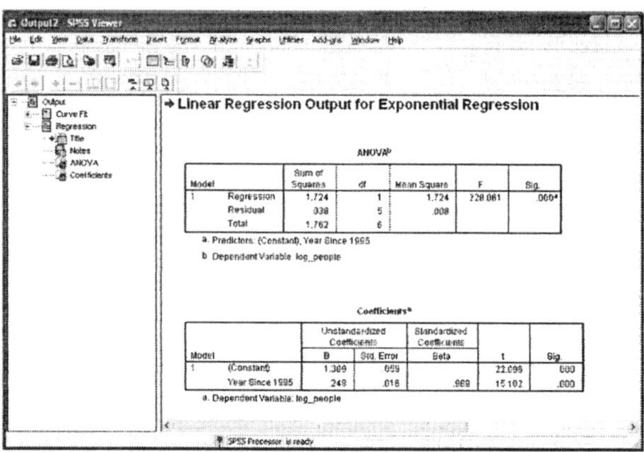

Linear Regression Output for Exponential Regression

ANOVA[b]

Model		Sum of Squares	df	Mean Square	F	Sig.
1	Regression	1.724	1	1.724	228.061	.000[a]
	Residual	.038	5	.008		
	Total	1.762	6			

a. Predictors: (Constant), Year Since 1995
b. Dependent Variable: log_people

Coefficients[a]

Model		Unstandardized Coefficients B	Std. Error	Standardized Coefficients Beta	t	Sig.
1	(Constant)	1.309	.059		22.096	.000
	Year Since 1995	.248	.016	.989	15.102	.000

a. Dependent Variable: log_people

As your text describes, the logarithm of the mean is a linear function of x. From our output we have $\hat{\log} y = 1.309 + .248x$. To transform this to an exponential model, the alpha coefficient is 10 raised to the power of the intercept and the beta coefficient is 10 raised to the power of the slope. Or, $\hat{\alpha} = 10^{1.309} = 20.37$ and $\hat{\beta} = 10^{.248} = 1.77$. So, the exponential regression model is $\hat{y} = 20.37 \times 1.77^x$, which is the same result we obtained above using the first approach.

Chapter 11 Multiple Regression

When we explored regression analysis in the previous chapter and in Chapter 3 we used examples of "simple" linear regression; that is, the regression of one Y (dependent) variable on one X (independent) variable. The "Regression" command in SPSS allows you to also do multiple regression with more than one independent variable. To demonstrate how to do this and how to read the output that SPSS produces we will use the data for Example 2 in Chapter 12 on the selling prices of houses. This data set is available on your text CD in the file "house_selling_prices" and a portion is shown in the table below.

Home	Selling Price	House Size	Number of Bedrooms	Number of Bathrooms	Lot Size	Real Estate Tax	NW
1	145,000	1240	3	2	18,000	1360	Yes
2	69,000	1120	3	2	17,000	830	No
3	163,000	1710	3	2	14,000	2150	Yes

Access this data set on your text CD and select **Analyze → Regression → Linear...** and enter "price" as the "Dependent" variable and both "size" and "lot" as "Independent(s)" variables.

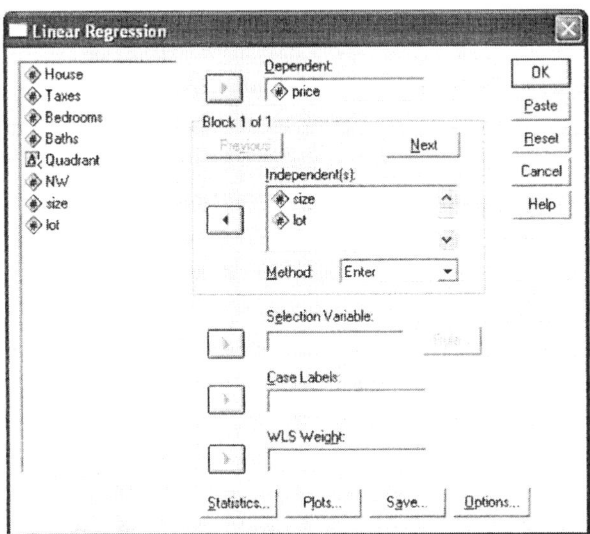

Click on **OK** to produce the output. The output is shown below. The "Coefficients" portion of the output provides us with the estimates of the parameters that describe the effects of house size and lot size on selling price. As shown in your text these values are 54.8 and 2.84, respectively.

Regression of Selling Price on House Size and Lot Size

Model Summary

Model	R	R Square	Adjusted R Square	Std. Error of the Estimate
1	.843a	.711	.705	30588.10044

a. Predictors: (Constant), lot, size

ANOVAb

Model		Sum of Squares	df	Mean Square	F	Sig.
1	Regression	223676226389	2	1.118E+011	119.532	.000a
	Residual	90756293211	97	935631888.8		
	Total	314432519600	99			

a. Predictors: (Constant), lot, size

b. Dependent Variable: price

Coefficientsa

Model		Unstandardized Coefficients		Standardized Coefficients	t	Sig.
		B	Std. Error	Beta		
1	(Constant)	-10536.0	9436.474		-1.117	.267
	size	53.779	6.529	.532	8.237	.000
	lot	2.840	.427	.430	6.656	.000

a. Dependent Variable: price

This portion of the output also includes T-tests for testing the significance of each of the independent variables, which will be discussed for another example below. The Model Summary portion of the output includes the values of the multiple correlation, R, and R-Squared. These values are .843 and .711, respectively. To see how to calculate R-Squared, the ANOVA portion of the output provides the value for the residual sum of squares and the total sum of squares. (This portion of the output is produced below with price recoded as the actual selling price divided by 1000, to match the output as given in the text. Transforming the dependent variable in this way does not affect the degree of association of selling price with the two independent variables.) The value of R^2 is $\dfrac{314{,}433-90{,}756}{314{,}433}=.711$.

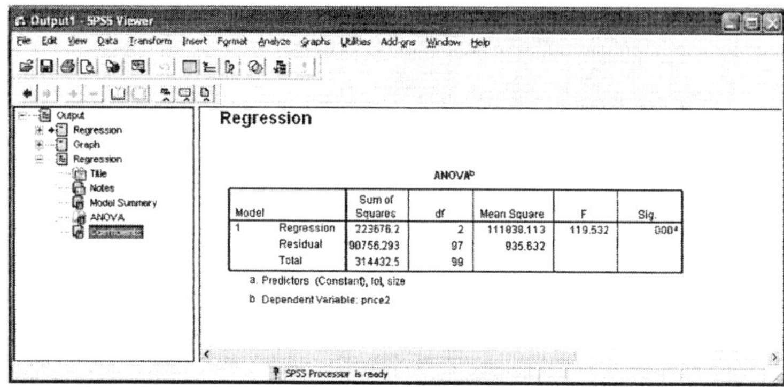

Regression

ANOVAb

Model		Sum of Squares	df	Mean Square	F	Sig.
1	Regression	223676.2	2	111838.113	119.532	.000a
	Residual	90756.293	97	935.632		
	Total	314432.5	99			

a. Predictors: (Constant), lot, size

b. Dependent Variable: price2

To see the associations between selling price with each of the independent variables select **Graphs → Scatter/Dot...** and select the "Matrix Scatter" option. Click on **Define**.

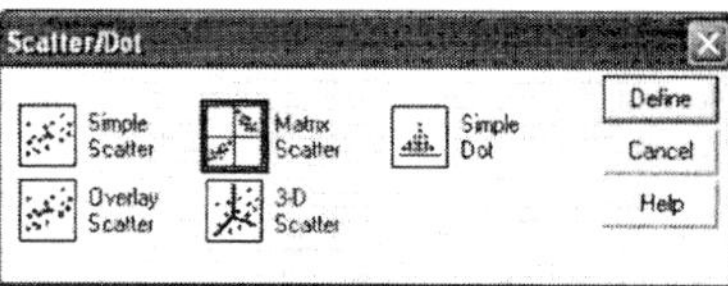

Enter the dependent variable and both independent variables in the “Matrix Variables” box.

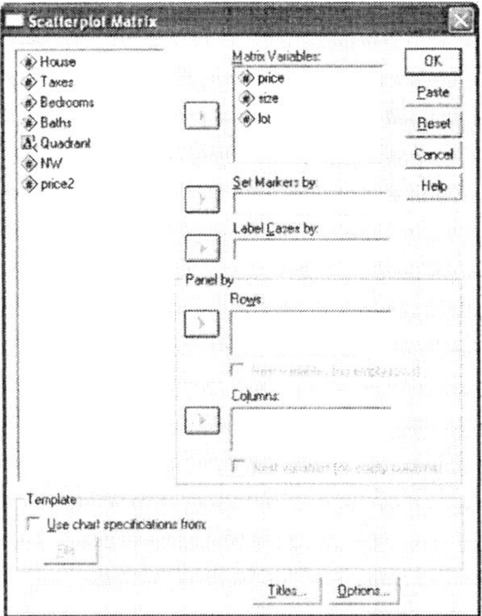

Click on **OK** to produce the output.

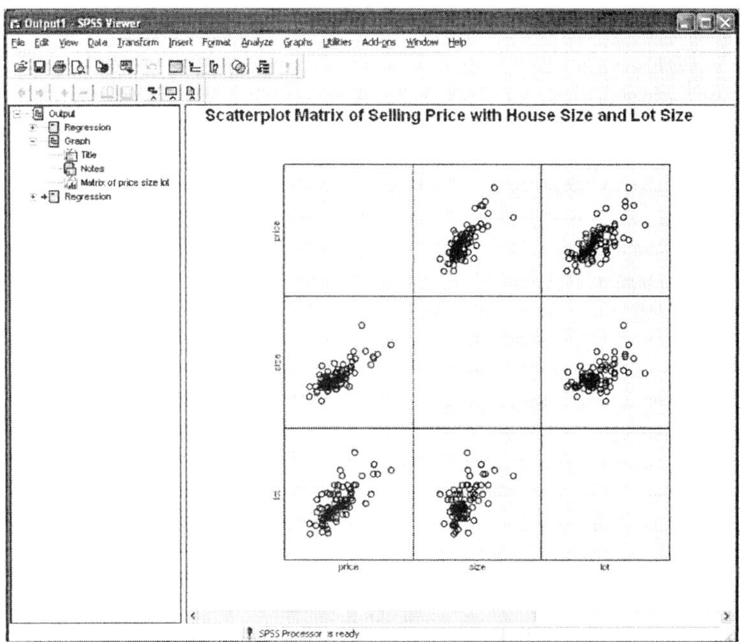

The top row of the matrix shows selling price on the Y-axis with house size on the X-axis in the second box and lot size on the X-axis in the third box. If you would like to see the corresponding correlation coefficients between each pair of variables in this matrix, select **Analyze → Correlate → Bivariate…** and enter all three variables in the Variables box.

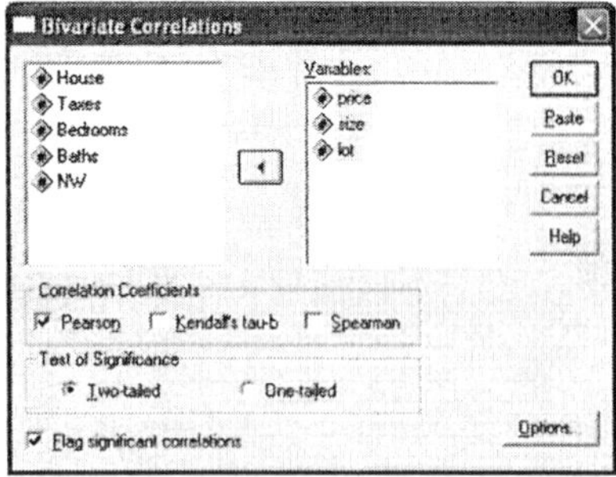

Click on **OK** to produce the output.

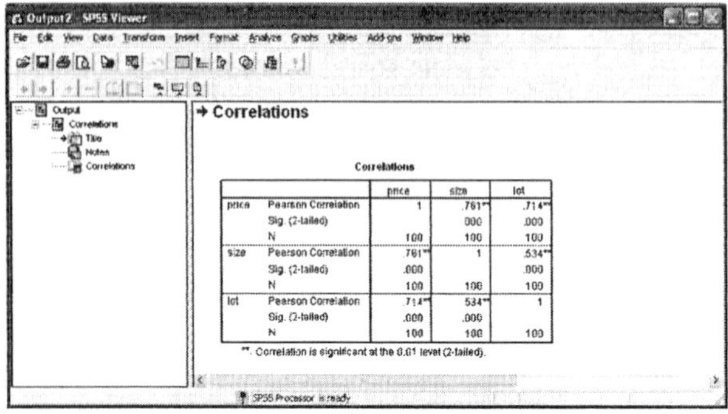

Correlations

		price	size	lot
price	Pearson Correlation	1	.761**	.714**
	Sig. (2-tailed)		.000	.000
	N	100	100	100
size	Pearson Correlation	.761**	1	.534**
	Sig. (2-tailed)	.000		.000
	N	100	100	100
lot	Pearson Correlation	.714**	.534**	1
	Sig. (2-tailed)	.000	.000	
	N	100	100	100

**. Correlation is significant at the 0.01 level (2-tailed).

Testing the Significance of the Regression Coefficients

To demonstrate how to use SPSS to test for the significance of the regression coefficients, we will use the "college_athletes" data set on your text CD. A portion of this data set is shown in the table below.

TBW	HGT	BF_%	AGE
96	62	13.0	23
130	65	16.3	17
107	64.5	17.3	21

Access this data set and select **Analyze → Regression → Linear...** Enter "TBW" as the "Dependent" variable and "HGT," "BF_Percent," and "AGE" as the "Independent(s)" variables.

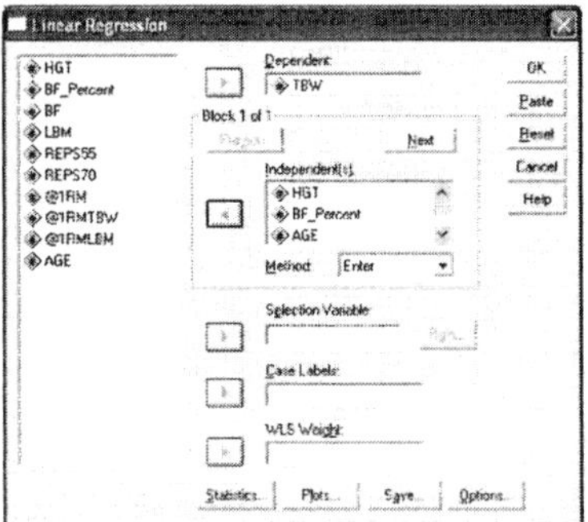

Click on **OK** to produce the output. The "Coefficients" portion of this output is shown below.

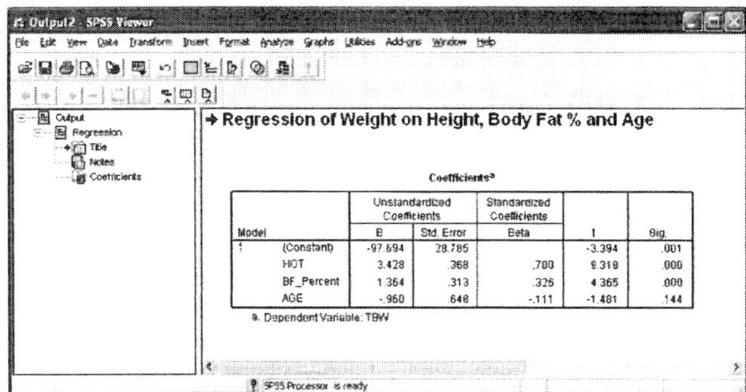

Regression of Weight on Height, Body Fat % and Age

Coefficients[a]

Model		Unstandardized Coefficients B	Unstandardized Coefficients Std. Error	Standardized Coefficients Beta	t	Sig.
1	(Constant)	-97.694	28.785		-3.394	.001
	HGT	3.428	.368	.700	9.319	.000
	BF_Percent	1.364	.313	.326	4.365	.000
	AGE	-.960	.648	-.111	-1.481	.144

a. Dependent Variable: TBW

The t-statistics in the "Coefficients" portion of the output, and their corresponding p-values (Sig.) indicate that both height and body fat percent are significant but age is not.

The Regression ANOVA

To determine if a set of independent variables collectively have an effect on a dependent variable you will use the "ANOVA" portion of the regression output. We will demonstrate this by also using the "college_athletes" data set. Select **Analyze → Regression → Linear…** and enter "TBW" as the "Dependent" variable and "HGT," "BF_Percent," and "AGE" as the "Independent(s)" variables. Click on **OK**. The "Model Summary" and "ANOVA" portions of the output are shown below.

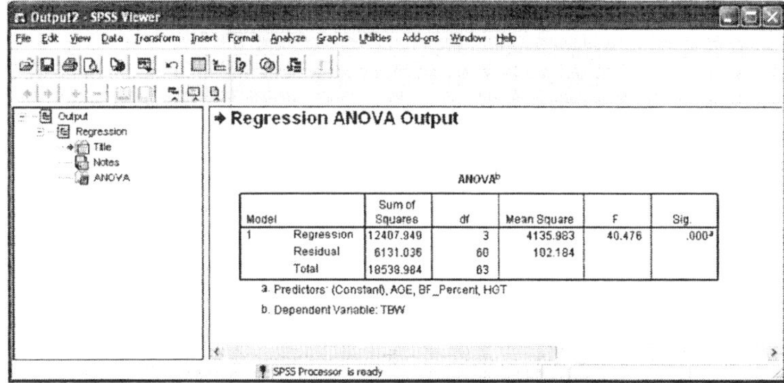

Regression ANOVA Output

ANOVA[b]

Model		Sum of Squares	df	Mean Square	F	Sig.
1	Regression	12407.949	3	4135.983	40.476	.000[a]
	Residual	6131.036	60	102.184		
	Total	18538.984	63			

a. Predictors: (Constant), AGE, BF_Percent, HGT

b. Dependent Variable: TBW

The F-statistic is for testing that at least one of the three independent variables is significantly different from zero. The corresponding p-value is given under "Sig." and, as with previous examples, this result tells us that the p-value is less than .00005. You will learn more about ANOVA tables in the next chapter.

Another Residual Plot

In Chapter 3 of this manual we looked at a histogram of the residuals as a diagnostic tool for assessing the fit of the best fit line and for looking for unusual observations. Another graph that is useful is to plot the residuals vs. the independent variable. The Regression command in SPSS has options under the "Plots" button for graphing the standardized residuals (*ZRESID) vs. the dependent variable (DEPENDNT) and for plotting the dependent variable vs. each of the independent variables separately (click on "Produce all partial plots), but to graph the standardized residuals vs. one of the independent variables you will have to save the residuals and produce the plot using the "Scatter/Dot" command. Let's demonstrate this using the "house_ selling_prices" data set. Access this data set and select **Analyze → Regression → Linear...** Enter "price" as the "Dependent" variable and "house size" and "lot size" as the "Independent(s)" variables. Click on the "Save" button and select "Standardized" under "Residuals." Click on **Continue** and then on **OK** to produce the regression output. A new column with name "ZRE_1" will be added to the Data Editor with the standardized residual of each observation.

To make the residual plot select **Graphs → Scatter/Dot...** and select the "Simple Scatter" option. Click on **Define** and enter the residuals ("ZRE_1") as the variable for the Y-axis and house size as the variable for the X-axis. Click on **OK** to produce the plot.

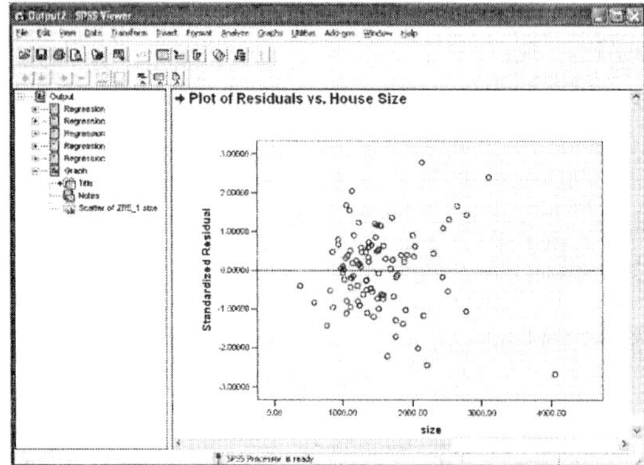

The horizontal line at zero on the Y-axis was added using the Chart Editor. Double click on the graph to open the Chart Editor, then select **Options → Y-axis reference line**. Then click on **Close** and close the Chart Editor. Make a similar graph with the other independent variable on the X-axis to check on its association with the residuals.

Categorical Predictor Variables

When a regression analysis is to be performed over groups identified by a categorical variable, the group to which a particular observation belongs can be identified in SPSS using an "indicator variable." An indicator variable takes just two values, 0 and 1, and so can be used when there are just two groups. If there are more than two groups, additional indicator variables can be used. The method for doing this is explained in your text.

Let's demonstrate the use of an indicator variable using the "high_jump2" data set on your text CD. This data set has three variables, "winning_height, "gender" and "year." The variable "gender" is an indicator variable, taking the value of "0" for females and "1" for males. A portion of this data set is shown below.

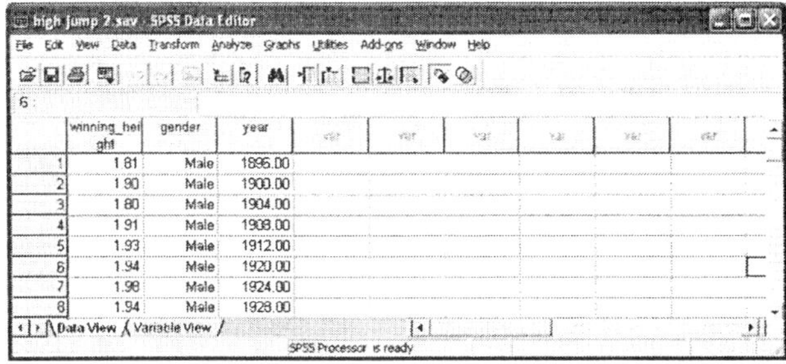

	winning_height	gender	year
1	1.81	Male	1896.00
2	1.90	Male	1900.00
3	1.80	Male	1904.00
4	1.91	Male	1908.00
5	1.93	Male	1912.00
6	1.94	Male	1920.00
7	1.98	Male	1924.00
8	1.94	Male	1928.00

To do the regression, select **Analyze → Regression → Linear...** and enter the "winning height" variable in the "Dependent" box and both "gender" and the "year" variable in the "Independent(s)" box. You do not need to do anything special to designate "gender" as an indicator variable.

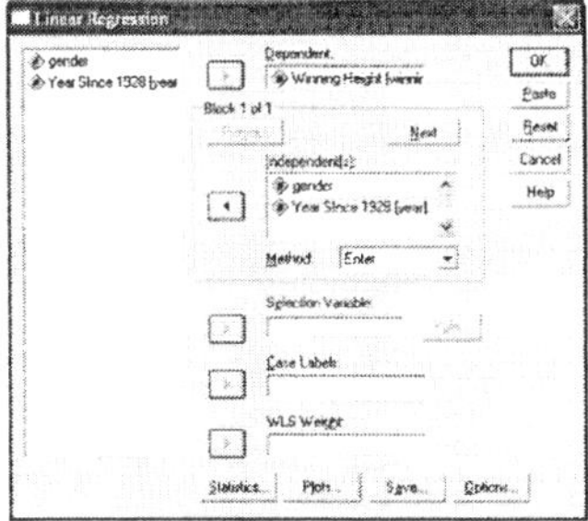

Click on **OK** to produce the output.

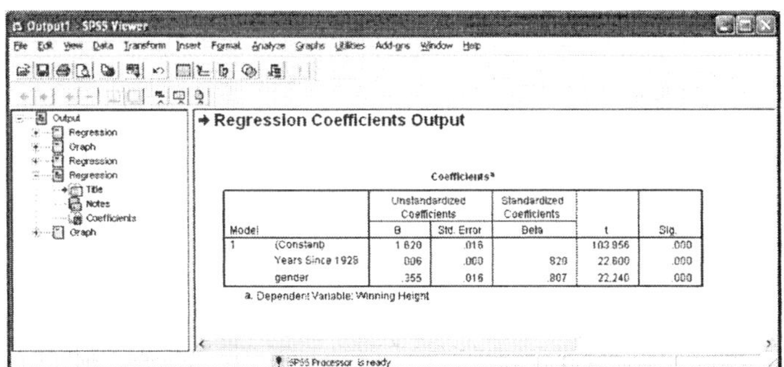

Regression Coefficients Output

Coefficients[a]

Model		Unstandardized Coefficients B	Std. Error	Standardized Coefficients Beta	t	Sig.
1	(Constant)	1.620	.016		103.956	.000
	Years Since 1928	.006	.000	.820	22.600	.000
	gender	.355	.016	.807	22.240	.000

a. Dependent Variable: Winning Height

The best-fit regression line is $\hat{y} = 1.62 + .006x_1 + .355x_2$. For females, where $x_2 = 0$, the best-fit line is $\hat{y} = 1.62 + .006x_1$ and for males, where $x_2 = 1$, the best-fit line is $\hat{y} = 1.62 + .006x_1 + .355(1) = 1.975 + .006x_1$. The result is two separate but parallel lines, one for the male data and the other for the female data.

To see if a model with two separate but parallel lines is a good "fit" to this data set, we can graph the data using a different symbol for the male and female data points. To do this select **Graphs → Scatter/Dot...,** and place the "height" variable in the "Y-axis" box and the "year" variable in the "X-axis" box. Then, enter the "gender" variable in the "Set Markers by" box.

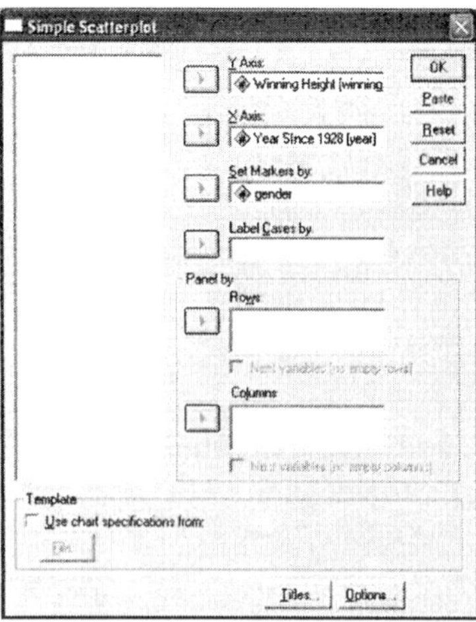

Click on **OK** to produce the plot.

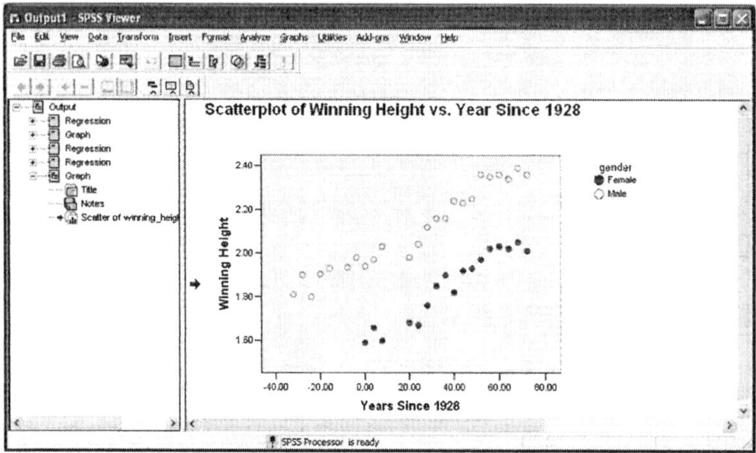

SPSS will use a different color for the two groups defined by the "gender" variable. You can use the Chart Editor to change the plot symbols and colors if you desire. In the plot above, the points for the female cases are filled in. We can see from this plot that two separate but parallel lines is a reasonable model for this data set.

Logistic Regression

SPSS can also do logistic regression on a binary response variable. To demonstrate this we will use the data from Example 12 in Chapter 12 on income (in 1000s of Euros) and whether or not the respondent has a travel credit card. The dependent variable Y is coded as 0 = "no" and 1 = "yes." Access the "credit_card_ and_income" data set from your text CD and select **Analyze → Regression → Binary Logistic...** Enter the variable "y" as the "Dependent" variable and "income" as the "Covariate."

Click on **OK** to produce the output.

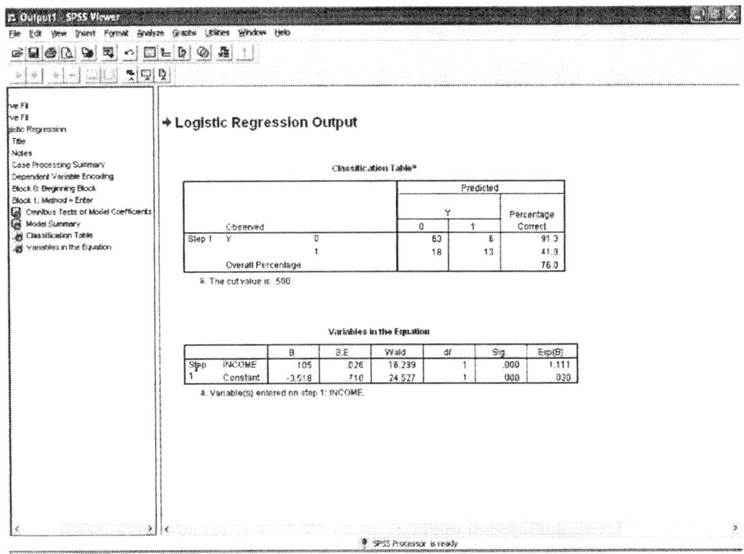

Logistic Regression Output

Classification Table[a]

Observed		Predicted Y 0	Predicted Y 1	Percentage Correct
Step 1 Y	0	63	6	91.3
	1	18	13	41.9
Overall Percentage				76.0

a. The cut value is .500

Variables in the Equation

		B	S.E.	Wald	df	Sig.	Exp(B)
Step 1[a]	INCOME	.105	.026	16.239	1	.000	1.111
	Constant	-3.518	.710	24.527	1	.000	.030

a. Variable(s) entered on step 1: INCOME.

The coefficients for the logistic model are given under "B" in the bottom portion of the output. The logistic equation would be $\hat{p} = \dfrac{e^{-3.52+0.105x}}{1+e^{-3.52+.105x}}$, which agrees with the result shown in your text. For example, for x=12 the estimate would be $\hat{p} = \dfrac{e^{-3.52+.105(12)}}{1+e^{-3.52+.105(12)}} = .0944$. So, the estimated proportion of those whose income is 12 thousand Euros that have a travel credit card is around 9%.

Chapter 12 Analysis of Variance

The extension of the independent samples T-test to compare more than two means is called Analysis of Variance. SPSS has two commands that perform ANOVA, "One-Way ANOVA" under the "Compare Means" command and "Univariate" under the "General Linear Model" command. We will demonstrate "One-Way ANOVA" in this section and "Univariate" for doing two-way and higher ANOVAs. Note that, when there are just two groups you can do either the independent samples T-test or a one-way ANOVA. Unlike the two samples situation, for ANOVA we have to assume that the variances of the groups being compared are the same. We will show how to check this assumption using SPSS.

One-Way ANOVA

To demonstrate how to compare the means of more than two independent samples, we will use the telephone holding data from Example 2 in Chapter 13. This data is shown in the table below.

Recorded Message	Holding Time Observations
Advertisement	5, 1, 11, 2, 8
Muzak	0, 1, 4, 6, 3
Classical	13, 9, 8, 15, 7

Enter this data in SPSS using one variable for the holding times and another for the type of message. The variable for the type of music must be coded numerically. To do the ANOVA, select **Analyze → Compare Means → One-Way ANOVA...** and enter the holding time variable in the "Dependent List" and the message type variable in the "Factor" list.

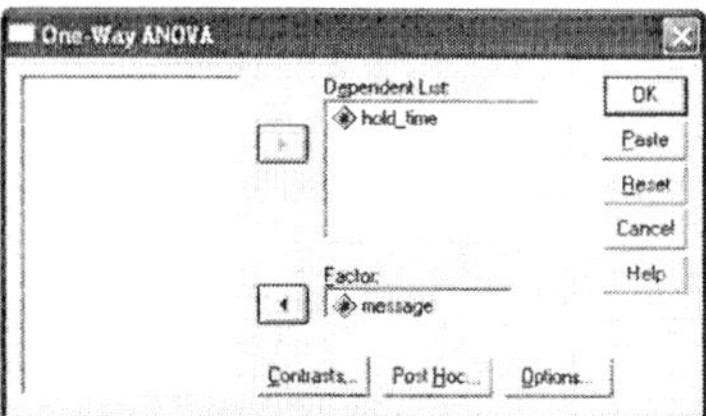

Click on **OK** to produce the output.

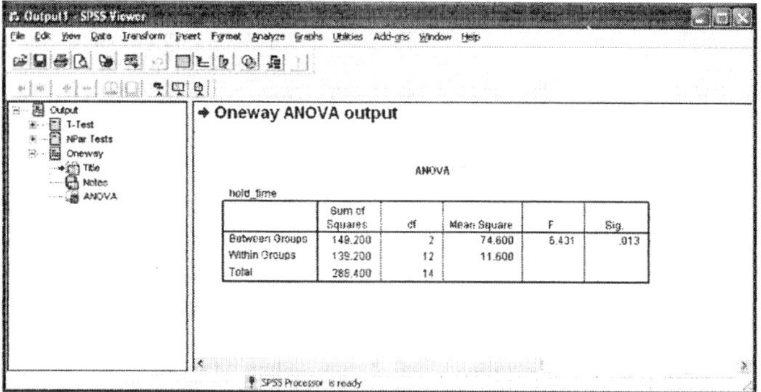

ANOVA

hold_time

	Sum of Squares	df	Mean Square	F	Sig.
Between Groups	149.200	2	74.600	6.431	.013
Within Groups	139.200	12	11.600		
Total	288.400	14			

The p-value is small so this suggests that there is a difference in the average amount of time subjects were willing to hold based on the type of message they heard.

To find out which groups differ and if the equal variances assumption is met, we can use some of the options of this command. Select **Analyze → Compare Means → One-Way ANOVA...** again and click on the "Post Hoc..." button. Click on "Tukey" and then on **Continue.**

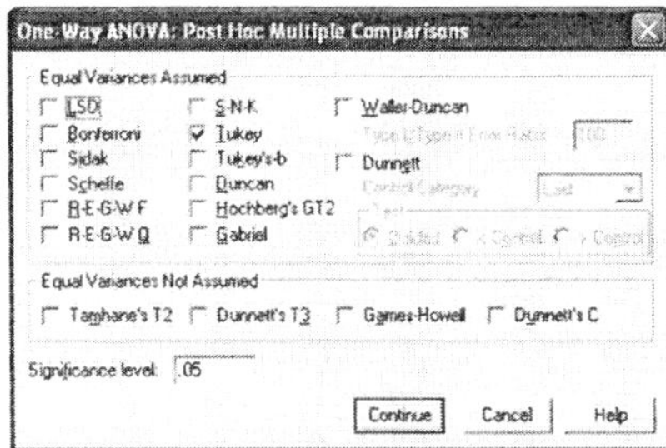

Click on the "Options" button and then on both "Descriptive" and "Homogeneity of variance test."

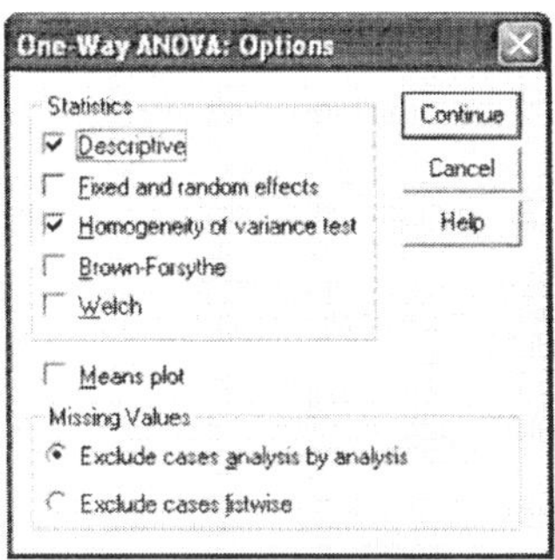

Click on **Continue** and then on **OK**. SPSS will produce a lot of additional output.

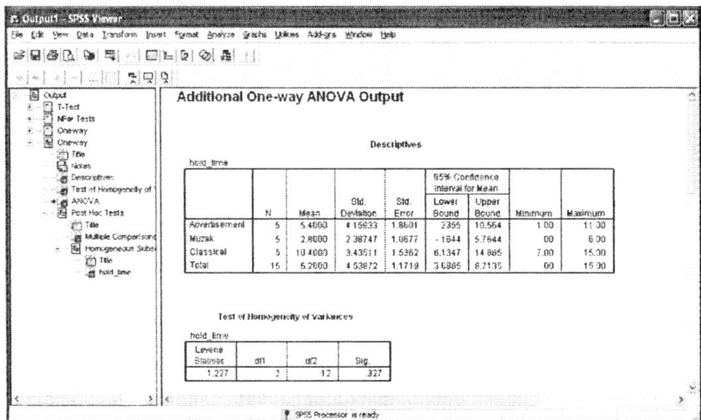

Additional One-way ANOVA Output

Descriptives

hold_time

	N	Mean	Std. Deviation	Std. Error	95% Confidence Interval for Mean Lower Bound	Upper Bound	Minimum	Maximum
Advertisement	5	5.4000	4.15933	1.8601	.2365	10.564	1.00	11.00
Muzak	5	2.8000	2.38747	1.0677	-.1644	5.7644	.00	6.00
Classical	5	10.4000	3.43511	1.5362	6.1347	14.665	7.00	15.00
Total	15	6.2000	4.53872	1.1719	3.6865	8.7135	.00	15.00

Test of Homogeneity of Variances

hold_time

Levene Statistic	df1	df2	Sig.
1.227	2	12	.327

The output above shows the group means and standard deviations along with individual confidence intervals for the mean of each group. Levene's test for equal variances is also produced. The p-value for this test is .327, indicating that there is no evidence to suggest that the variances are unequal.

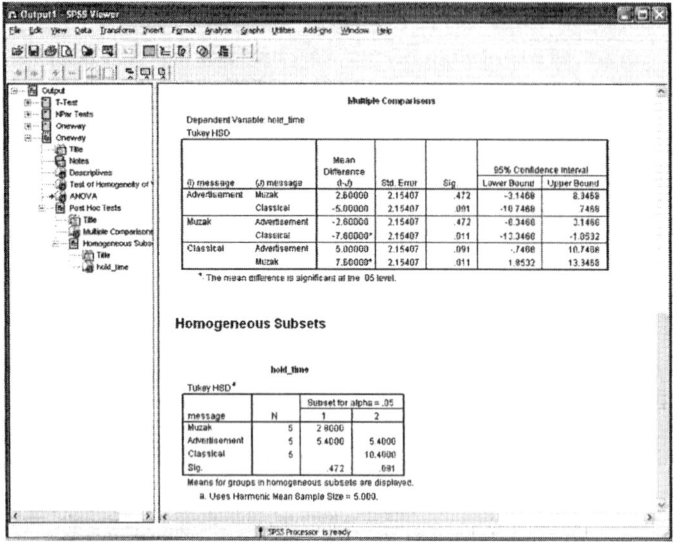

This output shows Tukey confidence intervals for simultaneous comparisons of each pair of groups. The last part of the output shows the conclusions of which groups differ based on the Tukey confidence intervals. If the Fisher-type confidence intervals are preferred, these can be produced using "Means" from the "Compare Means" command, as described in Chapter 8.

Two-Way ANOVA

As indicated at the beginning of this chapter, if you are doing a one-way ANOVA you can use either "One-Way ANOVA" under the "Compare Means" command or you can use "Univariate" under the "General Linear Model" command. To conduct an ANOVA with two or more factors you will need to use the "General Linear Model" command. To demonstrate this command we will use the agricultural data from Example 9 in Chapter 13. The data is provided below.

Fertilizer Level	Manure Level	Plot 1	2	3	4	5
High	High	13.7	15.8	13.9	16.6	15.5
High	Low	16.4	12.5	14.1	14.4	12.2
Low	High	15.0	15.1	12.0	15.7	12.2
Low	Low	12.4	10.6	13.7	8.7	10.9

Enter the data in SPSS in three columns, one for the crop yield and one each for the two factors, fertilizer level and manure level. Enter the value "1" for the "high" level of each factor and "0" for the "low" level. Use value labels under the Variable View to label the values appropriately. A portion of the data set is visible in the Data Editor shown below.

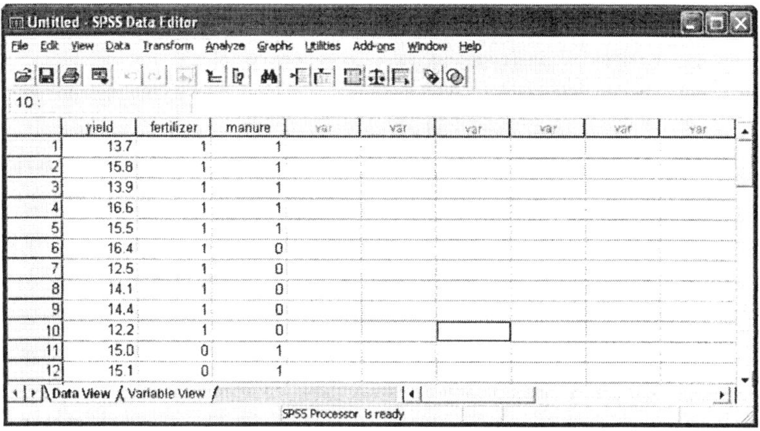

	yield	fertilizer	manure
1	13.7	1	1
2	15.8	1	1
3	13.9	1	1
4	16.6	1	1
5	15.5	1	1
6	16.4	1	0
7	12.5	1	0
8	14.1	1	0
9	14.4	1	0
10	12.2	1	0
11	15.0	0	1
12	15.1	0	1

To do the ANOVA select **Analyze → General Linear Model → Univariate…** and enter "yield" as the "Dependent Variable" and both "fertilizer" and "manure" as "Fixed Factor(s)."

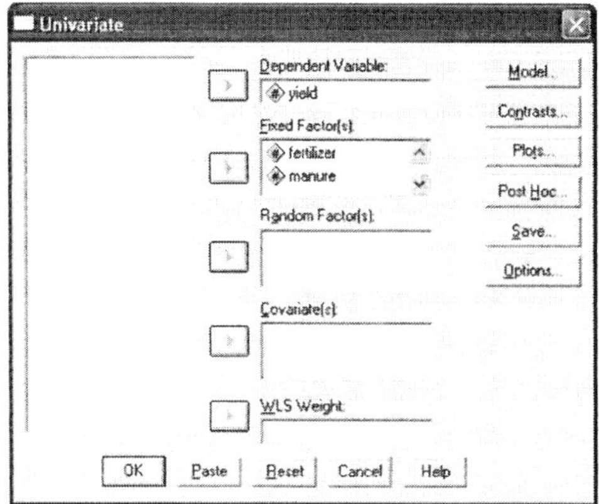

The default model in SPSS includes an interaction term. To run the ANOVA without this term, click on the "Model" button and then on "Custom." Click once on "fertilizer" under "Factors & Covariates" and then once on the arrow button underneath "Build Term(s)." Repeat this process with "manure." Your dialog box should look like the one shown below.

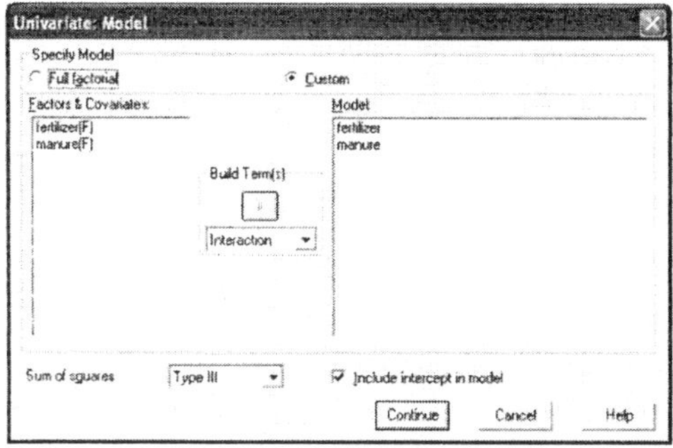

Click on **Continue** and then on **OK**.

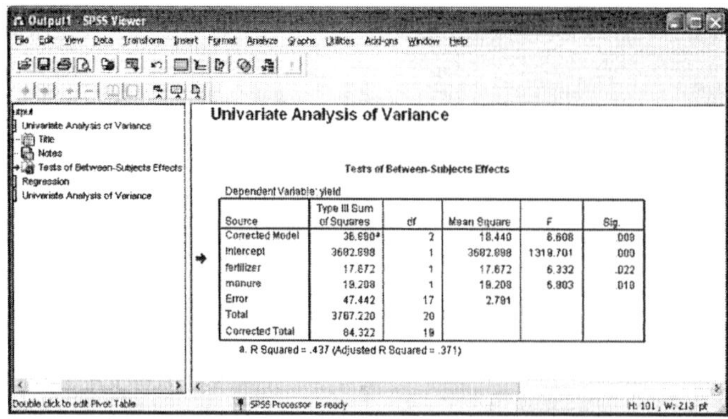

Univariate Analysis of Variance

Tests of Between-Subjects Effects

Dependent Variable: yield

Source	Type III Sum of Squares	df	Mean Square	F	Sig.
Corrected Model	36.880[a]	2	18.440	6.608	.008
Intercept	3682.898	1	3682.898	1319.701	.000
fertilizer	17.672	1	17.672	6.332	.022
manure	19.208	1	19.208	6.883	.018
Error	47.442	17	2.791		
Total	3767.220	20			
Corrected Total	84.322	19			

a. R Squared = .437 (Adjusted R Squared = .371)

This ANOVA output includes some additional terms under "Source" that are not in the ANOVA table in your text, but the results for "fertilizer" and "manure" are the same.

The "Options" button on the "Univariate" dialog box includes a number of useful options. For example, you can request the estimated marginal means, additional descriptive statistics and a test for equal variances (homogeneity).

The Regression Approach to ANOVA

For factors that have just two categories, such as "fertilizer" and "manure" in the previous example, you can also do ANOVA using the "Regression" command. For factors with more than two categories it is easier to just use the "General Linear Model" command (or the "Compare Means" command for a one-way ANOVA). Using the same agricultural data set select **Analyze → Regression → Linear...** and enter "yield" as the "Dependent" variable and both "fertilizer" and "manure" as "Independent(s)" variables.

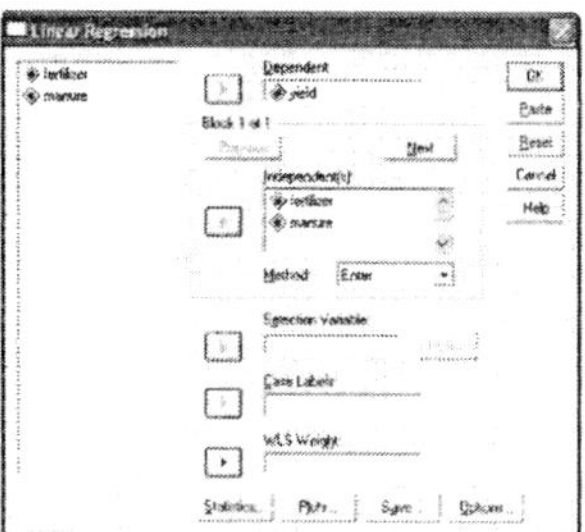

Click on **OK** to produce the output.

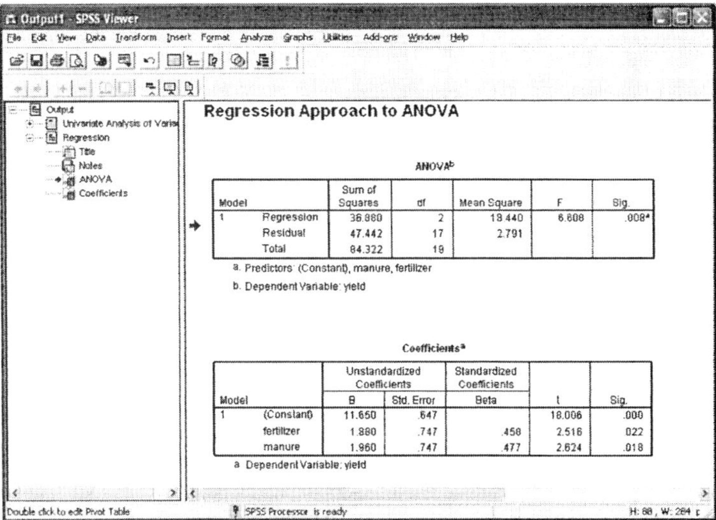

Regression Approach to ANOVA

ANOVA[b]

Model		Sum of Squares	df	Mean Square	F	Sig.
1	Regression	36.880	2	18.440	6.608	.008[a]
	Residual	47.442	17	2.791		
	Total	84.322	19			

a. Predictors: (Constant), manure, fertilizer

b. Dependent Variable: yield

Coefficients[a]

Model		Unstandardized Coefficients		Standardized Coefficients	t	Sig.
		B	Std. Error	Beta		
1	(Constant)	11.650	.647		18.006	.000
	fertilizer	1.880	.747	.458	2.516	.022
	manure	1.960	.747	.477	2.624	.018

a. Dependent Variable: yield

The ANOVA table from the "Regression" command does not separate out the effects of each factor separately, as in the first ANOVA table we produced for this data. But, the t-tests provided in the "Coefficients" portion of the output will allow you to test for the significance of each factor separately. The "slopes" associated with each factor under the "Coefficients" portion of the output provide the values you can use to predict the size of the yield for each combination of the levels of the two factors. For example, for a plot in which the high level of each factor was used we would predict an average yield of 11.65 + 1.88 + 1.96 = 15.49.

Two-Way ANOVA with Interaction

The default model for two-way and higher ANOVA in SPSS includes all interaction terms. For two-way ANOVA there is just one interaction term. To do the ANOVA with the interaction term for the agricultural data set, select **Analyze → General Linear Model → Univariate...** again and click on the "Model" button. This time, make sure that "Full factorial" is selected.

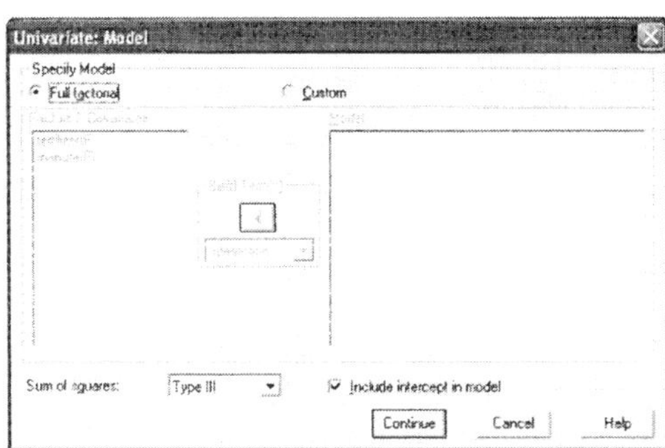

Click on **Continue** and then on **OK**.

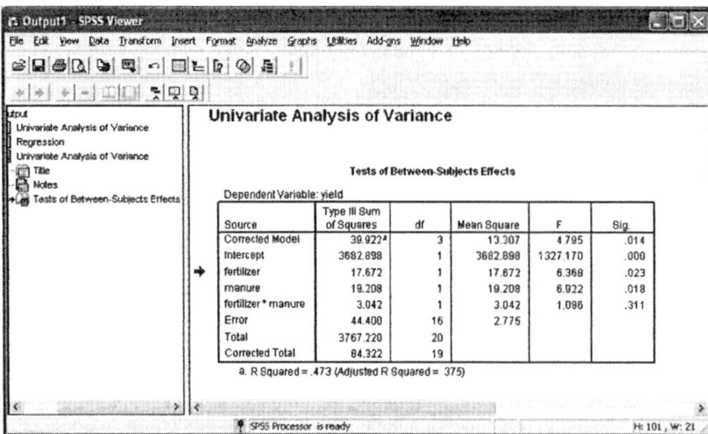

Univariate Analysis of Variance

Tests of Between-Subjects Effects

Dependent Variable: yield

Source	Type III Sum of Squares	df	Mean Square	F	Sig.
Corrected Model	39.922[a]	3	13.307	4.795	.014
Intercept	3682.898	1	3682.898	1327.170	.000
fertilizer	17.672	1	17.672	6.368	.023
manure	19.208	1	19.208	6.922	.018
fertilizer * manure	3.042	1	3.042	1.096	.311
Error	44.400	16	2.775		
Total	3767.220	20			
Corrected Total	84.322	19			

a. R Squared = .473 (Adjusted R Squared = .375)

Here the interaction term is not significant.

To include the interaction term using the regression approach, you will first have to create an interaction variable in the data set. Select **Transform → Compute…** and enter the variable name “interaction” under “Target Variable” and the expression “fertilizer*manure” in the “Numeric Expression” box. Click on **OK**. This new variable will be added to your data set.

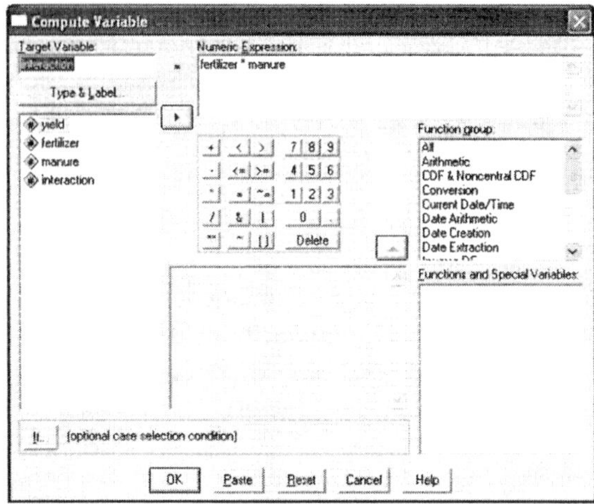

To do the ANOVA, select **Analyze → Regression → Linear…** and add the interaction variable to the “Independent(s)” variables list. Click on **OK**.

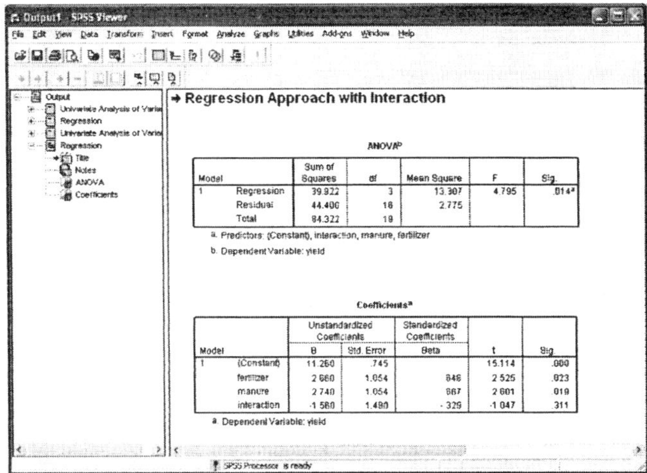

→ Regression Approach with Interaction

ANOVA[b]

Model		Sum of Squares	df	Mean Square	F	Sig.
1	Regression	39.922	3	13.307	4.795	.014[a]
	Residual	44.400	16	2.775		
	Total	84.322	19			

a. Predictors: (Constant), interaction, manure, fertilizer

b. Dependent Variable: yield

Coefficients[a]

Model		Unstandardized Coefficients		Standardized Coefficients	t	Sig.
		B	Std. Error	Beta		
1	(Constant)	11.260	.745		15.114	.000
	fertilizer	2.680	1.054	.848	2.525	.023
	manure	2.740	1.054	.867	2.601	.019
	interaction	-1.580	1.490	-.329	-1.047	.311

a. Dependent Variable: yield

As before, SPSS does not separate out the effect of the interaction term in the "ANOVA" portion of the output, but you can test for its significance using the T-test under "Coefficients."

Chapter 13 Nonparametric Statistics

All of the methods we've looked at so far for comparing means have an underlying normality assumption. It turns out that these methods are fairly "robust" when it comes to working with non-normal data; that is, the methods perform well in detecting a difference in means even when the data is not exactly normally distributed. For data from highly skewed distributions or from populations that differ substantially in the amount of variation present, these normal-based methods do not perform well and in these situations we need some alternate methods that do not involve a normality assumption.

The Wilcoxon Test

The nonparametric test for this situation is called the Wilcoxon test (or the Wilcoxon rank sum test or the Mann-Whitney test). As your text explains, it is based on replacing the actual data values with their associated ranks among all observations. To demonstrate this test, we will use the respiratory ventilation data from Exercise 14.4. This data set consists of respiratory ventilation measurements made on two groups, a control group and a treatment group that would undergo hypnosis. The data is shown in the following table.

Controls	3.99 4.19 4.21 4.54 4.64 4.69 4.84 5.48
Treated	4.36 4.67 4.78 5.08 5.16 5.20 5.52 5.74

Enter the data in SPSS using two variables, one for the ventilation measurements and one for the group. For this analysis, the grouping variable must be entered as a numeric variable. Recall that for the "Compare Means" and "One-way ANOVA" commands you could enter the grouping variable as either a numeric of string variable.

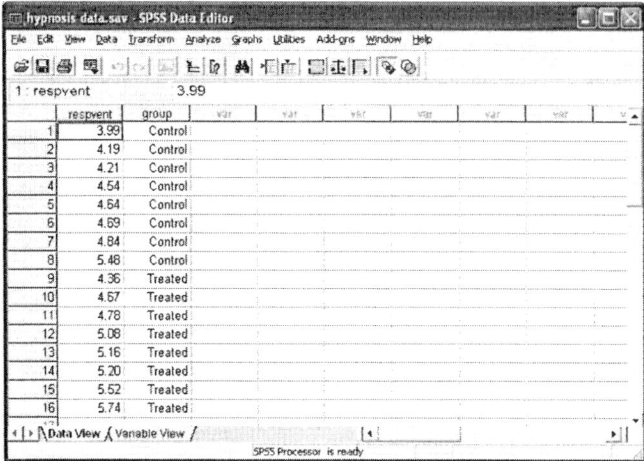

To do the nonparametric test, select **Analyze → Nonparametric Tests → 2 Independent Samples...** and enter the "respiratory ventilation" variable under "Test Variable List" and the "group" variable in the box under "Grouping Variable." Click on **Define Groups** and enter the values you used to define the two groups. In the example below you can see that the groups were defined using the values 1 and 2.

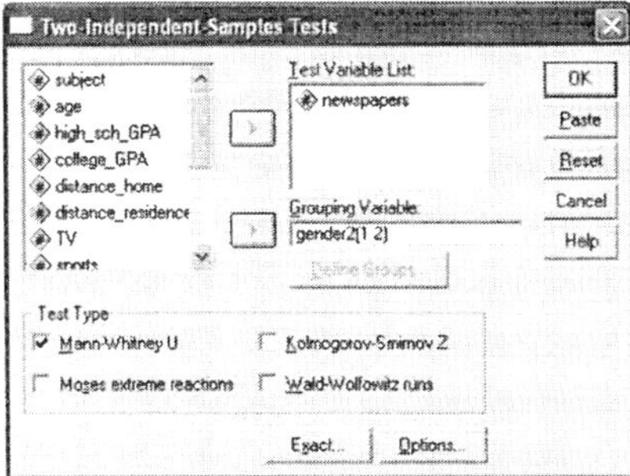

Click on **Continue** and then on **OK**.

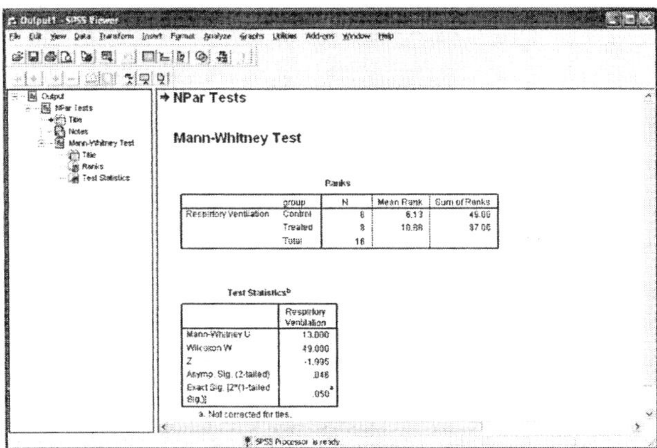

	group	N	Mean Rank	Sum of Ranks
Respiratory Ventilation	Control	8	6.13	49.00
	Treated	8	10.88	87.00
	Total	16		

	Respiratory Ventilation
Mann-Whitney U	13.000
Wilcoxon W	49.000
Z	-1.995
Asymp. Sig. (2-tailed)	.046
Exact Sig. [2*(1-tailed Sig.)]	.050[a]

SPSS will assign a rank to each observation but only the mean rank and sum of ranks are reported for each group. The results for this data set show an exact two-sided p-value of .050, but for this situation we need the one-sided p-value. Since we want to know if the treated group (those who would be hypnotized) ventilated more, and we can see that the mean ranks for the treated group is larger for the treated group, the p-value will be half of .050 or .025.

The Kruskal-Wallis Test

The extension of the Wilcoxon test for comparing the means of more than two samples is called the Kruskal-Wallis Test. In this test, as in the Wilcoxon test, the observations are replaced by their ranks among all observations. To demonstrate this test we'll use the data from Example 5 in Chapter 14. This data, shown in the table below, is from a study to investigate the impact of dating on GPA.

Dating Group	GPA observations						
Rare	1.75	3.15	3.50	3.68			
Occasional	2.00	3.20	3.44	3.50	3.60	3.71	3.80
Regular	2.40	9.25	3.40	3.67	3.70	4.00	

Enter this data in SPSS with one variable for the GPA and another for the dating group.

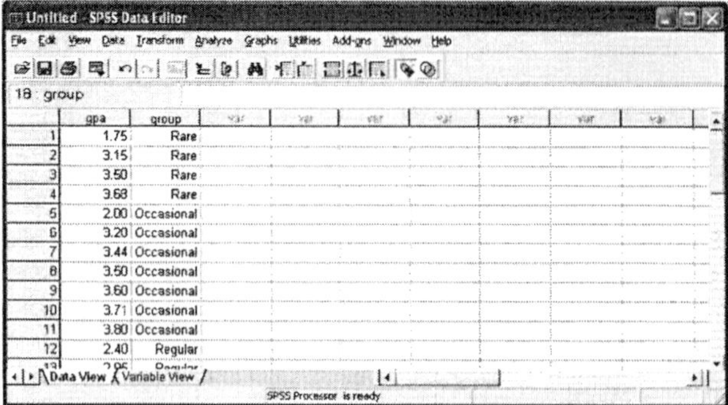

	gpa	group
1	1.75	Rare
2	3.15	Rare
3	3.50	Rare
4	3.68	Rare
5	2.00	Occasional
6	3.20	Occasional
7	3.44	Occasional
8	3.50	Occasional
9	3.60	Occasional
10	3.71	Occasional
11	3.80	Occasional
12	2.40	Regular

To conduct the Kruskal Wallis nonparametric test on this data set select **Analyze → Nonparametric Tests → K Independent Samples...** Enter "GPA" in the "Test Variable List" box and the "group" variable in the "Grouping Variable" box. The dialog box is similar to the ANOVA dialog box except that you need to click on the "Define Range..." button and enter the minimum and maximum values of the "Grouping Variable" that identify the groups to be compared.

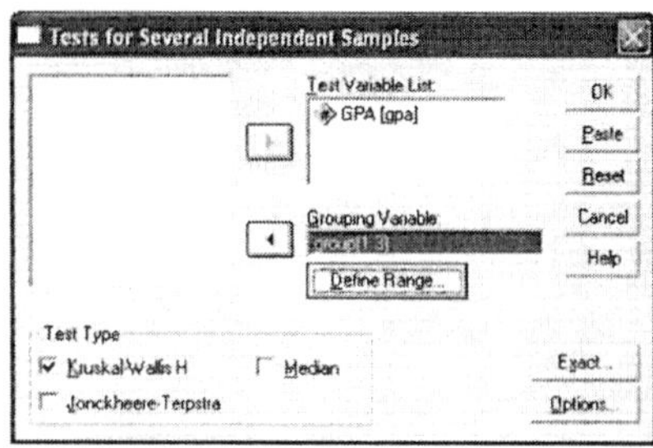

The Kruskal-Wallis test is the default option for this command. Click on **OK** to produce the output.

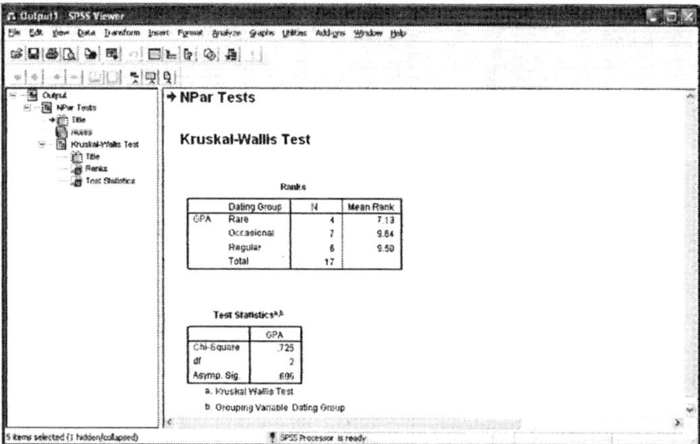

NPar Tests

Kruskal-Wallis Test

Ranks

	Dating Group	N	Mean Rank
GPA	Rare	4	7.13
	Occasional	7	9.64
	Regular	6	9.50
	Total	17	

Test Statistics[a,b]

	GPA
Chi-Square	.725
df	2
Asymp. Sig.	.696

a. Kruskal Wallis Test

b. Grouping Variable: Dating Group

As in the previous test, SPSS will assign a rank to each observation but the output reports just the mean rank for each group. The test statistic and its associated p-value are reported, as well. The large p-value for this data set suggests that GPA is independent of dating group.

The Sign Test

For analyzing data from paired or dependent samples, your text describes two separate tests, the Sign Test and the Wilcoxon Signed-Ranks Test. The difference between these two tests is that, in the sign test only the number of positive differences is recorded while in the Wilcoxon Signed-Rank the differences are ranked in absolute value and the sum of the positive ranks is recorded. To demonstrate the Sign Test we will use data from the "georgia_student_suvey" file from your text CD. This example is described in Example 6 from Chapter 14 of your text. The paired variables are the time (in minutes per day) spent watching TV and browsing the Internet. Access this data file and select **Analyze → Nonparametric Tests → 2 Related Samples...** and enter the paired variables in the same way as for the Paired T-test described in Chapter 8 of this manual. Click once on the "internet" variable and once on the "TV" variable and then click on the triangle button to enter the pair in the "Test Pair(s) List" box. Click on the "Sign" option to select it. The Wilcoxon Test (demonstrated in the next example) is the default option. You will have to click once on it to un-select it.

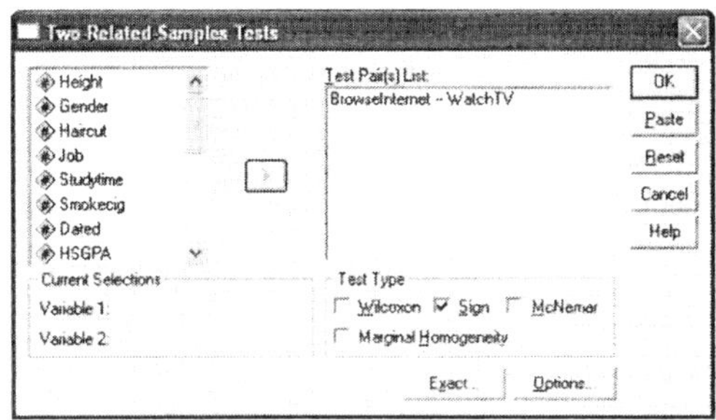

Click on **OK** to produce the output.

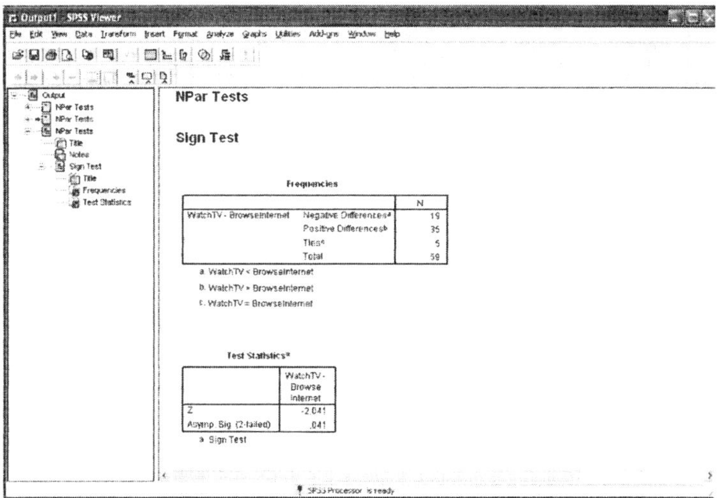

The output shows that there were 19 students who spent more time browsing the Internet (so the difference was negative) and 35 students who spent more time watching TV. All students who spent the same amount of time on these activities are not included in the analysis. SPSS reports both the exact value of the test statistic and is associated p-value.

The Wilcoxon Signed-Ranks Test

Let's use the same data set to demonstrate this test. Select **Analyze → Nonparametric Tests → 2 Related Samples...** again except this time select the Wilcoxon test (and un-select the Sign test).

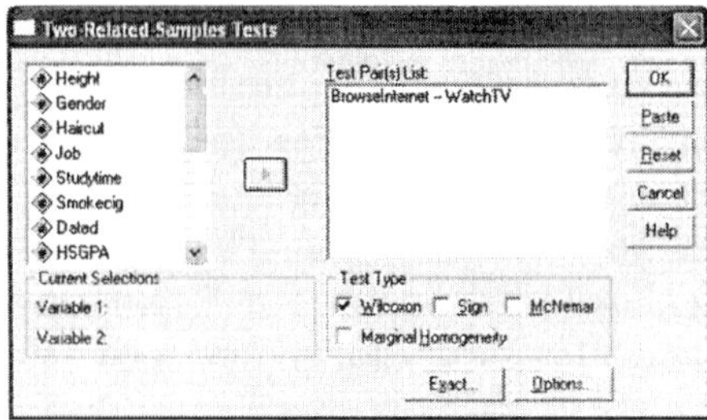

Click on **OK** to produce the output.

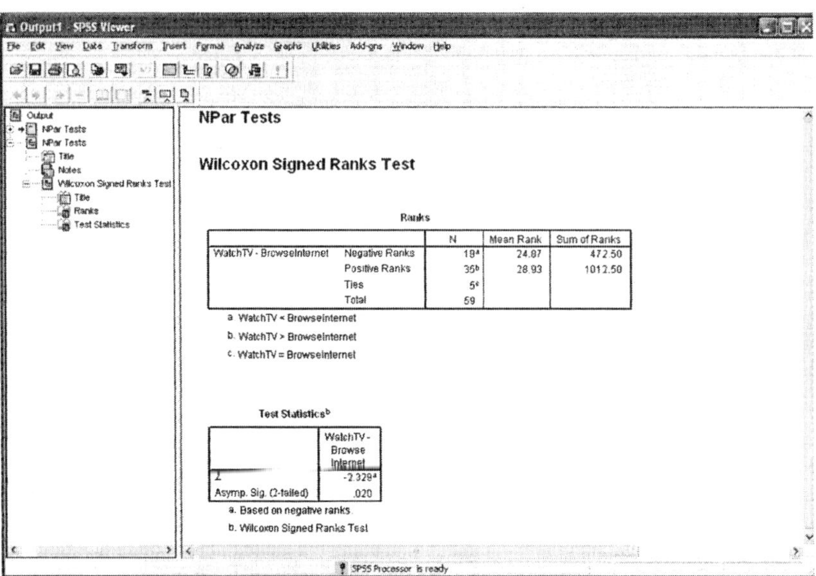

NPar Tests

Wilcoxon Signed Ranks Test

Ranks

		N	Mean Rank	Sum of Ranks
WatchTV - BrowseInternet	Negative Ranks	19[a]	24.87	472.50
	Positive Ranks	35[b]	28.93	1012.50
	Ties	5[c]		
	Total	59		

a. WatchTV < BrowseInternet

b. WatchTV > BrowseInternet

c. WatchTV = BrowseInternet

Test Statistics[b]

	WatchTV - BrowseInternet
Z	-2.328[a]
Asymp. Sig. (2-tailed)	.020

a. Based on negative ranks.

b. Wilcoxon Signed Ranks Test

SPSS reports both the mean ranks and sum of ranks for the negative and positive differences. The test statistic is the sum of the positive ranks for which SPSS reports a two-sided p-value.

Appendix A: Summary of SPSS Commands

The following is a list of frequently used SPSS procedures using the convention:

Select **Menu name→ Command name → Subcommand** or **option**

In the dialog box that appears, select one or response (dependent) variables and explanatory (independent) variables where appropriate. When you have entered enough information for SPSS to be able to complete the command, the "OK" button will become highlighted.

Operating SPSS

Creating a new data set

Start SPSS and select "Type in new data" or select **File → New → Data**. (Page 23)

Opening an existing data file

Select **File → Open → Data…** and locate the file from options under "Look in:" (Page 34)

Changing the window display

Select **Window→ SPSS Data Editor** or **SPSS Viewer** (Page 8)

Saving data

Select **File → Save as…** when the Data Editor is the active window. (Page 6)

Saving output

Select **File → Save as…** when the Output Viewer is the active window. (Page 13)

Printing output

Select the output to be printed, then select **File → Print.** (Page 12)

Exiting SPSS

Select **File→ Exit.** (Page 13)

Data Management

Naming a variable

Click on the **Variable View** tab and enter a new name under "Name." (Page 4)

Labeling a variable

Click on the **Variable View** tab and enter a label under "Label." (Page 4)

Selecting a variable type

Click on the Variable View tab, click on the box under "Type" and select "Numeric" or "String." (Page 3)

Labeling the values of a numerically coded qualitative variable

Click on the **Variable View** tab then on the box under "Values," enter a value in the "Value" box and a label in the "Label" box, then click on "Add." (Page 4)

Calculating a new variable from existing variables

Select **Transform → Compute…** and enter a variable name in the "Target Variable" box and an expression in the "Numeric Expression" box. (Pages 72, 90, 104, and 122)

Selecting a random sample

Select **Data → Select Cases…** then select "**Random sample of cases…**" (Pages 67 and 86)

Omitting specific cases

Select **Data → Select Cases…** then select "**If condition is Satisfied…**" and enter an expression that will eliminate specific cases; to analyze all cases for subsequent commands, Select **Data → Select Cases…** again and select "**All cases**." (Page 67)

Weighting cases for tabulated data

Select **Data → Weight Cases…** and enter a count variable in the "Frequency Variable" box. (Page 15)

Repeating an analysis of a quantitative variable for each value of a qualitative variable

Select **Data → Split File...** and select the option for comparing groups, enter the qualitative variable in the "Groups Base on" box; to analyze all cases for subsequent commands select **Data → Split File...** again, select "**Analyze all cases**..." (Pages 57 and 58)

Descriptive Statistics and Graphs

Listing data

Select **Analyze → Reports → Case summaries...** (Page7)

Frequency distributions

Univariate: Select **Analyze → Descriptive Statistics → Frequencies...** (Page 25)

Bivariate: **Analyze →Descriptive Statistics → Crosstabs...** (Pages 52 and 55)

Descriptive Statistics

Select **Analyze → Descriptive Statistics → Descriptives...** (Pages 9, 42 and 50)

Select **Analyze → Descriptive Statistics → Frequencies...** and request optional statistics. (Pages 41, 43, 44 and 50)

Select **Analyze → Compare means → Means...** (Page 56)

Select **Analyze → Descriptive Statistics → Explore...** and request optional statistics. (Pages 45, 49, 50, 56 and 81)

Univariate Graphs

Categorical Data

Pie Chart: Select **Graphs → Pie...** (Pages 16 and 20)

Bar Chart: Select **Graphs → Bar...** (Pages 23, 53 and 55)

Pareto Chart: Select **Graphs → Pareto...** (Page 23)

Quantitative Data

Dot Plot: Select **Graphs → Scatter/Dot...** (Page 26)

Histogram: Select **Graphs → Histogram...** (Pages 10 and 29)

Boxplot: Select **Graphs → Boxplot…** (Pages 46 and 47)

Bar Chart, Pie Chart or Histogram: Select **Analyze→ Descriptive Statistics → Frequencies…** and request optional Charts. (Page 25)

Boxplot, Histogram or Stem-and-Leaf Plot: Select **Analyze→ Descriptive Statistics → Explore…** and request optional Plots. (Page 28)

Line Plot: Select **Graphs → Line…** (Page 36) <u>or</u> select **Graphs → Sequence…** (Page 38) <u>or</u> select **Graphs → Scatter/Dot…** if data includes a time variable (Page 39) to which the regression line can be added. (Page 65)

Bivariate Graphs

One Categorical Variable and One Quantitative Variable

Select **Graphs → Boxplots…** and select **Summaries for groups of cases** (Page 47)

Select **Analyze → Descriptive Statistics → Explore…** (Page 56)

See **Data → Split File…** above under <u>Data Management</u> .

Two Quantitative Variables

Scatterplot: Select **Graphs → Scatter/Dot…** and select **Simple Scatter** (Pages 57 and 114) for a graph of Y vs. X or select **Matrix Scatter** (Page 109) for a graph of Y vs. several X variables.

Investigating Associations

Two Categorical Variables

Select **Analyze → Descriptive Statistics → Crosstabs…** (Pages 94, 96, 97 and 98)

Two Quantitative Variables

Select **Analyze → Correlate → Bivariate…** (Pages 59 and 109)

Select **Analyze → Regression → Linear…** (Pages 60, 100, 101, 105 and 107)

Select **Analyze → Regression → Curve Estimation…** (Page 103)

Select **Analyze → Regression → Binary Logistic…** (Page 115)

Comparing Means and Proportions

Comparing Means

One Sample

Select **Analyze** → **Compare Means** → **One-Sample T-Test…** (Page 86)

Two Independent Samples

Select **Analyze** → **Compare Means** → **Means…** (Page 56)

Select **Analyze** → **Compare Means** → **Independent Samples T-test…** (Page 88)

Select **Analyze** → **Nonparametric Tests** → **2 Independent Samples…** (Page 125)

Two Dependent Samples

Select **Analyze** → **Compare Means** → **Paired-Samples T-Test…** (Page 90)

Select **Analyze** → **Nonparametric Tests** → **2 Related Samples…** (Pages 128 and 129)

Two or More Independent Samples

Select **Analyze** → **Compare Means** → **One-Way ANOVA…** (Page 117)

Select **Analyze** → **General Linear Model** → **Univariate…** (Pages 120 and 122)

Select **Analyze** → **Nonparametric Tests** → **K Independent Samples…** (Page 127)

Select **Analyze** → **Regression** → **Linear…**(Page 122)

Comparing Proportions

One Sample

Select **Analyze** → **Nonparametric Tests** → **Binomial…** (Page

Select **Analyze** → **Nonparametric Tests** → **Chi-Square…** (Page 83)

Two Independent Samples

Select **Analyze** →**Descriptive Statistics** → **Crosstabs…** and select the Chi-square statistics under "Statistics." (Pages 97 and 98)

Two Dependent Samples

Select A**nalyze → Nonparametric Tests → 2 Related Samples...** (Pages 128 and 129)

Managing Output

Hiding Output

In the Output Navigator, double click on the icon of the output; double click on the icon again to "un-hide" (Page 9)

Making Copies of Output

Select **Edit → Copy** and then **Edit → Paste After** (Page 30)

Modifying Output

Titles (Page 11)

Text (Page 11)

Graphs

Resizing (Page 11)

Pie Charts – changing the items displayed and the color of the plot (Page 18)

Bar Charts – changing axis labels (Page 23)

Pareto Charts – removing counts and the cumulative percent line (Page 27)

Dot Plots – changing plot symbol (Page 27)

Histograms – changing the number of intervals and their width (Page 31)

Time Plots – changing the scale on the time axis (Page 37)

Boxplots – rotating the plot (Page 47)

Scatter Plots – connecting the points (Page 39); changing the plot symbol(Page 62); adding the regression line (Page 62)

PHStat 2.5 with Data Files for use with the Technology Manual to accompany Statistics: The Art and Science of Learning From Data
Alan Agresti and Christine A. Franklin
ISBN 0-13-149736-7
CD License Agreement

READ THIS LICENSE CAREFULLY BEFORE OPENING THIS PACKAGE. BY OPENING THIS PACKAGE, YOU ARE AGREEING TO THE TERMS AND CONDITIONS OF THIS LICENSE. IF YOU DO NOT AGREE, DO NOT OPEN THE PACKAGE. PROMPTLY RETURN THE UNOPENED PACKAGE AND ALL ACCOMPANYING ITEMS TO THE PLACE YOU OBTAINED THEM. THESE TERMS APPLY TO ALL LICENSED SOFTWARE ON THE DISK EXCEPT THAT THE TERMS FOR USE OF ANY SHAREWARE OR FREEWARE ON THE DISKETTES ARE AS SET FORTH IN THE ELECTRONIC LICENSE LOCATED ON THE DISK:

1. **GRANT OF LICENSE and OWNERSHIP:** The enclosed CD-ROM ("Software") is licensed, not sold, to you by Pearson Education, Inc. publishing as Pearson Prentice Hall ("We" or the "Company") in consideration of your adoption of the accompanying Company textbooks and/or other materials, and your agreement to these terms. You own only the disk(s) but we and/or our licensors own the Software itself. This license allows instructors and students enrolled in the course using the Company textbook that accompanies this Software (the "Course") to use and display the enclosed copy of the Software on up to one computer of an educational institution, for academic use only, so long as you comply with the terms of this Agreement. You may make one copy for back up only. We reserve any rights not granted to you.
2. **USE RESTRICTIONS:** You may not sell or license copies of the Software or the Documentation to others. You may not transfer, distribute or make available the Software or the Documentation, except to instructors and students in your school who are users of the adopted Company textbook that accompanies this Software in connection with the course for which the textbook was adopted. You may not reverse engineer, disassemble, decompile, modify, adapt, translate or create derivative works based on the Software or the Documentation. You may be held legally responsible for any copying or copyright infringement that is caused by your failure to abide by the terms of these restrictions.
3. **TERMINATION:** This license is effective until terminated. This license will terminate automatically without notice from the Company if you fail to comply with any provisions or limitations of this license. Upon termination, you shall destroy the Documentation and all copies of the Software. All provisions of this Agreement as to limitation and disclaimer of warranties, limitation of liability, remedies or damages, and our ownership rights shall survive termination.
4. **DISCLAIMER OF WARRANTY: THE COMPANY AND ITS LICENSORS MAKE NO WARRANTIES ABOUT THE SOFTWARE, WHICH IS PROVIDED "AS-IS." IF THE DISK IS DEFECTIVE IN MATERIALS OR WORKMANSHIP, YOUR ONLY REMEDY IS TO RETURN IT TO THE COMPANY WITHIN 30 DAYS FOR REPLACEMENT UNLESS THE COMPANY DETERMINES IN GOOD FAITH THAT THE DISK HAS BEEN MISUSED OR IMPROPERLY INSTALLED, REPAIRED, ALTERED OR DAMAGED. THE COMPANY DISCLAIMS ALL WARRANTIES, EXPRESS OR IMPLIED, INCLUDING WITHOUT LIMITATION, THE IMPLIED WARRANTIES OF MERCHANTABILITY AND FITNESS FOR A PARTICULAR PURPOSE. THE COMPANY DOES NOT WARRANT, GUARANTEE OR MAKE ANY REPRESENTATION REGARDING THE ACCURACY, RELIABILITY, CURRENTNESS, USE, OR RESULTS OF USE, OF THE SOFTWARE.**
5. **LIMITATION OF REMEDIES AND DAMAGES: IN NO EVENT, SHALL THE COMPANY OR ITS EMPLOYEES, AGENTS, LICENSORS OR CONTRACTORS BE LIABLE FOR ANY INCIDENTAL, INDIRECT, SPECIAL OR CONSEQUENTIAL DAMAGES ARISING OUT OF OR IN CONNECTION WITH THIS LICENSE OR THE SOFTWARE, INCLUDING, WITHOUT LIMITATION, LOSS OF USE, LOSS OF DATA, LOSS OF INCOME OR PROFIT, OR OTHER LOSSES SUSTAINED AS A RESULT OF INJURY TO ANY PERSON, OR LOSS OF OR DAMAGE TO PROPERTY, OR CLAIMS OF THIRD PARTIES, EVEN IF THE COMPANY OR AN AUTHORIZED REPRESENTATIVE OF THE COMPANY HAS BEEN ADVISED OF THE POSSIBILITY OF SUCH DAMAGES.** SOME JURISDICTIONS DO NOT ALLOW THE LIMITATION OF DAMAGES IN CERTAIN CIRCUMSTANCES, SO THE ABOVE LIMITATIONS MAY NOT ALWAYS APPLY.
6. **GENERAL:** THIS AGREEMENT SHALL BE CONSTRUED IN ACCORDANCE WITH THE LAWS OF THE UNITED STATES OF AMERICA AND THE STATE OF NEW YORK, APPLICABLE TO CONTRACTS MADE IN NEW YORK, EXCLUDING THE STATE'S LAWS AND POLICIES ON CONFLICTS OF LAW, AND SHALL BENEFIT THE COMPANY, ITS AFFILIATES AND ASSIGNEES. This Agreement is the complete and exclusive statement of the agreement between you and the Company and supersedes all proposals, prior agreements, oral or written, and any other communications between you and the company or any of its representatives relating to the subject matter. If you are a U.S. Government user, this Software is licensed with "restricted rights" as set forth in subparagraphs (a)-(d) of the Commercial Computer-Restricted Rights clause at FAR 52.227-19 or in subparagraphs (c)(1)(ii) of the Rights in Technical Data and Computer Software clause at DFARS 252.227-7013, and similar clauses, as applicable. Should you have any questions concerning this agreement or if you wish to contact the Company for any reason, please contact in writing: Pearson Education, Inc., One Lake Street, Upper Saddle River, New Jersey 07458 "AS IS" LICENSE

****
SYSTEM REQUIREMENTS
****

*Microsoft Windows, Windows 98, Windows NT, Windows 2000, Windows ME, or Windows XP
*In addition to the minimum processor requirements for the operating system your computer is running, this CD requires a Pentium II, 200 MHz or higher processor
*In addition to the RAM required by the operating system your computer is running, this CD requires 64 MB RAM for Windows 98, Windows NT 4.0, Windows 2000, Windows ME, and Windows XP
*Macintosh OS 9.x or 10.x
*In addition to the minimum processor requirements for the operating system your computer is running, this CD requires a PowerPC G3 233 MHz or better
*In addition to the RAM required by the operating system your computer is running, this CD requires 64 MB RAM
*Microsoft Excel 97, 2000, 2002, or 2003 (Excel 97 use must apply the SR-2 or a later free update from Microsoft in order to use PHStat2. Excel 2000 and 2002 must have the macro security level set to Medium) for Windows; MINITAB Version 14 or 12; JMP version 5.1; SPSS versions 13.0, 12.0, 11.0; or other statistics software.
*PHStat will not work on the Macintosh
*Microsoft Excel Data Analysis ToolPak and Analysis ToolPak VBA installed (supplied on the Microsoft Office/Excel program CD)
*CD-ROM or DVD-ROM drive; Mouse and keyboard; Color monitor (256 or more colors and screen resolution settings set to 800 by 600 pixels or 1024 by 748 pixels)
*Internet browser for Windows (Netscape 4.x, 6.x or 7.x, or Internet Explorer 5.x or 6.x) and Internet connection suggested but not required
*Internet browser for Macintosh (Internet Explorer 5.x or Safari 1.x) and Internet connection suggested but not required
*This CD-ROM is intended for stand-alone use only. It is not meant for use on a network.
****
CD-ROM CONTENTS
****

--Data Files : Included within that folder are:
--JMP Data Files (Folder: JMP_Agresti_Data)
JMP is required to view and use these files. Information about JMP can be found on the internet at http://www.jmp.com
--SPSS Data Files (Folder: SPSS_Agresti_Data)
SPSS is required to view and use these files. Information about SPSS can be found on the internet at http://www.spss.com
--MINITAB Data Files (Folder: minitab_Agresti_Data)
MINITAB is required to view and use these files. Information about MINITAB can be found on the internet at http://www.minitab.com/support/index.htm.
--Data Files (Folder: ASCII_Agresti_Data)
For use with SPSS or other statistics software.
--Excel Files (Folder: Excel_Agresti_Data)
Excel is required to view and use these files. Information about Excel can be found on the Internet at http://office.microsoft.com/en-us/default.aspx
--TI-83/84 Files (Folder: TI-8x_Agresti_Data)
TI-83/84 calculator is required http://education.ti.com/us/product/main.html.
--PHStat 2.5
Prentice Hall's PHStat statistical add-in system enhances Microsoft Excel to better support learning in an introductory statistics course. http://www.prenhall.com/phstat/
--Readme.txt
****
TECHNICAL SUPPORT
****

If you continue to experience difficulties, call 1 (800) 677-6337, 8 am to 8 pm Monday through Friday and 5 pm to 12 am Sunday (all times Eastern) or visit Prentice Hall's Technical Support Web site at http://247.prenhall.com/mediaform. Our technical staff will need to know certain things about your system in order to help us solve your problems more quickly and efficiently. If possible, please be at your computer when you call for support. You should have the following information ready:

- Textbook ISBN
- CD-Rom/Diskette ISBN
- Corresponding product and title
- Computer make and model
- Operating System (Windows or Macintosh) and Version
- RAM available
- Hard disk space available
- Sound card? Yes or No
- Printer make and model
- Network connection
- Detailed description of the problem, including the exact wording of any error messages.

NOTE: Pearson does not support and/or assist with the following:
- 3d-party software (i.e. Microsoft including Microsoft Office Suite, Apple, Borland, etc.)
- homework assistance
- Textbooks and CD-ROMs purchased used are not supported and are non-replaceable.

For assistance with third-party software, please visit:
JMP Support (for JMP): http://www.jmp.com/support/techsup/index.shtml
MINITAB Support: http://www.minitab.com/support/index.htm
SPSS Support: http://www.spss.com/tech/spssdefault.htm
TI-8x Support: http://education.ti.com/us/support/main.html
Excel Support: http://support.microsoft.com/
PHStat Support: http://www.prenhall.com/phstat/phstat2/phstat2(main).htm

----